STRUCTURAL SAFETY & RELIABILITY
VOLUME 2

Structural Safety & Reliability, Schuëller, Shinozuka & Yao (eds) © 1994 Balkema, Rotterdam, ISBN 90 5410 357 4

Table of contents

Inspection

Fatigue and fracture

Stability

Structural Safety & Reliability, Schuëller, Shinozuka & Yao (eds) © 1994 Balkema, Rotterdam, ISBN 90 5410 357 4

On the probabilistic stability analysis of tubular struts with initial imperfections

Andrea Benedetti
Institute of Civil Engineering, Faculty of Engineering, University of Ferrara, Italy

ABSTRACT : In this paper we considered the elasto-plastic analysis of simple layouts of imperfect tubular column in order to show how interactions between bar buckling and truss snapping modify the overall structural behaviour; in this context, we can introduce relations which allow a reliability analysis in a probabilistic formalism.
Starting from previous analyses where simple but accurate relations describing the behaviour of an imperfect strut are presented [1,2], it is possible to build up the collapse probability distribution functions for a simple two bar arch with elastic supports.
The discussion of the distribution found for a paradigmatic case helps to understand the influences of involved parameters on local failure of real single layer lattices.

1 INTRODUCTION

The structural reliability of single layer space trusses such as vaults and domes is strongly influenced by the imperfections arising from strut manufacturing, erection technique and, last but not least, human workmanship [Augusti (1984), Ditlevsen (1980)].

Since the limitation of imperfection statistics under a fixed (low) level is an engineering task with high specific cost, it seems important to perform qualitative analysis allowing the selection of the variables which affect the carrying capacity of the structural assemblage.

However, since the behaviour of real three dimensional frames is strongly sensitive to imperfections which affect both individual members and geometry of the structural arrangement, the relevant question is how to select main parameters and tolerances capable to improve the structural design. Moreover, this is particularly true for single layer vaults and shells, which have highly nonlinear responses even in the case of perfect structures [Rothert (1981), Papadrakakis (1983), Fuji (1989)].

Several solution techniques able to trace the load path also in the postbuckling range have been proposed; basically they use both finite element methodologies to spatially discretize the problem, and incremental-iterative strategies for the pointwise solution of the nonlinear problem [Wriggers (1990)].

A necessary ingredient of numerical analyses is a perturbation pattern of the perfect structure; in this respect, several studies concerning the evaluation of the "worst" initial imperfection were carried out [Ikeda (1990)], and their results appear very useful for practical applications.

Unfortunately, numerical solutions do not enable for a systematic sensitivity analysis of imperfection parameters; on the other hand, asymptotic expansion of the post critical path through a perturbation technique cannot deal with non linear material problems [Hansen (1974)].

In this paper, starting from a previous introductory analysis [Benedetti (1991)], the problem is reconsidered introducing a probabilistic context for the input data.

Firstly, a simplified but accurate solution for an hollow imperfect stocky strut with elasto-plastic material constitutive law is reviewed, as well as its validity limits [Benedetti (1992)]. In this respect, if we look for an analytical solution for imperfect bar assemblages, we can consider a simplified version of the method, making use only of the elastic and fully plastic parts of the load-deflection relationships: in fact this includes all the features of the more complex numerical solution.

Afterwards, assuming that the initial imperfections are scalar stochastic variables, the analytical model is written down in a form allowing the evaluation of the structural component reliability index; it appears worth remarking that the ongoing of collapse is

pointed out (in a probabilistic sense) from a sharp increase of the applied load variance.

This result, having large experimental evidence [Maier (1970)], can be used to arrive at a probabilistic definition of stability; in fact, it seems possible to draw a functional relation between deterministic and stochastic buckling load, linking tangent stiffness vanishing and variance raising[Roorda (1975), Konishi (1981)].

As a further step we consider a simple two bar arch that reproduces the behaviour of many axisymmetric assemblages which are locally present in most of single layer reticular shells. Assuming now an optimal design geometry, we can carry out the probabilistic analysis of the series system characterized by the bar buckling failure and/or the arch snapping failure. The comparison of the two event p.d.f.'s gets an insight into the strength of the assumed imperfections versus the outcome of possible stability losses.

2   THE IMPERFECT BAR BEHAVIOUR

Let us consider the imperfect hollow tubular bar of fig.1; as a further assumption, we set that the material can be described through an elastic plastic constitutive relation of holonomic type.

The general solution of the problem can be sought in a system of nonlinear differential equations [1,2], where the unknowns are the functions ε, χ, θ, w, u describing axis line strain, curvature, rotation and displacements respectively.

A first simplification can be obtained substituting the unknown functions with their mid span values interpolated by a form function which is set a priori

$$z(x) := z_o(x = L/2) \cdot F(x) \mid F(x) \text{ known.} \qquad (1)$$

A further simplification comes up considering only the usual range of dimensional ratios used in common practice engineering design and thus restricting the problem to a linearized elastic kinematics. Making reference to Benedetti (1991,1992) where the substitutions are described in full length, we obtain finally the nonlinear system:

$$\begin{cases} \int_A \sigma(\varepsilon_m + y\,\chi_m)\,dA = P\,, \\[2mm] \int_A \sigma(\varepsilon_m + y\,\chi_m)\,y\,dA = P\,w_m\,L\,, \\[2mm] w_m = w_e(P) + w_p(\chi_m - \chi_e\,,\,\ell_p)\,, \\[2mm] \dfrac{P}{A} + \dfrac{PLR}{J}\left(w_e(P) + w_p(\chi_m - \chi_e\,,\,\ell_p)\right)\Big|_{\bar{x}} = \bar{\varepsilon}\,. \end{cases}$$

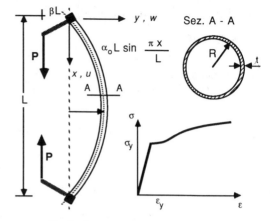

Fig. 1 : *Geometry of the imperfect column*

where w_m is assumed as path parameter, ℓ_p is the central zone with plastic spreading and $\bar{\varepsilon}$ is a strain value assumed as a limit of elastic behaviour.

As discussed in [1,2], the proposed solution method allows for an accurate solution of the probabilistic response of the imperfect bar. In fig. 2 and 3 the solutions obtained are compared with statistical data of the large experimental program of Maier (1970).

Even if the proposed solution can solve the imperfect bar problem it is far complex to use for a bar assemblage. A more simple solution can be composed starting from the load-displacement relations in elastic and fully plastic range:

$$W_{el} = \frac{\alpha L}{1-\rho^2} + \frac{\beta L}{\cos\dfrac{\pi\rho}{2}}\,, \qquad W_{pl} = \frac{M_p}{P}\left(\frac{1}{P} - \frac{1}{P_p}\right). \qquad (3)$$

where α and β are the imperfections, $\rho = \sqrt{P/P_{crE}}$, M_p is the fully plastic moment, P_p is the fully plastic axial force.

Replacing the trigonometric function with the suitable Taylor series, inverting the two relations and scaling them so that the load value at intersection point is preserved, we arrive at the final approximate load - dispacement relation:

$$P = \frac{8}{\pi^2}\,\frac{P_{crE}\,P_p}{P_{max}}\left(1 - \frac{\gamma}{\gamma + w}\right)\frac{M_p}{M_p + P_p\,(\gamma + w)} \qquad (4)$$

where $\gamma = (\alpha + \beta)\,L$ and w represents the incremental mid span displacement.

Similarly, starting from the general relation:

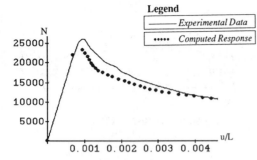

Fig. 2 : *Load - Shortening relationship*
experimental and computed

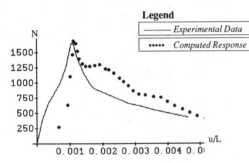

Fig. 3 : *Standard deviation of Load*
Experimental and computed

$$u = \int_0^L (1 + \varepsilon)(1 - \cos\theta(x))\, dx \qquad (5.a)$$

$$u \doteq L - \int_0^L \sqrt{1 - w_x^2}\, dx + \int_0^L \frac{P}{\tilde{E}A(x)}\, dx , \qquad (5.b)$$

and introducing some simplifications [Benedetti, 1991], we arrive at:

$$u = \frac{P\,L}{E\,A} + \frac{8\,w^2}{3\,L} + \frac{32\,w^4}{5\,L^3} \qquad (6)$$

where the load P must be considered as a function of displacement. As a consequence, the load-shortening relation of the bar can be constructed explicitly in a parametric way, assuming w as a parameter.

3. BAR ASSEMBLAGES

In order to assess the interaction phenomenon, we can take into account a simple bar subassemblage like those in fig. 4.

In this case, the local structure can be simulated through a simple two bar truss with elastic supports like in fig 5.

In the aforementioned cases the spring effect modelling the restraining outer ring cannot be left out; for instance, in the hexagonal layout, the ring compliance equals the inverse of the truss bar stiffness.

The analysis of the truss requires the solution of a non linear problem; however, because of the little value of the ratio $n = 2\ f_o/L_o$, we can introduce again a Taylor series expansion of the trigonometric quantities [Belluzzi (1954)]:

$$l_b \doteq L_o + \frac{f_o^2}{2\,L_o} \doteq L_o + \Delta L + \frac{f^2}{2\,L_o} \qquad (7)$$

After some algebra and manipulation we obtain:

$$V_{snap}(f) = \frac{A_s\,E_s}{L_o^3}\, f\,(f_o^2 - f^2)\, \frac{\xi}{(1+\xi)(1+n^2) - n^2} \qquad (8.a)$$

$$N_{snap}(f) = \frac{A_s\,E_s}{L_o^2}\,(f_o^2 - f^2)\, \frac{\xi\,(2+n^2)}{4(1+\xi)(1+n^2) - 4n^2} \qquad (8.b)$$

$$\Delta l_b(f) = V_{snap}(f)\, \frac{L_o^2 + f_o^2}{2\,f\,A_s\,E_s} \qquad (8.c)$$

where ξ holds for the ratio $k_{sp}\,l_b / (E_s\,A_s)$.

Setting to zero the derivative of the load with respect to f, we solve for f_{cr}; substituting its value to f we arrive at the critical load:

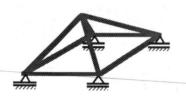

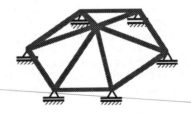

Fig. 4: *Bar Subassemblages*

747

$$V_{cr} = \frac{2}{3\sqrt{3}} A_s E_s \frac{\xi n^3}{(1+\xi)(1+n^2) - n^2} \tag{9.a}$$

$$N_{cr} = E_s A_s \frac{\xi n^2 (2+n^2)}{6 (1+\xi)(1+n^2) - 6 n^2} \tag{9.b}$$

From this last expression we can obtain the transition load, i.e. the load which causes both the snapping of the arch and the buckling of the bar simultaneously.

$$n = \sqrt{\frac{3 P_{max}}{E_s A_s} \frac{1+\xi}{\xi}} \tag{10}$$

In fact, the optimal geometry shows ratios lesser than 10 %; nevertheless this is the case when, for instance, a vault arch of 90° is made up of 8 or 12 bars.

4. PROBABILISTIC ANALYSIS

The definition of a procedure suitable for non linear reliability analysis of trusses can be done into the framework of the PFEM (Probabilistic Finite Element Method, see Liu & Belytschko (1989)).

However, a qualitative analysis intended to clarify the relevance of variables, do not necessarily need such a general tool; then, starting from the presented relations, and making the assumption that all distributions are of normal type, the moments of the target variable can be computed also by means of a Taylor series expansion [Madsen et al. (1986)]:

$$E[y] = f(E[x_1], ..., E[x_n]), \tag{11}$$

$$V[y] = D[y]^2 = \sum_{i=1}^{n} \left(\frac{\partial y}{\partial x}\right)^2 V[x_i]. \tag{12}$$

So, considering the vertical deflection of the two bar truss as a parameter, and introducing mean and standard deviation for all the geometrical and mechanical variables, it is possible to evaluate the axial force statistics.

More precisely, it is very interesting to compare the axial force distribution derived from the relations describing the behaviour of the truss and the result of the single imperfect bar analysis. The load level in the bar can be parametrized as a function of vertical truss displacement, making use of the inverse displacement - shortening relation.

With the data of Table I, we can derive the p.d.f. of the truss sag f_o from the chord and bar lengths; however, the small difference between the two

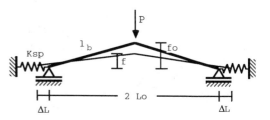

Fig. 5 : *Two Bar Arch with End Springs*

Table I

Variable	Mean	C.o.V.
L_o	1500 mm	0.001
l_b	1496 mm	0.001
α/L_o	0.001	0.05
β/L_o	0.001414	0.05
R	30 mm	0.02
t	3 mm	0.02
E_s	$2 \cdot 10^5$ MPa	0.05
σ_y	250 MPa	0.05
ξ	1	-

lengths causes a very high $V[f_o]$ value (the cov is nearly 26,2 %), but, in reality, it is damped from tolerances and assembling adjustements. So, it is interesting to introduce a $V[f_o]$ reduction factor φ that enables us to test different locking conditions.

In fig. 6 the coefficient of variation for both internal forces N_{snap} and P_{bar} are compared; moreover, N_{snap} was computed for the φ values: 1.0 and 0.5. It is apparent that, due to the geometrical fuzziness, $D[N_{snap}]$ is by far greater than $D[P_{bar}]$.

However, while N_{snap} can increase beyond the critical value (although the external load V cannot), P_{bar} shows an upper limit. So, when this limit is come over, the failure probability due to bar buckling increases rapidly, merging soon (up to exceed) to the probability of snapping.

The reliability index can be computed either from resistance and sollicitation distributions, or from Hasofer Lind method [Madsen et al. (1986)]; in the latter case the evaluation of the FORM design point requires the solution of a constrained minimization problem:

$$P_f = \Phi\left(-\frac{E[R-S]}{D[R-S]}\right) = \Phi(-\beta_c). \tag{13}$$

$$\beta_{HL} = Min\left\{ \| x \| := x <- \frac{z - E[Z]}{C[Z]} \right\},$$

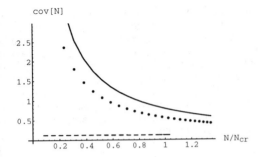
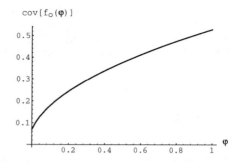

Fig. 6: *Axial force and sag coefficient of variation.*
(line → snapping $\varphi = 1.0$, dots → snapping $\varphi = 0.5$, dash → bar buckling)

subject to: $R(\mathbf{x}) - S(\mathbf{x}) = 0$ (14)

In the present case the two possible failure mechanisms constitute a series system; consequently the failure probability is the maximum value between the events:

$$P_{fb} = \text{prob}\left\{ P_{max} - P \le 0 \right\} \quad (15.a)$$
$$P_{fs} = \text{prob}\left\{ V_{cr} - V \le 0 \right\} \quad (15.b)$$

In the following analysis we set the pair $\{f,w\}$ as parameters describing a loading process with displacement control; since the bar shortening is linked to the crown displacement by the relationship:

$$f = \left[f_o^2 - u(w)\frac{(1+\xi)(1+n^2)-n^2}{\xi} \cdot \frac{L_o^3}{L_o^2+f_o^2} \right]^{\frac{1}{2}}, \quad (16)$$

we can use w as path parameter. Moreover, in this preliminary analysis, we can retain the limit values P_{max} and V_{cr} as deterministic.

So, it is a straightforward (despite lengthy) exercise to compute the first order Taylor series of the load and axial force probability distribution functions.

Fig. 7 shows the variation of the Cornell reliability index β_c for the two collapse mechanisms as a function of vertical displacement.

The computation of the Hasofer-Lind index requires further computational effort; in this case we have to solve the constrained minimization problem (14). Since the limit state functions are linear in P and V, to remove the constraint we can use a standard technique.

As a matter of fact, we can express P or V with relation to the other random variables; then, these last are mapped into the standard Gaussian multivariate space through a scaling operation. Finally, we can solve for the minimum:

$$\left\{ \mathbf{x_R} := \{x_2....x_n\} : \delta^2 = \text{Min}\{ \, \| \, x_1(x_2...x_n) + \mathbf{x_R} \, \| \} \right\}$$

the value δ is the reliability index β_{HL} we looked for.

In fig. 8 the obtained result is plotted once again as a function of vertical dispacement; a comparison between fig.s 7 and 8 reveals a substantial

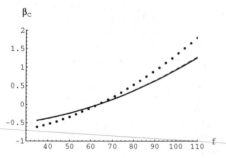
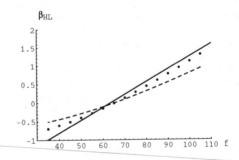

Fig. 7: *Variation of Cornell reliability indices*   **Fig. 8:** *Variation of Hasofer-Lind reliability indices*

(line → snapping $\varphi = 1.0$, dots → snapping $\varphi = 0.5$, dash → bar buckling)

agreement. In particular, when the displacement is small the critical (i.e. the lower) index arises from bar buckling while beyond a certain displacement value, the snapping probability becomes larger than the former.

5   DISCUSSION

It is somehow surprising that the randomness of the geometrical form does not yield to a very low level of reliability against snapping but, leaving out initial stress phenomena due to the faceting closure error, reliability against snapping comes out from the actual and critical crown displacement ratio which is indeed almost independent from initial geometry.

Moreover, during the displacement increase, the axial force is not bounded to the critical value (in fact, it can rise until the initial sag is vanished), whereas the limit load of the bar can be not exceeded anyhow.

As a consequence, although the snapping p.d.f. is more dispersed, the buckling p.d.f. is passed more quickly by the (displacement controlled) load value; as a result, until a nearby critical displacement is attained, buckling proneness overcomes the snapping one regardless of the dispersion.

Finally, it is interesting to discuss the evolution of a real collapse event on the basis of the computed behaviour.

When the structural design is carried out searching for a minimal weight, buckling and snapping probabilities are not very different; so, if in the life of the structure the vertical displacement takes a sufficiently large value, the two collapse mechanisms become nearly equally probable.

However, it is clear that the random nature of both global and individual resistances near the limit point yields to a strong coupling which can decrease considerably the overall reliability.

Once raised the limit point the stiffness turns suddenly to negative values and, if bars are beyond the elastic limit, a bifurcation phenomenon can break the displacement symmetry [Wu et al. (1987)].

On the other hand, if snapping takes place, in the motion from the critical point to the vanishing of the intial sag, the axial force in the bars rises up considerably, leading ultimately to a buckling instability again.

In this respect, it seems that an optimal design can lead to a non conservative design if the fuzziness of the geometrical behaviour amplifies the interaction effects of global and individual collapse modes.

More precisely, in the author's opinion, a correct member size would require that the bar buckling limit load was at least comparable to the largest bar axial force appearing in the whole snapping process of the assemblage under examination.

On the other hand, if the snapping limit load is adequately greater than the buckling one, no coupling phenomenon is expected and, as usual, the design can be worked out using standard optimization techniques.

6   CONCLUSION

In this analysis we considered some deterministic results enabling a direct treatment for the instability collapse of simple bar assemblages.

After, through the development of a first order probabilistic analysis we pointed out that, despite the variability of the geometrical parameters, both snapping and bar buckling instabilities can modify the structural system reliability with similar strength.

Finally, the computed reliability and the discussion of the evolution of a collapse event allowed to draw some useful criteria for a preliminary design of single layer pin jointed structures.

REFERENCES

Augusti G., Baratta A., Casciati F. 1984. *Probabilistic Methods in Structural Mechanics.* London: Chapman & Hall.

Belluzzi O. *"Scienza delle Costruzioni ",* I-IV, Zanichelli, Bologna, 1954.

Benedetti A. 1991. On the Reliability Analysis of Elastic Plastic Struts with Initial Imperfections. *IASS Symposium 91* III: 123-130.

Benedetti A. 1992. Sull'Analisi probabilistica di aste tubolari con imperfezioni iniziali. *AIMETA 92 - XI Congresso della Ass. It. di Mecc. Teorica e Appl.* :

Davister M., Britvec S. 1984. Evaluation of the most unstable dynamic buckling mode of pin jointed space lattices. *Third Int. Conf. on Space Struct.*: 468-473.

Ditlevsen O. 1980. Formal and real structural safety. Influence of gross errors. *Proceedings IABSE* 4: 185-204.

Elishakoff I. 1983. How to introduce the imperfection-sensitivity concept into design. *Collapse: the buckling of structures in theory and practice,* (Thompson and Hunt Ed.s): 347-357. Cambridge University Press.

Fujii F. 1989. Scheme for elasticas with snap-back and looping. *Jour. of Engrg. Mech.* 115 No. 10: 2166-2181.

Hansen J.S., Roorda J. 1974. On a Probabilistic Stability Theory for imperfection sensitive Structures. *Int. Jour. of Solids & Struct.* 10: 341-349.

Konishi I., Takaoka N. 1981. Some Comments on the Reliability of Civil Engineering Structures. *3rd Int. Conf. on Struct. Safety & Reliability* (Moan Shinozuka Ed.s). New York: Elsevier.

Ikeda K., Murota K. 1990. Critical initial imperfection of structures. *Int. Jour. Solids and Struct.* 26 No. 8: 865-886.

Little G.H. 1982. The collapse of steel model columns. *Int. J. of Mech. Sci.* 24 No. 5: 263-278.

Liu W.K., Belytschko T. Ed.s. 1989. *Computational Mechanics of Probabilistic and Reliability Analysis* Lausanne: Elmepress International.

Madsen H.O., Krenk S., Lind N.C. 1986. *Methods of Structural Safety.* Englewood Cliffs N.J.: Prentice Hall.

Maier G., Zavelani Rossi A. 1970. Sul comportamento di aste metalliche compresse eccentricamente: indagine sperimentale e considerazioni teoriche. *Costruzioni Metalliche* 4: 1-15.

Papadrakakis M. 1983. Inelastic post-buckling analysis of trusses *Jour. of Struct. Engnrg.* 109 No. ST9: 2129-2147.

Roorda J. 1975. The Random Nature of Column Failure. *J. Struct. Mech.* 3: 239-257.

Rothert H., Dickel T., Renner D. 1981. Snap-through buckling of reticulated space trusses. *Jour. of the Struct. Div.* 107 No. ST1: 129-143.

Wriggers P., Simo J.C. 1990. A general procedure for the direct computation of turning and bifurcation points. *Int. Jour. for Num. Meth.s in Engrg.* 30: 155-176.

Wu X.Q., Liu C., Yu T.,X. 1987. A Bifurcation phenomenon in an elastic-plastic symmetrical shallow truss subjected to a symmetrical load. *Int. J. Solids Struct.* 23 no.9: 1255-1233.

Structural Safety & Reliability, Schuëller, Shinozuka & Yao (eds) © 1994 Balkema, Rotterdam, ISBN 90 5410 357 4

Buckling reliability of laminated composite cylindrical shells

Hyo-Nam Cho, Seung-Jae Lee & Young-Min Choi
Department of Civil Engineering, Hanyang University, Seoul, Korea

Jae-Chul Shin
Department of Civil Engineering, Chungnam University, Taejon, Korea

ABSTRACT: In the paper, an attempt is made to suggest a buckling reliability model, and to investigate the characteristics of buckling reliability for laminated composite cylindrical shells.

Based on a comparison of theoretical and codified buckling strength, it is demonstrated that the practical ASME and JRPS strength formulas may be successfully used for the evaluation of buckling reliability. It is also demonstrated that this kind of reliability problems can be precisely evaluated by using the Importance Sampling Method.

Based on the numerical application of both theoretical and practical models, it may be stated that the buckling reliability of laminated composite cylindrical shells is sensitive to the variation of some critical parameters and uncertainties.

1. INTRODUCTION

One of the distinct features of laminated cylindrical shells is the fact that their mechanical behaviors are quite different from those of isotropic materials due to the coupling effects arising from fiber orientation and lamination. In general, the buckling of laminated composite cylindrical shells is not only sensitive to the variations of lamina strength/modulus, slenderness parameters and initial imperfections, but also significantly affected by volume content, lamination angle, stacking sequence and number of layers. As a result, the buckling reliability of laminated composite cylindrical shells is extremely difficult to assess precisely because of the complexity of analysis under generalized loading and highly complicated ramdom parameters.

This study as the first step toward more generalized model is intended to propose practically applicable model and method for the assessment of the buckling reliability of laminated composite cylindrical shells, and to investigate the reliabilities of the composite shells subject to external pressure designed by the current conventional design methods.

2. BUCKLING LIMIT STATE MODEL

When laminated cylindrical shells are subject to external pressure as shown in Fig.1, the buckling strength depending upon the external loading can be obtained analytically or numerically from an eigen–value analysis using the governing buckling equation of the Donnell type equilibrium(Kasuya and Uemura, 1985) and assuming the harmonic series form of displacement functions.

The buckling limit state function in general can be expressed in terms of random buckling strength $P_B(\cdot)$ as

$$g(\cdot) = P_B(\cdot) \cdot X_B - P \qquad (1)$$

where P = applied external pressure ; X_B = random variate of uncertainties associated

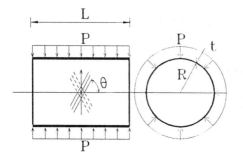

Fig.1. Configuration and coordinate of laminated composite cylindrical shell

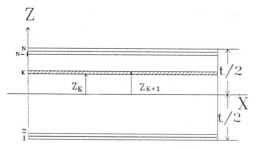

Fig.2. Geometry of a N-layered laminate

with imperfections(X_I), manufacturing /construction(X_M) and modelling(X_P).[X_B = $X_I \cdot X_M \cdot X_P$]

Theoretical random buckling strength $P_B(\cdot)$ can be derived in terms of random stiffness parameters $T_{ij}(\cdot)$ of laminated cylindrical shells(Kasuya and Uemura, 1985).

$$P_B(\cdot) = (D_n \cdot X_D/2n^2)[T_{33}(\cdot) + \{2T_{12}(\cdot) \\ T_{13}(\cdot)T_{23}(\cdot) - T_{11}(\cdot)T_{23}^2(\cdot) - \\ T_{22}(\cdot)T_{13}^2(\cdot)\}/\{T_{11}(\cdot)T_{22}(\cdot) - \\ T_{12}^2(\cdot)\}]$$
$$(2)$$

where D_n = nominal diameter of cylindrical shells ; X_D = basic random variate of geometric dimensional uncertainties.

It may be noted that $T_{ij}(\cdot)$ is function of random stiffness parameters of laminates $A_{ij}(\cdot)$, $B_{ij}(\cdot)$, $C_{ij}(\cdot)$, and X_D, as follows. For instance, in the case of angle-ply anti-symmetrical laminates

(Kasuya and Uemura, 1985),

$$T_{11}(\cdot) = A_{11}(\cdot)(m\pi/L)^2 + A_{66}(\cdot)(2n/D_n X_D)^2$$
$$T_{12}(\cdot) = (A_{12}(\cdot) + A_{66}(\cdot))(m\pi/L)(2n/D_n X_D)$$
$$T_{22}(\cdot) = A_{22}(\cdot)(2n/D_n X_D)^2 + A_{66}(\cdot)(m\pi/L)^2$$
$$T_{13}(\cdot) = -\{2A_{12}(\cdot)(m\pi/L)^2/D_n X_D + 3B_{16}(\cdot) \\ (m\pi/L)^2(2n/D_n X_D) + \\ B_{26}(\cdot)(2n/D_n X_D)^3\}$$
$$T_{23}(\cdot) = -\{2A_{22}(\cdot)(2n/D_n X_D)/D_n X_D + B_{16}(\cdot) \\ (m\pi/L)^3 + 3B_{26}(\cdot) \\ (m\pi/L)(2n/D_n X_D)^2\}$$
$$T_{33}(\cdot) = D_{11}(\cdot)(m\pi/L)^4 + 2\{D_{12}(\cdot) + 2D_{66}(\cdot)\} \\ (m\pi/L)^2(2n/D_n X_D)^2 + \\ D_{22}(\cdot)(2n/D_n X_D)^4 + 4A_{22}(\cdot)/(D_n X_D)^2 \\ + 8B_{26}(\cdot)(m\pi/L)(2n/D_n X_D)/D_n X_D$$
$$(3)$$

where m, n, respectively, represent the values of half/full-period indices in the x, y directions and L represents the length of composite cylindrical shell. Also, note that $A_{ij}(\cdot)$, $B_{ij}(\cdot)$, $C_{ij}(\cdot)$ which are stiffness parameters of force-strain equation of laminates, in turn, may be expressed as functions of random stiffness parameters of lamina $\overline{Q_{ij}}(\cdot)$ in the following way, i.e.,

$$A_{ij}(\cdot) = \sum_{k=1}^{n} \{\overline{Q_{ij}}(\cdot)\}_k (z_k - z_{k-1})$$

$$B_{ij}(\cdot) = \sum_{k=1}^{n} \{\overline{Q_{ij}}(\cdot)\}_k (z_k^2 - z_{k-1}^2) \qquad (4)$$

$$D_{ij}(\cdot) = \sum_{k=1}^{n} \{\overline{Q_{ij}}(\cdot)\}_k (z_k^3 - z_{k-1}^3)$$

where $A_{ij}(\cdot)$ = extensional stiffness parameter ; $B_{ij}(\cdot)$ = coupling stiffness parameter ; $D_{ij}(\cdot)$ = bending stiffness parameter ; z_k = distance to the kth layer(= ±(k-1)t·X_t/n), in which t is nominal thickness of laminate and X_t is basic random variate of uncertainty of t.

$\overline{Q_{ij}}(\cdot)$ depending upon fiber orientation can also be expressed as functions of reduced stiffness of orthotropic lamina (principal material elastic constants) $Q_{ij}(\cdot)$.

$$\overline{Q_{11}}(\cdot) = Q_{11}(\cdot)\cos^4\theta + 2[Q_{12}(\cdot) + 2Q_{66}(\cdot)] \\ \sin^2\theta\cos^2\theta + Q_{22}(\cdot)\sin^4\theta$$
$$\overline{Q_{12}}(\cdot) = [Q_{11}(\cdot) + Q_{22}(\cdot) - 4Q_{66}(\cdot)]\sin^2\theta$$

$$\overline{Q_{22}}(\cdot) = Q_{11}(\cdot)\sin^4\theta + 2[Q_{12}(\cdot)+2Q_{66}(\cdot)] \\ \cos^2\theta+Q_{12}(\cdot)(\sin^4\theta+\cos^4\theta) \\ \sin^2\theta\cos^2\theta+Q_{22}(\cdot)\cos^4\theta$$

$$\overline{Q_{16}}(\cdot) = [Q_{11}(\cdot)-Q_{12}(\cdot)-2Q_{66}(\cdot)]\sin\theta \\ \cos^3\theta+[Q_{12}(\cdot)-Q_{12}(\cdot)+2Q_{66}(\cdot)] \\ \sin^3\theta\cos\theta$$

$$\overline{Q_{26}}(\cdot) = [Q_{11}(\cdot)-Q_{12}(\cdot)-2Q_{66}(\cdot)]\sin^3\theta \\ \cos\theta+[Q_{12}(\cdot)-Q_{12}(\cdot)+2Q_{66}(\cdot)] \\ \sin\theta\cos^3\theta$$

$$\overline{Q_{66}}(\cdot) = [Q_{11}(\cdot)+Q_{22}(\cdot)-2Q_{12}(\cdot)- \\ 2Q_{66}(\cdot)]\sin^2\theta\cos^2\theta+Q_{66}(\cdot) \\ (\sin^4\theta+\cos^4\theta)$$

$$(5)$$

Finally, the random reduced lamina stiffness $Q_{ij}(\cdot)$ can be expressed in terms of basic random variables of material properties as follows.

$$Q_{11}(\cdot)=\frac{E_L X_{EL}}{1-\nu_{LT}\nu_{TL}X_{\nu_{LT}}X_{\nu_{TL}}}$$

$$Q_{12}(\cdot)=\frac{\nu_{LT}E_L X_{\nu_{LT}}X_{EL}}{1-\nu_{LT}\nu_{TL}X_{\nu_{LT}}X_{\nu_{TL}}}$$

$$Q_{22}(\cdot)=\frac{E_T X_{ET}}{1-\nu_{LT}\nu_{TL}X_{\nu_{LT}}X_{\nu_{TL}}}$$

$$Q_{66}(\cdot)=G_{LT}X_{GLT}$$

$$(6)$$

where E_L, E_T = modulus in longitudinal and transverse direction of fiber, respectively ; G_{LT},ν_{LT},ν_{TL} = nominal shear modulus and poisson's ratio ; $X_{EL},X_{ET},X_{GLT},X_{\nu_{LT}},X_{\nu_{TL}}$= basic random variate of uncertainties of E_L, E_T, G_{LT}, ν_{LT}, ν_{TL}, respectively.

In lieu of the complicated theoretical buckling strength $P_B(\cdot)$ shown as above, the simplified buckling strength formulas specified in the codes and standards(ASME, 1992; JRPS, 1985) are more preferable for practical application if these are appropriate for the prediction of buckling reliability. It has been found that the design formulas for ring–stiffened shells and unstiffened long shells specified in the ASME code provide about same results as those of the JRPS code. The randomized JRPS formulas for $P_B(\cdot)$ may be stated as follows;

for ring–stiffened shells,

$$P_B(\cdot)=E\cdot X_E\left\{\frac{\pi^4}{n^4(n^2-1)}\left[\frac{D_n\cdot X_D}{2L}\right]+ \\ (n^2-1)\frac{1}{12(1-\nu^2 X_\nu^2)}\left[\frac{2t\cdot X_t}{D_n\cdot X_D}\right]^2\right\} \\ \left\{\frac{2t\cdot X_t}{D_n\cdot X_D}\right\}$$

$$(7)$$

for unstiffened long shells,

$$P_B(\cdot)=\frac{E\cdot X_E}{4(1-\nu^2 X_\nu^2)}\left[\frac{2t\cdot X_t}{D_n\cdot X_D}\right]^3 \quad (8)$$

where E, ν=tensile modulus and poisson's ratio in the circumferential direction of laminates ; D_n, t, L=inner diameter, thick and length of cylindrical shell, respectively ; X_E,X_ν,X_D,X_t = basic random variate of uncertainties of E, ν, D_n, t, respectively.

Similarly, the randomized ASME formulas may be stated as follows;

for ring–stiffened shells,

$$P_B(\cdot) = 2.42E\cdot X_E\left[(\frac{t\cdot X_t}{D_n\cdot X_D})^{2.5}\right]/ \\ \left[(1-\nu_1\nu_2\cdot X_{\nu_1}X_{\nu_2})^{0.75}\times \\ \left\{(\frac{L}{D_L\cdot X_D})-0.45(\frac{t\cdot X_t}{D_n\cdot X_D})^{0.5}\right\}\right]$$

$$(9)$$

for unstiffened long shells

$$P_B(\cdot)=\frac{2E\cdot X_E}{(1-\nu_1\nu_2 X_{\nu_1}X_{\nu_1})}\left[\frac{t\cdot X_t}{D_n\cdot X_D}\right]^3$$

$$(10)$$

where ν_1, ν_2 = poisson's ratio in major and minor direction ; D_n, L = outside diameter and length between stiffeners of cylindrical shell, respectively.

It may be noted that in the case of unstiffened long shells, both ASME and JRPS code formulas become identical.

3. RELIABILITY ANALYSIS

As shown above, since the theoretical buckling limit state model of laminated composite cylindrical shells becomes extremely complicated function, it can not be expressed in analytical nonlinear functions but only in implicit forms. Therefore, obviously this kind of limit state function may not be effectively tackled by the conventional second moment methods such as MFOSM(Mean First Order Second Moment Method) or AFOSM(Advenced First Order Second Moment Method)(Ang and Tang, 1984) because of extreme nonlinearity and implicit nature of the limit state function. Therefore, more elaborate and some advanced reliability methods may have to be used(Schuëller, 1987; Shinozuka, 1983). Of course, when applied loads are also random variables under gernernalized loadings, the FERM(Finite Element–Based Reliability Method)(Liu and Kiureghian, 1989) may be the most effective approach which could handle complicated implicit limit state function. However, when the loading is considered as deterministic or when the response gradient is insensitive to the variation of resistance parameters, it seems that the ISM(Importance Sampling Method) is one of the most effective approach as a practical analysis tool. Also note that the safety index β can be practically evaluated as $\beta = -\varPhi^{-1}(P_F)$. In the paper, an improved IST algorithm (Cho and Kim, 1991) is used for the reliability analysis of the proposed model for the evaluation of the buckling reliability of laminated composite cylindrical shells.

4. NUMERICAL RESULTS AND DISCUSSION

The material used for the numerical application is a typical E–Glass/Epoxy. Unfortunately, the statistical data base for composite material properties are not available in public, and thus only limited data of composite material uncertainties can be obtained from the references (AFML, 1977; ASM, 1987; Lee, 1989). A set of material uncertainties of E–glass/Epoxy lamina is summarized in Table 1.

For the investigation of the sensitivity of the buckling reliability to the variation of various stiffness parameters and uncertainties, the illustrative GFRP laminated composite shells are made to be designed by using the ASME design formulas.

Table 1. Uncertainties of E–glass/Epoxy lamina

	Mean	COV(%)
E_L	$619.0 t/cm^2$	10
E_T	$252.9 t/cm^2$	4
G_{LT}	$122.6 t/cm^2$	5
ν_{LT}	0.23	10

(Fiber volume content, $V_f = 0.72$)

4.1 Applicability of Design Strength Formulas

As shown in the previous section, the theoretical appoach for the evaluation of the buckling strength and reliability of laminated composite cylindrical shells are too complicated to use in practice. Thus, the applicability of the codified strength formulas is investigated by comparing the theoretical buckling strength with the ASME and JRPS formulas. Fig.3. comparatively shows the plots of buckling strength P_{cr} vs. the slenderness parameter D/t evaluated, respectively, by the theoretical formula and the ASME and JRPS formulas for ring–stiffened shells and unstiffened long shells. It can be definitely seen that there exist no significant differences between the theoretical P_{cr} curve and the JRPS code strength curves in the case of stiffened shells although the ASME code strength curve shows some significant deviation in the range of D/t = 30~100. However, for unstiffened shells, the code formulas render identical results and, compared to the theoretical strength, show some minor deviation but provide consistently conservative results. This clearly indicates

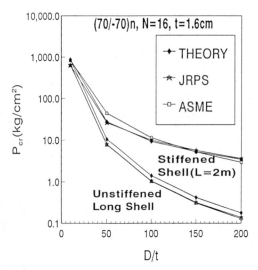

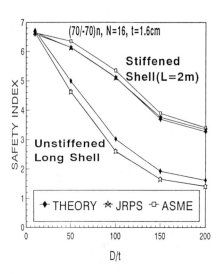

Fig.3. Comparison of critical buckling strength between different formulas : P_{cr} vs. D/t

Fig.4. Comparison of buckling reliability between different limit state models : β vs. D/t

that rather than the complicated theoretical formula, the practical ASME and JRPS code formulas may be preferably used for the practical evaluation of the buckling strength and reliability of laminated composite cylindrical shells in practice. This argument is evidently supported, as shown in Fig.4, by the comparative results of the buckling reliability evaluated by each different limit state function corresponding to the ASME and JRPS code strength formulas. Fig.4 graphically shows the deviations of β vs. D/t plots by the codified limit state models from the theoretical one. In the case of stiffened shells, it can be seen that the differences in β values between the theoretical and the JRPS are almost negligible, but also in the corresponding case with the ASME the differences are relatively insignificant with less than 0.15 at worst. However, in the case of unstiffened shells, it may be noted that both code formulas provide indentical results, as expected, and, when compared to the theoretical β, they show some conservative underestimated value with differences less tnan 0.3 at most.

In essence, it may be stated that in most cases the JRPS formula provide better results compared to the ASME, and shows almost identical or some minor conservative deviation compared to the theoretical formula, which strongly indicates that the JRPS formula may be successfully used in practice in the development of the reliability based design criteria or for the safety evaluation of composite cylindrical shells.

4.2 Comparison between FOSM and IST Methods

In order to demonstrate the computational efficiency of the ISM (Importance Sampling Mehtod) for the kind of highly nonlinear complicated reliability problems, the computational results obtained by using FOSM methods(Ang and Tang, 1984) are compared with those of the ISM(Cho and Kim, 1991). Fig.5 comparatively shows the calculated results obtained by the Lind−Hasofer's AFOSM, the Cornell's MFOSM and the ISM. It may be observed that the results by FOSM methods too widely deviate from those of the ISM,

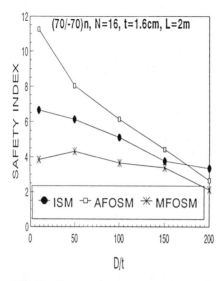

Fig.5. Comparison between FOSM and ISM: β vs. D/t

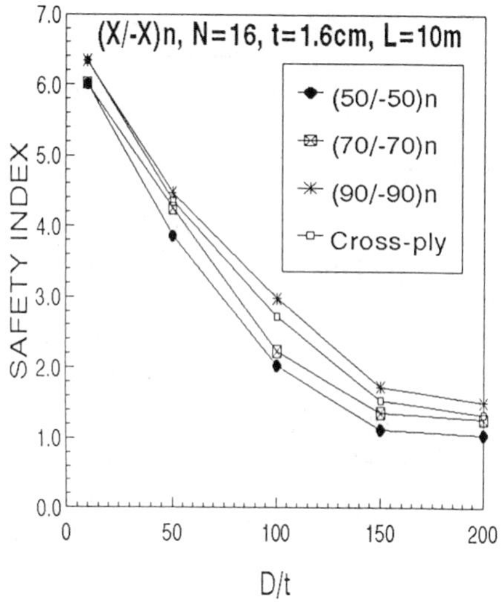

Fig.6. Buckling reliability to the variation of lamination : β vs. D/t

respectively. The accuracy of the results by the ISM was proven by crude Monte Carlo simulations. Therefore, it may be stated, as expected, that the conventional FOSM methods could not effectively tackle the extremely complicated nonlinear limit state functions for the buckling reliability analysis of laminated composite cylindrical shells. Therefore, more generalized advanced and elaborate first or second order reliability methods (Schuëller, 1987; Shinozuka, 1983) or some efficent sampling techniques such as the ISM proposed in the paper or some other sampling methods(Melchers, 1990; Moses and Fu, 1988; Ditlevsen and Bjerager, 1989) may have to be used for accurate reliability analysis.

4.3 Effect of Lamination and Fiber Orientation

The effects of the variations of lamination and fiber orientations on the reliability of the laminated composite cylindrical shells are comparatively shown in Fig.6 based on four different lamination with (50/−50)n, (70/−70)n, (90/−90)n angle−ply and cross ply. For those representative laminations, it may

be observed that the pattern of β variation with the increase of D/t does not change, but a great deal of differences ranging 0.2~1.0 arise in the reliability evaluated by each different lamination. It may be noted that one of the most popular type of (90/−90)n angle−ply lamination usually supplemented with woven fabrics in practice provides highest reliability, whereas (50/−50)n angle−ply lamination results in lowest reliability. On the other hand, the cross−ply lamination seems also very effective if no extra manufacturing effort is involved, but its reliability is lower by 0.1~0.15 than those of (90/−90)n angle−ply lamination. Therefore, as shown in Fig.7, it may be observed that the rate of increase of the safety index to the increase of fiber angle is different depending upon the slenderness parameter, but in any cases, it may be stated that if no extra manufacturing cost is involved, the highest angle−ply lamination that can be used in practice seems most effective with maximum reliability, but if the manufacturing cost varies with lamination and fiber angle, optimum angle−ply lamination may have to be obtained by a reliability−based

758

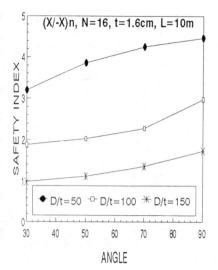

Fig.7. Sensitivity of reliability to the variation of fiber orientation : β vs. θ

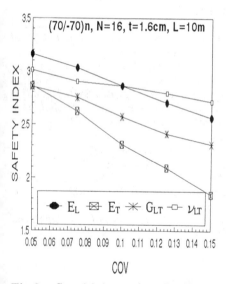

Fig.8. Sensitivity of reliability to the variation of uncertainties of several basic random variables : β vs. Ω

optimization. Although it is not shown herein, it has been also found that the effects of the number of layer or symmetry and anti symmetry of lamination on the buckling reliability are negligible.

4.4 *Effect of Uncertainties of Material Properties*

For the precise assessment of the buckling reliability, the senstivity of the reliability to the variation of the uncertainties of basic random variables E_L, E_T, G_{LT}, and ν_{LT}, ν_{TL} have to be identified. Fig.8 comparatively shows the results of β variations to the increase of the uncertainties of the basic variables. It can be observed that the degree of sensitivity of the reliability index to the increase of the coefficient of variation Ω of each basic random variate is different from each other. It is interesting to note that the rate of β reduction to the increase of Ω of the poisson ratio, ν_{LT}, results in the least slope by $\Delta\beta/\Delta\Omega \approx 0.3/0.1$, while that of the elastic modulus E_T of the matrix materials has the most sensitive slope by $1.0/0.1$. This clearly indicates that the uncertainty of elastic modulus E_T of the matrix materials is most important and very critical for the precise assessment of buckling reliability compared to other uncertainties.

5. CONCLUDING REMARKS

In this paper, the characteristics of the buckling reliability of laminated composite cylindrical shells under deterministic external pressure are investigated based on the proposed reliability model and method. The following conclusions may be drawn from this study :

The results of buckling load and reliability evaluated by the ASME and JRPS code formulas for stiffened shells, in general, shows insignificant deviation, but those for unstiffened shell result in some conservative deviation compared to those of the theoretical buckling model. However, the JRPS formula provides better results in most cases with relatively negligible differences. Thus, the JRPS code formulas may be more successfully used in practice for the evaluation of buckling reliability and in the development of reliability-based design code.

759

· The effect of the lamination and fiber orientation on the buckling reliability is found to be very significant. And the popular type of lamination using (90/−90)n angle−ply ideally provides most effective buckling reliability, but, in reality, the optimum fiber orientation minimizing the manufacturing cost may have to be searched based on a reliability−based optimization.

· Buckling reliability is also found to be very sensitive in varying degree to uncertainties of some material properties. It may be stated that the elastic modulus of matrix materials is the single most critical uncertainty that needs precise estimation for buckling reliability assessment.

Further research on the buckling reliability incorporating imperfections and stochastic loading is being undertaken and in progress.

ACKNOWLEDGEMENT

This reaserch was supported by the KOSEF(Korea Science and Engineering Foundation). The writers gratefully acknowledge this support.

REFERENCES

AFML 1977. *Advanced Composite Design Guide*. Air Force Materials Laboratory, Write−Patterson Air Force Base, Third Ed., Vol.1, p.1.21−3.

Ang, A.H−S & W.H. Tang 1984. *Probability Concepts in Engineering Planning and Design*. John Wiley & Sons, Inc., New York.

ASM 1987. *Engineered Material Handbook, Composites*. ASM International, Vol. 1., U.S.A.

ASME 1992. *Fiber-Reinforced Plastic Pressure Vessels*. ASME Boiler and Pressure Vessel Code, Sec. X., The American Society of Mechanical Engineers, New York, N.Y. 10017.

Cho, H.N. & Kim, I.S. 1991. Importance Sampling Technique for Practical System Reliability Analysis of Bridge Structures. *Proc. of the US-Korea-Japan Trilateral Seminar:* 87~100. Honolulu, 21−24 Oct. 1991.

Ditlevsen, O. and P. Bjerager 1989. Plastic Reliability Analysis by Directional Simulation. *Journal of Eng. Mech. ASCE* 115: 1347~1362.

JRPS 1985. *Fiber Reinforced Plastic Standards*. FRPS C001−1985, Japan Reinforced Plastic Society.

Kasuya, H. & M. Uemura 1985. Buckling of Laminated Composite Cylindrical Shells under External Pressure. *Journal of the Society of Materials Science* 34: 262~266.

Lee, G.S. 1989. *Materials of Composites*. Kwanghwa−Mon publishers: Seoul: Korea.

Liu, P.L. & A.D. Kiureghian 1989. Finite Element Reliability of Two Dimensional Continua with Geometrical Nonlinearity. *Proc. 5th ICOSSAR:* 1089~1096. San Francisco: U.S.A.

Melchers, R.E. 1990. Search−Based Importance Sampling. *Journal of Structural Safety* 9: 117~128.

Moses, F. & G. Fu 1988. Impotance Sampling in Sturctural System Reliability. *Fifth ASCE EMD/GTD/STD Specialty Conference on Probabilistic Mechanics*: 340~343. Blacksburg: Virginia.

Schuëller, G.I. & R. Stix 1987. A Critical Appraisal of Methods to Determine Failure Probabilities. *Journal of Structural Safety* 4: 293~309.

Shinozuka, M. 1983. Basic Analysis of Structural Safety. *Journal of Structural Engineering* 109: 721~740.

Structural Safety & Reliability, Schuëller, Shinozuka & Yao (eds) © 1994 Balkema, Rotterdam, ISBN 90 5410 357 4

Probabilistic and convex models of uncertainty in buckling of structures

I. Elishakoff
Center for Applied Stochastics Research & Department of Mechanical Engineering, Florida Atlantic University, Boca Raton, Fla., USA

G.Q. Cai
Center for Applied Stochastics Research, Florida Atlantic University, Boca Raton, Fla., USA

J.H. Starnes, Jr
NASA Langley Research Center, Hampton, Va., USA

ABSTRACT: Buckling of initial imperfection sensitive structure - column on a nonlinear elastic foundation - is investigated. A criterion based on the concept of "modal buckling load" is proposed to determine which modes should be included in the analysis when the weighted residuals method is utilized to calculate the buckling load for a given initial deflection. For stochastic analysis, a random field model is suggested for the uncertain initial imperfection, and Monte Carlo simulations are performed to obtain the probability density of the buckling load and the reliability of the column. Finally, a non-stochastic, convex model of uncertainty is employed to describe a situation when only limited information is available on uncertain initial deflection, and the minimum buckling load is obtained for this model. The results from both the stochastic and the non-stochastic approaches are derived and critically contrasted.

1 INTRODUCTION

It is known that an initial geometrical imperfection in a structure, inevitably present due to the nature of manufacturing process, has significant effects on buckling problem of a structure. To investigate such effects, a typical structure is a finite column on nonlinear elastic foundation with initial deflection and subject to an axial force (Fraser 1956, Fraser and Budiansky 1969, Elishakoff 1979, Videc and Sanders 1976, Hansen and Roorda 1973, 1974). The governing equation in non-dimensional form is given by

$$\frac{d^4 u}{d\eta^4} + \alpha\gamma \frac{d^2 u}{d\eta^2} + k_1 u - k_3 u^3 = -\alpha\gamma \frac{d^2 \bar{u}}{d\eta^2} \quad (1)$$

subjected to the boundary conditions

$$u = \frac{d^2 u}{d\eta^2} = 0, \quad at \ \eta = 0 \ and \ \eta = 1 \quad (2)$$

where $\bar{u}$ is the initial deflection, u is the additional deflection due to the axial load α, k_1

and k_3 are positive constants, representing, respectively, the linear and nonlinear spring constants of the foundation, and γ is a constant, obtained from

$$\gamma = \pi^2 m_*^2 + \frac{k_1}{\pi^2 m_*^2} \quad (3)$$

in which the integer m_* is determined by

$$\min_m \left[\pi^2 m^2 + \frac{k_1}{\pi^2 m^2} \right] = \pi^2 m_*^2 + \frac{k_1}{\pi^2 m_*^2} \quad (4)$$

It should be noted that all variables in equation (1) are non-dimensionalized.

In the present paper, the buckling problem of the column, represented by equation (1) and boundary conditions (2) is investigated. The concept of "modal buckling load" is defined for the column on a linear foundation. For the case of nonlinear foundation, a criterion based on the concept of "modal buckling load" is proposed to determine which modes are the most significant

and should be retained in calculating the buckling load. For reliability study, a random field model - Fourier series with truncated normally distributed coefficients is suggested to describe a random initial deflection. Monte Carlo simulations are performed to obtain the probability density of the buckling load and the reliability of the column. In the frequently encountered case where the sufficient knowledge about the initial imperfection is absent for substantiation of the stochastic analysis, an alternative non-stochastic approach (Ben-Haim and Elishakoff 1990, Elishakoff and Ben-Haim 1990, Elishakoff 1991) is applied to model the initial deflection and to obtain the minimum buckling load. In this study, results from the stochastic and non-stochastic approaches are critically contrasted.

2 BUCKLING LOAD FOR A GIVEN INITIAL DEFLECTION

Expand the initial deflection and the additional deflection as

$$\bar{u}(\eta) = \sum_{m=1}^{\infty} \bar{\xi}_m \sin(m\pi\eta)$$
$$(5)$$
$$u(\eta) = \sum_{m=1}^{\infty} \xi_m \sin(m\pi\eta)$$

Substitution of (5) into (1) results in (Fraser 1956, Fraser and Budiansky 1969)

$$\alpha_m \xi_m - \alpha(\xi_m + \bar{\xi}_m) - \frac{s\, m_*^2 I_m}{8 m^2} = 0 \quad (6)$$

where

$$s = \frac{2k_3}{k_1 + \xi^4 m_*^4} \quad (7)$$

$$\alpha_m = \frac{\pi^2 m^2 + k_1 / (\pi^2 m^2)}{\pi^2 m_*^2 + k_1 / (\pi^2 m_*^2)} \quad (8)$$

$$I_m = \sum_{p=1}^{\infty} \sum_{q=1}^{\infty} \sum_{r=1}^{\infty} \xi_p \xi_q \xi_r [\delta_{p+q,r+m} - \delta_{|p-q|,r+m}$$
$$- \delta_{p+q,|r-m|} + \delta_{|p-q|,|r-m|} + \delta_{p,q}\, \delta_{r,m}] \quad (9)$$

in which $\delta_{p,q}$ is the Kronecker function. It is noted that $\alpha_{m*} = \min(\alpha_m) = 1$ is the buckling load for the column without initial imperfection.

If the elastic foundation is linear, then $s = 0$. It can be seen from equation (6) that if the initial deflection contains the m_* mode, namely, $\bar{\xi}_{m*} \neq 0$, then the buckling occurs always in the m_* mode, and the buckling load $\alpha_{m*} = 1$, which is the same as that of the column without initial imperfection. However, if the initial deflection contains only the kth ($k \neq m_*$) mode, the additional deflection becomes unbounded not at unity load, but at α_k. Thus, we call α_m the mth "modal buckling load". The m_*th mode is called the classical critical mode.

For a nonlinear column with initial deflection and under an axial force, an explicit expression for the additional deflection is not obtainable, and a numerical algorithm has to be used to calculate the additional deflection. Depending on the nonlinearity, the system may exhibit initial imperfection sensitivity. Fraser et al. (1956,1969) defined the buckling load α^* as

$$[d\alpha/dF]_{\alpha=\alpha_*} = 0 \quad (10)$$

where function F is defined as the end shortening of the column. According to this definition, the buckling load α_* represents a maximum load the structure can sustain, i.e. the buckling load is defined as a limit load. If the initial and additional deflections are of the forms (5), the end shortening can be obtained as

$$F = \frac{\pi^2}{4} \sum_{m=1}^{\infty} m^2 \xi_m (\xi_m + 2\bar{\xi}_m) \quad (11)$$

Equation (10) cannot be solved either analytically or numerically since equations (6)-(9) and (11) have infinite number of terms. Approximate solutions were obtained by truncating these equations and retaining the "most important modes" (Fraser 1956, Fraser and Budiansky 1969, Elishakoff 1979). It is obvious that for "weak" nonlinearity, i.e., $k_3 << k_1$, the most significant contribution to the buckling load is expected to come from the m_*th mode which is the critical mode for the corresponding linear case. In this case, an explicit relation between the buckling load and the initial deflection has been obtained (Fraser,

1956). However, if the nonlinearity is not small and/or a high accuracy for the approximate solution is required, more modes must be taken into consideration. Thus, a criterion for determining which mode is more important is desirable.

Let us first consider a specific case where $k_1 = 16\pi^4$. It can be easily determined that m. =2 and that $\alpha_1 = 2.125$, $\alpha_2 = 1$, $\alpha_3 = 1.347$, $\alpha_4 = 2.125$ and $\alpha_5 = 3.205$. This specific example will be investigated in numerical calculations throughout this study. The approximate buckling loads are calculated for the case of $\xi_1 = \xi_3 = \xi_4 = \xi_5 = 0.3$ and $\xi_2 = 0.1$, and depicted in Fig.1. Different combinations of modes are selected to carry out numerical computations. The second mode is included in all cases since it is the critical mode. It can be observed from Figure 1 that (1) if the initial deflection is small and the nonlinearity is weak, one mode approximation may give acceptable results, (2) within two mode approximation, these two modes should be taken as the second mode and the third mode, and (3) four mode approximation gives very accurate results and the contribution of the fifth mode is negligible. These results well "correlate" with the fact that $1 = \alpha_2 < \alpha_3 < \alpha_1 = \alpha_4$, namely, the closer the modal buckling load is to unity, the more significantly the corresponding mode contributes to the accurate evaluation of the buckling load. For the simplest approximation, only the critical

mode needs to be considered. However, if more accurate results are required, the magnitude of the modal bucking load can be considered as a criterion to decide whether or not the corresponding mode should be included in calculating the buckling load by Galerkin method.

3 STOCHASTIC ANALYSIS

In some cases, we can obtain statistical properties of a initial deflection by measurements or past experience. In these cases, the initial deflection should be treated as a random field. Every sample in this random field can still be described by a Fourier series. However, different samples have different coefficients in their respective Fourier series. Thus, a random initial deflection can be described by the Fourier expansion with random coefficients $\bar{\xi}_m$ ($m = 1, 2, ...$). To proceed with the reliability analysis, a knowledge of the joint probability distribution for random variables $\bar{\xi}_m$ is necessary. The normal distribution is a popular choice due to its simplicity in analysis, but it may not be appropriate for practical cases since the initial deflection can be visualized as limited in a certain range (Indeed, quality control will discard "extremely imperfect" columns). To improve this model, we propose a *truncated* normal distribution for each random variable $\bar{\xi}_m$ as follows

$$p(\bar{\xi}_m) = \begin{cases} [2b_m \, erf(\dfrac{A_m}{b_m})]^{-1} \exp(-\dfrac{\bar{\xi}_m^2}{b_m^2}), & |\bar{\xi}_m| \le A_m \\ 0, & |\bar{\xi}_m| > A_m \end{cases}$$

(12)

where $p(\bar{\xi}_m)$ is the probability density of $\bar{\xi}_m$, each A_m is a maximum possible value for the random variable $\bar{\xi}_m$, b_m are parameters, and erf (.) is the error function defined as

$$erf(x) = \int_0^x \exp(-t^2)\,dt$$

(13)

The probability density functions of the random

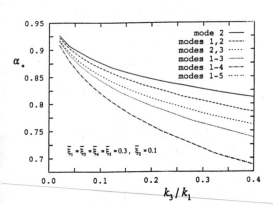

Fig.1 Buckling load α. computed by using different combination of modes

coefficients are shown in Figure 2 with $\bar{\xi}_m, A_m$ and b_m replaced by $\bar{\xi}, A$ and b for simplicity. With a given A, the probability density depends exclusively on b. A large b corresponds to a large deviation of $\bar{\xi}$. When $b^2 >> A^2$, $\bar{\xi}$ is nearly uniformly distributed, as shown by the case of $b=1$ in Figure 2.

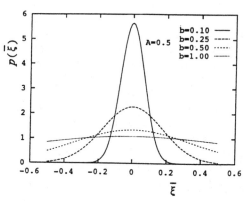

Fig.2 Probability density for truncated normally distributed random variable

The random field model is simple to simulate. For simplicity, we assume that the random coefficients $\bar{\xi}_m$ are jointly independent. The realizations of $\bar{\xi}_m$, denoted by $(\bar{\Sigma}_m)_k$, $k = 1, 2,$ ..., can then be generated by

$$(\bar{\Sigma}_m)_k = b_m\,erf^{-1}[(2\delta_k - 1)\,erf(\frac{A_m}{b_m})] \qquad (14)$$

where δ_k, $k = 1, 2, ...$ are independent random numbers uniformly distributed in $[0,1]$.

With given parameter A_m and b_m in the probability density functions $p(\bar{\xi}_m)$ for the initial deflection, Monte Carlo simulations can be carried out to obtain the probability density for the buckling load. For every sample, the Galerkin method can be applied by retaining finite modes in both initial and additional deflections according to the criterion proposed in the last section. Figure 3 shows the computed probability densities of the buckling load for the example column with $k_1 = 16\pi^4$ and $k_2 = 0.1k_1$.

Four modes were retained in computations and every mode was assumed to have the same probability distribution, namely, $A_1 = A_2 = A_3 = A_4 = A$ and $b_1 = b_2 = b_3 = b_4 = b$. Three different cases of $b = 0.1, 0.25$ and 1 were considered and 10^5 samples were calculated for each case. Figure 3 shows that with the same bound $A_m = A$ for the Fourier coefficients of the initial deflection, the distributions of the buckling load can be significantly different, depending on the distributions of the initial deflections. For a larger deviation of the initial deflection (the case $b=1$), the probability that a smaller buckling load occurs increases.

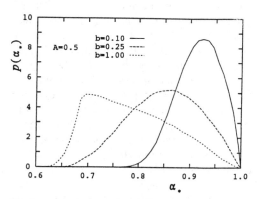

Fig.3   Probability density function of the buckling load

The reliability function for the column with a random initial imperfection, subject to a prespecified axial load α is defined as (Fraser 1956, Fraser and Budiansky 1969, Hansen and Roorda 1973 1974)

$$R(\alpha) = Prob\,[\,\alpha_* \geq \alpha\,] \qquad (15)$$

where α_* is the random buckling load.

4 NON-STOCHASTIC, CONVEX MODELING

In some cases, it is even difficult to estimate the probability distribution of the initial imperfection. In these circumstances, the stochastic approach is not applicable and a non-

probabilistic model for the initial imperfection must be adopted. We still expand the initial deflection as a Fourier series. A simple non-probabilistic model for the initial deflection is the Fourier series with coefficients varying in a hyper-cuboid set:

$$\left\{ Z(A): |\overline{\xi}_m| \le A_m \, (A_m \ge 0, m = 1, 2, ...) \right\} \quad (16)$$

where $A = \{A_1, A_2, ...\}$ is a constant vector. Our objective now is to find the minimum, least favorable limit load for all possible initial deflection $\overline{\xi} = \{\overline{\xi}_1, \overline{\xi}_2, ...\}$ belonging to the set $Z(A)$. If we design a column based on the non-stochastic approach, the minimum buckling load is the maximum value for the admissible axial load. Thus, we arrive at the alternative way of determining the admissible axial load, which can be applied to an ensemble of columns with bounded Fourier coefficients.

The buckling load is a function of $\overline{\xi}$, then the problem to find the minimum buckling load becomes an extreme value problem

$$\min_{\overline{\xi} \in Z(A)} (\alpha_*) = \min_{\overline{\xi} \in Z(A)} \psi(\overline{\xi}) \quad (17)$$

According to the method of Lagrange multipliers, we construct an auxiliary function

$$G(\overline{\xi}, A, x) = \psi(\overline{\xi}) + \sum_{m=1}^{N} \lambda_m(\overline{\xi}_m^2 - A_m^2 + x_m^2) \quad (18)$$

where N is the number of the most significant modes involved in computations, $x = (x_1, x_2, ..., x_N)$ is an auxiliary vector variable and λ_m $(m = 1, 2,..., N)$ are Lagrange multipliers. By letting $\partial G / \partial \overline{\xi}_m = 0$ and $\partial G / \partial x_m = 0$, we obtain

$$\frac{\partial \psi}{\partial \overline{\xi}_m} + 2\lambda_m \overline{\xi}_m = 0, \quad m = 1, 2, ..., N \quad (19)$$

$$\lambda_m x_m = 0, \quad m = 1, 2, ..., N \quad (20)$$

which, combined with

$$\overline{\xi}_m^2 - A_m^2 + x_m^2 = 0, \quad m = 1, 2, ..., N \quad (21)$$

constitute a set of nonlinear algebraic equations for unknown $\overline{\xi}_m$, x_m and λ_m. The solution points of equations (17)-(19) and the points at

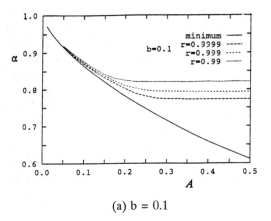

(a) b = 0.1

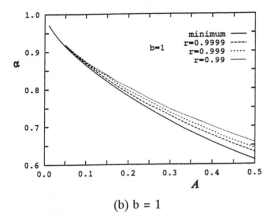

(b) b = 1

Fig.4   Comparison of admissible axial load compnted from stochastic and non-stochastic approaches

which any of the derivatives $\partial \psi / \partial \overline{\xi}_m$ does not exist have to be included as candidate points to seek the extreme values of the function $\psi(\overline{\xi})$.

Figures 4(a) and 4(b) show the minimum buckling load for the example column with the initial deflection bound $A_j = \text{constant} = A$ as a variable. Also, the admissible loads corresponding to different values of reliability are calculated from the stochastic approach and depicted in the same figures. For the case of b = 1 in Figure 4(b), namely, large deviation of the initial deflection, the minimum buckling load and the admissible load corresponding to different required reliability levels $r = 0.9, 0.99$ and 0.999 do not exhibit much difference

regardless of the magnitude of the boundary A. Therefore, design can be made based on the non-stochastic approach since it is much simpler than the stochastic one. The same situation can also be found for the case of $b = 0.1$ and small A ($A < 0.2$ in the present case), shown in Figure 4(a). However, if the deviation of the initial deflection is small and the boundary for the initial deflection is large ($b = 0.1$ and $A > 0.2$ in the present case, shown in Figure 4(a), the admissible value for the axial load obtained from the stochastic approach may be well above the minimum buckling load. In such circumstances stochastic analysis must be employed. To sum up, if probabilistic information on initial imperfection is fully specified, then the stochastic approach should be used; if the distribution is with relatively large standard deviation, non-probabilistic approach, due to its simplicity, is preferable the stochastic treatment; if the probabilistic information is unknown, then the convex modeling is one of the viable alternatives.

5 CONCLUSIONS

The comparison of the results of fully stochastic and non-stochastic analyses indicates that the design based on the non-stochastic approach is applicable for a initial deflection with a large deviation, even if probabilistic information is known. The new methodology, namely, the convex modeling must be used if the probabilistic information on initial imperfection is unavailable, for either small or large deviations.

6 ACKNOWLEDGEMENT

This study has been supported by the grant NAG-1-1310, from NASA Langley Research Center. This support is gratefully appreciated.

REFERENCES

Ben-Haim, Y. & Elishakoff, I. 1990. *Convex Models of Uncertainty in Applied Mechanics*. Elservier Science Publications, Amsterdam.

Elishakoff, I. 1979. Buckling of a stochastically imperfect finite column on a nonlinear elastic foundation - A reliability study. *Journal of Applied Mechanics* 46: 411-416.

Elishakoff, I. 1991. Convex versus probabilistic modeling of uncertainty in structural dynamics. In M. Petyt, H. F. Wolfe & C. Mei (eds.), *Structural Dynamics - Recent Advances*. Elsevier Applied Science Publishers, London.

Elishakoff, I. & Ben-Haim, Y. 1990. Dynamics of a thin cylindrical shell under impact with limited deterministic information on initial imperfections. *Journal of Structural Safety* 8: 103-112.

Fraser, W. B. 1956. Buckling of a structure with random imperfection. Ph.D Thesis, Division of Engineering and Applied Physics, Harvard University, Cambridge, Mass..

Fraser, W. B. & Budiansky, B. 1969. The buckling of a column with random initial deflections. *Journal of Applied Mechanics* 36: 232-240.

Hansen, J. S. & Roorda, J. 1973. Reliability of imperfection sensitive structures. In S. T. Ariaratnam & H. H. E. Leipholz (eds.), *Stochastic Problems in Mechanics*, 229-242. University of Waterloo, Canada.

Hansen, J. S. & Roorda, J. 1974. On a probabilistic stability theory for imperfection sensitive structures. *International Journal of Solids and Structures* 10: 341-359.

Videc, B. P. & Sanders, J. L. Jr. 1976. Application of Khas'minskii limit theorem to the buckling problem of a column with random initial deflections. *Quarterly of Applied Mathematics* 33: 422-428.

Structural Safety & Reliability, Schuëller, Shinozuka & Yao (eds) © 1994 Balkema, Rotterdam, ISBN 90 5410 357 4

An improved approach to the reliability-based design of structures subjected to bifurcation buckling

G. V. Palassopoulos
Military Academy of Greece, Athens, Greece

ABSTRACT: A new method is presented for the reliability-based design of structures which are subjected to bifurcation buckling and are sensitive to structural imperfections. The new method employs modern perturbation theory with a universal imperfection magnitude parameter as the control variable of the problem. Within this analytical framework, the well-known potential energy criterion for stability is reduced to the solution of a generalized eigenvalue problem for its dominant eigenvalue. The method is fundamentally different from Koiter's theory, on which the current practice is based, and surpasses most of its limitations. The suitability of the new method for the reliability-based design of imperfection sensitive structures in buckling is demonstrated via an example problem.

1 INTRODUCTION

The present paper addresses the problem of the reliability-based design of structures, which are subjected to bifurcation buckling and are sensitive to small structural imperfections. The interest in this problem stems mainly from two facts. First, very small imperfections, which are unavoidable in engineering practice, may cause quite significant deteriorations in the buckling strength of these structures, up to the order of 70% in certain exceptional cases (Yamaki 1984). Secondly, many structures, which are particularly important in engineering practice, fall into this category, for example most types of thin shells and many types of arches, frames and columns (Bazant and Cedolin 1991).

From the point of view of the design engineer, the problem poses two major difficulties. On the one hand, there is sensitive dependence of the response (buckling strength) on the initial data (structural imperfections), so that the response exhibits certain characteristics of chaotic systems (Thompson and Hunt 1984). On the other hand, the initial data (structural imperfections) are stochastic in nature, so that a reliability-based approach to the problem is necessitated (Shinozuka 1987).

The current practice of designing these structures is based on Koiter's theory (Budiansky 1974) and concentrates on the effect of one-mode imperfections

in the shape of the active buckling mode. However, there are two main reservations in this respect. From the analytical point of view, it is noted that Koiter's theory is strictly valid only asymptotically close to the bifurcation buckling load, while the results of the theory are frequently extrapolated in practical applications to buckling loads which are far away from this load (Palassopoulos 1991). From the experimental point of view, it is noted that the dominance of the imperfection in the shape of the active buckling mode has not been validated by experiments. On the contrary, the experimental results clearly indicate that the buckling strength does depend on a multitude of imperfection components (Arbocz et al 1987).

In the present paper, an improved method for the reliability-based design of these structures is presented, which is fundamentally different from Koiter's theory and surpasses most of its limitations. The method, which has been tentatively named "critical imperfection magnitude method", employs modern perturbation theory (Kevorkian and Cole 1981) with a universal imperfection magnitude parameter as the control variable of the problem. Within this analytical framework, the well-known potential energy criterion for stability is reduced to the solution of a generalized eigenvalue problem for its dominant eigenvalue (Gourlay and Watson 1973). The method has its roots in an old idea

(Palassopoulos and Shinozuka 1973), which has been recently fully reworked (Palassopoulos 1991, 1992), so as to cover all possible components of shape imperfections. In the present paper, the method is further extended to cover all possible sources of structural imperfections, i.e. all the geometric, material and load imperfections of the structure. The theoretical development of the method has been presented in detail in a companion paper (Palassopoulos 1993), while this paper concentrates on the application of the method to the reliability-based design of practical engineering structures.

Because of length limitations for the present paper, certain intermediate steps have been omitted from the presentation. However, it is hoped that enough detail has been included, so that the interested reader can understand the potentialities of the new method and readily rework those intermediate steps.

2 SOURCES OF IMPERFECTIONS

In what follows, the terms "perfect structure" and "actual structure" will be used repeatedly. The term "perfect structure" will denote the conventional deterministic structural model as it is routinely assumed for the purposes of structural analysis. The term "actual structure" will denote a more detailed structural model which simulates some or all of the unintended deviations of the real structure from the "perfect structure." These unintended deviations will be collectively denoted as "structural imperfections" and in engineering practice are usually related to:

(1). Small deviations of the structural shape from the intended one (shape imperfections).

(2). Small variations of other geometric properties of the structure from their nominal or assumed values (e.g. cross-sectional data, position of supports).

(3). Small variations of the material properties of the structure from their nominal values (e.g. modulus of elasticity, stiffness of supports).

(4). Small variations of the applied loads from their nominal values (e.g. change in the magnitude or the direction or the point of application of a concentrated force).

3 IMPERFECTION MAGNITUDE

It is intended to employ modern perturbation theory for the analysis of the buckling problem. Following the standard practice in such cases, the structural imperfections will be represented analytically as products of imperfection patterns, either measured or simulated, and a universal imperfection magnitude

parameter ε. Although it is not strictly necessary from a theoretical point of view, it is always possible to impose certain normalization conditions on the imperfection patterns (e.g. the root mean square magnitude of the pattern is equal to unity), so that ε is a well-defined measure of the magnitude of the imperfections. In the analysis of the buckling problem, the imperfection patterns will be considered fixed, while ε will serve as control variable.

4 ASSUMPTIONS

As already noted, the analysis of the problem will be based on modern perturbation theory and the well-known potential energy criterion for the stability of structures. Within this context, the following three assumptions are routinely made and will be also employed for the purposes of the present paper:

(1). The imperfection magnitude ε is sufficiently small, so that the first few terms of the relevant perturbation series in ε represent the exact solution with sufficient accuracy. This assumption is the usual one in all perturbation approaches and is expected to be satisfied in all but the most pathological cases of real structures.

(2). The actual structure, together with the applied loads, constitute a conservative elastic system with a single-valued total potential energy function. This assumption covers the great majority of engineering applications.

(3). The perfect structure exhibits a bifurcation buckling response, which has been fully analyzed by means of the classical stability theory. The bifurcation load ϱ_{cl} of the perfect structure will be denoted as classical buckling load and its buckling modes will be denoted as classical buckling modes. This assumption corresponds to the usual initial step in all similar analyses.

5 DISCRETIZATION OF THE POTENTIAL ENERGY

Although it is possible to present the analysis in continuum form, the discretized form is more advantageous for purposes of practical applications. With this in mind, it is further assumed that the potential energy V of the structure has been discretized in terms of a finite number M of the generalized coordinates q_j, $j = 1...M$:

$$V = v_o + a_j q_j + b_{jk} q_j q_k + c_{jkl} q_j q_k q_l + d_{jklm} q_j q_k q_l q_m + ... \tag{1}$$

In this expression and in what follows, the usual Cartesian tensor summation convention from 1 to M is implied over all the repeated subscripts j,k,l,m in the terms of a product, unless stated otherwise. All the coefficients v_o, a_j, b_{jk}, c_{jkl}, d_{jklm},..., j,k,l,m = 1...M, are, in general, functions of the applied loads and of the material and geometric properties of the structure. It is noted that these coefficients can always be selected, so as to be symmetric with respect to any permutation of their indices, i.e. so that:

$$b_{jk} = b_{kj}, \; c_{jkl} = c_{kjl} = c_{jlk} = ..., \text{ etc.}$$

The term v_o represents constant terms of the potential energy expression and has no effect on the results. The other coefficients can be assumed analytic in ε and can be expanded in the following form:

$$a_j = a_{oj} + \varepsilon a_{1j} + \varepsilon^2 a_{2j} + ... \quad (2a)$$

$$b_{jk} = b_{ojk} + \varepsilon b_{1jk} + \varepsilon^2 b_{2jk} + ... \quad (2b)$$

$$c_{jkl} = c_{ojkl} + \varepsilon c_{1jkl} + \varepsilon^2 c_{2jkl} + ... \quad (2c)$$

$$d_{jklm} = d_{ojklm} + \varepsilon d_{1jklm} + \varepsilon^2 d_{2jklm} + ... \quad (2d)$$

In general, it is a simple algebraic task to determine all the coefficients in the right-hand side of (2) for any given patterns of structural imperfections.

6 FORMULAE FOR THE CRITICAL IMPERFECTION MAGNITUDE

As already noted, the theoretical development of the "critical imperfection magnitude method" is presented in full generality in the companion paper (Palassopoulos 1993) and will not be repeated here. As proved in this paper, the resulting formulae attain their most simple form, if the space of the classical buckling modes is employed for the discretization of the potential energy. Employing this space, the well-known potential energy criterion for the inception of buckling can be reduced to the following generalized eigenvalue problem:

$$(\varepsilon \, \gamma_{1jk} + \varepsilon^2 \gamma_{2jk} + ...) (\delta q_k) = (\delta q_j) \quad (3)$$

where (δq_j), j = 1...M, are the buckling modes of the actual structure and $\gamma_{1jk}, \gamma_{2jk},...$ are readily computed components of square matrices:

$$\gamma_{1jk} = \frac{1}{\sqrt{b_{ojj} b_{okk}}} \left[-b_{1jk} + \frac{3c_{ojkl} a_{11}}{2b_{oll}} \right] \quad (4a)$$

$$\gamma_{2jk} = \frac{1}{\sqrt{b_{ojj} b_{okk}}} \left[-b_{2jk} + \frac{3c_{1jkl} a_{11}}{2b_{oll}} + \frac{3c_{ojkl} a_{21}}{2b_{oll}} \right.$$
$$- \frac{3c_{ojkl} b_{11m} a_{1m}}{2b_{oll} b_{omm}} + \frac{9c_{ojkl} c_{olmn} a_{1m} a_{1n}}{8b_{oll} b_{omm} b_{onn}}$$
$$\left. - \frac{3d_{ojklm} a_{11} a_{1m}}{2b_{oll} b_{omm}} \right] \quad (4b)$$

All that is required, both theoretically and analytically, is to compute the coefficients $\gamma_{1jk}, \gamma_{2jk},...,$ j,k = 1...M, from (4) and solve the eigenvalue problem (3) for its dominant (absolutely smallest) eigenvalue ε_{cr}. This dominant eigenvalue signifies the inception of the buckling of the actual structure and will be denoted as "critical imperfection magnitude."

7 EXAMPLE PROBLEM

The suitability of the "critical imperfection magnitude method" for the reliability-based design of structures will be illustrated by analyzing the buckling strength of a simply supported column on a linear elastic foundation. This problem has been selected because it is relatively simple and has been extensively studied, so that the new results of the method can be clearly recognized and easily verified. On the other hand, the problem is by no means trivially simple because, by adjusting the stiffness of the foundation, one can cover the cases of simple and compound bifurcation buckling and the cases of zero, low, medium and high imperfection sensitivity. Furthermore, the problem is interesting in its own right because it is widely considered as the archetype for a number of shell buckling problems.

Let L be the length of the column, E the modulus of elasticity of the material, I the moment of inertia of the cross-section, F the stiffness of the foundation per unit length, and P the applied axial force (Fig. 1). The column is assumed to be axially inextensional, and let X denote the axial coordinate of the column, while U denotes its lateral deflection. The analysis will be carried out in terms of the following non-dimensional variables:

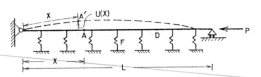

Fig. 1: Column on elastic foundation

$$x - \frac{\pi}{L} X, \quad u - \frac{\pi}{L} U, \quad \phi - \frac{L^4}{\pi^4 E_o I_o} F, \quad \varrho - \frac{L^2}{\pi^2 E_o I_o} P \quad (5)$$

where E_o and I_o are the mean values of E and I, respectively, along the axis of the column. In terms of these non-dimensional variables, the classical buckling modes of the structure are simply $\sin(jx)$, $j = 1,2,...$

For purposes of identifying all the significant sources of structural imperfections, the variability of all the field variables of the problem and the associated structural imperfections will be considered explicitly. Namely, it is assumed that E, I and ϕ vary along the axis of the column around their respective mean values E_o, I_o, and ϕ_o in the form:

$$E(x) - E_o \left[1 + \varepsilon.e(x) \right] \tag{6a}$$

$$I(x) - I_o \left[1 + \varepsilon.g(x) \right] \tag{6b}$$

$$\phi(x) - \phi_o \left[1 + \varepsilon.f(x) \right] \tag{6c}$$

where e(x), g(x) and f(x) are specified imperfection patterns for the modulus of the column material, the moment of inertia of the column cross-section and the stiffness of the foundation, respectively. At the same time, it is assumed that the axis of the column has shape imperfections in the form:

$$u_o(x) = \varepsilon.h(x) \tag{7}$$

where u_o is the value of u in the unloaded configuration and h(x) is a specified pattern for the shape imperfections.

The analysis of the problem starts with the expression for the total potential energy V of the actual structure in non-dimensional form (Palassopoulos 1991):

$$V - \frac{1}{2} \int_o^\pi \left[(1+\varepsilon e) \, (1+\varepsilon g) \left| \frac{u''}{\sqrt{1-u'^2}} - \frac{eh''}{\sqrt{1-\varepsilon^2 h'^2}} \right|^2 \right.$$

$$\left. + \phi_o (1+\varepsilon f) \, (u-\varepsilon h)^2 + 2\varrho\sqrt{1-u'^2} \right] dx \quad (8)$$

where the primes denote differentiation with respect to x.

Next, the total potential energy is discretized in the linear space of the classical buckling modes, i.e. by means of the following expansions:

$$u(x) - \sum_{j-1}^{M} q_j \sin(jx) \tag{9}$$

$$e(x) - \sum_{j-1}^{N} e_j \cos(jx), \quad f(x) - \sum_{j-1}^{N} f_j \cos(jx), \tag{10a}$$

$$g(x) - \sum_{j-1}^{N} g_j \cos(jx), \quad h(x) - \sum_{j-1}^{N} h_j \sin(jx) \tag{10b}$$

where M and N are the numbers of the modes which are taken into account in the numerical computations.

Substituting (9) and (10) into (8) and performing the algebraic operations, one obtains V, discretized in a form similar to (1) and (2), with the following coefficients:

$$a_{oj} - 0 \tag{11a}$$

$$a_{1j} - - \frac{\pi}{2} \, (j^4 + \phi_o) \, h_j \tag{11b}$$

$$b_{ojj} - \frac{\pi}{4} \left[j^4 - \varrho j^2 + \phi_o \right], \quad b_{ojk} - 0 \text{ if } j \neq k \tag{11c}$$

$$b_{1jk} - \frac{\pi}{8} \left[\phi_o f_1 + j^2 k^2 e_1 + j^2 k^2 g_1 \right] J_{jkl} \tag{11d}$$

$$b_{2jk} - \frac{\pi}{16} \, j^2 \, k^2 \, e_1 \, g_m \, K_{jklm} \tag{11e}$$

$$c_{1jkl} - - \frac{\pi}{48} \, h_m \, m^2 \, j \, k \, l \left[j \, K_{mjkl} \right.$$
$$\left. + k \, K_{mklj} + l \, K_{mljk} \right] \tag{11f}$$

$$d_{ojklm} - - \frac{\pi}{64} \, \varrho j k l m J_{jklm} + \frac{\pi}{96} j k l m \left[j k K_{jklm} \right.$$

$$\left. + j l K_{jlkm} + j m K_{jmkl} \right]$$

$$+ \frac{\pi}{96} \, j k l m \left[k l K_{kljm} \right.$$

$$\left. + k m K_{kmjl} + l m K_{lmjk} \right] \tag{11g}$$

where:

$$J_{jkl} = \delta(j-k+l) + \delta(j-k-l) + \delta(j+k+l) + \delta(j+k-l) \tag{12a}$$

$$J_{jklm} = \delta(j-k+l+m) + \delta(j-k+l-m) + \delta(j-k-l+m)$$
$$+ \delta(j-k-l-m) + \delta(j+k+l+m) + \delta(j+k+l-m)$$
$$+ \delta(j+k-l+m) + \delta(j+k-l-m) \tag{12b}$$

$$K_{jklm} = \delta(j-k+l+m) + \delta(j-k+l-m) + \delta(j-k-l+m)$$
$$+ \delta(j-k-l-m) - \delta(j+k+l+m) - \delta(j+k+l-m)$$
$$- \delta(j+k-l+m) - \delta(j+k-l-m) \tag{12c}$$

with $\delta(0) = 1$ and $\delta(j) = 0$ if $j \neq 0$.

Formulae (11) and (12) conclude the formal analysis of the problem. For any specific realization of the coefficients e_j, f_j, g_j, and h_j, $j = 1..N$, in (10), it is a simple numerical task to compute all the coefficients a_{oj}, a_{1j}, b_{1jk}, b_{2jk}, c_{ojkl}, c_{1jkl}, d_{ojklm}, $j,k,l,m = 1..N$, from (11) and (12). Then, one can readily compute γ_{1jk}, γ_{2jk},..., $j,k = 1..M$, from (4) and solve the generalized eigenvalue problem (3) for its dominant eigenvalue ε_{cr}.

8 SIGNIFICANT IMPERFECTION SOURCES AND COMPONENTS

A major problem of any design engineer, who attempts to take quantitatively into account the effect of the structural imperfections on the buckling strength, is the question of which imperfection sources and components to consider. Obviously, because of practical limitations, he cannot consider all the possible imperfection sources and components (their number is infinite).

This problem has been addressed in the companion paper (Palassopoulos 1993). Three are the main conclusions of the new theory:

a. The main result of Koiter's theory, i.e. that the imperfection in the shape of the active buckling mode is the only dominant form of imperfection, is validated only in the limit, i.e. for structures with almost zero imperfection sensitivity. Unfortunately, for all practical engineering structures with average or high imperfection sensitivity, this result of Koiter's theory cannot serve, even as rough first order approximation. Supporting evidence for this conclusion has been provided by experiments on carefully manufactured specimens. These specimens have imperfections in the shape of other buckling modes, besides the active one, and have demonstrated surprisingly high imperfection sensitivity.

b. The relative significance of the various imperfection sources and components varies between wide margins and depends strongly on the imperfection sensitivity of the structure. The higher the imperfection sensitivity of the structure, the more imperfection components must be taken into account. Indirect supporting evidence for this conclusion has been provided by a number of experiments, most notably on shells with very large intentional imperfections (and/or cut-outs), which have frequently demonstrated a surprisingly high buckling strength. It is noted that the new method provides a convenient working tool for identifying the most significant imperfection sources and components.

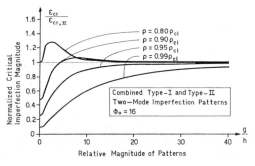

Fig. 2: Interaction between type-I and type-II imperfections

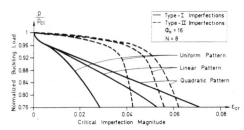

Fig. 3: Effect of the assumptions about the specific pattern of imperfections

c. The response of the structure to all the imperfections associated with the patterns $e(x)$, $f(x)$ and $g(x)$ is qualitatively similar. The imperfections associated with these patterns have been collectively denoted as type-II imperfections, in order to distinguish them from the imperfections associated with the $h(x)$ pattern, which have been denoted as type-I imperfections. It is noted that Koiter's theory considers only one-mode type-I imperfections. To the writer's knowledge, this is the first time that type-II imperfections are considered.

Figures 2 and 3 are representative of some new results, which have not been published yet. In Fig. 2, the interaction between the type-I and type-II imperfections is considered. It is observed that this interaction depends on the relative magnitude of the interacting imperfection sources, so that measurements and/or plausible assumptions for the relative magnitude of the imperfection patterns are required in applications.

In Fig. 3, the effect of the assumptions about the specific pattern of the imperfections is considered. It is noted that this effect is a mild one, even for structures with high imperfection sensitivity, so that

plausible assumptions can be readily introduced for the simulation of structural imperfections.

9 SIMULATION OF THE STRUCTURAL IMPERFECTIONS

Having identified the most significant imperfection sources and components, the next step is their simulation for the purposes of the design. The best approach, which is currently available for such a simulation, is undoubtedly the Shinozuka method of "Spectral Representation." (Shinozuka and Deodatis 1991). This line of research work is currently under implementation and the new results will be presented in a forth-coming paper.

For the limited purposes of the present paper, an alternative approach has been employed, which is based on the triangular decomposition of the covariance matrix of the imperfection components (Elishakoff 1983). Because of length limitations for the present paper, this approach will be presented, in what follows, only for the simulation of the shape imperfection pattern h(x) in (7). However, it is clear from (6) and (10), that the simulation of all the other imperfection patterns can proceed along entirely similar lines.

Following considerations presented in a previous paper (Palassopoulos 1989), the first step of this approach is the introduction of the following autocorrelation function for the imperfection pattern h(x):

$$K_h(x_1, x_2) = \overline{K}_h \exp(-\mu_h |x_1 - x_2|) \cos(2\omega_h |x_1 - x_2|) \quad (13)$$

where $\overline{K}_h$, μ_h and ω_h are suitable spectral parameters (Yaglom 1962). It is noted that the parameter $\overline{K}_h$ controls the relative magnitude of the imperfection source, the parameter μ_h the position of the maximum of $K_h(x_1,x_2)$ and the parameter ω_h its sharpness. Similar autocorrelation functions $K_e(x_1,x_2)$, $K_f(x_1,x_2)$ and $K_g(x_1,x_2)$ are employed for the other imperfection patterns e(x), f(x) and g(x), as well, with parameters $(\overline{K}_e, \mu_e, \omega_e)$, $(\overline{K}_f, \mu_f, \omega_f)$ and $(\overline{K}_g, \mu_g, \omega_g)$, respectively.

The second step of the approach is the evaluation of the elements H_{ij} of the covariance matrix of the imperfection components h_i, $i=1..N$, from the assumed autocorrelation function $K_h(x_1, x_2)$. To this end, it is noted that the following relation holds between $K_h(x_1, x_2)$ and H_{ij}, i,j=1..N, if the discretized form of the imperfection pattern h(x) in (10d) is taken into account:

$$K_h(x_1,x_2) = \sum_{i=1}^{N} \sum_{j=1}^{N} h_{ij} \sin(ix_1) \sin(jx_2) \quad (14)$$

Inverting (14), one obtains:

$$H_{ij} = \int_0^\pi \int_0^\pi K_h(x_1, x_2) \sin(ix_1) \sin(jx_2) \, dx_1 \, dx_2 \quad (15)$$

The double integral (15) can be readily evaluated numerically. A closed-form analytical expression for this integral has been given in a previous paper (Palassopoulos 1989).

The third step of the approach is the Cholesky decomposition of the matrix $[H_{ij}]$ into an upper triangular matrix $[H'_{ij}]$, i,j=1..N, and its transpose, i.e. so that:

$$H_{ij} = H'_{im} H'_{mj}, \quad i,j,m = 1..N \quad (16)$$

The forth and final step of the approach is the simulation of the structural imperfections in the form:

$$h_i = H'_{ij} z_j \quad (17)$$

where z_j, j=1..N, are standardized normally distributed independent random variables with zero mean and unit variance. It can be easily verified that the h_i from (17) have the required autocorrelation function (13).

10 RELIABILITY-BASED RESULTS

For the limited purposes of the present paper, the applied load ϱ will be considered deterministic and the reliability of the structure will be defined as the probability of the event that this deterministic load is less than the buckling load ϱ_{cr}. For any given values of the applied load ϱ and the imperfection magnitude ε, this probability is equal to the probability of the event that ε is less than the critical imperfection magnitude ε_{cr}. In order to estimate this probability, a large number of sample realizations of the imperfection patterns is produced, as described in the previous chapter, ε is computed for each such realization, and the estimated reliability function R is obtained as the ratio:

$$R = \frac{\text{Results with } e_{cr} < e}{\text{All Results}}$$

In what follows, the number of realizations at every level of ϱ and ε will be fixed to 512, which is

772

sufficient for the estimation of R with an accuracy of ±2% at a 95% confidence level (Harr 1987). Furthermore, following considerations presented earlier in chapter 8, paragraph c, the Type-II imperfections will be restricted to the imperfections of the foundation, which are described by the imperfection pattern f(x) in (6). It is understood that the imperfections associated with the patterns e(x) and g(x) do exhibit behavior qualitatively similar to f(x).

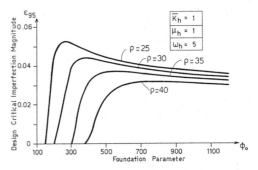

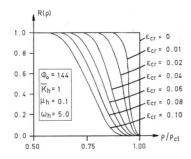

Fig. 4: Dependence of the reliability function R on the applied load ϱ and the critical imperfection magnitude ε_{cr}

Fig. 4 is representative of the dependence of the reliability function R on the applied load ϱ for various values of ε_{cr}. Such diagrams are helpful in assessing the efficiency of the simulation of the imperfections and the effect of the parameters $(\overline{K}_f, \mu_f, \omega_f)$ and $(\overline{K}_h, \mu_h, \omega_h)$ on the estimated reliability function. In this respect, it is helpful to note that the corresponding diagram for the deterministic case is the step function and that the area enclosed by the reliability function is equal to the average buckling load. It is observed that the dependence of R on ϱ is a smooth one and can be readily approximated for practical design purposes.

Figures 5 and 6 introduce the notion of the design imperfection magnitude, which is very useful for practical design purposes. Within the framework of a reliability-based design, the design imperfection magnitude ε_R may be conveniently defined as the critical imperfection magnitude at a pre-specified level R of reliability. For the purposes of the present paper, R=95% is selected and the corresponding design imperfection magnitude is denoted as ε_{95}. Figure 5 is representative of the dependence of ε_{95} on the foundation parameter ϕ_0, while Fig. 6 is representative of the dependence of ε_{95} on the

Fig. 5: Dependence of the design imperfection magnitude ε_{95} on the foundation parameter ϕ_0.

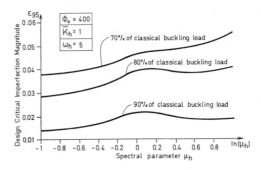

Fig. 6: Dependence of the design imperfection magnitude ε_{95} on the spectral parameter μ_h.

spectral parameters μ and ω. It is clear that such the dependencies are relatively smooth and it is possible to adopt certain simplifying assumptions for the purposes of the design. It is intended to introduce such assumptions and investigate their plausibility in a separate paper.

In closing this chapter, it is interesting to note a counter-intuitive result of the new theory for the present example problem. As Fig. 5 shows, increasing the foundation parameter ϕ_0, does not always result in an increase of the buckling strength, as predicted by classical stability theory and as is intuitively anticipated. This is due to the simultaneous increase of the imperfection sensitivity of the column. The experimental verification of the above result of the new theory would be an interesting task.

11 CONCLUSIONS

A new method, tentatively named "critical

imperfection magnitude method", has been presented for the analysis and design of imperfection sensitive structures in buckling. The new method has shed new light into this old problem, explaining and classifying the effect of the various imperfection sources and components, while demonstrating at the same time the significant limitations of Koiter's theory for practical design purposes. Furthermore, via an example problem, it was shown that the new method is well suited to the reliability-based design of imperfection sensitive structures in buckling.

Undoubtedly, much development is still necessary before the new method is introduced in practical applications. Questions like the radius of convergence of the perturbation expansions, the effect of the post-buckling response, the integration into the finite element method, the appropriate imperfection patterns and the measurement of the imperfections of real structures need to be addressed and answered before any such introduction.

However, it is firmly believed that the new method, if it is applied within the framework of the reliability-based design, provides the appropriate stepping stone for a more rational design and for the reduction of the very large empirical safety factors (of the order of 10), which are currently employed for the design of imperfection sensitive structures in buckling.

12 REFERENCES

Arbocz, J., M. Pottie-Ferry, J. Singer and V. Tvergaard 1987. *Buckling and Post-Buckling.* New York: Springer.

Bazant, Z.P. and L. Cedolin 1991. *Stability of Structures.* New York: Oxford University Press.

Budiansky, B. 1974. Theory of Buckling and Post-Buckling Behavior of Structures. *Advances in Applied Mechanics:* Chia-Shun Yih (Ed.). New York: Academic Press.

Elishakoff, I. 1983. *Probabilistic Methods in the Theory of Structures.* New York: Wiley.

Harr, M.E. 1987. *Reliability-Based Design in Civil Engineering.* New York: McGraw-Hill.

Gourlay, A.R. and G.A. Watson 1973. *Computational Methods for Matrix Eigenproblems.* New York: Wiley and Sons.

Kevorkian, J. and J.D. Cole 1981. *Perturbation Methods in Applied Mathematics.* New York: Springer.

Palassopoulos, G.V. 1989. Optimization of Imperfection Sensitive Structures. ASCE: *J. Engrg. Mech.:* Vol. 115, 8: 1663-1682.

Palassopoulos, G.V. 1991. Reliability-Based Design of Imperfection Sensitive Structures. ASCE: *J. Engrg. Mech.:* Vol. 117, 6: 1220-1240.

Palassopoulos, G.V. 1992. Response Variability of Structures Subjected to Bifurcation Buckling. ASCE: *J. Engrg. Mech.,* Vol. 118, 6: 1164-1183.

Palassopoulos, G.V. 1993. A New Approach to the Buckling of Imperfection-Sensitive Structures. ASCE: *J. Engrg. Mech.,* Vol. 119, 4: 850-869.

Palassopoulos, G.V. and M. Shinozuka 1973. On the Elastic Stability of Thin Shells. *J. Struct. Mech.,* Vol. 1 4: 439-449.

Shinozuka, M. (ed.) 1987. *Stochastic mechanics:* Vols. I, II, III. New York: Columbia University: Department of Civil Engineering and Engineering Mechanics.

Shinozuka, M. and G.Deodatis 1991. Simulation of stochastic processes by spectral representation. ASME: *Appl. Mech.* Rev., Vol. 44, 4:191-204.

Thompson, J.M.T. and G.W. Hunt 1984. *Elastic Instability Phenomena.* New York: Wiley.

Yamaki, N. 1984. *Elastic Stability of Circular Cylindrical Shells.* New York: North-Holland.

Yaglom A. M. 1962. *Stationary random functions.* New York: Dover.

13 ACKNOWLEDGMENT

This research work has been partially supported by the NATO Grant CRG 930176.

System identification

Structural Safety & Reliability, Schuëller, Shinozuka & Yao (eds) © 1994 Balkema, Rotterdam, ISBN 90 5410 357 4

Flow-induced vibrations and noises as useful signals for on-line internal state monitoring and estimating safety of PWR components

M. F. Dimentberg
Institute for Problems in Mechanics, Russian Academy of Sciences, Moscow, Russia (Presently: Worchester Polytechnic Institute, Mass., USA)

A. I. Menyailov & A. A. Sokolov
Institute for Problems in Mechanics, Russian Academy of Sciences, Moscow, Russia

V. H. Hayretdinov & A. I. Usanov
Institute Gydro Press, Podolsk, Russia

ABSTRACT: Flow-induced vibrations and noises in Pressurized Water Reactors (PWR) are usually of concern for engineers because of their potential detrimental effects such as fatigue, fretting wear that can lead to mechanical failure. In many cases however these effects may be insignificant, and moreover, measurable vibration and/or noise signals may after proper processing provide valuable information about internal state of structure. A brief survey of certain algorithms for such processing is presented, based on qualitative or so-called semiqualitative and parametrical identification, such as estimation of damping ratios structural modes estimation of structures stability margin and /or detection of restoring force non-linearity, e.g. those due to transition into impact regime of motion, from measured subcritical structure response to turbulence of the coolant flow. The results of processing vibration signals in certain PWR components are presented together with their interpretation.

1 INTRODUCTION

Development of reliable algorithms for parametric, qualitative an semiqualitative identification is of a great importance for internal state monitoring of PWR. The primary parameters of this structure change continuously during operation and it is necessary to have robust algorithms for obtaining data about these variations. The assumption of a broadband excitation permits to estimate the system's natural frequencies as those of the responses spectral densities peaks, whereas the bandwidth of any of these peaks is used commonly for estimating the corresponding modal damping ratio. This interpretation implies that linear model of structure is valid, otherwise even a small non-linearity of the structure's restoring force may lead to such a broadening of the respose spectral density peak, that the above damping estimate would become meaningless. Moreover, some additional peaks due to such a non-linearity may be misinterpreted as being due to response of other modes. So such algorithms sometimes cannot be used in monitoring diagnostics system.

New algorithms which are based on estimating of autocorrelation function of response's amplitude and exclude the above non-linear effects are considered in section 2.

The identification of restoring force non-linearity from structure's response is of importance not only by itself but for the detection of transition into impact regime of motion for system with rigid barrier(s) or due to apperance of gaps in mechanical joints. New identification algorithms of a modal restoring force nonlinearity are proposed in section 3.

In some cases there is no attempt to estimate all parameters of the assumed model of structure but only a few basic qualitative indexes are sought, e.g. estimation of a system's safety or stability margin from its measuared response. Diagnostic algorithms of determination of the change of "reserve" of stability of some structure of PWR with both external and parametric excitation, that always took place due to random oscillation of water pressure, from on-line response are presented in section 4.

2 MODAL DAMPING RATIOS ESTIMATES FOR STRUCTURE

For well separated spectral peaks in a MDOF-system response modal damping ratio may be estimated either from the spectral density of the

response,e.g. from its halfpower bandwidth, or from the decay rate of the envelope of the response autocorrelation function $K_{xx}(\tau)$ which is estimated from a response sample directly or using the "Randomdec algorithm" (Vandiver, Dunwoody, Campbell and Cook 1982):

$$D_{x_0}(\tau) = \langle x(t+\tau)|x(t) = x_0 \rangle = \left[\frac{K_{xx}(\tau)}{K_{xx}(0)} \right] x_0$$

$$(2.1)$$

From (2.1) the modal damping ratio is estimated directly from the decay rate of $D_{x_0}(\tau)$.

Another possible approach is to use a semilogarithmic plot of the envelope $\bar{K}_{xx}(\tau)$ of $K_{xx}(\tau)$. For an SDOF system, this should yield a straight line with a slope $-\alpha$, since $-\ln \bar{K}_{xx}(\tau)/K_{xx}(0) = \alpha\tau$. The value of α may be estimated not from the whole plot, but rather from that part which is actually linear, so as to exclude the influence of bandpass filtering. This procedure is not robust with respect to small non-linearity of the system's restoring force.

An appropriate procedure, which excludes this nonlinear effect almost completely from the analysis is based on exctracting the amplitude $A(t)$ (or its square $V = A^2$) of response. The correlation function of the zero-mean part $V_0(t) = V(t) - \langle V \rangle$ of the squared amplitude of the linear response is (Dimentberg, Frolov and Menyailov 1991):

$$K_{V_0 V_0}(\tau) = \langle V_0(t) \cdot V_0(t+\tau) \rangle$$
$$= K_{V_0 V_0}(0) e^{-2\alpha\tau}, \quad \alpha \geq 0, \quad (2.2)$$

So that the modal damping ratio may be estimated from the plot of $-\ln K_{V_0 V_0}(\tau)$, once again within its straight line part. Moreover, the verification of a basic linear time invariant SDOF model is also obtained if the slope of the $-\ln K_{V_0 V_0}(\tau)$ curve is twice as high as that of $-\ln \bar{K}_{xx}(\tau)/K_{xx}(0)$.

Results of this section were applied for estimating modal damping ratios of fuel bundle's elements of PWR in two orthogonal directions x and y and are presented in Table 1.

The values of modal damping ratio of a fuel element inside bundle were obtained from estimates of decay rates both of the envelopes of autocorrelation functions of the responses x(t), y(t) and of autocorrelation functions of squared amplitudes of responses (second and

Table 1. Results of estimates of the values of damping ratios of a fuel element for two cases: with one and two rigid constraints

Processes of vibration of an element a fuel bundle	Results by using $\bar{K}_{xx}(\tau)$, $\bar{K}_{yy}(\tau)$	Results by using $K_{V_{0x}V_{0x}}(\tau)$, $K_{V_{0y}V_{0y}}(\tau)$
1st mode - axis x	0.13	0.12
2nd mode - axis x	0.29	0.26
1st mode - axis y	0.19	0.17
2nd mode - axis y	0.29	0.28
1st mode - axis x	0.14	0.09
2nd mode - axis x	0.32	0.15
1st mode - axis y	0.18	0.12
2nd mode - axis y	0.21	0.16

third column in Table 1, respectively). Special control tests with a free fuel element gave results very close to values in the third column. So, the latter method, that may be named as $K_{V_0 V_0}$-algorithm, confirmed small sensitivity to the non-linearity of the system. The table shows clearly that whilst the upper bundle is linear (values in second and third columns are found to be almost the same), the lower one is not - the effect of broadening of responses' spectral densities due to non-linearity is obvious.

3 IDENTIFICATION NON-LINEARITY OF RESTORING FORCE FROM STRUCTU-RE'S RESPONSE TO RANDOM EXCITA-TION

Detection of restoring force non-linearity of a structure from its measured response is important to estimate essential changes in internal state of structure. The main model that will be under consideration is SDOF corresponding the one-mode approximation of motion of a structure:

$$\ddot{x} + 2\alpha\dot{x} + f(x) = \zeta(t) \qquad (3.1)$$

where $\zeta(t)$ is a stationary zero-mean Gaussian white noise with intensity D. In (Dimentberg and Sokolöv 1991) was considered the case of bilinear f(x) - suitable a model of a "flapping" crack. Here the case of transitions into impact

regime is considered, i.e. $f(x) = \Omega^2 (x - \Delta \, \text{sgn} \, x)$, $\text{sgn} \, x = \{+1, x>0; -1, x<0\}$.

The first approach of identification non-linearity is based on estimating the stationary probability density $w(x)$ of the measured response $x(t)$. Steady state solution of (3.1) one has (Dimentberg 1988):

$$w(x) = C \cdot \exp[-(2/D) \int f(x) dx] \quad (3.2)$$

In view of (3.2) the non-linearity of $f(x)$ may be detected most easily from that of the relation between $\ln w(x)$ and z, where $z = x^2 \text{sgn} x$, but results are usually too sensitive to the available finite sample length of $x(t)$.

Another identification method is based on the detection and analysis of the additional peaks of the response spectral density at integer multiples of the small-amplitude natural frequency Ω. This effect may described analytically by an asymptotic approach (Dimentberg 1988), provided that the so-called improved first approximation is used for $x(t)$ (Bogoliubov and Mitropolsky 1974). That is:

$$x = A \cdot \cos \Psi + \Omega^2 \sum_{k=2}^{\infty} [C_k(A)/(k^2-1)] \cos k\Psi$$

$$C_k(A) = (2\pi A)^{-1} \cdot \int_0^{2\pi} f(A \cos \Psi) \cos k\Psi d\Psi \quad (3.3)$$

We propose the following procedure of cross-correlational analysis, which is the key point of our approach. It can be seen that any coefficient of the series (3.3) with $k \geq 2$ should be correlated with the first one, which is simply A, since both contain the same slowly varying random function $A(t)$. Therefore, a non-zero cross-correlation between amplitudes of the principal harmonic and one of the higher ones is a necessary and sufficient condition for the restoring force non-linearity; the most sensitive, in general, should be the lowest-order harmonics: second one if $f(x)$ is not odd and third one if it is.

The corresponding procedure for processing measured signal $x(t)$ may be as follows. The signal is passed simultaneously through a pair of band-pass filters, one of which is tuned to Ω and the other one to 2Ω or 3Ω. The output signals of the filters are both narrow-band, and their slowly varying amplitudes are extracted. Then, the normalized cross-correlation factor g_{1k} is estimated for zero-mean parts of these amplitudes, where $k = 2$ or $k = 3$. If g_{1k} is not zero, then the restoring force of the system is non-linear indeed, whereas the value g_{1k}

provides a measure of the level of non-linearity. This factor according to equation (3.3) should be (Dimentberg, Frolov and Menyailov 1991):

$$g_{1k} = \frac{\langle (A - \langle A \rangle)(C_k(A) - \langle C_k(A) \rangle) \rangle}{\left[\langle (A - \langle A \rangle)^2 \rangle \langle (C_k(A) - \langle C_k(A) \rangle)^2 \rangle \right]^{1/2}} \quad (3.4)$$

This algorithm was applied to estimate quality of assembling mechanical joints of the reactor vessel with base after repairing works at Chmelnitskoy nuclear power plant. The "pendulums-type" oscillations of the vessel were investigated. The results are presented in Table 2.

Table 2. Results of estimates factors g_{13} for two cases: base joints of the vessel without and with defects

signals	g_{13}
1.1	0.079
1.2	0.035
2.1	0.715
2.2	0.375

On the reactor vessel two uniaxial accelerometers were located measuring the two horizontal components of the response. The signals (2.1; 2.2) were obtained originally, which implied presence of strong non-linearity in joints of the vessel with base due to gaps. After improving this assembling defects signals (1.1;1.2) were recorded. So the algorithm confirmed its very high sensitivity to non-linearity (in this case due to gaps in mechanical joints) of the system under consideration.

4 STABILITY MARGIN AND PARAMETRIC AMPLIFICATION ESTIMATES FOR STRUCTURES WITH EXTERNAL AND PARAMETRIC RANDOM OR PERIODIC EXCITATION

Starting with the case of periodic parametric excitation, consider equation of motion of a structure in one-mode approximation:

$$\ddot{x} + 2\alpha\dot{x} + \Omega^2 x(1 + \lambda \sin 2vt) = \zeta(t) \quad (4.1)$$

where $\zeta(t)$ is a stationary zero-mean broadband random Gaussian process with spectral density $\Phi_{\zeta\zeta}(\omega)$. The behavior of the system (4.1) was

considered in the vicinity of the so-called main parametric resonance, thus the detuning $\nu - \Omega$ is assumed to be small; $|\nu - \Omega|$, α, $\lambda \propto \varepsilon^2$, and $\zeta(t) \propto \varepsilon$. By rigorous mathematical derivation, using the Krylow-Bogoliubov averaging and method moments joint probability density $p(x_C, x_S)$ of the state variables ($x = x_C \cos \nu t + x_S \sin \nu t$) was obtained (Dimentberg 1988):

$$p(x_c, x_s) = \frac{1}{2\pi\sigma^2(1-\mu^2)^{1/2}} \cdot$$
$$\exp\left[-\frac{K_{ss}x_c^2 - 2K_{cs}x_cx_s + K_{cc}x_s^2}{2\sigma^4(1-\mu^2)}\right]$$
$$(4.2)$$

Here,

$$\sigma^2 = (K_{cc} + K_{ss})/2,$$
$$\mu = \left[K_{cs}^2 + (K_{cc} - K_{ss})^2/4\right]^{1/2}/\sigma^2$$

or

$$\sigma = \sigma_0^2/(1-\mu^2), \mu = 2r/(1+\delta^2)^{1/2}$$

$$r = \lambda\Omega^2/8\nu\alpha \approx \lambda\Omega/8\alpha, \quad \delta = \Delta/\alpha,$$
$$\Delta = (\nu^2 - \Omega^2)/2\nu \approx \nu - \Omega$$
$$(4.3)$$

Introducing in (4.2) two new variables - amplitude A and phase ϕ such that $x_C = A \cos \phi$, $x_S = A \sin \phi$ and integrating (4.2) over ϕ and A in the intervals $[0, 2\pi]$ and $[0, \infty]$ respectively, yields expressions for the one-dimensional probability densities of amplitude - p(A) and phase w(ϕ) :

$$p(A) = \frac{A}{\sigma^2(1-\mu^2)^{1/2}} \exp\left[-\frac{A^2}{2\sigma^2(1-\mu^2)}\right] \cdot$$
$$I_0\left(\frac{\mu A^2}{2\sigma^2(1-\mu^2)}\right)$$
$$(4.4)$$

$$w(\phi) = (1-\mu^2)^{1/2}/2\pi(1-\mu\cos 2(\phi-\Theta)) (4.5)$$

The periodic parametric excitation affects both amplitude and phase of the subcritical response, however, the major evolution of the phase properties with increasing parametric excitation amplitude is more suitable for identification purpose.

Several typical curves of the w(ϕ) are

presented in Fig.1, as result of numerical integration of equation (4.1) with some different values of the parametric excitation amplitude λ.

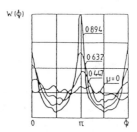

Fig. 1 Evolution of probability density of the phase response for diferent values of the parametric excitation amplitudes

It is clear from Fig. 1 that system's stability margin, as expressed in term of the nondimensional parametric excitation amplitude $\mu = \lambda/\lambda_*$, where λ_* - the critical value of the amplitude parametric excitation corresponding to the margin of parametric resonant - is closely related to the "nonuniformity" of the response phase probability density w(ϕ). This parameter can be estimated from the stationary phase probability density w(ϕ), that is defined by expression (4.5), so "phase nonuniformity factor" is:

$$\mu = \frac{W_{max} - W_{min}}{W_{max} + W_{min}} \qquad (4.6)$$

where W_{max} and W_{min} are the maximal and minimal values of w(ϕ) respectively. Moreover, this estimate also simultaneously provides a value for the parametric amplification factor from (4.6),(4.3):

$$\sigma_0^2 = \sigma^2(1-\mu^2) \qquad (4.7)$$

Here σ^2 is the averaged-over-the-period variance of the measured response, x(t), whereas σ_0 is the rms value of x(t) when $\lambda = 0$, i.e. in the absence of parametric excitation. Thus the quantity $1/(1 - \mu^2)$ is a parametric amplification factor of response, whereas its reciprocal $1 - \mu^2$ is a stability margin of a system (4.1) with near-resonant parametric excitation.

Alternative approach may be used, which doesn't require calculations of phase. In this, the inphase and quadrature components $x_C(t)$ and $x_S(t)$ of the response are extracted from the

available sample x(t). The second-order moments of x_C, x_S are now estimated, namely $K_{CC,SS} = <x_{C,S}^2>$, $K_{CS}=<x_Cx_S>$ and μ is calculated from formula (4.2):

$$\mu = \frac{[(K_{CC} - K_{SS})^2 + 4K_{CS}^2]^{1/2}}{K_{CC} + K_{SS}}$$

and

$$\sigma^2 = (K_{CC} + K_{SS})/2 \qquad (4.8)$$

Thus, an estimation of the phase may be avoided, although an accurate determination of $x_C(t)$, $x_S(t)$ may require many precautions. Nonzero mean values of x_C and/or x_S imply that some external periodic excitation is also present.

This approach has been used in the processing of vibration signal during a "cold" startup, or essentially hydraulic tests, of a reactor unit at the nuclear power plant. Spectral analysis of vibration signal from the exit pipe of a main coolant pumps (MCP) revealed several peaks. One of these peaks was very close to one-half of the shaft rotational frequency of the MCP. Formula (4.8) yielded the estimate $\mu \approx$ 0.06, which is definitely sufficiently close to zero, therefore it was deduced that the observed vibrational signal was solely due to purely random excitation (presumably a flow-induced one), and was unrelated to the MCP shaft ; the rapid decay of the filtered response autocorrelation functions also implies that no external periodic excitation was present at this frequency.

5 CONCLUSIONS

New algorithms of parametric and semiqualitative identification have been presented, which can be used effectively in practical diagnostics systems in machines and structures.

Some of the algorithms described were implemented in the diagnostic systems supplied for the Russian-built PWR-based nuclear power plants.

Other cases of semiqualitative identification with broadband random rather than periodic variations in the natural frequency and detailed discription of algorithms are described (Dimentberg, Frolov and Menyailov 1991).

6 ACKNOWLEGEMENTS

Great thanks are due to G.I. Schuëller, director of the Institute of Engineering Mechanics at the University of Innsbruck for the scientific discussion results of this work.

REFERENCES

Bogoliubov, N.N. and Mitropolsky, Ju.A. 1974. *Asymptotic method in the theory of non-linear oscillations* (in Russian). Moscow: Nauka.

Dimentberg, M.F. 1988. *Statistical dynamics of non-linear and time-varying systems.* Taunton: Research Studies Press.

Dimentberg, M.F., Frolov, K.V. and Menyailov, A.I. 1991. *Vibroacoustical diagnostics for machines and structures.* Taunton: Research Studies Press.

Dimentberg, M.F. and Sokolov, A.A. 1991. Identification of restoring force non-linearity from a systems response to a white noise excitation, *Intern. J. Nonlinear Mechanics* 26: 851-855.

Vandiver, J.K., Dunwoody, A.B., Campbell, R.B. and Cook, M.F. 1982. A mathematical basis for the random decrement vibration signature analysis technique, *J. Mech. Des.* 2: 135-147.

Structural Safety & Reliability, Schuëller, Shinozuka & Yao (eds) © 1994 Balkema, Rotterdam, ISBN 90 5410 357 4

A comparative analysis of system identification techniques for earthquake engineering applications

R. Ghanem
Civil Engineering Department, State University of New York at Buffalo, N.Y., USA

M. Shinozuka
Department of Civil Engineering and Operation Research, Princeton University, N.J., USA

ABSTRACT: System identification has long been the focus of intensive research activities. A number of excellent reviews describing the development of its theoretical and practical aspects continue to appear in the literature. In this paper, special emphasis will be placed on those system identification techniques that are suitable for on-line diagnostics. These techniques have the merit of being able to track time variation in the parameters being identified, and hence being suitable for implementation within a structural control strategy. These issues, associated with structural control, are expected to gain practical importance as more sophisticated hardware becomes available for building the control devices. Also, a comprehensive parametric study is used as the basis for a comparative review of the various system identification techniques.

1 Introduction

Associated with any system identification process, two major issues can be identified whose adequate treatment is crucial for the results to be of practical value. The first of these consists of identifying a mathematical model which is completely determined by a finite set of parameters. This model should be able to anticipate the behavior of the system within an acceptable tolerance. The second issue is to identify these parameters based on the observed behavior of the system. The identification of nonlinear relationships between the measured input and output of a given system is still in its infancy, particularly as related to structural engineering systems. This is in sharp contrast with the expected nonlinear behavior of structural systems, particularly those subjected to extreme loading conditions forcing them to sustain a certain amount of damage. Due to this differential in the state-of-the-art between identification and analysis, system identification has traditionally focused on estimating a linearized model of the structural system, which is equivalent, in some sense, to the original nonlinear system. Once a linear model for the system has been decided on, any of a number of algorithms can be used to identify the parameters which will completely determine this model. Each of these algorithms is the computational incarnation of a theoretical effort which minimizes a certain norm of a certain error. The specific error and the associated norm are, obviously, dependent on the

particular algorithm. In general, these algorithms fall into two categories, depending on whether they operate on the data in the time domain, or on the Fourier transform of the data, in the frequency, or more generally, the wave number domain. Frequency domain algorithms have been the most popular, mainly due to their simplicity, and also for historical reasons. These algorithms, however, involve averaging temporal information, thus discarding any of the details thereof. For structural systems, whose parameters are expected to degrade with time, this tradeoff of temporal information for frequency information is not always justifiable. Other, time domain algorithms, also aim at identifying models with time invariant parameters. These have the same shortcoming as the frequency domain approach mentioned above, albeit they make use of the details of the temporal variation in the measured data. Given the restriction of identifying a linear model of a given structure, it is apparent that the model which is capable of extracting the most information out of the available data, is one that identifies a linear model which is evolving in time. Such a model would be linear in its parameters which are themselves, time varying, and therefore have to be repeatedly updated through an identification process. Recent research efforts in structural system identification have aimed at further developing and validating the structural engineering applications of recursive and other time domain algorithms capable of tracking the variation in time of the parameters of a given system.

All of the above discussion relates to the identification of linear models of structural systems. As mentioned above, however, these systems can be highly nonlinear. A validation of this practice is highly desirable. This paper will report on the results from such an effort. A number of system identification algorithms are implemented in the processing of data collected in the course of laboratory experiments.

The class of structures that fall within the scope of the present investigation can be adequately modeled by the following N-dimensional equation which describes the motion of the structure,

$$\mathbf{M}\ddot{\mathbf{u}} + \mathbf{C}\dot{\mathbf{u}} + \mathbf{K}\mathbf{u} + \mathbf{g}[\mathbf{u}, \dot{\mathbf{u}}, t] = \mathbf{f}(t). \quad (1)$$

Here, $\mathbf{M}$ denotes the inertia matrix associated with the structure, $\mathbf{C}$ denotes the corresponding viscous damping matrix and $\mathbf{K}$ the stiffness matrix. Furthermore, the vector $\mathbf{f}(t)$ denotes the externally applied forces, and $\mathbf{g}[\mathbf{u}, \dot{\mathbf{u}}]$ is a vector whose components are nonlinear functions of the structural displacement $\mathbf{u}$ and its first derivative $\dot{\mathbf{u}}$. The measurement process has usually associated with it a certain level of noise, leading to the following observation equation which relates the observation vector at the ith observation time interval to the response vector at that instant,

$$\mathbf{y}_i = \mathbf{H}\mathbf{v}_i + \mathbf{e}_i, \quad \mathbf{v}_i^T = \{\mathbf{u}_i \;\; \dot{\mathbf{u}}_i \;\; \ddot{\mathbf{u}}_i\} \quad (2)$$

In the above equation, $\mathbf{H}$ is a matrix which reflects the location of the measurement devices in relation to the structural nodes, and the associated amplification or attenuation factors, and $\mathbf{e}_i$ is a vector denoting the measurement noise and is usually assumed to be a zero-mean Gaussian white noise. The vector $\mathbf{v}_i$ incorporates the displacement vector, as well as the velocity and the acceleration vectors. This is indicative of the fact that either of these response quantities may be monitored depending on the specific application. Also, the discrete form of the equation is commensurate with the form of data retrieval and storage used in practical applications. The structural system identification problem can then be stated as follows: to infer about the parameters of the model used to represent the system using noise corrupted observations of the response and the associated input.

Alternatively, the identification problem can be cast completely in the observation space, in terms of the observed input and output, without any reference to the underlying mechanics or the associated differential equation. This approach provides an algorithm which permits forecasts of the response of the structure that are compatible, in some sense, with measured past input and output. In the important case of a linear dependency, the functional relationship can be conveniently written as,

$$\mathbf{y}_i = \boldsymbol{\theta}_i^T \mathbf{x}_i + \mathbf{e}_i \quad (3)$$

where $\boldsymbol{\theta}_i$ is a matrix of the coefficients in the linear regression, and

$$\mathbf{x}_i^T = [\,\mathbf{y}_{i-1}, \ldots, \mathbf{y}_{i-k}, \mathbf{f}_i, \ldots, \mathbf{f}_{i-l}\,]. \quad (4)$$

From a knowledge of the coefficients in the difference equation (3), the modal parameters of an equivalent linear system can be recovered. An equivalence can thus be established between formulations based on equation (1) and equation (3). This equivalence, however, involves the implicit assumption that the motion of the structural system is governed by a linear differential equation of a particular form.

Of all the system identification techniques reported in this paper, only the extended Kalman filter deals directly with the differential equation model of the structural system. It also provides for the nonlinear behavior of the structure. All the other techniques start by identifying a linear prediction model as in equation (3), from which the modal parameters are subsequently obtained.

2 Experimental Results

All the identification techniques discussed in this paper have a sound theoretical basis and provide estimates that are optimal in a well defined sense. Important factors that differentiate those techniques from each others include the sensitivity of the convergence with respect to both the initial guess and the specific data being analysed, as well as the computational effort required to compute the estimated parameters. The order of importance of these factors depends to a large extent on the specific context to which system identification is being applied. For example, the computational effort may be crucial for on-line identification while being quite irrelevant for off-line applications. The experimental results detailed in this section are meant to tackle these specific issues by applying the various algorithms described in the previous section to experimental data.

Two sets of experiments provided acceleration time histories for the verification of the above parameter estimation algorithms. The experiments involved a three-story and a five story steel building models. The three-story building model was excited

by three different base motions. These consisted of the N-S component of the ground acceleration corresponding to the El-Centro earthquake, a sine sweep input, and a white noise input. The five-story model was excited with the El-Centro ground acceleration, and a white noise acceleration. In both experiments, accelerometers measured the structural response at all floor levels. Digital band-pass filters conditioned the acceleration time histories after digital data acquisition. Filtering the low frequency components is especially important in time-domain analyses since experimental acceleration bias errors are physically meaningless in structural vibrations. A detailed presentation and analysis of the results is presented elsewhere (Ghanem et.al 1991).

Except for the extended Kalman filter, all the parameter estimation techniques described in previous section involve two stages. In a first stage, the parameters of a linear prediction model are computed. These represent the regression coefficients of each new observation on previous observations. In the second stage, these coefficients are used to obtain approximations to the modal parameters of a linear differential equation model of the structure. It is emphasized that this second stage involves assumptions that cannot necessarily be inferred from the measured data.

Each of the estimation algorithms was implemented using each of the data sets obtained from the experiments. Each of the algorithms were run in turn on combinations of two measured records. The first one was the acceleration measured at the base of the structure, while the second one consisted of the acceleration at one of the floor levels. This way, the system parameters of the three-story model was identified using three different sets of data, while those of the five-story structure were identified using five sets.

2.1 Recursive Least Squares Algorithms

The recursive least squares algorithm was implemented on the data as described above. The modified least squares algorithms as described in (Ghanem et.al, 1991) were also implemented. These consisted of an exponential window and a truncated exponential window being superimposed on the observed records. In order to provide a comprehensive analysis of the effects of these windows on typical engineering data, a parametric study was carried out by varying the parameter controlling the exponential decay of the window.

Typical results for the five-story building model associated with the unmodified recursive least squares are shown in Figure (1). This figure shows the evolution of a typical estimated parameter

corresponding to the El-Centro input motion as more observations are being processed. The parameters estimated from some of the observed records failed to reach a steady state value by the end of the measurement period. The extent of the ensuing error can only be assessed by investigating the capability of the resulting model at predicting the behavior of the system. For the three-story building model, very good convergence was achieved by the coefficients associated with the white noise input, while very poor convergence was observed in connection with the sine-sweep input.

The exponential window algorithm was implemented on the same data. Values of the parameter α, controlling the rate of decay of the exponential window, equal to 0.7, 0.8, 0.9, and 0.99 were tried. Only the case corresponding to a value of α of 0.99 resulted in meaningful estimates. Other values of α resulted in estimated parameters that exhibited very large and frequent variation, and will therefore be omitted from the present discussion. An important observation can be made concerning the results associated with the exponential window. Specifically, it is noted that the effect on the first few observations is a desirable smoothing of the estimates, which deteriorates for later observations. With that in mind, a variant of the algorithm was implemented whereby the exponential window was used only for a fraction of the observations. In this case, one fourth of the data at the beginning of each record was processed through an exponential window with a value for the parameter α equal to 0.99.

The processing of only an initial block of the data through the exponential window had a substantial positive effect on the results as can be seen in Figure (2). The fluctuations have disappeared from all the estimates, except for the sine-sweep excitation in the three-story building model. Also, the monotic trend in the estimates has been reduced substantially, thus indicating that the bias associated with the least squares estimation technique has been substantially reduced.

2.2 Recursive Instrumental Variable Algorithms

The recursive instrumental variable algorithms described in (Ghanem et.al, 1991) were implemented. The first algorithm involved an unfiltered instrumental variable series. The coefficients of the linear prediction model identified in this fashion (Figure (3)) exhibited a pronounced transient behavior which was indicative of either a nonlinear relationship between the input and output series, or a deficient instrumental variable series which was incapable of identifying the parameters of the model.

The use of a filtered instrumental variable series in the identification algorithm resulted in a substantial improvement in the behavior of the coefficients. The algorithm consists of using as the instrumental variable series the series corresponding to the input motion after passing it through an auxiliary filter, with a parameter γ, so that it approximates the real output of the system, uncorrupted by measurement noise. A parametric study was performed by varying the value of the parameter γ in the auxiliary filter. A typical result from this analysis pertaining to the five story model is shown in Figure (4). A clear observation from this analysis related to the sensitivity of the estimation process to values of γ. Indeed, for certain combinations involving a specific value of γ and a set of measured records, the estimation process diverged. For other such combinations, the estimated parameters of the prediction model reached their stationary values at an early stage in the estimation process. Also, it was observed that the suitable value of γ was not the same for a given input motion. It depended both on the particular input motion used as well as on the particular floor level from which the measurements were obtained. Based on these observations, this parameter estimation technique does not seem fit for on-line identification, since it requires pre-tuning the auxiliary filter to the given data. However, in an off-line context, the results obtained using this technique feature a number of desirable properties such as the stability of the coefficients.

2.3 Maximum Likelihood Estimation

The program LINEARID was used in obtaining the results in this section. LINEARID is a program that implements parameter identification algorithms for multi-output systems. The program provides, in addition to the maximum likelihood technique, for least squares estimation and instrumental variable estimation. However, only results pertaining to the maximum likelihood estimation capability of the program are reported herein. LINEARID requires as many input records as the number of degrees of freedom to be identified. Therefore, only a single run was required on each of the two building models investigated. The output from the program consists of estimates of the matrices $\mathbf{M^{-1}K}$, $\mathbf{M^{-1}C}$, and $\mathbf{M^{-1}F}$, where $\mathbf{M}$, $\mathbf{K}$, $\mathbf{C}$, and $\mathbf{F}$ denote respectively, the mass matrix, the stiffness matrix, the damping matrix, and the load vector associated with the system being analysed. It should be noted, at this point, that the mass matrix $\mathbf{M}$, refers here to the real mass matrix, and not a lumped or otherwise simplified matrix. This fact can be expected to cause unsymmetric and full matrices to be associated with

the estimated stiffness and damping. Indeed, the resulting matrices associated with the three-story model excited by a white noise input were found to be equal to

$$\mathbf{M^{-1}K} = \begin{bmatrix} 21930 & -11650 & 1056 \\ -12300 & 23920 & -12990 \\ 953.5 & -13060 & 12390 \end{bmatrix}, \quad (5)$$

$$\mathbf{M^{-1}C} = \begin{bmatrix} 2.802 & 2.193 & 2.229 \\ -0.4784 & 0.3337 & -1.307 \\ 0.5817 & 0.3494 & 1.389 \end{bmatrix}. \quad (6)$$

The results for the three-story model corresponding to the El-Centro input motion were found to be

$$\mathbf{M^{-1}K} = \begin{bmatrix} 8222 & -11460 & 3297 \\ -12140 & 25740 & -13890 \\ 998 & -11290 & 10780 \end{bmatrix}, \quad (7)$$

$$\mathbf{M^{-1}C} = \begin{bmatrix} 0.07 & 4.317 & -7.291 \\ 3.452 & 3.476 & -1.149 \\ 0.6698 & 1.234 & 0.423 \end{bmatrix}. \quad (8)$$

Note the wide discrepancy in the results, indicating a poor performance of the program for the given data. Moreover, the program failed to converge in the case of the five-story building model. Furthermore, the results obtained from this estimation procedure are not compatible with the results obtained from the other techniques used in the investigation. Specifically, the stiffness matrix cannot be directly related to the natural frequencies of the system, nor can the damping matrix be related to the modal damping ratios. However, the structure of the resulting matrices indicate the extent of cross-modal correlation and can therefore be used as an indication of the significance of an uncoupled modal analysis of the system. In addition to the above results, LINEARID was utilized to identify the dominant mode of the system present in each of the floor accelerations. Thus, the program was implemented in a single-input single-output mode, using the ground motion as input, and one of the floor accelerations as output. This was done for each of the floor accelerations, and for both the three-story model and the five-story model. In this case, the results from LINEARID were interpreted as representing the square of the natural frequencies, ω_i^2 and the damping quantity $2\xi_i\omega_i$, respectively. Accordingly, the modal parameters could be calculated from the output of the program. The results associated with the two building models are shown in Tables (1) and (2) for various input motions. The results in this case are much more

Table 1: Estimated Modal Parameters for the Three-Story Building Model using LINEARID in Single Input Mode

El Centro Input Motion		
Input Record	Frequency (Hz.)	Damping Ratio (%)
First Floor	12.87	0.18
Second Floor	14.91	0.17
Third Floor	5.02	0.44

Table 2: Estimated Modal Parameters for the Five-Story Building Model using LINEARID in Single Input Mode

El Centro Input Motion		
Input Record	Frequency (Hz.)	Damping Ratio (%)
First Floor	5.31	0.45
Second Floor	5.46	0.39
Third Floor	6.43	0.27
Fourth Floor	7.57	0.05
Fifth Floor	7.56	0.12

Table 3: Identification of the Five-Story Building Model From El-Centro Input; Extended Kalman Filter Algorithm.

El Centro; 1st Floor		
Mode	Frequency (Hz)	Damping Ratio
1	3.17	0.375
2	10.2	0.059
3	18.7	0.129
4	28.1	4.88
5	30.1	5.81

El Centro; 3rd Floor		
Mode	Frequency (Hz)	Damping Ratio
1	3.17	0.382
2	10.2	0.058
3	18.7	0.160
4	-	-
5	-	-

Table 4: Comparison of System Identification Algorithms

Identification Techniques	Required Expertise	Numerical Convergence	On-Line Potential	Initial Guess	Reliabili of Resul
Maximum Likelihood	substantial	sometimes	low	close	good
Extended Kalman Filter	substantial	sometimes	low	close	good
Recursive Least Squares	minimal	always	high	anywhere	medium
Recursive Least Squares with Exponential Window	minimal	always	high	anywhere	good
Recursive Instrumental Variable	medium	always	high	anywhere	medium
Recursive Instrumental Variable with Filter	substantial	sometimes	high	anywhere	medium

consistent than those obtained in the multi-output mode. It is observed that the results from the estimation algorithm are in the range of the two lowest natural frequencies of the structure. It is also obvious that the dominant frequency in a given measured record depends to a great extent on both the particular input motion and the particular floor level on which the measurements were obtained.

2.4 Extended Kalman Filter Estimation

The program EXKAL2 (Maruyama et.al, 1989) was used to estimate the parameters the model buildings using the extended Kalman filter algorithm. It was observed that the behavior of the estimates depended to some extent on the excitation type. Within each test, frequency and damping estimates were consistently estimated. These estimates did, however, vary between tests. Large initial covariances allowed the parameters to deviate from their initial values before converging on the values reported in Table (3). In the table, entries with a − indicate that EXKAL2 could not identify the corresponding quantity. In some cases values for a particular frequency were repeated and in other cases the values were clearly in error.

Considering the reliance of EXKAL2 upon the linear acceleration method for estimation of dynamic properties, it fared remarkably well when applied to the data from the five-story building model. In these tests the sample rate was only three times the highest natural frequency. The accuracy of the linear acceleration method deteriorates rapidly as the number of points per sinusoidal oscillation decreases. In fact, sample rates of at least five times the highest response frequency are recommended for numerical integration. Errors associated with the numerical integration of the fourth and fifth modes may have prevented accurate estimation of those modes using data from floors in which those modes do not contribute strongly to the overall response. In some cases, (the 1st and 4th floors of the white noise excitation case) the slow sample rate resulted in meaningless parameters for all floors or failure of the program to converge at all. These results are not reported. Nevertheless, the fourth and fifth modes were identified from the 1st and 2nd floors of the El Centro excitation case. Also, lower modes could be identified in a consistent fashion using data from any of the floors.

3 Conclusions

This paper reported the results from a research effort whose aim was a comprehensive treatment of system

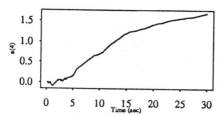

Figure 1: Recursive Least Squares; Sixth Coefficient; Output Measured at 3rd Floor.

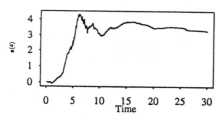

Figure 2: Recursive Least Squares with Partial Exponential Window; Sixth Coefficient; Output Measured at 3rd Floor.

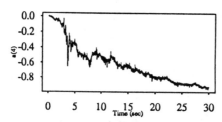

Figure 3: Recursive Instrumental Variable; Sixth Coefficient; Output Measured at 3rd Floor.

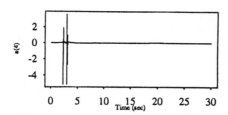

Figure 4: Filtered Recursive Instrumental Variable; $\gamma = 0.1$; Sixth Coefficient; Output Measured at 3rd Floor.

identification techniques in structural engineering applications.

The emphasis placed throughout the investigation on time domain techniques for system identification is justified by the desire to monitor the evolution in time of the identified parameters. This capability has the potential of permitting the synthesis of more meaningful damage assessment indices, as well as enhancing the reliability of adaptive schemes that may be used for on-line control of structural systems.

Table (4) summarizes the recommendations from this research while highlighting the issues that were deemed important in assessing the worthiness of each of the identification algorithms.

4   Acknowledgement

This work was partially supported by contracts No. NCEER-90-3006 under the auspices of the National Center for Earthquake Engineering Research under NSF Grant No. ECE-86-07591.

Bibliography

[1] Ghanem, R., Gavin, H., and Shinozuka, M., "Experimental Verification of a Number of Structural System Identification Algorithms," NCEER Technical Report 91-0024, 1991.

[2] Maruyama, O., Yun, C-B., Hoshiya, M., and Shinozuka, M., *Program EXKAL2 for Identification of Structural Dynamic Systems,* Technical Report NCEER-89-0014, National Center for Earthquake Engineering Research, Buffalo, NY, 1989.

[3] Yun, C-B., and Shinozuka, M., *Program LINEARID for Identification of Linear Structural Systems,* Technical Report NCEER-90-0011, National Center for Earthquake Engineering Research, Buffalo, NY, 1990.

Structural Safety & Reliability, Schuëller, Shinozuka & Yao (eds) © 1994 Balkema, Rotterdam, ISBN 90 5410 357 4

Global and local damage detection of existing structures

T. Hamamoto
Department of Architecture, Musashi Institute of Technology, Tokyo, Japan

I. Kondo
Engineering Research Institute, Sato-kogyo Co., Atugi, Japan

ABSTRACT: Structural damage may alter the stiffness and change the modal properties of structural system, such as natural frequencies, damping ratios and mode shapes. The change of structural system depends on both the sverity and location of the damage. In this study, a two-stage damage detection scheme is presented to detect structural damage using system identification techniques. The approach breaks the damage detection problem into a global damage detection at the system level and a local damage detection at the element level. The global damage detection is performed to identify the modal properties of structural system by using a multivariate autoregressive moving average (ARMA) model Then, if the global damage is judged to be significant, the damage detection of structural element is carried out to estimate the location and extent of the damage by using an inverse modal perturbation method. To show the applicability of the scheme, a series of shaking table tests on a simple three-bar truss system have been performed. Based on the analytical and experimental results, the efficiency and effectiveness of the scheme are discussed.

1 INTRODUCTION

Most building systems continuously accumulate damage during their service life. The damage of building structures may be due to natural hazards such as earthquakes and wind-storms and due to long-duration aging. For the purpose of assuring safety, it is necessary to monitor the damage as to its occurrence, its location and as to the extent of damage. Undetected damage may potentially cause more damage and eventually catastrophic structural failure. In addition, the functional requirements of recent building structures become more complex and critical. Hence, a rapid structural damage detection will be more essential. Information on the damage may be utilized to make a decision on whether repairs, partial replacement or demolition should be done after severe natural hazards or long-duration usage.

Periodic monitoring of existing building structures using vibration measurements is one of the effective nondestructive tests to identify their damage states. Especially, vibration measurements under natural ambient excitations, such as microtremors and winds, seem to be preferable. The reason is mainly due to the daily acquisition of measurement data, the easy installation of measurement instruments and the cost of artificial excitations such as a vibration generator test.

The damage detection using vibration measurements has been investigated by many investigators(Caridis and Mozakis, 1986). For the purpose of damage detection, a variety of system identification techniques have been proposed and applied to building structures(Liu and Yao, 1978). The system identification may be carried out in the frequency domain or in the time domain. The time domain approach is known to be superior to the frequency domain approach because of a good discrimination of modal properties between different modes of vibration. An autoregressive moving average (ARMA) model belongs to the time domain approach and has been successfully used to identify modal properties of structures (Gersh et al., 1976). On the other hand, an inverse perturbation method has been often used to detect the

location and extent of damage by making use of the change in modal properties between damaged and undamaged states in the fields of offshore and space technology (Sandstrom et al., 1982; Chen, 1988).

In this study, a two-stage system identification scheme (Hamamoto and Kondo, 1992) is presented and applied to detect the damage of a simple three-bar truss system. The first stage is the identification of modal properties such as natural frequencies, damping ratios and mode shapes by using a multivariate ARMA model at the system level. The second stage is the detection of the location and extent of structural damage by using an inverse modal perturbation method at the element level. Based on the analytical and experimental results, the efficiency and effectiveness of the scheme are discussed.

2 DAMAGE DETECTION SCHEME

The schematic representation of the proposed damage detection methodology for existing building structures is shown in Fig.1. The determination of a mathematical model being in undamaged state is needed to update the model in accordance with the evolution of damage. The method for improving the model is based on a parameter estimation using modal properties. The updating procedure leads to an equivalent linear model. If the damage affects a structural system to such an extent that the equivalent linear model is no longer applicable for the prediction of damage, it may still be used for the detection of damage.

Once an initial mathematical model is determined, the subsequent procedure may consist of periodical and nonperiodical monitoring. At this

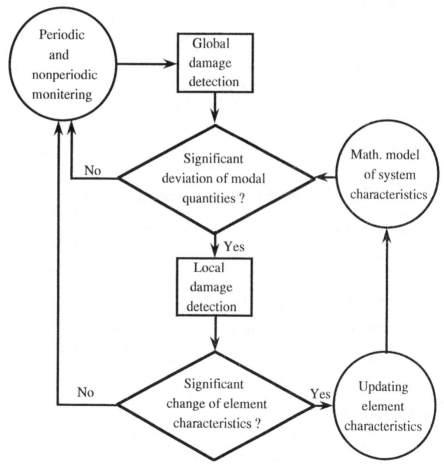

Fig.1. Damage detection scheme

stage, the question arises as to which modal property is suitable to compare it with the initial one. Natural frequencies may be the first candidate as a global damage indication, although damping ratios and mode shapes are also needed at the stage of local damage detection. A significant deviation in natural frequencies from the undamaged state indicates the possible occurrence of damage. If the significant deviation is not observed, the monitoring is repeatedly continued. If the significant deviation is observed, the change in modal properties containing damping ratios and mode shapes are translated into the change of element stiffness. This leads to a local damage detection, i.e., the determination of the location and extent of the damage. Based on the result, an initial mathematical model is replaced by the updated one. The above procedure is iterated until the damage state of structural system violates a specified damage limit state.

3 GLOBAL DAMAGE DETECTION

A multivariate autoregressive moving average (ARMA) model is used to identify the global damage of building structures. The global damage detection is divided into four steps.

Step 1. The multivariate autoregressive model, AR(p), is fitted to the time series of structural response under an ambient excitation. The multivariate AR(p) model is described by

$$\{Y_t\} - \sum_{k=1}^{p}[\Psi_k]\{Y_{t-k}\} = \{X_t\}, \qquad (1)$$

in which $\{X_t\}$ is the discrete time-series of a white noise excitation, $\{Y_t\}$ is the discete time-series of a stationary response, and $[\Psi_k]$ is the autoregressive coefficient vector. Writing down Eq.(1) from $p+1$ to M with respect to t and using a linear least squares method, $[\Psi_k]$ may be obtained.

Step 2. The multivariate AR(p) model is converted to a multivariate ARMA(n,m) model by applying an inverse function method (Pandit and Wu, 1983). The multivariate ARMA(n,m) model is described by

$$\{Y_t\} - \sum_{k=1}^{n}[\Phi_k]\{Y_{t-k}\} = \{X_t\} - \sum_{k=1}^{m}[\Theta_k]\{X_{t-k}\}, \qquad (2)$$

in which $[\Phi_k]$ is the autoregressive coefficient vector and $[\Theta_k]$ is the moving average coefficient vector.

Step 3. The transfer function of the multivariate ARMA (n,m) model is put into a parallel form realization by using a partial fraction expansion. Using Z-transform operator, the transfer function of the multivariate ARMA(n,m) model is obtained as

$$H(Z) = [[I] - \sum_{k=1}^{m}[\Theta_k]Z^{-k}][[I] - \sum_{k=1}^{n}[\Phi_k]Z^{-k}]^{-1}$$

$$= adj([I] - \sum_{k=1}^{n}[\Phi_k]Z^{-k})[[I] - \sum_{k=1}^{m}[\Theta_k]Z^{-k}]/det([I] - \sum_{k=1}^{n}[\Phi_k]Z^{-k}).$$

(3)

We first determine the poles, i.e., the roots of denominator polynomial in Eq.(3). For a stable system, the poles are all in complex-conjugate pairs and have modulus less than one, i.e., the poles are all located inside the unit circle in the complex plane. By combining the pair of terms corresponding to the pair of complex-conjugate poles, we can rewrite the transfer function of the multivariate ARMA(n,m) model as

$$H(Z) = \sum_{i=1}^{N}\left[\frac{[R_i]Z}{Z - \alpha_i} + \frac{[R_i^*]Z}{Z - \alpha_i^*}\right], \qquad (4)$$

in which N is the number of degrees-of-freedom, α_i and α_i^* are the i-th pair of complex-conjugate poles, and $[R_i]$ and $[R_i^*]$ are the i-th pair of complex conjugate residues given by

$$[R_i] = \frac{H(Z)(Z - \alpha_i)}{Z}\Big|_{Z=\alpha_i}, \qquad (5a)$$

$$[R_i^*] = \frac{H(Z)(Z - \alpha_i^*)}{Z}\Big|_{Z=\alpha_i^*}. \qquad (5b)$$

The form of Eq.(4) is called a parallel form realization of the transfer function and may be described by the corresponding mechanical model. The highest power of Eq.(3) is equal to twice the number of modes. Eq.(4) shows that the transfer function, $H(Z)$, is equal to the sum of multiple second order filters. Each second order filter represents a single-degree-of-freedom damped oscillator and corresponds to a mode of the structure. The relationships between the i-th damped natural frequency, ω_i, and damping ratio, h_i, and the i-th pair of complex-conjugate poles are given by

$$\omega_i = \sqrt{\log \alpha_i \log \alpha_i^*}/\Delta t, \qquad (6a)$$

$$h_i = -(\log \alpha_i + \log \alpha_i^*) \cdot (2\omega_i \Delta t), \qquad (6b)$$

where Δt is the time interval of discrete time-

series. The pairs of complex-conjugate residues correspond to mode shape coefficients. The absolute value of complex residues is the amplitude of mode shape coefficient. The argument of complex residues is used to judge on whether the mode shape coefficient is positive or negative.

Step 4. The change in modal properties between undamaged and damaged states are obtained. The change in the i-th natural frequency, damping ratio and mode shape between damaged and undamaged states may be given by

$$\Delta\omega_i^2 = \omega_{0i}^2 - \omega_{ti}^2, \qquad (7a)$$

$$\Delta h_i = h_{ti} - h_{0i}, \qquad (7b)$$

$$\{\Delta\psi_i\} = \{\psi_{ti}\} - \{\psi_{0i}\}, \qquad (7c)$$

in which the subscripts 0 and t denote the undamaged and damaged states, respectively, and Δ represents the change in modal properties between undamaged and damaged states. The change of mode shapes may be simply expressed in terms of the undamaged mode shapes as follows:

$$\{\Delta\psi_i\} = \sum_{j=1,j\neq i}^{n} \beta_{ij}\{\psi_{0j}\}, \qquad (8)$$

in which β_{ij} denotes the participation of the j-th mode to the change in the i-th mode.

4   LOCAL DAMAGE DETECTION

An inverse modal perturbation method is applied to detect the location and extent of the damage. The modal properties identified by a multivariate ARMA model are used to obtain the change of element stiffness as an indication of local damage. The local damage detection is divided into three steps.

Step 1. To determine the change of system characteristics in such a way that a small change in modal properties is satisfied, a linearized dynamic model of structural system is constructed at every monitoring time. A structural system in an undamaged state is described by the system of equation

$$[M_0]\{\ddot{X}\} + [C_0]\{\dot{X}\} + [K_0]\{X\} = 0, \qquad (9)$$

in which $[M_0]$, $[C_0]$ and $[K_0]$ represent the mass, damping and stiffness matrices, respectively, and $\{X\}$ is a vector of structural displacement. At some later time, the structure is damaged in one or more locations and the resulting equation of motion becomes

$$[M_t]\{\ddot{X}\} + [C_t]\{\dot{X}\} + [K_t]\{X\} = 0. \qquad (10)$$

During the intervening period, the stiffness and damping matrices are changed as follows:

$$[K_t] = [K_0] - [\Delta K], \quad [C_t] = [C_0] + [\Delta C]. \qquad (11)$$

The mass matrix is assumed to be unchanged throughout the service life, i.e., $[M_0] = [M_t]$.

Step 2. The relationship between the change in modal properties and the change of system stiffness is derived. Substituting Eqs.(7a,b,c) and (11) into Eq.(10), making use of usual orthogonal restrictions, and disregarding more than second order terms and damping effect, the change in modal properties may be given by the equations

$$\Delta\omega_i^2 = -\frac{1}{M_{0i}}\{\psi_{0i}\}^T[\Delta K]\{\psi_{0i}\}, \qquad (12a)$$

$$\{\Delta\psi_i\} = -\sum_{j=1,j\neq i}^{N} \frac{\{\psi_{0j}\}}{M_{0j}(\omega_{0i}^2 - \omega_{0j}^2)}\{\psi_{0j}\}^T[\Delta K]\{\psi_{0i}\}, \qquad (12b)$$

in which M_{0i} is the i-th generalized mass in the undamaged state.

Step 3. The relationship between the change in modal properties and the fractional change of element stiffness is derived. Decomposing the change of system stiffness, $[\Delta K]$, into multiple changes of element stiffness, $[\Delta K_e]$, i.e.,

$$[\Delta K] = \sum_{e=1}^{M}[\Delta K_e], \qquad (13)$$

in which M is the total number of element. The stiffness change of each element can be expressed as

$$[\Delta K_e] = [K_{e0}]\alpha_e, \qquad (14)$$

in which $[K_{e0}]$ is the undamaged stiffness of element e and α_e represents the fractional change. Substituting Eqs.(13) and (14) into Eqs.(12a,b), we can write

$$\Delta\omega_i^2 = -\frac{1}{M_{0i}}\sum_{e=1}^{M}[\{\psi_{0i}\}^T[K_{e0}]\{\psi_{0i}\}]\alpha_e, \qquad (15a)$$

$$\{\Delta\psi_i\} = -\sum_{e=1}^{M}\left[\sum_{j=1,j\neq i}^{N} \frac{\{\psi_{0j}\}}{M_{0j}(\omega_{0i}^2 - \omega_{0j}^2)}\{\psi_{0j}\}^T[K_{e0}]\{\psi_{0i}\}\right]\alpha_e. \qquad (15b)$$

Thus, the change in modal properties is expressed by the stiffness change of each element. Eqs.(15a,b) may be solved for the fractional change, α_e, by using a least squares method.

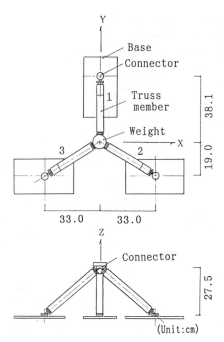

Fig.2. Experimental model.

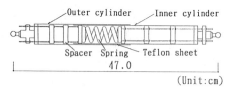

(Unit:cm)

Fig.3. Details of truss member.

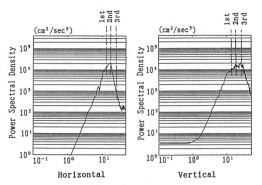

Fig.4. Power spectra of input time series and natural frequencies of experimental model

Table 1. Undamaged and damaged states.

Member		1	2	3
Case0	F.C.	0.0	0.0	0.0
	kg/cm	40.3	19.3	70.0
Case1	F.C	0.0	0.52	0.0
	kg/cm	40.3	9.25	70.0
Case2	F.C.	0.52	0.52	0.0
	kg/cm	19.3	9.25	70.0
Case3	F.C.	0.52	0.52	0.42
	kg/cm	19.3	9.25	40.3

Table 2. Modal properties of analytical model.

	Mode	Natural freq.(Hz)	Mode shape coef.		
			x	y	z
Case0	1st	12.14	1.000	0.586	0.781
	2nd	17.54	-0.039	1.000	-0.700
	3rd	23.13	-1.164	0.654	1.000
Case1	1st	8.40	1.000	0.580	0.793
	2nd	17.54	-0.032	1.000	-0.691
	3rd	23.13	-1.172	0.654	1.000
Case2	1st	8.40	1.000	0.590	0.786
	2nd	12.14	-0.028	1.000	-0.715
	3rd	23.12	-1.188	0.682	1.000
Case3	1st	8.40	1.000	0.587	0.783
	2nd	12.14	-0.034	1.000	-0.706
	3rd	17.55	-1.174	0.665	1.000

5 EXPERIMENTAL VERIFICATION

To illustrate and verify the damage detection scheme, a series of shaking table tests on a simple three-bar truss system, as shown in Fig.2, have been performed. The dimensions and details of a truss member are shown in Fig.3. A weight of 3.25kg is attached to the apex. An undamaged state(Case0) and three different damaged states(Cases 1, 2 and 3) of the truss system are assumed as shown in Table 1 in terms of the fractional change of each member stiffness. Sample time-series are generated in the horizontal and vertical directions and used as the excitations to the truss system at the base. The input time-series were originally intended as a limited white noise sequence whose frequency range extends from 0 to $50Hz$. However, the real simulated excitations are contaminated due to the limitation of shaking table performance, as shown in Fig.4. The sampling

Table 3. Effect of non-whiteness on the accuracy of identified frequencies and damping ratios.

	Mode	Target	Ideal white	Non-white
Natural	1st	8.403	8.345	8.285
freq.	2nd	12.141	12.071	12.965
(Hz)	3rd	23.124	23.043	22.868
Damping	1st	0.054	0.060	0.021
ratio	2nd	0.055	0.051	0.038
	3rd	0.076	0.072	0.042

time interval is 0.01sec. For each simulated time-series, the change in modal properties is identified by a global damage detection and the location and extent of the damage are detected by a local damage detection.

Modal properties of analytical model are shown in Table 2 for each case. The natural frequency in the first mode is only changed in Case1, those in the first and second modes are changed in Case2, and those in all modes in Case3. The change in mode shapes is not clearly observed in all cases.

The order of the AR model is taken to be 19 in all cases. The AR(19) model is converted to the ARMA(10,9) model. The natural frequencies, damping ratios and mode shapes of the trusse are identified by using two orthogonal components of acceleration response which contain the predominant response direction.

In the formulation of damage detection, the excitation was assumed to be an ideal white noise time series. However, the spectral density function is not uniform in the frequency domain in real situation. Therefore, the damage detection scheme should be still effective as long as the excitation can be regarded as a nearly white noise time series in the neighborhood of predominant modes of vibration. To test the robustness of the damage detection scheme, modal properties of analytical model are identified by using an ideal white noise time series and a sample time series (non-white noise) observed on the shaking table. Table 3 and Fig.5 compare both results identified by using both excitations. It is observed that the accuracy of natural frequencies and mode shapes is not so disturbed by a non-white noise input. On the other hand, the accuracy of damping ratios is much influenced by a non-white noise input. Therefore, for the purpose of local damage detection, it seems to be reasonable to use only the information on natural frequencies and mode

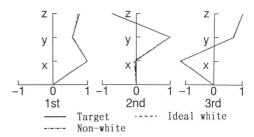

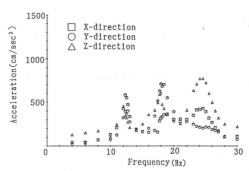

Fig.5. Effect of non-whiteness on the accuracy of identified mode shapes.

Fig.6. Resonance curves by harmonic excitation.

shapes and disregard the information on damping ratios.

6 RESULTS AND DISCUSSION

An example of the resonance curves derived from harmonic excitation tests is shown in Fig.6. The natural frequencies and damping ratios identified by using an experimental model subjected to non-white random excitations are shown in Table 4 together with measured values when subjected to harmonic excitations. The accuracy of natural fre-

Table 4. Identification of natural frequencies and damping ratios.

	Mode	Natural freq. (Hz)		Damping ratio	
		Identified	Measured	Identified	Measured
Case0	1st	11.92	12.00	0.048	0.038
	2nd	17.50	17.74	0.037	0.043
	3rd	23.69	24.20	0.169	0.062
Case1	1st	9.04	9.05	0.085	0.050
	2nd	17.80	17.50	0.038	0.034
	3rd	23.78	24.40	0.097	0.111
Case2	1st	8.97	9.05	0.067	0.044
	2nd	12.17	12.25	0.053	0.090
	3rd	23.87	24.00	0.081	0.071
Case3	1st	8.78	9.05	0.060	0.055
	2nd	11.86	12.25	0.038	0.045
	3rd	18.47	18.30	0.059	0.036

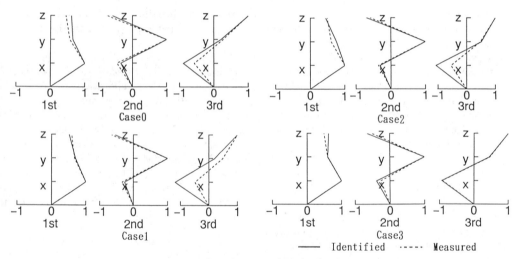

Fig.7. Identification of mode shapes.

quencies is quite well in comparison with that of damping ratios and mode shapes. It can be seen that the change in natural frequencies between un-damaged and damaged states depends on which member is damaged. The damage of members 1, 2 and 3 affects the second, first and third modes of vibration, respectively. The identified and mea-sured mode shapes are shown in Fig.7. The dif-ference may be due to the backlash and play at the joint of experimental model. More elaborate experiment seems to be necessary, because the

change in mode shapes between undamaged and damaged states is quite small.

The local damage detection is carried out by using different combinations of modal properties for each damaged state. The fractional change of member stiffnesses between undamaged and dam-aged states is shown in Table 5 for both cases in-cluding and neglecting mode shapes. Even when only natural frequencies are available, a local dam-age detection seems to be possible. The location and extent of damage may be detected by using

Table 5. Identification of member stiffness reduction

Damaged state	F.C.	Settled value	Selected modes (incl.& negl. mode shapes)							
			1		1.2		1.3		1.2.3	
			Incl.	Negl.	Incl.	Negl.	Incl.	Negl.	Incl.	Negl.
Case1	α_1	0.000	———	0.012	-0.035	-0.035	0.313	0.012	-0.028	-0.028
	α_2	0.520	———	0.459	0.456	0.456	0.448	0.456	0.457	0.457
	α_3	0.000	———	0.011	0.040	0.049	-0.086	-0.085	-0.085	-0.085

			1		2		1.2		1.2.3	
			Incl.	Negl.	Incl.	Negl.	Incl.	Negl.	Incl.	Negl.
Case2	α_1	0.520	-2.351	0.096	0.547	0.547	0.545	0.567	0.573	0.573
	α_2	0.520	0.458	0.463	0.850	0.850	0.406	0.415	0.521	0.521
	α_3	0.000	-1.090	0.096	-0.191	0.174	0.296	0.089	-0.065	-0.065

			3		1.3		2.3		1.2.3	
			Incl.	Negl.	Incl.	Negl.	Incl.	Negl.	Incl.	Negl.
Case3	α_1	0.520	———	0.045	0.571	0.370	0.506	0.494	0.431	0.431
	α_2	0.520	———	0.069	0.389	0.421	1.015	-0.005	0.459	0.459
	α_3	0.420	———	0.505	0.468	0.495	0.367	0.543	0.471	0.471

solely the first mode information for the case with a single damaged member (Case1), both the first and second mode information for the case with two damaged members (Case2), and all mode information for the case with three damaged members (Case3). It is occasionally observed that inclusion of mode shape information disturbs the accuracy of a local damage detection. More accurate identification of mode shapes will be required to achieve satisfactory results.

7 CONCLUSIONS

A damage detection scheme has been developed for existing structures. The scheme is comprised of two stages: a global damage detection and a local damage detection. The global damage is identified in terms of modal properties of structural system by using a multivariate ARMA model. The local damage is detected in terms of the stiffness reduction of each member by using an inverse modal perturbation method. Based on the analytical and experimental results, the following conclusions are obtained.

1. The proposed two-stage damage detection scheme is a promising method to detect the global and local damage of existing structures systematically.

2. The identification of damping ratios is sensitive to the frequency content of input motion, although the identification of natural frequencies and mode shapes is robust against the frequency content.

3. The identificaion of natural frequencies is not so affected by joint mechanism of experimental model, although the identification of mode shapes is very sensitive to joint mechanism.

4. Careful selection of the modal properties identified at the stage of global damage detection is essential to accurately detect the local damage.

5. More than one location of damage may be detected by using natural frequencies only, although the extent of damage will be more accurately predicted if we can improve the identification of mode shapes.

REFERENCES

Carydis, P. and Mouzakis, H.P. 1986 Small amplitude vibration measurements of buildings undamaged, damaged and repaired after earthquakes. Earthquake Spectra, Vol.2, No.3:515-535.

Chen, J.C. 1988. On-orbit damage assessment for large space structures. AIAA Journal, Vol.26, No.9:1119-1126.

Gersch, W., Taoka, G.T. and Liu, R. 1976. Structural system parameter estimation by two-stage least-squares method. J. of Engineering Mechanics, ASCE, Vol.102, No.5:883-889.

Hamamto, T. and Kondo, I. 1992. Damage detection of existing building structures using two-stage system identification. Theoretical and Applied Mechanics, Univ. of Tokyo Press, Vol.41:147-157.

Liu, S.C. and Yao, J.T.P. 1978 Structural identification concept. Proc.ASCE, Vol.104, No.ST12:1845-1858.

Pandit, S.M. and Wu, S.M. 1983. Time series and system analysis with applications. John Wiley and Sons.

Sandstrom, R.E. and Anderson, W.J. 1982. Modal perturbation methods for marine structures. Trans. of the Society of Naval Architects and Marine Engineers, Vol.90:41-54.

Structural Safety & Reliability, Schuëller, Shinozuka & Yao (eds) © 1994 Balkema, Rotterdam, ISBN 90 5410 357 4

Extended Kalman filtering for identification of damping and stiffness parameters in finite element models of structural systems

T. Herrmann & H. J. Pradlwarter
Institute of Engineering Mechanics, University of Innsbruck, Austria

ABSTRACT: Extended Kalman filtering is utilized to identify stiffness and especially damping parameters in large finite element models. In the presented approach numerical problems induced by the large number of DOF are avoided by introducing complex modal analysis in the prediction and filtering equations. Thereby the algorithm is advantageous from the viewpoint of computational efficiency and storage effort. Provisions to obtain convergence will be addressed and discussed. The scheme is applied to determine the parameters of a generalized Maxwell model for describing local damping in a finite element model.

1 INTRODUCTION

The Extended Kalman filter (Jazwinski 1970) (EKF) is capable to estimate physical parameters of a mechanical system directly from measured time series of the excitation and the response quantities. This feature is important in case the identification of parameters in finite element models is the goal. EKF has been widely applied to many linear and nonlinear structural dynamic problems (see e.g. Hoshiya 1988, Loh & Tsaur 1988, Imai, Yun, Maruyama & Shinozuka 1989, Ghanem, Gavin & Shinozuka 1991). In these publications, structural systems with less than 10 degrees of freedom are treated. However, utilizing EKF for realistic structural models described by finite elements, many more degrees of freedom are required. A significant increase of the number of DOF entails generally serious problems not encountered when dealing with few DOF. The most serious one is certainly the problem of convergence.

Several concepts are available to treat larger system models by the EKF. The main idea is always to reduce the number of estimated quantities. One way is to identify only modal parameters like modal damping ratios and eigenfrequencies of the system after a modal decomposition of the system equations (Hoshiya & Saito 1984). In further

existing approaches the coefficients of ARMA representations of the system are estimated. Then modal parameters or the coefficients of the underlying differential equation (Maruyama, Aizawa & Hoshiya 1990) are recovered. Another way is to decompose the system in substructures and apply the EKF on these resulting smaller systems (Koh, See & Balendra 1991).

In this paper complex modal analysis is utilized in the Extended Kalman filter for reducing the computational effort for large systems. The resulting approach has numerically well conditioned equations which contributes to the improvement of the convergence properties of the whole scheme. As an example, the identification of damping and stiffness parameters respectively of a generalized Maxwell model, included as a local damping element in a 60 DOF finite element model, is given.

2 STATEMENT OF THE PROBLEM

The EKF estimates the optimal values of the state variables of the system based on the measured time series of all external forces and some response quantities. In EKF the unknown parameters of a model are treated formally as state variables. The result is a new state vector which includes the parameters itself. A consequence of this approach

is that the model equations are nonlinear also in case of a physical linear system. Consider a linear system

$$M\ddot{u} + C\dot{u} + Ku = g(t) \qquad (1)$$

in which M, C and K are the mass, damping and stiffness matrices of the system with dimensions $[n \times n]$ and u the associated displacement vector of dimension $[n \times 1]$. The function $g(t)$ is a deterministic excitation vector of dimension $[n \times 1]$. The matrices C and K depend upon the constant parameters Θ which shall be identified.

The state space formulation

$$B\dot{x}(t) = f[x(t)] + c(t) \qquad (2)$$

where

$$x^T = (u^T, \dot{u}^T, \Theta^T) \quad , \quad c^T(t) = (0, g^T(t), 0) \qquad (3)$$

and

$$B = \begin{pmatrix} I & 0 & 0 \\ 0 & M & 0 \\ 0 & 0 & I \end{pmatrix} \qquad (4)$$

can be established. The nonlinear system function $f[x(t)]$ which reads

$$f[x(t)] = \begin{pmatrix} 0 & I & 0 \\ -K(\Theta) & -C(\Theta) & 0 \\ 0 & 0 & 0 \end{pmatrix} \begin{pmatrix} u \\ \dot{u} \\ \Theta \end{pmatrix} \qquad (5)$$

must be linearized since the Kalman filter equations require a linear system description. For this purpose the function f will be expanded by a Taylor series of the form

$$f[x(t)] = f[\hat{x}(t)] + \left.\frac{\partial f}{\partial x}\right|_{\hat{x}(t)} (x(t) - \hat{x}(t)) \qquad (6)$$

in which $\hat{x}(t)$ is the mean value of x at the time t. Insertion of this expansion in the system equation leads to

$$B\dot{x} = A(t)x + h(t) + c(t) \qquad (7)$$

where

$$A(t) = \left.\frac{\partial f}{\partial x}\right|_{\hat{x}(t)} \qquad (8)$$

and

$$h(t) = f[\hat{x}(t)] - \left.\frac{\partial f}{\partial x}\right|_{\hat{x}(t)} \hat{x}(t) \quad . \qquad (9)$$

This equation can be solved iteratively by assuming the system matrix A to be constant in the considered time step. The discrete measurement

equation which describes the relationship between the measured values and the system state variables reads

$$y(t_k) = Tx(t_k) + n(t_k) \qquad (10)$$

in which T is a transformation matrix of dimension $[k \times n]$. $n(t_k)$ is a vector of uncorrelated white noise processes of dimension $[k \times 1]$ with covariance matrix N at time t_k.

3 MODAL PREDICTION OF STOCHASTIC RESPONSE

The prediction as well as the filtering procedure is accomplished with modal coordinates of the associated important m modes (Pradlwarter & Li 1991). Starting from equation (7) the eigenvalue problems

$$A\Phi = B\Phi\Lambda \quad \text{and} \quad A^T\Psi = B^T\Psi\Lambda \qquad (11)$$

with normalization conditions

$$\Lambda = \Psi^T A\Phi \quad \text{and} \quad I = \Psi^T B\Phi \qquad (12)$$

are solved. Φ and Ψ are the right and left hand sided complex modal matrices. The diagonal matrix Λ contains the complex eigenvalues. Then, modal coordiantes z with

$$x = \Phi z \qquad (13)$$

can be introduced in equation (7). The resulting decoupled system equations read

$$\dot{z}(t) = \Lambda z(t) + q(t) \qquad (14)$$

where

$$q(t) = \Psi^T (h(t) + c(t)) \quad . \qquad (15)$$

From equation (14) the mean $E[z]$ and covariance P_z of the response can be calculated as

$$E[z_k(t)] = e^{\lambda_k(t-t_0)}[E[z_k(t_0)] + \int_0^{(t-t_0)} q_k(t_0 + \tau)e^{-\lambda_k\tau}d\tau] \qquad (16)$$

and

$$\begin{aligned} P_{z,ij}(z(t)) &= P^\star_{z,ji}(z(t)) \\ &= e^{(\lambda_i + \lambda_j^\star)(t-t_0)} P_{z,ij}(z(t_0)) \quad .(17) \end{aligned}$$

The backward transformation of the response from modal coordinates into real space reads

$$E[x] = \Phi E[z] \quad \text{and} \quad P_x = \Phi P_z \Phi^{T\star} \quad . \qquad (18)$$

The backward transformation of covariances can be expressed in an alternate form which will be used in the filtering procedure. After solving the eigenvalue problem

$$P_z \Gamma = \Gamma \Sigma \tag{19}$$

the relation

$$P_x = S \Sigma S^T \tag{20}$$

where

$$S = \Phi \Gamma \tag{21}$$

can be established. Since the covariance matrix P_z is Hermitian, the diagonal matrix Σ contains only real eigenvalues which are identical with the variances σ_k^2 of all z_k in the uncorrelated space. Furthermore the matrix S is real valued.

4 KALMAN FILTERING EQUATIONS

In the filtering procedure, the measurements taken at discrete time steps are utilized in the EKF procedure. The underlying filtering equations can be formulated also in modal coordinates. For completeness, the equations of the EKF shall be stated briefly. They are describing the mean $\hat{x}$ and the covariance matrix P_x of the state variables after filtering:

$$\hat{x}(t_{k+1}|t_{k+1}) =$$
$$\hat{x}(t_{k+1}|t_k) + K\left[y(t_{k+1}) - T\hat{x}(t_{k+1}|t_k)\right] \tag{22}$$

where

$$K = P_x(t_{k+1}|t_k)T^T \left(T P_x(t_{k+1}|t_k)T^T + N\right)^{-1} \tag{23}$$

and

$$P_x(t_{k+1}|t_{k+1}) = (I - KT) P_x(t_{k+1}|t_k) \quad . \tag{24}$$

Equation (24) can be written also in the form (Jazwinski 1970)

$$P_x(t_{k+1}|t_{k+1}) =$$
$$(I - KT) P_x(t_{k+1}|t_k) (I - KT)^T + K N K^T \tag{25}$$

which avoids indefinitness of the filtered covariance matrix. For minimizing the numerical effort the decomposition

$$P_x(t_{k+1}|t_k) = S_0 \Sigma_0 S_0^T \tag{26}$$

in equation (20) is introduced in equation (23) and (25). The index 0 indicates the state before the

filtering procedure. The results for the so called gain matrix K and the filtered covariance matrix $P_x(t_{k+1}|t_{k+1})$ are

$$K = S_0 \Sigma_0 S_0^T T^T \left(T S_0 \Sigma_0 S_0^T T^T + N\right)^{-1} \tag{27}$$

and

$$P_x(t_{k+1}|t_{k+1}) =$$
$$S_0 \Sigma_0^{\frac{1}{2}}[(I - F^T E F)(I - F^T E F)^T]\Sigma_0^{\frac{1}{2}} S_0^T +$$
$$S_0 \Sigma_0^{\frac{1}{2}}[F^T E N E^T F]\Sigma_0^{\frac{1}{2}} S_0^T \tag{28}$$

with

$$F = T S_0 \Sigma_0^{\frac{1}{2}} \tag{29}$$

and

$$E = (T S_0 \Sigma_0 S_0^T T^T + N)^{-1} \quad . \tag{30}$$

For computing equation (28) the new eigenvalue problem

$$D\Omega = \Omega \Sigma_1 \tag{31}$$

with

$$D = (I - F^T E F)(I - F^T E F)^T$$
$$+ F^T E N E^T F \tag{32}$$

is solved. Then, the relation for the filtered covariances reads

$$P_x(t_{k+1}|t_{k+1}) = S_0 \Sigma_0^{\frac{1}{2}} \Omega \Sigma_1 \Omega^T \Sigma_0^{\frac{1}{2}} S_0^T$$
$$= S_1 \Sigma_1 S_1^T$$
$$= (S_1 \Sigma_1^{\frac{1}{2}})(S_1 \Sigma_1^{\frac{1}{2}})^T \quad . \tag{33}$$

The index 1 denotes the state after the filtering process. The resulting updating form for the covariance matrix of the state variables reduces the storage effort and ensures the symmetry and positive definitness of the matrix. This fact is important for obtaining convergence. From the computational point of view it is also advantageous, since the updating is carried out only in m modal coordinates and the involved expressions are real valued.

5 NUMERICAL RESULTS

5.1 Example

As an example the proposed approach is applied to the composite structure, shown in fig.1. The system was designed to investigate the material properties of the gasket materials used in combustion

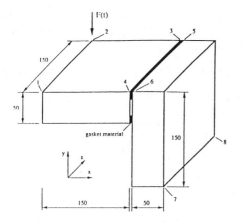

Figure 1. Composite structure (length unit [mm]).

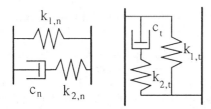

Figure 2. Continuous stiffness distribution.

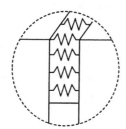

Figure 3. Generalized Maxwell models.

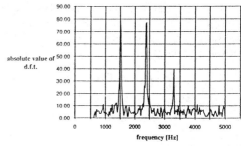

Figure 4. Discrete Fourier transformation of excitation.

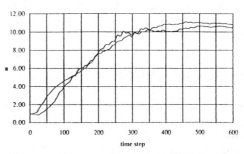

Figure 5. Factor a of c_n, c_t / 10 % RMS noise.

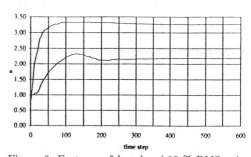

Figure 6. Factor a of $k_{n,1}, k_{t,1}$ / 10 % RMS noise.

engines. It consists of two steel plates connected by screws. Between them two strips of gasket material are located.

Each steel plate is modeled by a 24 DOF condensed finite element model. The material between the surfaces has a continuous distribution of stiffness and damping properties (s.fig.2). It is modeled by a generalized Maxwell model (s.fig.3) (Flügge 1975). This model is capable to describe a frequency dependent stiffness and a damping characteristic which shows a maximum of energy dissipation at one frequency. In each DOF of the contact surfaces a Maxwell model is introduced.

The parameters of these models are formulated in terms of four stiffness and two damping parameters in normal and tangential direction of the contact surface $(k_{n,1}, k_{n,2}, k_{t,1}, k_{t,2}, c_n, c_t)$ respectively (s.fig.3). The parameters $k_{n,2}$ and $k_{t,2}$ describe the frequency dependent parts of the stiffness. For the identification procedure the ratios $\kappa_1 = k_{n,2}/k_{n,1}$ and $\kappa_2 = k_{t,2}/k_{t,1}$ are fixed. Including the internal DOF of the Maxwell models the total number of DOF of the model is 60.

The measured time series are simulated numerically (time step length $\Delta t = 2.0 \cdot 10^{-5} s$) since the properties of the algorithm shall be investigated.

802

Table 1. Simulated and initial values of damping and stiffness parameters.

	$c_n[Ns/m^3]$	$c_t[Ns/m^3]$
$x_{i,sim}$	$2.621 \cdot 10^8$	$2.674 \cdot 10^6$
$x_{i,0}$	$2.621 \cdot 10^7$	$2.674 \cdot 10^5$
	$k_{n,1}[N/m^3]$	$k_{t,1}[N/m^3]$
$x_{i,sim}$	$1.451 \cdot 10^{11}$	$4.315 \cdot 10^{11}$
$x_{i,0}$	$6.667 \cdot 10^{10}$	$1.333 \cdot 10^{11}$

For this purpose the excitation vector was located at the first steel plate (s.fig.1). It consists of three sinoidals with fixed amplitudes and a linear time varying frequency around a center value (f_i= 1510, 2390 and 3295 Hz). The resulting time series are filtered such that it contains only components in the frequency domain between 600 and 5000 Hz (s.fig.4).

Measurements of the stationary acceleration response of the structure with 10.0% RMS noise level at the points 1-8 in x- , y- and z-direction (s.fig.3) are assumed. The measured accelerations are utilized to evaluate the corresponding displacement and velocity response due to a forward and backward Fourier transformation in the frequency domain. The properties of the measurement noise covariance matrix is chosen according to the RMS value of each reconstructed time series.

The simulation is performed utilizing the values x_{sim} for $k_{n,1}$, $k_{t,1}$, c_n and c_t listed in table 1. The κ_1 and κ_2 are fixed to $\kappa_1 = 0.411$ and $\kappa_2 = 0.220$. In the simulation and identification procedure the first five eigenmodes of the composite structure are considered. For the parameter values used for simulation the first five eigenfrequencies and modal damping ratios are $f_i = 1375, 1722, 2989, 3324, 3588$ Hz and $\xi_i = 2.39, 2.01, 1.81, 2.55, 1.67$ % respectively.

The initial values for the mean of the state variables are $u_{i,0} = 0.0$ and $\dot{u}_{i,0} = 0.0$. The starting values $x_{i,0}$ of the damping and stiffness parameters are given in table 1. The initial covariance matrix is set in modal coordinates as $P_{z,ii} = 0.01$ except the coordinates which correspond to the parameters. They are chosen as $P_{z,ii} = 1.0$.

5.2 Convergence Properties

The most critical issue in context with EKF is to obtain convergence of the algorithm (Ghanem,

Gavin & Shinozuka 1991). The convergence properties of the proposed concept was significantly influenced by the following steps.

- The scheme for the filtered covariance matrix in equation (28) and (33) ensures the positive definitness and symmetry of the covariance matrix. Especially numerically induced unsymmetry can cause divergence of the filter (Verhaegen & Van Dooren 1986).

- It is a severe disadvantage of the EKF that the filter has to reach the domain of the true parameters already within the first time steps. As a consequence the filter can diverge, because the state covariance matrix decreases very rapidly (Kumar, Yadav & Srinivas 1991). This statement holds especially in case of measurement signals which are almost exact, i.e. only slightly noisy. The behavior of the filter improves by increasing the elements of the measurement noise covariance matrix. The measurement information is seen then less reliable and the decay of variances decreases. Therefore the parameter estimation is carried out slowly and "smoother". The best results are obtained in case the noise covariances are decreased exponentially from the artificially increased level at the beginning of the filtering procedure to the level from simulation. In the considered numerical example the measurement noise level was multiplied by a factor of 10^3 in the filtering equations. Then is was decreased exponentially till the time step 300.

- Very important is an appropriate scaling of the identified parameters. Great differences between the magnitudes of parameters and state variables (in this example greater than 10^{10} due to the high frequency region) cause numerical problems in calculating the eigenvalues of covariance matrix P_z. If k_1 is a parameter which shall be identified and $\boldsymbol{K}_1$ the associated element stiffness matrix a decomposition of the form

$$\boldsymbol{K}_1 \cdot k_1 = \boldsymbol{K}_1 \cdot k_{1,0} \cdot a$$

was performed. $k_{1,0}$ is the starting value for the filtering algorithm and a is the identified factor.

Table 2. Identified correcting factors.

	c_n	c_t	$k_{n,1}$	$k_{t,1}$
$a_{i,sim}$	10.000	10.000	2.176	3.237
$a_{i,0}$	1.0	1.0	1.0	1.0
$a_{i,end}$	10.443	10.802	2.193	3.267
$r.e.[\%]$	4.4	8.0	0.8	0.9

5.3 Results of Identification

Even for a rather high noise contamination of the measured acceleration response, the relative errors (r.e.) of the identified parameters are rather small (range of 1 % for stiffness and 10 % for damping parameters) (s.tab.2). The stiffness parameters are reaching the true values very rapidly (s.fig.6). For determining the damping parameters, naturally more time steps are required to approach the values from simulation (s.fig.5).

6 CONCLUDING REMARKS

- A concept for applying the Extended Kalman filter for identification of parameters in finite element models is proposed. Numerical problems caused by the great number of DOF are avoided by introducing complex modal analysis. The identified parameters are parameters of stiffness and damping matrices, i.e. a direct identification of physical parameters is possible.

- In this paper only linear models are considered. The concept is, however, extendable to physical nonlinear models.

- Convergence is the most important requirement for Extended Kalman filtering procedures. The best convergence properties have been obtained by a measurement noise controlling strategy. In the beginning of the filtering procedure the measurement noise covariances are increased artificially. Then this level is decreased with an exponential decay to the one known from simulation.

ACKNOWLEGDGEMENT

This work was partially supported by the Austrian Industrial Research Promotion Fund (FFF) under project No. 6/636, which is gratefully acknowledged by the authors.

REFERENCES

Flügge,W. 1975. Viscoelasticity. Berlin, Heidelberg, New York: Springer Verlag.

Ghanem,R.G., Gavin,H. & Shinozuka,M. 1991. Experimental verfication of a number of structural system identification algorithms. National Center of Earthquake Engineering Research. Technical Report NCEER-91-0024.

Hoshiya,M.& Saito,E. 1984. Structural identification by Extended Kalman filter. Journal of Engineering Mechanics. Proc. of ASCE. 110,2: 1757-1770.

Hoshiya,M. 1988. Application of the Extended Kalman filter - WGI method in dynamic system identification. Stochastic structural dynamics - Progress in theory and applications. Edi. Ariaratnan et al. Elsevier Applied Science. 103-124.

Imai,H., Yun,C.B., Maruyama,O. & Shinozuka,M. 1989. Fundamentals of system identification in structural dynamics. Probabilistic Engineering Mechanics. 4,4: 162-173.

Jazwinski,A.H. 1970. Stochastic processes and filtering theory. New York, London: Academic Press.

Koh,C.G., See,L.M. & Balendra,T. 1991. Estimation of structural parameters in time domain : A substructure approach. Earthquake Engineering and Structural Dynamics. 20: 787-801.

Kumar,K., Yadav,D & Srinivas,B.V. 1991. Adaptive noise models for Extended Kalman filter. Journal of Guidance Control and Dynamics. 14,2: 475-477.

Loh,C.H. & Tsaur,Y.H. 1988. Time domain estimation of structural parameters. Engineering Structures. 10: 95-105.

Maruyama,O., Aizawa,J. & Hoshiya,M. 1990. Identification of dynamic properties of existing structures by multivariante ARMA representations. Proc. of JSCE. 416,13: 439-447.

Pradlwarter,H.J.& Li,W.L. 1991. On the computation of the stochastic response of highly non-

linear large MDOF-systems modeled by finite elements. Probabilistic engineering mechanics. 6,2:109-116.

Verhaegen,M.& Van Dooren,P. 1986. Numerical aspects of different Kalman filter implementations. IEEE Transactions on Automatic Control. AC-31,No. 10: 907-917.

Yun,C.B. & Shinozuka,M. 1980. Identification of nonlinear structural dynamic systems; Journal of Structural Mechanics, Proc. of ASCE, 1980, Vol.8, No.2, pp.187-203

Structural Safety & Reliability, Schuëller, Shinozuka & Yao (eds) © 1994 Balkema, Rotterdam, ISBN 90 5410 357 4

Parameter identification of structural systems with unmodeled dynamics

H. Imai
Department of Mechanical Engineering, Setsunan University, Osaka, Japan

H. Zui
Department of Civil Engineering, Setsunan University, Osaka, Japan

ABSTRACT: This paper deals with the problem of structural systems identification. A major difficulty in this problem is caused by the difference in model structure between the true system and the nominal model; i.e., higher modes are neglected in building the nominal model. Such an unmodeled (or neglected) dynamics results in a bias error. An identification algorithm is proposed which minimizes the effect of the unmodeled dynamics. The ellipsoidal bounding or the set membership algorithm is applied to obtain an upper bound for the model error, which can be computed based on *a priori* information including the bound for the measurement noise and the truncated impulse response sequence. An example of multi-story building is also given for illustrating the result.

1. Introduction

The problem of parameter identification of structural systems has been treated by many authors. Among them we refer to Shinozuka et al. (1982), Hac and Spanos (1990), Fassois et al. (1990), and Moustafa (1992). A major difficulty in this problem is caused by the difference in model structure between the true system and the nominal model; the true system has so many vibrational modes that higher modes are, in practice, neglected in building the nominal model. Therefore, the transfer function of the nominal model with estimated parameters will not converge, at least at some frequencies, to that of the true system even if the number of data tends to infinity. Such an error is called a bias error. It is possible that such a bias error causes spillover instability in control systems of flexible structures. The recently developed robust control theory seems most promising to overcome the difficulty, which requires not only the nominal model but also a hard bound for the bias model error.

Existing approaches to characterizing the hard bound can be classified into the following two groups.

(1) A worst case $/H_\infty$ identification: Helmicki et al. (1991) proposed an algorithm which provides estimates of frequency transfer function and their hard bound based on a finite number of noisy measurements of the plant frequency response and a hard bound for the measurement noise sequence. Several works in this direction have been reported by Gu and Khargonekar (1992), Makila (1991), and Partington (1991).

(2) Set membership or ellipsoidal bounding algorithm: Set membership algorithm computes the set of parameters consistent with the measured data, the hard bound for measurement noise and the assumed model structure (Fogel and Huang (1982), Norton (1987), Belforte et al. (1990)). Extensions of these algorithms are made to deal with the identification of unmodeled dynamics by several authors (Younce and Rohrs (1992), Kosut et al. (1992), and Wahlberg and Ljung (1992)).

This paper attempts to apply the set membership algorithm to the identification of structural systems. Especially, we follow Wahlberg and Ljung (1992). They assume the nominal model to be linear in parame-

ters, such as Laguerre model or Kautz filter, which makes it possible to apply the linear regression technique and to estimate the nominal model and the model error simultaneously. However, it seems that the Laguerre model or the Kautz filter is not convenient for the identification of structural systems because the relationship between parameters of Laguerre model or Kautz filter and those of structural systems such as damping and stiffness is not obvious. On the other hand, the parameters of ARX model can be easily transformed into structural parameters (Shinozuka et al. 1982). Therefore, in the current paper the parameters of the nominal model is identified first using ARX model, then a bound for the model error is obtained via Wahlberg and Ljung's approach.

The organization of this paper is as follows. The problem of parameter identification of single input multi output structural systems is stated in Section 2, where it is shown that the problem is amount to the identification of single input single output (SISO) systems. In Section 3, we discuss the relationship between unmodeled dynamics and bias model error. Brief review of the robust stability theory is presented in Section 4. In Section 5, an estimation problem is formulated as that of minimizing a weighted integral of squared model error, and it is shown that the problem can be solved by the output error method applied to filtered input and output data. It is shown in Section 6 that a bound for the model error can be obtained by means of Wahlberg and Ljung's approach applied to the data consisting of the output error and the simulated output. An example of multistory building is given in Section 7.

2. Mathematical model of structural systems

Most of the structural systems can be modeled by the following finite dimensional vector second order differential equation:

$$\ddot{y}(t)+J\dot{y}(t)+Ky(t)=Lu(t) \qquad (1)$$

where $y(t)$ is an n-dimensional displacement vector, $u(t)$ is a forcing function which is assumed to be a scalar, and J, K, L are coefficient matrices of appropriate dimen-

sions. The problem dealt with in this paper is to estimate the parameters J, K, and L on the basis of the sampled input-output data:

$$Z_N = \{y(i),u(i),i=1,2,...,N\} \qquad (2)$$

where

$$y(i)=y(iT) \quad \text{and} \quad u(i)=u(iT)$$

with T denoting the sampling interval.

Since the data are given at discrete time instants, it is convenient to rewrite the nominal model (1) in a discrete form appropriate for parameter identification. It is shown by Shinozuka et al. (1982) that (1) can be written in the following ARX model:

$$y(i+1) = -F_1 y(i)-F_2 y(i-1)$$
$$+G_1 u(i)+G_2 u(i-1). \qquad (3)$$

Let q denote the shift operator such that

$$q^{-1}y(i)=y(i-1).$$

Then, the transfer function of the nominal model can be written as

$$G(q, \theta) = (I+F_1 q^{-1}+F_2 q^{-2})^{-1}(G_1 q^{-1}+G_2 q^{-2})$$
$$= [G_1(q, \theta), G_2(q, \theta), ..., G_n(q, \theta)]' \quad (4)$$

where θ denotes the set of parameters to be identified, and $G_i(q, \theta)$ is the transfer function from the forcing function to the i-th output, which is written as

$$G_i(q, \theta) = \frac{N_i(q, \theta)}{D(q, \theta)} \qquad (5)$$

where

$$D(q, \theta) = |I+F_1 q^{-1}+F_2 q^{-2}|$$
$$= 1+d_1 q^{-1}+d_2 q^{-2}+...+d_{2n}q^{-2n} \qquad (6)$$

and

$$N_i(q, \theta) = n_{i1} q^{-1}+n_{i2}q^{-2}+...+n_{i,2n}q^{-2n}. \qquad (7)$$

Once the $2n\times(n+1)$ parameters $d_j, n_{ij}, i=1,2,...,n, j=1,2,...,2n$, have been identified, the parameters J, K, L can be recovered by means of the following scheme:

1. Obtain the controller canonical realization (F, G, C_i) of $\dfrac{N_i(q, \theta)}{D(q, \theta)}$ as follows:

$$F = \begin{bmatrix} 0 & 1 & 0 & \cdots & 0 \\ 0 & 0 & 1 & \cdots & \cdot \\ \cdot & \cdot & 0 & \cdots & 0 \\ \cdot & \cdot & \cdot & \cdots & 0 \\ \cdot & \cdot & \cdot & \cdots & 1 \\ \cdot & \cdot & \cdot & \cdots & \cdot \\ -d_{2n} & -d_{2n-1} & -d_{2n-2} & \cdots & -d_1 \end{bmatrix}$$

$$G = \begin{bmatrix} 0 \\ 0 \\ \cdot \\ \cdot \\ \cdot \\ \cdot \\ 1 \end{bmatrix} \quad C_i = [n_{i,2n}, \; n_{i,2(n-1)}, \; \ldots, \; n_{i1}],$$

and let

$$C = [C_1', C_2', \ldots, C_n']'$$

2. Find the continuous counterpart (A, B) of the pair (F, G). An algorithm for this is given in Shinozuka et al. (1982).

3. A coordinate transformation

$$A_o = TAT^{-1}, \; B_o = TB, \; C_o = CT^{-1}$$

with

$$T = \begin{bmatrix} C \\ CA \end{bmatrix}$$

results in

$$A_o = \begin{bmatrix} 0 & I_n \\ A_{21} & A_{22} \end{bmatrix} \quad B_o = \begin{bmatrix} B_1 \\ B_2 \end{bmatrix}$$

$$C_o = [I_n \; 0_{n \times n}].$$

It should be noted that the realization is unique, and hence we can identify

$J = -A_{21}$, $K = -A_{22}$, $L = B_2$, and $B_1 = 0$,

provided no measurement noise exists and the measured data are generated by the nominal model (1).

Our task is now to identify the parameters d_j, n_{ij}, $j=1, 2,.., n$, $i=1, 2,..., 2n$, or transfer functions $G_i(q, \theta)$. We note that the denominator polynomial is common to all $G_i(q, \theta)$'s. So, we first estimate the denominator coefficients $\{d_j, j=1,2,...,2n\}$ based on the data consisting of the input $\{u(i)\}$ and the first element $\{y_1(i)\}$ of the output; then estimate the coefficients of each numerator polynomial $N_i(q, \theta)$ based

on the data consisting of the input and the i-th output.

3. Unmodeled dynamics and bias error

It is shown in the previous section that the problem amounts to identifying each SISO transfer function $G_i(q, \theta)$ which has $4n$ unknown parameters. In view of this, we restrict our attention to SISO systems in what follows. We assume that the measured data Z_N are generated by the true system so that

$$y(i) = G^o(q)u(i), \tag{8}$$

where $G^o(q)$ denotes the transfer function of the true system. No measurement noise is assumed in this paper. Also, we define the nominal model set according to

$$\mathbf{M} = \{G(q, \theta): \theta \in \mathbf{R}^{4n}\}.$$

where

$$\theta = [d_1, d_2, , \ldots, d_{2n}, n_1, n_2, , \ldots, n_{2n}]'.$$

Let $\hat{\theta}_N$ denote an estimate for θ obtained by an appropriate identification algorithm, e.g., the output error method or the maximum likelihood method, based on Z_N. It is well known that under reasonable conditions the estimate $\hat{\theta}_N$ is consistent, i.e.,

$$\lim_{N \to \infty} \hat{\theta}_N = \theta^o$$

provided $G^o \in \mathbf{M}$, where θ^o is the true parameter such that $G^o(q) = G(q, \theta^o)$.

It is always the case, however, that the true transfer function $G^o(q)$ is more complex than the nominal model $G(q, \theta)$, so that some modes are neglected in building the nominal model. Hence, the true system G^o does not belong to the set $\mathbf{M}$. In such case, we may not expect that

$$G(q, \hat{\theta}_N) \to G^o(q) \quad \text{as} \quad N \to \infty,$$

and some bias error due to undermodeling is expected to exist.

4. Model uncertainty and robust stability

We briefly review the theory of robust stability. Let us consider the standard feedback control system illustrated in Fig. 1, where $K(q)$ denotes a controller. Let the

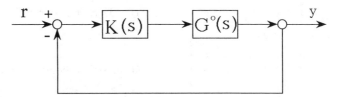

Fig. 1. A feedback control system.

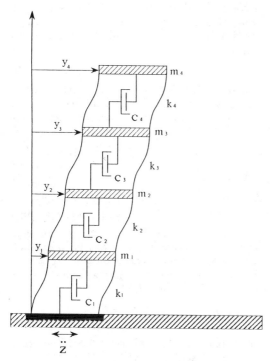

Fig. 2. A multistory building.

model error is given in the following multiplicative form:

$$G^o(q) = G(q, \hat{\theta})(1+\Delta(q)). \qquad (9)$$

Suppose that the feedback system with the nominal model is stable, and that the model error satisfies the following inequality:

$$|\Delta(e^{j\omega T})| < |1+(G(e^{j\omega T})K(e^{j\omega T}))^{-1}| \quad \forall \, \omega.$$

Then, it can be shown under some reasonable conditions that the perturbed system is also stable (see, for example, Vidyasagar 1985).

5. Weighted output error identification

In view of the robust stability theory, it is reasonable that we seek an estimate such that

$$\hat{\theta} = \arg \min_{\theta} \int_{-\pi/T}^{\pi/T} |Q(e^{j\omega T})\Delta(e^{j\omega T})|^2 d\omega \qquad (10)$$

where $Q(e^{j\omega T})$ denotes a weighting function which is introduced in order to emphasize the frequency range where higher model accuracy is needed.

The estimate minimizing (10) can be obtained by the following algorithm:

1. Let $\Sigma_u(\omega)$ denote the power spectrum of $\hat{u}(i)$. Then, by means of spectral decomposition theorem there exists a rational function $U(q)$ such that

$$\Sigma_u(\omega) = U(e^{j\omega T})U(e^{-j\omega T}) \qquad (11)$$

2. Filter the measured data according to

$$\tilde{y}(i) = \frac{Q(q)}{G(q, \theta)U(q)}y(i) \qquad (12a)$$

$$\tilde{u}(i) = \frac{Q(q)}{G(q, \theta)U(q)}u(i) \qquad (12b)$$

3. Apply the output error method to estimate parameters $\{d_i, n_i, i=1,...,2n\}$ using the filtered data $\{\tilde{y}(i), \tilde{u}(i)\}$ in place of $\{y(i), u(i)\}$.

To see that the above algorithm provides an estimate having the required property, it is enough to note that

$$\tilde{e}(i) \equiv \tilde{y}(i) - G(q, \theta)\tilde{u}(i)$$

$$= [G^o(q) - G(q, \theta)]\frac{Q(q)}{G(q, \theta)U(q)}u(i)$$

$$= \Delta(q)\frac{Q(q)}{U(q)}u(i) \qquad (13)$$

and hence we have

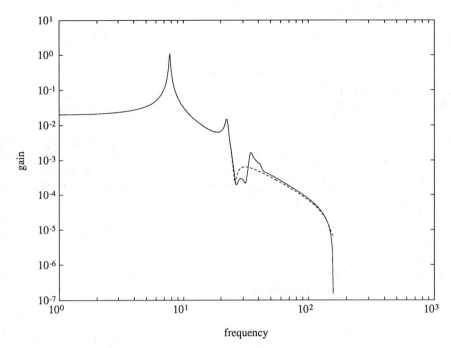

Fig. 3. True(solid) and estimated(dashed) frequency responses.

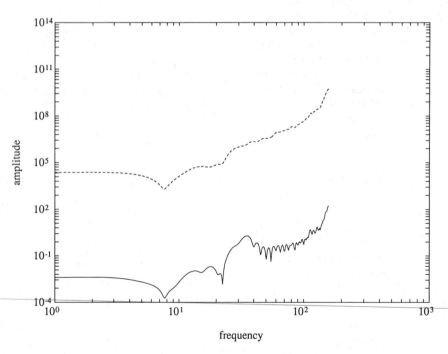

Fig. 4. Frequency plots of estimated model error (solid) and its
bound(dashed).

$$\sum_{i=1}^{\infty} \bar{e}(i)^2 = \frac{T}{2\pi} \int_{-\pi/T}^{\pi/T} |Q(e^{j\omega})\Delta(e^{j\omega T})|^2 d\omega.$$

$$(14)$$

We remark that we need the value of θ in (12), so that recursive computation may be necessary.

6. Estimation of unmodeled dynamics

In this section a bound for model error $\Delta(q)$ will be obtained via the ellipsoidal bounding algorithm. First we note from (9) that

$$G^o(q) - G(q, \hat{\theta}) = \Delta(q)G(q, \hat{\theta})$$

We see from this that the output error $e(i) \equiv y(i) - G(q, \hat{\theta})u(i)$ can be written as follows

$$e(i) = \Delta(q)\hat{y}(i) \qquad (15)$$

where

$$\hat{y}(i) \equiv G(q, \hat{\theta})u(i) \qquad (16)$$

denotes the simulated output sequence. This reveals that the unmodeled dynamics $\Delta(q)$ can be modeled by a linear system with the output error $e(i)$ as its output and the simulated output $y(i)$ as its input.

In order to estimate $\Delta(q)$, we suppose that it can be approximated by a finite impulse response model such that

$$\overline{\Delta}(q) = \sum_{k=0}^{m-1} g_k q^{-k}. \qquad (17)$$

Then, we have

$$\Delta(q) = \overline{\Delta}(q) + \tilde{\Delta}(q) \qquad (18)$$

where

$$\tilde{\Delta}(q) \equiv \sum_{k=m}^{\infty} g_k q^{-k} \qquad (19)$$

denotes the error term due to truncation of impulse response sequence.

With this definition, (15) can be written as follows.

$$e(i) = \overline{\Delta}(q)\hat{y}(i) + \varepsilon(i)$$
$$= \gamma' \phi(i) + \varepsilon(i) \qquad (20)$$

where

$$\gamma = (g_0, g_1, ..., g_{m-1})'$$
$$\phi(i) = (\hat{y}(i), \hat{y}(i-1), ..., \hat{y}(i-m+1))'$$

and

$$\varepsilon(i) = \tilde{\Delta}(q)\hat{y}(i) = \sum_{k=m}^{\infty} g_k \hat{y}(i-k).$$

The next thing for us to do before applying the ellipsoidal bounding algorithm is to evaluate the bound for the truncated term $\varepsilon(i)$. To do this, we assume that the truncated impulse response sequence satisfies

$$\sum_{k=m}^{\infty} |g_k| \leq \beta_g. \qquad (21)$$

Also we write

$$\max_i |\hat{y}(i)| = \beta_y. \qquad (22)$$

Then, we obtain

$$|\varepsilon(i)| \leq \beta_y \beta_g \equiv \beta_\varepsilon. \qquad (23)$$

Also, we assume that an *a priori* knowledge on the parameter γ is given in the following ellipsoidal form:

$$[\gamma - \hat{\gamma}_0]' P_0^{-1} [\gamma - \hat{\gamma}_0] \leq 1 \qquad (24)$$

Applying the set membership algorithm of Wahlberg and Ljung (1992), after processing the data $\{ \hat{y}(i), e(i) , 1=1, 2, .., N\}$, the ellipsoid (24) is updated as follows:

$$[\gamma - \hat{\gamma}_N]' P_N^{-1} [\gamma - \hat{\gamma}_N] \leq 1 \qquad (25)$$

where

$$P_N = \sigma_N \overline{P}_N. \qquad (26)$$

$$\overline{P}_N^{-1} \equiv P_0^{-1} + \sum_{i=1}^{N} \phi(i)\phi(i)' \qquad (27)$$

$$\hat{\gamma}_N = \overline{P}_N \left[P_0^{-1} \hat{\gamma}_0 + \sum_{i=1}^{N} \phi(i) e(i) \right] \qquad (28)$$

$$\sigma_N = 1 + N\beta_\varepsilon^2 - \sum_{i=1}^{N} \delta(i)^2 - \left[\hat{\gamma}_N - \hat{\gamma}_0 \right]' P_0^{-1} \left[\hat{\gamma}_N - \hat{\gamma}_0 \right]$$

$$(29)$$

$$\delta(i) \equiv e(i) - \phi(i)' \hat{\gamma}, \qquad (30)$$

The algorithm is also given in recursive form (Wahlberg and Ljung (1992)).

To obtain the corresponding bound in frequency domain, we define

$$W(e^{j\omega T}) \equiv [1, \ e^{-j\omega T}, \ e^{-j\omega 2T}, ..., \ e^{-j\omega(m-1)T}].$$

It is easily checked that

$$\overline{\Delta}(\omega, \gamma) = W(\omega)\gamma \qquad (31)$$

and

$$\overline{\Delta}(\omega, \hat{\gamma}_N) = W(\omega)\hat{\gamma}_N \qquad (32)$$

By Lemma 3.1 of Wahlberg and Ljung (1992) with some modification we can prove from (25) that

$$|\overline{\Delta}(\omega, \gamma) - \overline{\Delta}(\omega, \gamma_N)|^2 \leq R(\omega) \qquad (33)$$

where

$$R(\omega) \equiv [W(\omega)PW^*(\omega)]$$

in which [ ]* denotes the conjugate transpose of a matrix.

Our aim in this section is to obtain a bound for $|\Delta(\omega)|$. To this end, we rewrite (18) as follows:

$$\Delta(\omega) = [\overline{\Delta}(\omega, \gamma) - \overline{\Delta}(\omega, \hat{\gamma})] + \overline{\Delta}(\omega, \hat{\gamma}) + \tilde{\Delta}(\omega), \qquad (34)$$

so that

$$|\Delta(\omega)| \leq |\overline{\Delta}(\omega, \gamma) - \overline{\Delta}(\omega, \hat{\gamma})| + |\overline{\Delta}(\omega, \hat{\gamma})| + |\tilde{\Delta}(\omega)|. \qquad (35)$$

A bound for the first term of the above equation has been obtained by (33); the second term can be easily computed from (32); a bound for the last term is obtained from the *a priori* knowledge given by (21) as follows:

$$|\tilde{\Delta}(\omega)| \leq \sum_{k=m}^{\infty} |g_k||e^{-j\omega kT}| = B_g. \qquad (36)$$

7. Example

We consider a multi-story building illustrated by Fig. 2 which is subject to ground excitation $\ddot{z}$. We would like to model this structure by a two-degrees of freedom system. The measured data are generated by simulating a 4-degrees of freedom model given by

$$m_i\ddot{\xi}_i + c_i(\dot{\xi}_i - \dot{\xi}_{i-1}) - c_{i+1}(\dot{\xi}_{i+1} - \dot{\xi}_i)$$
$$+ k_i(\xi_i - \xi_{i-1}) - k_{i+1}(\xi_{i+1} - \xi_i) = -m_i\ddot{z},$$
$$i = 1, 2, 3, 4$$

where the mass, damping, and stiffness coefficients are taken to be such that

$$\frac{c_i}{m_j} = 0.5(\frac{1}{sec}), \qquad \frac{k_i}{m_j} = 2000(\frac{1}{sec^2})$$

$$i, j = 1, 2, 3, 4.$$

The spectrum of the ground acceleration is assumed to be

$$\Sigma_{\ddot{z}} = \frac{1 + 4\beta^2(\frac{\omega}{\omega_0})^2}{\{1 - (\frac{\omega}{\omega_0})^2\}^2 + 4\beta^2(\frac{\omega}{\omega_0})^2}$$

where

$$\beta^2 = 0.410, \qquad \omega_0^2 = 242$$

The parameters of the two-degrees of freedom model is estimated based on the data $y = (\xi_2, \xi_4)$ and $u = \ddot{z}$. No measurement noise is assumed in this example.

Fig. 3 illustrates the bode diagram of the true system and the low order model with identified parameters. In order to evaluate the model error, we need some *a priori* knowledge on parameters γ_0, P_0, and the bound B_g for the truncated impulse response sequence. Those parameters are evaluated via a preliminary analysis of data $Y_N = \{e(i), \hat{y}(i), i = 1, 2, ..., N\}$, i.e., an ARX model is fitted to Y_N and γ_0 and B_g is estimated from the ARX model. P_0 is taken to be $10^5 I$. The model error and its bound are plotted in Fig. 4, in which the order m of the finite impulse response model is taken to be 64. It is observed that the bound is fairly loose while the shape of the bound resembles the estimated model error.

8. Conclusion

The problem of structural system identification with unmodeled dynamics is considered. An algorithm is proposed which minimizes the effect of the unmodeled dynamics. An upper bound for the model error is also given, which can be computed

based on *a priori* information including the bound for the measurement noise and the truncated impulse response sequence. It is seldom the case that such an information is available *a priori*. Therefore, it is desirable that the algorithm is improved so that those bounds can be estimated based on the measured data.

The numerical example shows that the obtained bound is very loose. The looseness may be resulted from the autocorrelation of $\varepsilon(i)$, and the much conservative bound for $\varepsilon(i)$ given by (23), which may be also expanded in transforming the bound in parameter space into frequency domain. More work should be made to remove those sources for the looseness to improve the algorithm.

REFERENCES

Belforte, G., B. Bona and V. Cerone 1990. Parameter estimation algorithms for a set-membership description of uncertainty: *Automatica*, Vol. 26, No. 5, 887-898.

Fassois, S. D., K. F. Eman and S. M. Wu 1990. A linear time-domain method for structural dynamic identification: *ASME Journal of Vibration and Acoustics*, Vol. 112, 98-106.

Fogel, E. and Y. F. Huang 1982. On the value of information in system identification- bounded noise case: *Automatica*, Vol. 18, No. 2, 229-238.

Hac, A., and P. D. Spanos 1990. Time domain method for parameter system identification: *ASME Journal of Vibration and Acoustics*, Vol. 112, 281-287.

Helmicki, A. J., C. A. Jacobson and C. N. Nett 1991. Control oriented system identification: a worst-case / deterministic approach in H_∞: *IEEE Trans. on Automat. Contr.*, Vol. 36, No. 10, 1163-1176.

Gu, G. and P. P. Khargonekar 1992. Linear and nonlinear algorithms for identification in H_∞ with error bounds. *IEEE Trans. on Automat. Contr.*, Vol. 37, No. 7, 953-963.

Kosut, R. L., M. K. Lau and S. P. Boyd 1992. Set-membership identification of systems with parametric and nonparametric uncertainty: *IEEE Trans. on Automat. Contr.*, Vol. 37, No. 7, 929-941.

Makila, P. M. 1991. On identification of stable systems and optimal approximation: *Automatica*, Vol. 27, No. 4, 663-676

Moustafa, K. A. F. 1992. Time-domain structural identification using free response measurements: *International Journal of Control*, Vol. 56, No. 1, 51-65.

Norton, J. P. 1987. Identification and application of bounded-parameter models: *Automatica*, Vol. 23, No. 4, 497-507.

Partington, J. R. 1991. Robust identification and interpolation in H_∞: *Int. J. Control*, Vol. 54, No. 5, 1281-1290.

Shinozuka, M., C.-B. Yun and H. Imai 1982. Identification of linear structural dynamic systems: *ASCE Journal of Engineering Mechanics*, Vol.108, No.EM6, 1371-1390.

Vidyasagar, M. 1985. *Control system synthesis: A factorization approach*, The MIT Press, Cambridge, Massachusetts.

Wahlberg, B. and L. Ljung 1992. Hard frequency-domain model error bounds from least-squares like identification techniques: *IEEE Trans. on Automat. Contr.*, Vol. 37, No. 7, 900-912.

Younce, R. C. and C. E. Rohrs 1992. Identification with nonparametric uncertainty: *IEEE Trans. on Automat. Contr.*, Vol. 37, No. 6, 715-728.

Zhu, Y. C. 1989. Estimation of transfer functions: asymptotic theory and a bound of model uncertainty: *International Journal of Control*, Vol. 49, No. 6, 2241-2258.

Structural Safety & Reliability, Schuëller, Shinozuka & Yao (eds) © 1994 Balkema, Rotterdam, ISBN 90 5410 357 4

An integrated approach for the identification of mechanical systems with nonlinear force elements

R. Krause
Daimler Benz Research Institute AEG, Frankfurt, Germany

ABSTRACT: The parameter identification of multy body systems with nonlinear force elements like friction, back lash, cubic stiffness and prestress has been discussed. An extended covariance method including the measured data of a technical system and a proper model are used to determine the kind of nonlinear force elements and to identify the parameters.

1 INTRODUCTION

Modeling dynamic systems requires assumptions and idealizations resulting in mathematical equations which determine the dynamic behavior of the model. In general, such models include characteristic parameters with physical meaning, like mass or geometrical data. The values of these parameters can be found by direct measurement. Other parameters result from idealized physical laws, e.g. stiffness and damping elements, and cannot be determined directly. It is the task of identification methods to evaluate these parameters indirectly by comparing the dynamic behavior of both dynamic system and the model.

The covariance method was developed for parameter identification of linear dynamic systems subject to stochastic excitation. The parameters of the system can be identified using second moments of the response of a linear filter added to the system's measurement devices /1,2/.

The proposed paper shows an extension of the covariance method for the detection of nonlinear damping and stiffness elements and the parameter identification of the equivalent nonlinear forces. Fig. 1 shows the scheme of parameter identification.

It begins with a nonlinearity test, which is based on the parameter identification of a linear model with respect to different excitation levels. If one of the parameters changes the value depending on different excitation levels, the system is nonlinear. It can be shown that the

variation of the parameters with respect to the excitation level is characteristic to certain nonlinearities.

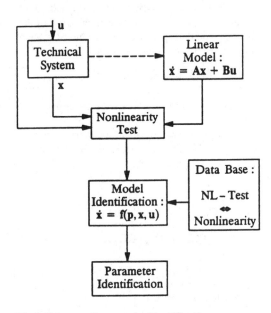

Fig.1. Scheme of parameter identification

The relationship between these nonlinearities and the parameter curve is stored in a data base and can be used to find nonlinear models for unknown technical systems. At least, the parameters of

these models can be identified with the extended covariance method.

2 NONLINEARITY TEST

In the first step a linear time-invariant model is chosen for the detection of nonlinearities of a mechanical system

$$\dot{x} = Ax + Bu \quad ,x \in IR^n , \qquad (1)$$

instead of using higher moments /3/. Matrices A and B comprise known and unknown parameters of the model. The measured data of the excitation $u(t)$ and response $x(t)$ of the technical system are considered as the excitation and response of the linear model. The stochastic excitation $u(t)$ is supposed to be ergodic, Gaussian and a stationary process which can be described by a time-invariant form filter

$$\dot{u} = Ku + Lw \quad ,u \in IR^m , \qquad (2)$$

where $w(t)$ is a gaussian white noise with zero mean. For detection or identification purposes the excitation and response of the technical system have to be passed through a linear filter

$$\dot{y} = Fy + Hx + Gu \quad ,y \in IR^l , \qquad (3)$$

with known matrices F, H and G and $l>m+n$, see Fig. 2.

Using (1), (3) and the measured data, stationary of $C_{yx} = E\{yx^T\}$ will yield algebraic relations for the system parameters:

$$\frac{d}{dt}C_{yx} = C_{yx}A^T + C_{yu}B^T +$$

$$FC_{yx} + HC_{xx} + GC_{ux} = 0 . \qquad (4)$$

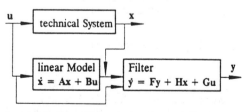

Fig. 2. Nonlinearity test

After comprising all unknowns of A and B in a parameter vector p and collecting all non-trivial equations of (4) we end up with an overdetermined set of linear algebraic equations

$$Cp = c \qquad (5)$$

for identifying the unknown parameters. These parameters may be estimated in the sense of least squares by premultiplying (5) with C^T nd solving the equations for the parameters

$$p = (C^TC)^{-1}C^Tc . \qquad (6)$$

If the system is linear, the parameters are independent of the excitation intensity C_{uu}. In case the system is nonlinear, the parameters will change with respect to the excitation intensity and kind of nonlinearity.

3 MODEL IDENTIFICATION

Some applications of a single degree of freedom system

$$m\ddot{x} + \delta\dot{x} + kx + F(x,x) = u(t) \qquad (7)$$

with constant coefficients m,δ,k and nonlinear force elements $F(x,x)$ like prestress, back lash, cubic stiffness and coulomb friction show for a linear model

$$m\ddot{x} + d\dot{x} + cx = u(t) \qquad (8)$$

characteristic parameter curves, see Fig. 3.

Numerical applications to mechanical systems up to 4 degree of freedom and 7 nonlinearities yield in good results /4/. Depending on the excitation level, only the parameters of the force elements change their values, which replace the nonlinear elements. All parameters of linear force elements from the system remain constant by the detection of nonlinearities. If multiple nonlinearities belong to one force element, the single characteristic curves in Fig. 3 must be added. Problems occur only if a nonlinear force element includes two nonlinearities of *inverse* behavior like prestress and back lash.

Nonlinear Force Element	Linear Parameter

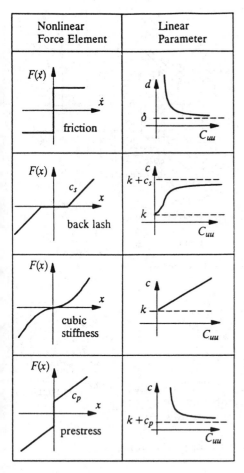

Fig. 3. Characteristic parameter curves for
certain nonlinearities

4 PARAMETER IDENTIFICATION

After the detection of nonlinearities it is possible
to define a nonlinear model

$$\dot{x} = f(x,u,p) = \sum_{i=1}^{q} p_i \, f_i(x,u) \qquad (9)$$

with the model vector $f(x,u,p)$ and the parameter
vector $p \in \mathbb{R}^q$. The parameters must be a linear
combination with respect to all other variables for
the parameter identification. Now, it is not
sufficient to use the covariance method (4) for
linear models. A system with a prestress element
e.g. can have the same covariances for low
prestress and high stiffness or high prestress and

low stiffness. Different excitation levels must be
used to distinguish systems with different
parameter values. In this case an extended model
is defined

$$\dot{z} = \sum_{i=1}^{q} p_i \, g_i(z,u_1 \, ... \, u_r) \qquad (10)$$

$$\text{with } z = \begin{bmatrix} x_1 \\ \vdots \\ x_r \end{bmatrix} \quad , g_i = \begin{bmatrix} f_{i\,1}(x_1,u_1) \\ \vdots \\ f_{i\,r}(x_r,u_r) \end{bmatrix}$$

where r is the maximal number of parameters to
describe a stiffness or damping element of the
nonlinear model. It is important here, that the r-
different excitation levels are chosen carefully to
have a different influence on the nonlinearities.
Using (10), (3) and the measured data, the
extended covariance analysis is given by

$$\frac{d}{dt} C_{yz} = \sum_{i=1}^{q} p_i \, C_{ygi} +$$

$$F C_{yz} + H C_{zz} + G C_{uz} = 0 . \qquad (11)$$

Analogous to solve (4) with respect to the
parameters eq. (11) is an overdetermined set of
equations which can be solved as shown in (5)
and (6).

5 PARAMETER IDENTIFICATION WITH
UNCOMPLETE STATE MEASUREMENT

The parameter identification with uncomplete
measurement is more difficult, because unknown
parameters may be multiplied with not
measurable covariances. An extension of the
identification scheme is necessary as shown in
Fig. 4.

The linear model (19) must be transformed to
the observability form for the nonlinearitiy test. In
that case all unknown covariances can be
eliminated and the elements of the state form can
be identified. The relationship between the
coefficients of the state form and the physical
parameters is given by a set of algebraic nonlinear
equations. They can be solved numerically to get
the characteristic parameter curves and the
nonlinear model. Usually, the parameter
identification of the nonlinear model is not
possible by using a nonlinear observability form if

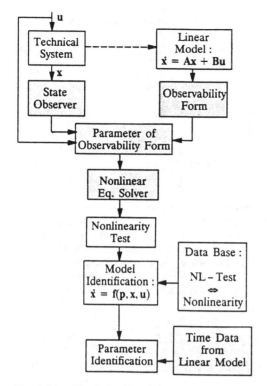

Fig. 4. Identification with uncomplete state measurement

the system has nondifferentiable force elements. Therefore the unknown covariances will be replaced by numerical data of the linear model. The covariances of the linear model are equivalent to the covariances of the technical system for certain excitations.

6 IDENTIFICATION OF A ROBOT DRIVE

The method is applied by experimental data of a robot drive. Fig. 5 shows the harmonic drive gear with a manipulator arm. The harmonic drive gear is a planetary gear with a high reduction. The robot is controlled to keep the arm in a horizontal position. The complete experiment is described in /5/.

The mechanical model is a two degree of freedom system, see Fig. 6. The manipulator arm and the rotor of the electronic motor are modelled as rigid bodies. The harmonic drive is idealized by a damper and a spring element. For the identification only the displacement of the

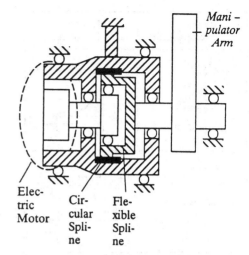

Fig. 5. Harmonic drive gear

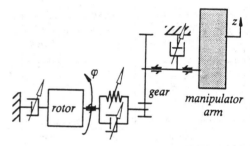

Fig. 6 Mechanical model

manipulator arm or the angular velocity of the rotor can be measured.

The model identification shows that friction of gear wheels and back lash of the gear have an essential influence on the dynamic.

The result of the nonlinear parameter

	Measured Data	Data Sheet
Stiffness [Nm/rad]	19400	23900
Back Lash [min]	4	3
Friction [Nm]	7.4	–

Fig. 7. Parameter of the nonlinear elements

identification is shown in Fig. 7. Back lash and stiffness can be compared with the values of the harmonic drive data sheet. The identified parameter of back lash is lower than the values on the data sheet.

The experimental results demonstrate the efficiency of the identification method.

7 CONCLUSION

It is shown that the covariance analysis is an efficient tool for the detection of nonlinear force elements. The identified parameter curves with respect to the excitation are characteristic for the above mentioned nonlinearities, also if multiple nonlinearities occur in one force element. The parameter identification is also very accurate in case the nonlinearities have an influence on the dynamic behavior. Nonlinear force elements, which have no influence on the interesting excitation level can not be identified.

REFERENCES

/1/ Weber, H.I. and Schiehlen, W.O.: A filter technique for parameter identificatin. Mech. Res. Comm. 1983, 10, pp 259-265.

/2/ Kallenbach, R.: Kovarianzmethoden zur Parameteridentifikation zeitkontinuierlicher Systeme. Fortschrittberichte VDI, VDI-Verlag, Düsseldorf, 1987.

/3/ Krause, R. and Schiehlen, W.O.: Detection of System Nonlinearity. Proceeding of ICOSSAR'89, 5th. Int. Conf. on Structural Safety and Reliability, Vol. II, pp 1382-1397, San Francisco. New York: Amer. Soc. of Civil Eng., 1990.

/4/ El-Dessouki, N.: Beschreibung nichtlinearer stochastisch angeregter Systeme. Stuttgart: Diplomarbeit DIPL 35. Institut B für Mechanik, Universität Stuttgart, 1990.

/5/ Krause, R.:Analyse und Parameteridentifikation stochastisch angeregter Mehrkörpersysteme mit nichtlinearen Kraftgesetzen. Fortschritt-Berichte VDI, Reihe 11, Nr. 177. Düsseldorf: VDI-Verlag 1992.

Structural Safety & Reliability, Schuëller, Shinozuka & Yao (eds) © 1994 Balkema, Rotterdam, ISBN 90 5410 357 4

Problems and related countermeasures in mathematical model improvement: A survey

H.G.Natke, N.Cottin & U.Prells

Curt-Risch-Institute, University of Hannover, Germany

ABSTRACT: Updating of parametric mathematical models with measured data leads to a verified and validated mathematical model. The prior mathematical model is uncertain, and the measured data are erroneous and incomplete. If the irregular measurement errors are modelled statistically, then the estimators to be applied lead to linear or nonlinear systems of equations for the parameters to be sought. These resulting equations are often ill-conditioned, in consequence the solution, if exists, is unstable or in the nonlinear case, the iterative procedure does not converge. Problems to be solved are to optimize the test design, to choose suitably the parameters to be estimated, and to assure the uniqueness of the problem. It is discussed how to overcome these problems.

1 INTRODUCTION

System here is heuristically defined as a part of the real world, that is an object, a technical construction. Within system identification a distinction is made between the subject areas experimental modal analysis and estimation of physical model parameters. In the related literature the latter is often named as mathematical model correction (updating, improvement, reconciliation calibration). That means a prior parametrical mathematical model is given, test data are available, and if the model structure is adequate to the purpose (of modelling), then system identification is reduced to parameter estimation. The result will be a validated mathematical model.

The prior mathematical model is uncertain. This uncertainty concerns

1. the model structure, i.e. for linear systems

   - the type of damping
   - the effective number of degrees of freedom (Natke 1968)

and

2. the parameters.

Correction of this mathematical model requires the modelling of the uncertainties and a data set independent from that used in (theoretical)

modelling. The best data set is formed by the measured data from the existing system under consideration. However, these data are erroneous and incomplete. The errors can be deterministic and irregular, and the latter are often modelled additively and stochastically. Deterministic errors should be avoided, detected otherwise and theoretically corrected. Consequently, if the system identification problem is formulated as an inverse problem, then estimators have to be applied. An erroneous model structure will lead to biased parameter estimates. Therefore the model structure has to be verified first for the purpose-equivalent model. This, for example, can be done with some prior knowledge (from physics and experience) simultaneously with the parameter estimation and an additional parametrized structure well-known in the observer theory (Mook 1992), (Natke, Yao 1993) and a posteriori validation. Validation here is understood as homomorphy between the system and the model, which means using a system data set independent of that used for model correction. This validated model, of course, is only a model valid within the class of excitation signals used in system identification.

Parameter estimation with the classical procedures leads to ill-posed (mathematical) problems. In the case of finite dimensions the operators representing matrices are ill-conditioned. In consequence, if a solution exists it is unstable, otherwise

the solution domain has to be extended for a generalized solution. For this purpose regularization methods have to be applied. However, the first attempt must be a well-posed modelling process (Natke 1992c), and minimum uncertainties in the prior mathematical model as well as in the measurements. The latter can be achieved by suitable test planning and an optimum test design. Additionally, the maximum information available must be used for system identification.

2 OPTIMUM TEST DESIGN

The first problem to be solved is to design the test optimally. This test design concerns the pickup and excitation locations, the type of excitation, and if working in the frequency domain, the choice and number of excitation frequencies to be used in the updating procedures. The problem can be solved, for example, by maximizing the value of the determinant of the information matrix while taking the measuring constraints into account.

2.1 Excitation

To accomplish the test-supported modelling the type of excitation has to be chosen dependent on the purpose of the model and the environmental conditions. If all the natural modes within a pre-specified frequency interval are required, then harmonic multi-point excitation is optimal, assuming that the high test and time expenditure is acceptable. Otherwise random excitation is recommended. The reader can find discussions of the advantages and disadvantages of various types of excitation, for example, in (Natke 1992a).

The frequency range, the force shape expressed by the force vector, the required amplitude and, of course, the system properties determine the choice of exciter types. For example, step relaxation is not suitable for exciting heavy diesel engines. Electro-hydraulic exciters enable a pure harmonic excitation with large amplitudes in the lower frequency range (roughly up to $150 Hz$). For a detailed discussion see, for instance, (Natke 1988a).

The exciter locations can be determined theoretically by the maximum work done through the force in line with the interesting eigenmodes. Let us look, for example, at the spectral decomposition of the dynamic response of a finite-dimensional model with viscous damping in the image domain with the Laplacian variable s (Natke 1992a), Eq.(5.133),

$$U(s) = \sum_{l=1}^{2n} \frac{a_l}{s - \lambda_{Bl}} \tag{1}$$

with the vector a_l equal to the work done by the force multiplied by the eigenvector $\hat{u}_{Bl}$ of the model corresponding to the damped system. λ_{Bl} is the eigenvalue corresponding to $\hat{u}_{Bl}$. As can be seen, the statement made above holds true. It is disadvantageous that one has to know the modes within a given frequency interval from, for example, the results of system analysis. However, a lot of experience is available for particular systems. The available number of exciters to be applied is often restricted, otherwise one should apply as many as possible in connection with the maximum work.

Attempts to find the optimum exciter locations are rare, and the authors are aware only of the recent paper (Niedbal, Klusowski 1991), where several methods are described for localizing optimum exciter positions and determing adequate exciter forces when applying phase resonance and phase separation techniques.

Optimum test design with respect to the choice of the input signal is treated in (Mehra 1991); fundamentals of optimum experimental design are described in (Pázman 1986), and some results concernig the optimum experimental design for the parametric identification of linear elastomechanical systems are published in (Cottin 1990).

2.2 Measurements

Measurement technology is outside the range of this paper, since this refers to sensor choice, sensor fixation and the calibration of sensors, which can be a problem if many sensors have to be installed ($\rightsquigarrow$ computer aided self-calibration) etc. It should be mentioned that expert systems are under development for the application of measurement technology for non-experts within civil engineering (NN 1992).

The number of sensors and their locations are also determined by the goal being aimed at. The dynamic response (displacements, velocities, accelerations, strains etc.) to be measured must be approximated by the measurements. This means that the shape, the slopes and the curvature must be reconstructable with the required accuracy. In other words, prior knowledge of the mathematical model is needed or pre-tests must be performed in order to find the required positions and their number.

For optimal sensor locations see, for example, (Shah, Udwadia 1978), (Lallement, Cogan 1991). A survey concerning recent developments in this topic is given in (Kirkegaard, Brincker 1992), where the field of the optimum location of sensors is treated for parametric identification of linear structural systems by means of the Fisher information matrix. There it is also indicated that the optimum sensor location problem for parametric identification is closely related to the problem of the optimum location of sensors and controllers for control systems and for the failure detection of systems by vibration monitoring.

2.3 Adaptive excitation

We know the required "appropriate" excitation in the phase resonance testing in order to isolate one natural mode of the system (Natke 1992a). If one is interested in the identification of a few parameters of the model, then an adaptive excitation is required which can be determined using selective sensitivity (Ben-Haim 1992),(Ben-Haim, Prells 1993), (Prells, Ben-Haim 1993). If we recall the description of a stationary state in the frequency domain:

$$\psi(\omega, p) = HF(\omega, p)G\phi(\omega) \in \mathbb{C}^{nr}, \; \phi \in \mathbb{C}^{ne}$$

with the frequency response matrix

$$F^{-1}(\omega, p) = -\omega^2 A_2(p) + j\omega A_1(p) + A_0(p) \in \mathbb{C}^{n \times n}$$

$$= \sum_{\alpha=1}^{N_p} B_\alpha(\omega)p_\alpha, \quad N_p \leq \tfrac{3}{2}n(n+1).$$

Depending on the frequency ω, on the excitation $\phi(\omega)$ and on the model parameters p the nonnegative scalar

$$S_\alpha := \|\frac{\partial \psi}{\partial p_\alpha}\|^2 = \|HF(\omega, p)B_\alpha(\omega)F(\omega, p)G\phi(\omega)\|^2$$

$$=: \phi^\dagger D_\alpha \phi \quad (B_\alpha := \frac{\partial Q}{\partial p_\alpha})$$

is called sensitivity of the response $\psi = \psi(\phi(\omega), p)$ with respect to parameter p_α, where $\alpha \in \{1, ..., N_p\} =: \mathcal{I}$

Let $\mathcal{I} = \mathcal{K} \cup \mathcal{J}, \mathcal{K} \cap \mathcal{J} = \emptyset$. The model is then called selectively sensitive (selectively insensitive) (of rate ϵ) with respect to $\mathcal{J}$ ($\mathcal{K}$) at frequency ω, if an excitation exists $\phi(\omega, p)$, such that

$$S_\alpha(\phi) \begin{cases} = 0 \;\; (\leq \epsilon), & \text{if } \alpha \in \mathcal{K} \\ > 0 \;\;\;\; (\epsilon), & \text{if } \alpha \in \mathcal{I}. \end{cases}$$

By applying these adaptively chosen excitations to the system it is possible to estimate only a few (selective sensitive) parameters at a time.

2.4 Data set for identification

System identification can be understood as test- and computer-aided modelling (TACAM). Therefore some requirements must be fulfilled. We are interested in an improved model, which means that for the improvement we will use a data set which is not used for the prior mathematical model. In addition the improved model must be verified, which means that the data used for improvement can obviously be reconstructed (with respect to the applied quality criterion) by this model. Additionally, the improved model must be validated, i.e. homomorphy between the system and model for a data set independent of that used for the improvement. This identified model is one possible representation of the system within the class of excitation signals which have generated the data sets for improvement. Finally, this validated model must be usable. This means its errors must be as small as necessary to provide a sufficiently great number of measurements of sufficient accuracy and a correction of systematic errors if there are any.

3 REGULARIZATION OF ILL-CONDITIONED PROBLEMS OF LINEAR SYSTEMS

Inverse problems of system identification are generally ill-posed in the sense of Hadamard (Baumeister 1987), (Louis 1989). First one has to look for the existence of a solution. It can be necessary to enlarge the solution domain ⤳ the generalized solution. Even if the finite-dimensional (mathematical) operator possesses a continuous inverse it may be ill-conditioned. In consequence, the application of regularization is required (Baumeister 1987), (Louis 1989), (Natke 1992b), (Baumeister 1993).

3.1 Classical methods

The numerical handling of finite-dimensional mathematical problems first requires the application of suitable methods. This is to avoid amplifying the errors contained. However, this is not sufficient: the operator itself must be modified in order to obtain an (adjacent) stable solution. This can be achieved, for example, with the SVD by introducing weighting (regularization parameters) and truncation of those singular values which correspond to the measurement noise. The difficulty

is, of course, how to distinguish between the SV with system information and those existing due to disturbances. This procedure can be interpreted as the well-known filtering.

Another method should be mentioned here: that is Tikhonov- regularization (Tikhonov, Arsenin 1977). Let

$$Aa = g \qquad (2)$$

an overdetermined system of algebraic equations. The extended weighted least squares (EWLS) method (Natke 1992a) with the a priori knowledge $a_0 = 0$ leads to the loss function

$$J_\gamma(a) = (Aa - g)^* G_v(Aa - g) + \gamma^2 (Ba)^* Ba, \quad (3)$$

with the compositable weighting matrices G_v and $B^* B$ with required properties, γ the regularization parameter, and $(...)^*$ the conjugate transposed quantity $(...)$. Minimization of Eq. (2) yields

$$(A^* G_v A + \gamma^2 B^* B) a_\gamma = A^* G_v g. \qquad (4)$$

As can be seen, the solution a_γ is dependent on the regularization parameter γ. Two questions arise concerning regularization, and these have to be answered. The first is how one can estimate the regularization influence on the generalized solution, and the second is how to determine the regularization parameter for given weightings. The first problem can be solved by applying triangular inequality (see, for example (Louis 1989)). Several methods for prior and posterior determination are available for solving the second problem. Here, only cross-validation (Golup, van Loan, 1983) and data error sensitivity (Prells 1991) should be mentioned.

For further methods see the related references and the citations there.

3.2 Parameter topology

Regularization can be achieved for a given data set by introducing a coarser parameter topology. The result will be a relatively greater number of equations for estimating a small number of parameters, and therefore a relative enlargement of the information content.

In the field of updating this was already introduced in 1974 (Natke, Collmann, Zimmermann, 1974). For the finite-dimensional equation of motion

$$M\ddot{u}(t) + C\dot{u}(t) + Ku(t) = f(t) \qquad (5)$$

of n degrees of freedom and with the common interpretation of the used symbols, the parameter matrices are split into submodel matrices of the same order:

$$M = \sum_{r=1}^{R} M_r, \ C = \sum_{i=1}^{I} C_i, \ K = \sum_{j=1}^{J} K_j. \quad (6)$$

How this should be done depends on the physical subsystems, and on the assembling of parameters of the same error magnitude etc. (Natke 1992a). Then the parameters to be estimated are introduced as global factorial corrections:

$$M^c = \sum_{r=1}^{R} a_{Mr} M_r, \ C^c = \sum_{i=1}^{I} a_{Ci} C_i, \ K^c = \sum_{j=1}^{J} a_{Kj} K_j.$$
$$(7)$$

Estimates of a_{Mr}, a_{Ci}, a_{Kj} inserted into the above equations yield estimates $\hat{M}, \hat{C}, \hat{K}$. These $R+I+J$ parameters to be estimated can be a much smaller number than $3n^2$. A side-effect is the economics of this method. However, in the limit all the elements of the parameter matrices can be theoretically corrected (without regularization effect).

3.3 Bayesian approach: Extended weighted least squares

The EWLS method has already been introduced with the Tikhonov- Philipps procedure, but without mentioning the statistical weighting. In order to obtain the posterior most probable parameter estimates, the weighting matrices $G_v, G_0 = B^* B$ should be equal to the inverse covariance matrices with respect to the residuals and the parameters of the prior mathematical model, assuming normal distributions both for the measurement errors and for the prior parameters. G_v can be estimated or approximated, but G_0 is often unknown. Here, regularization with a matrix of regularization parameters or with an estimate of G_0 and one regularization parameter has the advantages of

1. substituting missing measurements

2. restricting the distances from the estimates to the prior values

3. making the problem convex

4. modifying the operator (adjacent stable solution)

5. providing the possibility of constructing an always convergent iteration procedure.

3.4 Additional information

Generally, additional measurements are excluded. Therefore, other available additional information should be used for regularization. This can be

- symmetry properties of the matrices

- sparse configuration of the matrices

- orthonormal properties of the eigenvectors

- resonances as well as anti-resonances

- force paths of the structural system

- band-limitations due to band-restricted excitations

- energy restrictions from practice

- measurement restrictions as described, for example, by the measuring equation

- inequalities from restrictions

- ...

Additional masses or/and stiffnesses (theoretically applied) will modify the system (model) properties and also give additional information to be used. At least a re-formulation should be considered, e.g. as described in (Nalitolela 1992).

4 REGULARIZATION WITHIN STRUCTURE IDENTIFICATION USING POLYNOMIALS

Non-linear systems need first structure (of the mathematical model) identification. No general systematic procedure for this exists up to now due to the diversity of existing nonlinearities. If one chooses the model class of polynomials for an additively separable model with respect to restoring and damping forces, then structure identification can be reduced by parameter estimation: estimating the degree of a polynomial simultaneously with its parameters (Natke 1988b), (Natke, Prells 1991), (Zamirowski 1992).

4.1 Consequences from the use of polynomials

Bearing in mind the Stone-Weierstrass theorem, the approximation of descriptions of nonlinear behaviour by polynomials includes a wide class of systems. However, multi-valued functions such describing hysteretic and deteriorating behaviour cannot be directly approximated by polynomials.

But if a transformation is done into the state space domain with the state variables $\dot{x}(t), \dot{f}(t)$ (Benedettini, Capecchi, Vestroni 1990) then polynomials will approximate this type of nonlinearities relatively well.

4.2 Regularized solutions

The first attempt used the economization formula from approximation theory and Chebychev polynomials combined with multi-hypotheses testing. The introduction of penalties in order to influence the complexity etc. failed (Natke, Zamirowski 1990). Subsequently, regularization methods were applied succesfully (Natke, Prells 1991). The economization formula in (Natke, Prells 1991) was included in the loss function as a penalty term without and with regularization parameter. The results show as an example that the regularization parameter is almost independent of the measurement noise, and that the regularized results for the degree are smaller than those obtained by multihypotheses decision.

5 REMARKS ON APPLICATIONS

The applications of system identification clearly show the need of regularization for identification problems. Applications are: the direct improvement of mathematical models which are necessary for optimization and for system qualification (safety proofs), and the localization of model uncertainties, which is equivalent to damage detection and location. A validated model is also needed for diagnosis and damage assessment (model supported diagnostics).

Our own applications concern only simulations with a non-proportional damped vibration chain and the polynomial approximations. Information is provided about the latter in (Natke, Prells 1991). The vibration chain example is used for demonstration purposes in (Prells 1991) and (Cottin, Prells, Natke, 1992).

6 CONCLUSIONS

The main problems in experimental modal analysis concern the test design (excitation and sensoring) and the error estimation (it is mainly systematic errors that are difficult to detect and to determine). In the correction of mathematical models with the use of measured data the test design requires attention. Test design means optimizing

the excitation by means of the optimum selection of exciter positions, type of force, and the type of force vector to be applied. It also means the measurement, and especially the choice of sensor locations. The aim of optimization is to identify unbiased estimates with minimum variances and covariances. It is assumed that the model structure is adequate to the model purpose. First results are mentioned here. A new approach uses adaptive excitation. What is sought is that excitation which gives a dynamic response that is sensitive with respect to a pre-chosen subset of parameters to be estimated: selective sensitivity. It is a displacement excitation (or, more generally, a response excitation) instead of a force excitation. Here the dynamic response of the system is excited sensitively only with respect to this pre-chosen subset of parameters. Consequently, by estimating only a few parameters now, it is shown that the inverse problem of sytem identification leads to fewer difficulties than that of estimating many parameters (magnitude of more than 10).

The discrete inverse problems of identification are ill-conditioned. This seems to be the reason for many non-convergent procedures and for erroneous estimates. It is assumed that the structure identification results in a purpose equivalent parametrization; otherwise the parameter estimates are biased. The choice of a suitable numerical algorithm is necessary (to avoid an amplification of numerical errors), but regularization is required in addition: looking for a generalized and stable solution. Several approaches are discussed here.

Dynamic systems with nonlinear behaviour which are modelled by the class of polynomials can be validated (identified) in such a way that the model structure, expressed here by the maximum power of the polynomial, and the coefficients can be estimated simultaneously.

REFERENCES

Baumeister, J., 1987. *Stable Solution of Inverse Problems. Vieweg Advanced Lectures in Mathematics.* Friedr. Vieweg & Sohn Braunschweig/Wiesbaden

Baumeister, 1993. Identification of parameters, ill-posedness and adaptive systems. In: Natke, H.G., Tomlinson, G.R., Yao, J.T.P. (Eds.); 1993. *Safety Evaluation Based on Identification Approaches Related to Time-variant and Nonlinear Structures.* Vieweg Verlag Braunschweig, Wiesbaden

Benedettini, F., Capecchi, D., Vestroni, F., 1990. *Nonparametric models in identification of hysteretic oscillators.* DISAT - Pubbl. n. 4, Dic., Dip. di Ing. delle Strutture, delle Acque e del Terreno, Univ. dell'Aquilla

Ben-Haim, Y., 1992. Adaptive diagnosis of faults in elastic structures by static displacement measurement: the method of selective sensitivity. *Mechanical Systems and Signal Processing,* Vol. 6, No. 1. January 1992, 85-96, Academic Press

Ben-Haim, Y., Prells, U., 1993. *Selective sensitivity in the frequency domain, Part 1: Theory. Mechanical Systems and Signal Processing.* to appear

Cottin, N., 1990. On the optimum experimental design for the parametric identification of linear elastomechanical systems. *Proc. of the European Conf. on Structural Dynamics, EURO-DYN '90.* Bochum, Germany, 5-7 June 1990, 347-354

Cottin, N., Prells, U., Natke, H.G., 1992. A parameter identification technique for elastomechanical systems using modal quantities. *Int. J. Analytical and Experimental Modal Analysis,* July 1992, 197-212.

Golup, G.H., van Loan, C.F., 1983. *Matrix Computations.* North Oxford Academic

Kirkegaard, P.H., Brincker, R., 1992. On the optimal location of sensors for parametric identification of linear structural systems. *Proc. of the 17th Int. Conf. on Experimental and Numerical Methods in Structural Dynamics.* 21-25 Sept. 1992, Leuven, Belgium

Lallement, G., Cogan, S., 1991. Optimal selection of the measured degrees of freedom and application to a method of parametric correction. *Proc. of the Int. Forum on Aeroelasticity and Structural Dynamics.* June 3-6, 1991, Aachen, Germany (Preprint)

Louis, A.K., 1989. Inverse und schlecht gestellte Probleme. Teubner Studienbücher, Mathematik. B.G. Teubner Stuttgart

Mehra, R.K., 1981. Choice of input signals. In: P. Eykhoff (Ed.). *Trends and Progress in System Identification.* Pergamon Press Oxford, New York, Toronto, Sydney, Paris, Frankfurt

Mook, J., 1992. *Minimum model error estimation.* Lecture held at the "Oberseminar für Mechanik". University of Hannover (unpublished)

Nalitolela, N.G., 1992. *A new approach to update model parameters using the frequency response data.* IMAC X, San Diego, CA, USA, 1267-1273

Natke, H.G., 1968. *Ein Verfahren zur rechnerischen Ermittlung der Eigenschwingungsgrößen aus den Ergebnissen eines Schwingungsversuches in einer Erregerkonfiguration.* Dissertation Technische Hochschule München. Engl. translation: A method for computing natural oscillation magnitudes from the results of vibration testing in one exciter configuration. NASA - TT - F - 12446, 1969

Natke, H.G. (Ed.), 1988a. *Application of System Identification in Engineering.* CISM Courses and Lectures No. 296. Springer-Verlag, Wien, New York

Natke, H.G., 1988b. *On methods of structure identification for a class of nonlinear mechanical systems.* IFAC, Beijing, China, 916-921

Natke, H.G. 1992a. *Einführung in Theorie und Praxis der Zeitreihen- und Modalanalyse - Identifikation schwingungsfähiger elastomechanischer Systeme.* Vieweg Braunschweig, Wiesbaden, 3. Auflage

Natke, H.G., 1992b. *On regularization methods within system identification.* Keynote Lecture, IUTAM Symposium, 11-15 May 1992, Tokyo, Japan. will appear

Natke, H.G., 1992c. *The future of structural system analysis.* Opening address of the 17th Intern. Conf. on Experimental and Numerical Methods in Structural Dynamics, 21-25 Sept. 1992. Leuven, Belgium

Natke, H.G., Collmann, D., Zimmermann, H., 1974. *Beitrag zur Korrektur des Rechenmodells eines elastomechanischen Systems anhand von Versuchsergebnissen.* VDI-Berichte Nr. 221, 23-32

Natke, H.G., Prells, U., 1991. *Contribution to structure identification of nonlinear mechanical systems.* Symposium on identification of nonlinear mechanical systems from dynamic tests. EUROMECH 280, Oct. 29-31, 1991

Natke, H.G., Yao, J.T.P., 1993. Detection and location of damage causing non-linear system behaviour. In: Natke, H.G., Tomlinson, G.R., Yao, J.T.P. (Eds.); 1993. *Safety Evaluation Based on Identification Approaches related to Time-variant and Nonlinear Structures.* Vieweg Verlag Braunschweig, Wiesbaden

Natke, H.G., Zamirowski, M., 1990. On methods of structure identification for the class of polynomials within mechanical systems. *ZAMM, Z. f. angew. Math. Mech.* 70, 1990, 10, 415-420

Niedbal, N., Klusowski, E., 1991. *Optimal exciter placement and force vector tuning required for experimental modal analysis.* Proc. Int. Conf. Spacecraft Structures and Mechanical Testing. Noordwijk, The Netherlands, 24-26 April 1991 (ESA SP-321, Oct. 1991)

NN, 1992. *Knowledge-based system "Measuring Techniques".* Research Contract: (VW-Foundation). Curt-Risch-Institute, University of Hannover

Pázman, A., 1986. *Foundations of optimum experimental design.* D. Reidel Publ. Comp. Dordrecht, Boston, Lancaster, Tokyo

Prells, U. 1991. *Regularisierte Modellfehlerlokalisierungen.* CRI F2/91, Research Report of the Curt-Risch-Institute of the University of Hannover

Prells, U., Ben-Haim, Y., 1993. *Selective sensitivity in the frequency domain, Part 2: Applications; Mechanical Systems and Signal Processing.* to appear

Shah, P.C., Udwadia, F.E., 1978. *A methodology for optimal sensor locations for identification of dynamic sytems.* Transactions of ASME, Vol. 45, March 1978, 188-196

Tikhonov, A.N., Arsenin, V.Y., 1977. *Solution of Ill-posed Problems.* Wiley, New York

Zamirowski, M., 1992. *Einige zeitdiskrete Parameterschätzmethoden zur Identifikation nichtlinearer mechanischer Systeme.* Dissertation, Universität Hannover, Curt-Risch-Institut, CRI-F-1/92

Structural Safety & Reliability, Schuëller, Shinozuka & Yao (eds) © 1994 Balkema, Rotterdam, ISBN 90 5410 357 4

Safety evaluation of structures using system identification approaches

H.G. Natke
Curt-Risch-Institute, University of Hannover, Germany

G.R. Tomlinson
Department of Civil Engineering, University of Manchester, UK

J.T.P. Yao
Department of Civil Engineering, Texas A&M University, College Station, Tex., USA

ABSTRACT: It is reported on results of the international workshop on Safety Evaluation Based on Identification Approaches Related to Time-variant and Nonlinear Structures held in Lambrecht, Pfalz, Germany on 6 - 9 September 1992, which was organized by the authors of this paper. During the workshop, several new methods for instrumentation, parameter identification, decision-making and damage assessment were discussed and explored. There seems to exist variability in several new methods for studying nonlinear and time-varying systems. In most studies, simulations were used and catagorised the effects of faults and there is a lack of application to real data. Work has been done to find procedures not requiring parametrization and into which a variety of engineering damage models can be utilized. This is a positive step toward the development of a practical and implementable damage assessment procedure.

1 INTRODUCTION

The paper deals with safety evaluation of existing and real systems, in mechanical, civil, aeronautical, naval, and other engineering disciplines. The following topics are relevant to this subject matter:

- signal processing including signal transforms,

- symptoms as sensitive quantities due to damage/faults and their combinations in the form of patterns and features,

- identification methods formulated as inverse problems which are ill-posed,

- model-based knowledge inference,

- damage assessment using fuzzy logic and neural networks, and

- decision-making for possible actions to be followed (policy for repair and system replacement).

This wide range of topics was discussed by invited participants of an international workshop on Safety Evaluation Based on Identification Approaches Related to Time-variant and Nonlinear Structures (Natke, Tomlinson, Yao, 1993) held in Lambrecht, Pfalz, Germany on 6 - 9 September 1992, which was organized by the authors of this paper. Results of this workshop are summarized in this paper. Fig. 1 indicates the topics covered, and Table 1 lists the participants and relates their work to the subject matter.

2 OVERVIEW OF THE WORKSHOP TOPICS

Starting with informal opening papers on "Review on Identification Methods" (Natke) and on "Choice of Training Variables for Neural Networks Used in Modelling a Class of Nonlinearities" (Tomlinson), the following categories of topics were treated:

- Test design, monitoring of structural behaviour (mainly advanced methods such as intensity measurements, spatially continuous measurements),

- damage detection and localisation by studying time-variant and nonlinear behaviour,

- methods of system identification (with respect to time-variant and nonlinear sytems) including mathematical solutions of the generally ill-posed problems in the inverse formulations, parameter identification using

SUMMARY

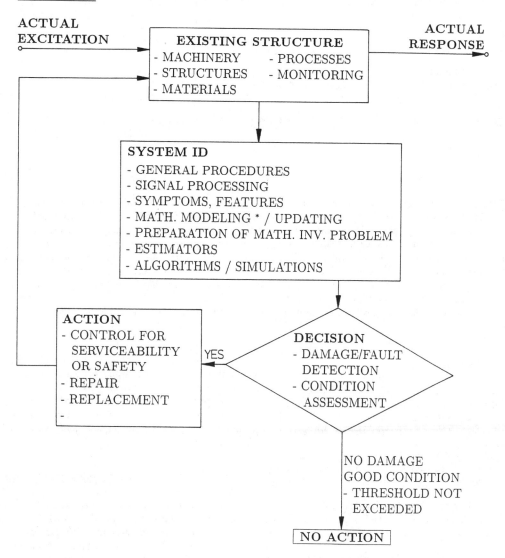

* INCLUDES DAMAGE EVOLUTION, UNCERTAINTY, ETC.

FIGURE 1

neural networks, treating the associated uncertainties, and their effects on structural reliability,

• procedures for safety evaluation with emphasis on time-variant and nonlinear systems (including neural nets and fuzzy logic).

An attempt at evaluating several classes of diagnostic methods was also presented.

3 STATE OF THE ART

This workshop was organized with the following objectives

1. to focus attention on the problem of damage assessment and safety evaluation of time-varying and nonlinear systems using identification approaches, and

2. to exchange information among a group of invited participants with expertise in a variety of subject areas ranging from mathematics via acoustics to engineering.

Four general subject areas were addressed in this workshop: The utilisation of new measurement techniques and of new sensors for damage detection in time-variant and nonlinear systems, damage detection and localisation methods including sensitivity investigations and regularisation methods, the identification of nonlinear systems, and damage evaluation.

Signal processing is essential in fault detection. In addition to the classical methods producing signatures and probabilistic-based functions in the time and frequency domain wavelets, the Hilbert transform and the Wigner distribution have been discussed.

Guided stress waves, propagating energy measurements, intensities as directional quantities can be used to localise sources and sinks (damage). These results from acoustics studies already are applied for structural waves in linear systems. However, methods for nonlinear structures are under further research and development.

Sensoring is a major topic in monitoring and inspection. Local contacting and non-contacting sensors are in common use, while spatially continuous (field) measuring (distributed sensors) devices represent a relatively new topic. It can be distinguished between fully, segmented, shaped and convolved designs of distributed sensors. Applications are in active structures (sensing and control) and for high-precision systems. Monitoring needs knowledge of the quantities to be measured. It is difficult to decide generally (independent on the system under measurement and independent on the method applied) which symptoms and features should be taken. In this context a user-friendly monitoring system with respect to screening and interpretation is important.

Several papers deal with damage detection and localisation based on system identification methods. Useful approaches exist for specific time-variant and nonlinear problems to be solved. System identification methods can be using forward strategy and an inverse problem formulation which is ill-posed. Ill-posed problems require regularisation for its stable solution. Dimensionality plays a role. The regularisation theory for linear (operator) equations is almost complete from a mathematical point of view, which is not true for nonlinear problems including parameter identification. Identification methods exist for special problems. The mostly missing solution is that of (model)

structure identification. This is substituted in many cases by parametrisation through polynomials (restoring surface method). First steps to apply neural networks are accomplished, neural network algorithms and training seems to be well established.

Model-based fault diagnosis with parameter estimation and knowledge-based inference is applied for electrical and structural processes and systems. The mathematical model used must be validated. Dynamic mathematical models may be used for fault detection and localisation. However, if the damage assessment is based on (local) maximum stresses, then a more detailed (static) model is needed.

In addition, damage measures and prediction/detection of damage evolution are important and are under development. This a step in the direction to include environmental effects. This research is closely connected with damage measures transformation into symptoms and reverse, and with sensitivity regarding the damage.

Specific damage detection and localisation schemes were applied for nonlinear composites using stationary and progressive waves, and for electrical towers using vibration signatures. In addition, faults and their causes of a general nature were described using finite element calculations in order to explain their effect on the response of nonlinear systems.

4 IMPACT OF THE WORKSHOP

Because of the diverse backgrounds of participants, each learned a lot from others. Meanwhile, much interaction and group dynamics enabled cross fertilisation and led to potential collaboration among several participants.

During the workshop, several new methods for instrumentation, parameter identification, decision-making and damage assessment were discussed and explored. There seem to exist variability in several new methods for studying nonlinear and time-varying systems. In most studies, simulations were used and catagorised the effects of faults and there is a lack of application to real data. In fact, there exists a huge communication gap between researchers and practitioners. A significant result of this workshop was to sensitise all participants as to the need for improved communication between researchers and practicing engineers.

Work has been done to find procedures not requiring parametrization and into which a variety of engineering damage models can be utilized.

TABLE 1

	Machinery	Materials	Structures	Robot	Gen. Procedures	Signal Processing	Symptoms / Features	Math. Model	Inv. Problem	Estimators	Algor./ Simulation	Decision	Uncertainty	Damage Assess.
Baumeister									X					
Ben - Haim			X			X	X		X			X		
Ben - Haim/Cempel Natke / Yao					X	X	X					X		
Cempel / Natke	X		X				X	X				X	X	X
Cowley	X	X	X		X						X	X		
Heckl			X		X								X	
Isermann	X			X	X		X	X	X	X	X	X		
Jezequel			X		X			X						
Luong			X				X						X	X
Masri	X	X	X	X	X	X		X		X	X			
Natke / Yao			X		X			X	X			X		X
Soong			X		X			X	X	X	X			X
Stubbs		X	X		X		X	X		X	X			X
Tomlinson					X	X	X	X						
Tzou	X	X	X	X	X	X	X	X					X	
Valente			X		X	X	X	X				X	X	X
Vinh			X				X						X	X
Zadeh					X	X	X	X	X	X	X	X	X	
Zhang	X						X	X						

This is a positive step toward the development of a practical and implementable damage assessment procedure.

5 FUTURE DEVELOPMENT AND TREND: MODEL-BASED DETECTION, DIAGNOSIS AND ASSESSMENT

Due to the diversity of nonlinear phenomena there is no universal and unique approach for safety evaluation of general systems. Detection and localisation of nonlinearities in systems, for example, due to damage/faults, needs more detailed investigation for each particular application (symptoms, patterns, features). Guided stress waves, intensity measurements may be developed for detection and localisation as possible methods for practical implementation. Active systems should be included and they will impact the investigation by new possibilities, for example, by using sensor/actuator signals, e.g. smart structures.

Improved simulations with validated models are necessary for studying possible faults and summarizing them in a catalogue/diagram as patterns for dignosis purposes. Precision in its meaning is not and can not be the aim. However, uncertainty considerations (uncertainty modelling) and error estimation are important and should be studied.

From a methodological point of view inverse formulations have to be applied carefully due to their ill-posedness. Regularisation methods for nonlinear problems, and sensitivity measures for parameter identification are needed and should include convergence considerations. Other alternative procedures such as neural nets and fuzzy sets are under investigation.

Structure (of the mathematical model) identification is an unsolved problem. Patterns of nonlinear behaviour obtained from special transformations but also from phase-plane diagrams should be used more intensively. Parametrical but physically not interpretable models are the first step within the procedures discussed, however, they must be developed in a direction which relates to physical interpretability.

In addition to model-based assessment, diagnosis and assessment based on fuzzy logic and neural nets deserves further attention and development. Neural networks have been used by several investigators dealing with nonlinear and time-varying systems with promising results. It is expected that more studies will be conducted using this relatively new tool.

There exist Type 1 and Type 2 errors (e.g., failure to replace the faulty system when it needs to be replaced, and replacement of the system when it is not necessary). A bigger problem along this line is to find the optimal policy which requires the evaluation of probabilities. There are too many factors involved in this problem to make precise analysis for making the optimal policy. High precision is incompatible with high complexity (the principle of incompatability). Damage is a matter of degree (e.g., slight damage or moderate damage) which may involve fuzziness.

It is clear that in order to combine identification procedures for nonlinear and time-variant systems with safety evaluation formulations some attention must be given to the diversity of the techniques and approaches used, as evident from the workshop. An accepted basis for evaluating the effectiveness and choice of a given methodology is needed which would provide a sub-grouping and hence a first step in the reduction of the diversity problem.

Generally, a holistic investigation which includes environmental effects as causes of damage as well as damage evolution is one of todays requirements.

REFERENCE

Natke, H.G., Tomlinson, G.R., Yao, J.T.P. (Eds.), 1993 *Safety Evaluation Based on Identification Approaches Related to Time-variant and Nonlinear Structures.* Vieweg Verlag Braunschweig, Wiesbaden

Structural Safety & Reliability, Schuëller, Shinozuka & Yao (eds) © 1994 Balkema, Rotterdam, ISBN 90 5410 357 4

Techniques for parameter estimation of structural systems

A. Nishitani & S. Suganuma

Department of Architecture, Waseda University, Tokyo, Japan

ABSTRACT: System identification in structural dynamics is of great significance in discussing structural reliability as well as structural response control. In performing structural analysis, a structural system is converted into certain mathematical model. Although it is not an easy task either to select or build suitable model for an existing or designed structure, appropriate estimation of those parameters involved in such a properly-selected or newly-established model is also a nontrivial issue. Parameter estimation from the observed data of inputs and outputs for a structure leads to the treatment of a nonlinear filtering technique even for a linear system. This paper demonstrates the usefulness and benefit of the statistical processing filter in comparison with the extended Kalman filter. The statistical processing filter is to estimate the parameters in corporation with the Monte Carlo simulation.

1 INTRODUCTION

System identification in structural dynamics has obtained considerable attention by many researchers and professional engineers in discussing not only structural reliability and safety but also structural response control.

In performing structural analysis, a structural system is converted into certain mathematical model. Although it is not an easy task either to find suitable model or to build new model for an existing or designed structure, appropriate estimation of those parameters involved in such a properly-selected or newly-established model is also a nontrivial issue. If accurate parameter estimation of the structural model can be completed, it will be really possible to assess the accuracy of the performed structural design and construction and also to evaluate the structural safety under certain external disturbance, say, earthquake excitation. Therefore, parameter estimation is one of the most important issues of system identification. The accumulation of these data for various types of civil engineering structures may provide useful and beneficial information to the structural design methodologies and also may be helpful for improving the design theory and philosophy. In addition, with these data available, accurate damage assessment can be possible for future big earthquake.

Parameter estimation from the observed data of inputs and outputs for a structure is usually performed by means of nonlinear filtering. In the structural engineering and engineering mechanics field, the extended Kalman filter has been often utilized to deal with these nonlinear filtering problems (Ljung 1979). However, there remain some problems in making practical application of this filter technique (Hoshiya & Saito 1983). This paper demonstrates that the improvement of parameter estimation is performed by using the statistical processing filter (Nakamura 1972 and Tsuji & Nakamura 1973). The technique presented in this paper provides a practical and useful method of the parameter estimation.

2 EXTENDED KALMAN FILTER AND UNKNOWN PARAMETER ESTIMATION

Consider a system described by the following discrete-time state equation:

$$x_{t+1} = f(x_t, u_t) + w_t, \qquad w_t \sim N(0, Q) \qquad (1)$$

in which

$\qquad t$: discrete time with constant sampling interval τ

$\qquad x_t$: n-dimensional state vector

$\qquad u_t$: r-dimensional input vector

$\qquad f(\cdot)$: state transition function

$\qquad w_t$: n-dimensional system noise vector

$N(\mu, V)$: Gaussian random variable vector with mean vector μ and covariance matrix V

By assuming that only certain part of the state vector is observed, the observation equation is written as:

$$y_t = Hx_t + v_t, \qquad v_t \sim N(0, R) \qquad (2)$$

in which

y_t : m-dimensional observation vector

H : $m \times n$ matrix specifying the states to be observed

v_t : m-dimensional observation noise vector

In the above equations, the noises w_t and v_t are assumed to be white noises independent of the past states prior to the time t. Of course, it would be desirable to represent the dynamic system without any noises included. But it is unfortunately impossible and hence the above expression is employed.

Supposing the input u_t and the output y_t are obtained, the states are given, on the basis of Bayesian estimation, as the following conditional expectations:

$$\hat{x}_{t+1-} = E[x_{t+1} \mid Y_t] \qquad (3)$$

$$\hat{x}_{t+} = E[x_t \mid Y_t] \qquad (4)$$

in which

$E[\cdot \mid Y_t]$: conditional expectation given Y_t

$\qquad Y_t$: information obtained from the observed data

In the above expression, $\hat{x}_{t-}$ and $\hat{x}_{t+}$ are referred to as "priori estimate" and "posteriori estimate" at the time t, respectively. The process of the state estimation, corresponding to Eq. 3, is called "prediction" and the process corresponding to Eq. 4 is referred to as "filtering". The covariance matrices of the estimate errors are

$$P_{t-} = E[(x_t - \hat{x}_{t-})(x_t - \hat{x}_{t-})^T] \qquad (5)$$

$$P_{t+} = E[(x_t - \hat{x}_{t+})(x_t - \hat{x}_{t+})^T] \qquad (6)$$

In general, however, Eqs. 3 and 4 cannot be easily completed. Only providing the linear state and linear observation equations, the conditional expected values can be described by the mean vector and covariance matrix alone. Through the repetition of prediction and filtering, x_t can be estimated along with the measurement of u_t and y_t. This estimation process is the Kalman filter (Kalman 1960 and Kalman & Bucy 1961), which is depicted schematically in Fig. 1.

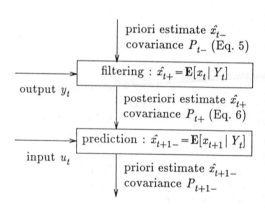

Figure 1. Filtering and prediction.

By assuming that the estimate obtained from the processes of filtering and prediction can be represented by the mean vector and covariance matrix even for a non-linear state equation, similar filter formulation as a linear state equation is constructed. By linear-expanding the non-linear state equation (Eq. 1) with respect to the expected value of the posteriori estimate, $\hat{x}_{t+}$, and applying the Kalman filter technique to the

linear-expanded non-linear system, the extended Kalman filter formulation is obtained in the following fashion:

$$\hat{x}_{t+1-} = f(\hat{x}_{t+}, u_t) \tag{7}$$

$$P_{t+1-} = F_t P_{t+} F_t^T + Q \tag{8}$$

$$\hat{x}_{t+} = \hat{x}_{t-} + K_t(y_t - H\hat{x}_{t-}) \tag{9}$$

$$P_{t+} = P_{t-} - K_t H P_{t-} \tag{10}$$

$$F_t = \frac{\partial}{\partial x^T} f(x, u_t)\Big|_{x=\hat{x}_{t+}} \tag{11}$$

$$K_t = P_{t-} H^T (HP_{t-} H^T + R)^{-1} \tag{12}$$

in which K_t is a $(n \times m)$ Kalman gain matrix, the superscript "-1" denotes a inverse matrix and the superscript T represents a matrix or vector transpose. Basically, the extended Kalman filter formulation involves some errors resulting from two approximations: (i) the states described by only the mean and covariance and (ii) linear expansion. However, these approximations are not necessarily reasonable. Then, the extended Kalman filter formulation based on them does not provide the conditional expectation i.e. the Bayesian estimate in the strict sense.

In the case of a system with some unknown parameters, Eq. 1 is rewritten as:

$$x_{t+1} = f(\theta; x_t, u_t) + w_t \tag{13}$$

in which θ denotes unknown parameter vector (l-dimensional). In dealing with the unknown parameters, the subscript t is artificially added to θ in similar way to the other states and noises, though the parameter vector θ is time-invariant. Considering the combination of the state and unknown parameter vectors, the system is described by

$$\begin{bmatrix} x_{t+1} \\ \theta_{t+1} \end{bmatrix} = \begin{bmatrix} f(\theta_t; x_t, u_t) \\ \theta_t \end{bmatrix} + \begin{bmatrix} w_t \\ 0 \end{bmatrix} \tag{14}$$

$$y_t = [H \ 0] \begin{bmatrix} x_t \\ \theta_t \end{bmatrix} + v_t \tag{15}$$

Equations 14 and 15 are now in the same form as Eqs. 1 and 2 with the elongated state vector $[x_t^T \ \theta_t^T]^T$. Therefore, simultaneous estimation of the state and unknown parameters can be conducted. In other words, the parameter estimation problem leads to the state estimation for a nonlinear system.

3 PARAMETER ESTIMATION BY STATISTICAL PROCESSING FILTER

The extended Kalman filter is formulated for a nonlinear system based on two approximations as mentioned above. However, this filter in some cases provides undesirable results.

In the statistical processing filter, on the other hand, the conditional expectations are evaluated through numerical simulations such as the Monte Carlo simulation in the following manner:

$$\hat{x}_{t+1-} = \frac{1}{N} \sum_{i=1}^{N} {}_i x_{t+1} \tag{16}$$

$$P_{t+1-} = \frac{1}{N} \sum_{i=1}^{N} ({}_i x_{t+1} - \hat{x}_{t+1-})$$
$$\times ({}_i x_{t+1} - \hat{x}_{t+1-})^T + Q \tag{17}$$

$${}_i x_{t+1} = f({}_i x_t, u_t) \tag{18}$$

in which
N : number of simulated samples
${}_i x_t$: i-th sample of x_t
Since the statistical processing filter technique evaluates the conditional expectations based on the ensemble average, it is particularly appropriate for the case where certain constrainted conditions required for the states and parameters should be taken into account. Specifically, those samples which do not satisfy the required condition can be omitted from the calculation of the ensemble average.

4 NUMERICAL EXAMPLES

For the purpose of demonstrating the effectiveness of the statistical processing

filter in estimating the unknown parameters, numerical examples are presented.

First of all, consider a single-degree-of-freedom (SDOF) structural system with unknown stiffness shown in Fig. 2.

mass : $m = 1$
stiffness : $k = 1000$
damping : $c = 3$

Figure 2. SDOF model.

By applying the linear acceleration method with the sampling interval $\tau = 0.01$ to the structural system subjected to ground acceleration, the equation of motion is converted into a discrete-time state equation. In this example, the ground acceleration is assumed to be a discrete-time Gaussian white noise expressed as $N(0, 1 \times 10^6)$. Because of its irregularity or randomness, the white noise is less favorable input as an excitation than realistic and practical input such as microtremor records. This example, then, deals with an unfavorable situation as an unknown parameter estimation problem. Suppose that measurement is to be made for the ground acceleration and the acceleration of the mass. System and measurement noises are assumed to be in the following expressions:

$$Q = 1 \times 10^{-4} \cdot Q_0 \tag{19}$$

$$R = 1 \times 10^{-4} \cdot R_0 \tag{20}$$

in which
Q_0 : covariance matrix of responses without system noise involved
R_0 : matrix consisting of only the diagonal elements of the covariance matrix of outputs without observation noise involved

Under these conditions, the observed ground acceleration with 500 discrete-time intervals is simulated and then the response acceleration of the mass can be computed. Both record samples are shown in Fig. 3.

Then, the unknown stiffness is to be estimated from both the observed records of

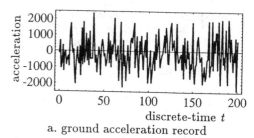

a. ground acceleration record

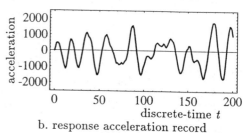

b. response acceleration record

Figure 3. Simulated observed records.

the ground and response accelerations. The initial estimates are assigned by

$$\begin{bmatrix} \hat{x}_{0-} \\ \hat{\theta}_{0-} \end{bmatrix} = \begin{bmatrix} 0 \\ \tilde{\theta} \end{bmatrix} \tag{21}$$

in which
$\tilde{\theta}$: Gaussian random sample expressed as $N(0, \{\alpha_\theta \cdot \text{diag}(\theta)\}^2)$
α_θ : constant coefficient for determining the initial estimated value of θ
$\text{diag}(\theta)$: diagonal matrix whose elements are the elements of the vector θ

In addition, the estimation error covariance matrix of the initial values is assumed to be idealized as

$$E\left[\begin{bmatrix} x_0 - \hat{x}_{0-} \\ \theta_0 - \hat{\theta}_{0-} \end{bmatrix} \begin{bmatrix} x_0 - \hat{x}_{0-} \\ \theta_0 - \hat{\theta}_{0-} \end{bmatrix}^T \right]$$
$$= \begin{bmatrix} Q_0 & 0 \\ 0 & \{\alpha_\theta \cdot \text{diag}(\theta)\}^2 \end{bmatrix} \tag{22}$$

Under the above condition, the extended Kalman filter (EKF) and processing filter (SPF) techniques are compared. Figure 4 illustrates how the unknown stiffness k is

being estimated by EKF, presenting a typical example with peculiar convergence of the EKF estimation. This is because the error resulting from the linear expansion inappropriately underestimates the covariance matrices of the estimation error.

improves the convergence of the estimation in early stage. In the following discussion, then, the SPF is to always reflect the constraints of non-negative values such as stiffness.

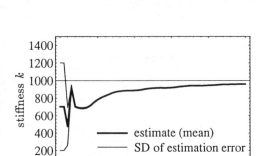

Figure 4. Convergence of estimate by EKF.

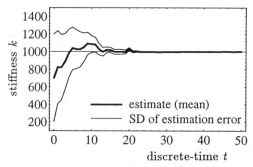

Figure 6. Convergence of estimate by conditional SPF.

In Fig. 5, SPF represents the result based on the ensemble average of the sample number $N=512$ with the same initial estimates under the same condition as Fig. 4. From the comparison between Fig. 4 and 5, superiority of SPF technique can be recognized. The estimation error becomes zero in the earlier stage for SPF (in Fig. 5) than for EKF (in Fig. 4).

In order to judge the superiority of estimation, the following index is introduced.

$$S_t = \log_{10} \| \operatorname{diag}^{-1}(\theta) \cdot (\hat{\theta}_{t+} - \theta) \| \qquad (23)$$

in which $\| \cdot \|$ denotes the Euclidian norm. For the purpose of inspecting the effect of the initially assigned values, i.e., the effect of the values of α_θ, the values of S_{500} are compared between SPF and EKF in Fig. 7. From this figure it is observed that EKF with the initial values $\alpha_\theta = 0.3$ or more does

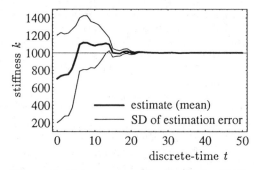

Figure 5. Convergence of estimate by SPF.

The results shown in Fig. 6 is obtained by considering the constraint that every simulated sample of the stiffness should be positive. The comparison between Fig. 6 and Fig. 5 indicates that the constraint

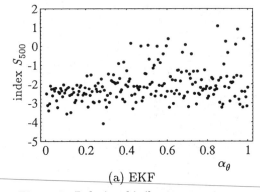

(a) EKF

Figure 7. Relationship between estimate error and initial parameter.

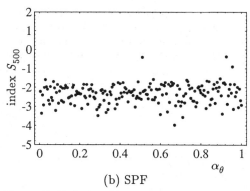

(b) SPF

Figure 7. Relationship between estimate
error and initial parameter.

not provide satisfying estimation (this is because of the errors resulting from the linearization), whereas SPF can manage to estimate the unknown parameters with dexterity even in starting with undesirable initial values.

Although the SPF carries advantageous benefits as mentioned above, it requires large amount of computation time in nature. The required time can be considered to be proportional to the sample number. Figure 8 shows the effect of the employed sample number on the estimation accuracy in the case of $\alpha_\theta = 0.5$. The increment of the sample number is found to be produce a stable estimation. In employing, however, the sample number of over 100, significant reduction of the accuracy is not recognized. The used computer program for

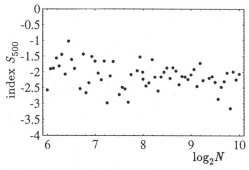

Figure 8. Relationship between estimate
error and number of samples.

SPF does not aim at the fastest efficiency. The computation time of EKF roughly amounts to that of SPF with 2 sample simulation.

More realistic and practical application is presented in the next example with a damped two-degree-of-freedom (2-DOF) system as shown in Fig. 9.

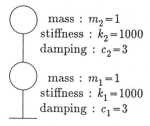

mass : $m_2 = 1$
stiffness : $k_2 = 1000$
damping : $c_2 = 3$

mass : $m_1 = 1$
stiffness : $k_1 = 1000$
damping : $c_1 = 3$

Figure 9. 2-DOF model.

In this example, all the stiffnesses k_1, k_2 and dampings c_1, c_2 are to be estimated, assuming that the mass values m_1, m_2 are known. Since the dampings are to be estimated in this case, the non-negative constraints is accounted for both dampings and stiffnesses through SPF. The other conditions are same as the SDOF model example.

Under these conditions the estimates are given by setting $\alpha_\theta = 0.1$ for a initial condition. The processes of estimation for the top storey damping c_2 and stiffness k_2 through EKF and SPF are compared in Fig. 10. SPF is accomplished with the sample number $N = 256$. It is observed from Fig. 10, that EKF do not provide the successful results, while, SPF produces much more satisfying results.

As a closing presentation, the relationship between the value of α_θ and the index S_{500} is shown in Fig. 11. EKF for this example cannot keep the accuracy of estimation at smaller values of α_θ than for SDOF model. In Fig. 11(a), the values of the index $S_{500} \geq 2$ appears along the ordinate of $S_{500} = 2$. This is induced by the errors due to the linear expansion of the complex structural system with respect to the inappropriate values at the early stage. On the contrary, SPF provides the stable and successful results, handling the inappropriate initial values.

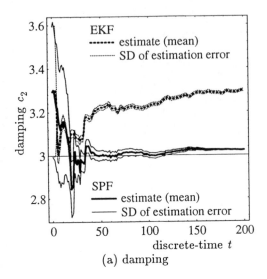

(a) damping

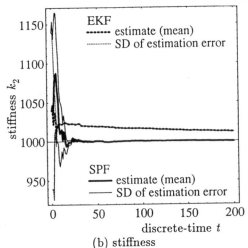

(b) stiffness

Figure 10. Convergence of estimator.

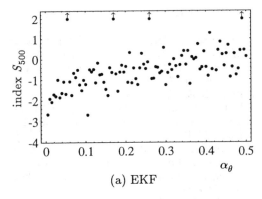

(a) EKF

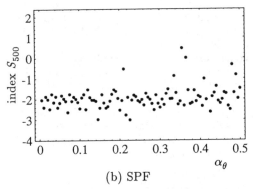

(b) SPF

Figure 11. Relationship between estimate error and initial parameter.

5 CONCLUDING REMARKS

Parameter estimation leads to the treatment of a nonlinear filtering technique even for a linear system. Among several nonlinear filtering techniques, the extended Kalman filter has been often utilized in many cases because of simple formulation. However, it is known that successful results are not always obtained from the extended Kalman filter.

In this paper, both the statistical processing filter and the extended Kalman filter are employed in evaluating the unknown parameters involved in the mathematical models of structural systems. On the basis of the comparison between the results provided by these two filters, it is concluded that the statistical processing filter can manage the initial errors with much more dexterity and efficiency, i.e., it can provide more successful estimates in the early stage even under such a severe condition as dealing with several parameter estimation.

In the statistical processing filtering, the complexities associated with nonlinear problems are handled by large amount of computation. However, considering the employment of suitable sampling techniques and also the development of high-speed computer, the statistical processing filter is expected to be a practical and efficient tool in performing the parameter estimation for civil engineering structures.

REFERENCES

Hoshiya,M. & Saito,E. 1983. Identification problem of some seismic systems by extended Kalman filter. *Proc. JSCE*, No.339: 59-67. (in Japanese)

Kalman,R.E. 1960. A new approach to linear filtering and prediction problems. *Trans. ASME, J. Basic Eng.*, Vol.82D, No.1: 35-45.

Kalman,R.E. & Bucy,R.S. 1961. New results in linear filtering and prediction theory. *Trans. ASME, J. Basic Eng.*, Vol.83D, No.1: 95-108.

Ljung,L. 1979. Asymptotic behavior of the extended Kalman filter as a parameter estimator for linear system. *IEEE Trans. on Automatic Control*, Vol.AC-24, No.1: 36-50.

Nakamura,M. 1972. Nonlinear estimation using a statistical processing technique. *Trans. IEEJ*, Vol.92-C, No.2: 81-89. (in Japanese)

Tsuji,S. & Nakamura,M. 1973. Nonlinear filter using a statistical processing technique. *Trans. IEEJ*, Vol.93-C, No.5: 109-116. (in Japanese)

Structural Safety & Reliability, Schuëller, Shinozuka & Yao (eds) © 1994 Balkema, Rotterdam, ISBN 90 5410 357 4

Parameter identification and probabilistic prediction of settlement in geotechnical engineering

Makoto Suzuki & Kiyoshi Ishii
Ohsaki Research Institute, Shimizu Corporation, Tokyo, Japan

ABSTRACT: The safety and stability of proposed soil structures depend upon accurate predictions of the ground displacement that will result from construction. The finite element method is an effective tool for predicting ground displacements but, the analytical solution produced is only as accurate as the soil properties used in calculation. However, reliable soil properties are often difficult to obtain. To overcome this parameter-estimation problem, random field theory was employed to introduce back analysis, using the Kalman filter technique, into a finite element solution approach. Credible predictions, based on the stochastic finite element analysis, can be performed using the results. This paper describes the formulation of the method and discusses the applicability of the proposed method using an embankment example.

1 INTRODUCTION

The reliable assessment of soil-structure safety is an important issue in geotechnical engineering. Methods for predicting structure behavior can be divided into three categories with respect to their time of application in the construction process (Lambe 1973). "Type A" predictions are deterministic numerical solutions that provide information prior to construction. "Type B" predictions are numerical evaluations based on field observations that are made during construction and, "type C" predictions are those made after construction has been completed. All types of assessment are essential in the design and construction of soil structures.

The finite element method has been a successful analytical tool for the prediction of ground settlement. However, the reliability of finite element analysis is dependent upon accurate estimation of soil properties. Unfortunately, it is not always easy to obtain precise soil property values through site investigation. The randomness and uncertainty, inherent to soil properties, produce an array of test and statistical errors. The significance of the spatially varying soil properties makes it natural to treat them as random functions in spatial coordinates (Vanmarcke 1977). The application of probability theory or random field theory to soil modeling has recently enabled the solution of various geotechnical engineering problems by stochastic finite element analysis (Baecher & Ingra 1981; Ishii & Suzuki 1989).

When ground settlement is observed during the construction process soil parameters are identified using a technique known as "back analysis". This technique presents a rational method to numerically interpret field performance, such as final settlement (Arai et al. 1983; Goida 1985; Goida & Sakurai 1987). Back analysis using the Kalman filter technique (Hoshiya & Saito 1984) was incorporated into the finite element method to determine the optimal location of field observation points (Murakami 1988).

The soil properties at two neighboring locations are rarely identical. Because of this, additional observed values should be used to interpret the spatial variation of the soil properties. If soil properties are allocated to each element of the finite element method, the number

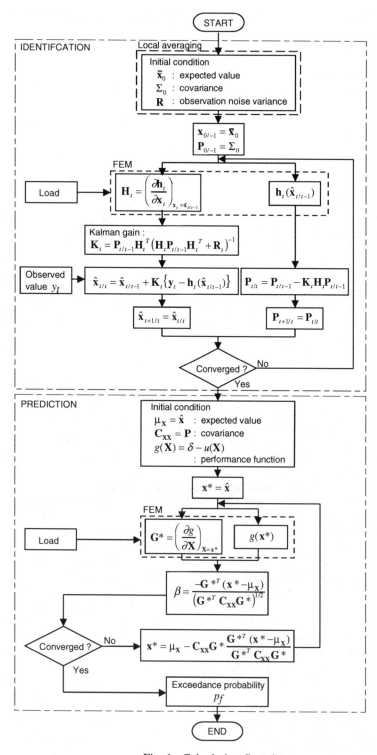

Fig. 1   Calculation flowchart

of unknown parameters is generally larger than the number of observed values. The formulation of the approach is therefore similar to that of Bayes and, the estimation of the unknown parameters can be performed using the extended Kalman filter algorithm (Cividini et al. 1983). Furthermore, the soil properties estimated can also be used to provide a continuous probabilistic prediction of settlement when an area is subjected to additional loading.

2 FORMULATIONS OF IDENTIFICATION AND PREDICTION METHODS

2.1 Extended Kalman Filter Approach using FEM

In the extended Kalman filter approach, the following observation and state equations of the nonlinear system dominate the field relationship governing the geomechanical problem, the Taylor expansion is adopted for linearization (Kalman & Bucy 1961) as follows:

$$y_t = h_t(x_t) + v_t, \tag{1}$$
$$x_{t+1} = f_t(x_t) + G_t w_t, \tag{2}$$

where x_t is a state vector, y_t is an observation vector, h_t is a nonlinear vector function of observation, v_t is an observation noise, f_t is a nonlinear vector function of state transition, G_t is a driving matrix, and w_t is a system noise . Equation 2 represents a discrete formulation where subscript t denotes the time or iteration step number. The stochastic characteristics of the observation and system noises are assumed to be independent Gaussian white noises.

The formulation of the extended Kalman filter approach in conjunction with finite elements will be given below for the purpose of estimating the spatially varying soil property x_t using the observed values. To formulate the extended Kalman filter described above, the stiffness equation is rewritten in the inverted form

$$u = K^{-1}f . \tag{3}$$

where K is a stiffness matrix formulated by the state vector of soil properties, u is a displacement vector, and f is a load vector. An observation equation is defined as

$$y_t = u + v_t$$
$$= h_t(x_t) + v_t . \tag{4}$$

Since soil properties are constant, the following stationary condition of parameters defines the state equation for the Kalman filter:

$$\hat{x}_{t/t+1} = \hat{x}_{t/t} . \tag{5}$$

This condition should be deterministic so that the state equation does not include system noise.

In a Bayesian approach (Cividini et al. 1983), initial estimate values and a covariance matrix of estimate errors should be given for the mean values and a covariance matrix of prior distribution. This approach is effective when the number of unknown parameters is greater than the number of measurements.

Due to the nonlinearity of the observation equation as seen in Eq. (4), an iterative procedure is adopted for fixed observed displacements until the convergence of the identified parameters can be achieved. In this case, the subscript t in Eqs. (4) and (5) does not denote the time axis but rather the "iterative axis."

2.2 Stochastic FEM using identified results

An analytical approach to the stochastic finite element method has been developed by introducing the advanced first-order approximation at a design point in a space consisting of random variables (Ang & Tang 1984). The estimate values and error covariance of x, calculated by the former procedure, are applied to the stochastic finite element analysis. When a critical displacement δ is given, its exceedance probability can be incorporated into a cumulative distribution function. In the calculation, a performance function g(x) is defined as follows :

$$g(x) = \delta - u(x) \tag{6}$$

where u(x) is a settlement obtained by this method. In Eq.(6), an exceedance probability means a probability of negative g(x).

Figure 1 illustrates the computation flowchart for both the identification and prediction procedures.

3 NUMERICAL EXAMPLE

3.1 Identification of soil properties

To examine the validity of the above mentioned

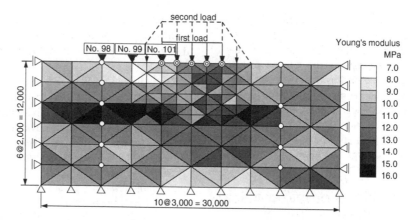

Fig. 2   Analytical model

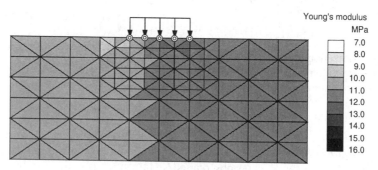

Fig. 3   Estimate values from vertical displacements
(case 1: 5 observation points)

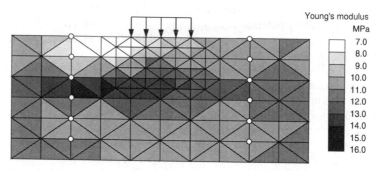

Fig. 4   Estimate values from horizontal displacements
(case 2: 12 observation points)

procedure, the spatially varying Young's modulus in a soil stratum is to be identified for the vertically and horizontally displacements observed when a load acts on the surface. Figure 2 shows the analytical model and provides a finite element discretization under the two-dimensional plane strain condition. In this figure, symbol ◎ and symbol ○ denote the observation points of the vertical and horizontal displacements respectively. Young's moduli are assumed to constitute a random field having a mean value of 10.0 MPa, a coefficient of variation of 0.2, and an

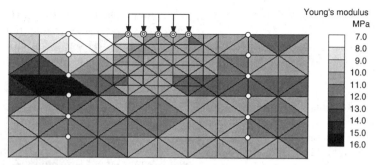

Young's modulus
MPa

| | 7.0 |
| 8.0 |
| 9.0 |
| 10.0 |
| 11.0 |
| 12.0 |
| 13.0 |
| 14.0 |
| 15.0 |
| 16.0 |

Fig. 5   Estimate values from vertical and horizontal displacements
(case 3: 17 observation points)

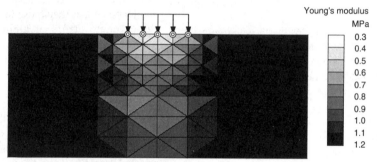

Young's modulus
MPa

| 0.3 |
| 0.4 |
| 0.5 |
| 0.6 |
| 0.7 |
| 0.8 |
| 0.9 |
| 1.0 |
| 1.1 |
| 1.2 |

Fig. 6   Estimate errors from vertical displacements
(case 1: 5 observation points)

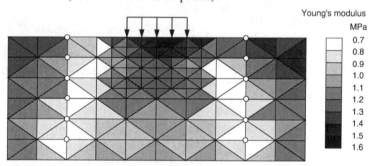

Young's modulus
MPa

| 0.7 |
| 0.8 |
| 0.9 |
| 1.0 |
| 1.1 |
| 1.2 |
| 1.3 |
| 1.4 |
| 1.5 |
| 1.6 |

Fig. 7   Estimate errors from horizontal displacements
(case 2: 12 observation points)

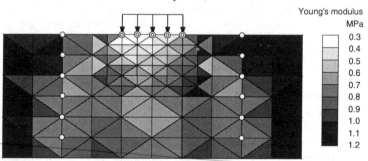

Young's modulus
MPa

| 0.3 |
| 0.4 |
| 0.5 |
| 0.6 |
| 0.7 |
| 0.8 |
| 0.9 |
| 1.0 |
| 1.1 |
| 1.2 |

Fig. 8   Estimate errors from vertical and horizontal displacements
(case 3: 17 observation points)

auto-correlation function given by

$$\rho(\Delta x, \Delta y) = \exp\left[-\left\{\left(\frac{\Delta x}{5.0}\right)^2 + \left(\frac{\Delta y}{1.0}\right)^2\right\}\right] \quad (7)$$

The first load is produced by an embankment of 0.24 MPa. The Poisson's ratio is treated as a constant of 0.3. A sample field of the spatially varying Young's moduli is first generated by the simulation method. This field is to be used for comparison with the results from the identification procedure. The Young's modulus of each element is assessed by a local averaging technique.

Figure 2 also illustrates the targeted values of the elements in relation to Young's modulus. In the analysis, the Young's moduli of all elements are regarded as state values and will be identified from observed values. The darker shades of the elements correspond to the higher values of Young's moduli while the lighter shades correspond to the lower values. The statistical properties of Young's moduli are assumed to be known.

In order to evaluate the applicability of this procedure, three cases are examined. In case 1, the observed data is given only for vertical displacements (settlements). In case 2, the observed data is given only for horizontal displacements. In case 3, the observed data is given for both vertical and horizontal displacements (case 1 + case 2). Using the back analysis and assuming that the same Young's modulus value exists in all elements, an initial value is estimated to be approximately 11.0 MPa.

The results of the estimate values in cases 1, 2 and 3 are shown in Figs. 3, 4 and 5 respectively. Figure 3 estimates the values using vertical displacements. This results in a high deviation from the analytical model as a whole but the values near the observation points closely match those at the top of the analytical model. Figure 4 uses horizontal displacements to estimate the values of the elements. By doing so, the overall values closely resemble those of the analytical model, especially around the observation points. These results indicate that the values surrounding the observation points tend to be more consistent with the analytical model then those values which are not. Figure 5 estimates the values by using both vertical and horizontal displacements. The

values obtained by this method most accurately match those of the analytical model. The patterns in Figs. 3, 4 and 5 show that the horizontal correlation is stronger than the vertical correlation because the initial spatial correlation defined by Eq. (7) obviously affects the estimate values.

Also, the results of estimate errors for cases 1, 2 and 3 are shown in Figs. 6, 7 and 8 respectively. In Fig. 6, the estimate errors under the embankment are below 0.5 MPa, which is approximately one quarter of the initial values (2.0 MPa). In Fig. 7, the estimate errors as a whole are greater than those in Fig. 6. However, the estimate errors of elements around the observation points in Fig. 7 are much smaller than the same element's values in Fig. 6. The result in case 3 is a combination of the case 1 and 2 results, which is shown in Fig. 8.

Since the results depend on the number and/or location of observation points, the procedure may not generate the uniqueness of the results. However, when the observed locations are selected carefully, the estimate errors of Young's moduli can be reduced greatly.

3.2 Prediction of settlement

When an additional loading acts on the same ground model, by which the repaired embankment is approximately twice as high as the original embankment, continuous probabilistic prediction is performed by the stochastic finite element method using the estimate values and errors of the Young's moduli derived in section 3.1. Three sets of the cumulative probabilities of settlements at Nos. 98, 99 and 101 are shown in Figs. 9, 10 and 11 respectively. In Fig. 9, the variations of the predicted settlements are rather small and a precise prediction can be performed. The horizontal displacement is only observed at No. 98. In Fig. 10, variations for cases 1 and 2 are same and their expected values are different when the observed data is not given at No. 99. In Fig. 11, variations for cases 1 and 3 are small when compared to case 2. However, the coefficients of variation for cases 2 and 3 are less than 5% as shown in Fig. 9. Small variations in the predicted second-stage settlements can be evaluated by comparison with the initial values in each case.

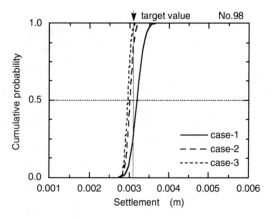

Fig. 9 Cumulative probability of settlement at the point No. 98

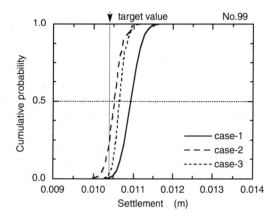

Fig. 10 Cumulative probability of settlement at the point No. 99

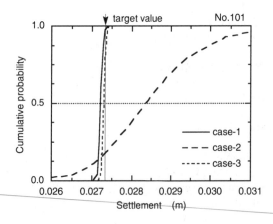

Fig. 11 Cumulative probability of settlement at the point No. 101

4 CONCLUSION

The back analysis procedure using the extended Kalman filter technique, which is incorporated into the finite element method, was proposed to identify the spatially varying soil properties from the observed data. As the derivation of soil properties is assumed to be done in a sample field (a realization of the random field), the initial condition of estimate values and a covariance of estimate errors are adopted by using the expected value and the covariance of random field. A numerical example of an embankment, which was modeled by the two-dimensional plane strain elements, was applied and the results were confirmed. Probabilistic prediction was also applied using the results thus identified.

Four specific points produced by this study can be summarized as follows. First, the proposed method can be performed when the number of unknown parameters is larger than the number of observed values, which is usually the case. The resulting Young's moduli can suitably represent the target values. Second, estimate errors can be evaluated simultaneously, since the concept is based on the Bayes' theory. Third, the observation points should be carefully selected, as the observed values affect the estimated result. Fourth, accurate prediction based on the stochastic finite element analysis can be performed using the estimate values and errors.

REFERENCES

Ang, A.H.-S.& W.H. Tang 1984. Probability Concepts in Engineering Planning and Design, Vol. II - Decision, Risk, and Reliability, John Wiley & Sons.

Arai, K., H. Ohta & T. Yasui 1983. Simple optimization techniques for evaluating deformation moduli from field observation. Soil and Foundations 23: 107-113.

Baecher, G.B. & T.S. Ingra. 1981. Stochastic FEM in settlement prediction. Jour. Geotech. Eng. ASCE, 107 (GT4): 449-463.

Cividini, A., G. Maier & A. Nappi 1983. Parameter estimation of a static geotechnical model using a Bayes' approach. Int. Journal of Rock Mechanics Mining Sciences and Geomech. 20: 215-226.

Gioda G. 1985. Some remarks on back analysis and characterization problems in geomechanics. Proc. 5th ICONMG: 47-61. Nagoya.

Gioda G. & S. Sakurai. 1987. Back analysis procedures for the interpretation of field measurements in geomechanics. Int. Jour. Numer. Anal. Methods Geomech. 11: 555-583.

Hoshiya, M. & E. Saito 1984. Structural identification by extended Kalman filter. Jour. Eng. Mechanics, ASCE, 110 (EM12): 1757-1770.

Ishii, K. & M. Suzuki 1989. Stochastic finite element analysis for spatial variations of soil properties using kriging technique. Proc. 5th ICOSSAR: 1161-1168. SanFrancisco.

Kalman, R.E. & R.S. Bucy 1961. New results in linear filtering and prediction theory. Jour. Basic Eng. Trans. ASME, 83: 95-108.

Lambe, T.W. 1973. Predictions in soil engineering. Geotechnique 23:149-202.

Murakami, A. & T. Hasegawa 1988. Back analysis by Kalman filter-finite elements and optimal location of observed points. Proc. 6th ICONMG: 2051-2058, Innsbruck.

Vanmarcke, E.H. 1977. Probabilistic modeling of soil profiles. Jour. Geotech. Eng. ASCE, 103 (GT11): 1227-1246.

Structural Safety & Reliability, Schuëller, Shinozuka & Yao (eds) © 1994 Balkema, Rotterdam, ISBN 90 5410 357 4

A system identification technique with unknown input information

Duan Wang & Achintya Haldar
Department of Civil Engineering and Engineering Mechanics, University of Arizona, Tucson, Ariz., USA

ABSTRACT: An element level time domain system identification procedure is proposed to evaluate a structural system. This procedure does not need any information on input excitation forces. Also, this method has no restriction on types of input forces and output responses, and no information requirement on modal properties of the structures. Only a small number of observation time points are required. The unknown exciting forces can also be applied at ground level representing seismic excitation. This method is verified using two examples. For verification purposes, both the noise-free and noise-included output responses are considered. In all cases, the results of examples indicate that the proposed method identifies the structural parameters very well.

1. INTRODUCTION

Nondestructive evaluation (NDE) methods to study the damaged states of existing structures have recently received increasing attention in our profession. The axiom that the extent of degradation will be reflected in the changes in the behavior of the structure is applicable, and in turn is dependent on the changes in the structural parameters at the element level in terms of stiffness and damping characteristics and their variations with time. Based on this concept, the most attractive among many promising NDE methods is a method based on the structural dynamic response measurements to identify a suitable mathematical model to represent the structure under consideration. This mathematical model should change to reflect changes in the physical state of the structure. One approach that can be used very effectively for this purpose is the use of system identification (SI) techniques.

The use of SI for NDE problems has expanded in recent years. Although this approach is very appealing, it has several limitations which reduce its practical applications. In most system identification approaches, the information on input loading and output responses must be known. In many cases, measuring the input information may take most of the resources, and it is very difficult to accurately measure the input information during actual vibrations of practical importance, e.g., earthquakes, winds, microseismic tremors, mechanical vibration, etc. However, the desirability and application potential of SI to real structures could be highly improved if an algorithm is available which can estimate structural parameters based on the response data alone without the input information. The development of such an algorithm is the subject of this paper.

Since damage detection at the element level is a major concern, time-domain SI techniques are more desirable for this purpose. The available time-domain SI approaches applicable to civil engineering structures can be divided into two

groups: (1) methods where input excitation information is necessary, and (2) methods where input excitation information is not necessary. When input excitation information is available, some of the commonly used methods are the recursive Least-Squares algorithm, the Instrument Variable method, the Maximum Likelihood method, the Kalman Filter technique and the extended Kalman Filter technique.

However, as stated earlier, time-domain SI without input excitation information is desirable for wider applications and is the subject of this paper. An attempt has been made to summarize the state of the art in the area of system identification without input information and the deficiencies in the available methods. After an extensive literature survey, the following five techniques have been identified where input information is not required. These are: (1) Fourier Analysis; (2) Adaptive Filtering Techniques; (3) Free-Decay Curve Analysis; (4) Stochastic Approach; and (5) Random Decrement Method. The details of these methods can not be discussed here due to lack of space.

To justify the desirability of the method proposed in this paper, the major deficiencies of these techniques are identified below. For most of these techniques, the modal properties are identified first and the system parameters are identified from the modal properties (Fourier Analysis, Adaptive Filtering Techniques, and Random Decrement Method). Because the modal property is a global property and a reliable estimation of only a few lower modes of vibration are generally available, this is not desirable for complicated structures. It will fail to detect the exact location of the damaged structural element or elements. In addition, in all these methods, the requirements for input excitations and output measurements are very restrictive in nature, e.g., the input is impulsive (Yang, 1985), or it has a zero mean (Stochastic Approach) and the output is free-decay type [Random Decrement Method and Free-Decay Curve Analysis (Yang 1985, Toki et al. 1989)].

Rather than representing actual operating conditions in real structures, it may not be possible to record the free-decay type response, thus making these techniques inappropriate to identify existing structures.

From this brief discussion it is clear that a robust time-domain SI technique needs to be developed to identify a structure at the element level without information on input excitation and without any restriction on input excitation and output response types.

2. PROPOSED METHOD

A method is proposed here in order to eliminate some of the deficiencies in the currently available time-domain SI techniques without input information. The proposed method is based on the linear finite element-based algorithm. This method is expected to be very economical, simple and easy to implement since input information is not required. This procedure is discussed very briefly below.

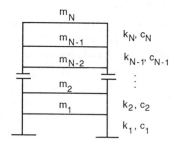

Fig 1. N-story shear building

Without losing any generality, the governing equation of motion of a N-story shear building shown in Fig. 1 can be written as:

$$M \ddot{X}(t) + C \dot{X}(t) + K X(t) = f(t) \qquad (1)$$

where M is the diagonal mass matrix and can be expressed as:

$$M = \text{diag} (m_1, m_2, \cdots, m_N) \qquad (2)$$

K is the damping matrix and C is the stiffness matrix and can be shown to be:

$$K = \begin{bmatrix} k_1+k_2 & -k_2 & \cdots & 0 & 0 \\ -k_2 & k_2+k_3 & \cdots & 0 & 0 \\ \cdots & \cdots & \cdots & \cdots & \cdots \\ 0 & 0 & \cdots & k_{N-1}+k_N & -k_N \\ 0 & 0 & \cdots & -k_N & k_N \end{bmatrix} \qquad (3)$$

$$C = \begin{bmatrix} c_1+c_2 & -c_2 & \cdots & 0 & 0 \\ -c_2 & c_2+c_3 & \cdots & 0 & 0 \\ \cdots & \cdots & \cdots & \cdots & \cdots \\ 0 & 0 & \cdots & c_{N-1}+c_N & -c_N \\ 0 & 0 & \cdots & -c_N & c_N \end{bmatrix} \qquad (4)$$

and $\ddot{X}(t)$, $\dot{X}(t)$, and $X(t)$ are acceleration, velocity and displacement vectors; and $\mathbf{f(t)}$ is the force vector.

Assuming M to be known, Eq. 1 can be rewritten as:

$$[C \vdots K] \begin{bmatrix} \dot{X}(t) \\ X(t) \end{bmatrix} = f(t) - M \ddot{X}(t) \qquad (5)$$

For an N-story shear building, this equation can be rearranged as:

$$[A]_{N\times2N} [P]_{2N\times1} = [F(t)]_{N\times1} \qquad (6)$$

where [A] is an $N\times2N$ matrix composed of the response vectors of velocity and displacement; [P] is a $2N\times1$ vector composed of the unknown system parameters, which are damping and stiffness; and [F(t)] is an $N\times1$ vector composed of input and inertia forces at any time t.

Suppose the response of the structure is measured for a duration of $m\cdot\Delta t$ at all stories, where m is the number of sample points, and Δt is the constant time increment. Then Eq. 6 can be rewritten as:

$$[A]_{(m\times N)\times2N} [P]_{2N\times1} = [F]_{(m\times N)\times1} \qquad (7)$$

The matrix [F] in Eq. 7 can be shown to be:

$$[F] = [F(t_0), F(t_1), \cdots, F(t_m)]^T \qquad (8)$$

where $F(t_i)$ at any time t_i for all N stories can be shown to be:

$$F(t_i) = \begin{bmatrix} f_1(t_i)-m_1\ddot{x}_1(t_i) \\ f_2(t_i)-m_2\ddot{x}_2(t_i) \\ \cdots \\ f_N(t_i)-m_N\ddot{x}_N(t_i) \end{bmatrix} \qquad (9)$$

Similarly, the matrix [A] in Eq. 7 can be expressed as:

$$[A] = [A(t_0), A(t_1), \cdots, A(t_m)]^T \qquad (10)$$

where $A(t_0)$, $A(t_1)$,$\cdots$, $A(t_m)$ are the response quantities at time t_0, t_1,$\cdots$, t_m, respectively; and

$$A(t_i) = \begin{bmatrix} \dot{x}_1 & \dot{x}_1-\dot{x}_2 & \cdots & 0 & \vdots & x_1 & x_1-x_2 & \cdots & 0 \\ 0 & \dot{x}_2-\dot{x}_1 & \cdots & 0 & \vdots & 0 & x_2-x_1 & \cdots & 0 \\ \cdots & \cdots & \cdots & \cdots & \vdots & \cdots & \cdots & \cdots & \cdots \\ 0 & 0 & \cdots & \dot{x}_{N-1}-\dot{x}_N & \vdots & 0 & 0 & \cdots & x_{N-1}-x_N \\ 0 & 0 & \cdots & \dot{x}_N-\dot{x}_{N-1} & \vdots & 0 & 0 & \cdots & x_N-x_{N-1} \end{bmatrix} \qquad (11)$$

and the unknown system parameters [P] in Eq. 7 can be shown to be:

$$[P]= \begin{bmatrix} c_1 & c_2 & \cdots & c_N & \vdots & k_1 & k_2 & \cdots & k_N \end{bmatrix}^T \qquad (12)$$

For mathematical convenience, Eq. 7 can also be expressed as:

$$\sum_{s=1}^{2N} A_{rs} P_s = F_r \qquad r=1,2,\cdots,m\times N \qquad (13)$$

Suppose $\hat{P}_s$ is the predictor of the sth system parameter P_s that needs to be evaluated, i.e., the stiffness and damping. Then the total error E of the system can be shown to be:

$$E = \sum_{r=1}^{mxN} (F_r - \sum_{s=1}^{2N} A_{rs} \hat{P}_s)^2 \qquad (14)$$

To minimize the total error, Eq. 14 can be differentiated with respect to each one of the $\hat{P}_q$ parameters as:

$$\frac{\partial E}{\partial \hat{P}_q} = \sum_{r=1}^{mxN} (F_r - \sum_{s=1}^{2N} A_{rs} \hat{P}_s) A_{rq} = 0 \qquad (15)$$

where $q = 1, 2, \cdots, 2N$.

Eq. 15 gives 2N simultaneous equations. The solution of Eq. 15 will give all 2N unknown parameters. For a structure with known input information, if the responses can be measured at all stories, then all the 2N system parameters can be evaluated easily using Eq. 15. However, if the input information is unknown, Eq. 15 can not be solved directly.

The identification of the system parameters without input information is under consideration in this paper. In general, the structure can be excited at any or all stories, including at the base. The exciting force can be anything including impulsive, sinusoidal, random or very irregular as in earthquakes.

The proposed procedure addresses all these issues. It is an iterative procedure. To start the iteration process, it is necessary to have information on input excitation. Since the input excitation is not available at any time, the iteration can be started by assuming it is zero at time t_i, $i=1, 2, \cdots, p$, where p is the number of time points to be determined as discussed below, and $p \leq m$. From a practical point of view, since it is not correct to assume zero input excitation at all time points, p must be kept to a minimum without compromising the convergence or the accuracy of the proposed method. It will be elaborated further that p can be only 2 points if the structure is excited at any floor; and only 4 points if the structure is excited at the base representing seismic motion. The proposed algorithm is not sensitive to this initial assumption. The basic concept of the proposed algorithm can be described in the following steps.

Iterative Steps

1. Case a - For a case with the input forces at the ith floor, the input forces $f_i(t_0)$ and $f_i(t_1)$ need to be assumed for at least 2 time points t_0 and t_1 in order to obtain a non-singular solution of Eq. 15. In this study, since the information is not available, they are assumed to be zero.
Case b - If seismic loading is under consideration, i.e. the exciting force is applied at the base of the structure, the input forces need to be known for at least at four time steps t_0, t_1, t_2, and t_3. Again, since the information is not available, it is assumed to be zero.

2. Since the response quantities are available at all stories at m points, $F(t_0)$ and $F(t_1)$ can be obtained at the first two time steps using Eq. 9. For seismic motion, it will be evaluated at the first four time steps. Considering Eq. 8, it can be shown that the vector [F] will be of dimensions 2N x 1 for Case a and 4N x 1 for case b.

3. Referring to Eq. 15, the first estimation of the system parameter matrix [$\hat{P}$] can be obtained by solving 2N simultaneous equations since the output response information A_{rs} is available and the required F_r values are estimated in Step 2.

4. Using the information on the system parameters [$\hat{P}$] obtained in Step 3, the unknown input force $f_i(t)$ at all m points now can be generated using Eq. 1.

5. Using the generated input time history at the total m observation points obtained in Step 4, the next estimation of [$\hat{P}$] can be obtained by solving Eq. 15.

6. The information on [$\hat{P}$] obtained in Step 5 can be used to again solve Eq. 1 and the input forces at t_0 and t_1 can be obtained for Case a. For Case b, i.e., seismic excitation, the information on seismic input excitation at 4 points can be similarly obtained.

7. Steps 2 through 6 need to be reiterated until

the convergence of input forces at t_0 and t_1, or t_0, t_1, t_2 and t_3 can be obtained with a predetermined accuracy. Once the algorithm converges, the updated $[\hat{P}]$ will give the unknown system parameters.

The proposed algorithm needs to be verified at this stage. Two examples are considered in the following section to clarify and amplify several desirable features of the proposed method.

3. NUMERICAL EXAMPLES

Example 1: A two-story shear building is considered first, as shown in Fig. 1 with N = 2. A proportional damping matrix is assumed for this example. The actual values of the parameters are:

$$m_1 = 136, \quad c_1 = 307, \quad k_1 = 30700$$
$$m_2 = 66, \quad c_2 = 443, \quad k_2 = 44300$$

The structure is assumed to be excited by a sinusoidal force $f(t) = 10000 \sin(20t)$, applied horizontally at the top floor level. The theoretical responses are calculated in terms of displacement, velocity, and acceleration of the structure at the two mass points. To describe the proposed methodology, these response quantities are assumed to be measured at both locations of the structure. The input sinusoidal exciting force is completely ignored. The task is to estimate the system parameters as accurately as possible with the proposed algorithm using the measured output responses only.

Although the responses are available for a long duration, only the responses from 0.2 to 0.7 sec are considered here to establish the robustness and efficiency of the proposed method. Furthermore, it is assumed that the responses are available at 0.01 sec time intervals providing 50 sample points. Initially, all the responses are assumed to be noise free. The structure is identified using the proposed procedure; the actual and the predicted values of K and C are shown in Table 1. The maximum

error in the stiffness estimation is only 0.014%, and for damping the maximum error is 0.12%. As a practical consideration, some noise is expected in the output measurements. To address the issue of noise in the response quantities, numerically generated white noise with intensities of 1% and 5% of the mean-square values of the responses observed at the first story are added to the output data. A similar procedure was followed by Toki et al. (1989). The structure is identified again using this contaminated output data. The estimation of K and C values for the two noise level cases are also shown in Table 1. The maximum errors in the stiffness estimation are 0.365% for 1% noise and 1.814% for 5% noise. The corresponding maximum errors for damping are 1.47% and 7.6%, respectively. It also needs to be pointed out at this stage that the structure is identified using only 0.5 sec of measured output responses. A long duration record is not necessary for the successful implementation of the proposed method.

Example 2: To show the applicability of the proposed method in solving seismic excitation, a three-story shear building considered by Toki et al. (1989) is studied here, as shown in Fig. 1 with N = 3. The actual structural parameters considered by Toki are:

$$m_1 = 10.0, \quad c_1 = 40.0, \quad k_1 = 3000.0$$
$$m_2 = 10.0, \quad c_2 = 40.0, \quad k_2 = 2000.0$$
$$m_3 = 10.0, \quad c_3 = 30.0, \quad k_3 = 1000.0$$

The structure is excited by a seismic load at the ground level. The input motion was simulated by multiplying the white noise with an evolutionary function (Toki et al., 1989). The input time history is shown in Fig. 2. The theoretical responses are calculated in terms of displacement, velocity, and acceleration of the structure.

Again, to show the applicability of the proposed method, these response quantities are assumed to be measured at all the stories of this structure and the input seismic data are

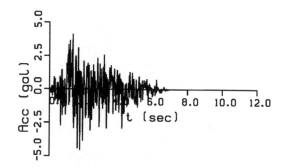

Fig 2. Seismic input motion for example 2

stiffness and damping values and the associated estimation errors are shown in Table 2 for both cases. For the noise free case, the maximum error is only 0.08% in the stiffness estimation and 0.88% in the damping estimation. For the 1% noise case, the corresponding maximum errors are 1.75% and 5.43%, respectively. Toki (1989) reported a maximum error of about 10%.

considered to be unavailable. Although the response quantities are available for a long duration, to increase the efficiency of the algorithm, a very small duration, from 0.8 to 1.8 sec, is considered here. The time interval of the output measurements is considered to be 0.02 sec. The structure is identified using the proposed procedure for noise free and 1% noise cases. The noise is added to the output measurements in a similar way to that discussed in Example 1. The actual and the identified

4. CONCLUSIONS

An element level time domain linear system identification procedure that does not require input loading information is proposed here. The unknown exciting forces can be of any type and can be applied at any location, including at ground level representing seismic excitation. This makes the procedure extremely robust. The method is verified using two examples. The errors in the estimation of the parameters are considerably smaller than those in the other methods currently available in the literature. For verification purposes, both the noise-free and

Table 1. The result of Two D.O.F. system
(0.2-0.7 sec sampling time at 0.01 sec time interval)

| Floor | K | | | | C | | | |
	1	Error %	2	Error %	1	Error %	2	Error %
Exact	30700.0		44300.0		307.0		443.0	
No Noise	30700.4	0.001	44293.6	0.014	306.63	0.12	442.87	0.03
1% Noise	30693.7	0.021	44461.8	0.365	311.10	1.34	436.48	1.47
5% Noise	30686.0	0.046	45103.6	1.814	329.55	7.35	409.31	7.60

Table. 2 Results obtained for Example 2
(Δt=0.02, sample points=50, sampling time 0.8-1.8 sec)

| Floor | K | | | | | C | | | | |
	Exact	No noise	Error %	1% Noise	Error %	Exact	No noise	Error %	1% Noise	Error %
1	3000.	2997.53	0.08	2988.69	0.38	40.0	40.02	0.05	42.17	5.43
2	2000.	2000.99	0.05	1964.93	1.75	40.0	39.65	0.88	41.74	4.35
3	1000.	999.94	0.01	998.39	0.16	30.0	30.14	0.47	29.98	0.07

noise-included output responses are considered. In all cases, the proposed method identified the structural parameters very well. The proposed method is extremely economical and efficient in identifying actual existing structures. Since the input exciting forces are not required, there is no restriction on the type of exciting force, only a small number of observation points are required and no information is required on the modal properties of the structure.

ACKNOWLEDGMENTS

This paper is based upon work partly supported by the National Science Foundation under Grant No. MSM-8896267. Any opinions, findings, conclusions and recommendations expressed in this paper are those of the writers and do not necessarily reflect the views of the sponsor.

REFERENCES

Caravani, P., Watson, M.L., and Thomson, W.T. (1977). "Recursive least-squares time domain identification of structural parameters." Journal of Applied Mechanics, ASME, 44(2), 135-140.

Davis, P., and Hammond, J.K. (1984). " A comparison of Fourier and parametric method for structural system identification." Journal of Vibration, Acoustics, Stress, and Reliability in Design, ASME, 106(1), 40-48.

Haldar, A., and Reddy, R.K. (1991). " A new iterative algorithm for structural identification," 13th Canadian Congress of Applied Mechanics, Vol. 1, 290-291.

Toki, K., Sato, T., and Kiyono, J. (1989). "Identification of structural parameters and input ground motion from response time histories. " Journal of Structural Engrg. / Earthquake Engrg., 6(2), 413s-421s.

Yang, J.S.C., Tsai, T., Tsai, W.H., and Chen, R.Z. (1985). "Detection and identification of structural damage from dynamic response measurements. " Proc. 4th Int. Symp. on Offshore Mech. and Arctic Engrg., Vol. 2., 496-504.

Wang, D., and Haldar, A. (1993). "An element level system identification with unknown input ", Journal of Engineering Mechanics, ASCE (accepted for publication).

Structural Safety & Reliability, Schuëller, Shinozuka & Yao (eds) © 1994 Balkema, Rotterdam, ISBN 90 5410 357 4

Improved frequency domain identifications of structures

Chung-Bang Yun
Korea Advanced Institute of Science and Technology, Taejon, Korea

Kyu-Seon Hong
Samwoo Engineering Company, Seoul, Korea

ABSTRACT : An improved method for estimating frequency response functions (FRF) of structural systems is presented. The new FRF estimator $H_4(f)$ takes a weighted average of two conventional estimators, $H_1(f)$ and $H_2(f)$, utilizing the characteristics that $H_2(f)$ gives more accurate estimate at resonance, while $H_1(f)$ yields better result at antiresonance. An exponential weighting function is used, and the shape of the weighting function is determined by way of minimizing the resolution bias error over the significant frequency range. The effectiveness of the present estimator is investigated through numerical and experimental studies. The estimated results indicate that the new estimator gives more accurate results than other estimators.

1 INTRODUCTION

The structural identification has become an important subject in the area of structural engineering, particularly in connection with the prediction of structural responses to adverse environmental loads, such as earthquakes, wind and wave forces, and also with respect to the estimation of existing conditions of structures for the assessment of damages and deteriorations. Frequency response functions (FRF) are the most fundamental data for the frequency domain identifications of the dynamic characteristics of structural systems.

Several techniques have emerged for the estimation of FRF's in such a way to minimize the effect of measurement noise. By converting many samples of measured time history records into the frequency domain, the conventional FRF's, which usually designated as $H_1(f)$ and $H_2(f)$, are obtained from the relations between the averaged power spectra and the cross spectra of the excitation and response measurement records(see Bendat and Piersol (1971) and Mitchell (1982)). In a recent paper, Fabunmi and Tasker (1988) introduced a FRF estimator $H_3(f)$, which is based on the minimization of the measurement noises.

More recently, an improved FRF estimator $H_4(f)$ was proposed by the current authors (Hong and Yun (1993)). The new FRF estimator also takes the weighted average of $H_1(f)$ and $H_2(f)$ as $H_3(f)$. Considering that $H_2(f)$ is more accurate at resonances while $H_1(f)$ is better at antiresonances, $H_4(f)$ is designed to have the characteristics that it approaches $H_2(f)$ at resonances and $H_1(f)$ at antiresonances, by selecting the weighting function appropriately. An exponential weighting function is used, and the shape of the weighting function is determined by way of minimizing the resolution bias error over the significant frequency range. The effectiveness of the proposed estimator is investigated through numerical and experimental studies. The estimated results indicate that the new estimator gives better estimates than other estimators.

2 CONVENTIONAL FREQUENCY RESPONSE FUNCTIONS

Using the measurement data, the frequency response functions are usually calculated as

$$H_1(f) = S_{xy}(f)/S_{xx}(f)$$
$$= H(f)/(1 + S_{mm}(f)/S_{uu}(f)) \quad (1)$$
$$H_2(f) = S_{yy}(f)/S_{yx}(f)$$
$$= H(f)(1 + S_{nn}(f)/S_{vv}(f)) \quad (2)$$

where $H(f)$ is the true frequency response function; u and v denote the true input and output; x and y denote the measured excitation and response; m and n denote the measurement noise of input and output(see *Figure 1*); $S_{uu}(f)$, $S_{vv}(f)$, $S_{xx}(f)$, $S_{yy}(f)$, $S_{mm}(f)$ and $S_{nn}(f)$ are the power spectral density(PSD) functions; $S_{xy}(f)$ and $S_{yx}(f)$ are the cross-spectral density functions.

During experiments on real structural systems, it is frequently observed that the input PSD function, $S_{uu}(f)$, drops drastically near resonance, particularly for the case with random excitations(see Mitchell (1982)). The low input power spectrum in the resonance region causes $H_1(f)$ much less than the true FRF. On the other hand, $H_2(f)$ gives an excellent estimate of $H(f)$ in the resonance region, since the response PSD function, $S_{vv}(f)$, at resonance is considerably larger than that of output noise, $S_{nn}(f)$. However, $H_2(f)$ is a much poorer estimator in the antiresonance region than $H_1(f)$, since the true output power spectrum at antiresonance drops drastically to the level of the output noise spectrum.

Noise is not the only source of error in the FRF estimation. The resolution bias error causes underestimation of spectral peaks and overestimation of spectral troughs. Cawley(1984) investigated the effects of the resolution bias error on the FRF estimators, $H_1(f)$ and $H_2(f)$. Schmidt (1985) analyzed the resolution bias errors in spectral density, frequency response and coherence function measurements. It has been shown that, due to the resolution bias error, $H_1(f)$ produces a

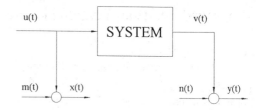

Figure 1. System model for measuring FRF

result of underestimation of the FRF at resonances, but it gives good results at antiresonances. On the other hand, $H_2(f)$ yields more accurate estimates at resonances, but gives poor estimates at antiresonances.

In a recent paper by Fabunmi and Tasker (1988), a new estimator for the FRF was proposed, which was based on the joint minimization of the measurement input and output noises as

$$H_3(f) = \frac{H_2(f)[\gamma_{xy}^2(f)F(f)/|H_1(f)|^2 + 1]}{F(f)/|H_1(f)|^2 + 1} \quad (3)$$

where $\gamma_{xy}(f)$ is coherence function and $F(f)$ is a normalizing function. $H_3(f)$ of equation (3) was intended to approach $H_1(f)$ at antiresonances and $H_2(f)$ at resonances. For that purpose, $F(f)/|H_1(f)|^2$ should be significantly smaller than unity at resonance, whereas $\gamma_{xy}^2(f)F(f)/|H_1(f)|^2$ should be larger than unity at antiresonances. For displacement and acceleration FRF's, the normalizing function was obtained as

$$F(f) = -((S_{xy}(f) - S_{yx}(f))/2S_{yy}(f)^2 \quad (4)$$

and for the velocity FRF, it was obtained as

$$F(f) = ((S_{xy}(f) + S_{yx}(f))/2S_{yy}(f)^2 \quad (5)$$

Since the normalizing function, $F(f)$, depends on the power and the cross spectra of the measured records, it may not be possible that the intended characteristics for the normalizing

intended characteristics for the normalizing function are achieved; namely, depending on the measurement data, $F(f)/|H_1(f)|^2$ may not be significantly smaller than unity at resonance, and $\gamma_{xy}^2(f)F(f)/|H_1(f)|^2$ may not be much larger than unity at antiresonances. Therefore, $H_3(f)$ may give undesirable results over some frequency ranges, as shown in the first example analysis.

3 IMPROVED FREQUENCY RESPONSE FUNCTION

A new FRF estimator $H_4(f)$ proposed by the current authors is taken as a weighted average of $H_1(f)$ and $H_2(f)$,

$$H_4(f) = (1 - W(f))H_1(f) + W(f)H_2(f) \quad (6)$$

where $W(f)$ is the weighting function. The new FRF estimator in equation (6) is designed to have the characteristics that it approaches $H_1(f)$ at antiresonances and $H_2(f)$ at resonances. Thus, the weighting function, $W(f)$, is to have unity at resonances and zero or near zero at antiresonances. An exponential function is selected for the weighting function, which is given as

$$W(f) = \exp\left(-\left(\frac{f/f_0 - 1}{\alpha}\right)^2\right) \quad (7)$$

where f_0 is the natural frequency, and α is the parameter of the exponential weighting function. The weighting function, $W(f)$, has unity correctly at the resonance and approaches zero as the frequency being far apart from the resonance frequency. The shape of the weighting function depends on the parameter α, which may be determined by way of minimizing the resolution bias error as described later(see equation (10) and Figure 2).

The frequency response function of a SDOF system can be written as

$$H(f) = a/(1 - \hat{f}^2 + j2\zeta\hat{f}) \quad (8)$$

where $\hat{f} = f/f_0$, ζ is the damping ratio, a is the system constant, and j is complex unit.

The resolution bias error of the amplitude of the new estimator, $H_4(f)$, can be obtained as

$$e[|H_4(f)|] = |H_4(f)| - |H(f)|$$
$$= (1 - W(f))e[|H_1(f)|] + W(f)e[|H_2(f)|] \quad (9)$$

where $e[|H_1(f)|]$ and $e[|H_2(f)|]$ are the bias errors of the amplitudes of $H_1(f)$ and $H_2(f)$. The expressions for the resolution bias error of $H_1(f)$ and $H_2(f)$ can be found for rectangular and Hanning windows in References by Schmidt (1985).

The parameter of the weighting function, α, is determined by way of minimizing the integral of the square of $e[|H_4(f)|]$ over the significant frequency range, which is taken from zero to two times of the natural frequency as

$$\min_{\alpha} J = \int_0^{2f_0} e[|H_4(f)|]^2 \, df \quad (10)$$

The integrals of the squared error are calculated for various values of α and damping ratios. Then, the value of α is determined at the point of the minimum J for each damping ratio. Through the successive applications to various values of damping ratios, the value of α is determined as a function of the damping ratio, and the results are shown in Figure 2. It has been observed that α is nearly linear to damping ratio for both cases of windowing techniques; i.e., rectangular and Hanning windows.

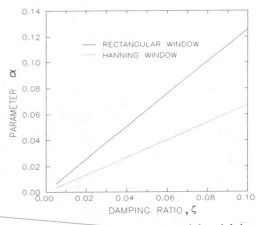

Figure 2. Parameter of exponential weighting function vs Damping ratio

861

For a SDOF system, a single weighting function is used over the entire frequency range. However, for a MDOF system, series of different weighting functions may be defined for individual spectral peaks. In the latter case, the frequency range of the weighting function in the neighborhood of each spectral peak may be taken as the range between two points of the troughs of $H_1(f)$ adjacent to each peak as shown in *Figure 3*.

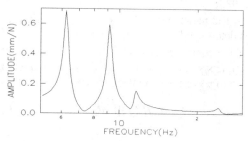

(a) Frequency Response Function

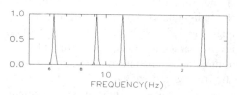

(b) Weighting Function

Figure 3 Example frequency response function and weighting function for MDOF system

4 EXAMPLE CASES

The effectiveness of the new FRF estimator was investigated through a series of example analyses. At first, numerical simulation analysis is performed for a MDOF system with additive noises. Then, experimental study is also carried out for a rotor system to which random input are applied by a magnetic exciter.

4.1 Simulation study for a bridge model

To investigate the effectiveness of $H_4(f)$ for MDOF systems, numerical studies were carried out for a 3-span continuous bridge model which has the span lengths of 35, 42 and 35 *m* (see *Figure 4*). The excitation point(A) is assumed to be at 15 *m* from the left abutment, and the response measuring points(B and C) are at 30 *m* and 45 *m* from the left abutment, respectively. The moment of inertia of the cross section of the girder is assumed to be 0.2 m^4. Damping ratios are all assumed to be 0.02. Sample functions of the displacement measurement records at Point B and C are simulated for the random excitations at Point A based on the finite element model of the bridge structure. The input and the output noises are added respectively. The intensities of the noises are taken as 10 % of those of the original signals in the root mean square(RMS) level.

In *Figure 5(a)*, the amplitudes of the exact FRF, $H_1(f)$, $H_2(f)$ and $H_4(f)$ at Point B are shown. The Hanning window has been employed. From the figure, it is ascertained that $H_1(f)$ gives substantially underestimated results

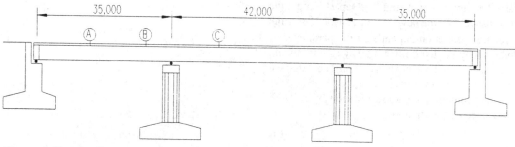

Figure 4. Sketch of a continuous 3-span bridge model (in mm)

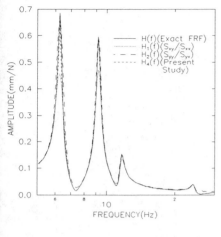

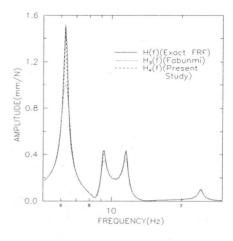

(a) FRF's at Point B

(b) FRF's at Point C

Figure 5 . Frequency response functions of a 3-span bridge model at Point B and Point C
(using Hanning Window)

in the resonance region, while $H_2(f)$ gives overestimated ones in the off-resonance region. On the other hand, the proposed FRF estimator, $H_4(f)$, yields more accurate results either at resonance or at antiresonance. In *Figure 5(b)*, $H_4(f)$ is compared with $H_3(f)$ at Point C. It can be observed that, for the present example, $H_3(f)$ gives considerably poor results even in the resonance region compared with those by $H_4(f)$. $H_3(f)$ gives significantly underestimated result at resonance, like $H_1(f)$.

4.2 Experiments for a rotor system

The schematic diagram of a rotor system and the experimental set-up are shown in *Figure 6*. The rotor consists of a shaft of 0.85 *m* long and 3 *cm* in diameter, three rigid discs and two high-speed ball bearings of self-aligning type. The random excitations are applied to the overhung disc of the rotor by a magnetic exciter, and the horizontal displacement of the shaft at a station is measured. The sampling frequency is 345 *Hz*. Each record consists of 2048 data points, and 24 sets of records are used for ensemble averaging.
To determine the shape of the weighting

function, the first two natural frequencies and damping ratios are approximately estimated as : $f_1 = 50$ *Hz*, $f_2 = 106$ *Hz*, $\zeta_1 = 0.25\%$,   and $\zeta_2 = 1.5\%$. Then, the weighting functions are constructed based on the damping ratios for individual spectral peaks. The amplitudes of the estimated $H_1(f)$, $H_2(f)$, $H_3(f)$ and $H_4(f)$ over a wide frequency range are shown in *Figure 7(a)*, and those near the first spectral peak are shown in *Figure 7(b)*. Hanning window has been used. Although the differences among the FRF's may not be appreciated in *Figure 7(a)*, considerable discrepancies can be observed between $H_1(f)$ and $H_2(f)$ in the frequency region near the first peak, as shown in *Figure 7(b)*. The small peak at 60 *Hz* is considered as the noise effect due to the current of 60 *Hz*. Based on the estimated FRF's, the modal parameters are evaluated by the complex exponential method. The results from four different FRF's are compared in *Table 1*. It can be observed that the results of $H_4(f)$ approach $H_2(f)$ at resonance and $H_1(f)$ at antiresonance, as intended. However, $H_3(f)$ is almost the same as $H_2(f)$ over the whole frequency band including the first peak frequency.

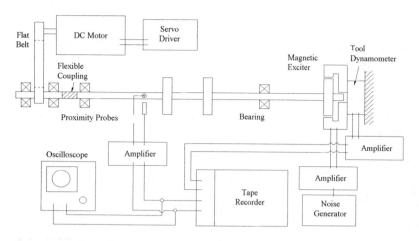

Figure 6. Schematic diagram of a rotor system and experimental setup

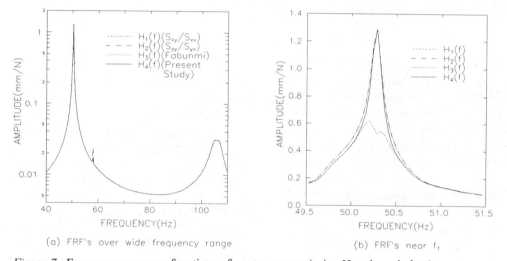

(a) FRF's over wide frequency range

(b) FRF's near f_1

Figure 7. Frequency response functions of a rotor system (using Hanning window)

Table 1. Estimated modal parameters of a rotor system

		$H_1(f)$	$H_2(f)$	$H_3(f)$	$H_4(f)$
1st Peak	$f_1\ (Hz)$	50.23	50.24	50.24	50.24
	$\zeta_1\ (\%)$	0.42	0.20	0.19	0.21
	$A_1(1/kg)$	1.213	1.368	1.361	1.322
2nd Peak	$f_2\ (Hz)$	105.84	105.82	105.84	105.83
	$\zeta_2\ (\%)$	1.55	1.48	1.55	1.55
	$A_2(1/kg)$	0.717	0.693	0.718	0.716

A_1 and A_2 are the modal constants, which is similar to the system constant a in Eq. 8.

5 CONCLUSION

An improved estimator for the frequency response functions of structural systems is presented. The new FRF estimator, $H_4(f)$, takes a weighted average of two conventional estimators, $H_1(f)$ and $H_2(f)$. It utilizes the characteristics that $H_2(f)$ is more accurate at the resonances than $H_1(f)$, and vice versa at the antiresonances. An exponential weighting function is used, and the shape of the weighting function is determined by way of minimizing the resolution bias error over the significant frequency range.

The numerical results of the simulation studies for a 3-span continuous bridge model indicate that the new FRF estimator gives more accurate results than other FRF estimators, such as $H_1(f)$, $H_2(f)$, and $H_3(f)$. From the experimental studies, it has been also found that $H_4(f)$ gives the estimate which approaches $H_2(f)$ at resonances and $H_1(f)$ at antiresonance, as intended.

ACKNOWLEDGMENT

This work was supported by the Korea Science and Engineering Foundation under the grant number 91-07-00-15 during the period of 1991 - 1993. The financial support is gratefully acknowledged.

REFERENCES

Bendat, J. S. and Piersol, A. G., 1971, *Random Data: Analysis and Measurement Procedures*, John Wiley & Sons, New York

Cawley, P., 1984, "The reduction of bias error in transfer function estimates using FFT-based analyzers", *J. Vib. Acous. Stress Reliability in Design*, ASME, Vol.106, pp.29-35

Fabunmi, J. A. and Tasker, F. A., 1988, "Advanced techniques for measuring structural mobilities", *J. Vib. Acous. Stress Reliability in Design*, ASME, Vol.110, pp.345-349

Hong, K-S. and Yun, C-B., 1993, "Improved method for frequency domain identifications of structures", *Eng. Struct.*, Vol.15, No.3, pp.179-188

Mitchell, L. D., 1982, "Improved methods for the Fast Fourier Transform(FFT) calculation of the frequency response function", *J. Mech. Design*, ASME, Vol.104, pp.277-279

Schmidt, H., 1985, "Resolution bias errors in spectral density, frequency response and coherence function measurements, III: Application to second-order systems (white noise excitation)", *J. Sound Vib.*, Vol.101, No.3, pp.377-404

Schmidt, H., 1985, "Resolution bias errors in spectral density, frequency response and coherence function measurements, V: Comparison of different frequency response estimators", *J. Sound Vib.*, Vol.101, No.3, pp.413-418

Bridge reliability

Structural Safety & Reliability, Schuëller, Shinozuka & Yao (eds) © 1994 Balkema, Rotterdam, ISBN 90 5410 357 4

Reliability analysis of the cable stayed bridge in construction and service stages

A. Florian & J. Navrátil
Technical University of Brno, Czech Republic

ABSTRACT: The time-dependent behavior of the cable stayed bridge with respect to the uncertainties in mechanical, geometrical, environmental and loading characteristics is analyzed in the paper. The combination of time-discretization method and Finite Element Analysis is used. For the purpose of reliability analysis the technique Latin Hypercube Sampling and "Curve-Fitting" method are utilized. The detail reliability analysis provides better understanding of the structural behavior.

1 INTRODUCTION

The usual deterministic analysis of modern light-weight structures does not predict the dispersion of the structural response caused by the uncertainties in mechanical, geometrical, environmental and loading characteristics, and also by the uncertainties due to imperfect modeling and gross errors.

The main target of this paper is to describe the reliability analysis of the cable stayed bridge in construction and service stages. The following problems are analyzed in the paper:

1. The sensitivity of the structure on
- the random variability of some material and mechanical input characteristics,
- the size of initial imperfections caused by the manufacturing and construction process,
- the uncertainties in environmental and loading characteristics.

2. Estimates of some statistical parameters of structural response quantities.

3. The time dependency of the statistical parameters.

4. The probability of exceeding the ultimate values of internal forces and deformations.

5. The effects of the redistribution of internal forces caused by creep and shrinkage on the magnitude of influence of random variability of input variables.

2 DETERMINISTIC MODEL

A method and a computer program have been developed (Navrátil 1991) for the time-dependent analysis of composite plane frames with the fully acting cross-section. The linear aging viscoelastic theory is applied.

Various operations used in the construction, such as addition or removal of structural elements and prestressing cables, changes of boundary conditions, loads and prescribed displacements may be modeled. The creep and shrinkage effects are taken into account.

The method is based on a step-by-step computer procedure in which the time domain is subdivided by discrete time nodes into time intervals. The finite element analysis is performed in each time node. The plane frame element with eight degrees of freedom (6 external and 2 internal) is used. Its stiffness matrix and load vector terms include the effect of axial, bending and shear deformations. The axial and transverse displacements are approximated by the polynomial function of order 2 and 3 respectively. The centroidal of each element can be placed in a linear eccentricity relating to the reference axis.

The creep prediction model is based upon the assumption of linearity between stresses and strains to assure the applicability of linear superposition. The rheological behavior of structural members is predicted through the mean properties of a given cross-section taking into account the average relative humidity and member size.

The method has been proved to be an efficient tool for the analysis of the concrete structures with respect to their rheological behavior (Navratil 1991).

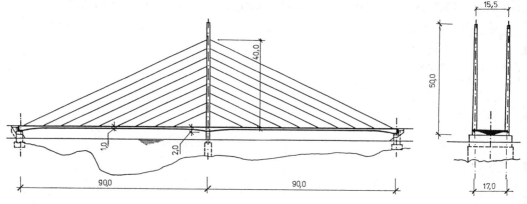

Fig. 1 Cable stayed bridge in Lycksele (Sweden)

3 RELIABILITY ANALYSIS

Reliability analysis of the problem analyzed consists of three types of analyses - statistical, sensitivity and probability analysis.

3.1 *Statistical analysis*

The Latin Hypercube Sampling technique is used for the purpose of statistical analysis (McKay et al. 1979, Iman and Conover 1980). It theoretically provides estimates of some statistical parameters - mean value, statistical moments, cumulative distribution function (CDF) - which are unbiased and whose variance is reduced. As proved by numerical studies, the features mentioned above still hold for estimates of other widely used statistical parameters - standard deviation, coefficient of variation, skewness, excess, minimal or maximal value of sample. For some of these statistical parameters the bias is negligible but for all of them the variance is substantially reduced in comparison with Simple Random Sampling or other sampling techniques. The number of simulations needed for reliable estimates of commonly used statistical parameters is generally within the range 20 - 50. It is a real advantage, because other sampling techniques need hundreds or thousands of simulations. The number of simulation equals 20 in the present study.

3.2 *Sensitivity analysis*

In the sensitivity analysis, the measure based on the Spearman rank correlation coefficient r_k^s is used. It is defined as

$$r_k^s = 1 - \frac{6.\sum_i (m_{ik} - n_i)^2}{N.(N-1).(N+1)} \qquad (1)$$

where m_{ik} - rank of interval in the i-th simulation for k-th input variable, n_i - rank of the output variable in sample arranged from the smallest to the largest values, N - number of simulation, and r_k^s - Spearman rank correlation coefficient between the input variable X_t and the output. The measure has the following desirable features:

1. It can be used for the description of any monotonic statistical correlation between two random variables.

2. The CDFs of random variables can be non-normal and different each other.

3. It is defined with the help of the ranks of two samples and thus it is a meaningful measure to define statistical correlation between the individual input variables and the output in the case of Latin Hypercube Sampling.

4. The higher the Spearman coefficient, the higher the sensitivity of the output to the random variability of an appropriate input variable.

5. Results of sensitivity analysis are obtained as a part of statistical and/or probability analysis. No additional simulations are necessary.

3.3 *Probability analysis*

Probability analysis in the present study is based on the "curve fitting" method. In this method, the most suitable theoretical distribution model is selected (Florian and Novák 1988) for the given random sample obtained from statistical analysis. Based on the selected theoretical distribution model with the cumulative distribution function G(y), the theoretical failure probability p_t (e.g. the probability of exceeding ultimate values of internal forces or deformations) is calculated as a value of the distribution function

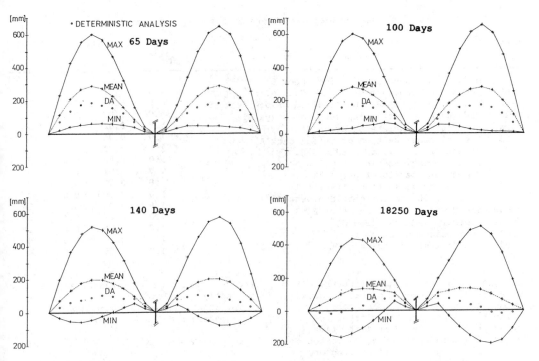

Fig. 2 Random variability of deflections of the bridge deck

$$P_f = G(y) \qquad (2)$$

The following theoretical distribution models (from which the best one is selected) are included in competition in this study: normal, lognormal, truncated normal, Weibull, Pearson III. Also, another measure of the theoretical failure probability is used. It is based on the reliability index β defined for $y = const$ as follows

$$\beta = \frac{\bar{y} - y}{s_x} \qquad (3)$$

where $\bar{y}$ - mean value, s_x - standard deviation.

4 STRUCTURAL ANALYSIS

4.1 *Description of structure analyzed*

The reliability analysis of time-dependent behavior of the cable stayed bridge (see Fig. 1) designed by Stráský, Hustý and Partners (Hustý and Hubík 1991) was performed. The bridge was designed as a cable stayed symmetric structure consisting of two 90 m long spans. The 16 m wide deck was suspended in two lateral planes from two concrete pylons 50 m high.

The stays were of a harp arrangement, were symmetric with the tower and were assembled of 42 strands 15.7 mm each. The strands were placed in steel ducts and grouted with cement mortar. The depth of the cast-in-situ concrete deck varied in the transverse direction from 1.0 m in the middle of the cross-section to 0.55 m at the edges. In the longitudinal direction the deck was assumed to be rigidly connected to the pylons and simply supported at the abutments. To avoid the negative reactions at the ends of the bridge the massive deck over the abutments was 3.0 m deep. Both pylons of "I" shape were designed of reinforced concrete (class K 50, according Swedish Code Bronorm 88) and were connected the concrete block of the intermediate support. Their dimensions varied from 3.5 x 1.5 m at the deck to 2.5 x 1.5 m at the top. The deck was prestressed both in longitudinal and transverse directions.

4.2 *Structural model*

The structure described above is in fact a highly complicated system of plate and beam elements, loaded both in plane and transversely. To decrease the requirements for computer capacity, the simplified static model was used. The structure was

871

Table 1. Statistical parameters of input random variables

Number	Input Variable	Mean Value	COV	Skewness	CDF
1	Modulus of concrete of bridge deck E_c^d	1.0	0.06	0.0	N
2-21	Density of concrete of bridge deck ρ_c^{1-20}	25.0	0.02	-0.4	LN
22	Modulus of prestressing steel in bridge deck E_p^d	195.0	0.02	0.0	N
23	Modulus of prestressing steel of stays E_p^s	195.0	0.02	0.0	N
24	Modulus of concrete of pylon E_c^p	1.0	0.06	0.0	N
25	Area of cross-section of bridge deck A^d	1.0	0.015	0.0	N
26	Moment of inertia of bridge deck I^d	1.0	0.02	0.0	N
27	Area of cross-section of pylon A^p	1.0	0.01	0.0	N
28	Moment of inertia of pylon I^p	1.0	0.015	0.0	N
29	Area of cross-section of stays A^s	1.0	0.005	0.0	N
30	Prestressing force in bridge deck P_p^d	1.0	0.02	-0.8	LN
31-48	Prestressing force in stays P_p^{66-83}	1.0	0.01	-0.4	LN
49-52	Superimposed dead load G^{1-4}	1.0	0.05	0.4	LN
53	Coefficient of creep ε_c	1.0	0.2	0.0	N
54	Coefficient of shrinkage ε_s	1.0	0.25	0.0	N

analyzed as a plane frame structure. It consists of 83 elements of bridge deck, pylon, stays and prestressed cables.

The time-dependent behavior of the structure was analyzed in fifty time steps. The main time nodes are selected:

0, 10, 20 days: time of casting one third of the pylon,

30 days: casting of the bridge deck on the centring, introducing the structural dead load of the bridge deck,

65 days: initiation of the stiffness of the deck elements into stiffness matrix of the structure, installing and prestressing of the stays, removing of the centring, initiation of the stiffness of the stays,

100 days: installing and prestressing of prestressed cables, initiation of the stiffness of prestressed cables,

140 days: introducing the superimposed dead load (pavements, carriage way etc.)

1000, 10000 and 18250 days: evaluation of the structural response quantities.

4.3 Statistical parameters of input variables

Every input variable is described by its CDF - e.g. normal (N), truncated normal (TN), lognormal (LN), Weibull (W) or Pearson III (P3) - and by its statistical parameters, see Tab. 1. These statistics are based on data taken from technological manuals, from the appropriate Swedish code of practice, from Florian 1989, and are also based on professional judgment. For simplicity, all input variables are supposed to be mutually statistically independent in this study, with the following exception: the area and the moment of inertia of the cross-section of bridge deck and pylon (random variables 25

and 26, and 27 and 28 respectively) are perfectly correlated.

The input random variables number 1 and 24 - 54 are modeled as

$$X = K \cdot X_{nom} \qquad (4)$$

where K is the random variable with the mean value equal to 1, and X_{nom} is the nominal value of the appropriate variable.

The influence of imperfect modeling is incorporated into the reliability analysis by the following formula

$$Z = R \cdot Y \qquad (5)$$

where Y - output from a deterministic model, R - random variable describing the influence of imperfect modeling, Z - output from a model, in which the influence of imperfect modeling is included. The statistical parameters and CDF of the random variable R can be obtained from calibration of the appropriate deterministic model with respect to the results obtained from laboratory tests or a more precise model.

For the sake of simplicity the creep and shrinkage coefficients were taken directly as the random variables. Thus the uncertainties due to different factors influencing the creep and shrinkage, and also due to imperfect modeling of these phenomena are incorporated in the random variables No. 53 and 54 respectively, see Tab. 1.

5 RESULTS

For the sake of brevity only some illustrative results of effects of the long-time loading are shown in Figs. 2 - 7. For details see Navrátil and Florian 1992. The

following notation is used: MEAN - mean value, MIN - minimal value in the sample, MAX - maximal value in the sample, DA - deterministic analysis, 1, 5, 95, 99 - quantiles 1%, 5%, 95%, and 99% respectively, s_x - standard deviation, v_x - coefficient of variation, a_x - skewness, r_k^s - sensitivity measure based on Spearman rank correlation coefficient.

5.1 Deflections of the bridge deck

The bridge is designed according to the Swedish bridge standard Bronorm 88. Maximal deflection caused by live load, which must not exceed 1/400 of a span (225 mm in this case), is a decisive requirement in the cross-section proposal. Mean values of hogging of the bridge deck caused by the long-time loading are decreasing with respect to time, see Fig. 2, and are coming closer to zero. They are symmetrically distributed along the bridge deck, in spite of the fact that the structure is nonsymmetrical due to the uncertainties in input variables. They differ remarkably from the deflections obtained from the deterministic analysis.

Standard deviation of deflections is rather high, see Fig. 3, and varies along the bridge deck. The structure is highly sensitive to the uncertainties in input variables. The interval of possible values of deflections of the bridge deck is rather wide, e.g. hogging in the middle of the bridge deck at the time of 65 days varies from 50 mm to 600 mm, see Fig. 2. Skewness shows the tendency to hogging of the bridge deck.

Random variability of deflections in the middle of the bridge deck is dominantly influenced by random variability of modulus of prestressing steel of stays. It is highly influenced by the dead load of some elements of the bridge deck, by the

prestressing force in some stays, and by the superimposed dead load. No significant changes in the influence of individual input variables on the random variability of deflections are observed over the time.

5.2 Bending moments of the bridge deck

Mean values of bending moments of the bridge deck are symmetrically distributed along the bridge deck and are decreasing with respect to time. Random variability of bending moments in construction stages is rather high, but it quickly decreases in time, especially during the period before the prestressing of the cables, see Fig. 4. The strong influence of rheological effects of concrete is obvious. The mean values of bending moments in the part of the deck with constant cross-section are close to zero in service stages. The interval of possible values of bending moments of the bridge deck is rather wide, e.g. from 0 to -28000 kNm in the middle of the bridge deck after 65 days. The extreme values of bending moments are not symmetrically distributed along the bridge deck. The maximal values decrease very quickly in time, but the minimal values decrease slowly.

Random variability of bending moments is dominantly influenced by the modulus of prestressing steel of stays. No significant changes in the influence of different input variables on bending moment in the middle of the bridge deck are observed with respect to time.

5.3 Normal forces of the stays

Normal forces in individual stays change in time. This may be due to the redistribution of internal forces. During the construction the forces in all stays

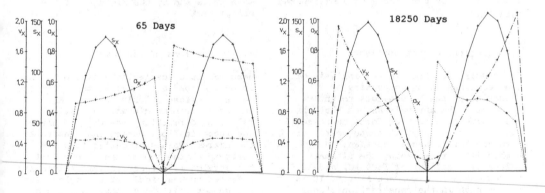

Fig. 3 Statistical characteristics of deflections of the bridge deck

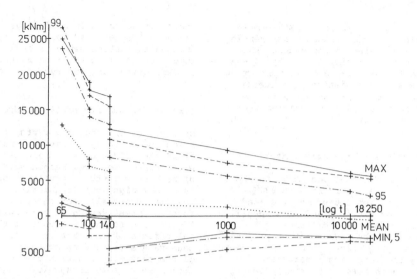

Fig. 4 Time dependency of random variability of bending moments of the bridge deck in the
middle of span

are approximately at the same level.
After introducing the superimposed dead
load, normal forces in stays near both
ends of the pylon are decreasing in time,
and those near the middle of the pylon are
increasing. Standard deviation is not too
high. The highest standard deviation is
observed for the stays near both ends of
the pylon and it slowly increases with
respect to time.

Random variability of normal forces in
stays in the middle of the pylon is
dominantly influenced by the random
variability of the modulus of prestressing
steel of stays and also by rigidity of the
bridge deck, see Fig. 5. It is highly
influenced by the dead load of some
elements of the bridge deck, by
prestressing force in some stays, by area
of the cross-section of stays, by creep of
concrete and by some parts of superimposed
dead load.

5.4 Reactions at the abutments

The crucial question in the analysis of
the reactions is the necessity of
anchoring of the bridge deck to the
abutments. The analysis showed that the
mean value of anchoring force decreases in
time. After 140 days it drops nearly to
zero and then changes the sign. The
probability of exceeding the ultimate
values of reactions of 1500 kN, and 1000
kN decreases in time, see Fig. 6. It is
rather high in the construction stages
(0.05, and 0.5 respectively). After

introducing the superimposed dead load the
probability drops to the level of 10^{-6}, and
$5*10^{-5}$ respectively. Thus the anchored
bearings at the abutments, or the ballast
load on both ends of the bridge deck, are
necessary during the whole period
analyzed. The minimum anchoring force
during the construction can be assumed to
be the value of 1500 kN.

Standard deviation of the reactions
decreases in time. Thus creep and
shrinkage of concrete suppress the random
variability of the reactions of the
structure analyzed.

The random variability of the reactions
is dominantly influenced by the random
variability of the modulus of prestressing
steel of stays and highly influenced by
the dead load of some elements of the
bridge deck, by the prestressing force in
some stays and by the rigidity of the
bridge deck.

5.5 Deflections of the pylon

Mean values of deflections caused by the
long-time load are equal to zero. Extreme
values increase quickly in time. Also
curves describing minimal and maximal
values of deflection, see Fig. 7, change
the shape in time. During the construction
the extreme values of deflections are
observed at the top of the pylon. During
the service life these extreme values are
observed in the middle of the pylon. Thus
the geometrically nonlinear analysis of
the pylon should include both the

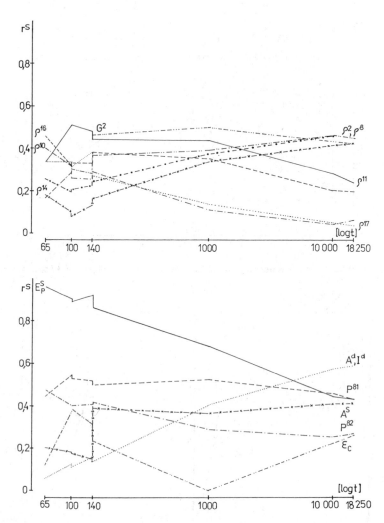

Fig. 5 Influence of some input variables on the random variability of normal forces of the stay in the middle of pylon

Table 2. - The most suitable CDF for appropriate structural response

Structural Response	CDF	Note
Deflection of the bridge deck	LN, TN	near pylon
	N	near abutments
Deflection of the pylon	N	at 65 days
	TN	at 100 days
	W	at 140 days
	LN	at 18250 days
Bending moments of the bridge deck	W, P3	near pylon
	N, LN	near abutments
Bending moments of the pylon	W, TN	construction
	W	service
Normal forces of the stays	TN	construction
	W	service, ends of pylon
	N, LN, TN	service, middle of pylon
Reactions in abutments	N	construction
	LN, W	service

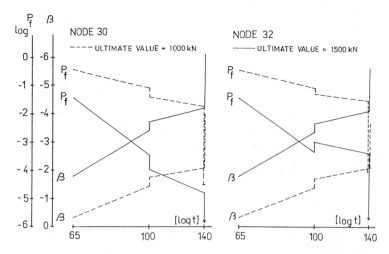

Fig. 6 Probability of exceeding of ultimate values of the reactions in supports

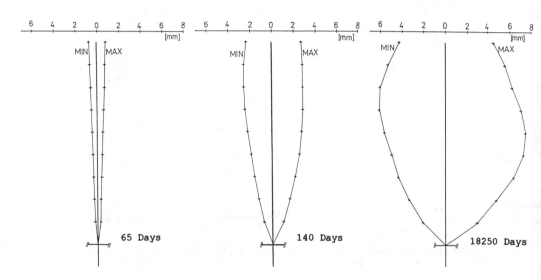

Fig. 7 Random variability of deflections of the pylon

influence of uncertainties in input variables and creep of concrete.

5.6 *The most suitable CDF*

The most suitable CDFs for describing the random behavior of some structural response quantities are summarized in Tab. 2. Generally, various CDFs are suitable for describing the random behavior at different time points and at different parts of the structure.

6 CONCLUSIONS

The uncertainties in input variables cause a significant statistical scatter both in the predicted internal forces and deflections. The statistical spread of the results depends on the accuracy of geometrical, material and mechanical properties and on the proper choice of model. Reduction of this scatter may be achieved by the consistent observation and measurement of input variables, top discipline in technology and utilization of adequate methods for the analysis. The random behavior of various response quantities (deflections, bending moments, normal

forces in stays, support reactions etc.) differs each other. The statistical distribution of the structural response is clearly nonsymmetrical with generally a non-zero coefficient of skewness. The influence of individual input variables on the random variability of structural response is different at different time nodes. Some input variables are dominant, others have a very small influence.

The results of this study show, that the analysis of the complicated light-weight structures should incorporate both the influence of rheological effects and the iffluence of the uncertainties and imperfections in input variables. The reliability analysis provides not only the mean values but also the upper and lower confidence limits of the structural response in which we may consider the project to be without significant errors. At the same time the designer may estimate the critical development possibilities of the structural behavior and prepare the means rectification of the stays etc. to control all variants of the development.

ACKNOWLEDGMENT

The research was performed at the Technical University of Brno as a part of its research program. The authors would like to express their sincere thanks to Stráský, Hustý and Partners for the support, interest and collaboration.

REFERENCES

Florian, A. 1988. *Nondeterministic analysis of prestressed concrete beams.* Ph.D. thesis, Dept. of Struct. Mech., Technical University of Brno, Czecho-slovakia (in Czech)

Florian, A. and Novák, D. 1988. The statistical model selection for random variables. *Software for Engineering Workstations.* 4(3), pp. 158 - 162

Hustý, I. and Hubík, P. 1991. *Preliminary design of the cable stayed bridge across the river Ume in Lycksele.* Stráský, Hustý and Partners, Czechoslovakia, Jacobson and Widmark, Sweden.

Iman, R. L. and Conover, W. J. 1980. Small sample sensitivity analysis techniques for computer models, with an application to risk assessment. *Commun. Statist.* A9, pp. 1749 - 1842

McKay, M. D., Beckman, R. J. and Conover, W. J. 1979. A comparison of three methods for selecting values of input variables in the analysis of output from a computer code. *Technometrics.* 2, pp. 239 -245

Navrátil, J. 1991. *Numerical models for time-dependent analysis of modern concrete structures.* Ph.D. thesis, Dept. of Concrete and Masonry Struct., Technical University of Brno, (in Czech)

Navrátil, J. and Florian, A. 1992. *The influence of stochastical and rheologi-cal phenomena on the behavior of a cable stayed bridge in construction and servi-ce stages.* Research Rep., Technical University of Brno, Czechoslovakia

Structural Safety & Reliability, Schuëller, Shinozuka & Yao (eds) © 1994 Balkema, Rotterdam, ISBN 90 5410 357 4

Reliability based proof load factors for short and medium span bridges

Gongkang Fu & Jianguo Tang
Structures Research, New York State Department of Transportation, Albany, N.Y., USA

ABSTRACT: This paper proposes proof load factors for highway bridge evaluation by proof load testing, to determine target proof load and load rating. This formula is based on a target structural safety index β of 2.3, which is consistent with current bridge evaluation practice and an evaluation method of load and resistance factors under development in the United States. It is demonstrated that the proposed factors will assure a relatively uniform level of bridge structural safety, and that possible changes in input data and probability distribution assumptions in the reliability models will not affect the results obtained here. The proposed method can be applied to highway bridge evaluation by proof load testing, and the resulting rating can be directly input to the current US national bridge rating inventory. It may be included in specifications for highway bridge evaluation by proof load testing.

1. INTRODUCTION

According to the US Federal Highway Administration, about 40 percent of highway bridges in US are considered either structurally deficient or functionally obsolete. Funds necessary to replace and rehabilitate them will not be available in the foreseeable future. These statistics are based on current evaluation technology in US. For existing bridge structures, AASHTO's Manual for Maintenance Inspection of Bridges [1], referred to here as "the AASHTO Manual," provides technical guidelines for routine analytical evaluation based on data supplemented by field inspection. However, such an analytical evaluation is often not applicable because of inaccessibility of bridge structural components due to their locations or protection methods, and/or lack of such detailed information on bridges as design plans, etc. When a bridge is evaluated by physical testing, many assumptions critical to analytical evaluation methods may be unnecessary. In addition, in many cases

physical testing may be the only way to obtain a reliable rating. The testing results have also demonstrated higher load carrying capacity than predicted by conventional analytical methods. Despite these advantages of physical testing in bridge evaluation, the current AASHTO Manual does not include provisions for field testing to evaluate highway bridges.

This paper presents partial results of a study to develop a proof load testing program for bridge rating, based on evaluation of current practice in this area. This study was undertaken based on recognition that proof load testing is one of the most effective approaches to examine structural load carrying capacity. On the other hand, basis of proof test load requirements in many current codes has not been well documented. Development of a consistent method for determining both target proof load and load rating is a major focus here, based on a criterion of target structural reliability. Proof testing has been studied in the context of structural reliability [7,11,13,14,15,16]. The proof load factors proposed here are intended to

apply to evaluation of short- and medium-span highway bridges whose response is governed by vehicular loads.

2. PROOF LOAD FORMULA FOR BRIDGE RATING

A proof load formula is proposed, in the load and resistance factor format:

$$\phi \, Y_p = \alpha_L \, L_n \, g_n \, I_n \qquad (1)$$

where Y_p is target proof load effect. L_n is nominal static live load effect, g_n is nominal load distribution factor, and I_n is impact factor accounting for dynamic effect of vehicular loading. They are specified by AASHTO [4] respectively based on the current AASHTO rating vehicles with a lane load, empirical estimations, and road surface roughness. ϕ and α_L are resistance reduction and live-load factors, respectively. The load and resistance factors are to be determined in this study based on a structural reliability criterion. In addition to determining target proof load, the proposed formula is also intended to be used for rating through proof load testing:

$$\begin{aligned} \text{Rating Factor} &= \phi \, R_p \, / \, \alpha_L \, L_n \, g_n \, I_n \\ &= R_p \, / \, Y_p \end{aligned} \qquad (2)$$

where R_p is proved capacity for live load, equal to or lower than the target value Y_p. This rating methodology is consistent with the current rating method given by AASHTO manual [1] in both concept and format. Note that only bending moment as load effect is considered in this paper, although Eqs. 1 and 2 are in general forms.

3. STRUCTURAL RELIABILITY MODELS FOR BRIDGE PROOF TESTING

Consider a limit state function Z for a typical primary bridge structural member:

$$Z = R - D - L = R' - L \qquad (3)$$

where R, D, and L are true values of resistance,

dead load, and live load effects, respectively, and $R' = R - D$ is resistance margin for live load. Considering uncertainties and random variation associated with these quantities, R' and L are modeled by independent lognormal random variables. The live-load effect is further modeled by a combination of the following factors [2,3,6]:

$$L = a \, HW_{.95} \, m \, g \, I \qquad (4)$$

where all variables are modeled by independent lognormal random variables except a, which is a deterministic coefficient correlating truck weight to load effect based on the AASHTO rating vehicles [1]. H is a factor accounting for multiple presence of vehicles on the bridge, and $W_{.95}$ is a characteristic value of the vehicle weight spectrum. Their product is treated here as a single variable. m covers effect of vehicle configuration variation on the load effect, g is lateral distribution factor, and I is impact factor for dynamic effect.

Two application cases of proof load testing are considered here: 1) verifying or enhancing an existing rating obtained by analytical methods, and 2) establishing a rating for those bridges not suitable for analytical rating. For Case 1, R' in Eq.3 is described by its mean and standard deviation obtained by the existing rating [2]. For Case 2, $R' = Y_p$ is assumed conservatively, where Y_p is the target proof load effect. Figs.1a and 1b schematically illustrate how the target proof load is determined based on a target failure probability P_f that $L > R'$ using Eq.3 (or equivalently a target safety index β [2,3,6]), respectively for Cases 1 and 2. In Fig.1a the original probability density distribution f of R' is truncated by the proof load effect and then normalized for its area under f, with residual total probability being 1. The posterior failure probability for $L > R'$ is then computed using the updated probability density. The transformation method [7] was used to compute the reliability index β, which has been further verified of its accuracy by a direct integration method. In Fig.1b the posterior failure probability for $L > R'$ is calculated by setting $R' = Y_p$. The computation is much simpler than the other case.

Table 1 contains a database input into the

reliability models in Eq.3. Its sources are identified in [9]. This database was established by a variety of techniques and considered typical in covering the statistical variation of current practice and traffic loadings in the United States. It has been found [2,3,4] that the average safety level assured by the current US bridge evaluation code [1] is equivalent to a reliability index $\beta=2.3$. Thus this value was used as the target level for proof load factor determination here.

4. PROOF LOAD FACTORS BASED ON TARGET STRUCTURAL RELIABILITY

To be consistent with bridge evaluation practice by current analytical methods [1] and the recently developed method of load and resistance factors [4], the resistance reduction factor ϕ is proposed to be 0.95, 0.9, and 0.95, respectively, for steel, reinforced concrete, and prestressed concrete materials. Thus α_L is the only factor to be determined in the proof load formula, to reach the target safety index of 2.3. Given an α_L, Y_p determined by Eq.1 is used in limit state functions for Eq.3 for the two application cases of proof load testing. Their safety indices are then calculated to be compared with the target value of 2.3. This mechanism allows selection of α_L to satisfy the requirement for structural reliability. It is noted that for a given α_L, β varies with traffic loading condition, span length, and material type. Thus for each traffic loading condition, α_L is selected by minimizing β's variation due to other factors.

For application of proof load testing when a rating factor (RF) exists, variation of required α_L with RF was found to be less significant when RF<0.7, regardless of traffic loading conditions [9]. Thus, RF equal to 0.7 is selected as a threshold whether to take into account the existing rating in determining the target proof load. In other words, when an existing rating factor is higher than 0.7, it is considered an important piece of information to be included in selecting the target proof-load level, but lower than 0.7 is not worth considering. By this criterion, proposed proof

load factor α_L is listed in Table 2 for this case. Fig.2 shows safety index β using the proposed load and resistance factors. It is seen that they produce a relatively uniform reliability level of safety index equal to 2.3. Note that reinforced concrete bridges have significantly different live-to- dead load ratios than steel and prestressed concrete bridges. This is a major factor causing safety indices higher than the other two types of bridge, especially for longer spans. Its reliability assessment is not performed for spans longer than 100 ft, since the available empirical ratio of dead-to-live load is considered valid only up to this span length and few R/C highway bridges in the United State exceed this span length.

For the application case when a rating factor is not available, Table 2 contains also proposed proof load factor α_L for the four categories of live-load traffic. They are higher than those for the previous application case, because less information is required to reach the same target safety level. Fig. 3 shows safety index assured by the proposed proof load factor for this case. It is seen that a relatively uniform safety level 2.3 is realized with respect to span lengths. In this application case, differences in β for different materials are lower than in the previous case. This is because the dead-load influence on β is eliminated as it no longer appears in the limit function. Note that when an existing rating factor is lower than 0.7 these live-load factors can be used to determine required target proof load and load rating by proof load testing. They are the maximum proof load levels needed for bridge rating to ensure the target reliability level, which do not depend on any a priori information about the bridge's capacity.

5. SENSITIVITY ANALYSIS

It is well recognized that the input data and certain assumptions in modeling and calculation just described may influence the obtained safety index, and in turn may affect the proposed proof load factors. A sensitivity analysis thus was undertaken to examine if reasonable changes in input data and assumptions would

881

affect uniformity in reliability level and satisfaction of appropriate target safety levels by the proposed proof load formula. It was found that the proposed proof load factors are not sensitive to the lognormal distribution assumption used and input statistical parameters [9].

6. CONCLUSIONS

Reliability-based proof-load factors are proposed for bridge rating by proof-load testing. They are included to be used to determine the target proof load and resulting bridge load rating. The factors of the proposed formula are prescribed based on a target safety index of 2.3, which is consistent with the current practice in the United States. Two cases of proof load testing application are covered: 1) an analytical rating is available but unsatisfactory and 2) no analytical rating can be obtained analytically. A relatively uniform reliability level is assigned by the proposed proof-load formula for these cases over a practical range of span length, material type, and traffic condition. A comprehensive sensitivity analysis has been performed to examine implications of the input data. Its results show that the proposed proof load formula is not sensitive to the input statistical parameters and probability distribution assumptions.

7. ACKNOWLEDGEMENTS

This study is partially supported by the Federal Highway Administration. Discussions with Dr.D.Verma of Altair Engineering and Dr.F.Moses at University of Pittsburgh are acknowledged. The authors also thank Messrs. D.B.Beal and G.A.Christian of New York State Department of Transportation for their valuable comments and suggestions during the course of this study.

8. REFERENCES

[1] Manual for Maintenance Inspection of Bridges, Washington, DC: American Association of State Highway and Transportation Officials, 1983

[2] Moses,F. and Verma,D. "Load Capacity Evaluation of Existing Bridges: Phase II", Draft Final Report on National Cooperative Highway Research Program Project 12-28(1), July 1989

[3] Moses,F., and Verma,D. " Load Capacity Evaluation of Existing Bridges", Report 301, National Cooperative Highway Research Program, Transportation Research Board, 1987

[4] Guide Specifications for Strength Evaluation of Existing Steel and Concrete Bridges, Washington, DC: American Association of State Highway and Transportation Officials, 1989

[5] Ontario Highway Bridge Design Code, Downsview, Ontario, Canada, Ministry of Transportation and Communication, 1983

[6] Fu,G., and Moses,F. "Overload Permit Checking Based on Structural Reliability", Transportation Research Record No.1290, 1991, pp.279-89

[7] Fujino,Y., and Lind,N.C. "Proof-Load Factors and Reliability", Journal of the Structural Division, American Society of Civil Engineers, Vol.103, No.ST4, 1977, pp.853-870

[8] Fu,G., Saridis,P, and Tang,J. "Proof Load Testing for Highway Bridges", Interim Report for Research Project 209-1, Engineering R & D Bureau, New York State Department of Transportation, July 1991

[9] Fu,G. and Tang,J. "Proof Load Formula for Highway Bridge Rating", Transportation Research Record No.1371, 1992, p.129

[10] Ellingwood,B., Galambos,T.V., MacGregor,J.G., and Cornell,C.A. "Development of a Probability Based Load Criterion for ANSI A58", Report NBS 577, National Bureau of Standards, June 1980

[11] Grigoriu,M. and Hall,W.B. "Probabilistic Models for Proof Load Testing", ASCE J.Stru.Eng. Vol.110, No.2, 1984, pp.260-274

[12] Ladner,M. "In Situ Load Testing of Concrete Bridges in Switzerland", American Concrete Institute Symposium on Strength Evaluation of Existing Concrete Bridges, SP-88, Detroit, MI, 1985, pp.59-79

[13] Shinozuka,M. and Yang,J.-N. "Optimum Structural Design Based on Reliability and Proof-Load Test", Annals of Assurance Sci.,

Proc. of the Reliability and Maintainability Conference, V.8, pp.375-391, 1969

[14] Rackwitz,R. and Schrupp,K. "Quality Control, Proof Testing and Structural Reliability", Structural Safety, Vol.2, 1985, pp.239-244

[15] Veneziano,D., Meli,R., and Rodriguez,M. "Proof Loading for Target Reliability", ASCE J.Stru.Div. Vol.104 No.ST1, 1978, pp.79-93

[16] Yang,J.-N. "Reliability Analysis of Structures under Periodic Proof Tests in Service", AIAA Journal, Vol.14, No.9, p.1225, 1976

Table 1   Statistical Database

	Mean	COV(%)
m Span(ft):		
30	1.0	11
50	.95	11
70	.90	11
90	.92	7.5
120	.95	5.4
160	.96	3.4
200	.97	3.2

$HW_{.95}$

Traffic	Com.	Single	Com.	Single
Enforced				
Light	170kips	92kips	5	8
Heavy	180kips	100kips	6	8
Unenforced				
Light	210kips	120kips	10	10
Heavy	225kips	125kips	10	10

I Surface		
Smooth	1.1	10
Medium	1.2	10
Rough	1.3	10

	Bias			COV(%)		
	Steel	R/C	P/C	Steel	R/C	P/C
R	1.05	1.05	1.00	16	14	11
D	1.00	1.00	1.00	10	10	10
g	0.90	0.97	0.96	13	11	8

Note:Com.=Combination; Bias=Mean/
Nominal; COV=Standard Deviation/
Mean.

Table 2 Proposed α_L According
to Traffic Condition

Existing Rating Factor>0.7
Category:	Enforced	Unenforced
Volume:		
Light	1.35	1.80
Heavy	1.45	1.90

Existing Rating Factor<0.7 or
 No Rating Factor Available
Category:	Enforced	Unenforced
Volume:		
Light	1.45	1.90
Heavy	1.55	2.00

a) Case 1:
An Analytical Rating Available

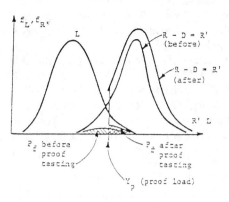

b) Case 2:
No Analytical Rating Available

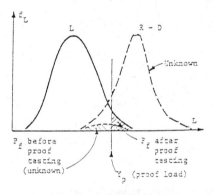

Fig.1   Structural Reliability
Model: Failure Probability
P_f and Target Proof Load Y_p

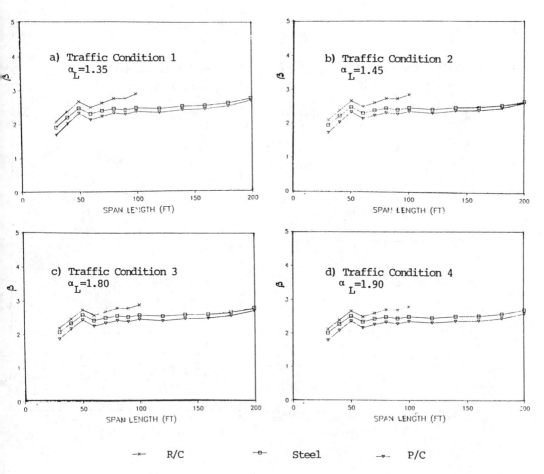

Fig. 2 Structural Reliability Based on Propsed Proof Load Formula - with Analytical Rating

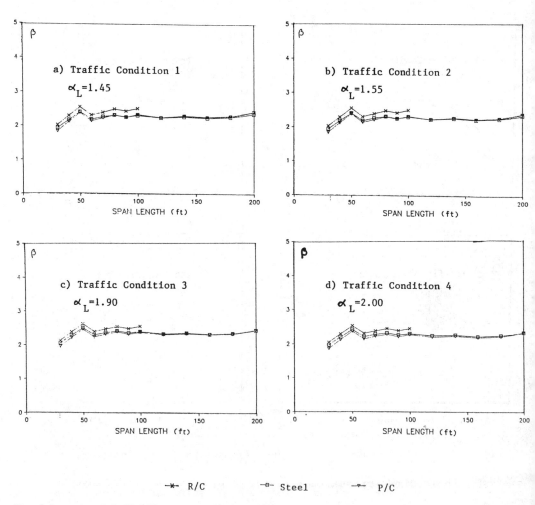

Fig. 3 Structural Reliability Based on Proposed Proof Load Formula – without Analytical Rat…

Structural Safety & Reliability, Schuëller, Shinozuka & Yao (eds) © 1994 Balkema, Rotterdam, ISBN 90 5410 357 4

Reliability and redundancy of highway bridges

Michel Ghosn & Jian Ming Xu
Department of Civil Engineering, The City College of New York, N.Y., USA

Fred Moses
Department of Civil Engineering, The University of Pittsburgh, Pa., USA

ABSTRACT: This paper proposes a framework for implementing the concept of redundancy and system safety criteria in the design and evaluation of typical concrete bridges. The proposed format consists of providing tables of partial system factors that could be used as measures of the level of redundancy for typical bridges. The proposed system factors depend on the capability of the system geometry to redistribute the applied load following either overloading of the intact bridge or a sudden failure of one main member. These system factors are derived by examining different combinations of numbers of girders, spans, and member ductility parameters so that serviceability and collapse conditions are avoided with reasonably high safety margins.

1 INTRODUCTION

Structural redundancy is defined as the ability of a structure to redistribute its load after the failure of one or more of its members. Despite the fact that they are normally designed on a component by component basis, many structures have some levels of redundancy that are often unaccounted for. Concerns about the safety of the deteriorating bridge infrastructure are offset by the limited resources available for rehabilitation. In order to optimize the use of these resources, we need to reconsider the current structural safety criteria by accounting for the benefits of redundancy in the evaluation process. This paper presents an overall framework for considering system effects and redundancy in the design of new bridge superstructures and the load capacity evaluation of existing bridges.

To analyze the redundancy of bridges, the ideal situation would be for the engineer to have available a structural model and a finite element analysis package that considers elastic, inelastic, instability and dynamic behavior. Such a program could evaluate both overloads and the consequences of assumed hazard conditions. The engineer would then check the structure to verify whether a) acceptable behavior, b) unserviceable conditions or, c) collapse states occurred. Even in the ideal situation with a comprehensive analysis

program available, there are still many safety related decisions that must be made. These include: 1) The level of overloads that must be carried before either unserviceable or collapse states have occurred, 2) the type of hazards and magnitude of accident conditions that must be borne by the structure, 3) the inclusion of uncertainties in the strength analysis model and, 4) the reliability level that is required for each of the overload and hazard conditions and their corresponding serviceability and collapse damage states. The goal of this paper is to develop a reliability-based procedure that will account for these factors in evaluating the redundancy of bridges.

Some bridges can be easily classified for both geometry and member ductility. Bridges of the same material that have similar geometries and have members designed according to the same specification can be expected to have similar levels of redundancy. To minimize the level of effort of the evaluating engineer, these bridges can be studied as a group and a specifications format can be developed. Other bridges are unique because of their geometry and non-typical member properties, these will have to be studied individually with general guidelines provided to help the engineer perform the evaluation. This paper addresses typical concrete T-beam bridges and proposes a framework for implementing system safety criteria in their design and evaluation. The

Table 1. Load Factors for Concrete T-Beam Bridges with Two-Lane Loading.

Span Length		4 Beams			6 Beams			8 Beams	
		S=4 ft.	S=6 ft.	S=8 ft.	S=4 ft.	S=6 ft.	S=8 ft.	S=4 ft.	S=6 ft.
45 ft.	Ultimate	2.603	3.797	4.972	3.901	5.647	7.150	5.058	7.020
	Serviceability	2.351	3.429	4.490	3.469	5.102	5.224	4.449	5.755
60 ft.	Ultimate	2.791	4.022	5.378	4.165	5.965	7.937	5.436	7.610
	Serviceability	2.490	3.633	4.939	3.731	5.469	6.408	4.980	6.520
80 ft.	Ultimate	2.846	4.169	5.200	4.268	6.243	7.741	5.684	8.060
	Serviceability	2.718	3.796	4.882	4.065	5.633	7.288	5.429	7.000
100 ft.	Ultimate	3.039	4.174	5.301	4.559	6.252	7.939	6.077	8.290
	Serviceability	2.889	3.984	5.102	4.330	6.001	7.618	5.765	7.440

Table 2. Load Factors for Concrete T-Beams Bridges with One-Lane Loading.

Span Length		4 Beams			6 Beams		
		S=4 ft.	S=6 ft.	S=8 ft.	S=4 ft.	S=6 ft.	S=8 ft.
45 ft.	Ultimate	5.193	7.579	9.747	7.779	10.970	13.220
	Serviceability	4.670	6.780	8.500	7.001	8.500	8.750
60 ft.	Ultimate	5.540	7.952	10.650	8.273	11.910	15.820
	Serviceability	5.017	7.250	9.833	7.550	10.500	11.500
80 ft.	Ultimate	5.691	8.331	10.390	8.536	12.480	15.450
	Serviceability	5.410	7.540	9.680	8.080	11.280	14.320
100 ft.	Ultimate	6.079	8.346	10.600	9.118	12.520	15.880
	Serviceability	5.775	7.920	10.170	8.675	11.920	15.230

Table 3. Input Data for Reliability Analysis.

Span Length	Total Live Load					Dead Load(per beam)			
	One Lane		Two Lanes		C.O.V.	Beam Spacing			C.O.V.
	2 Years	75 Years	2 Years	75 Years		4 ft.	6 ft.	8 ft.	
45 ft.	1.604	1.745	2.726	3.030	0.246	0.386	0.475	0.565	0.100
60 ft.	1.653	1.790	2.810	3.120	0.225	0.628	0.738	0.848	0.100
80 ft.	1.745	1.890	2.966	3.280	0.197	0.950	1.047	1.195	0.100
100 ft.	1.838	2.000	3.124	3.440	0.188	1.184	1.359	1.533	0.100

All bridges are designed to satisfy AASHTO's LFD criteria
Random variables are: Resistance, Dead Load, Total live load including impact
All random variables are assumed to be lognormal

Loads given in table are factors of HS20 load effect
Resistance: Bias=1.13 COV=13.5%
Impact factor: for one lane=1.15 , for two lanes=1.10
COV for impact included in live load table

proposed format consists of providing tables of partial system factors that would be used to modify component strengths. These factors will depend on the structural properties of the component. The system factors also depend on the capability of the system geometry to redistribute loading following either an overload or a sudden failure of one main member. These system factors are derived by examining different combinations of numbers of girders, spans, and member ductility parameters. The output consists of tabulated system factors that

modify the component strengths so that serviceability and collapse conditions are avoided with reasonably high safety margins.

2 DUCTILITY OF CONCRETE MEMBERS

The behavior of concrete bridge members is a function of member ductility. In a special finite element program developed in this study, the nonlinear behavior and ductility of bridge members is represented by their moment

versus rotation curves. These curves are obtained based on the results of analytical and empirical equations developed by previous research studies. The first step consists of developing a moment versus curvature relationship based on the properties of the concrete section being analyzed. This is achieved using typical stress versus strain curves for the concrete and the reinforcing steel. An example of a moment versus curvature plot for a typical concrete section is given in figure 1. As a member is subjected to higher moments, the section will undergo plastic rotations. These are related to the curvatures by the length of the plastic hinge formed. Empirical formulas are available to estimate the length of the plastic hinge. The member will keep deforming until a maximum plastic hinge rotation is reached. At that point the member is assumed to loose its capacity to carry any load. Formulas to estimate the maximum possible plastic hinge rotation that a section can sustain are given in terms of the reinforcement index, the type of steel used, the concrete strength, the confinement ratio and the geometric properties of the section. The proposed model is valid for both reinforced concrete and prestressed concrete members. To verify the validity of the method used in this study, figure 2 compares the analytical results to the published experimental results for two reinforced concrete beams. The results show good agreement for the whole range of behavior including the determination of the failure point.

3 ANALYSIS OF CONCRETE T-BEAM BRIDGES

A large number of concrete T-beam bridges were designed for the purpose of this study. All the T-beam bridges discussed herein are designed using AASHTO's LFD approach. Simple span bridges of 45, 60, 80 and 100 ft lengths are analyzed. These are assumed to have 4 beams at 4, 6 or 8 ft spacings; 6 beams at 4, 6 or 8 ft spacings; or 8 beams at 4 and 6 ft spacings. The sections are assumed to have just enough steel to satisfy the LFD requirements. The slab thickness is assumed to be 7" in all the cases considered. The 45 and 60 ft spans are assumed to have three diaphragms at the ends and the midspan. The 80 and 100 ft spans are assumed to have 5 diaphragms equally spaced at each quarter point. Concrete strength is assumed to be 3000 psi and grade 40 steel is used for the base case. A sensitivity analysis however, confirmed that neither the steel grade nor the concrete strength were important as long as the required member capacity is maintained.

3.1 Base Cases

The analysis of the reserve strength of these bridges was performed using the program developed in this study. The first set of analyses is performed assuming a minimum level of shear reinforcement i.e. the main members are assumed to be unconfined. The slab is modeled by linear elastic plate elements. The behavior of the diaphragms and the longitudinal members is modeled using the moment versus rotation curves developed for each member according to its section properties. Two side-by-side HS-20 vehicles are placed on the bridge at the point of maximum moment. These vehicle loads are then incremented until the bridge ultimate capacity is reached. This is defined as the load at which one of the longitudinal members reaches its maximum plastic hinge rotation. The other limit state noted is the load factor at which a maximum displacement equal to span length/300 is reached. Figure 3 shows a plot of load factor versus maximum displacement for one example bridge.

Table 1 gives the results of the analysis for the cases studied in this section. As expected for AASHTO bridges, the following simple observations can be immediately made: a) The load factor increases as the span length increases, b) the load factor increases as the beam spacing increases, and c) the load factor increases as the number of beams increases. These observations are valid for both the ultimate load condition and the serviceability limit state i.e. the load factor at L/300. More specific observations are also noted, these include:

1. There is a strong relationship between the load factor at ultimate and the load factor at L/300. The ratio of the load factor at ultimate to the load factor at L/300 is on the average equal to 1.10 with a coefficient of variation equal to 6.3%.

2. There is a strong relationship between the load factors for the beams with 8 ft , 6 ft and 4 ft spacings. The ratio of the load factors for the 6 ft spacings to those with 4 ft is on the average equal to 1.43. The ratio of the load factors of beams with 8 ft spacings to those with 4 ft spacings is 1.85.

3. There is a strong relationship between the load factors for the bridges with 8 beams, 6 beams and 4 beams. The ratio of the load factor for the 6 beam bridge to that of the 4 beam bridge is on the average 1.5. The ratio of the load factor of the 8 beam bridge to that of the 6 beam bridge is 1.3.

These observations are made for concrete T-

Table 4. Safety Indices.

No. of Beams		4			6			8	
	Span Length	Beam Spacing			Beam Spacing			Beam Spacing	
		S=4 ft.	S=6 ft.	S=8 ft.	S=4 ft.	S=6 ft.	S=8 ft.	S=4 ft.	S=6 ft.
2 Lane Loading (Ultimate)	45 ft.	1.954	3.318	4.332	3.367	4.752	5.637	4.256	5.461
	60 ft.	2.107	3.419	4.504	3.391	4.676	5.716	4.148	5.299
	80 ft.	1.948	3.288	4.025	3.058	4.378	5.022	3.707	4.875
	100 ft.	1.937	2.936	3.687	2.941	3.870	4.553	3.510	4.365
2 Lane Loading (Serviceability)	45 ft.	1.955	3.306	4.308	3.284	4.706	4.674	4.097	4.965
	60 ft.	2.039	3.366	4.495	3.286	4.623	5.085	4.082	4.913
	80 ft.	2.081	3.218	4.048	3.125	4.203	4.986	3.740	4.505
	100 ft.	2.026	3.011	3.770	2.969	3.904	4.562	3.493	4.115
1 Lane Loading (Ultimate)	45 ft.	3.748	5.184	6.137	5.134	6.396	6.979	5.210	6.620
	60 ft.	3.723	5.012	6.067	4.804	6.039	6.982	5.055	5.940
	80 ft.	3.311	4.628	5.274	4.131	5.396	5.931	4.557	5.652
	100 ft.	3.142	4.057	4.724	3.853	4.692	5.287	4.221	4.976
1 Lane Loading (Serviceability)	45 ft.	3.624	5.007	5.819	4.961	5.528	5.385	4.726	5.978
	60 ft.	3.569	4.845	5.923	4.621	5.664	5.729	4.751	5.169
	80 ft.	3.309	4.401	5.137	4.063	5.111	5.722	4.460	5.096
	100 ft.	3.134	4.014	4.700	3.800	4.614	5.217	4.105	4.575
One Member	45 ft.	1.648	2.964	3.835	2.570	3.931	4.110	3.585	3.527
	60 ft.	1.756	3.030	3.889	2.624	4.062	4.457	3.748	3.847
	80 ft.	1.898	2.989	3.791	2.667	3.965	4.684	3.632	4.051
	100 ft.	1.871	2.862	3.567	2.573	3.721	4.288	3.422	3.769

Table 5. System Factors.

No. of Beam		4			6			8		Average
Span Length	Target Index	Beam Spacing			Beam Spacing			Beam Spacing		
		S=4 ft.	S=6 ft.	S=8 ft.	S=4 ft.	S=6 ft.	S=8 ft.	S=4 ft.	S=6 ft.	
45 ft.		0.675	0.910	1.112	0.909	1.194	1.181	1.063	1.241	Average
60 ft.	3.77	0.723	0.929	1.138	0.918	1.159	1.251	1.055	1.212	1.009
80 ft.		0.730	0.911	1.044	0.888	1.074	1.219	0.990	1.127	
100 ft.		0.736	0.871	0.986	0.873	1.016	1.135	0.956	1.057	
45 ft.		0.643	0.868	1.063	0.869	1.143	1.132	1.018	1.191	
60 ft.	4.00	0.693	0.891	1.092	0.882	1.114	1.203	1.014	1.166	0.969
80 ft.		0.702	0.876	1.004	0.855	1.034	1.175	0.953	1.086	
100 ft.		0.708	0.838	0.950	0.841	0.979	1.094	0.921	1.019	
45 ft.		0.577	0.782	0.963	0.787	1.040	1.033	0.927	1.088	
60 ft.	4.50	0.631	0.813	0.999	0.807	1.021	1.105	0.931	1.072	0.888
80 ft.		0.644	0.805	0.923	0.787	0.952	1.083	0.878	1.001	
100 ft.		0.651	0.772	0.875	0.775	0.903	1.009	0.850	0.940	

beam bridges with high levels of transverse distribution capability (represented by elastic deck and high moment capacity of transverse beams). According to these observations, the load factors for all the cases analyzed can be deduced from the load factors for the 4-beam bridges with 4 ft beam spacing. High transverse moment capacity is represented by a ratio of ultimate transverse beam moment to ultimate moment of main members higher than 25%.

3.2 Effect of loading pattern

The results given in table 2 show that for one-vehicle loading in one lane of concrete bridges with linear deck behavior or "high" transverse capacity, the results are very similar to those of the two-vehicle loading. In general, the load factors shown in table 2 are double the values given in table 1 indicating that the bridges are able to carry the same total load whether they are loaded by one vehicle or two vehicles. A sensitivity analysis has shown that this conclusion does not hold for the case of nonlinear deck behavior or "low" transverse capacity. Most concrete bridges however, are expected to have "high" transverse capacity and the observation made here should be valid for most cases.

4 RELIABILITY ANALYSIS

A simple reliability analysis is performed for all the bridges analyzed. The safety index β was first calculated for the most critical longitudinal member assuming an elastic distribution of load. The safety index for one member is normally used to calibrate new codes as currently done for the development of AASHTO's LRFD specifications. System safety indices were also calculated for the ultimate capacity of the complete bridge structure and for the serviceability limit state defined as a maximum displacement of L/300. The input data for the reliability analysis is given in Table 3. The safety index calculations for the ultimate capacity use a maximum expected live load moment corresponding to a bridge life of 75 years. For the serviceability limit state, a 2 year exposure period is assumed which corresponds to the mandatory inspection period for US bridges. The calculations are performed for one lane or two-lane loadings. The results of the safety index calculations performed in this study are given in table 4. The results obtained show that in addition to the span length and beam spacing, the number of beams affects the distribution of load to the individual members and thus the safety indices.

The safety indices obtained for one member vary between 1.65 and 4.68. The lowest value corresponds to a short (45' span) and narrow bridge with 4 beams at 4 ft spacings (we are assuming that there is enough overhang to carry two lanes of traffic). The average member safety index obtained for all the cases considered is 3.35.

The system safety indices for ultimate capacity vary between 1.95 and 5.72. Here again, the lowest value corresponds to the short narrow bridge. The highest values correspond to the short but wide bridges. Indicating that a transverse redistribution of load is effective in improving the system safety for bridges with high number of beams at relatively wide spacings. The average safety index is 3.89.

The serviceability limit state gives safety indices roughly equal to those of ultimate capacity except for wide bridges. In the latter case, the serviceability limit state gives consistently lower safety indices. This indicates that as the number of beams and the beam spacing increases, a bridge might exhibit high safety levels against collapse, the large deformations that it will have to sustain however, will make it unusable way before collapse is reached.

Therefore, in this study we propose to use the lower system safety index value as a measure of the system capability of a bridge (i.e. for a given bridge, take the lower of the system index at ultimate and the system index at L/300 as the final index that will control the bridge's safety). This will produce an average system safety index of 3.77 for all the bridges analyzed. This indicates that on average, the bridge system will give a safety index higher than the member safety index by 0.42. In general, adding 1.0 to the safety index decreases the risk or probability of failure by about a factor of 10.

The calculations for one lane loading produced higher safety indices than for the two-lane loadings indicating that for typical concrete bridges the two-lane loadings are most critical.

5 CALIBRATION OF SYSTEM FACTORS

Specifications, and design in general, deal with individual components and are likely to continue to be done on a component basis. Thus, for typical bridges, this study proposes to use system factors which consider the system behavior in the routine design or strength evaluation of individual components. These system factors can be introduced into the checking equations with either a Working Stress Design (WSD), Load Factor Design (LFD) or Load and Resistance Factor Design (LRFD) format. For example, for AASHTO's LFD format for concrete members, the checking equation will become:

$$\phi_s \, \phi \, R_n = 1.3 \, (D_n + 5/3 \, L_n) \qquad (1)$$

ϕ_s is the system factor. R_n is the nominal resistance. D_n is the dead load. L_n is the design live load including lateral distribution and impact. ϕ is the strength reduction factor which is 0.9 for flexure. 1.3 is the dead load factor and 5/3 is the additional live load factor.

ϕ_s should be calibrated such that when bridge members are designed according to equation 1, the bridge should produce a system safety index value β_{sys} equal to a predetermined target value. In this case, we propose that all bridges should satisfy a target system index equal to the average system index which is 3.77. For example, the 45 ft bridge with 6 beams at 4 ft center to center produced a system index of 3.37 when designed according to AASHTO's LFD approach (equation 1 without ϕ_s). In order to raise its system index to the target value of 3.77, each one of its members should be increased by a factor of 1.08. This means that for this case, ϕ_s is 0.92 (1/1.08).

Table 5, gives the system factor ϕ_s required

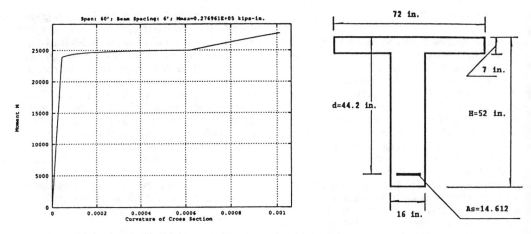

Figure 1. Typical Cross Section and Moment Curvature Plot.

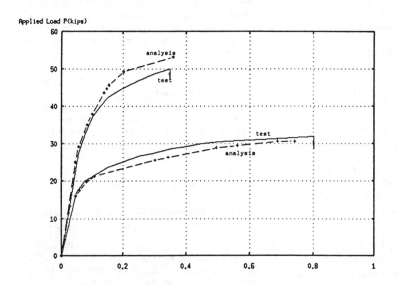

Figure 2. Comparison of Load vs. Deflection for 2 Tests performed by Mattock et. al.

for each bridge analyzed herein. The lowest observed ϕ_s factor is 0.68. If a bridge is associated with a value less than 1.0 it indicates that the bridge has unacceptable levels of redundancy. To improve the design, member strengths should be multiplied by a factor of $1/\phi_s$. Thus ϕ_s can be defined as a "system reduction factor" which penalizes designs that produce low levels of redundancy and rewards bridges with high redundancy. The highest factor obtained is 1.25 for the 60ft bridge with 6 beams at 8 ft. Since ϕ_s in this case is greater than 1.0, this means that the particular bridge

will provide about 25% more capacity than the minimum required.

The bridges with 4 beams at 4 ft spacings seem to provide the most critical cases. Since such geometry is rather unusual for two lane bridges, the calibration process is repeated herein without considering these extremely narrow configurations. The target system safety index then becomes 4.0, and the system reduction factors become as shown in table 5. Here, the system factors vary from a value of 0.64 to a value of 1.20. If the target system index is increased to 4.5 the system factors will vary from 0.58 to 1.1. On the average, a change in

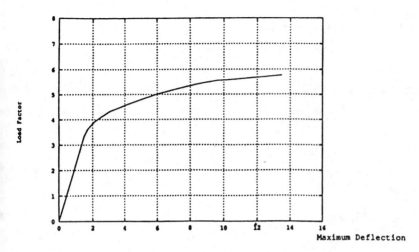

Figure 3. Typical Load vs. Deflection Curve.

the target index on the order of 0.5 will reduce
the system factors by about 0.08.

6 CONCLUSIONS

A method to develop system factors that consider
bridge redundancy in the design of concrete
bridges is presented. System factors provide a
simple technique for assuring that some
minimum level of redundancy exists after the
most critical member reaches its limit. The
proposed factors can be easily incorporated in
the checking equations used for the design and
load capacity evaluation of bridge components.
The calibration of these factors uses reliability
theory, however, once developed, the engineer
will be able to use these factors in a
deterministic approach which is a simple
extension of the current methods recommended
in the specifications.

ACKNOWLEDGEMENT

This work is sponsored by the American
Association of State Highway and Transportation
Officials (AASHTO), in cooperation with the
Federal Highway Administration (FHWA), and is
conducted in the National Cooperative Highway
Research Program (NCHRP) which is
administered by the Transportation Research
Board (TRB) of the National Research Council
(NRC). The opinions and conclusions expressed
or implied in this paper are those of the authors
and are not necessarily those of the
Transportation Research Board, the National
Research Council, or the U.S. Government.

Structural Safety & Reliability, Schuëller, Shinozuka & Yao (eds) © 1994 Balkema, Rotterdam, ISBN 90 5410 357 4

Simulation for limit state analysis of box girder bridges

T. Kitada
Osaka City University, Japan

H. Ikeda
Japan Bridge, Japan

M. Dogaki
Kansai University, Japan

Y. Takeda
Kawada Industry, Japan

S. Ishizaki
Sakai Iron Works, Japan

ABSTRACT: Safety factors of three steel box girder bridge models designed according to the current JSHB are evaluated at their ultimate and serviceability limit states. Simulations for calculating live load effect are carried out, based on statistical data on live loads surveyed in highway bridges in Japan. The two limit states of steel box girder bridges are defined. Then a method proposed by Kitada et al. (1992) is used to predict the limit states. The accuracy of the method is verified by 15 experimental results. The characteristic values of ultimate and serviceability stress resultants at the cross sections under consideration are decided by considering the probabilistic variation of live load and yield stress of steel. The inherent safety factors of the box girder bridge models are at least 1.44 and 1.52 at the ultimate and serviceability limit states, respectively.

1 INTRODUCTION

In the near future, a limit state design method will also be adopted in design specifications for highway steel bridges in Japan. For this purpose, it is necessary, (1)to clearly define the serviceability and ultimate limit states for all the bridge members, (2)to derive the design criteria for checking these limit states, and (3)to investigate the inherent safety of each bridge member designed by the current Japanese Specifications for Highway Bridges (JSHB) in order to determine a target level for the safety.

Three 2−spans continuous steel box girder bridge models designed by the JSHB are used for a case study in this research, in order to examine the three issues mentioned above for development of a limit state design method for steel box girder bridges. A calculation formula proposed by Kitada et al. (1992) is adopted for predicting the ultimate stress resultants of box girders subjected to the combination of bending, shear and torsion. In the method, the ultimate strength of the web plates in the box girders is estimated using effective thickness based on the elastic buckling theory, and the ultimate strength of the stiffened compression flanges is evaluated by a column model approach considering initial deflection and residual stress. Only a serviceability limit state decided by the yielding of a cross section is considered as a dominant serviceability limit state for these bridge models in this study, although other serviceability limit states concerning vibration, fatigue strength, excessive deflection due to live load, etc. can be considered.

The characteristic values of the ultimate and serviceability stress resultants at some typical cross sections of the box girder models are decided by considering the probabilistic variation of yield stress of steel. The error of the calculation formula is evaluated through 15 experimental data.

The live load effect, that is, applied bending moment, shear force and torsional moment, is calculated through the Monte Carlo simulation method using three types of traffic data surveyed in Japan. The characteristic values of the live load effect are defined by fitting the simulation results to a type−I asymptotic extreme distribution.

Finally, the inherent safety factors of the box girder bridge models are evaluated by comparing the characteristic values of resistance with those of load effect at the ultimate and serviceability limit states.

2 LIMIT STATES OF BOX GIRDER BRIDGES

2.1 *Ultimate limit state*

It can be considered that the ultimate limit state of steel box girders subjected to combined bending, shear and torsion is mainly governed by the buckling of stiffened flange plates under compression, the buckling of stiffened web plates under bending and/or shear, or the interactive buckling of these plates.

In this paper, a level of safety at the ultimate limit state is evaluated by the following equation (Kitada et al. 1992).

$$\phi_{ULT} = \sqrt[4]{(\frac{M}{M_{ULT}})^4 + (\frac{S+S_t}{S_{ULT}})^4} < \Phi^*_{ULT}, \quad (1)$$

where ϕ_{ULT} is defined as an applied stress resultant index for the ultimate limit state, Φ^*_{ULT} is an ultimate stress resultants index concerning the error of ϕ_{ULT} (see Eq.(3) and Fig.1), M_{ULT} is the ultimate pure bending moment of a steel box girder section subjected to pure bending, and S_{ULT} is the ultimate pure shear force of the steel box girder section subjected to predominant shear force. M, S and S_t are the applied bending moment, shear force, and pseudo−shear force decided with shear stress caused by torsional moment, respectively.

2.2 Serviceability limit state

Applied stresses due to all the loads and their combination which occur often during bridge life should be kept in elastic region with proper reliability. Then a serviceability limit state is defined by yielding of a flange plate taking shear lag phenomenon into consideration. The yielding criterion of steel box girders subjected to bending, shear and torsion is given by the following equation (Kitada et al. 1992).

$$\bar{\sigma} = \sqrt{\sigma^2 + 3 \cdot \tau^2} < \sigma_y, \quad (2)$$

where $\bar{\sigma}$ is the equivalent stress, and σ_y is the yield stress of steel. σ and τ are the normal and shear stresses occurring on the effective cross sections considering the effective width due to shear lag phenomenon, and they are the sum of normal and warping stresses, and the sum of shear stresses due to shear and torsion, respectively.

2.3 Verification of predicted ultimate stress resultants by experimental results

In order to compare the ultimate stress resultants predicted by the formula (1) with experimental results, the experimental ultimate stress resultants index ϕ^*_{ULT} is defined as follows:

$$\phi^*_{ULT} = \sqrt[4]{(\frac{M^*}{M_{ULT}})^4 + (\frac{S^*}{S_{ULT}})^4}, \quad (3)$$

where M^* and S^* are the ultimate bending moment and shear force of the experimental models, respectively.

Substituting the experimental ultimate bending moments M^* and shearing forces S^* into the Eq.(3), the values of ϕ^*_{ULT} can be obtained as plotted in Fig.1 (Mikami et al. 1980, Niwa et al. 1979, and Dowling et al. 1973).

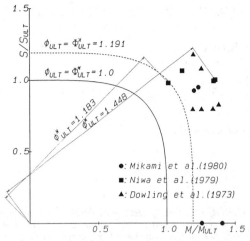

Fig.1 Experimental results and interaction curve for ultimate stress resultants

It can be seen from these results that this method predicts the conservative ultimate strengths.

3 SIMULATION ANALYSIS FOR LIVE LOAD EFFECT

3.1 Simulation analysis

Dead and live loads can be considered as the major loads for box girder bridges. The variation of dead load is, however, very small. Therefore, it is neglected in this study.

The live load effect is evaluated by the Monte Carlo simulation method. Three traffic flow models which mean rows of vehicles are generated based on the existing data obtained from the traffic surveys at an urban expressway (Kameta 1986), a national road (Shinohara 1988) and a rural road (Matsui & Kanbara 1989) in Japan. For example, the statistical parameters of the heaviest truck in each traffic model are shown in Table 1.

Table 1 Heaviest truck models

	Total weight W (KN)		Axle weight ratio (%)			P. D. F.
	μ	σ	W1	W2	W3	
Urban expressway	316.2	21.97	18.0	30.0	52.0	L. N.
National road	196.1	96.11	26.1*	36.9*	37.0*	L. N.
Rural road	208.8	66.68	22.3*	38.8*	38.9*	L. N.

The values noted by * are mean axle weight ratios.
μ :Mean value
σ :Standard deviation
P. D. F:Probabilistic Density Function
L. N. :Log-Normal

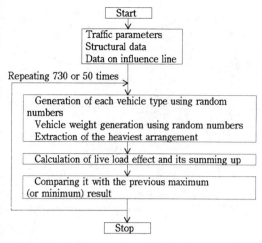

Fig.2　Simulation procedure for live load effect

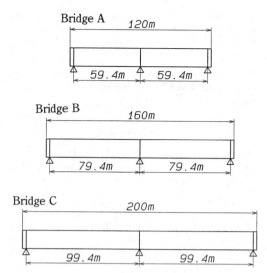

Fig.4　Side elevation of model bridges

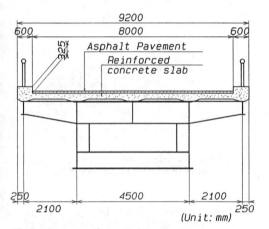

Fig.3　Cross section of bridge models

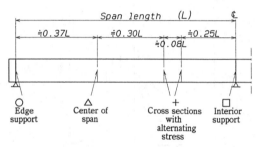

Fig.5　Cross sections under consideration

As the severest conditions of vehicle loading, two types of traffic jam models are taken into consideration, that is, a normal traffic jam which occurs during commuter rush hours in every morning and evening, and an unexpected traffic jam which takes place incidentally due to road repairing or traffic accident in midnight. The characteristics of these traffic jams are assumed referring to an existing investigation (Kameta 1986) as follows, for the lack of surveyed data. Namely, as for the mixing ratio of each type of vehicle, the statistical parameters of the traffic flow models mentioned above are used for the normal traffic jam, and those for the unexpected traffic jam are created modifying the statistical parameters of the normal traffic jam by proportional allotment so that the summation of mixing ratio of heavy weight vehicles becomes about 60%. The distance between vehicles is

fixed to 5 meters for the normal traffic jam, and 2.5 meters for the unexpected one. Frequency of the occurrence of traffic jam is assumed to be 730 times per year for the normal traffic jam, and 50 times per year for the unexpected one. The simulation procedure for live load effect is shown in Fig.2.

3.2 Model bridges and cross sections under consideration

Three 2–spans continuous steel box girder bridges are chosen as the bridge models in order to investigate the effect of combined action of bending and shear on the ultimate and serviceability limit states. The standard cross section of the model bridges is shown in Fig.3, and their span lengths and the locations where the inherent safety factors are calculated are shown in Figs.4 and 5. These are in the vicinity of the edge support with predominant shear force, at the center of span with large bending moment,

near the interior support subjected to the combination of bending and shear, and near the points with alternating stress due to live loading.

4 ULTIMATE STRESS RESULTANTS INDEX AND EQUIVALENT STRESS FOR SERVICEABILITY LIMIT STATE

4.1 Ultimate stress resultants index

The yield stress of steel should also be considered as a probabilistic variable. The ultimate pure stress resultants, M_{ULT} and S_{ULT} in Eq.(1) are calculated considering the probabilistic variation of yield stress. It is assumed that the variation of the yield stress is subject to a Log–Normal distribution function (Fukumoto 1980) with the mean value $\mu =1.148$ (normalized with the nominal yield stress) and the

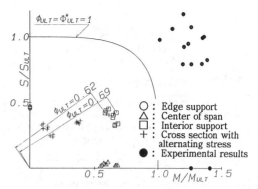

Fig.8 Simulation results and interaction curves for ultimate stress resultants (Normal traffic jam)

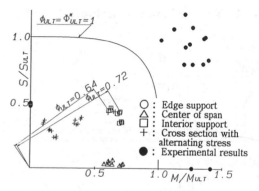

Fig.9 Simulation results and interaction curves for ultimate stress resultants (Unexpected traffic jam)

coefficient of variation $\delta =0.118$.

The maximum (or minimum) values of applied bending moment and corresponding shear force and torsional moment are decided through the simulation analysis throughout 2 years. Then, these maximum (or minimum) stress resultants, the dead load effect, and M_{ULT} and S_{ULT} calculated with a yield stress arbitrarily took out of the population of the yield stress as a random number are substituted into Eq.(1) to obtain ϕ_{ULT}. By repeating such simulation 50 times for all the combination of three traffic flow models and three bridge models, the histograms of ϕ_{ULT} are generated. It is found that the distribution of ϕ_{ULT} well fits a type–I asymptotic extreme value distribution. The characteristic values of ϕ_{ULT} are defined with the exceedance probability of 0.4% for the ultimate limit state. This exceedance probability for the ultimate limit state corresponds to 500 years return period for bridge life of 50 years. The representative examples of the distribution of ϕ_{ULT} are shown in Figs. 6 and 7 in case of the cross section at the center of span.

The applied stress resultants corresponding to

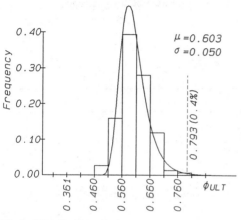

Fig.6 Distribution of ϕ_{ULT}
(Normal traffic jam, center of span)

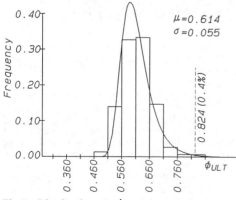

Fig.7 Distribution of ϕ_{ULT}
(Unexpected traffic jam, center of span)

500 years return period are nondimensionalized with the ultimate pure bending moment or pure shear force of the cross sections, and are plotted on the interaction diagrams in Figs. 8 and 9 together with the experimental results and the interaction curve for ultimate stress resultants.

It can be seen from Figs. 8 and 9, that ϕ_{ULT} takes the maximum characteristic values at the interia support where bending moment and shear force are predominant, and ranges from 0.62 to 0.69 for the normal traffic jam and from 0.64 to 0.72 for the unexpected one.

4.2 Equivalent stress for serviceability limit state

The characteristic values of live load effect in the serviceability limit state are determined with 50 maximum (or minimum) values of live load effect in every year.

The histograms of $\bar{\sigma}$ also well fit a type-I asymptotic extreme value distribution. In case of the serviceability limit state, the characteristic values of $\bar{\sigma}$ are defined with the exceedance probability of 5% as shown in Fig.10.

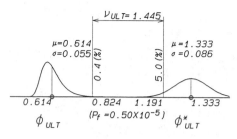

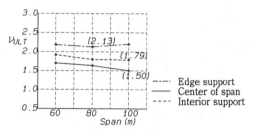

Fig.11 Probabilistic distributions of ϕ_{ULT} and ϕ^*_{ULT} (Unexpected traffic jam, center of span)

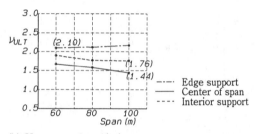

(a) Normal traffic jam

(b) Unexpected traffic jam

Fig.12 Safety factors

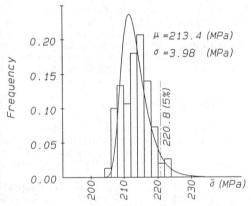

Fig.10 Distribution of $\bar{\sigma}$ for serviceability limit state (Unexpected traffic jam, interior support)

5 EVALUATION OF SAFETY FACTORS FOR ULTIMATE AND SERVICEABILITY LIMIT STATES

A probability of failure can be calculated by Eq.(1) with a numerical integration or other procedures. The number of experimental data and simulation results for live load effect are, however, not enough to evaluate the accurate probability of failure of the box girder bridge models.

5.1 Safety factors for ultimate limit state

In this study, a safety factor ν_{ULT} is, therefore, calculated which is defined as the ratio of the characteristic value of exceedance probability of 0.4% for ϕ_{ULT} to that of non−exceedance probability of 5% for ϕ^*_{ULT} as shown in Fig.11 in case of the unexpected traffic jam. The probabilistic distribution functions of ϕ^*_{ULT} is assumed to fit a Normal distribution. These safety factors for the typical cross sections of the model bridges, as shown in Figs. 11 and 12, are at least 1.50 for the normal traffic jam and 1.44 for the unexpected traffic jam.

For reference, the maximum probability of failure P_f in all the cases, which are calculated by changing the probabilistic distribution of ϕ_{ULT} for two years to that for the bridge life of 50 years by power method, is shown in fig. 11, although it is so notional.

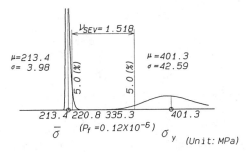

Fig.13 Probabilistic distributions of $\bar{\sigma}$ and σ_y
(unexpected traffic jam, interior support)

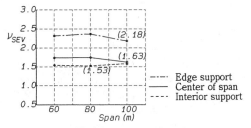

(a) Normal traffic jam

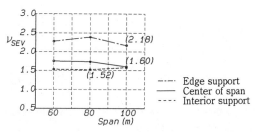

(b) Unexpected traffic jam

Fig.14 Safety factors

5.2 Safety factors for serviceability limit state

A safety factor ν_{SEV} is also calculated which is defined as the ratio of the characteristic value of exceedance probability of 5% for $\bar{\sigma}$ to that of non–exceedance probability of 5% for σ_y as shown in Fig.13. The probabilistic distribution function of σ_y is assumed to fit a Log–Normal distribution. It can be seen from Fig. 14 that these safety factors are at least 1.53 for the normal traffic jam and 1.52 for the unexpected traffic jam.

For reference, the maximum probability that $\bar{\sigma}$ exceeds σ_y in case of the interior support and unexpected traffic jam is shown in Fig.13, although it is notional.

6 CONCLUSION

The safety factors of three box girder bridge models designed according to the current Japanese Specifications for Highway Bridges have been examined through the simulation analysis for live load effect and by using the method for predicting the ultimate strength of box girders subjected to bending, shear and torsion, on condition that only the parameters on live load and the yield stress of steel are probabilistic variables. The following main conclusions have been obtained in this study.

1) The safety factors for the ultimate limit state for three box girder bridge models are at least 1.50 for the normal traffic jam and 1.44 for the unexpected one.
2) The severest safety factors for the serviceability limit state is 1.52 at the interior support for the unexpected traffic jam.

ACKNOWLEDGMENT

Part of the study was carried out as a research work in the committee of "Limit States Design Method for Ultimate and Fatigue Strengths of Steel Bridges" in Kansai Branch of JSCE. The authors would like to express their gratitude to the chairman Prof. Fukumoto, the secretary general, Prof. Matsui and the other members in the committee for their helpful discussions.

REFERENCES

Dowling,P.J. , Chatterjee, S. , Frieze,P.A. & Moolani,F.M. 1973. Experimental and Predicted Collapse Behaviour of Rectangular Steel Box Girders : Steel Box Girder Bridges : ICE, London

Fukumoto,Y. 1980. (Chairman for Study Group on Steel Structures in Tokai): Evaluation of Resistance of Structural Members and its Application to Reliability Design, Bridge and Foundation Engineering, Vol.14, Nos. 11 and 12, 33−41 and 8−44

Kameta,Y. (Chairman of HDL Committee). 1986. A Study and Investigation on Design Load System in Hanshin Expressway, Hanshin Expressway Public Corporation

Kitada,T. , Dogaki,M. , Ishizaki,S. , Ikeda,H. & Takeda,Y. 1992. : Safety Factor of Steel Box Girder Bridges at Ultimate and Serviceability Limit States : Proc. of 3rd Pacific Structural Steel Conference : Tokyo, Japan : 221−228

Matsui,S. & Kanbara,Y. 1989. Study on Vehicle Loads for Highway Bridges, Journal of Structural Engineering , JSCE , Vol. 35A, 419−431

Mikami,I., Dogaki,M. & Yonezawa,H. 1980. Ultimate Load Test on Multi−Stiffened Steel Box Girders: Technology Report of Kansai Univ., NO.21:157−169

Niwa,Y., Watanabe,E. & Nishigome,A. 1979. A Study on Shear Capacity of Box Girder Webs with Multi-Stiffeners, Preliminary Reports of Annual Meeting, Kansai Branch of JSCE

Shinohara,Y. 1988. Public Works Research Institute, Ministry of Construction: A Study on Design Load for Limit State Design Method, Technical Memorandum, No.2539

Structural Safety & Reliability, Schuëller, Shinozuka & Yao (eds) © 1994 Balkema, Rotterdam, ISBN 90 5410 357 4

Evaluation of load carrying capacity of bridge deck from vibrational response

S.S.Law
Department of Civil and Structural Engineering, Hong Kong Polytechnic, Hong Kong

G.B.Shi & R.Z.Chen
Guangdong Communication Science Research Institute, Guangzhou, People's Republic of China

ABSTRACT: This paper presents a relationship between the flexural stiffness and the load carrying capacity of a reinforced concrete beam-slab structure. A method is proposed to assess the load carrying capacity of a simply supported reinforced concrete beam-slab structure with no design information from its vibrational response. The only information required are the fundamental modal frequency and the geometrical dimensions of the structure. Assumptions for this method are given. This method is evaluated against the test results of a model reinforced concrete deck slab and a database of thirteen prototype bridge decks. Good results are obtained in the assessment of steel percentage in the main beams of the prototype bridge decks with a maximum error of +16.2%. A reliability-based approach for the analysis of the test results is also discussed.

1 INTRODUCTION

The most common approach to evaluate the load carrying capacity (LCC) of a bridge deck is to employ the same calculations for evaluation as for design. If a component is found to be weaker than required by design calculations, the bridge is declared substandard. This philosophy has been adopted in the reanalysis of the structure basing on measured damage, and in the more sophisticated proof load testing.

The method using "performance index" was proposed (Cabrera 1988) to evaluate concrete bridges which is based on the observation of signs of distress and their quantification using weightings based on frequency and extent. Numerical values are obtained by weighting the three main signs of distress, i.e., leaks, cracks and surface defects and a tentative scale of weights was proposed.

In United States of America, a rating system of 0 to 9 (Lauer 1991) is used as a damage classification system for concrete structures. It involves adjectivial ratings summarizing the condition of individual bridge components into four general categories of good, fair, poor or critical. A similar system is also used in China for the inspection and classification of concrete bridges. It involves four classes with adjectivial terms along with crack width limitations, (Ministry of Communication 1986).

The evaluation of bridges is an increasingly important topic in the effort to deal with the deteriorating infrastructure in the country. To avoid the high costs of replacement or repair, the evaluation must accurately reveal the present LCC of the structure and predict loads and any further changes in the capacity in the applicable time span.

The type of structure under studied is a simply supported reinforced concrete Tee-beam and slab bridge deck which is most common in the People's Republic of China. A number of these bridges are deteriorating due to ageing, inadequate maintenance and increasing load spectra, and some of the bridges built in the fifties do not have complete design information. The structural

behaviour and the load carrying capacity of the structure are usually calculated from an estimated value of the basic properties of the structure. A more deterministic approach is required however to assess more accurately the actual load carrying capacity.

A method is developed in this research using the basic configuration of the reinforced concrete bridge deck, i.e. the geometry and dimension, and the vibrational response of the structure to assess its LCC. The LCC is defined by the different limit states in the design standard, of which the most popular ones are the strength limit, the crack width limit and the deformation limit. Each of these limits can be expressed as a function of the percentage of reinforcement in the main beams of the structure. Thus the LCC is related to only one single design parameter, i.e. the percentage of reinforcement of the main beam cross-section which is required to be identified in the following investigation.

A relation between the cracked moment of inertia of the bridge beams and its steel content is established throughout its working range. A reasonable finite element model formed for the structure is used to simulate the stiffness of the bridge deck from its vibrational responses, and the LCC of the bridge beam is thus obtained. With sufficient information on the transverse load distribution property of the structure, the LCC of the whole bridge deck can then be found.

The type of bridge deck under study fails primarily in tension failure due to loading only, with sufficient detailing and design provision to prevent other types of failure, like shear and compression failures.

The approach on reliability analysis of the estimated value of steel content for damage assessment of the bridge decks is also discussed with reference to the different types of random variables included in the method.

2 THE STRUCTURES UNDER STUDY

2.1 *The Small Scale Bridge Deck*

The structure was 3.2m long and 1.68m wide consisting of five precast main beams

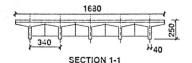

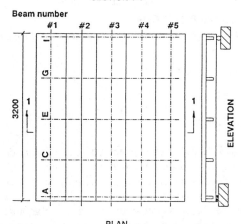

PLAN

Fig 1 Grid System and Dimensions of Model Bridge Deck

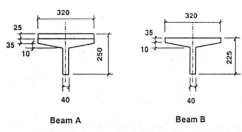

Beam A   Beam B

Fig. 2 Cross-section of Beams A and B

connected transversely by five diaphragms. The precast beams were placed side by side and the reinforcement projecting from the edge of the flanges were hooked together. An additional layer of reinforcement was placed on top and concrete was cast in-situ to form the integral deck. The diaphragms were precast together with the beams. Reinforcement of the diaphragms were welded together both at the top and bottom to simulate actual practice on site. A plan view and sectional elevations of the model are shown in Fig 1.

Two additional precast Tee-beams were also cast. One was cast with the in-situ deck slab whereas the other was not. They are denoted Beam A and Beam B respectively. The

cross-sections are given in Fig 2. The beams in the model bridge deck have a flange 20mm wider than for Beam A due to fabrication requirements and is called Beam C for convenience.

The bridge deck and the two beams were supported on rigid concrete blocks well-founded on the concrete slab of the laboratory. Static point load was applied at midspan of the beams and on the second beam of the deck, and vibrational response to ambient excitation was recorded after the removal of load in each loading stage.

2.2 *The Prototype Bridge Decks*

A total of thirteen prototype reinforced concrete Tee-beam and slab bridge decks have been measured for their dynamic response to traffic-generated vibration in the People's Republic of China from 1984 to 1987. They are of different span length and width consisting of precast Tee-beams at 1.4m or 1.6m spacing with a layer of in-situ concrete cast on top and were designed according to Standards (Standard JTJ023-85 1985; Standard JT/GQB011-73 1973). The bridge decks are of the same type of construction as the model bridge deck with diaphragms equaly spaced along the structure.

Yuan Dun Bridge consists of 7 spans. The first, third and fifth span of Yuan Dun Bridge were tested before and after rehabilitation works (Law et al 1992a).

3 THEORETICAL CRACKED MOMENT OF INERTIA

The theoretical cracked moment of inertia are plotted against the neutral axis depth for section of Beams A, B and C in Fig 3. There are three stages in the variation of the cracked moment of inertia with neutral axis depth. When a bending moment is applied, tensile crack develops at the bottom of the beam together with a decrease in the moment of inertia. The crack grows very rapidly under a small load until the tensile stress in the main steel increases to form a balancing moment with the concrete in compression to resist the applied bending moment. The second stage has a relatively constant moment of inertia of about 71.0×10^{-6} m^4 for Beams A and C and 53.0×10^{-6} m^4 for Beam B. This relatively constant value of moment of inertia is considered to be a property of the beam. It is called I_{con} denoting a relatively constant cracked moment of inertia of the cracked section. The last stage is an unstable stagewhere the moment of inertia decreases rapidly with further reduction in the neutral axis depth until the beam section is fully cracked over its depth.

4 THEORETICAL MOMENT CAPACITY

In an under-reinforced concrete beam section, the strength and serviceability requirements of crack width and deflection (Standard JTJ023-85 1985) can be expressed in terms of the steel percentage. The theoretical moment capacity of two Beams, A and B are thus plotted in Figs 4 and 5. The steel percentage is calculated based on the gross cross-sectional area of the Tee-beam section as specified in the Standard with an allowable range of 0.6% to 2.0%. Details of the calculations on the moment capacity of a beam section are referred to (Law et al 1992b).

It is noted that if the percentage of reinforcement of a beam is known, the moment capacity of a bridge beam can be estimated from graphs similar to Figs 4 to 5. The LCC of a bridge deck can then be estimated with information on the transverse load distribution properties of the structure.

Similar limiting requirements for the beams from the British Standard (BS5400:Part 4

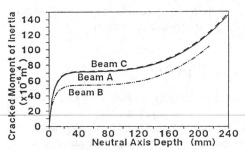

Fig 3 Theoretical Cracked Moment of Inertia of Beams A, B and C

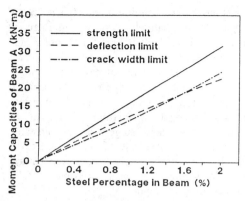

Fig 4 Moment Capacity of Beam A (JTJ023-85)

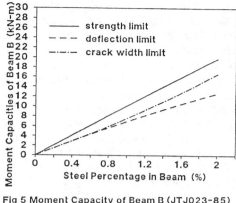

Fig 5 Moment Capacity of Beam B (JTJ023-85)

1984) can also be expressed as a relationship between the percentage of reinforcement and the moment of resistance of the beam section.

5 ASSUMPTIONS FROM EXPERIMENTAL OBSERVATIONS OF THE MODEL BRIDGE DECK

1. The transverse load distribution properties of the bridge deck are relatively constant over the whole loading range before any yielding of reinforcement in the main beams occurs;

2. The neutral axis of a Tee-beam section goes near to the soffit of the slab after cracking first appears. Under further loading, the variation is small;

3. A cracked Tee-beam section resists the applied bending moment with a relatively constant cracked moment of inertia I_{con} throughout most of its working range of loading;

4. All beams in a bridge deck have approximately the same longitudinal distribution of cracked moment of inertia throughout the loading range;

6 THE STRATEGY

The first bending modal frequency is less affected by changes in the transverse load distribution properties of the bridge deck and is more sensitive to cracks near midspan of the structure. It is therefore possible to estimate

the flexural stiffness, with an assumed longitudinal distribution in the major members of the bridge deck by developing the best match to the measured first modal frequency with a FEM. The percentage ofreinforcement in the longitudinal members of the bridge deck and hence the LCC of the structure is then estimated. The steps of implementation are:

Step 1. Measure the fundamental modal frequency of the structure in free vibration.

Step 2. Select a FEM which best represents the geometrical configuration of the bridge deck.

Step 3. The moment of inertia of the beams in the FEM is adjusted to have the best match in the fundamental modal frequency in free vibration.

A uniform distribution of the cracked moment of inertia is assumed over the length of the main beams, and all the main beams have the same value and distribution of inertia in the FEM. The value of inertia that gives the closest fit of the measured first modal frequency is the best estimate of the moment of inertia of the beams, I_{est}.

Step 4. Modify the estimated moment of inertia I_{est} by dividing it with a Reference Factor β_{LS}, which is a statistical average of the Factor β, defined as the ratio of I_{est} over I_{con} of many similar test samples. The result is an improved estimate of I_{con}.

Step 5. Calculate I_{con} of the bridge beam for different percentages of reinforcement, and plot the graph of I_{con} of the cracked section versus percentage of reinforcement in the

beam. Find the percentage of reinforcement in the beam from the graph using the estimated I_{con} obtained from Step 4.

Step 6. Calculate the moment of resistance of the beam according to the different limiting criteria, and plot the theoretical moment capacity of the beam against different steel percentages.

Step 7. Find the moment resistance of the beam from the graph plotted in Step 6 with the percentage of reinforcement obtained from Step 5.

Step 8. Obtain the transverse load distribution property of the bridge deck from test or from the literature. Calculate the LCC of the structure according to the required load configuration.

Factor ß is a measure of the deviation of I_{est} from the I_{con} of the beam, and Reference Factor $ß_{LS}$ is a statistical average of ß with which the percentage of reinforcement of the beam can be estimated from the measured I_{est}. Hence Factor ß is called the Load Carrying Capacity (LCC) Factor.

As I_{est} and I_{con} are functions of the steel percentage of the beam, Factor ß can be assumed to be independent of the steel percentage. However since I_{est} is calculated from the FEM and an assumed uniform distribution of moment of inertia along the length of the beams, Factor ß is therefore assumed to be dependent on the length dimension of the structure.

7 FINITE ELEMENT MODELLING

The SAP IV software package was used in the modelling of the individual beams and the bridge deck. Stiffness arising from fencing and pedestrian pavements on the edges was not considered, but the masses of these elements were added on to the nodes along the edge beams in the FEM. The bridge deck was modelled as a grillage of beams resting on rigid supports. The experimental density and modulus of elasticity were used.

The actual reinforcement detail along the beam, (i.e. part of the tensile reinforcement being bent up near the ends as shear reinforcement and the precise position of the

centroid of tension reinforcement) were not taken into account in calculating the stiffness of each beam element in the FEM.

8 APPLICATION

8.1 *The Model Bridge Deck*

1. The results shown here are for beams with a constant percentage of reinforcement throughout their length.

2. The Load Carrying Capacity (LCC) Factor ß is calculated, and it varies from 1.467 in the bridge deck after first cracking to 1.213 in the failed bridge Beam A. The magnitude of ß indicates the appropriateness of assuming I_{con} for the beams in the FEM. The larger ß is the less appropriate the assumption. However, the small range of ß indicates that it might have a potential use as an invariant factor in the different cracked states of the structure.

3. Reference factor $ß_{LS}$ is taken to be 1.31 for illustration of the method. The individual beams and the bridge deck exhibit a variation of between -6.29% to +15.73% in the estimate of reinforcement in the beams, with the larger value for the bridge deck after first cracking and the smaller value for Beam A close to failure. This suggests that the error in the estimate is little affected by the width of the deck and the steel percentage in the structure. In fact the fundamental frequency has been checked (Law et al 1992b) to be little dependent on the width of the structure provided there is sufficient stiff transverse connection.

4. As elastic flexural stiffness is defined as the product of elastic modulus and the moment of inertia, a reduction in the value of elastic modulus used in the FEM would give a proportional increase in the estimate of I_{est}. Therefore an accurate assessment of the elastic modulus is essential which can be obtained from the literature, a design document or from site measurements. However the LCC Factor ß are all greater than unity. This may possibly be due to the non-uniformity of cracks along the beam and the inappropriateness of the linear computer model in modelling the behaviour of a cracked reinforced concrete

907

member. If a factor of 1.35 is used to take account of these effects, the estimated dynamic moment of inertia would be closer to I_{con} to within +8.7% and -8.0% in all the cracked working states.

8.2 *The Prototype Bridge Decks*

Curves plotting the I_{con} value versus steel percentage of the 16m long bridge beams are shown in Fig. 6. The LCC Factors ß are calculated and are listed in Table 1. Reference Factor $ß_{LS}$ is obtained for bridges of the same span length with minimum least-squares error on the estimated steel percentages of the bridge beams. The errors on the estimates of the percentage of reinforcement are shown in the last column of Table 1. The maximum error in the 16m span group is only -12.33 % whereas that in the 22m and 30m group is +16.17% and -14.74% respectively. It is noted that the first group of 16m long bridge decks consists of relatively new structures whereas the last two groups are combinations of damaged and rehabilitated bridge decks.

It is also noted that if the Reference Factor $ß_{LS}$ used in the calculation is obtained from a set of undamaged structures, assessment in a badly damaged structure gives a negative error which is on the safe side.

The accuracy on the estimate of steel percentage in the bridge beams is not sensitive to the proportion of non-structural stiffness contribution from fencing, support etc. in each bridge deck as seen from Table 1. This is probably due to the reduction of the random error in the estimate by including the Reference Factor $ß_{LS}$ in the calculation.

Reference Factor $ß_{LS}$ varies with the span length of the structure as shown in Table 1. In fact, less tensile reinforcement towards the ends of the beams in practice would give a smaller moment of inertia of the beam section. This actual distribution of moment of inertia along the beam length would deviate from a straight line assumption. However, if all the bridge decks are constructed to the same steel detailing practice, the Reference Factor $ß_{LS}$ obtained should be relatively constant for a particular span length.

Table 1. Estimate of percentage of reinforcement of bridge beams.

Span (m)	Bridge Deck	LCC Factor ß	Reference Factor $ß_{LS}$	Percentage Error in Estimate
16	Ni Zi (1st span)	2.082		+2.73%
	Ni Zi (2nd span)	1.960		-4.08%
	ZTYQ (1st span)	2.198	2.034	+9.53%
	ZTYQ (8th span)	2.131		+5.67%
	NDZT	1.818		-12.33%
22	Yuan Dun (1st span) (before rehab.)	1.782		+16.17%
	Yuan Dun (6th span) (before rehab.)	1.416	1.593	-13.94%
	Yuan Dun (1st span) (after rehab.)	1.609		+1.45%
	Yuan Dun (7th span) (after rehab.)	1.506		-7.54%
30	Yuan Dun (3rd span) (before rehab.)	0.977		-14.74%
	Yuan Dun (5th span) (before rehab.)	1.045	1.097	-6.53%
	Yuan Dun (3rd span) (after rehab.)	1.118		+2.86%
	Yuan Dun (5th span) (after rehab.)	1.186		+12.26%

9 DISCUSSIONS

1. The Load Carrying Capacity Factor ß is little dependent on the percentage of reinforcement in the main beams and width ofthe bridge deck;

2. A Reference Factor $ß_{LS}$ can be chosen for the different cracked states of the structures with a minimum least-square error in the estimate of steel percentage of the beam;

3. The Factors ß and $ß_{LS}$ are smaller for longer span groups, and the Factor $ß_{LS}$ is related to the span length by the following formula from the limited data obtained,

$$ß_{LS} = 3.09 - 0.07 * Span \qquad (1)$$

908

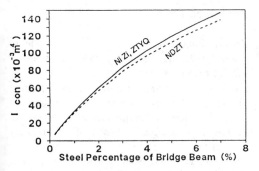

Fig 6 Theoretical I_{con} of 16m Bridge Beams

4. A factor of 1.35 is suggested to account for the errors due to an assumed uniform distribution of the cracked moment of inertia along the member and errors of the linear computer model in modelling a cracked reinforced concrete structure.

10 RELIABILITY ANALYSIS APPROACH

The approach of reliability analysis of the test data is discussed in the following paragraphs with no numerical results on the reliability indices of the structures.

10.1 *The Limit States*

For a typical bridge girder, the limit state functions can be formulated for various conditions, including:

1. Bending moment capacity;
2. Shear capacity;
3. Buckling capacity; overall and local;
4. Deflection;
5. Vibrations;
6. Accumulated damage conditions, including corrosion (lapping of reinforcement, prestressing tendions, structural steel sections,) fatigue, cracking and other forms of material deterioration.

Each limit state is associated with a set of limit state functions which determines the boundaries of the acceptable performance. For the accumulated damage, the functions are difficult to formulate. For the type of bridge deck under study and using the same design method as in the assessment, the limit state functions for deflection, cracking and strength of the girder are simplified into one single performance function

$$Z = \rho_e - \rho_a \qquad (2)$$

where ρ_a is the actual percentage of reinforcement in the beam and ρ_e is the estimated value. The performance function can be used for the reliability analysis of damage in the bridge decks. The larger the accumulated damage due to cracking and corrosion, etc., the smaller the value of ρ_e, and a comparison with ρ_a gives the distance of its condition (the three limit states) from the true design moment capacity. The range for an acceptable (safe) estimate is for $Z \geq 0$.

Usually reaching a limit state by one of the members of the whole structure does not mean that the bridge deck also reached its limit state. Traditional design and analysis of bridges is based on identification of the governing limit state in each member. Safety provisions are applied to ensure an adequately low probability of occurrence for these limit states. However, the analysis of the whole bridge deck will not be discussed in this paper.

10.2 *Random Variables*

The random variables shown below can be selected for use in the analysis.

1. variables related to the support conditions of the bridge deck;
2. variables related to the material properties of reinforced concrete;
3. variables related to the uniformity of crack distribution along the member length and the use of a linear FEM for this problem;
4. variables related to the error of structural dimensions due to construction.

The variables can be selected through sensitivity analysis in terms of a wider range of variation and a relatively significant effect on the estimated steel content. The sensitivity of random variables can be estimated by the relative magnitude of deviation Δ between ρ_c and ρ' as

$$\Delta = \left| \frac{\rho_e - \rho'}{\rho_e} \right| \cdot 100\% \qquad (3)$$

where ρ_e defines the steel content when randomness of all variable are taken into account and ρ' defines the steel content when the randomness of variables other than each variable are considered.

The statistical characteristics of the random variables are obtained directly from available field data. In cases where no field data is available, they can be selected from the published data.

By assuming ρ_a and ρ_e are uncorrelated, the reliability index for each structure can be computed as (Madsen et al 1986):

$$\beta_c = \frac{\mu_{\rho e} - \mu_{\rho a}}{\sqrt{\sigma_{\rho e}^2 + \sigma_{\rho a}^2 + \sigma_m^2}} \qquad (4)$$

where $\mu_{\rho a}$ and $\mu_{\rho e}$ are the mean values of ρ_a and ρ_e, and $\sigma_{\rho a}$, $\sigma_{\rho e}$ and σ_m are the standard deviations of ρ_a, ρ_e and random variables considered respectively.

11 CONCLUSIONS

The proposed strategy provides a method of estimating the steel percentage and hence the load carrying capacity of a bridge deck from the first modal frequency. A Reference Factor β_{LS} has been adopted to modify the measured moment of inertia I_{est} to an estimate of the cracked moment of inertia I_{con} of the beam section. The maximum percentage error in the estimation of steel percentage in the prototype bridge decks is +16.2%. A reliability-based approach for the analysis of the test results with the statistical properties of different random variables in the problem is presented.

12 ACKNOWLEDGEMENTS

The authors are indebted to Mr. X. Li of South China University of Technology for their contribution in the laboratory tests and their useful discussions. The authors would also like to thank Dr. P. Waldron and Dr. C. Taylor of University of Bristol for their advice in this research.

REFERENCES

British Standards Institution. 1984. *Code of practice for the design of concrete bridges. BS5400:Part 4: 1984.*

Cabrera, J. 1988. Towards a Performance Index for Concrete Bridges. *Proc of Seminar on Assessment of Reinforced and Prestressed Concrete Bridges*, Inst of Structural Engineers.

Lauer, K. R. 1991. State of the Art Report: The Use of Damage Classification Systems for Concrete Structures. *Proc of the Int RILEM-IMEKO Conf on Diagnosis of Concrete Structures*, ed. T. Jávaor, Expertcentrum Czechoslovakia.

Law, S. S., Shi, G. B., Chen, R. Z., and Ward, H. S. 1992a. The Vibrational Response of Reinforced Concrete Bridge Deck Before and After Rehabilitation. *Proc of the Federation Internationale de la Precontrainte FIP'92 Symposium*, Vol.1, May 11-14 1992, 339-346.

Law, S. S., Ward, H. S., Shi, G. B., Chen, R. Z., Waldron, P. and Taylor, C. 1992b. Load Carrying Capacity of Bridge Deck from Vibrational Response. *J of Structural Engrg.* ASCE, (under review)

Madsen, H.O., Krenk, S. and Lind, N.C. 1986. *Methods of Structural Safety.* Prentice-Hall, Inc.

Ministry of Communications 1986. *Technical Specification for Highway Maintenance. JTJ073-85, 1986.* (in Chinese), The People's Communication Publishing House, Beijing

Standard JTJ023-85 1985. *Design Standard on Reinforced and Prestressed Concrete Highway Structures.* (in Chinese), Ministry of Transport, PRC, May 1985.

Standard JT/GQB011-73 1973. *Standard Drawing on Highway Structures - Precast Concrete Tee-Beam Bridge.* (in Chinese), The First Highway Engineering Design Office, Ministry of Transport, The People's Republic of China, September 1973.

Structural Safety & Reliability, Schuëller, Shinozuka & Yao (eds) © 1994 Balkema, Rotterdam, ISBN 90 5410 357 4

Evaluation of residual life of steel bridges concerning corrosion deterioration

M. Matsumoto, N. Shiraishi, H. Miyake & T. Takanashi
Department of Civil Engineering, Kyoto University, Japan

ABSTRACT: This study aims to evaluate the residual life in terms of corrosion deterioration of painted-steel plate-girder bridges.
Based on the detected corrosion data for up to 300 steel bridges in Japan, the corrosion damage index, X, was introduced and the sequential evaluation on the residual life of each bridge was developed.

1 INTRODUCTION

The assessment on corrosion deterioration and the related residual life of steelbridges are evidently important for the evaluation of bridge safety and reliability. It is known that the paint-film is effective to protect steel bridges from corrosion during some period depending on the paint-film material and the environmental factors. However, as a matter fact, for many cases it should be noted that the repainted-cycle is not so sufficiently short that the corrosion deterioration unavoidably occurs. The environmental factors, paint-film materials, water leakage from bridge deck, the usual maintenance condition and other factors cause the complicity of the corrosion deterioration of steel bridges.

The authors have tried to establish an easy system for the evaluation of the paint-film damage level and the corrosion damage level based on the sight observation data for up to 300 corroded bridges in Japan.(reference 3) Furthermore, the corrosion deterioration characteristics of the various bridge-parts were modelized as a function of environmental factors and the location of each bridge-part. Then thecritical corroded state, which means the termination of bridge life because of corrosion, was related to the certain value of damage index from previous practical experiences of bridge-replacement. This procedure enables the evaluation of residual life of bridges.

2 SIGHT OBSERVATION AND DAMAGE LEVEL CLASSIFICATION OF CORRODED BRIDGES

Damage state of paint-film and corrosion of steel at 26 various parts, from P1 to P26, such as shoe and web-plate, upper/lower surface of lower flange at mid/quarter/end parts of span, of external/internal girder, for up to 300 steel bridges was investigated by sight observation. Based on these observed results, the damage states of paint-film and corroded steel were classified into four levels, i.e., PDL1, PDL2, PDL3 and PDL4, according to the standard of former Japan National Railway Bureau (reference 5), and eight levels, i.e., CDL – A, CDL – B, CDL – C, CDL – D, CDL – E, CDL – F, CDL – G and CDL – G', as shown in Table 1(1) and 1(2), respectively. PDL1 and CDL – A in these levels correspond to integrity state, in another words non-damage state, of paint-film and corrosion, respectively. Figure 1 shows the examples of these damage states. The damage level of paint-film and corroded steel have been classified by the standard photographies, provided by authors.

3 QUANTIFICATION OF SIGHT – OBSERVED CORROSION STATE

The corrosion damage levels classified by sight

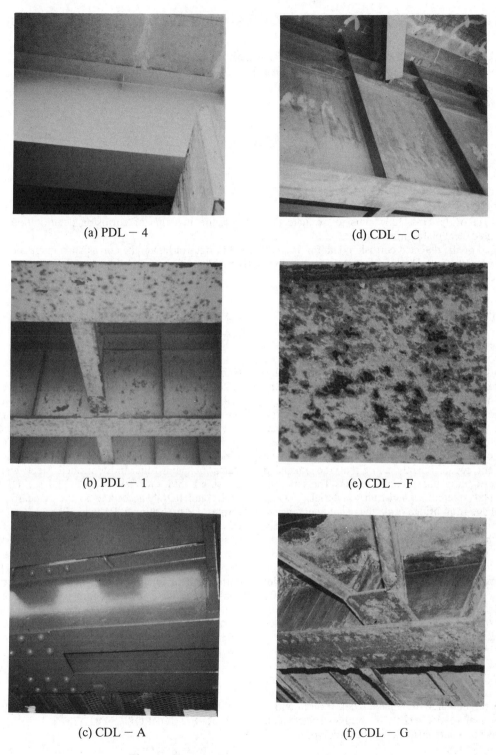

(a) PDL − 4

(d) CDL − C

(b) PDL − 1

(e) CDL − F

(c) CDL − A

(f) CDL − G

Fig. 1   Damage Level of Paint or Steel Surface

Table 1   Definition for Damage Levels

(1) Paint Damage Levels

PDL	Damage State
4	No defect or Defect < 0.03% of area
3	0.03% ≤ Defect < 0.3% of area
2	0.3% ≤ Defect < 5% of area
1	Defect ≥ 5% of area

Note PDL : Defect means rust, blister, crack or
peeling

(2) Corrosion Damage Levels

CDL	Damage State
A	No rust or Spot rust < 0.5% of area
B	0.5% ≤ Spot rust < 10% of area
C	10% ≤ Spot rust < 50% of area
D	50% ≤ Spot rust < 90% of area
E	Spot rust ≥ 90% of area or Uniform corrosion over the surface
F	Inital pitting corrosion
G	Severe pitting corrosion
G'	Very severe pitting corrosion

Note CDL : Corrosion Damage Levels

Table 2   Corrosion depth quantification

Corrosion levels	A	B	C	D
Corrosion depth (mm)	0.00	0.10	0.23	0.40

Corrosion levels	E	F	G	G'
Corrosion depth (mm)	0.60	0.90	3.00	10.0

observation is qualitative, therefore this should
be quantified in order to judge the member
strength deterioration or the residual life. The
quantification was performed by the direct
measurement of corrosion depth of corroded
steel plate by ultra-sonic plate thickness meter
after removing paint film for various damage
levels, from CDL‐A to CDL‐G'. 43 points of
one road bridge, 32 points of two railway
bridges and 5 points of the removed bridge
members were investigated. Based on the

Corrosion(mm)

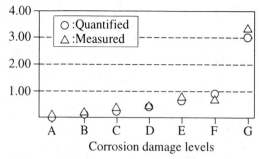

Fig. 2   Corrosion depth for each corrosion damage level

measurement, the corrosion depth corresponding
each damage level was quantified as shown in
Table 2 and in Figure 2.

4 CORROSION DAMAGE INDEX

In this study, two different corrosion damage
indices were introduced, that is the global
corrosion damage index, X, and the local
corrosion damage index, Y, to characterize the
corrosion damage level of each bridge. These
indices are expressed as follows ;

$$X= (0\times N_A+0.1\times N_B+0.23\times N_C+0.4\times N_D+0.6\times \\ N_E+0.9\times N_F+3\times N_G+10\times N_{G'})/(N_A+N_B+N_C+ \\ N_D+N_E+N_F+N_G+N_{G'}) \quad (1)$$

and

$$Y= 1\times N_F+3\times N_G+10\times N_{G'}, \quad (2)$$

in general,

$$N_A+N_B+N_C+N_D+N_E+N_F+N_G+N_{G'}=26, \quad (3)$$

where, N_i shows the respective number of i‐
ranked corrosion level, CDL‐i, and the water
leakage part is excluded in counting the number
in equation (1) and included in equation (2).
Damage index, X, in consequence, means the
average corrosion depth for each bridge, and this
index, X, is useful for the evaluation of corrosion deterioration. The histogram of corrosion
damaged 300 bridges is shown in Figure 3. In
which, 63 bridges were non-damaged.

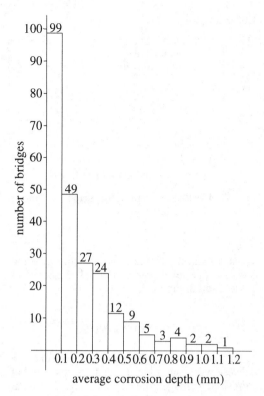

Fig. 3 Histogram of Corrosion Damaged Bridges

5 PAINT FILM DETERIORATION

The paint film damage (PDL) was classified into four levels, as described before, according the maintenance rule by former Japan National Railway Bureau. PDL4 corresponds to non-damage or integrity state, and PDL2 and PDL1 correspond to damage state where paint film cannot already protect its beneath steel surface from corrosion. It means when paint damage reaches PDL2 corrosion occurs. The linear-regressive method was applied to the ranked data from PDL4 to PDL1, characterized as a function of the time after the latest former repaint, for individual part of bridges which locate at the same environmental category, such as rural, country side or sea side, and are painted by the same material, such as alkyd resins, chlorinated rubber and so on. Thus the mean value, μ_{PDL}, and the standard deviation, δ_{PDL}, were obtained, then the paint-life could be determined the period which makes the value of ($\mu_{PDL} - 3\delta_{PDL}$) 2. An example, for the paint film deterioration

characteristics of the span-end of alkyd resins painted external girder located at rural category, is indicated in Figure 4. Furthermore, the paint film deterioration was characterized by two environmental factors, that is temperature, $X_1(^{\circ}C)$, and sea-salt particle, $X_5(10^{-4}g/cm^2/year)$, as follows;

$$\mu_{PDL} - 3\delta_{PDL} = 4 + (a_0 + a_1 X_1 + a_5 X_2) \times t \quad (4)$$

where, t:exposed time since the latest former repaint.

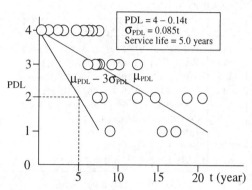

Fig. 4 Evaluation of paint life
(Span end of main external girder
Lower surface of lower flange (city A : rural category)
Paint type : alkyd resins
PDL : Paint damage levels)

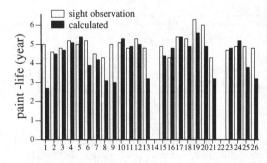

Fig. 5 Paint –Life for various bridge parts (City B : urban category, alkyd–resin paint film)

In general, the paint life of chlorinated rubber is a little bit longer than the one of alkyd resin, and the shoe part or the lower surface of lower flange tends to show a shorter life. The ob-

served/evaluated alkyd resin paint life at the urban category of city B for various part of bridge is shown in Figure 5.

6 CORROSION DETERIORATION MODEL BASED ON EXPOSED TEST

Using the data based on exposed test (reference 4 and 1), the corrosion depth for various exposed periods from 1 year to 5 years were modelized by the following equation in terms of five environmental factors ;

$$y_i = a_{0i} + a_{1i}X_1 + a_{2i}X_2 + a_{3i}X_3 + a_{4i}X_4 + a_{5i}X_5 \qquad (5)$$

where, y_i : corrosion depth during i–years, X_2 : humidity (%), X_3 : precipitation (mm/year), X_4 : sulphur-dioxide, SO_2, $(10^{-3}ppm)$

The coefficients of a_{ji} (j=0 to 5) in equation (5) were determined as shown in Table 3 by multi-regressive analysis.

The corrosion deterioration is characterized by the following formula ;

$$y(t^*) = k \times (t^*)^n \qquad (6)$$

where, $y(t^*)$ is the corrosion depth in mm during t^* years. t^* is corrosion related accumulated exposure time (year), and k and n are parameters depending on the environmental factors. The accumulated exposed time, t^*, is obtained from knowing the past history of repainting and the life time of paint film.

Table 3   Corrosion depth of non-painted steel

exposed to rain
$y_1 = 34.14X_1 - 2.48X_2 - 0.040X_8 + 12.25X_4 + 8.46X_5 - 30.92$
$y_2 = 38.65X_1 - 10.00X_2 + 0.072X_8 + 10.95X_4 + 16.21X_5 + 444.41$
$y_3 = 42.63X_1 - 5.11X_2 + 0.122X_8 + 24.30X_4 + 27.57X_5 - 185.14$
$y_4 = 34.83X_1 - 10.77X_2 + 0.260X_8 + 32.42X_4 + 32.88X_5 + 74.85$
$y_5 = 20.24X_1 - 22.39X_2 + 0.397X_8 + 32.22X_4 + 34.83X_5 + 962.18$

non-exposed to rain
$y_1 = 6.03X_1 + 5.25X_2 + 5.15X_4 + 27.76X_5 - 406.30$
$y_2 = 11.86X_1 + 17.62X_2 + 16.73X_4 + 53.28X_5 - 1427.35$
$y_3 = 17.68X_1 + 30.76X_2 + 29.31X_4 + 77.46X_5 - 2523.99$

Note X_1:Temperature(℃) X_2:Humidity(%)
X_3:Precipitation(mm/year)
X_4:SO_2concentration(10^{-3}ppm)
X_5:Sea–salt particles(10^{-4}g/cm²/year)
$y_1 - y_8$:Corrosion depth in 1−5 year - exposure time, respectively(10^{-4}mm)

7 CORROSION RATIO OF 26 VARIOUS PARTS

The corrosion level differs in dependence on the parts of a bridge, therefore the corrosion ratio was defined as the average ratio of corrosion depth at each part to the standard one obtained from equation (7), in taking the environmental factors at each bridge site into account, for surveyed 300 bridges. The corrosion ratio of each part is shown in Figure 6.

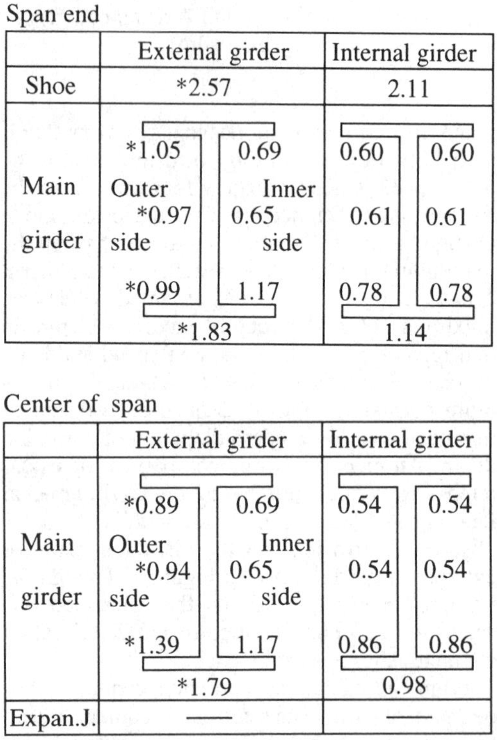

Span end

	External girder	Internal girder
Shoe	*2.57	2.11
Main girder	*1.05 / 0.69 — Outer side *0.97 / Inner side 0.65 — *0.99 / 1.17 — *1.83	0.60 / 0.60 — 0.61 / 0.61 — 0.78 / 0.78 — 1.14

Center of span

	External girder	Internal girder
Main girder	*0.89 / 0.69 — Outer side *0.94 / Inner side 0.65 — *1.39 / 1.17 — *1.79	0.54 / 0.54 — 0.54 / 0.54 — 0.86 / 0.86 — 0.98
Expan.J.	—	—

Note *:Members exposed to rain

Fig. 6   Corrosion ratio for 26 various parts

8 PREDICTION OF CORROSION DETERIORATION OF BRIDGE

Using the accumulated exposed time, t^*, which is obtained from the data of repaint cycle and paint life, the environmental factors at bridge site and corrosion ratio at each part, the corrosion depth at each part can be predicted as a function of time (in year). It has been reported that the corrosion speed changes drastically

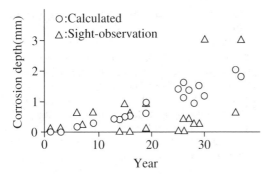

Fig. 7 Corrosion Deterioration of shoe part of external girder (City A :rural category, alkyd resin paint film)

Table 4 Modification Factors (α)

City A : rural	1.415
City B : seaside	0.961
City B : urban	0.909
City C : rural	0.780
City F : Mountainous	1.242
City I : seaside	1.112

Table 5 Evaluation of residual life
(city B : seaside)

Average paint cycle for each bridge
18years (Alkyd resins)
17years (Chlorinated rubber)

No.	Paint type	Age (year)	Xo	Residual life (year)
1	C	51	0.276	62
2	A	51	0.505	41
3	A	24	0.463	43
4	A	25	0.333	52
5	C	7	0.000	85
6	A	50	0.511	41
7	A	33	0.327	53
8	A	21	0.613	23
10	A	26	0.675	18
11	A	16	0.378	49
12	C	12	0.075	79
13	A	31	0.176	46
14	A	26	0.326	36
15	A	30	0.710	24
16	A	8	0.148	67
17	C	6	0.000	85
18	C	25	0.351	57

Note A : Alkyd resins C : Chlorinated rubber
Xo : Observed corrosion damage index

when the corrosion mode changes from "uniform corrosion" to "pitting corrosion", that is the latter speed is approximately four times than the former one.(reference 2) This corrosion mode change has been also reported to occur when the maximum corrosion depth reaches 0.7 to 1.4mm. Thus, in this study it is assumed when the maximum corrosion depth exceeds 0.7mm in one repaint-cycle, the corrosion speed suddenly accelerates by four times. In which the maximum corrosion depth is defined as the sum of the mean and the standard deviation of corrosion depth. An example of the comparison of calculated corrosion depth based on environmental factors and corrosion ratio with the sight-observed corrosion depth for shoe part of external girder is shown in Figure 7. Though the observed corrosion scatters, the calculated one seems to realize an actual corrosion deterioration characteristics to certain extent.

From the calculated corrosion depth of various part of bridge, the corrosion damage index, given by equation (1), X_c, can be obtained for arbitrary exposed time.

9 RESIDUAL LIFE TO CORROSION DETERIORATION

The global damage index, X, must be a scale of the residual life of bridge to corrosion. On the other hand, three from 300 bridges showed the larger value of X than 1.0 and these have been/will be soon reconstruced. Therefore, it is, in this study, assumed that exceeding unity of corrosion damage index, X, means the termination of life of a steel bridge.

Gaining acquaintance with environmental factors at bridge site, used paint film material, repaint cycle and corrosion ratio, the global corrosion damage index, X_c, can be calculated, then the residual life can be, in consequence, evaluated. However, as a matter of fact, the corrosion mechanism of an actual bridge is generally so complicated that the evaluated corrosion depth based on the proposed model in this study would have the discrepancy from the actual value, more or less. In order to increase

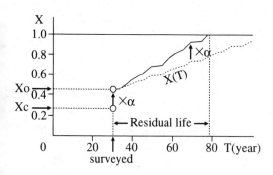

Fig. 8   X − T diagram

⎛ X : Corrosion damage index
⎜ T : Lapse after construction
⎜ Xo : Observed X
⎜ Xc : Calculated X
⎜ α : Modification factor
⎝ X(T) : Future prediction of X ⎞⎠

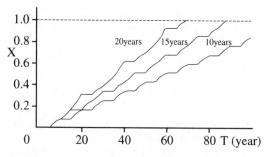

Fig. 9   Relationship between Paint cycle and
Corrosion damage index

(city B : seaside)

X : Corrosion damage index
T : Lapse after construction
Paint–service life = 5 years
Paint cycle = 10, 15, 20 years

the evaluation accuracy, the modification factor, α, is introduced here.

This factor is determined as the ratio of the observed global damage index, X_o, to the calculated one, X_c, and it changes in dependence on the environmental factors at bridge site and on the bridge management institution/authority. Since the misjudgment of the corrosion damage level possibly occurs if the paint film damage level is PDL4 or PDL3, the modification factor was obtained for only bridges whose lapse of time after the latest repaint exceeds the respective paint life. The modification factors as shown in Table 4 vary from 0.780 to 1.415.

Using the calculated global corrosion damage index, X_c, and the modification factor, , the residual life can be evaluated as illustrated in Figure 8. The residual life of bridges belongs to city B local government is shown in Table 5 with their age and the observed global corrosion damage index. In this study, the residual life of 118 bridges was investigated, then it was cleared 10 bridges had less than 20 years residual life.

10 ELONGATION OF BRIDGE LIFE

The elongation of bridge life subject to corrosion is more and more important on increasing decrepit bridges. More frequent repaint or the improvement of paint film material is one of decreasing the corrosion speed. Figure 9 shows

an example of the effect of paint cycle or paint film material on the bridge life.

11 CONCLUSION

Corrosion depth was quantified from the non-destructive and simple sight observation on corrosion damage state of steel surface of bridges, furthermore, the global corrosion damage index, X, was introduced to express the damage level, and was used to evaluate the residual life of a bridge subject to corrosion damage. When the average corrosion depth, which corresponds to X value, reaches 1.0mm, the bridge life was assumed to terminate based on the practical fact that the three bridges with X>1 among 300 researched bridges had just been replaced or would be reconstructed soon. As a result, approximately 10 % bridges among the investigated ones have less than 20 years residual life. Furthermore, it is cleared that the shorter repaint cycle or the improvement of paint film material are effective to elongate the residual life and their elongated life can be quantitatively evaluated.

12 ACKNOWLEDGMENT

The authors would like to acknowledge to Messrs. S.Rungthongbaisuree of King Monk University, Bangkok, Tailand, and T.Okamura

of Hitachi Zosen, Japan, for their great contribution of the sight observation of bridges at the site.

REFERENCES

1. Hanshin Expressway Public Corporation, 1980. *Anti-corrosion study group of bridge structure report*, (in Japanese)
2. Ito, H. 1976. *Life estimation of corroded materials using the statics of extremes*, Proc. of the 41th Symposium on Corrosion Protection, (in Japanese)
3. Matsumoto, M., Shiraishi, N., Rungthong baisuree, S., and Okamura, T. 1991. *On assessment of corrosion damages in steel bridges* Proc. of The Third East-Pacific Conference on Structural Engineering & Construction.
4. Public Works Research Institute, Ministry of Construction, 1988. *Results of the second measurement of corrosion depth from steel exposure test*, Research Report No.10, (in Japanese)
5. Sato, Y. and Hashimoto, T. 1974. *Investigation on corrosion of steel bridges and the method of maintenance painting*, Railway Technical Research Report No.392, (in Japanese)

Structural Safety & Reliability, Schuëller, Shinozuka & Yao (eds) © 1994 Balkema, Rotterdam, ISBN 90 5410 357 4

Structural reliability assessment of steel girder stiffeners on urban expressway bridges

A. Nanjo & H. Sekimoto
Hanshin Expressway Public Co., Japan

M. Kawatani
Osaka University, Japan

W. Shiraki
Tottori University, Japan

H. Furuta
Kyoto University, Japan

H. Okada & Y. Murotsu
University of Osaka Prefecture, Japan

H. Ishikawa
Kagawa University, Japan

ABSTRACT: This paper is concerned with the structural reliability assessment methods for fatigue, buckling and plastic strength of steel girder stiffeners on urban expressway bridges. At first, the stress analyses are performed to obtain influence coefficients of stress in the stiffeners with regard to roadway surface for traffic loads. Traffic loads are practically simulated by means of Monte Calro technique, and stress ranges for fatigue analyses are counted by means of the rainflow method. The probabilistic modeling of some uncertainties is made byutilizing field data and laboratory testing data together with the simulation results. The reliabilities of fatigue, buckling and plastic failure of steel girder stiffeners are evaluated by using the advanced first order second moment method. Finally, the features of the former and current design standard bridges are investigated through comparison between both results.

1 INTRODUCTION

In design of highway bridges, fatigue and buckling failures of steel members are important problems. Many studies for these problems of highway bridges have been performed during the last decade(Fisher 1977, British Standards Institution 1980, Nyman and Moses 1985, Hanshin Expressway Public Coporation 1990, 1991, 1992 and et al.). Recently in Japan, traffic volume and weight of large-size vehicles gradually increase, and fatigue cracks occur in steel girder bridges, especially on urban expressways(HDL Committee 1986 et al.).

Structural response of bridge members due to actual traffic loads should be accurately examined for preventing fatigue and future maintenance.

This paper is concerned with the structural reliability assessment methods for fatigue, buckling and plastic strength of steel girder stiffeners on urban expressway bridges. Two typical types of steel girder bridges are chosen from existing bridges designed by the old and current standards of Hanshin Expressway Public Corporation.

At first, the stress analyses are performed for vertical stiffeners connected to the web plate of each main girder at the connections of a central transverse girder and influence coefficients of stress in the stiffeners are obtained with regard to roadway surface for traffic loads.

Traffic loads are practically simulated by means of Monte Calro technique based on stochastic data of traffic flows measured on the Hanshin Expressway.

Time history of stress in the stiffeners is precisely calculated, which is attributed to the traffic loads, and stress ranges for fatigue analyses are counted by means of the rainflow method. S-N curve of welded members recommended by Japanese Society of Steel Construction is used. Fatigue life of the stiffeners is estimated on the basis of Miner's rule, and equivalent stress range is also calculated.

The probabilistic modeling of some uncertainties is made by utilizing field data and laboratory testing data together with the simulation results. By using these probabilistic models, the reliabilities of fatigue, buckling and plastic failure of steel girder stiffeners are evaluated by using the advanced first order second moment method.

Finally, the features of the old and current design standard bridges are investigated through comparison between both results from a viewpoint of reliability.

2 STRESS ANALYSIS FOR STEEL GIRDER STIFFENERS ON URBAN EXPRESSWAY BRIDGES

2.1 Modeling of urban highway bridges

Two steel girder bridges with five or seven I-section main girders on the Hanshin Expressway bridges are chosen and are modeled for analysis. The scheme and principal dimensions are shown in Fig. 1, where G1, G2, ... , G7 are main girders, CTG central transverse girder, and integer j joint number of nodal point.

2.2 Evaluation of load influence coefficients of stiffeners

Interest in this study is focused on the vertical stiffeners which connect the central transverse girder to the main girders. Load influence coefficients for stresses in the stiffeners are evaluated by the following method. First, stress ratio is defined by

$$K_{tc} = \sigma_{dc}/\sigma_{bc} \tag{1}$$

where σ_{dc} is the stress evaluated by the FEM in the stiffener connected to specified girder[1], and σ_{bc} the stress evaluated by the plane framework model illustrated in Figs. 1(a) and (b). The loading condition in stress analysis is shown in Fig. 1(c). The stress ratio K_{tc} is used as a correction factor to estimate the stresses in the stiffener σ_{st} for various loading conditions. Bending stress σ_{bt} for the framework model under a loading condition, stress σ_{st} is approximated by

$$\sigma_{st} = K_{tc} \cdot \sigma_{bt} \tag{2}$$

The general expression of Eq. (2) is obtained as

$$\alpha_{ij} = K_{tc} \cdot \sigma_{ij} \tag{3}$$

where α_{ij} and σ_{ij} are the load influence coefficient and the bending stress at the point i due to unit load acted on the point j, respectively.

The influence coefficients are calculated for each stiffener connected to Gi (i = 1, 2, ..., 7 and 1, 2, ..., 5) girders. Calculation results for the G4 side stiffener of G3 girder on the bridge designed by the current standard, and G3 side stiffener of G2 girder on the bridge designed by the old standard are shown in Fig. 2.

It is seen from this figure that influence factors due to loads acting on the girder with the stiffener attached are remarkably large.

3 FATIGUE LIFE EVALUATION OF STEEL GIRDER STIFFENERS

Fatigue analysis is performed for the vertical stiffeners connecting the central transverse girder and main girders. The rainflow method and Monte Calro simulation technique are used for the fatigue analysis.

3.1 Traffic loads models

The traffic load models used in this study are based on the research results by the Committee on Design Loads Development for Hanshin Expressway (HDL Committee 1986). Table 1 presents the vehicle classifications, the statistical characteristics of vehicle weights for each category, the numbers and ratios of axles, and the proportion relative to the total traffic volume. By using these traffic loads models, it is possible to generate the sequences of vehicles withthe aid of a simulation technique. In the simulation, the weights of vehicles and the distances between vehicles are assumed to be random, whereas the remaining quantities are deterministic. The traffic volume is considered to be 10500 per a day and a lane. As a total, 42000 vehicles are generated, which correspond to the traffic volume for four lanes and for a day.

Time history of stress in the stiffeners is precisely calculated based on stochastic data of traffic flows. Examples of time histories of stresses are shown in Figs. 3(a) and (b). For the current standard bridge, stress ranges are very smaller than those for the former standard bridge.

3.2 Fatigue analysis

In order to estimate the fatigue life, the following S-N curve is used, which is recommended by the Japanese Society of Steel Construction(JSSC 1989). Since attention is paid to the fatigue of vertical stiffeners of webs, the curve D in the JSSC recommendation for fillet welding is used.

$$S^m \cdot N = C_0 \tag{4}$$

where C_0 is $2.0 \times 10^6 \times (S_t)^m$, S_t 100 MPa and m 3, respectively. The fatigue limit is 39.0 MPa. The fatigue analysis is performed by using the rainflow method. Then,

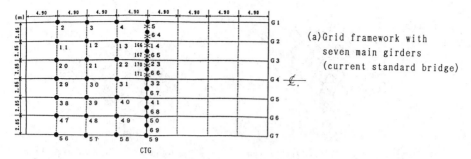

(a) Grid framework with
seven main girders
(current standard bridge)

● : Loading points for calculating influence surfaces
×: Locations of target stiffeners

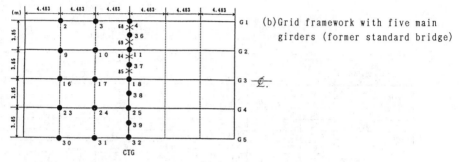

(b) Grid framework with five main
girders (former standard bridge)

● : Loading points for calculating influence surfaces
× : Locations of target stiffeners

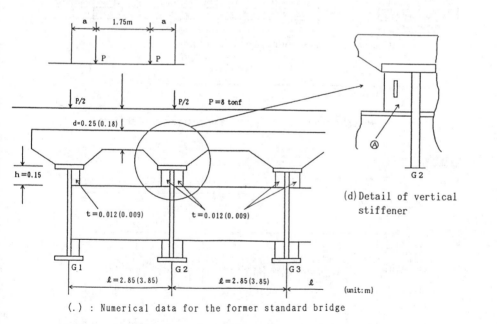

(d) Detail of vertical
stiffener

(.) : Numerical data for the former standard bridge

(c) Load condition and cross-section of span units

Fig. 1 Scheme of steel girder bridges

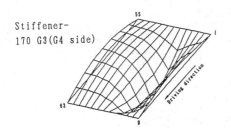

Stiffener-
170 G3(G4 side)

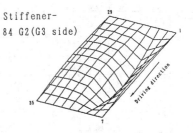

Stiffener-
84 G2(G3 side)

(a)Current standard bridge with seven   (b)Former standard bridge with five
   main girders   main girders

Fig.2 Influence surfaces of stiffeners

Table 1 Probabilistic models of traffic loads

vehicle classification		propor-tion (%) ashiya	total vehicle weight			vehicle length (m)	ratios of axle weight			
			dist.	μ(t)	σ(t)		first axle	second axle	third axle	fourth axle
2-axles large size trucks	unloaded	1.52	NOR	7.36	1.90	7.20	0.489	0.511		
	loaded	1.13	LOG	14.05	2.32		0.313	0.687		
	overloaded	0.01	EXP	23.14	3.14		0.238	0.762		
3-axles large size trucks	unloaded	1.80	NOR	11.21	1.89	7.40	0.436	0.206	0.358	
	loaded	6.19	LOG	19.77	3.18		0.262	0.270	0.468	
	overloaded	0.05	EXP	32.24	2.24		0.180	0.300	0.520	
3-axles large size trucks	unloaded	2.32	NOR	11.21	1.89	7.75	0.236	0.255	0.509	
	loaded	1.40	LOG	19.77	3.18		0.179	0.193	0.628	
trailer	unloaded	0.94	NOR	14.60	3.50	13.10	0.359	0.300	0.155	0.186
	loaded	1.46	LOG	26.98	8.63		0.189	0.325	0.221	0.265
mid size trucks		18.14	LOG	5.11	3.01	4.35	0.419	0.581		
passenger cars		65.04	LOG	1.31	0.34	4.10	0.527	0.473		

μ : Mean value, σ : Standard deviation
NOR : Normal distribution, LOG : Log-normal distribution, EXP:Exponential distribution.

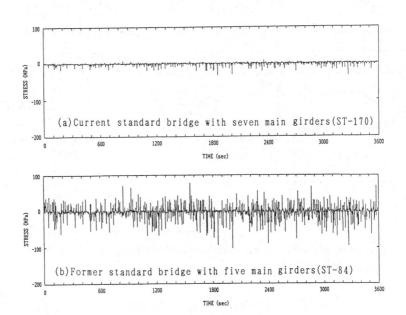

(a)Current standard bridge with seven main girders(ST-170)

(b)Former standard bridge with five main girders(ST-84)

Fig. 3 Examples of stress time histories of stiffeners

922

the equivalent stress range S_e is derived as follows:

$$S_e = (\Sigma S_i^m \cdot n_i / \Sigma n_i)^{1/m} \qquad (5)$$

Table 2 provides the equivalent stress ranges and the fatigue lives obtained for the vertical stiffeners.

From this table, it is concluded that fatigue life of the stiffeners in the bridges with five main girders designed by the old specifications practically corresponds to experimental data of fatigue occurrence. The fatigue life of the bridges with seven main girders designed by the current specifications becomes considerably large, and this means that the new bridges maintain the high level safety for fatigue damages.

4 RELIABLITY ANALYSIS OF FATIGUE, BUCKLING AND PLASTIC FAILURE OF STEEL GIRDER STIFFENERS

4.1 Probabilistic models of working stresses of stiffeners

Probabilistic distribution of stress range S is assumed to be of Weibull type, and then the cumulative distribution function $F_S(S)$ of stress range S is given by

$$F_S(S) = 1 - \exp[-(S/q)^h] \qquad (6)$$

where q and h are scale and shape parameters of Weibull distribution function, respectively (Wirsching and Chen 1987 and Takashima and Machida 1988 and et al.).

By using Miner's rule, S-N curve of Eq. (4) and c.d.f given by Eq. (6), fatigue damage D_f for number of cycle N_0 and equivalent stress range S_e are expressed as

$$D_f = (N_0/C_0) \cdot q^m \cdot \Gamma (m/h+1) \qquad (7)$$

$$S_e = q \cdot \{\Gamma (m/h+1)\}^{1/h} \qquad (8)$$

where $\Gamma(.)$ is gamma function.
The values of q and h are determined by using the simulation results. Numerical results are shown in Table 2.

4.2 Reliability analysis for fatigue of stiffeners

The reliability assessment model for 50 year fatigue life N_0 of the stiffeners is taken in the following form:

$$M = \ln(D_{cr}) - \ln(D_f) < 0 \qquad (9)$$

where D_{cr} is critical fatigue damage, and

D_f is the above mentioned fatigue damage, and their probabilistic properties are assumed as follows:

D_{cr} = LN(mean value = 1.0, coefficient of variation = 0.15)

D_f = $(N_0/C_0) \cdot q^m \cdot \Gamma (m/h+1)$

C_0 = LN(mean value = 2×10^{12}, coefficient of variation = 0.02)

m = 3.0(deterministic value)

q = LN(mean value = the value listed in Table 2, coefficient of variation = 0.1)

h = 1.2(deterministic value)

N_0 = 1×10^7 (for current standard bridge)

$= 5 \times 10^7$ (for old standard bridge)

$LN(.)$ = logarithmic normal distribution with parameters given in parentheses

Numerical results for the reliability assessment for 50 year fatigue life are given in Table 2, which draw similar conclusions mentioned in section 3.2

4.3 Reliability analysis for buckling and plastic failure of stiffeners

The reliability assessment model for the buckling and plastic strength corresponding to a 50-year maximum stress of stiffeners is given in the following form:

$$M = \phi_l \cdot \sigma_{Yl} - \sigma_l \qquad (10)$$

where ϕ_l is buckling and plastic strength factor(Johnston 1976), and its value is assumed to be $\phi_l = 0.98$ (deterministic value). σ_{Yl} is the yield stress of stiffeners, and σ_l a 50 year maximum stress of each stiffener. Their probabilistic properties are assumed as follows:

σ_{Yl} = N(mean value = 280 MPa, coefficient of variation = 0.08)

σ_l = N(mean value = $(S_{max})_{50}/2$, coefficient of variation = 0.10)

$(S_{max})_{50}$ = expected value of a 50-year maximum stress range given in Table 3

Table 2(a) Numerical results of reliability assessment for fatigue failure of
stiffeners (The current standard bridge with seven main girders)

| Stiffener number name | Simulation results | | Weibull parameters | | Expected value of 50 year cumulative damage (D_f) | Reliability index for 50 year fatigue life β |
	Fatigue life (year)	Equivalent stress range S_e (MPa)	q (MPa)	h		
G2(G3 side)(166)	268	39.2	26.3	1.2	0.302	1.06
G3(G2 side)(167)	355	38.1	25.7	1.2	0.282	1.12
G3(G4 side)(170)	134	39.9	26.7	1.2	0.316	1.01
G4(G3 side)(171)	1627	36.5	24.5	1.2	0.244	1.26

Table 2(b) Numerical results of reliability assessment for fatigue failure of
stiffeners (The former standard bridge with five main girders)

| Stiffener number name | Simulation results | | Weibull parameters | | Expected value of 50 year cumulative damage (D_f) | Reliability index for 50 year fatigue life β |
	Fatigue life (year)	Equivalent stress range S_e (MPa)	q (MPa)	h		
G1(G2 side)(68)	235.6	39.2	26.2	1.2	1.49	−0.36
G2(G1 side)(69)	25.4	46.0	30.7	1.2	2.40	−0.75
G2(G3 side)(84)	6.17	58.6	39.2	1.2	5.01	−1.30
G3(G2 side)(85)	14.7	52.8	35.2	1.2	3.62	−1.07

Table 3 Numerical results of reliability assessment for buckling and plastic
failures of stiffeners
(The current standard bridge)   (The former standard bridge)

Stiffener number name	Expected value of 50 year maximum stress $(S_{max})_{50}/2$ (MPa)	Reliability for buckling and plastic strengths β	Stiffener number name	Expected value of 50 year maximum stress $(S_{max})_{50}/2$ (MPa)	Reliability for buckling and plastic strengths β
G2(G3 side)(166)	154.0	4.49	G1(G2 side)(68)	130.0	5.66
G3(G2 side)(167)	149.0	4.73	G2(G1 side)(69)	168.0	3.85
G3(G4 side)(170)	161.0	4.17	G2(G3 side)(84)	220.0	1.75
G4(G3 side)(171)	136.0	5.36	G3(G2 side)(85)	194.0	2.74

N(.) : Normal distribution function with parameters given in parentheses

Numerical results for the reliability assessment are given in Table 3. It is seen from this table that both standard bridges maintain the sufficient safety for buckling and plastic failures.

5 CONCLUDING REMARKS

This paper presents the structural reliability assessment for fatigue, buckling and plastic strengths of steel girder stiffeners on urban expressway bridges. The main results are summarized as follows:

(1) The load influence coefficients for stresses in the stiffeners are evaluated by the approximation method.

(2) Time history of stress in the stiffeners is precisely calculated by the traffic load simulations, and the stress ranges for fatigue analyses are counted by means of the rainflow method.

(3) Analytical results of fatigue life of the stiffeners in the bridges designed by the old standard practically correspond to experimental data of fatigue occurrence. The fatigue life of the bridges designed by the current standard becomes considerably large, and the new bridges are reliable for fatigue damages.

(4) From the reliability assessment for 50 year fatigue life of stiffeners by assuming the Weibull distribution of stress ranges in service life based on the simulation results, similar conclusions are also drawn.

(5) From the reliabilty assessment for buckling and plastic failure of stiffeners, it is pointed out that both standard bridges maintain the sufficient safety for these static failures.

ACKNOWLEDGEMENTS

The authors would like to acknowledge financial support for their studies on the development of practical reliability-based design methods of bridge structures from the Society of Materials Science, Japan and Hanshin Expressway Public Corporation.

REFERENCES

BS5400, Brtish Standards Institution 1980 :Steel, Concrete and Composite Bridges, Part10, Code of Practice for Fatigue.

Fisher, J. W. 1977 : Bridge Fatigue Guide, American Institute of Steel Construction

Hanshin Expressway Public Corporation and Society of Materials Science, Japan 1990, 1991, 1992 : Report on Investigation for the Development of Practical Reliability-Based Design Methods of Bridge Structures (in Japanese).

HDL Committee 1986 : Report on Investigation of Design Load Systems on Hanshin Expressway Bridges, Hanshin Expressway Public Corporation (in Japanese).

Johnston, B. G. 1976 : Guide to Stability Design Criteria for Metal Structures, John Wiley & Sons, New York

Japanese Society of Steel Construction (JSSC) 1989 : Guide Line of Fatigue Design(Preliminary edition)(in Japanese)

Nyman, W.E. and Moses, F. 1985 : Calibration of Bridge Fatigue Design Model, J. of Structural Engineering, ASCE, 111-6, pp.1251-1266.

Takashima, H. and Machida, S. 1988 : Reliability Analysis of Fatigue Damage of Offshore Structure, J. of the Soc. of Naval Architects, Japan, Vol. 163, pp. 415-424(in Japanese).

Wirsching, P.H. and Chen, Y.N. 1987 : Considerations of Probability-Based Fatigue Design for Marine Structures, SNAME Marine Structural Reliability Symposium, Arlington, USA., pp.31-43

Structural Safety & Reliability, Schuëller, Shinozuka & Yao (eds) © 1994 Balkema, Rotterdam, ISBN 90 5410 357 4

Calibration of LRFD bridge design code

Andrzej S. Nowak
Department of Civil and Environmental Engineering, University of Michigan, Ann Arbor, Mich., USA

ABSTRACT: The paper summarizes the code development procedures used for the new LRFD (load and resistance factor design) bridge design code in the United States. The new code is based on a probability-based approach. Structural performance is measured in terms of the reliability (or probability of failure). Load and resistance factors are derived so that the reliability of bridges designed using the proposed provisions will be at the predefined target level. The paper describes the calibration procedure (calculation of load and resistance factors). The major steps include selection of representative structures, calculation of reliability for the selected bridges, selection of the target reliability index and calculation of load factors and resistance factors. The reliability indices for bridges designed using the proposed code are compared to the reliability indices corresponding to the current specification. The proposed code provisions allow for a consistent design with a uniform level of reliability.

1 INTRODUCTION

The objective of the paper is to summarize the procedures used in the development of a new LRFD bridge design code in the United States. The allowable stress method and even load factor design, do not provide for a consistent and uniform safety level for various groups of bridges. One of the major goals set for the new code is to provide a uniform safety reserve. The main parts of the current AASHTO (1992) specification were written about 5s0 years ago. There were many changes and adjustments at different times. In the result there are many gaps and inconsistencies. Therefore, the work on the LRFD code also involves the re-writing of the whole document based on the state-of-the-art knowledge in various branches of bridge engineering. This paper is focused on the reliability analysis of bridges designed using the current AASHTO (1992) provisions and proposed LRFD code.

The fundamental load combination for highway bridges includes dead load, D, live load, L, and dynamic load, I. Therefore, the basic LRFD formula is,

$$\phi R > \gamma_D D + \gamma_L (L + I) \qquad (1)$$

where R = resistance; ϕ = resistance factor; γ = load factor.

The objective of this calibration is to determine ϕ, γ_D and γ_L so that reliability of bridges designed using the new code is at the preselected target level.

2 CALIBRATION PROCEDURE

The development of the new code involves the following steps:

1. Selection of representative bridges. About 200 structures were selected from various geographical regions of the United States. These structures cover materials, types and spans which are characteristic for the region. Emphasis is placed on current and future trends, rather than very old bridges. For each selected bridge, load effects (moments, shears, tensions and compressions) were calculated for various components. Load carrying capacities are also evaluated.

Table 1. Statistical Parameters of Loads

Load Component	Bias	COV	Load Factors AASHTO 1992	LRFD
Dead load	1.03-1.05	0.08-0.10	1.3	1.25
Asphalt surface	90 mm	0.25	1.3	1.50
Live load (moment)	.75-2.10	0.18	2.17	1.6-1.7
Live load (shear)	1.50-1.90	0.18	2.17	1.6-1.7
Dynamic load	0.1 L	0.80	2.17	1.6-1.7

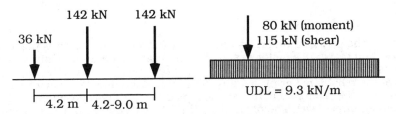

Fig. 1. HS20 Truck and Lane Loading (AASHTO 1992).

2. Establishing the statistical data base for load and resistance parameters. The available data on load components, including results of surveys and other measurements, is gathered. Truck survey and weigh-in-motion (WIM) data were used for modeling live load. There was little field data for dynamic load therefore a numerical procedure was developed for simulation of the dynamic bridge behavior. Statistical data for resistance include material tests, component tests and field measurements. Numerical procedures were developed for simulation of behavior of large structural components and systems.

3. Development of load and resistance models. Loads and resistance are treated as random variables. Their variation is described by cumulative distribution functions (CDF) and correlations. For loads, the CDF's were derived using the available statistical data base (Step 2). Live load model includes multiple presence of trucks in one lane and in adjacent lanes. Multilane reduction factors were calculated for wider bridges. Dynamic load is modeled for single trucks and two trucks side-by-side. Resistance models were developed for girder bridges. The variation of the ultimate strength was determined by simulations. System reliability methods are used to quantify the degree of redundancy.

4. Development of the reliability analysis procedure. Structural performance is

measured in terms of the reliability, or probability of failure. Limit states are defined as mathematical formulas describing the state (safe or failure). Reliability is measured in terms of the reliability index, β. Reliability index is calculated using an iterative procedure. The developed load and resistance models (Step 3) are part of the reliability analysis procedure.

5. Selection of the target reliability index. Reliability indices were calculated for a wide spectrum of bridges designed according to the current AASHTO (1992). The performance of existing bridges was evaluated to determine whether their reliability level is adequate. The target reliability index, β_T, was selected to provide a consistent and uniform safety margin for all structures.

6. Calculation of load and resistance factors. Load factors, γ, are calculated so that the factored load has a predetermined probability of being exceeded. Resistance factors, ϕ, are calculated so that the structural reliability is close to the target value, β_T.

3 LOADS AND RESISTANCE PARAMETERS

The load models are summarized in other

Table 2. Statistical Parameters of Resistance.

Type of Structure	Bias	COV
Steel Girder		
Moment	1.12	0.10
Shear	1.14	0.105
Reinforced Concrete T-Beams		
Moment	1.14	0.13
Shear	1.20	0.155
Prestressed Concrete Girders		
Moment	1.05	0.075
Shear	1.15	0.14

papers (Nowak and Hong 1991; Hwang and Nowak 1991). The statistical parameters of load components used in this study are presented in Table 1. Bias is a ratio of mean to nominal and COV is a coefficient of variation.

The current design live load is based on HS-20 truck or lane loading, as shown in Fig. 1. The proposed new live load is a superposition of HS-20 truck and a uniformly distributed load of 9.3 kN/m. The bias factors for the proposed live load are from 1.25 to 1.35 for moment and 1.15 to 1.25 for shear.

In the present AASHTO (1992), girder distribution factors (GDF) are specified as function of girder spacing, s, only. Recent studies indicate that GDF's are too conservative for larger spacings and spans, but too low for short s. New provisions rectify this problem and relate GDF to girder spacing and span length.

Current AASHTO(1992) specifies impact, I, as a function of span length only. In the LRFD code, it is proposed to use the dynamic load equal to 0.30 applied to the truck portion of live load only.

The reliability indices were calculated for bridges designed using the current AASHTO (1992). The design equation is

$$\phi R > 1.3 \, D + 2.17 \, (1+I) \, L \qquad (2)$$

where $I = 50/(125 + L_S)$; L_S = span in ft (1 ft = 0.305 m);
L = HS-20 load (truck or lane load whichever governs);
ϕ = 1.0 for steel girders;
ϕ = 1.0 for prestressed concrete girders;
ϕ = 0.9 for reinforced concrete T-beams.

The statistical parameters of live load (bias factor) vary depending on girder spacing and span length. This is reflected in the reliability indices. The results are shown in Fig. 2 for steel girders, reinforced concrete T-beams and prestressed concrete girders, for spans up to a 60 m.
Resistance models were obtained for girder bridges including steel girders, reinforced concrete T-beams and AASHTO-type prestressed concrete girders. The resulting statistical parameters of R are shown in Table 2.

4 RELIABILITY INDICES FOR AASHTO (1992)

The available reliability methods are presented in several publications (e.g. Thoft-Christensen and Baker 1982; Melchers 1987). In this study the reliability index, β, is determined by iterations. The calculations were carried out for the ultimate limit states (moment and shear) for girder bridges.

5 RELIABILITY INDICES FOR THE PROPOSED LRFD CODE

Reliability indices were also calculated for the proposed LRFD code. The considered design formula is,

$$\phi R > 1.25 \, D + 1.5 \, D_A + \gamma_L \, (1+I) \, L \qquad (3)$$

where D_A = dead load due to asphalt;
I = 0.30 (applied to truck load only);

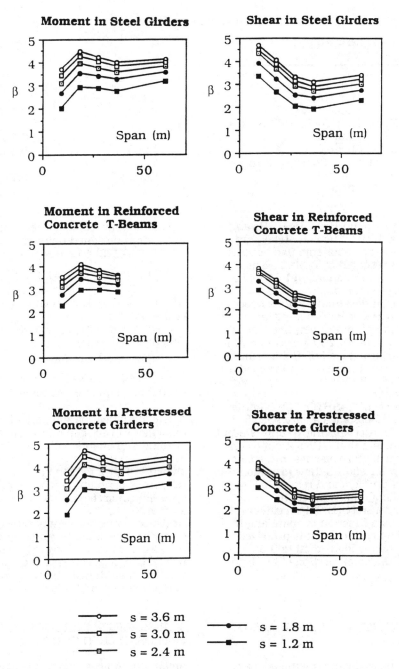

Fig. 2. Reliability Indices for Bridge Girders Designed According to AASHTO
(1992).

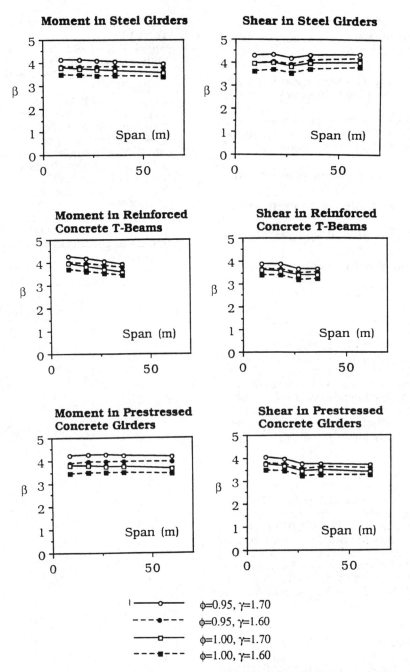

Fig. 3. Reliability Indices for Bridge Girders Designed According to the Proposed LRFD Code (γ is live load factor).

L = new proposed live load;
γ_L = live load factor, two value are considered: 1.60 and 1.70.

Various ϕ factors were considered. The resulting reliability indices are shown in Fig. 3.

The target reliability index was selected equal to 3.5. Consequently, the proposed resistance factors are as follows:

ϕ = 1.00 for steel girders, moment and shear;

ϕ = 1.00 for prestressed concrete girders, moment;

ϕ = 0.90 for reinforced concrete T-beams, moment;

ϕ = 0.90 for shear in concrete components.

6 CONCLUSIONS

Reliability indices for bridge girders designed according to the current AASHTO (1992) vary with girder spacing and span length. Probability-based calibration of the bridge design code provided a set of new load and resistance factors. Bridges designed using these factors have a uniform reliability level.

7 ACKNOWLEDGMENTS

The presented work on the new LRFD code for bridge design has been carried out in conjunction with NCHRP Project 12-33. Thanks are due to current and former research assistants at the University of Michigan for their help, in particular, Y-K. Hong, E-S. Hwang, S.W. Tabsh, H. Nassif, T. Alberski and A. Yamani.

8 REFERENCES

AASHTO, 1992, "Standard Specifications for Highway Bridges", American Association of State Highway and Transportation Officials, 15th edition, Washington, DC.

Hwang, E-S. and Nowak, A.S., "Simulation of Dynamic Load for Bridges," ASCE Journal of Structural Engineering, Vol. 117, No. 5, May 1991, pp. 1413-1434.

Melchers, R.E., 1987, "Structural Reliability Analysis and Prediction," Ellis Horwood Limited, Chichester, England.

Nowak, A.S. and Hong, Y-K., "Bridge Live Load Models," ASCE Journal of Structural Engineering, Vol. 117, No. 9, September 1991, pp. 2757-2767.

Tabsh, S.W. and Nowak, A.S., "Reliability of Highway Girder Bridges," ASCE Journal of Structural Engineering, Vol. 117, No. 8, August 1991, pp. 2373-2388.

Thoft-Christensen, P. and Baker, M.J., 1982, Structural Reliability Theory and Its Applications, Springer-Verlag, p. 267.

Structural Safety & Reliability, Schuëller, Shinozuka & Yao (eds) © 1994 Balkema, Rotterdam, ISBN 90 5410 357 4

Simulation of traffic loads for dynamics analysis

Jorge D. Riera
UFRGS, Porto Alegre, RS, Brazil

Ignacio Iturrioz
CPGEC, UFRGS, Brazil

ABSTRACT: An approach to determine rational loading criteria for highway bridges is herein proposed. Traffic loads are simulated taking into consideration traffic composition, vehicle characteristics, mean velocity and level of service. Relevant design quantities, such as midspan bending moment or torsion are obtained by numerical integration of the equations of motion of typical bridge structures, to generate random processes than can be characterized, for any given design situation, in terms of bridge span and damping. The prediction of peak values for predefined times of exposure, or of expected number of cycles for fatigue verification may be made on that basis.

1 INTRODUCTION

The specification of loads for the design of highway bridges was subjected to extensive scrutiny in recent years. This resulted from the need for an improved definition of bridge excitation, which in turn stems from two concurring factors: a) the trend of most codes and regulations on structural design towards a probabilistic, or at least semi-probabilistic format, and b) the noticeable evolution of the transportation industry towards heavier and faster vehicles which, in conjunction whith ligther structural systems is leading to dynamic effects that can no longer be regarded as a marginal influence.

Recent contributions in the area include surveys on truck loading and evaluations of peak loads obtained from exceedance limit analyses of random processes representative of traffic loading. Alternatively, efforts were likewise devoted to the determination of dynamic amplification factors in coupled models of vehicle-girder systems. The authors are not aware of formulations that incorporate both the random load amplitude and spatial distribution, as well as the ensuing dynamic structural response in the determination of loading criteria applicable to modern bridge design. With such purpose in mind, traffic loads per lane were generated by simulation, for any given mean lane velocity, by randomly

selecting the type of vehicle, its velocity and all relevant characteristics, introduced by means of appropriate probability distribution functions. The discrete random process thus simulated is next used to evaluate the dynamic response of typical bridge structures. Important design quantities, such as the midspan bending moment, are then characterized as stationary random processes, for which exceedance limits may be readily obtained for any prescribed exposure time, furnishing a sound basis for the development of rational bridge loading criteria.

2 REVIEW OF RECENT STUDIES RELATED TO BRIDGE LOADING

Information on traffic loads in Switzerland was recently presented by Bez 'et al' (1987). The study was motivated by the need to revise the swiss code on loading SIA 160 (1970) and includes result of measurements conducted between 1975 and 1976 by the Swiss Federal Laboratories for Testing Materials (EMPA), jointly whith ETH (Zürich). Nowak & Hong (1990) report on similar studies conducted by the Canadian Ministry for Transportation, between 1970 and 1980 . The results are referred to the AASHTO Standart Specification (1989) HS20 load and allow the evaluation of the expected value of the peak bending moment in the specified

life time.

Jacob 'et al'(1990) present valuable statistics on truck loading in Europe, based on measurements conducted on more than 10000 trucks. This work is part of current efforts within Eurocode 3, Part 12 "Traffic Loads on Bridges". Loriggio (1991) discusses truck loads measured in Brasil. At the other end of the problem, Kameda & Kubo (1990) propose a computational procedure to determine the peak excitation in bridges, during the specified lifespan. Whithin such context, Nowak (1990) observes that additional studies are still necessary to stablish load models applicable to different regions, variable periods and multiple presence of vehicles in each lane.

An intermediate but relevant aspect of the problem, namely the dynamic amplification of the traffic loads, will not be discussed in this paper for reasons of space. It is germaine to quote, though, that Gross & Rackwitz (1989) warn that the surface roughness can produce non-negligible additional loads. This effect is accounted for in the proposed model by means of the scheme described in Section 3.

3 FLUCTUATING LOAD COMPONENTS: VEHICLE-STRUCTURE INTERACTION

When a vehicle is in motion, the vertical load resultant at the wheel-road surface interface is no longer constant, but fluctuates as a result of the vehicle vibrations. The latter are induced by (a) pavement irregularities, (b) breaking or acceleration of the vehicle, (c) the vertical motion of the bridge structures under load. It was found that the first group, which includes pavement discontinuites, such as those encountered at construction joints, or common detritus, like broken bricks or wooden planks, are the most significant vibrations sources. Large loaded trucks have a fundamental vertical mode frequency in the range between 2 and 3 Hz, with a critical damping ratio of about 2%. Previous experimental and theoretical studies showed that the average amplitude of the fluctuating component is close to 10% of the mean wheel load at average cruising speeds, increasing with vehicle velocity V. While further studies on this subject are necessary, it was herein assumed that the expected value of the vertical vibration amplitudes, referred to the static wheel load, is given by: $a = 0.0067$ V (V in m/s).The excellent

correlation between theoretical results obtained employing this equation whith measured 'in loco' response (Riera 'et al', 1992) confirms that accurate predictions can be made when the loading induced by the vehicle is especified in terms of the following variables:

Table 1. Vehicle basic properties.

a- Vehicle total mass.
b- Number of axles.
c- Static load distribution per axle.
d- Mean forward velocity.
e- Fluctuating vertical load component.
f- Frequency of vibration.

As it travels,along the way the vehicle is excited by road irregularities or other sources, effect that was introduced in the global model described in section 4 by modulating the high frequency (2-3Hz) vibration components with a low frequency time function, as shown in figure 1. The amplitude of the fluctuating component depends on rms the mean velocity.

4 TRAFFIC LOAD REPRESENTATION

In order to represent traffic loads accurately and with great generality, highway vehicles were divided in five categories, namely:

Table 2. Vehicle categories.

1- Light vehicles (2 axles).
2- Buses (2 axles).
3- Pick-ups and light trucks (2 axles).
4- Three-axles trucks (3 axles).
5- Trailers and composite heavy trucks.

Each group is characterized by a fixed number of axles but all the other properties listed, indicated in Table 1, are considered random variables with known probability distributions, as described in Section 5. Now, traffic consists of a composition of vehicles, in which each

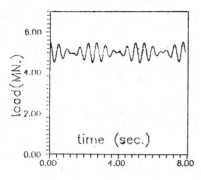

Fig. 1- Axle Load vs. Time.

Table 4. Levels of service.

LEVEL	DENSITY [*.km/h]	AVERAGE SPEED [km/h]	[m/s]	FLOW RATE/LANE [*/h]
A	≤ 9.5	≥ 97	27	700
B	≤ 12.5	> 89	25	1100
C	≤ 18.75	> 72	20	1550
D	≤ 45	> 56	15.6	1850
E	≤ 42	> 48	13.3	2000
F	> 42	< 48	13.3	(unstable)

* :equivalents automobiles.

Table 5. Traffic composition in simulated samples (percentage of vehicles).

Sample	C2					F2				
Categories	1	2	3	4	5	1	2	3	4	5
1 lane	61	5	11	11	12	61	5	11	12	15
2 lanes	61	5	11	11	12	61	5	11	12	15
	91	5	1	1	2	61	5	11	12	35

Sample	C3					F3				
Categories						1	2	3	4	5
1 lane	5	26	27	27	15	5	5	26	27	27
2 lanes	5	26	27	27	15	5	5	26	27	27
	15	20	20	20	35	35	5	26	27	27

category enters in a given proportion, that depends on the design situation as defined below:

Table 3. Design situations.

1-LIGTH TRAFFIC: Predominance of automobiles and buses, access to resort areas or access to city centers, from residential areas, in early morning hours.

2-NORMAL TRAFFIC: Average conditions in intercity, interstate highways.

3-HEAVY TRAFFIC: Access to large ports or harbours, export-import or load distribution centers. Special or abnormal situations in intercity highways.

Naturally, the traffic composition varies with the Design Situation, the number of lanes, and the level of service. The latter is correlated with the total flow of vehicles. In the Highway Capacity Manual (1985) , six levels of service are established, see Table 4.

For instance, C2 orresponds to normal traffic at level of service C. Note that the number of vehicles per hour indicated in table 4 refers to equivalent automobiles. In the evaluations conducted so far, only a limited number of simulations were considered, as indicated in Table 5.

5 RANDOM PROCESS SIMULATION

The load, per lane, especified for any given traffic configuration,is simulated as follows :the vehicle category (see table 2) for the first vehicle in the train is selected according to the relative participation in the total composition (Table 5). All properties indicated in Table 1 are then generated from the probability distributions assigned to each vehicle category. The separation distance to the following vehicle is also a random variable with expected value: $E[x] = \mu x = (V/F) - Cl$ in which V is the mean flow velocity, F the number of equivalent automoviles per hour and per lane and Cl the mean distance between vehicles. The latter two values depend on level of service (see Table 4).

It is assumed that x is uniformly distributed between $0.9\mu x$ and $1.1\mu x$. The simulation proceeds with the generation of the category to which the second vehicle in the group pertains, and so on. Thus, a detailed loading scheme that retains all recognized influencing factors can be generated for the analysis of straight or curved bridge structures.

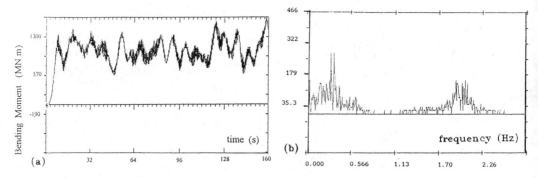

Fig. 2- Bending Moment vs. time and Spectral Density at midspan of 40m long. concrete bridge, for F3 traffic conditions.

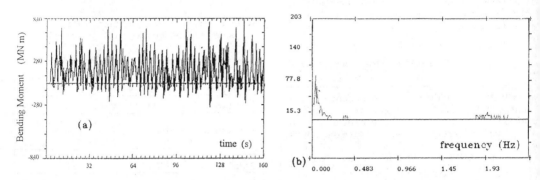

Fig. 3- Bending Moment vs. time and Spectral Density at midspan of 40m long. concrete bridge, for C3 traffic conditions.

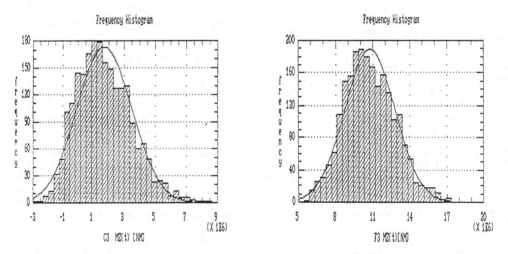

Fig. 4- Probability density function of stationary part of midspan bending moment under C3 and F3 traffic conditions (Histogram and Normal density).

6 DETERMINATION OF STRUCTURAL RESPONSE

A numerical procedure for evaluating the dynamic response of curved or straight, open or closed section viscously damped bridges was next employed to compute the structural response of typical prestressed concrete bridge structures to the simulated loading. The methods is based on the simultaneous numerical integration in the space and time domains of Vlasov's equations, as described by Carbonari 'et al' (1987). It was tested against the experimentally determined dynamic response of prestressed concrete bridges under truck loading by Riera 'et al' (1992). Since the loading obviosly represents a stationary random process, a stable (damped) linear system will present a likewise stationary response. The solution is obtained by numerical integration in the time domain, for null initial conditions (zero displacements and velocities). Thus, the first 20 seconds of the response correspond to a transient condition, as the traffic train enters into the bridge, and are disregarded in the following. The computed response betweeen 20 and 160 seconds was then used to evaluate the response quantities for single span structures indicated such as midspan bending moment and torsional moment.

If the estructure does not present significant dynamic response, the above functions should depend on the span l. When important vertical or torsional vibrations are induced, the natural frequencies and damping ratios become relevant variables as well. It has been shown, however, that for reinforced or prestressed concrete bridges, the fundamental vibration frequency associated to vertical motion is closely correlated with the span (Cantiene,1984; Riera 'et al' 1992).Therefore, it may be further assumed that the response for a given loading train is governed by the span l and first mode damping ζ_1 (straight bridges), and also by the curvature radius R in curved bridges.

Figures 2 - 4 show representative results for a 40m prestressed box-section concrete bridge whith fundamental frequency equal to 2.37 Hz and damping ratio 0.03. The variation with time of the bending moment Mz at midspan for condition C3(20m/s) and F3(5m/seg) can be seen in Figs. 2(a) and 3(a). The corresponding spectral density and probability density functions are also shown. Finally, Fig 5 presents sample evaluations of the variation of the midspan bending moment

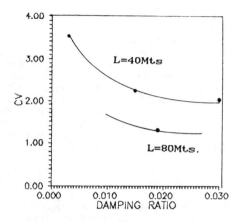

Fig. 5- Coef. of variation of midspan bending moment for C2 traffic conditions.

for the traffic condition (C2) in terms of bridges span and damping ratio. It may be seen that the approach permits an accurate quantification of decrease in variability of the response with span and damping, which in practice reflects in lower dynamic amplification factors.

7 EVALUATIOM OF EXTREME RESPONSE AND FATIGUE CONSIDERATIONS

According to the results presented in the preceding sections, all basic design variables, such as bending moment, torsion moment, etc can be characterized as stationary random processes with probability and spectral densities defined, for any given span, in terms of traffic conditions and damping, as well as radius whenever appropiate. If the lifespan of the bridge is denoted by T, and TA represents the total time that the bridge is expected to be under service level A, etc, then: T=TA+TB+....+TF, in which TA=TA1+TA2+TA3, etc.

Thus the problem at hand constitutes a classical situation in theory of random processes, namely the determination of the statistics of the extreme values, given the frequency distribution of the process and the time of exposure. By defining other excursion limits, rational criteria to evaluate the risk of fatigue failure can be determined. The condensation of the resulting information in rules useful for design is a complex problem, which will be subject of a forthcoming paper.

8 CONCLUSIONS

A procedure to develop rational design criteria for highway bridges based on a complete simulation of traffic loading, coupled to the evaluation of the dynamic response of typical bridge structures, has been outlined and representative examples given. The approach allows the consideration, at different stages in the process, of all factor that significantly influence the response statistics.

ACKNOWLEDGEMENTS

This research was partly supported by CNPQ, Brasil valuable information on traffic engineering furnished by J. Lindau is kindly acknowledged. The work is also partly related to current research efforts within CLAES (Latin American Commitee on Structures).

REFERENCES

Bez, R.; Rato & Jacquemound J. 1987. Modeling of Highway Traffic Loads in Switzerland. IABSE Procedings p.117/87; Periódica 3/1987,pp. 153-168.

Carbonari, G; Riera, J. D. & Awruch, A. M. 1987. Análise dinámica de vigas de eixo curvo e seçäo aberta submetidas a cargas móveis, Colloquia 97, Ed by UFRGS, Porto Alegre, RS, Brasil, Vol 5,17-34.

Catiene, R. 1984. Dinamic load testing of highway bridges. IABSE Proceedings, Zürich, (75):57-72, Aug. .

Gross, P. & Rackwitz, R. 1989. On the Dynamic Response of Bridges by Oscillating Vehicles on Rough Surfaces. 5th. Int. Conf. on Structural Safety and Reliability, ICOSSAR '89, San Francisco, USA, Aug., Proceedings, ASCE 1990, 2247-2250.

Highway Capacity Manual 1985. Special Report 209, Transport Research Board, National Research Council, Washington, D.C.

Jacob, B; Mailard, J.B. & Gorse, J.F. 1989. Probabilistic Traffic Load Models and Extreme Loads on a Bridge. 5th. Int. Conf. on Structural Safety and Reliability, ICOSSAR '89, San Francisco, USA, Aug. 1989, Proceedings, ASCE 1990, 1973-1981.

Kameda, H. & Kubo, M. 1989. Lifetime – maximum load effect for highway bridges based on stochastic combination of typical traffic loading. 5th. Int. Conf. on Structural Safety and Reliability, ICOSSAR'89, San Francisco, USA, Aug., Proceedings, 1783-1791.

Loriggio, D. D. 1991. Carregamentos e solicitações de serviço em lajes de pontes.", XXV Jornadas Sul-Americanas de Engenharia Estrutural, Porto Alegre, Brasil, Nov., Anais, Vol 5,(46-60) .

Norme SIA 160 1970 . Norme Concernent les charges, la mis en service et la surveillance des constructions, Zurich, SIA.

Nowak, A. S. 1989. Probabilistic basis for bridge design codes. 5th. Int. Conf. on Structural Safety and Reliability, ICOSSAR '89, San Francisco, USA, Aug. 1989, Proceedings, Publ. by ASCE, 1990, 2019-2026.

Nowak, A. S. & Hong, T. 1991. Bridge Live-Load Models. J. of Structural Engineering, ASCE, vol. 117, no 9, Sept., 2757-2767.

Standard Specification for Highway Bridges 1989: American Association of State Highway Transportation Officials (AASHTO), Washington, DC, USA.

Riera, J. D.; Doz, G. N. & Carbonari, G. 1992. Determinación numérica de la respuesta dinámica de puentes en planta curva. Jornadas Argentinas de Ingenieria Estrutural, Buenos Aires, Argentina, Ed. by AIE. Vol.I Set. 239-249.

Structural Safety & Reliability, Schuëller, Shinozuka & Yao (eds) © 1994 Balkema, Rotterdam, ISBN 90 5410 357 4

Optimal strategy for maintenance of concrete bridges using expert systems

Palle Thoft-Christensen & Henriette I. Hansen

CSR, Aalborg, Denmark

ABSTRACT: Expert systems for optimal reliability-based inspection and maintenance of reinforced concrete bridges are described. The deterioration of the reinforcement is assumed to be corrosion of the reinforcement due to carbonation and chloride. Expert system module BRIDGE 1 is used at the bridge site to assist during the inspection. BRIDGE 2 is used after an inspection during the detailed analysis of the bridge when testing in the laboratory has taken place.

1 INTRODUCTION

The research project 'Assessment of Performance and Optimal Strategies for Inspection and Maintenance of Concrete Structures using Reliability Based Expert Systems' supported by CEC within the BRITE/EURAM research programme is presented in this paper.

The main objective of the project is to optimize strategies for inspection, maintenance and repair of reinforced concrete bridges by developing improved methods for modelling the deterioration of existing as well as future structures using reliability based methods and expert systems.

Results from this research project are presented with special emphasis on reliability assessment, updating, and the software modules. Further, the functionalities and the structure of the expert systems and the optimal strategy is briefly discussed.

2 MODELLING OF REINFORCEMENT CORROSION

In the present version of the expert systems only one type of deterioration is included namely corrosion of the reinforcement due to carbonation and chloride. The reliability of a corroded reinforced structural element is dependent of a number of quantities like the area of the corroded reinforcement.

In the project the area of the reinforcement $A(t)$ as function of time t, the depth of the carbonation front d_c, and the chloride concentration C are es-

timated on basis of well known techniques (Thoft-Christensen 1992).

It is assumed that the area of the n reinforcement bars $A(t)$ as function of time t can be determined by the following formula

$$A(t) = \begin{cases} n \cdot D_i^2 \cdot \pi/4 & \text{for } t \le T_i \\ n \cdot (D(t))^2 \cdot \pi/4 & \text{for } T_i < t \le T_1 \\ 0 & \text{for } t > T_1 \end{cases} \quad (1)$$

where $T_1 = T_i + D_i/(0.023 \cdot i_{corr})$. The diameter $D(t)$ of a single bar (in mm) at time t is modelled by $D(t) = D_i - 0.023 \cdot (t - T_i) \cdot i_{corr}$. D_i is the initial diameter of a single bar. i_{corr} is the rate of corrosion in terms of a mean corrosion current density in $\mu A/cm^2$. It is assumed that a corrosion current of 1 A/m^2 is equivalent to an average oxidation or dissolution of approximately 1.6 $mm/year$ from the surface of steel.

T_i (in years), the corrosion initiation time, due to carbonation (as function of the carbonation front d_c in mm) and due to chloride (as function of the chloride diffusion coefficient D_c in cm^2/sec) is estimated by the following formulas.

For carbonation:

$$T_i = \left(\frac{d}{K}\right)^2 \quad (2)$$

For chloride:

$$T_i = \frac{d^2}{4 \cdot D_c} \left(erf^{-1} \left(\frac{C_{cr} - C_0}{C_i - C_0} \right) \right)^{-2} \quad (3)$$

where d is the concrete cover in cm, K is a constant depending of the concrete cover depth d, concrete mix proportions, water/cement ratio, cement con-

tent etc. C_0 is the equilibrium chloride concentration (in % by weight of cement) on the concrete surface and C_{cr} is the critical chloride concentration at the reinforcement level.

3 ASSESSMENT OF THE RELIABILITY

The reliability of the bridge is measured using the reliability index β for a single failure element or for the structural system (the bridge) (Thoft-Christensen & Baker 1982, Thoft-Christensen & Morutsu, 1986). The reliability is assumed to decrease in time due to the deterioration. The overall requirement is that the expected reliability index should never be smaller than some minimum reliability index. The failure modes can e.g. be stability failure of columns, yielding or shear failure in a number of critical cross-sections of the bridge. If a systems modelling is used then it is assumed that the structure fails if any one of these failure modes fails, i.e. a series system modelling is used.

In the present version of the expert systems only 3 failure modes are implemented, namely

- 'Positive' bending failure of a reinforced T-beam (see figure 1a).

- 'Negative' bending failure of a reinforced T-beam (see figure 1a).

- Compression failure of a rectangular reinforced column; two models, namely with and without bending moments (see figure 1b).

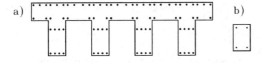

Figure 1: a) reinforced T-beam; b) rectangular reinforced column.

It is assumed that uncertain quantities like loading, strength and inspection results can be modelled by N stochastic variables $\vec{X} = (X_1, X_2, \ldots, X_N)$.

At present the stochastic variables shown in table 1 are used.

Table 1: Distribution types for stochastic variables.

Stochastic variable		Distribution
X_1	Concrete cover	Normal
X_2	Height of beam	Normal
X_3	Height of deck	Normal
X_4	Initial diameter of reinforcement	Normal
X_5	Width of column	Normal
X_6	Depth of column	Normal
X_7	Compression yield stress of concrete	Normal
X_8	Yield stress of reinforcement	Normal
X_9	Uniformly distributed dead load	Normal
X_{10}	Uniformly distributed traffic load	Gumbel
X_{11}	Point traffic load	Gumbel
X_{12}	Chloride concentration on the concrete surface	Normal
X_{13}	Chloride diffusion coefficient	Lognormal
X_{14}	Coefficient rate of carbonation	Normal
X_{15}	Rate of corrosion	Normal
X_{16}	Measurement uncertainties	Normal

4 UPDATING OF FAILURE PROBABILITIES

Two main types of updating of the probability of failure estimates are considered, namely

- Updating of stochastic variables based on measured samples of the stochastic variables, e.g. measurements of the yield strength of the reinforcement.

- Updating based on general information, e.g. the observation that the structure is not failed or that a corrosion degree smaller than a certain value is measured.

Basic variable updating is performed within the framework of Bayesian statistical theory (Lindley 1976, Aitchison & Dunsmore, 1975). The updating based on general information is mainly based on the Bayesian methods suggested by (Madsen 1987) and (Rackwitz & Schrupp 1985).

Let the density function of a stochastic variable X be given by $f_X(x, \Theta)$, where Θ is a parameter defining the distribution of X. The parameters Θ are treated as uncertain parameters (stochastic variables). $f_X(x, \Theta)$ is therefore a conditional den-

sity function $f_X(x|\theta)$. The initial (or prior) density function for Θ is called $g'_\Theta(\theta)$.

When an inspection is performed n realisations $\vec{x}^\star = (x_1, x_2, \ldots, x_n)$ of the stochastic variable X are obtained. The inspection results are assumed to be independent. An updated density function for Θ taking into account the inspection results is then defined by

$$g''_\Theta(\theta|\vec{x}^\star) = \frac{f_n(\vec{x}^\star|\theta)g'_\Theta(\theta)}{\int f_n(\vec{x}^\star|\theta)g'_\Theta(\theta)\,d\theta} \qquad (4)$$

where $f_n(\vec{x}^\star|\theta) = \prod_{i=1}^n f_X(x_i|\theta)$.

The updated density function of X taking into account the realisations $\vec{x}^\star$ is then obtained by

$$f_X(x|\vec{x}^\star) = \int f_X(x|\theta)g''_\Theta(\theta|\vec{x}^\star)\,d\theta \qquad (5)$$

In the expert systems the functions $g'_\Theta(\theta)$, $g''_\Theta(\theta)$, and $f_X(x|\vec{x}^\star)$ are implemented for normal distributions with unknown mean and standard deviations, for the Gumbel distribution with unknown location parameter, for the Weibull distribution with unknown location parameter, for the Frechet distribution with unknown location parameter and for the exponential distribution with unknown location parameter.

5 FUNCTIONALITIES OF THE EXPERT SYSTEM

The expert system is divided into two expert system modules which are used in two different situations, namely by the inspector of the bridge during the inspection at the site of the bridge and after the inspector has returned to his office.

During the inspection the expert system will supply information on: the causes of observed defects, appropriate diagnosis methods, and related defects. Further the inspector will be asked to record the inspection results so that they can be used later for e.g. assessment of the reliability of the bridge and in the decision whether a detailed structural assessment is needed.

A detailed analysis of the state of the bridge after an inspection takes place when the inspector has returned to his office and after testing in the laboratory has taken place. The output of the analysis includes an updated estimation of the reliability of the bridge, decision whether a structural assessment should take place, decision whether repair should take place, relevant repair procedures and the time for repair. Expert knowledge is used to improve the quality of the decisions.

6 ARCHITECTURE OF THE EXPERT SYSTEMS

The architecture of the expert system is shown in figure 2. It consists of two main modules called BRIDGE 1 and BRIDGE 2, a number of dBASE IV databases, several FORTRAN programmes, and INPUT and OUTPUT modules.

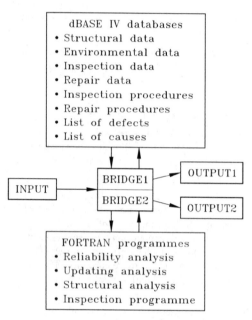

Figure 2: General Architecture of the expert systems.

The BRIDGE 1 expert system is used on the site of the bridge. BRIDGE 2 is used by the engineer in the office. All four FORTRAN programmes indicated in figure 2 are needed to run the BRIDGE 2 part of the expert system. The reliability, the updating, and the inspection modules are integrated parts of the expert systems, while any structural analysis programme can be used.

7 APPLICATION OF THE EXPERT SYSTEMS

The general inspection, maintenance, and repair model from inspection no. i at time t_i to inspection no. $i+1$ at time $t_{i+1} = t_i + \Delta t$ is indicated in figure 3, where also the application of the modules BRIDGE 1 and BRIDGE 2 is shown.

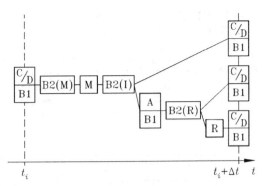

Figure 3: The general inspection, maintenance, and repair model.

The symbols used in figure 3 are:

C: Current inspections are performed with a fixed time interval, e.g. 15 months. The inspection is mainly visual and only the most critical areas are inspected.

D: Detailed inspections are also periodical with a fixed time interval which is a multiple of the current inspection time intervals, e.g. 5 years. The detailed inspections are visual inspections, but structural elements with potential problems are inspected more carefully. The inspections can also include non-destructive in-situ tests.

A: Structural assessments are only performed when a current or detailed inspection show some serious defects which require a more detailed investigation. Structural assessments are thus not periodical. They can include laboratory tests, in-situ tests with non-portable equipment, static and dynamic load tests.

M: Maintenance and repair of minor defects.

R: Structural repair.

B1: Application of BRIDGE 1 during the inspections.

B2(M): The maintenance subsystem in BRIDGE 2 assists in selection of maintenance work and on repair of minor structural defects to be performed. This subsystem is always used after a current or detailed inspection.

B2(I): The inspection module in BRIDGE 2 assists in selecting the next type of inspection. This module is always used after a current or detailed inspection.

B2(R): The repair subsystem in BRIDGE 2 assists

in selecting the best repair technique (including no repair, upgrading and replacement of an element or of the bridge). The selection is based on economic considerations and expert knowledge. This subsystem is used after a structural assessment.

After a current or a detailed inspection BRIDGE 2 is used to rate the maintenance and minor repair work needed and to decide if a structural assessment has to be performed. The decision is based partly on estimates of the reliability of the bridge and partly on expert knowledge. The decision does not include economic considerations.

After a structural assessment BRIDGE 2 is used to decide if a repair has to be performed and also to give the optimal point of time for the repair. Expert knowledge as well as numerical algorithms are used. The decisions are partly based on a cost-based optimization where different repair possibilities (selected by expert knowledge) and no repair are compared. The total expected costs are minimized using the FORTRAN inspection module.

8 DECISION MODEL WITH REGARD TO STRUCTURAL ASSESSMENT

Let t_i be the time of a periodic inspection and let the updated reliability index at time t be $\beta(t, t_i)$. The general decision model with regard to the structural assessment can then be formulated as:

If $\beta(t_{i+1}, t_i) > \beta^{min}$ then the inspection at time t_{i+1} should be a current or detailed inspection unless the damage is so serious that a structural assessment is needed. This decision is based on expert knowledge.

If $\beta(t_{i+1}, t_i) \leq \beta^{min}$ then a structural assessment should be performed before the next periodic inspection.

β^{min} is the minimum acceptable reliability index (e.g. 3.72).

9 REPAIR MODEL

After a structural assessment it must be decided whether the bridge should be repaired and if so, how the repair is performed. Solution of this problem requires that all future inspections and repairs are taken into account. However, the numerical calculations can then easily become very compli-

cated and also very time consuming.

The decision problem is therefore in the present version of the expert system solved approximately. After each structural assessment the total expected benefits minus expected costs in the remaining lifetime are maximized considering only repair events in the remaining lifetime. In this way an adaptive method is obtained where after each structural assessment the stochastic model is updated and the optimal decision is taken.

In order to decide which repair type is optimal after a structural assessment, the following optimization problem is considered for each repair technique:

$$\max_{T_R, N_R} \quad W(T_R, N_R) = B(T_R, N_R) - C_R(T_R, N_R)$$
$$- C_F(T_R, N_R) \qquad (6)$$
$$\text{s.t.} \quad \beta^U(T_L, T_R, N_R) \geq \beta^{min} \qquad (7)$$

where the optimization variables are the expected number of repair N_R in the remaining lifetime and the time T_R of the first repair. W is the total expected benefits minus costs in the remaining lifetime of the bridge. B is the expected benefits in the remaining lifetime of the bridge. C_R is the repair cost capitalized to time $t = 0$ in the remaining lifetime of the bridge. C_F is the expected failure cost capitalized to time $t = 0$ in the remaining lifetime of the bridge. T_L is the expected lifetime of the bridge. β^U is the updated reliability index. β^{min} is the minimum reliability index for the bridge (related to critical element or to the total system).

The repair decision is then based on the results of solving this optimization problem and on expert knowledge.

10 EXPERT SYSTEM MODULE BRIDGE 1

The expert system module BRIDGE 1 is as mentioned earlier used at the bridge site during a current inspection, a detailed inspection, and a structural assessment. This expert system module contains useful information concerning the bridge being inspected and the defects being observed. The information includes: general information about the bridge, appropriate diagnosis methods for each defect, probable causes for each defect, and other defects related to a defect. It is also possible to create a provisional defect report.

The general information about the bridge stored in the database for the selected bridge can be reviewed. The database contains information about:

bridge site, design, budget, traffic, strength, load, deterioration, factors that model the costs, and the cross-sections entered for the bridge.

New cross-sections can be entered for the selected bridge. The information stored in the database for each cross-section contains: cross-section identification, geometry of cross-section (detailed description of the reinforcement layers for cross-sections in the deck), failure mode, and load data.

Technical support can be provided for a defect, see figure 4.

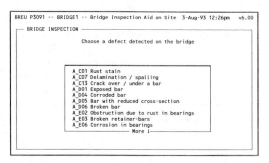

Figure 4: List of defects included in the expert systems.

The technical support include:

- List of diagnosis methods that can be used to observe a selected defect. The list is divided in high and low correlated diagnosis methods for the selected defect, see figure 5.

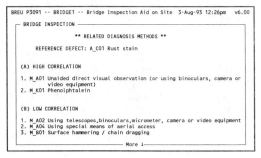

Figure 5: List of diagnosis methods related to the defect 'rust stain'.

- List of probable causes of a selected defect. The list is divided in high and low correlated causes for the selected defect, see figure 6.

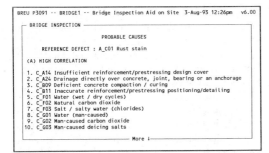

Figure 6: List of probable causes for
the defect 'rust stain'.

- List of defects associated with the selected defect. This list is very useful as the defects that with high probability can be found if the selected defect is observed can be reviewed. Measures for the correlations between the selected defect and the related defects are shown, see figure 7.

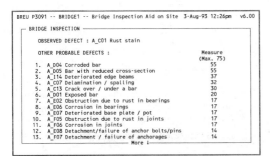

Figure 7: List of defects associated to
the defect 'rust stain'.

A provisional defect report can be recorded for an inspection. The date and type of inspection is recorded together with information about all defects observed. The information to be recorded for the defects depends on the type of subsystem for the defects.

- For defects related to the repair subsystem the following information is recorded: the cross-section where the defect is observed, one or more diagnosis methods used to conclude the detected defect, one or more of the causes of the defect, and comments. Only diagnosis methods and causes correlated with the observed defect can be selected.

- For defects related to the maintenance subsystem the following information is recorded: the cross-section where the defect is observed, defect classification (in terms of rehabilitation urgency, importance of the structure's stability, and affected traffic), and comments. The defect classification is based on pre-fixed rules that are shown on the screen.

11 EXPERT SYSTEM MODULE BRIDGE 2

The expert system module BRIDGE 2 is used to

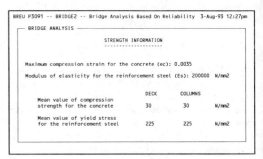

Figure 8: Example of strength data.

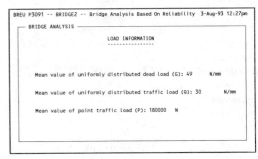

Figure 9: Example of load data.

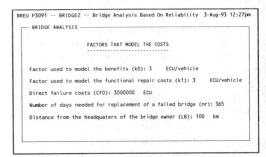

Figure 10: Example of cost data.

make a detailed analysis of the bridge after an inspection when testing in the laboratory has taken place. New bridges and cross-sections can be entered in the database and existing bridges and cross-sections can be edited. For the bridges in the database the following options are available: review provisional defect reports, enter inspection results, estimate the reliability index, plan maintenance work and estimate costs, plan structural repair work and estimate costs, and review the agenda of inspection for one bridge or all bridges. Further, the database can be updated after repair.

New bridges can be entered and existing bridges can be edited. The general information about the bridges stored in the database contains information about: bridge site, design, budget, traffic, strength, load, deterioration, factors that model the costs, and the cross-sections entered for each bridge. In figures 8–10 are shown examples of strength, load, and cost data.
New cross-sections can be entered and existing cross-sections can be edited. The information stored in the database for each cross-section contains: cross-section identification, geometry of cross-section (detailed description of the reinforcement layers for cross-sections in the deck), failure mode, and load data.

After an inspection the provisional defect reports recorded at previous inspections can be reviewed. A description of the detected defects and measurements of diagnosis methods can be entered. After a repair the databases can be updated. In figure 11 is shown a description corresponding to the observed defect 'rust stain'.

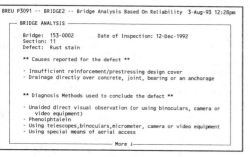

Figure 11: Defect 'rust stain'. Causes and used diagnosis methods.

The reliability index for the bridge can be estimated by the integrated FORTRAN program RELIAB. Both the reliability index where no inspection results are taken into account and the updated

reliability index where all inspections performed for the bridge are taken into account can be estimated.

The following submodules are integrated in BRIDGE 2:

- BRIDGE 2(M) is the maintenance/small repair submodule. This submodule is always used after a current or detailed inspection. It assists in selecting the maintenance work and repair of minor structural defects to be performed and estimate the maintenance costs.

The defects are rated based on the defect classification in terms of rehabilitation urgency, importance of the structure's stability, and affected traffic recorded during the inspection, see figure 12.

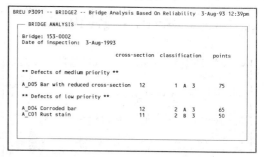

Figure 12: Rating of defects in the maintenance subsystem.

- BRIDGE 2(I) is the inspection strategy submodule. This submodule is always used after a current or detailed inspection. It assists in the decision whether a structural assessment is needed before the next periodic inspection. The decision taken in BRIDGE 2(I) is mainly based on the updated reliability index for the

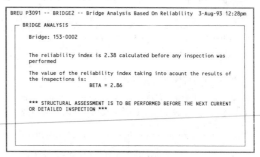

Figure 13: Decision tool related to structural assessment.

bridge calculated by RELIAB (see figure 13). If the value of the updated reliability index for the bridge is acceptable then each of the defects detected at the latest periodic inspection and the combination of defects are investigated. Based on expert knowledge it is investigated whether from a structural point of view a defect or combinations of defects require a structural assessment.

- BRIDGE 2(R) is the repair submodule. This submodule is always used after a structural assessment. It assists in selecting the optimal structural repair technique (including no repair) to be performed, when the repair should be performed, and the number of repairs in the remaining lifetime of the bridge. Further, the expected benefits minus costs are estimated. The repair plan is optimized based on a cost-benefit analysis by the FORTRAN program INSPEC (see figure 14).

```
BREU P3091 -- BRIDGE2 -- Bridge Analysis Based On Reliability  3-Aug-93 12:38pm

 ┌─ BRIDGE ANALYSIS ─────────────────────────────────────────────

   Bridge: 153-0002
   Date of inspection: 12-Dec-1992
   Section: 11
   Defect: A_C01 Rust stain

   Repair technique           Time   Number   Benefits-costs   Repair Cost
   -----------------          ----   ------   --------------   -----------

   R_C02 Concrete Patching    1995     1         26431713          5228
   R_D02 Concrete Patching    1995     1         26303962        145988
   R_D01 Concrete Patching    1995     1         26118570        366800
```

Figure 14: Optimized repair plan for the defect 'rust stain'.

ACKNOWLEDGEMENT

This paper presents results of the CEC supported research project BREU P3091 'Assessment of Performance and Optimal Strategies for Inspection and Maintenance of Concrete Structures Using Reliability Based Expert Systems'.

The partners in the project are:

- CSR, Aalborg, Denmark

- University of Aberdeen, Aberdeen, UK / Sheffield Hallam University, Sheffield, UK

- Jahn Ingenieurbureau, Hellevoetsluis, Holland

- Instituto Superior Técnico, Lisboa, Portugal

- LABEIN, Bilbao, Spain

The expert systems BRIDGE 1 and BRIDGE 2 are implemented by LABEIN. RELIAB and INSPEC are CSR software.

REFERENCES

Aitchison, J. & I.R. Dunsmore 1975. *Statistical Prediction Analysis*, Cambridge University Press, Cambridge.

Lindley, D.V. 1976. *Introduction to Probability and Statistics from a Bayesian Viewpoint*, Vol. 1+2, Cambridge University Press, Cambridge.

Madsen, H.O. 1987. *Model Updating In Reliability Theory*, Proc. ICASP5, pp. 564–577.

Rackwitz, R. & K. Schrupp 1985. *Quality Control, Proof Testing and Structural Reliability*, Structural Safety, Vol. 2, pp. 239–244.

Thoft-Christensen, P. & M.J. Baker 1982. *Structural Reliability Theory and Its Applications*, Springer Verlag.

Thoft-Christensen, P. & Y. Murotsu 1986. *Application of Structural Systems Reliability Theory*, Springer Verlag.

Thoft-Christensen, P. 1992. *A Reliability Based Expert System for Bridge Maintenance*. Proceedings from the Tekno Vision Conference on Road and Bridge Maintenance Management Systems, Copenhagen, No. 25.

Bridge reliability (ongoing research)

Structural Safety & Reliability, Schuëller, Shinozuka & Yao (eds) © 1994 Balkema, Rotterdam, ISBN 90 5410 357 4

Reliability of highway suspension bridge with respect to fatigue

D. Bryja, R. Sieniawska & P. Śniady
Institute of Civil Engineering, Technical University of Wrocław, Poland

ABSTRACT: The life expectancy of a single-span suspension bridge under stochastic excitation caused by traffic flow is studied. The traffic flow is modelled by a stationary Poisson process. The damage process of the bridge is determined by the Palmgren-Miner rule. The expected fatigue life of the bridge is calculated with respect to the reliability of cables and beam. Numerical results for a particular bridge with some practical load cases are presented.

1 BASIC ASSUMPTIONS AND SOLUTIONS

The loading of highway bridges is characterized by the occurrence of millions of repetitive random load events. This type of load causes material fatigue and ultimately damage of the structure (Fryba 1980, Lin 1967, Sieniawska, Śniady 1990). Some problems of fatigue of suspension bridges have been presented by Rakwitz and Faber (1991) and the first crossing problem by Bryja and Śniady (1991b). In this paper the life expectancy of suspension highway bridge with respect to fatigue of the cables and beam subjected to traffic flow is studied. The bridge is modelled by a single-span prismatic girder which is simply supported and underslung by means of vertical hangers to two whipped cables (Fig.1).

The cables are anchored at their ends and movable at their supporting points on the undeformable pylons. The dead-load curve of the cable within the span of the beam forms a parabola $z(x) = 4x(1-x)f/l^2$, the other segments being rectilinear. The mass of the bridge is assumed to be uniformly distributed, constant along the length of the span. In the paper the vertical vibrations of the bridge are considered and estimated with respect to the fatigue reliability problem. The geometrical nonlinearity of the cables is omitted because the effect of non-linear components on the response of the bridge under highway traffic is negligible (Bryja, Śniady 1991).

The linear integro-differential equation of vertical vibrations $w(x,t)$ of the bridge under the live-load $p(x,t)$ has the following form

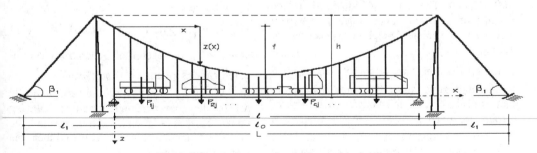

Fig. 1. Scheme of suspension bridge and its loading

$$E_b J_b w^{IV} - 2H_0 w'' + \frac{16k_c f}{l^2}\int_0^l w\,dx + \qquad (1)$$

$$+ c\dot{w} + m\ddot{w} = p,$$

as was detaily derived by Bryja and Śniady (1991a). $E_b J_b$ is the flexural rigidity of the beam, m is the mass of the bridge, c denotes the damping coefficient and $(\cdot)' = \partial/\partial x$, $(\dot{\cdot}) = \partial/\partial t$. The stiffness factor of a single cable is defined as

$$k_c = (8f/l^2)(E_c A_c/L_c), \quad L_c = \int_{(L)} \cos^{-3}\beta\,dx, \quad (2)$$

where E_c is the cable's modulus of elasticity and A_c is the cable's cross-section area. The all-horizontal component of cable tension H(t) is the sum of the dead-load tension $H_0 = mgl^2/16f$ and the vibrational increment $\Delta H(t)$, which can be expressed as

$$\Delta H(t) = k_c \int_0^1 w\,dx, \quad H(t) = H_0 + \Delta H(t). \quad (3)$$

Therefore, the normal stress of cable $S_c(x,t) = S_0(x) + \Delta S(x,t)$ is described by the relationships

$$S_0(x) = \frac{H_0}{A_c \cos\beta}, \quad \cos\beta = 1/\sqrt{1+z'^2},$$
$$\Delta S(x,t) = \frac{\Delta H}{A_c \cos\beta} = \frac{k_c}{A_c \cos\beta}\int_0^1 w\,dx. \quad (4)$$

The bending moment and equivalent normal stress of the beam cross-section x are defined as

$$M = -E_b J_b w'', \quad S_b = -z_w E_b w'' \quad (5)$$

where z_w denotes the level of the cross-section fibres.

The live-load p(x,t) is caused by the random highway traffic taking place in several separate traffic lanes. The traffic is idealized as the passage of trains of concentrated forces with values P_{ij} (vehicle weights) appearing at random times t_{ij}, $j = 1,2,...,n_p$ is the number of the traffic lane. The weights P_{ij} are assumed to be random variables, the probabilistic parameters of the forces, constant for each force of a train, are given as $E[P_{ij}^k] = E[P_j^k]$, where k is the natural exponent and E[ ] denotes the expected value. All forces of a single train are

travelling at the same deterministic constant speed v_j. It is assumed that a stationary Poisson process $N_j(t)$ with the parameter $\lambda_j(t) = \lambda_j$ is the appropriate model for the occurrence times t_{ij} of the random forces moving one by one in the same traffic lane "j". The excitation function for the loading process described above has the following form

$$p(x,t) = \sum_{j=1}^{n_p}\int_{t_0}^t P_j(\tau)\delta[x - v_j(t-\tau)]dN_j(\tau) \quad (6)$$

in which $dN_j(\tau)$ is the increment of the process $N_j(t)$ and δ denotes the Dirac delta function. Employing the theory presented by Bryja and Śniady (1991a), the normal stress response of cables and beam and the speed of their stress variations can be obtained from the formulae, respectively

$$\Delta S(x,t) =$$
$$= \frac{k_c}{A_c \cos\beta}\sum_{j=1}^{n_p}\int_{t_0}^t P_j(\tau)\int_0^1 H_{wj}(x,t-\tau)dx dN_j(\tau), \quad (7)$$

$$\Delta \dot{S}(x,t) =$$
$$= \frac{k_c}{A_c \cos\beta}\sum_{j=1}^{n_p}\int_{t_0}^t P_j(\tau)\int_0^1 \dot{H}_{wj}(x,t-\tau)dx dN_j(\tau), \quad (8)$$

$$S_b(x,t) =$$
$$= -z_w E_b \sum_{j=1}^{n_p}\int_{t_0}^t P_j(\tau)H_{wj}''(x,t-\tau)dN_j(\tau), \quad (9)$$

$$\dot{S}_b(x,t) =$$
$$= -z_w E_b \sum_{j=1}^{n_p}\int_{t_0}^t P_j(\tau)\dot{H}_{wj}''(x,t-\tau)dN_j(\tau). \quad (10)$$

The influence function of displacement $H_{wj}(x,t-t_i)$ is satisfying the equation (1) when the loading process $p(x,t) = \delta[x - v_j(t-t_i)]$ for $0 \le t - t_i \le 1/v_j$ and $p(x,t) = 0$ for $t - t_i > 1/v_j$. The function H_{wj} is approximated by sine series

$$H_{wj}(x,t-t_i) = \sum_n q_n(t-t_i)\sin(n\pi x/l) = q^T s, \quad (11)$$

where $\bar{q}$ is the vector of generalized coordinates and $\bar{s} = [\sin(\pi x/l), \sin(2\pi x/l),...]^T$ is the vector of kinetically and kinematically acceptable approximating functions. Having employed the variational Galerkin method, the equation of motion can be derived and then the generalized solutions $\bar{q}(t-t_i)$, $\dot{\bar{q}}(t-t_i)$ may

be obtained by means of the Newmark numerical method. After integrating in the space domain with taking into account the series (11), the two-dimensional cumulants of stresses and their speeds can be expressed as

$$\kappa_{n,k} = \sum_{j=1}^{n_p} \lambda_j E\left[P_j^{n+k}\right] \int_{t_0}^{t} H_{S_j}^n(x,t-\tau) H_{\dot{S}_j}^k(x,t-\tau) d\tau,$$

(12)

where k, n are the number of cumulants and the influence functions for the cables are equal

$$H_{S_j} = H_{\Delta S_j} = \frac{2k_c l}{A_c \cos\beta} \overline{g}^T \overline{q}_j,$$

$$H_{\dot{S}_j} = H_{\Delta \dot{S}_j} = \frac{2k_c l}{A_c \cos\beta} \overline{g}^T \dot{\overline{q}}_j,$$

(13)

and for the beam

$$H_{S_j} = H_{S_{bj}} = -z_w E_b H_{wj}'' = -\frac{z_w E_b}{l^2} \overline{q}_j^T \{d^2\} \overline{s},$$

$$H_{\dot{S}_j} = H_{S_{bj}} = -\frac{z_w E_b}{l^2} \dot{\overline{q}}_j^T \{d^2\} \overline{s},$$

(14)

in which $\quad \overline{g} = col(\frac{1}{\pi}, 0, \frac{1}{3\pi}, 0, ...)$

and $\{d\} = diag(\pi, 2\pi, ...)$. The cumulants (12) make possible the calculation of two-dimensional joint probability density function $p_2(S, \dot{S}, x, t)$ from the series (see Sieniawska, Śniady 1990). This function will be used to calculating the fatigue life of bridge (Lin 1967, Sieniawska, Śniady 1990). The expected damage per unit time can be derived (Lin 1967) from expression

$$E[D(x,t)] = C^{-1} E[M_T(x,t)] \int_{-\infty}^{\infty} \xi^b f_z(\xi,x,t) d\xi,$$ (15)

where $f_z(S,x,t)$ is the probability density function of the peak magnitude of the stresses, $E[M_T(x,t)]$ is the expected value of the total number of peaks per unit time regardless of their magnitudes and C and b are material constants. For a narrow-band process the above functions can be obtained from such approximate formulae

$$f_z(S,x,t) \cong -E[M_T(x,t)]^{-1} \frac{d}{dS} \int_0^{\infty} \dot{S} p_2(S,\dot{S},x,t) d\dot{S}$$ (16)

and

$$E[M_T(x,t)] \approx \int_0^{\infty} \dot{S} p_2(0,\dot{S},x,t) d\dot{S}.$$ (17)

The expected value of fatigue life for the steady state solution $(t_0 \to -\infty)$ and evaluating fatigue damage accumulation by the linear damage law (Palmgren-Miner's rule) is given by

$$T = E^{-1}[D]$$ (18)

and can be calculated with respect to the reliability of cables or the beam.

2 NUMERICAL RESULTS AND CONCLUSIONS

To illustrate the algorithm presented above the expected fatigue life of a suspension bridge under traffic flow is calculated versus vehicle speed and the number of traffic lanes. The calculation was made for the bridge span l=300 m and for the cross-section of cables x=0, of beam x=0,25l and x=0,5l. Other design parameters of the bridge are given in work of Bryja and Śniady (1991); material properties needed to fatigue live calculation are equal $b = 4$, $C_c = 4 \cdot 10^{12}(N/mm^2)^b$, $C_b = 1,59 \cdot 10^{12}(N/mm^2)^b$. The probabilistic parameters of forces moving in a single lane are taken from Fujino and Bhartia (1987) for the "C" traffic composition and the arrival rate λ_j is chosen respectively for each speed v_j. Two or four traffic lanes with the same load are considered. The effect of the vehicle speed on the mean value and the standard deviation of the cable stress is shown in Fig. 2. From Fig.3 one can see, that the bridge's life at first increases very quickly and becomes nearly fixed when the vehicle speed is up 50 m/s. This phenomenon corresponds to the shape of the curve evaluated for the standard deviation of stress. The results in Fig.3 are obtained for one and three terms of the series, which is employed to determine two-dimensional joint probability density function. It is noticeable, that the use of only one series term can be well accepted and for this case the expected fatigue life of the bridge with respect to the reliability of the beam is shown in Fig. 4.

951

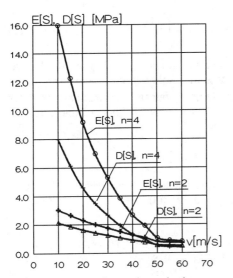

Fig.2 Expected value and standard
deviation of the cable stress

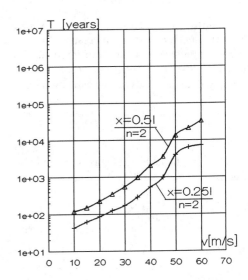

Fig.4.Expected fatigue life of bridge beam.

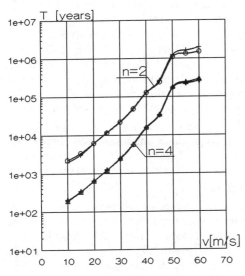

Fig.3. Expected fatigue life of bridge cables

One can see that the cross-section x=0,25l is
more competent in the reliability analysis of the
bridge as the midspan cross-section though the
expected value of stresses is greater at the
midspan. From the comparison of the results
shown in Figs 3 and 4 it follows that the
expected fatigue life of the cables is more
sensitive on the vehicle speed changes as the
expected fatigue life of the beam.

REFERENCES

Bryja, D. & Śniady, P. 1991. Spatially coupled
vibrations of a suspension bridge under
random highway traffic, *Earth. Eng. and
Struct. Dyn.* 20: 999-1010.
Bryja, D. & Śniady, P. 1991. Spatial vibrations
and reliability of a suspension bridge under
random traffic flow. *Proc. CERRA ICASP6:*
857-864. Mexico City: Ed. Inst. of Eng.,
UNAM.
Fryba, L. 1980. Estimation of fatigue life of
railway bridges under traffic loads. *Journ. of
Sound and Vibration* 70: 527-541.
Fujino, Y. & Bhartia, B.K. 1987. Effect of
multiple presence of vehicles on fatigue
damage of highway bridges. *Proc. CERRA
ICASP5:* 1157-1164.
Lin, Y.K. 1967. *Probabilistic theory of struc-
tural dynamic.* New York: McGraw-Hill.
Rakwitz, R. & Faber, M.H. 1991. Reliability of
parallel wire cable under fatigue. *Proc.
CERRA ICASP6 :* 166-175. Mexico City: Ed.
Inst. of Eng., UNAM.
Sieniawska, R .& Śniady, P. 1990. Life expec-
tancy of highway bridges due to traffic load.
Journ. of Sound and Vibration 140: 31-38.

Structural Safety & Reliability, Schuëller, Shinozuka & Yao (eds) © 1994 Balkema, Rotterdam, ISBN 90 5410 357 4

Repair service life evaluation based on imprecise information

K.C.Chou
Department of Civil Engineering, The University of Tennessee, Knoxville, Tenn., USA

Paul C.Hoffman
Department of Civil Engineering, Villanova University, Pa., USA

ABSTRACT: The service life reassessment of a repaired concrete deck is accomplished generally through a visual inspection and, in some cases, nondestructive testing. While traditional statistical analysis provides a means to consider the quantitative result of a nondestructive test, it cannot be applied directly to the qualitative reliability rating of a nondestructive test nor the linguistic rating from a visual inspection. A combination of Bayes' theorem and fuzzy set theory is proposed for the reduction of both qualitative and quantitative information to an overall service life prediction of a concrete deck repair.

1. INTRODUCTION

The updating of the service durability and reliability (service life) of a repair for even one component of a concrete bridge is dependent on imprecisely defined variables. For example, the quality of repair may be evaluated through a visual inspection as "good". The evaluation of such a linguistic term with respect to service life prediction is not possible through classical statistical procedures.

Recent research results [Chou and Yuan, 1993; Hoffman, et al., 1990] suggest that a combination of Bayes' theorem and fuzzy set theory can be used to analyze both crisp and imprecise information. Herein, an algorithm integrating fuzzy set and Bayesian method is proposed for the assessment of the renewed service life due to a repair.

2. PROBLEM SPECIFICS

First, bridge deck performance is evaluated typically through visual inspection of workmanship according to some checklist format. The final statement with respect to performance is a linguistic rating which is vague and imprecise. For example, after a deck is repaired, an inspector may rate, according to some checklist, the expected performance of the deck with respect to fair, good or very good categories.

Second, a nondestructive test (NDT) may be used to determine the concrete compressive strength, f'_c, as a surrogate variable for quality. The probe penetration test is one of the more popular in situ NDT techniques for determining f'_c [ACI, 1988]. Typically, a linear regression relationship between penetration depth and compressive strength is available, Figure 1. The result of the probe penetration test is a crisp ordered pair (depth of penetration and f'_c). However, the capability (or reliability) of the probe penetration test to accurately measure f'_c is vague. In an American Concrete Institute Committee 228 report, the reliability of the penetration test was listed as good [ACI, 1988].

The problem is how does one modify the probabilities of occurrences from a visual inspection with nondestructive test (NDT) information which consist of crisp and imprecise information.

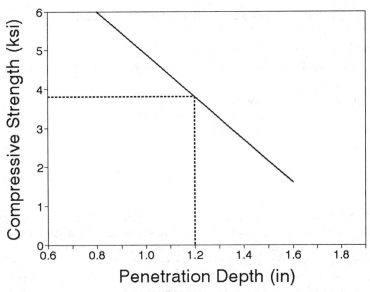

Figure 1. Typical relationship between penetration depth and compressive strength.

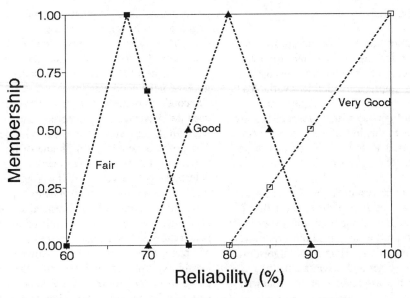

Figure 2. Typical membership function for reliability.

3. BAYESIAN ANALYSIS

For simplicity, the expected performance of a repaired deck with respect to the classification of fair F, good G, or very good VG is considered as a crisp event. The corresponding probability of occurrence would be the apriori probability based on the site specific conditions such as annual average daily traffic, and meteorological conditions.

Next, conditional probabilities need to be developed for the compressive concrete

strength f'_c in each category of expected repair service life. These probabilities can be developed through laboratory and field experiments. The posterior probability can then be obtained by incorporating the crisp event, f'_c, measured from probe penetration test. A basic application of Bayes' theorem yields

$$P[S \mid F_c] = \frac{P[F_c \mid S]P[S]}{\sum\limits_i P[F_c \mid S_i]P[S_i]} \tag{1}$$

in which S = expected service life; F_c = measured concrete compressive strength.

As an illustration, consider the situation where the inspector's evaluation of a repaired deck's expected service life is classified as good with a probability of occurrence of 0.6; while P[S=F] = 0.1 and P[S=VG] = 0.3. An in-situ probe penetration test indicates a penetration depth of 1.2 inches which corresponds to F_c = 3.8 ksi (Fig. 1). Now let D = ratio of measured concrete strength to design strength. If the design strength is 4 ksi, then D = 0.95. For illustrative purpose, the conditional probability P[D|S], which would be developed experimentally, are assumed (Table 1). The posterior probability of having a good service life that the measure strength is 3.8 ksi becomes

$$
\begin{aligned}
P[S{=}G \mid F_c{=}3.8ksi] &= P[S{=}G \mid D{=}0.95] \\
&= \frac{P[D{=}0.95 \mid S{=}G]P[S{=}G]}{\sum\limits_i P[D{=}0.95 \mid S_i]P[S_i]} \\
&= \frac{0.15(0.6)}{0.2(0.1){+}0.15(0.6)} \\
&= 0.818
\end{aligned}
\tag{2}
$$

Thus, with the knowledge of in-situ concrete strength, the probability that repaired deck will have a good service life is updated.

4. FUZZIFIED BAYESIAN APPROACH

If the probe penetration test results were abso-

lutely accurate in measuring the in-situ concrete strength, Bayes' theorem alone would be adequate to compute a posterior probability. As discussed previously, the probe penetration test has a reliability (capability of measuring the in-situ strength accurately) of good (ACI, 1988). Thus the posterior probability of 0.818 computed in Eq. 2 needs to be modified. Since the reliability rating of the test is imprecise, Eq. 1 is not applicable here.

Fuzzy set theory is a tool that allows one to quantify imprecise information. Membership functions for the linguistic terms for the reliability of NDT techniques need to be developed (Fig. 2). For illustration convenience, the membership functions can be formulated in discrete terms. The values are denoted by symbols while the lines are used to show the variation of the membership functions. Research has shown that such functions are elicited readily from experts [Hoffman, et al., 1990]. For computational convenience, the membership function can be formulated in discrete terms, for instance, "good" can be defined as

$$\mu_R(good) = 0.5 \mid 75\% + 1.0 \mid 80\% \atop + 0.5 \mid 85\% \tag{3}$$

(Note that the symbol "+" in Eq. 3 does not denote addition). If the reliability of the NDT test is 75%, then the posterior probability given by Eq. 2 would be modified to 0.75(0.818) = 0.614. According to Eq. 3, there is only a 50% belonging of having a 75% reliability for medium, this implies that there is also a 50% belonging that the posterior probability for good service life is 0.614. Implementing Eq. 3 into Eq. 2 yields a posterior membership function for expected good repair service life

$$\mu_S(good) = 0.5 \mid 0.614 + 1.0 \mid 0.655 \atop + 0.5 \mid 0.695 \tag{4}$$

Consequently, the vague rating, "good", from the visual inspection has been quantified with the imprecise NDT results.

Table 1. Conditional probability of D given service life

Service	D = measured f'_c / design f'_c					
Life	0.85	0.90	0.95	1.0	1.05	1.10
Fair	0.10	0.15	0.20	0.5	0.05	0.0
Good	0.0	0.05	0.15	0.6	0.15	0.05
Very Good	0.0	0.0	0.0	0.7	0.2	0.1

5. REMARKS

The implementation of imprecise as well as quantitative data in the reassessment of servicelife due to a repair is discussed. This was for illustration purpose. The authors are pursuing the fuzzification of the service life categories and the apriori probabilities. Combination of NDT results with additional imprecision are being considered in integrating fuzzy set with Bayes' theorem.

REFERENCES

ACI (1988) "In-place methods for determination of strength of concrete", ACI 228.1R-89, American Concrete Institute, Detroit, MI.

Chou, K.C. and Yuan, J. (1993) "Fuzzy-Bayesian approach to reliability of existing structures", Journal of Structural Engineering, ASCE, Nov. (in print).

Hoffman et al. (1990) "Fuzzy assessment at bridge deck repair techniques", Proc. of NAFIPS'90, Vol.1, Univ. of Toronto, June.

Structural Safety & Reliability, Schuëller, Shinozuka & Yao (eds) © 1994 Balkema, Rotterdam, ISBN 90 5410 357 4

Predicting critical chloride levels in concrete bridge decks

P.C. Hoffman
Villanova University, Pa., USA

R.E. Weyers
Virginia Polytechnic Institute and State University, Blacksburg, Va., USA

ABSTRACT: The transport of chloride through concrete is modeled as apparent diffusion. The statistics for apparent diffusion and surface chloride exposure are presented. Lognormal density functions for apparent diffusion and surface exposure were used in Monte Carlo simulations. Chloride concentrations at specific time periods are described by gamma probability density functions.

INTRODUCTION

Once the chloride concentration in a reinforced concrete structural component reaches a critical level at the reinforcement depth, corrosion may commence. If corrosion is initiated, the volume of the corrosion product produced is greater than that originally occupied by the steel. As a consequence, eventually spalling occurs due to the ever increasing tensile stresses.

In order to provide timely maintenance and repair, a reasonable estimate of the chloride content in the concrete is needed. It is recognized that field measurements of chloride transport include numerous factors such as ubiquitous microcracking. Nevertheless, field measurements seem to follow a diffusion like distribution with depth. Consequently, for the research project described herein, an apparent diffusion process was accepted as an appropriate model of the actual transport phenomenon.

The analysis of numerous historical records was undertaken for the expressed purpose of summarizing representative apparent diffusion constants and corrsponding surface chloride exposures (Hoffman, 1992). The records were compiled from several geographical areas (sixteen states) in the United States of America. Over twenty seven hundred samples taken from three hundred twenty one bridges were analyzed. Generally, each sample consisted of chloride concentrations measured at one half inch increments to an overall depth of two and a half to three inches.

COMPUTATIONAL APPROACH

While the top surface of any bridge deck is subjected to a continual changing chloride exposure, the near-surface chloride concentration, approximately the one half inch depth, was assumed to be virtually constant (Cady & Weyers, 1983, and Hoffman, 1992). Consequently, a closed form solution to the fundamental differential equation of diffusion for an isotropic semi-infinite medium was deemed applicable (Crank, 1975),

$$C_{x,T} = C_o \left[1 - erf\left(\frac{X}{2\sqrt{D_c T}}\right)\right] \qquad (1)$$

where $C_{X,T}$ = Chloride Concentration with depth and time.
C_o = Constant Surface Concentration
erf = Error function
X = Depth
D_c = Diffusion Constant
T = Time of exposure to C_o

As long as the chloride concentration at the bottom surface remained unchanged, Equation 1 was determined to be applicable also to decks of finite thickness.

A least squares fit of Equation 1 to each of the three hundred and twenty one bridge deck data sets was conducted. Each data set consisted of several samples, typically five to ten samples per bridge. The best single diffusion constant for

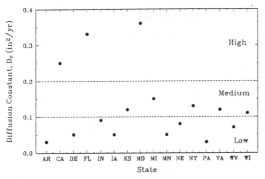

Figure 1: Mean Apparent Diffusion Constants

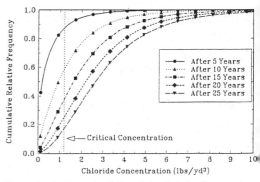

Figure 4: An Example of Cumulative Relative Frequencies

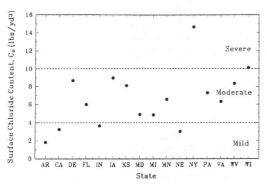

Figure 2: Mean Surface Chloride Concentrations

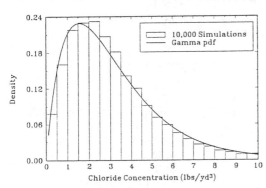

Figure 3: An Example of Chloride Concentrations at 2 Inch Depth After 25 years

each data set was selected on the basis of the cumulative sum of squares error.

A set of D_c values over a fixed range was defined. The sum of squares error (SSE) for each sample in a data set for each D_c value was computed. Then the cumulative SSE (sum of the sample SSE values) was totalled for each D_c.

Subsequently interpolation with a cubic spline fit was used to determine the D_c value that would result in the minimum cumulative SSE, that is the best fit for the data set under study.

COMPUTED STATISTICS

The first objective of the research was the determination of mean apparent diffusion constants for the states included in the study, Figure 1. The results indicated substantial differences in mean values between states. Likewise, the coefficients of variation of the computed apparent diffusion constants for the states ranged from 0.3 to 1.6 with a mean of 0.75. These variations were expected in that such variables as water-cement ratio (Herald, 1989) and temperature (Page, 1981) have been found to cause large variations in the mass transport of chloride through concrete. Unfortunately, the research database did not include water-cement ratios and temperature readings. Nevertheless, the apparent diffusion constant approach was found to be realistic in that the computed values are representative of the many variables that influence the transport of chloride through concrete. The results seemed to indicate three general categories or ranges of values for apparent diffusion constants, Figure 1.

During the initial phases of the research project, several models for the near-surface chloride concentrations increasing with time were considered. In particular, near-surface chloride concentration was assumed to vary with the square root of time, (Uji, Matsujka, and Maruya, 1990). However, with the acquisition of some time series data, it was concluded that assuming a constant near-surface concentration

958

was the best practical and realistic approach (Hoffman, 1992). The mean near-surface chloride contents, C_o, were grouped according to severity by state. First, the average C_o was computed for each bridge by state. Then an average near-surface chloride content for each state was computed. Comparisons of the magnitudes of the mean surface chloride contents by state with a Vehicle Corrosion Environment Map of the United States of America and road salt usage data by state during 1981-1982 and 1982-1983 resulted in the identification of three surface exposure categories, Figure 2.

MONTE CARLO SIMULATION

With the assumption that both the apparent diffusion constant and surface exposure are lognormal random variables, the statistics were the basic input to Monte Carlo simulations, Equation 1. For example, the mean values for D_c and C_o were set to 0.1 in²/yr and 10 lbs/yd³ respectively. Likewise the coefficients of variation were set to 0.75 and 0.5 respectively based on the least squares fit results. These values are presented solely for illustrative purposes and are not intended to be representative of any specific state.

The simulation results of distribution of chloride at two inches below the surface after twenty five years of exposure, C_o, support a gamma probability density function, Figure 3. Indeed, the gamma function was found to describe the distribution of chloride at all depths and time periods simulated.

The cumulative relative frequency distributions for the aforementioned illustration provide an example of the use of computed statistics for D_c and C_o, Figure 4. Given that 1.2 lbs/yd³ of chloride initiates corrosion (Cady & Weyers, 1983), the percent of a concrete deck with reinforcement at two inches subjected to corrosion is ascertained readily.

CONCLUSIONS

For the particular states included in the study, the corresponding computed statistics provide for realistic decision models. The estimate for the percent of bridge decks in which a critical chloride concentration is exceeded can be determined. For other states the statistics provide best estimate categories, Figure 1 and Figure 2. A comparison of the state being evaluated with the list of states included in the research should lead to a reasonable estimate of the appropriate statistics.

One area requiring additional research is the development of the statistics for the critical chloride concentration that may initiate corrosion. Based on a review of the appropriate technical literature it seems that the critical chloride concentration is between 1.2 lbs./cu.yd. and 2.0 lbs./cu.yd. On the other hand, it has been suggested that a critical chloride concentration specification is erroneous (Vaysburd, 1993).

REFERENCES

Cady, P.D. & R.E. Weyers 1983. Chloride Penetration and Deterioration of Concrete Bridge Decks. *Cement, Concrete and Aggregates* 5(2): 81-87.

Crank, J. 1975. *The Mathematics of Diffusion, (Second Edition), Oxford University Press.* New York: Oxford Univ. Press.

Herald, S.E. 1989. The Development of a Field Procedure for Determining the Chloride Content of Concrete and an Analysis in the Variability of the Effective Diffusion Constant. Master's Thesis: Virginia Polytechnic and State University.

Hoffman, P. C. 1992. *Statistics for Apparent Diffusion Constants and Chloride Surface Concentrations.* Internal Final Report, VPI&SU subcontract: Strategic Highway Research Project C103.

Page, C.L., N.R. Short & A. El Tarras 1981. Diffusion of Chloride Ions in Hardened Cement Pastes. *Cement and Concrete Research* 11(3): 395-406.

Uji, K., Y. Matsudka & T. Maruya 1990. Formulation of an Equation for Surface Chloride Content of Concrete due to Permeation of Chloride. *Corrosion of Reinforcement on Concrete*: 258-267. New York: Elsevier Applied Science.

Vaysburd, A.M. 1993. Some Durability Considerations for Evaluating and Repairing Concrete Structures. *Concrete International Design and Construction:* 15 (3): 29-35.

Structural Safety & Reliability, Schuëller, Shinozuka & Yao (eds) © 1994 Balkema, Rotterdam, ISBN 90 5410 357 4

Distribution of the fatigue life of prestressed concrete bridges

Mohammad A. Khaleel
Battelle Pacific Northwest Laboratories, Richland, Wash., USA

Rafik Y. Itani
Washington State University, Pullman, Wash., USA

ABSTRACT: This paper presents a method for calculating the fatigue life of partially prestressed concrete girder bridges under stochastically varying traffic loads. Under random loading with Poisson arrivals, a distribution for the cumulative damage is derived. The bridge is assumed to fail in fatigue due to the initiation of observable cracks in the girders. A girder is assumed to be a series system consisting of prestressing strands, reinforcing bars, a cast-in-place slab and a precast girder. The probabilities of survival for the structural components and bridge system are calculated. The contributions to cumulative damage are obtained for each traffic stream composed of a single category of vehicle.

1 INTRODUCTION

Partially prestressed concrete (PPC) girders are allowed to crack under service loads. Therefore, the stresses in the constituent materials fluctuate due to the opening and closing of the cracks posing a potential fatigue problem. Experimental data on the fatigue life of the bridge components show considerable scatter, due to both the stochastic nature of the imposed loading and the variability in their strength. Therefore, probabilistic concepts are employed to analyze and design partially prestressed concrete girders to ensure adequate fatigue resistance.

The following assumptions are made:

1. All permutations of load order are equally likely with incremental damage always being non-negative.

2. A finite number of different load categories are possible on the bridge. They are obtained by considering trucks with different numbers of axles, weights, and configurations.

3. The counting process for the random number of loads of each category of trucks is a stationary Poisson process.

4. The material fails due to the accumulation of an incremental damage past some critical level ω.

2 STOCHASTIC DAMAGE MODEL

The total cumulative damage for truck category i after $N_i(t)$ cycles is $\sum_{j=1}^{N_i(t)} X_{ij}$. The incremental damage of truck j of category i is X_{ij}. If Y_t is the total cumulative damage over all truck categories (k truck categories are possible), then

$$Y_t = \sum_{i=1}^{k} \sum_{j=1}^{N_i(t)} X_{ij} \qquad (1)$$

where the expected value of the incremental damage is $EX_{ij} = \mu_i$ and the variance is $var(X_{ij}) = \sigma_i^2$ (i=1,..k). Using Wald's lemma (Wald 1944), the expected total cumulative

damage and its variance are given by

$$EY_t = t \sum_{i=1}^{k} \lambda_i \mu_i = t \ \mu \qquad (2)$$

and

$$Var(Y_t) = t \sum_{i=1}^{k} \lambda_i (\sigma_i^2 + \mu_i^2) = t \ \sigma^2 \qquad (3)$$

Since fatigue is measured in millions of cycles, $\sum_{j=1}^{N_i(t)} X_{ij}$ can be approximated very well by the normal distribution. If T is a random time (or number of cycles) at which a critical level of damage is exceeded for the first time then

$$P[T \geq t] = P[Y_t \leq \omega] =$$

$$P \left[\frac{Y_t - t\mu}{\sqrt{t} \ \sigma} \leq \frac{\omega - t\mu}{\sqrt{t} \ \sigma} \right] \qquad (4)$$

If $\beta = \dfrac{\omega}{\mu}$ and $\alpha = \dfrac{\sigma}{\sqrt{\omega \mu}}$, then

$$P[T \leq t] = 1 - \Phi \left[\frac{1}{\alpha} \left(\sqrt{\frac{t}{\beta}} - \sqrt{\frac{\beta}{t}} \right) \right] \qquad (5)$$

This result provides an exact expression for fatigue life under stationary stochastic load. This distribution is the B-S distribution (Birnbaum and Saunders 1969). The scale parameter β, and shape parameter α, are given by (for derivation (Khaleel 1992)):

$$\beta = \frac{1}{\displaystyle\sum_{i=1}^{k} \frac{\lambda_i}{\beta_i}} \qquad (6)$$

$$\alpha = \beta \sum_{i=1}^{k} \frac{\lambda_i}{\beta_i} \left[\alpha_i + \frac{1}{\beta_i} \right] \qquad (7)$$

Eq. 6 provides an exact expression for the resulting distribution of fatigue-life under

stationary stochastic loads and gives the widely used Miner-Palmgren rule as the median life for that distribution while the shape parameter (coefficient of variation) is expressed in Eq. 7.

3 PROBABILISTIC DISTRIBUTIONS OF BASIC VARIABLES

Experimental data on the fatigue life of plain concrete, ordinary steel and prestressing steel from constant and variable amplitude loading is collected. The Weibull, Lognormal and B-S distributions were compared by computing statistical estimates of their parameters for each material under each stress regime. The goodness-of-fit of each distribution was measured (Khaleel 1992). It was found that the B-S distribution is appropriate for describing the fatigue life of plain concrete, steel, and prestressing strands. The parameters of the B-S distribution were derived based on complete and censored samples. Then, the parameters were expressed as functions of the imposed stress regime by using log-linear combinations of the actual stresses (see Al-Zaid 1986). The scale parameters for prestressing steel, concrete, and reinforcing steel are respectively:

$$\log \beta = 11.383 - 3.487 \log[max\varsigma - min\varsigma] \qquad (8)$$

$$\log \beta = 13.410 - 12.038 \frac{max\varsigma}{f_c'} \qquad (9)$$

$$\log \beta = 10.235 - .1641 \ max\varsigma + .1497 \ min\varsigma \qquad (10)$$

where $\varsigma(t)$ is a cyclic loading function.

Data on the variations in truck configurations, gross weight, load impact, and lane position were collected from weight-in-motion studies (Moses and Ghosn 1983). The truck weights were considered to be well described by the inverse Gaussian distribution (Chhikara and Folks 1989). Load-impact factors developed by Hwang and Nowak (1991) were found to follow the Lognormal distribution and were used in this study. Khaleel and Itani (1990) investigated the allocation of the live-load to the girders in skew bridges using finite element analysis. Empirical relations

for the lateral distribution and skew reduction factors were derived and assumed to follow the Lognormal distribution. The weight of the precast girders, concrete slab, and asphalt are all assumed to be normally distributed. Only one truck is allowed on the bridge at time t and its lateral location follows the lognormal distribution with a 2ft mean distance from the curb. The spacings between the axles of a truck are assumed to be uniformly distributed.

4 CALCULATING β_i AND α_i

To calculate the scale and shape parameters for each traffic stream composed of a single category of vehicles, a truck configuration from that category is selected at random. A PPC bridge of specified design and geometry is chosen. Distributions of the gross weight, the allocation of load axles, the impact, and lane position of that vehicle are determined based on the weight-in-motion information. Based on linear finite element analysis the mean values of the lateral distribution (D) and skew reduction (K) factors are calculated (Khaleel and Itani 1990). It was found that D and K follow the Lognormal distribution with a coefficient of variation of 8%. Using nonlinear stress analysis of PPC girders (Al-Zaid 1986) for the chosen bridge structure and the vehicle information, the stress response vector for all components are computed simultaneously. This gives the maximum stress fluctuation within each structural component. Minimum stress fluctuations are found by considering the dead load of the structure. Using Eqs. 8, 9 and 10, β_{ij} (scale parameter for truck j in category i) are calculated for prestressing steel, concrete, and reinforcing bars, respectively. Similar equations are used to calculate α_{ij} (Khaleel 1992). Based on Eqs. 6 and 7 β_i and α_i are calculated for each truck category (i.e.

$$\beta_i = 1/\sum_{j=1}^{N_i(t)} 1/\beta_{ij},$$

$$\alpha_i = \beta_i \sum_{j=1}^{N_i(t)} 1/\beta_{ij}[\alpha_{ij} + 1/\beta_{ij}]). \text{ Thus for the}$$

entire traffic burden the composite β and α are computed for each structural component. The probability of failure based on a specified number of cycles is computed.

5 SYSTEM RELIABILITY

PPC girders fail in fatigue by crushing of concrete or rupture of steel. This is a brittle failure. Therefore, redistribution of stresses within the section after failure of any component will result in a subsequent failure of other component. A PPC girder is assumed to be a series system. The components of a series system in this case are the prestressing strands, cast-in-place slab, precast beam, and reinforcing bars. The system probability of failure is given by:

$$\max_{i=1}^{4} P[E_i] \le P_f \le 1 - \prod_{i=1}^{4} (1 - P[E_i]) \qquad (11)$$

Where E_i is the event of failure of the ith structural component (i.e., rupture of prestressing strands, rupture of reinforcing bars, crushing of cast-in-place slab or crushing of precast girder in compression).

6 RESULTS AND CONCLUSIONS

Extensive parametric studies have been con-

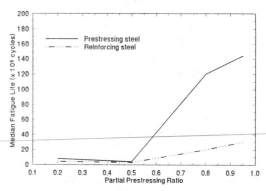

Fig. 1. The effect of PPR on median fatigue-life of prestressing and reinforcing steel

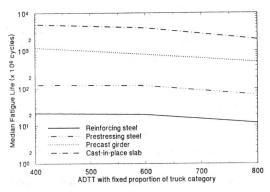

Fig. 2. The effect of ADTT on the median life of the girder components

ducted on several bridges with the same moment resistance but different partial prestressing ratios (PPR). Bridges with low PPR have a much shorter fatigue life than those designed with a high PPR as shown in Fig 1. PPC bridges should be designed for PPR higher than 0.75. The probability of failure of reinforcing bars is the highest among the girder components as shown in Fig. 2. Increasing the average daily truck traffic (ADTT) with fixed proportions of truck categories reduces the composite median life as shown in Fig. 2.

REFERENCES

Al-Zaid, R. 1986. Fatigue reliability of prestressed concrete girders" Ph.D. Thesis, The University of Michigan.

Birnbaum, Z.W. & S.C. Saunders. 1969. A new family of life distributions," *Journal of Applied Probability*, 6, 319-327.

Chhikara, R.S. & J.L. Folks. 1989. *The inverse gaussian distribution: theory, methodology, and applications*," New York: Marcel Dekker, Inc.

Hwang, E.-S. & A. Nowak. 1991. Simulation of dynamic load for bridges" *J. Struc. Div., ASCE*, 117(5), 1413-1434.

Khaleel, M.A. & R.Y. Itani. 1990. Live-load moments for continuous skew bridges," *J. Struc. Div., ASCE*, 116(9), 2361-2372.

Khaleel, M.A. 1992. Reliability-based analysis, sensitivity and design of partially prestressed concrete systems," Ph.D. Thesis, Washington State University, Pullman, Washington.

Moses, F. & M. Ghosn. 1983. Instrumentation for weighing trucks-in-motion for highway bridge loads," Report No. FHWA/OH-83/001, Case Western Reserve University, Cleveland, Ohio.

Wald, A. 1944. On cumulative sums of random variables, *Ann. Math. Statist.*, 15, 283-296.

Structural Safety & Reliability, Schuëller, Shinozuka & Yao (eds) © 1994 Balkema, Rotterdam, ISBN 90 5410 357 4

Probabilistic models of bridge live loads

Sang-hyo Kim & Hak-joo Hwang
Yonsei University, Korea

Hung-seok Park
Korea Institute of Construction Technology, Seoul, Korea

ABSTRACT: The most important loading acting on bridges is traffic loading, especially for bridges with short spans. Because of the variety of vehicle types, their weights, and traffic patterns, the estimation of lifetime load history appears to be a critical problem that calls for probability-based load modeling and simulation techniques.

To establish a framework of probability-based safety evaluation of highway bridges, this paper develops a methodology dealing with probabilistic traffic load modeling technique as well as simulation techniques of traffic flow and their effects. Various paramaters, such as heavy vehicle proportion, consecutive vehicle composition, speed of traffic flow, headway distribution, etc., are included and their sensitivities are investigated.

1 INTRODUCTION

Traffic loads show highly uncertain nature in individual vehicle weight and traffic flow. In recent years, an increasing number of overloaded heavy trucks were reported from many countries. This causes serious problems in ultimate load carrying capacities as well as fatigue failures of highway bridges. Therefore, it is required to evaluate the performance of bridges constructed with current and past design codes.

Because of the variety of vehicle types, their weights, and traffic patterns, the estimation of lifetime load history appears to be a critical problem that calls for probability-based load modeling and simulation techniques(Ghosn 1984, Jacob 1989, Kameda 1989 & Miller 1983).

To establish a framework of probability-based safety evaluation of highway steel bridges, this paper investigates a methodology dealing with probabilistic traffic load modeling. The proposed traffic load modeling techniques has been applied to investigate lifetime maximum traffic load effects on bridges as well as traffic-induced fatigue damage of steel bridges.

2 TRAFFIC LOAD MODELING

Traffic load models adopted in this study consist of models of individual vehicle gross weight, vehicle configuration, axle weight distribution, traffic volume, traffic composition, headway and heavy vehicle arrival modes, which are assumed to play important roles in bridge safety evaluation.

Five vehicle models have been selected to represent the commercial vehicle types and their axle configurations. P-type represents small vehicles including passenger cars and small trucks, B-type for buses, T-type for trucks with single axle, TT-type for trucks with tandom axle, and finally ST-type for semi-trailers.

Vehicle gross weights have been modeled in terms of probability distributions based on the axle weight survey data collected on domestic roads. About 22,000 vehicles were recorded, among which heavy trucks(T, TT, ST) are more than 19,000 units.

In Table 1, either uni-modal or bi-modal distributions has been recommended to represent properly the real situations. In general, the first mode stands for either empty or light-loaded vehicles, and the second for heavy-loaded vehicles. In cases of heavy trucks, such as T, TT, and ST-types shown in Fig.1, another distribution has been used to model the upper tail region. Weight models are limited with both lower and upper bounds.

According to the survey data on traffic flow, consecutive vehicles arriving at a bridge show high correlation in vehicle type with each other. In particular, the characteristics of heavy vehicles travelling in group have been clearly examined. This characteristic is caused by different driving abilities of vehicles and assumed to have significant effects to safety margins reserved in existing bridges, especially with medium or long span.

Based on the survey data 3 traffic patterns are modeled in terms of transition probability matrix depending on proportion of heavy vehicles (T, TT, and

ST-type). Models of 3 representative heavy vehicle proportions are tabulated in Table 2 . The compositions in Table 1 stand for the one-step transitional probability from vehicle type in the extreme left column to vehicle type in the upper row. Whereas, compositions of the extreme right column are those counted without considering the type of preceding vehicle.

3 FATIGUE DAMAGE ESTIMATION

This section presents a study of traffic-induced fatigue damage analysis based on the proposed traffic models. To demonstrate numerical examples, fatigue strength models proposed by JSSC(Japanese Society of Steel Construction) have

been used(JSSC 1989). The models have different endurance limits for material categories as well as stress characteristics(such as constant stress range and variable stress range).

Among many methods of fatigue damage accumulation proposed, Miner's rule appears to be a logical choice applicable to the type of random stress history. According to Miner's rule, the fatigue limit state for variabe amplitude loading is stated as $\Sigma\ C_i/C = 1$, where C is material constant and C_i represents fatigue damage caused by a group of i-th traffic pattern. In this study, the lifetime traffic flow is subdivided into groups with different levels of fatigue damage accumulated under same traffic volum. the parameter used for subdividing is the speed of moving flow, that is, headway distribution and impact effects.

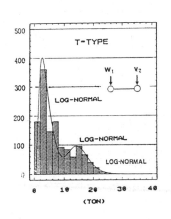 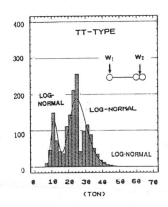 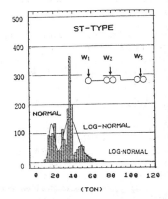

Fig.1 Vehicle weight distribution (1ton=9.8KN)

Table 1. Vehicle gross weight models (1ton=9.8KN)

vehicle type	mode	distribution type	parameters		lower bound (ton)	upper bound (ton)	modification factor
			μ, λ	σ, ζ			
P	1	L-N	0.398	0.317	0.7	5.0	1.0
B	1	Normal	4.1	1.02	1.4	17.1	0.098
	2	Normal	11.6	1.54	4.0	24.0	0.902
T	1	L-N	1.338	0.620	1.25	24.1	0.733
	2	L-N	2.721	0.221	1.25	24.1	0.277
	3	L-N	2.490	0.260	23.5	40.0	1.0
TT	1	L-N	2.467	0.178	7.3	41.3	0.219
	2	L-N	3.253	0.203	7.3	41.3	0.781
	3	L-N	3.240	0.210	41.3	65.0	1.0
ST	1	Normal	18.5	3.00	11.3	63.4	0.260
	2	L-N	3.650	0.202	11.3	63.4	0.742
	3	L-N	3.420	0.260	59.7	105.0	1.0

Table 2. Traffic flow models

unit : %

heavy vehicle proportion	Type	consecutive vehicle composition					simple composition
		light vehicles		heavy vehicles			
		P	B	T	TT	ST	
15	P	83.68	3.10	6.67	4.83	1.72	81.59
	B	77.95	7.48	6.65	6.17	1.75	3.25
	T	71.90	3.22	14.41	7.91	2.56	7.54
	TT	70.90	3.01	10.27	12.84	2.98	5.65
	ST	68.96	3.36	10.69	9.39	7.60	1.97
25	P	75.74	2.77	10.91	7.79	2.79	71.27
	B	69.92	6.63	10.77	9.86	2.82	2.83
	T	60.02	2.66	21.71	11.78	3.82	12.90
	TT	58.77	2.46	15.37	18.98	4.42	9.64
	ST	56.55	2.72	15.83	13.74	11.16	3.36
35	P	69.46	2.48	14.33	10.14	3.59	62.88
	B	63.71	5.89	14.05	12.74	3.61	2.46
	T	51.83	2.24	26.87	14.44	4.62	17.31
	TT	50.51	2.06	18.93	23.16	5.34	12.92
	ST	48.30	2.27	19.37	16.66	13.40	4.43

2-lane bridges of various span lengths with 4 steel plate girders are designed for demonstration. Fig.2 shows the results of the sensitivity study on proportion of heavy vehicles and the effect of considering the correlation of vehicle types of consecutive vehicles. (60m span, material category c). It is found that neglecting the characteristic of consecutive vehicle type results in an unconservative estimation of fatigue damage parameter, especially when the bridge has medium or long span and the level of endurance or fatigue limit is relatively high. This error of underestimation becomes significant when the proportion of heavy vehicles is low. The composition of heavy vehicles is the most significant parameter. This significance is remarkable for high material categories.

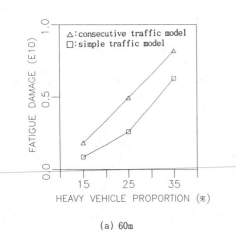

(a) 60m

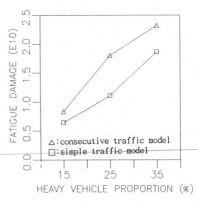

(b) 30m

Fig.2 Cumulative fatigue damage due to 10,000 vehicles

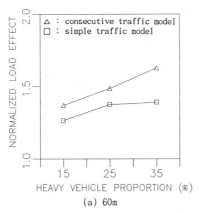

(a) 60m

(b) 30m

Fig.3 Extreme load effects

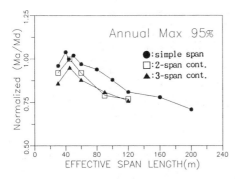

Fig.4 Variation of extreme load effects
due to effective span length

4 LIFETIME MAXIMUM TRAFFIC LOAD EFFECTS

The lifetime maximum load effects induced by traffic loads have been also investigated, for which an efficient methodology has been proposed. The 95% extreme values of annual maximum load effects on exterior girder are summarized in Fig.3, and 4. The values are normalized by the design load effects due to the first-class Korean standard design truck load(or lane load), in which the truck load has the total weight of 43.2ton(1ton = 9.8KN).

The effects of consecutive traffic model and heavy vehicle proportion are demonstrated well, in Fig.3. In Fig.4, the sensitivity of lifetime extreme load effects on effective span lengths are also summarized. The effective span length is the total length along which the influence line has positive effect. The ordinates in Fig.4 are normalized with respect to the result of 44m simple span bridge.

5 CONCLUSION

From the investigations on cumulative fatigue damage as well as lifetime maximum load effects, it can be said that traffic models and simulation process proposed in this study is quite useful to estimate fatigue damage and extreme load effects of bridges under traffic loadings.

The characteristics of consecutive vehicles arriving at a bridge have considerable effects on load effects. Neglecting this aspect underestimates considerably the cumulative fatigue damage as well as extreme load effects and this error becomes significant especially when the bridge has medium or long span. The proportion of heavy vehicles is also found to be important parameters.

ACKNOWLLEDGEMENT

This research work is supported in part by NON DIRECTED RESEARCH FUND, Korea Research Foundation, 1991.

REFERENCE

Ghosn, M.J. 1984. Bridge load modeling and realibility analysis, Ph.D. Thesis. Case Western Reserve Univ.

Jacob,B., J.B.Maillard & J.F.Gorse 1989. Probanilistic traffic load models and extreme loads on a bridge. Proc. ICOSSAR'89. : 1973-1989.

Japanese Society of Steel Construction 1989. Fatigue design manual.

Kameda,H. & M.Kubo 1989. Lifetime-maximum load effect for highway bridges based on stochastic combination of typical traffic loadings. Proc. ICOSSAR'89. : 1783-1790. San Francisco.

Miller,D. & W.H.Munse 1983. Mathematical modeling of loading histories for steel beam or girder highway bridges. : Civil eng. studies SRS508. Univ. of Illinois at Urbana-Champaign.

Structural Safety & Reliability, Schuëller, Shinozuka & Yao (eds) © 1994 Balkema, Rotterdam, ISBN 90 5410 357 4

Analysis of cable-stayed bridge subjected to stochastic earthquake loading

S. Malla & M. Wieland
Division of Structural Dynamics and Structural Reliability, Electrowatt Engineering Services Ltd, Zürich, Switzerland

ABSTRACT: When a lifeline structure is located in a seismically active zone, it is important to know about its dynamic behaviour and safety under earthquake loads. In this paper, a simple method of evaluating the seismic reliability of a cable-stayed bridge based on a stochastic dynamic analysis is presented. A probabilistic description of the earthquake hazard is prepared by means of a statistical analysis of the historical earthquake data. The case of the Karnali cable-stayed bridge in Nepal is investigated.

1. INTRODUCTION

The stochastic dynamic response of a cable-stayed bridge under seismic loading is computed in the frequency domain using the Kanai-Tajimi ground acceleration spectrum. The reliability of the structure is then estimated from its stochastic response and the probability density functions of magnitude and epicentral distance.

As a case study, the earthquake response of the Karnali cable-stayed bridge (Fig. 1) in Nepal is analysed. This bridge has a single pylon 120 m high and two spans comprising a main span of 325 m and a side span of 175 m. The cables are arranged in two approximately vertical planes with thirty cables in each plane. A composite girder, consisting of a 225 mm concrete slab over two trusses, each 3 m deep, has been used. The bridge is under construction at present. The bridge is modelled as a plane frame system and only the in-plane behaviour is considered.

2. EARTHQUAKE HAZARD

The number of earthquakes N exceeding a magnitude M can be expressed by the following empirical relationship:

$$log_{10} N = a - bM \qquad (1)$$

where a and b are constants which can be determined from the past earthquake records. The corresponding distribution function is

$$F_M(M) = \frac{1 - e^{-\beta(M-M_o)}}{1 - e^{-\beta(M_1-M_o)}} \qquad M_o \le M \le M_1 \quad (2)$$

where $\beta = b\,ln\,10$ and M_o and M_1 are the lowest magnitude of interest and the upper bound magnitude respectively (assumed as 4 and 8.7 respectively in the present study). The probability density function of the epicentral distance is also estimated by a statistical analysis.

3. POWER SPECTRA OF GROUND ACCELERATION AND RESPONSE

The Kanai-Tajimi power spectral density function (PSDF) of the ground acceleration can be written as

$$S_{\ddot{u}_g}(\bar{\omega}) = \frac{1 + 4\xi_g^2(\bar{\omega}/\omega_g)^2}{\left[1 - (\bar{\omega}/\omega_g)^2\right]^2 + 4\xi_g^2(\bar{\omega}/\omega_g)^2} S_o$$

$$(3)$$

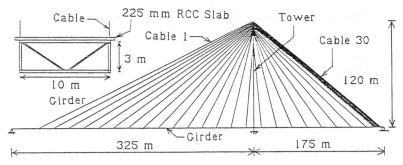

Figure 1   Karnali cable-stayed bridge

where S_0 is the white noise intensity, and ω_g and ξ_g are the characteristic ground frequency and damping ratio respectively. The variance of the ground acceleration is

$$\sigma_{\ddot{u}_g}^2 = \int_o^\infty S_{\ddot{u}_g}(\overline{\omega})d\overline{\omega} = \frac{\pi\, S_0 \omega_g}{4\xi_g}\left(1 + 4\xi_g^2\right) \quad (4)$$

The expected peak value of the ground acceleration a_o of duration T assuming a stationary zero-mean Gaussian random process can be approximated as (Davenport, 1961)

$$a_o = \sigma_{\ddot{u}_g}\left\{\sqrt{2\ln(2fT)} + \gamma / \sqrt{2\ln(2fT)}\right\} \quad (5)$$

where $\gamma = 0.5772$ and f is the average frequency of zero-crossing of the ground acceleration at a positive slope. For a cut-off frequency of 20 Hz in the Kanai-Tajimi spectrum and $\xi_g = 0.6$, it can be shown that $f = 0.3215\omega_g$. The peak ground acceleration a_o and the duration of strong ground motion T of an earthquake of magnitude M and epicentral distance r can be estimated by empirical relationships. Then, S_o can be found from eqs. (4) and (5). The PSDF of response $S_z(\overline{\omega})$ (e.g. Fig. 2) is computed by a dynamic analysis (Clough & Penzien, 1975). The total response is calculated assuming that the responses due to the horizontal and vertical components of earthquake are independent.

4.   STATISTICS OF RESPONSE

The j-th spectral moment of response $z(t)$ is defined as

$$\lambda_j = \int_o^\infty \overline{\omega}^j S_z(\overline{\omega})d\overline{\omega} \quad j = 0,1,2,... \quad (6)$$

The zeroth moment is equal to the variance σ_z^2 of the response (Fig. 3); other parameters of interest, such as the mean rate ν_o of zero-crossing at a positive slope and band width q of the response, can be computed from the spectral moments (Vanmarcke, 1976). The mean rate of crossing a level A at a positive slope by a stationary zero-mean Gaussian process is given by

$$\nu_a = \nu_o e^{-\psi^2/2} \quad (7)$$

where $\psi = A/\sigma_z$.

The probability of response remaining below a barrier A over the duration T of stationary response is

$$L(T) = L_0 e^{-\alpha T} \quad (8)$$

where L_0 is the probability of starting below the threshold and α is the decay rate.

In the present study, the cable stress is used as a criterion for reliability estimation. The stress barrier is essentially one-sided because of the relatively high initial cable tension caused by the dead loads. For a one-sided high barrier level,

970

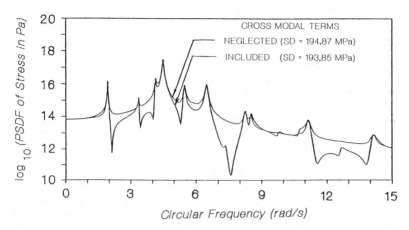

Figure 2 Power spectral density functions (PSDF) and standard deviations (SD) of stress in cable 11 with and without cross-modal terms (vertical earthquake component; $M = 8$; $r = 40$ km; damping ratio $\xi = 0.75\%$)

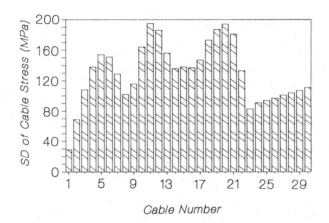

Figure 3 Standard deviations of cable stress due to vertical component of ground motion ($M = 8$; $r = 40$ km; $\xi = 0.75\%$)

L_O and α are approximately equal to one and v_a respectively.

For given values of M and r, the conditional probability that the cable stress exceeds the threshold can be calculated from eq. (8). Assuming that magnitude and epicentral distance are independent random variables, the probability of the cable stress exceeding the barrier per event can be expressed as (Malla, 1988)

$$p_E = P(z > A) =$$

$$\int_{M=4}^{8.7} \int_{r=0}^{\infty} P(z > A|M,r) f_M(M) f_r(r) dM dr \qquad (9)$$

where $f_M(M)$ and $f_r(r)$ are the probability density functions of M and r respectively.

Then, the probability of the cable stress exceeding the barrier in m years, with n events per year, is

$$p_f = 1 - (1 - p_E)^{mn} \qquad (10)$$

5. RESULTS AND CONCLUSIONS

In the present study, the stress barrier beyond which a cable is assumed to fail is taken as

971

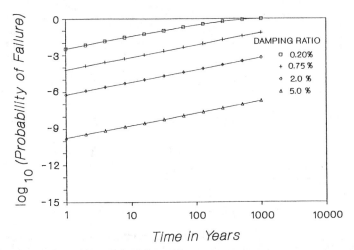

Figure 4   Probability of first-passage failure of one cable as a function of service life of bridge and damping ratio

1.52 GPa, which is equal to the specified ultimate strength. The probability of failure of one cable is shown as a function of service life of the bridge in Fig. 4. For a damping ratio of 0.75%, the annual probabilities of failure of one cable and two cables are estimated as 6.6E-5 and 6.9E-10 respectively. These annual probabilities are equal to 6.2E-7 and 5.6E-14 respectively for a damping ratio of 2%.

As the dynamic response and reliability estimates are sensitive to damping (Fig. 4), especially when it is low as in a cable-stayed bridge, the damping value should be very accurately known. However, in practice, damping is never accurately known, even in built bridges. Further, the probability density functions used should be more accurate for high magnitudes and short epicentral distances, as these ranges are predominantly significant when estimating the reliability.

6. ACKNOWLEDGEMENT

The research work presented in this paper was carried out while both authors were at the Asian Institute of Technology in Bangkok, Thailand.

7. REFERENCES

Clough, R.W. and Penzien, J. 1975. Dynamics of structures. McGraw-Hill, New York.

Davenport, A.G. 1961. The application of statistical concepts to the wind loading of structures. Proceedings, Institution of Civil Engineers, London, Vol. 19, pp. 449-472.

Malla, S. 1988. Safety and reliability of the cable system of a cable-stayed bridge under stochastic earthquake loading. M. Eng. thesis, Asian Institute of Technology, Bangkok, Thailand.

Vanmarcke, E.H. 1976. Structural response to earthquakes. Seismic risk and engineering decisions, Developments in geotechnical engineering, Elsevier Scientific Publishing Company, Amsterdam, Vol. 24, pp. 287-338.

Structural Safety & Reliability, Schuëller, Shinozuka & Yao (eds) © 1994 Balkema, Rotterdam, ISBN 90 5410 357 4

Reliability based assessment procedure for highway bridge decks

T. Micic & M. K. Chryssanthopoulos
Department of Civil Engineering, Imperial College of Science, Technology and Medicine

M. J. Baker
Department of Engineering, King's College, University of Aberdeen, UK

ABSTRACT : An assessment procedure based on structural system reliability principles has been developed for highway bridges with concrete or composite decks. The approach takes account of the fact that the failure of the structure is dependent on the combination of applied loads, their variability, both spatially and in amplitude, and on the variability of geometric and material properties. It requires the use of optimization techniques in conjunction with structural reliability analysis in order to establish the configuration of the critical failure pattern and associated probability of failure. By modelling a bridge as a virtual series system and analysing a finite number of basic failure patterns the total failure probability can be estimated.

INTRODUCTION

Until recently, methods for assessment of existing structures have not been distinguished from methods of initial structural design. Thus, the problem of introducing structure-specific data remains, to a large extent, unsolved. Typically, issues that have to be addressed in assessment guidelines are:

(i) development of methods to overcome analytical difficulties in strength prediction in existing structures where structure-specific data indicate non-conformity with assumed design tolerances;

(ii) development of methods for incorporating structure-specific data into assessment procedures, rather than relying solely on characteristic values assumed in design; availability of this data will not eliminate the uncertainty associated with loading, geometric and material variables, but will influence the *a priori* models assumed in design.

In the UK, such procedures are required in connection with the large number of existing highway bridges that need assessment, and possible repair or strengthening, mainly, as a result of:

- increased frequencies of Heavy Goods Vehicles
- increased weight of Heavy Goods Vehicles
- natural deterioration of existing bridge structures.

Developments in structural reliability have initiated a number of studies in the probabilistic approach to assessment. Moses and Nowak, among others, have carried out extensive studies in developing appropriate probability distributions for bridge resistance and loading [1,2]. Other studies [3-6] have dealt in more detail with probabilistic modelling of the resistance, particularly in relation to plastic slabs. It has been pointed out [5,6] that the critical yield line pattern from a deterministic analysis is not necessarily identical to the one relevant in a probabilistic analysis. Thus, in order to estimate an upper bound to the reliability of plastic slabs, a yield line pattern, which is described by generalised geometric parameters, needs to be considered and the minimum reliability index must be determined parametrically or by optimization with respect to these geometric parameters. It has been demonstrated that, in some simple cases [5], a good estimate of the minimum reliability index can be obtained by using the yield line pattern which is critical in deterministic analysis, i.e. implying that the uncertainty in various basic variables does not affect the configuration of the critical pattern. However, if a general reliability based procedure is to be developed that can be used in the assessment of various bridge structures with random variability in loading as well as geometric and material properties, adopting deterministically optimized yield pattern(s) may lead to unconservative estimates of the failure probability.

OVERVIEW OF THE PROPOSED METHOD

A typical bridge deck is considered as a structural system, which may form different kinematically admissible yield line patterns depending on the combination of applied loads, their variability, both

spatially and in amplitude, and the variability of geometric and material properties. The number of such patterns for large and/or complex structures can be very high. Thus, an important step towards developing a reliability based assessment procedure is to determine a finite set of yield patterns from which a sufficiently accurate estimate of the failure probability can be obtained [11]. In practice, for a specific structure and for any distinct load case, there is only a limited number of "basic yield patterns" that will contribute to the failure probability.

Thus, when such a set is identified, FORM/SORM analysis can be carried out to determine the failure probability associated with each of the basic yield patterns within the set and well known methods for reliability analysis of series systems can be used to obtain the system failure probability, as pointed out in [5,6]. As mentioned, due to random variability in the basic variables, for each basic yield pattern, there is an infinite number of basic variable combinations that may lead to failure, each associated with a fairly different configuration of the basic yield pattern.

Here, the required safety margin is obtained from yield line analysis [7,8]. Within the reliability analysis undertaken for each basic yield pattern, the safety margin expression is optimized with respect to some generalized geometric parameters which describe the geometry of the yield pattern. By repeating the analysis for all basic yield patterns, the system failure probability is estimated by modelling the structure as a virtual series system, where each element represents the event that one particular pattern has developed. The flowchart of the method is given in Fig. 1.

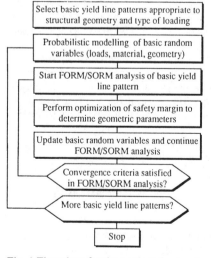

Fig. 1 Flowchart for the proposed procedure

So far, emphasis has been placed in developing the appropriate safety margins for bridges with concrete or composite deck under static loading, whose behaviour could be classified as that of a solid slab.

DEVELOPMENT OF SAFETY MARGINS

It is assumed that the behaviour of the structure is governed primarily by the reinforcement, whose properties are those of perfectly plastic material (under-reinforced concrete).The main assumption in rigid-plastic structural analysis is that the structure loses its integrity when sufficient hinges (or yield lines) have formed to allow rigid segment motions to take place without any further increase in load.

The yield line method, an upper bound plastic analysis method, has been adopted, as it is relatively simple to apply to many types of bridge decks. The first step is to propose kinematically admissible patterns consisting of lines along which yield of reinforcement takes place and rigid regions between yield lines [7,8].

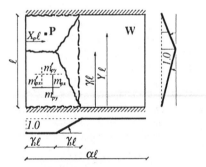

Fig. 2 General layout for the basic failure mode I

Displacements are assumed to be small enough for the plastic analysis to be based on the undeformed geometry of the structure prior to the loading. Further assumptions are that shear forces have no significant effect on the failure of the slab, and in-plane membrane forces are neglected. For the purpose of this study the virtual work method has been applied, as the safety margin expressions can be easily obtained for slabs with complex out-of-plane boundary conditions.

As mentioned, each yield pattern, is described by a number of geometric parameters γ_i (e.g. Fig. 2) and, assuming a unit deflection at an arbitrary point, the deflections at any point of the deck and the rotations of the yield lines are expressed as functions of γ_i. Thus, the governing equation, has the following general form:

$$\sum_{\substack{each \\ yield\ line}} m_p\theta\ell - \lambda \sum_{\substack{each\ rigid \\ region}} \iint w\delta\,dxdy = 0$$

where m_p represents ultimate moment of resistance/unit length, ℓ length of the yield line, θ rotation of the yield line which is calculated from the geometry of the yield pattern, w external load per unit area, δ displacement of the slab, and λ is the non-negative loading proportionality factor. In deterministic analysis with proportional loading, the critical yield line position is obtained from the minimisation of the load factor with respect to yield line geometry parameters for a particular set of input parameters (i.e. loading and resistance variables).

However, even for simple statically indeterminate structures, the number of kinematically admissible collapse mechanisms can be very large and unless a mathematical programming approach is used the problem can be extremely difficult to assess. A set of possible basic yield patterns for a slab bridge subject to typical highway loading has been presented in [11]. By describing each of these with generalised geometric parameters, a very large number of yield patterns is covered. From this set, a subset relevant to a particular loading case may be formed and by implementing the optimization techniques, the critical values of the geometric parameters for each of the basic modes may be found. For example, for the pattern shown in Fig. 2, the following safety margin equation is developed assuming rigid-perfectly plastic stress-strain relationship.

$$M = (\gamma_2 + \gamma_3)\frac{m_{px} + m'_{px}}{\gamma_1(1-\gamma_1)} + \frac{m_{py} + m'_{py}}{\gamma_3}$$
$$- w\ell^2(\frac{\gamma_2}{2} + \frac{\gamma_3}{3}) - P\frac{1-Y_P}{1-\gamma_1}$$

where γ_i's represent generalised geometric parameters and m_{px}, m'_{px}, m_{py}, m'_{py} ultimate moments of resistance/unit length in x and y directions and in the bottom and top layers; w represents uniformly distributed load, P point load intensity, (X_p, Y_p) spatial co-ordinates of the point of application. It is worth noting that the function is expressed in general terms so that it is possible, for various combinations of applied loads to calculate their contribution to the work equation as well as the energy dissipated within yield lines.

PROBABILISTIC METHODOLOGY

For the probabilistic modelling of the resistance, the procedure developed here assumes that parameters such as yield strength of reinforcement (f_y), concrete compression strength (f_{cu}), effective depth to reinforcement (d_{eff}), area of reinforcement =(A_r) etc. are random variables. Therefore, the plastic moment of resistance per unit length at any position on the slab is also a random variable and can be expressed in terms of a number of basic random variables,
$$M_p = M_p(\overline{X}) = M_p(A_r, f_y, d_{eff}, k, f_{cu}...)$$

Furthermore, the amplitude of uniformly distributed loads ($\mathbf{W}$), point loads ($\mathbf{P_i}$) and knife edge loads are assumed to be basic random variables, as well as some of the points of load application, namely those associated with concentrated (X_p), knife edge or strip loading (X_{HA}).

Thus, the safety margin associated with each of the basic yield line patterns, is given by

$$\mathbf{M} = min(\ \sum_{\substack{each\ yield \\ line}}^{m} M_p\ell\theta(\overline{\gamma}) - \sum_{\substack{each\ load \\ type}}^{n} W_i\delta(\overline{\gamma})\ )$$

subject to constraints: $g_j(\overline{\gamma}) \geq 0\ \ j=1,...,k$ and is a function of random variables and generalized geometric parameters, where the minimisation is performed at each step within the FORM/SORM analysis with respect to the geometric parameters using the method given in [9]. The constraints are such that the various patterns do not violate the overall geometry. For mode I, Fig. 2 they are:

$$0 < \gamma_1 < 1.0;\ \ 0 \leq \gamma_2 \leq \alpha;\ \ 0 < \gamma_3 \leq \alpha;\ \ \gamma_2 + \gamma_3 \leq \alpha$$

Clearly, for each basic pattern i

$$P_f = P[M_i \leq 0],\ i=1,2,...,N$$

where N is the number of basic yield patterns applicable to any load case.

Finally, the total probability of failure may be estimated using a series system model from

$$P_{f_{sys}} = P(\cup E_i)\text{ where } P(E_i) = P[M_i(\overline{X}) \leq 0]$$

and, as is well known, by linearizing the safety margins at the so-called design points, P_{fsys} is approximately obtained from $P_{f_{sys}} = 1 - \Phi_m(\beta, R)$ where Φ_m is the multi-normal distribution function, β is the vector of reliability indices for the various basic yield line patterns and R is the correlation matrix of the linearized safety margins at the design points.

EXAMPLE

In this example, the reliability index for one particular yield pattern, given in Fig. 2, is estimated, assuming that the loading consists of two lane loads and two symmetrically placed point loads (Fig. 3).

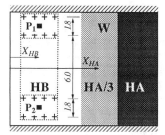

Fig.3 Plan of the load case considered in the example

The basic random variables are for simplicity, assumed to be normally distributed (Table 1).

Table 1. Probabilistic modelling used in the example.

Basic Variable		μ	σ
A_r	(m^2)	1.612 10^{-3}	0.25 10^{-4}
d_{eff}	(m)	0.510	0.05
f_y	(kN/mm^2)	322.000 10^{+3}	32.00 10^{+3}
f_{cu}	(kN/mm^2)	32.600 10^{+3}	4.89 10^{+3}
W	(kN/m^2)	15.800	1.42
P_i	(kN)	500.000	25.00
HA	(kN/m)	30.000	3.60
X_{HA}	(m)	7.75/10.24	0.50
X_{HB}	(m)	3.100	0.50

Using the method described above, a value of β_{min} = 5.1 is obtained. By assuming that the yield line pattern is optimized prior to reliability analysis by setting all random variables equal to their mean value, a value of β=5.9 was obtained. Fig. 4 shows the variation of β with respect to two geometric parameters.

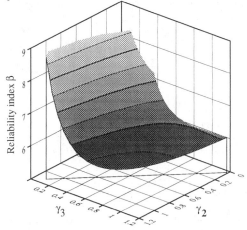

Fig. 4 Change in reliability index with variation of geometric parameters.

CONCLUSION

By combining FORM/SORM reliability analysis and non-linear optimization it is possible to estimate upper bounds on the reliability of solid (pseudo solid) bridge decks and, moreover, to estimate system effects. The proposed procedure relies on defining a relatively small set of "basic yield line patterns", which describes the virtual series system, but identifies the critical configuration associated with each of these in a probabilistic analysis A simple example has been presented to illustrate its use. At present, further work is carried out on typical bridge decks using this method and will be reported in [11]. Preliminary findings indicate that, depending on the load case considered, the difference between stochastic and deterministic critical failure pattern may be significant both in terms of geometric parameters and estimated reliability index.

REFERENCES

1   Rashedi, R. & F. Moses 1988. Identification of Failure Modes in System Reliability: *J. Stru. Eng.*, ASCE 114: 292-313.
2   Nowak, A. & J. Zhou 1990. System Reliability Models for Bridges, *Struct. Safety* 7: 247-254.
3   Middleton, C. & A. Low, 1990, The Use of Reliability in the Assessment of Existing Bridges, in *Proc. 1st Int. Conf. on Bridge Management*, ed. J.E. Harding et al., University of Surrey.
4   Simoes L.M.C. 1990. Reliability Assessment of Plastic Slabs, *Comp. and Struct.* 35: 689-703.
5   P. Thoft-Christensen & G.B. Pirzada 1988: Upper Bound Estimate of the Reliability of Plastic Slabs, *Probabilistic Methods in Civil Engineering,* ed. P. D. Spanos, ASCE, NY.
6   Thoft-Christensen, P. 1989. Reliability of plastic slabs, in *Proc. 5th ICOSSAR,* ed. A.H.S. Ang et al., San Francisco, USA.
7   Nielsen M.P. 1984. Limit Analysis and Concrete Plasticity, *Prentice-Hall Inc:* Englewood Cliffs, NJ.
8   Clark, L.A. 1983. Concrete Bridge Design to BS 5400, *Construction Press,* London.
9   Schittkowski, K. 1983. NLPQL: A FORTRAN Subroutine Solving Constrained Non-linear Programming Problems, *Annals of Operations Research.*
10 BS 5400 Concrete, Steel and Composite Bridges, 1984, *British Standards Institution.*
11 Micic, T. M.J. Baker, M.K. Chryssanthopoulos, Reliability Analysis for Highway Bridge Assessment, *submitted for publication,* 1993.

Structural Safety & Reliability, Schuëller, Shinozuka & Yao (eds) © 1994 Balkema, Rotterdam, ISBN 90 5410 357 4

Surveillance of structural properties of large bridges using dynamic methods

R.G. Rohrmann & W. Rücker

Federal Institute for Materials Research and Testing (BAM), Berlin, Germany

ABSTRACT: Permanently working monitoring systems will be used to observe and control the behavior of bridges. These monitoring systems are necessary for the detection of faults in bridges at an early stage, the localisation of these faults and their assessment. Dynamic monitoring quantities give information about the dynamic operational state and the eigenvalues of the bridges under inspection. The present study gives advices for the installation of dynamic monitoring systems, based on researches of a big reinforced concrete highway brigde in Berlin.

1 INTRODUCTION

The sudden interruption of traffic on main roads can lead to an enormous economical loss. If the interruption has been caused by an unpredicted fault in a bridge, the repair works will continue over a longer period. To minimize such risks permanently working, intelligent monitoring systems will be installed on bridges for the early detection of faults inside the bridge structure and the information about these. The successful operation of such systems, using modal parameters for the building assessment, requires a profound dynamic analysis of the structure during its preparation phase, consisting of:
1) an experimental modal analysis of the inspected building
2) an adjusted finite-element-model and
3) an examination of the present excitation sources

The following discussion is based on the concept of Natke and Yao (1989). Herein the changes of the natural frequencies ω_i and natural mode shapes V_i are considered as the representative qualities to measure changes of the structural parameters. Considering the sensitivity of these quantities, the following is valid for the natural frequencies in a first approximation

$$\Delta \omega_i^2 = \frac{1}{I} \alpha B(\hat{x}) [V_i''(\hat{x})]^2 \Delta x$$

with
I = kinetic energy, B = bending stiffness,
$0 \leq \alpha \leq 1$ damage parameter,
V_i = eigenfunctions
$[x_1, x_2]$ = damaged area, $\hat{x} \epsilon [x_1, x_2]$,
$\Delta x = \| x_2 - x_1 \|$

It becomes obvious that the greatest sensitivity for structural changes is reflected in those natural frequencies ω_i where the curvatures V_i'' of the natural mode shape reaches a maximum in the damaged area $[x_1, x_2]$.

2 EXCITATION OF LARGE STRUCTURES

The continuous vibration measurement of certain parts of structures by a monitoring system requires a permanently active excitation source. This cannot be realized by an artificial excitation. Wind, microseismic and service loads can be used as natural excitation sources. As e. g. in damaged concrete structures the modal parameters depend on the level of the static load, it makes sense to use dynamic traffic loads as an excitation source. Various measurements of eigenvalues on brigdes under traffic loads are made (Brownjohn 1989, Luz 1992). However, it has never been fully clearified, if the so gained frequencies and mode shapes can really be considered as eigenvalues and are not for example specific excitation frequencies, resulting from the natural frequencies of cars or of the particular interaction of the system car-roadsurface-bridge. Here the question arises if traffic in general can be the excitation source for reproducable natural mode shapes. Clarification is only possible on the basis of a classical experimental modal analysis with determined excitation forces. An other problem is the amount of exciteable natural frequencies traffic induced. Their clarification is important to produce higher vibration modes of the inspected bridge for they are in general more sensitive towards faults because of the

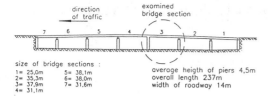

size of bridge sections :
1= 25,0m	5= 38,1m
2= 35,3m	6= 38,0m
3= 37,9m	7= 31,6m
4= 31,1m	

average heigth of piers 4,5m
overall length 237m
width of roadway 14m

Figure 1: View of the Highway Bridge Westend

greater curvature V" of their eigenfunctions. Furthermore the adjustment of the analytical model towards the changing eigenvalues during the monitoring process requires a sufficient amount of eigenvalues.

3 SURVEY OF THE HIGHWAY BRIGDE

A survey of a P/C highway bridge in Berlin (see figure 1) has been conducted to clarify the above stated questions and to prepare the installation of a permanently working monitoring system. Figure 2 shows one result of a longterm observation under traffic loads over the period of eight days and nights, starting on friday 0.00 a. m.

The ordinate shows at every time point the belonging frequencies of 10 accelerationmaxima respectively, taken from a response spektrum averaged out of 50 measurements with a frequency resolution of $3 \cdot 10^{-2}$ Hz. Figure 2 shows the daily and weekly rhythm of the traffic loads as well as accumulations of frequency points. The mean values resulting from the frequency concentrations in narrow frequency ranges were interpreted as natural frequencies excited by traffic loads. Figure 2 also shows that between approximately 7 and 13 Hz a high density of accelerationmaxima can be found, which cannot be integrated in narrow ranges. An explanation

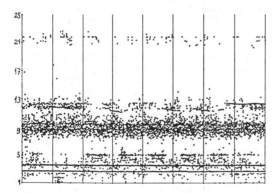

Figure 2: Results of Longtime Observations : Measured Frequencies Versus Time

will be given furtherdown. One realises, that from these averaged response data due to traffic loads frequency results higher than 13 Hz rarely can be found.

During the next step the bridge has been investigated with impact and snap-back excitation forces and without traffic loads. With these results a modal analysis of the bridge was performed. In the frequency range between 0 and 25 Hz more than 80 frequencies were found which could be considered as eigenfrequencies. The corresponding global mode shapes of the multispanbridge result in broad frequency ranges as combinations of basic and higher order mode shapes of a single span structure, with different phase relations respectively. That means traffic loads can excite different but similar mode

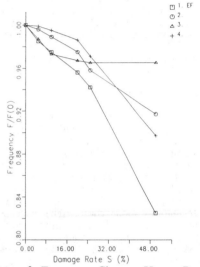

Figure 3: Frequency Changes Versus Damage Rate

shapes in a broad frequency range. The mode shapes frequently differ from each other only in the extent of coupling between bending and torsion. These all is the reason for the scattering of the frequency points in Figure 2 between 7 and 13 Hz. A more sensitive averaging method can be helpful here. Comparing of the natural mode shapes resulting from artificial excitation with the vibration modes, which belong to the accumulation frequencies in Figure 2, a good correspondence can be found up to 13 Hz.

Figure 4 shows the frequency spectra of 3 measurement points due to a normal traffic event. If such events are measured and averaged 50 times the results are as shown in Figure 4. The effect is that the basic frequencies dominate in the averaging process. The response signals in the higher frequency range are averaged comparatively against zero. Hereby information about the dynamic behavior in the

higher frequency range, that are very important for a damage analysis, get lost. Therefore a continuous monitoring system requires the use of averaging methods, which are working in certain frequency sections averaging due to the magnitude of the measured response signal.

4 ANALYTICAL MODEL AND SENSITIVITY ANALYSIS

An analytical model based on beam elements has been developed. The adjustment of the analytical model due to the measuring results was carried out according to the substructure method (Natke 1988). The first 4 bending eigenfrequencies were chosen as comparative quantities for the residuum. These eigenfrequencies were also the same which could be measured definitly under traffic loads. The comparison of analysed and measured eigenfrequencies are shown below.

	calculated	measured
1	2.56	2.73
2	3.63	3.61
3	8.72	8.89
4	13.63	13.77

With this adjusted model sensitivity calculations were carried out. The aim was to find out which eigenfrequencies and which natural mode shapes can be used to show local faults best. For the investigated bridge one can expect that span number 3 will show faults in future. The results in figure 3 show that from the investigated eigenfrequencies the first is in particular sensitive towards a global damage in span number 3. However, the change of this eigenfrequencies compared to the damage is relatively low. In figure 5 the changes of the global eigenfunctions caused by damages in span number 3 were examined. As one can see, in particular those natural mode shapes react sensitive towards damages, which corresponding natural frequencies are not remarkable. As one also can see in figure 5 the natural mode shapes of number 3 and 4 are in particular valid to be observed by a permanently working monitoring system. This result will affect the choise for sensor localisations during the installation of the monitoring system.

5 CONCLUSIONS

The successful operation of a monitoring system requires individuell adjustment to the specific building. In general it is possible to use the traffic loads as a suitable excitation source. Profound preperation phase with following steps is necessarry:

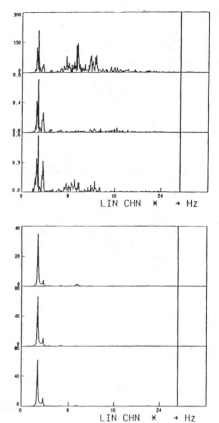

Figure 4: Response Spectra at Three Measurement Points in Span 3 from Traffic Loads : One Event (above), 50 Events Averaged (below)

1. Investigation of the frequency content of the structural response excitated by traffic
2. Experimental modal analysis to verify the eigenvalues excited by traffic
3. a first analytical model.
   The result of modal adjustment is an
4. updated analytical model.
   By means of sensitivity analysis a decision will be made
5. which eigenvalues are in particular suitable to be observed by a monitoring system.
   This affects the localisation of the sensors and other hardware equipment.
   According to figure 6 the operation phase of the monitoring requires:
6. a monitoring system for the measurement of the changes of the observed eigenvalues
7. an updated analytical model for the quantity identification of faults and localisations
8. a method for the identification and assessment of damage.

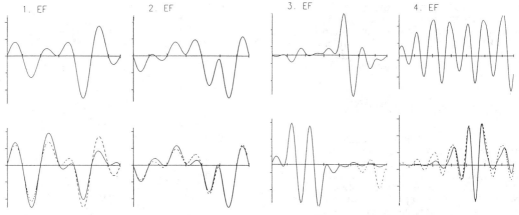

Figure 5: Changes of Modal Shapes Due to Stiffness Loss in Span 3
(above: 0%, below: --- 25%, ——50%)

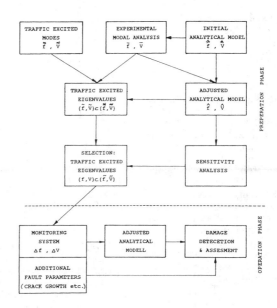

Figure 6: Longtime Monitoring System in the Preparation and Operation Phase

REFERENCES

Brownjohn, J.M.W., A.A. Dumanogm, R.T. Severn & A. Blakeborough 1989. Ambient vibration survey of Bosporus Suspension Bridge. *Earthquake engineering and structural dynamics* 18.

Luz, E. & J. Wallaschek 1992. Experimental modal analysis using ambient vibration. *International yournal of analytical and experimental modal analysis.*

Natke, H.G. 1988. Application of system identification in engineering. *CISM* 269. Springer Verlag.

Natke, H.G. & J.T.P. Yao 1989. System identification methods for fault detections and diagnosis. *Proc.5th ICOSSAR:* 1387-1393

Rohrmann, R.G., W. Rücker, G. Möller & W. Nitsche 1992. Dynamische Untersuchungen und Langzeitmessungen an der Autobahnbrücke Westend in Berlin (in German). *Research report,* BAM Berlin.

Structural Safety & Reliability, Schuëller, Shinozuka & Yao (eds) © 1994 Balkema, Rotterdam, ISBN 90 5410 357 4

Bridge seismic retrofit decisions

R.G.Sexsmith
University of British Columbia, B.C., Canada

ABSTRACT: Bridge seismic retrofit decisions are ideally suited to treatment by principles of expected value decision making. This paper illustrates the decision situation in terms of minimizing present expected value of the sum of damage consequences and retrofit construction costs. For a group of bridges, retrofit priorities are established by ranking benefit/cost ratios, where benefit is net present value of damage consequences mitigated. An ongoing investigation at UBC for improvement of damage probability estimates, and an application example involving a major bridge in the City of Vancouver, are discussed.

1 INTRODUCTION

The design of bridges did not include significant attention to seismic design until the San Fernando, California, earthquake of 1972. Early design codes in North America required only that a lateral force of 2%, 4%, or 6% (depending on the foundation conditions) of the weight of the superstructure be applied. This force is usually less than wind or the effects of live load, and was therefore insignificant. The past 20 years has seen a major increase in attention to details, force levels, seismicity, and subsurface conditions, with the result that modern bridges are designed to higher seismic standards. Unfortunately the vast majority of bridges that will be in service for many decades to come were built prior to the mid-1970's. We therefore have a problem of inconsistent safety in the overall bridge stock, and a risk of major bridge failures in the event of an earthquake. Bridges are typically an important part of the lifeline system that must survive in an earthquake, and the consequences of failure can be severe.

Bridge engineers are therefore faced with major retrofit decisions for existing bridges. Code-based assessments of existing bridge structures usually show that expensive modifications have to be made to reach code prescribed safety levels. Budget limitations preclude automatic adherence to new criteria, however, and tradeoffs between safety and cost are necessary. This is especially true in the case of retrofits of existing bridges for seismic effects, since the structure may otherwise be quite safe and serviceable. The application of decision analysis principles can provide an appropriate alternative that will provide more safety for less money in the short term, while maintaining the long term objective of reaching some consistent high safety standards.

2 DECISION ANALYSIS

The application of decision analysis to the bridge retrofit situation is illustrated in Figure 1. The independent variable is the annual probability of failure (or failure rate) u, where u_0 is the value before the retrofit. Note that in the figure u is decreasing to the right. The present expected value of consequences C_p, based on continuous compounding, is

1) $$C_p = C_f \, u \, / \, (i + u)$$

where C_f is the estimated consequences due to damage in the seismic event, i the real interest rate (excluding inflation), and u the annual probability of the damage. The occurrence of seismic events causing the damage is assumed to be in accordance with the Poisson process. C_p is plotted in the figure as a function of u. Construction retrofit cost C_0 appears in the same figure. It increases as better safety (reduced u) is achieved. The total cost $C_T = C_0 + C_p$ shows a minimum at the optimum failure rate $u_{optimum}$. This analysis assumes that the future life of the retrofitted bridge is reasonably long. If a fixed future life is assumed, the probability model on the time to the seismic event has to be modified from the exponential pdf assumed here.

The maximum expected value decision maker will choose to spend funds on the retrofit to bring the structure to reliability represented by $u_{optimum}$. Where this is less safe than modern criteria would demand, this can be taken as an interim measure until funds are available to bring the structure to the prescribed code safety level.

Bridges tend to be owned by governmental bodies that are responsible for a large number of bridges. Typically there will not be sufficient funds available at the outset to bring all structures to the optimum point. The total cost curve of Figure 1 provides the basis for the choice of priorities among the stock of deficient bridges. Each retrofit to $u_{optimum}$ involves spending a construction cost C_i. The net gain is R_i, the reduction in total cost C_T. The optimal strategy is to get the most reduction in total cost for a fixed available total sum of the C_i values for all bridges retrofitted. For each bridge considered for retrofit, C_i and R_i are determined. The highest priority is then the alternative with the greatest ratio R_i/C_i. This is simply a ranking by benefit/cost ratio, where the benefit is the reduction in net present expected cost due to the increased safety.

The decision process provides a framework for action involving retrofit strategy, but at the present state of knowledge, estimates for use in the model are very crude. The estimation of damage levels and corresponding probabilities requires further analysis. The consequences corresponding to the estimated damage levels require more detailed economic studies. Finally, predictions of the effectiveness of retrofit designs has to be improved. These difficulties, however, do not preclude decision making under the current state of knowledge, as subjective estimates are preferable to avoidance of rational action.

The value of formal decision making as outlined herein is primarily in providing an organized focus on the important parameters, rather than in obtaining numerical values. The numbers are crude, but the focus helps to counter the tendency of designers to require code criteria as a substitute for actual decision making. Formal decision analysis also helps to justify the codes and standards that apply to retrofits, by demonstrating the appropriateness of lower safety levels where retrofits are involved, due to the high marginal cost of increased safety.

The State of California embarked on a retrofit program for bridges after the San Fernando Earthquake. They identified numerous problems in existing bridges, but they chose to embark on a program of span restraints, in which cable or bar restrainers are installed to limit movement of sliding bearings to ensure spans could not fall from supports. This is clearly an intuitively reasonable strategy, consistent with the above decision prescription, because span restraints provide significant protection against collapse for relatively small construction costs.

3 DAMAGE PROBABILITY

Among the several approaches to estimation of damage levels from seismic events, the IDARC program (Park, Reinhorn, and Kunnath 1987, Kunnath, Reinhorn, and Lobo 1992) appears promising. It carries a cumulative summation of hysteretic energy as a fraction of maximum energy corresponding to "full damage", in terms of moment-rotation of concrete cross-sections, for input time histories. Where a time history can be associated with probability levels, damage probabilities can be estimated. Currently at University of British Columbia we are testing a number of replicates of bridge portal piers at a large scale with slow cyclic loads. The piers are 0.45 scale models of actual piers from Vancouver's Oak Street Bridge, currently under

detailed retrofit design. It is intended that the model piers and actual piers be instrumented for ambient vibration testing, which will determine mode shapes and periods. On the model piers this will be done at several stages of cyclic loading as damage increases, and on the unretrofit as well as retrofit specimens.

Analysis with the IDARC program combined with the experimental results during degradation of the specimens will provide improved means to establish criteria for damage probabilities.

4 APPLICATION EXAMPLE

In a recent seismic upgrade project in Vancouver, a major 40 year old bridge (Granville Bridge) was assessed using decision analysis techniques to establish priorities for retrofit. One aspect of the project involved six major portal piers that support large deck trusses (see fig. 2). The capacity/demand ratio for lateral strength of the portal piers was calculated assuming demand was from a seismic event with peak ground velocity (pgv) 0.2m/s, having an annual probability of exceedence of 0.002. At this pgv, design rules aim at a damage threshold of minor repairable damage. Corresponding pgv for other probability levels at Vancouver are: 0.074, 0.115, 0.197, and 0.292, for probability of exceedence per annum of 0.010, 0.005, 0.002, and 0.001, respectively.

Where the force level is proportional to the peak ground velocity, it is now possible to estimate the peak velocity corresponding to the damage threshold assuming that capacity/demand of unity corresponds to the design damage threshold of 0.2 m/s. Table 1 presents the corresponding v and annual probability based on the above Vancouver data. Because the design criteria presumes that minor damage is expected at the design event, we can assume that the probability of damage is less than the probability of v. Based on experience at Loma Prieta and similar events, it was assumed for the Granville portal piers that the damage probability would be about half the probability of v. This is tabulated in the last column of Table 1.

The Granville bridge portal piers were estimated to have capacity/demand ratios about

0.4. Damage probability was therefore taken as 0.005 per annum. Damage cost was the collapse of the span due to pier failure, with an estimated cost of $50 million. With an interest rate of 5%, the present expected cost of damage (eqn 1) is $5 million. The construction cost to achieve capacity/demand of 1.0 was estimated at 0.4 million, based on preliminary retrofit designs involving longitudinal post-tensioning of the portal cap beam and construction of a surround to increase the beam width and depth, and to add shear reinforcing. This retrofit brings the damage probability to a negligible value. Although figure 1 shows a continuous curve, in practice it is more likely that a few reasonable retrofit concepts will lead to a few discrete points on the curve. Where only one retrofit alternative is being considered, it will be accepted when the retrofit cost is less than the present expected value of the consequences without retrofit, as it is in this example.

Table 1. Annual Probability of Damage

cap/dem	v	probability of v	u
0.2	0.04	-	0.010
0.4	0.08	0.010	0.005
0.6	0.12	0.005	0.003
0.8	0.16	-	0.002
1.0	0.20	0.002	0.001
1.2	0.24	-	0.0008
1.4	0.28	0.001	0.0005

There were many other aspects of the Granville Bridge that required assessment and possible retrofit. As outlined above, the net reduction in expected cost R_i provides an indicator for setting priorities. In the first phase retrofit, span restrainers for the truss spans and the portal pier cap retrofit gave the greatest benefit. Further work is currently under way. Priorities were set with due account of the R_i/C_i values, but adjusted to suit appropriate contract arrangements and timing constraints.

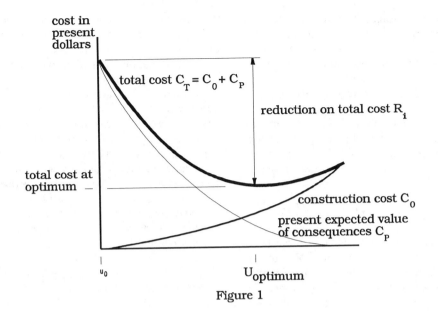

Figure 1

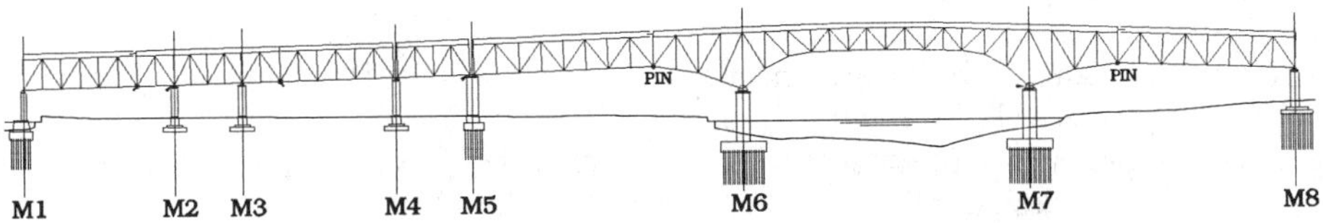

Figure 2

REFERENCES

Kunnath, Reinhorn, and Lobo 1992, "IDARC version 3.0: A Program for the Inelastic Damage Analysis of Reinforced Concrete Structures". Technical Report NCEER-92-0022, National Center for Earthquake Engineering Research, SUNY, Buffalo NY.

Park, Reinhorn, and Kunnath 1987, "IDARC: Inelastic Damage Analysis of Reinforced Concrete Frame-Shear Wall Structures, National Center for Earthquake Engineering Research, Technical Report NCEER-87-0008, SUNY, Buffalo NY.

Sexsmith 1992, "Seismic Risk Management for Existing Structures", submitted to the Canadian Journal of Civil Engineering, National Research Council of Canada, Ottawa.

ACKNOWLEDGEMENTS

The principals and staff at Buckland and Taylor Ltd, and the staff at the City of Vancouver were responsible for the technical development of the Granville bridge retrofit. Thanks are also due to the University of British Columbia Department of Civil Engineering and the Natural Sciences and Engineering Research Council of Canada for research support.

Structural Safety & Reliability, Schuëller, Shinozuka & Yao (eds) © 1994 Balkema, Rotterdam, ISBN 90 5410 357 4

Random system response of reinforced and prestressed concrete bridges

Juan A. Sobrino & Juan R. Casas
Technical University of Catalunya, Civil Engineering Department, Barcelona, Spain

ABSTRACT: In this paper, an alternative to evaluate the bridge response is presented. The proposed method is used to obtain the response for Ultimate Limit States based on the moment-curvature relationship and ultimate cross-section response, taking into account the uncertainties in the parameters involved, and the internal forces redistribution for each mode of failure. Each failure state is obtained with a nonlinear concrete bridge analysis, using a finite element formulation. The proposed method implies an important computer effort reduction and is perfectly applicable to evaluate accurately both the probabilistic behaviour and the structural capacity of an existing bridge. The method is compared with other analyses (elastic and plistic).

1 INTRODUCTION

In order to evaluate the real structural capacity of Reinforced and Prestressed Concrete Bridges, or other linear members, in terms of flexural capacity of a system, a non-linear analysis should be made, taking into account the possible redistribution of moments in statically indeterminate systems. Most of the classical reliability analysis for existing structures do not consider the ductility available in the structures (**elastic analysis**). In other cases, a **plastic analysis** is performed, considering a total redistribution of moments at failure.

The scope of this study is to present a method to evaluate the probabilistic ultimate resistance of concrete bridges and its correlation with load effects. The proposed method is compared when an elastic or a plastic analysis is used in the reliability analysis.

2 STRUCTURAL RELIABILITY ANALYSIS

The proposed method versus other classical analyses (elastic analysis and plastic analysis) to evaluate the reliability of existing bridges is presented with an example included in [1]. A continous prestressed concrete slab, designed to have a high ductility, is analyzed, Figure 1. Only a mode of failure and one case of load is considered, live load in second span. The load models used are summarized in Table 1.

The probabilistic ultimate response of the criti-

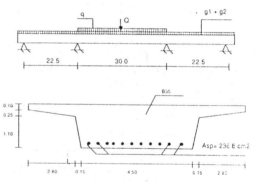

Figure 1. Prestressed Concrete Bridge. Static model, external loads and typical cross-section at mid span.

cal cross sections has been evaluated considering the uncertainties in materials and geometry and non-linear behaviour of materials, using a large data bank recently collected in Spain [2] [3]. The results are shown in Table 2 and in Figures 3 and 4

2.1. Elastic Analysis:

An elastic analysis of the bridge is performed for each load. The moments may be written in terms of moments in a reference simply supported beam (Figure 4) as:

$$M^i = \lambda^i_j \bar{M}^j \qquad (1)$$

Table 1. Load Models used in the reliability analysis

Type of Load	Notation	Mean	C.O.V.	Type of Distribution
Self weigth	g_1 (kN/m)	214.37	0.05	Normal
Permanent load	g_2 (kN/m)	39.32	0.15	Normal
Uniform. Live load	q (kN/m)	41.60	0.15	Gumbel
Concent. Live load	Q (kN)	600.0	0.15	Gumbel

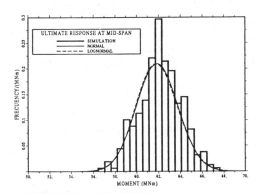

Figure 2.– Ultimate response of mid-span section

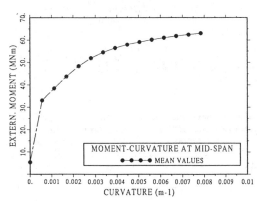

Figure 3.– Moment-Curvature Relationship at mid-span section

Where: i= No. Section, j= Case of Load and $\bar{M}^j$ = Moment in the reference static beam. The failure function, in terms of moments in the critical sections is:

Table 2.- Probabilistic ultimate response of Critical sections

Ultimate Moment	Mean	C.O.V.	Type of Distribution
M_u (MN m) Support	44.92	0.048	Lognormal
M_u (MN m) Mid Span	63.43	0.050	Lognormal

$$M_u^i - \lambda_{g_1}^i \bar{M}_{g_1} - \lambda_{g_2}^i \bar{M}_{g_2} - \lambda_q^i \bar{M}_q - \lambda_Q^i \bar{M}_Q + M_{hyp}^i = 0 \tag{2}$$

Where: M_u^i = Ultimate bending moment in section i. M_{hyp} = Hyperstatic (redundant) moment due to prestressing (including its sign), assumed Normal variable with a C.O.V.= 5 %.

2.2. Plastic Analysis:

The ultimate flexural response of the bridge, if a perfect elasto-plastic behaviour is considered, can be evaluated considering plastic hinges, Figure 5. In this cases the value of λ may be written as:

$$\lambda = \frac{M_u^2}{\frac{M_u^1 + M_u^3}{2} + M_u^2} \tag{3}$$

The failure function will be:

$$M_u^2 - \lambda \ [\bar{M}_{g_1} + \bar{M}_{g_2} + \bar{M}_q + \bar{M}_Q] = 0 \tag{4}$$

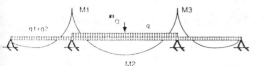

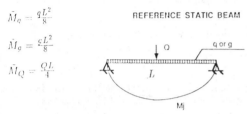

$$\bar{M}_g = \frac{qL^2}{8}$$

$$\bar{M}_g = \frac{gL^2}{8}$$

$$\bar{M}_Q = \frac{QL}{4}$$

REFERENCE STATIC BEAM

Figure 4. Elastic analysis

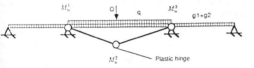

Figure 5. Plastic analysis

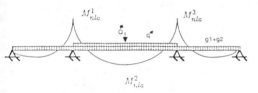

Figure 6. Non-linear analysis. Internal forces at the failure state.

The hyperstatic moment due to prestressing may be ignored since a mechanism occurs at failure, implying zero structural stiffness.

2.3. Non-Linear Analysis:

In order to evaluate the real capacity of the structure, a great computational effort has been made to evaluate the final redistribution of the internal forces at failure, using Monte-Carlo simulations, taking into account the uncertainty in mechanical properties, based on a large data collected in concrete bridges in Spain [3].

The failure function may be written as equation 5, (Figure 6):

$$M_u^i - \lambda^i [\bar{M}_{g_1} + \bar{M}_{g_2} + \bar{M}_q + \bar{M}_Q] = 0 \qquad (5)$$

The values of λ^i, for each section, in this case of failure, can be evaluated as :

$$\lambda^i = \frac{M_{nla}^i}{\frac{M_{nla}^1 + M_{nla}^3}{2} + M_{nla}^2} \qquad (6)$$

Where M_{nla}^i are the final redistributed moments in the critical sections obtained in the Non-Linear Analysis, Figure 6.

The values of the λ^i parameters obtained are shown in Table 3.

Table 3. Values of λ, Moment redistribution.

λ	Mean	C.O.V	Type of Distribution
Support (1)	0.4213	0.016	Shifted Lognormal
Mid-Span (2)	0.5787	0.012	Lognormal

2.4 Proposed Method with Non-Linear Analysis:

The failure fuction is the same as equations 5 and 6. In this case the Hyperstatic Moment due to prestressing can be neglected because it has been already included in the Non Linear Analysis.

The proposed method is based on:

1) The C.O.V. of the moment response for each section is practically constant with the curvature after yielding [2].

2) Only one Non-Linear Analysis is made [4], using the mean values of parameters, to evaluate the moment redistribution.

3) The variability of the mechanical properties and geometrical uncertainties is such that does not change the failure mode.

4) The geometrical and mechanical models are based on a large data base recently collected in Spain for concrete bridges.

5) Incremental live load steps to achieve failure is performed.

2.5. Reliability Analysis:

The values of the Hasofer-Lind Reliability Index, β obtained using the 4 methods exposed and the load models of Table 1 and ultimate response of sections, Table 2, are summarized in Table 4.

Table 4.- Reliability Index

Elastic Analysis	Plastic Analysis	NLA + Simul.	Proposed Method
13.8	14.2	12.8 - 13.9 (*)	14.2

(*) The range of variability of β is due to different assumed correlations between λ and M_u.

3 CONCLUSIONS

1.- The proposed method takes into account the real structural behaviour (system response) at failure and the uncertainties in geometry and materials.

2.- The computational effort to evaluate the probabilistic response of the system is reduced to only one Non-Linear Analysis versus other methodologies (Monte-Carlo simulations, etc).

3.- The exposed example shows the Reliability Index values from the proposed method and other classical analyses. In this particular case, the differences between plastic analysis and the proposed method are not significant because of the considerable ductility of the structure. Because of this ductility, almost total moment redistribution is produced at failure.

4.- Due to the high ductility of the system the values of β are very close with the four different structural analysis. The results derived from the deterministic analysis [1], amplification live load factors until failure, are not realistic compared with those provided by the probabilistic analysis.

4 ACKNOWLEDGEMENTS

This study is included in a research program, supported by the Education Department of the Goverment of Catalonia for Post-Graduated Students.

5 REFERENCES

[1] Aparicio, A.C.;*Some examples of non-linear analysis of prestressed concrete continuous decks under increasing loads.* Bull. Information du C.E.B No. 153, pp 9-22, 1982.

[2] J.A. Sobrino y J.R. Casas; *Probabilistic response of reinforced and prestressed bridge cross-sections.* IABSE Colloquium "Remaining Structural Capacity", Copenhagen, 1993.

[3] J.A. Sobrino; *Evaluation of Structural Safety and Serviceability of existing prestressed and reinforced concrete bridges.* Doctoral Thesis (in development), Departamento de Ingeniería de la Construcción, ETSICCPB, UPC, Barcelona.

[4] Ramos, G.; *Behaviour of externally prestressed concrete bridges in Service and failure* Doctoral Thesis (in development). Departamento de Ingeniería de la Construcción, ETSICCPB, UPC, Barcelona.

Inspection

Structural Safety & Reliability, Schuëller, Shinozuka & Yao (eds) © 1994 Balkema, Rotterdam, ISBN 90 5410 357 4

X-ray inspection reliability for welded joints

C. M. Chang, I. K. Chen, H. K. Shee & H. Y. Chen
Quality Assurance Center, Chung-Shan Institute of Science and Technology, Lungtan, Taoyuan, Taiwan

J. N. Yang
Department of Civil Engineering, University of California, Irvine, Calif., USA

ABSTRACT: Nondestructive inspection (NDI) is very important for ensuring the reliability and safety of mechanical and structural components in their design service lives. It is well-known that the detection of defects involves considerable statistical uncertainties. As a result, the capability of an NDI system is defined by the probability of detection (POD) as a function defect size, referred to as the POD curve. The POD curve is determined from experimental test results. In this paper, an experimental program was conducted using welded joint specimens of SAE 4130 steel plate. The X-ray inspection method was used to measure the diameter of natural defects (pores) in the welded joints. Then, the method of metallography was used to determine the exact defect (pore) size. Statistically meaningful experimental data were generated and analyzed using the statistical method to establish the POD curves associated with different confidence levels. The resulting POD curves indicate that the X-ray inspection method is quite accurate and it is appropriate for quality assurance of welded joints of many types of mechanical and structural components used in practice.

1 INTRODUCTION

Current nondestructive inspection (NDI) systems are not capable of repeatedly producing correct indications when applied to defects of the same size. The chance of detecting a given defect depends on many factors, such as the location, orientation and shape of the defect, materials, inspectors, inspection environments, etc. As a result, the probability of detection (POD) for all defects of a given size has been used in the literature to define the capability of a particular NDI system in a given environment. The probability of detection as a function of the defect size "a" is referred to as the POD(a) curve [e.g., Berens, Hovey 1981, 1984, Packman 1976] as shown in Fig. 1.

It follows from Fig. 1, that an NDI system may result in two types of incorrect indications: (i) failure to give a positive indication in the presence of a crack whose length is greater than a_{NDE}, referred to as the Type I error, and (ii) given a positive indication when the crack length is smaller than a_{NDE}, referred to as the Type II error. The Type I error allows components containing a crack length longer than a_{NDE} to remain in service, thus greatly increasing the potential safety hazard. For safety critical components, the Type I error is of primary concern. The Type II error rejects good components and, hence, has an adverse effect on the cost of repair/replacement and the life cycle cost. In applications, such as the retirement-for-cause (RFC) life management, both Type I and Type II errors are important, because the criterion used in RFC life management is the minimization of the life-cycle-cost [e.g., Yang, Chen 1985, 1986]. For a given NDI system with a single inspection, it is impossible to reduce the Type II error without increasing the Type I error and vice versa [e.g., Yang, Donath 1982, 1983(a), 1983(b)]. It is obvious that the ideal inspection capability of an NDI system is a unit step function. Figure 1 shows schematically an ideal and a realistic POD curve. The ideal inspection system would detect all flaws larger than a_{NDE} and none smaller than a_{NDE} as indicated by a unit step function in Fig. 1 in which both Type I and Type II errors are zero. Unfortunately, such an ideal NDI system is far from reality. Methodologies to reduce both types of errors using multiple inspection procedures were proposed by Yang et al [Yang, Donath 1982, 1983(a), 1983(b)].

In aerospace applications, a nondestructive inspection limit, a_{NDE}, is frequently used as the initial flaw length [e.g., Gallagher, et al 1982]. The fracture mechanics crack propagation life, N_f, is the life for the flaw length, a_{NDE}, to propagate to the critical flaw length, a_c, under expected usage environments. The nondestructive inspection limit, a_{NDE}, is the flaw length corresponding to a 90% detection probability and 95% confidence level for a particular NDI system used. The required inspection interval τ is equal to N_f divided by a safety factor S_f, i.e., $\tau = N_f/S_f$. The information on the POD curve for an NDI system is important in quality assurance and reliability analysis of mechanical

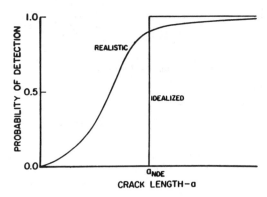

Fig. 1: Schematic of probability of detection curves

and structural components in service. It is also important for the determination of the inspection interval for scheduled maintenance [e.g., Yang, Trapp 1974(a), 1974(b), 1975, Shinozuka 1976, Yang, Chen 1985, 1986].

In this paper, an experimental test program was conducted using SAE 4130 steel plate specimens with a 1 cm TIG welded joint in the center along the specimen length. This type of specimen was used to simulate the welded joints in many mechanical components, such as pressure vessels, rocket motor cases, etc. The X-ray inspection method was used to detect and measure the diameter, â, of the natural defects (pores) in the welded joints. Statistically meaningful data were generated using four different X-ray epuipments (systems) inspected by three inspectors independently. Then, the method of metallography was used to determine the exact diameter of the defect (pore) size, denoted by "a".

A statistical methodology using the experimental results to establish the POD curve associated with different confidence levels is described. The experimental data sets (â, a) were analyzed and the probability of detection (POD) as a function of the defect (pore) size "a" was established. The POD curves associated with different levels of confidence were also constructed. The resulting POD curves indicate that the X-ray inspection method is quite accurate and appropriate for quality assurance of welded joints for many types of mechanical and structural components currently used in practice.

2 STATISTICAL ESTIMATION OF POD CURVE

The POD is a measure of inspection uncertainty, not the cause. The causes of uncertainty for the inspection results, â (detected defect size or diameter of pores), comes from two sources of variations: (1) the material properties, the defect location, geometry and orientation, etc., which are strictly associated with the individual defects, and (2) factors that change from inspection to inspection, including inspectors and equipment factors. Consequently, the inspection result â can be expressed by either

$$\hat{a} = f(a) + c + e \tag{1}$$

or

$$\ln \hat{a} = \ln f(a) + c + e \tag{2}$$

where f(a) represents the overall mean trend in â as a function of the actual defect size "a", c represents the defect to defect variation, and e represents the variation from inspection to inspection of the same defect. The function f(a) is fixed while the variables c and e are random with means of 0. Equation (1) or (2) provides the basic model for the analysis of the experimental NDI data of â versus a. The methodology for the determination of the POD(a) function from the â versus a data sets is described in the following [Berens, Hoveg 1984].

The defect related and defect independent terms c and e are random variables with means equal to 0 and variances equal to s_c^2 and s_e^2, respectively. It follows from Eq. (1) that the mean and variance of â for a single inspection of a defect of size "a" picked at random are: $E(\hat{a}|a) = f(a)$ and $Var(\hat{a}|a) = s_c^2 + s_e^2$.

Suppose the model of Eq. (2) is used and â and a are transformed into Y and X through

$$Y = \ln \hat{a}, \qquad X = \ln a \tag{3}$$

Then, Eq. (2) becomes

$$Y = \alpha + \beta X + c + e \tag{4}$$

in which f(a) is considered a power function of a, i.e., $f(a) = e^{\alpha} a^{\beta}$ with α and β being parameters to be determined from NDI experimental data of â versus a. If the error variables c and e are assumed to have normal distributions with zero means, it follows from Eq. (3) that Y is a normal random variable with a mean value of $\alpha + \beta X$ and a standard deviation $S = (s_c^2 + s_e^2)^{1/2}$.

The POD(a) function is the probability that the detected defect size, â, is greater than the threshold value a_{th}. It is, therefore, given by

$$POD(a) = P[\hat{a} > a_{th}] = P\{\ln(\hat{a}) > \ln(a_{th})\} =$$
$$P(Y > Y_{th}) = 1 - \Phi \left[\frac{Y_{th} - (\alpha + \beta X)}{S} \right] \tag{5}$$

in which

$$S = \sqrt{s_c^2 + s_e^2} \quad ; \quad Y_{th} = \ln a_{th} \tag{6}$$

and Φ is the standard normal distribution function. Using the symmetry properties of $\Phi(x)$, Eq. (5) becomes a lognormal distribution function

$$POD(a) = \Phi\left(\frac{X - \left(\frac{Y_{th} - \alpha}{\beta}\right)}{S/\beta}\right) = \Phi\left(\frac{\ln a - \bar{\mu}}{\bar{\sigma}}\right)$$

(7)

in which the mean value, $\bar{\mu}$, and standard deviation, $\bar{\sigma}$, of the log defect size are given by

$$\bar{\mu} = (\ln(a_{th}) - \alpha)/\beta \quad ; \quad \bar{\sigma} = S/\beta$$

(8)

Since Eq. (4) is linear, the method of linear regression can be used conveniently to estimate the parameters α, β and S appearing in the POD(a) function of Eq. (7) as follows. The NDI experimental data are expressed in n $(\hat{a}_i, a_i)$ pairs. These n pairs of data are transformed into a set of n (Y_i, X_i) pairs through the transformation of Eq. (3), i.e., $Y_i = \ln \hat{a}_i$ and $X_i = \ln a_i$. By use of the linear regression analysis, the estimates $\hat{\alpha}$, $\hat{\beta}$, and $\hat{S}$ for the parameters α, β and S are obtained as follows.

$$\hat{\beta} = [(\Sigma X_i Y_i) - n\bar{X}\bar{Y}] / SSX ;$$

$$\hat{\alpha} = \bar{Y} - \hat{\beta}\bar{X} \quad ; \quad SSX = (\Sigma X_i^2) - n^{-1}(\Sigma X_i)^2$$

$$S^2 = (n-2)^{-1} \sum (Y_i - \hat{a} - \hat{\beta} X_i)^2 ;$$

$$\bar{X} = n^{-1}\Sigma X_i \quad ; \quad \bar{Y} = n^{-1}\Sigma Y_i$$

(9)

in which the summation is from 1 to n. After estimating α, β and S, the parameters $\bar{\mu}$ and $\bar{\sigma}$ can be computed from Eq. (8) in which α, β and S are replaced by $\hat{\alpha}$, $\hat{\beta}$ and $\hat{S}$, respectively. Hence, the POD(a) function, Eq. (7), is completely defined.

It is noticed from Eq. (7) that the POD(a) function depends on the specified detection threshold a_{th}. The effects of the detection threshold a_{th} on POD(a) are described in the following. First, the median inspection defect size increases with the detection threshold. Second, the slope of the POD(a) function decreases as the detection threshold increases.

The POD(a) function derived in Eq. (7) is the lognormal cumulative distribution function of the defect size a. It represents the POD(a) with 50% confidence level. To determine the POD(a) function with any arbitrary percent of confidence level, say $\gamma\%$, a method described by Chang and Iles (1983) for calculating confidence bounds on the lognormal cumulative distribution can be used; with the result

$$POD(a, \gamma) = \Phi(z_\gamma)$$

(10)

in which

$$z_\gamma = \hat{z} - \sqrt{\frac{\lambda}{n}\left(\frac{\hat{z}^2}{2} + \frac{(X - \bar{X})^2}{SSX} + 1\right)}$$

(11)

where

$$\hat{z} = \frac{X - \bar{\mu}}{\bar{\sigma}} \quad and \quad X = \ln a$$

(12)

In Eq. (11), n is the sample size, λ is the γth percentile of the Chi-Square distribution with two degrees of freedom. SSX and $\bar{X}$ are given by Eq. (9) and X is defined by Eq. (3).

3 EXPERIMENTAL PROGRAM

Test specimens used were SAE 4130 steel plates with dimensions of 250mm long, 150mm wide, and 6mm thick. The middle of the plate is a TIG welded joint of 10mm wide in the longitudinal direction as shown in Fig. 2. This type of specimen was used to simulate the welded joints for many types of mechanical and structural components used in practice. Three different X-ray equipments (systems) were used to inspect the welded joint and to measure the diameter of the defects (pores) in the joint under the same inspection conditions as follows: voltage = 200kvp; current = 8MA; exposure time = 80 seconds; Kodak film M type; Kodak X-OMAT automatic development equipment; distance between film and focus lens = 1,000mm. The negatives developed for each X-ray equipment (system) were examined and defect sizes were measured by four inspectors independently. As a result, there are twelve (12) data points of the NDI results, $\hat{a}$, associated with each detected defect (pore), "a".

After all X-ray inspections were conducted, the method of metallography was used to determine the exact diameter, a, of the detected defects for all specimens. Note that the use of the method of metallography is very time consuming. Portions of the

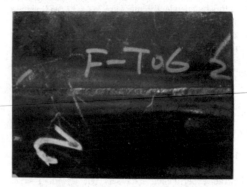

Fig. 2: SAE 4130 welded joint specimen

993

Table 1. Some results of metallography and x-ray inspections

| Metallo-graphy | X-Ray Inspection Results, â(mm) | | | | | | | | | | | |
a(mm)	#1	#2	#3	#4	#5	#6	#7	#8	#9	#10	#11	#12
0.250	0.00	0.25	0.25	0.25	0.25	0.25	0.00	0.25	0.25	0.25	0.25	0.25
0.280	0.25	0.25	0.25	0.25	0.25	0.25	0.25	0.25	0.25	0.25	0.25	0.25
0.375	0.00	0.30	0.30	0.30	0.00	0.30	0.30	0.30	0.25	0.30	0.30	0.30
0.395	0.40	0.30	0.30	0.30	0.40	0.30	0.30	0.30	0.40	0.30	0.30	0.30
0.412	0.60	0.70	0.70	0.70	0.60	0.70	0.70	0.63	0.60	0.70	0.70	0.63
0.500	0.50	0.40	0.40	0.40	0.50	0.40	0.51	0.40	0.50	0.40	0.50	0.40
0.587	0.30	0.30	0.30	0.30	0.30	0.30	0.30	0.50	0.40	0.30	0.50	0.30
0.628	0.50	0.50	0.50	0.50	0.50	0.50	0.50	0.64	0.50	0.50	0.60	0.50
0.692	0.50	0.50	0.50	0.50	0.50	0.50	0.51	0.50	0.50	0.50	0.51	0.50
0.797	0.75	0.50	0.00	0.75	0.75	0.75	0.75	0.77	0.80	0.75	0.70	0.75
0.859	0.60	0.50	0.50	0.00	0.60	0.50	0.63	0.60	0.60	0.50	0.64	0.50
0.921	0.70	0.75	0.75	0.75	0.70	0.75	0.70	0.76	0.60	0.75	0.70	0.70
0.922	0.90	0.70	0.70	0.70	0.90	0.70	0.70	0.76	0.90	0.70	0.70	0.60
1.021	1.00	0.80	0.85	0.80	1.00	0.80	0.80	0.89	1.00	0.80	0.80	0.80
1.296	1.00	1.00	1.00	1.00	1.00	1.00	1.02	1.00	1.00	1.00	1.02	1.00
1.406	1.50	1.20	1.20	1.20	1.50	1.20	1.20	1.20	1.50	1.20	1.20	1.20
1.790	1.10	1.00	1.00	1.00	1.10	1.00	1.14	1.00	1.10	1.00	1.14	1.00
3.343	2.50	2.50	2.50	2.50	2.50	2.50	2.50	2.50	2.50	2.50	2.50	2.50

resulting experimental data (â, a), i.e., the actual diameter, a, versus the measured diameter, â, by the X-ray inspection technique, are shown in Table 1. Since each defect is inspected using three X-ray systems and the result (film) from each X-ray system is read by three inspectors independently, there are 3x4=12 readings for â for each defect of size "a" obtained using the metallography as shown in Table 1. Since the thickness of the specimens is 6mm and the sensitivity of the X-ray instrumentation is approximately 4%, defect sizes smaller than 0.2mm cannot be found. Hence, â is set to be 0.2mm if a defect was not detected. A total of 540 (â, a) data points were obtained, i.e., n=540 in Eq. (9).

4 ANALYSIS OF EXPERIMENTAL RESULTS

With the experimental data sets $(â_i, a_i)$ for $i=1,2, ...,n=540$ and the analysis procedures described in the previous section, the regression parameters were obtained as $\alpha=-0.2586$, $\beta=0.9038$ and $S=0.2385$, Eqs. (5) to (9). The data sets in log scale are presented in Fig. 3, in which the solid line represents the regression results. The POD curve depends on the specified detection threshold level, a_{th}, see Eqs. (5) to (9). Based on Eq. (7), the POD curves for three different detection threshold levels, a_{th}, are presented in Fig. 4. The solid curves shown in Fig. 4 correspond to a 50% confidence level, Eq. (7), whereas the dashed curves correspond to a 95% confidence level, Eq. (10). As observed from Fig. 4, the POD curve is shifted to the right-hand side as the detection threshold a_{th} increases. This indicates that a less stringent requirement (or larger a_{th}) will allow components (or welded joints) with larger defects to pass the inspection. The inspection limit, a_{NDE}, that is the defect size corresponding to a 90% detection probability and a 95% confidence level is obtained from Fig. 4 as follows: $a_{NDE}=0.55mm$ for $a_{th}=0.3mm$; $a_{NDE}=0.95mm$ for $a_{th}=0.5mm$; and $a_{NDE}=2.0mm$ for $a_{th}=1.0mm$.

As mentioned previously, the capability of an NDI

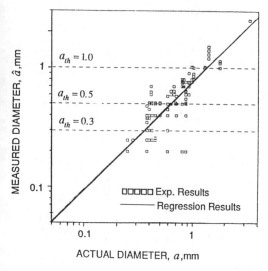

Fig. 3: Measured defect size, â, vs. actual defect size, a, and regression analysis results

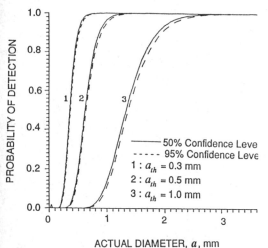

Fig. 4: POS curves for x-ray inspection of welded joints

system is defined by the corresponding POD curve. The median point of the POD curve, i.e., the defect size with a 50% detection probability and 50% confidence level, is a measure of the average (or general trend) of the NDI capability. The range or the slope of the POD curve indicates the uncertainty of the NDI system, which is a most important factor in determining the appropriate use of a particular NDI system. Since the range of the POD curves presented in Fig. 4 is small, the X-ray inspection method is quite accurate and appropriate for applications to the quality assurance of welded joints.

5 CONCLUSION

An experimental test program has been conducted to generate statistically meaningful data for establishing the probability of detection curves associated with different levels of confidence for X-ray inspection systems. Welded joint specimens of SAE 4130 steel plates were used to simulate natural defects (pores) for various types of mechanical components used in practice, such as pressure vessels, rocket motor cases, etc. A statistical analysis methodology is used to construct the probability of detection curve using experimental test results. It is demonstrated that the X-ray inspection method is quite accurate and appropriate for detecting defects (pores) in the welded joints. The establishment of the POD curves for nondestructive inspection systems is very important for ensuring the reliability and safety of mechanical and structural components in their design service lives.

REFERENCES

Berens, A.P. and Hovey, P.W., Evaluation of NDE reliability characterization, AFWAL-TR-81-4160, Air Force Wright Aeronautical Laboratories, WPAFB, Ohio, 1981.

Berens, A.P. and Hovey, P.W., Flaw detection reliability criteria, vol. 1 - method and results, AFWAL-TR-84-4022, Air Force Wright Aeronautical Laboratories, WPAFB, Ohio, 1984.

Berens, A.P., NDE reliability data analysis, in Metals Handbook, Vol. 17, 9th edition: Nondestructive Evaluation and Quality Control, pp. 689-701, 1988, ASM International.

Cheng, R.C.H. and Iles, T.C., Confidence bands for cumulative distribution functions of continuous random variables, Technometrics, Vol. 25, No. 1, 1983, pp. 77-86.

Gallagher, J. P. et al, U.S.A.F. damage tolerant design handbook: guidelines for the analysis and design of damage tolerant aircraft structures, AFWAL-TR-82-3073, Air Force Wright Aeronautical Laboratories, Wright-Patterson Air Force Base, Ohio, 1982.

Packman, P.F., Klima, S.J., Davies, R.L., Malpani, J., Moyzis, J., Walker, W., Yee, B.G.W. and Johnson, D.P., Reliability of flaw detection by nondestructive inspection, ASM Metal Handbook, Vol. 11, 8th Edition, Metals Park, Ohio, 1976, pp. 214-224.

Yang, J.N. and Donath, R.C., Improving NDE capability through multiple inspections with applications to gas turbine engine, AFWAL-TR-82-4111, Air Force Wright Aeronautical Laboratories, WPAFB, Ohio, 1982.

Yang, J.N. and Donath, R.C., Improving NDE reliability through multiple inspections, in Review of Progress in Quantitative NDE, edited by D.O. Thompson and D.E. Chimenti, Plenum Press, NY, Vol. 1, 1983(a), pp. 69-78.

Yang, J.N. and Donath, R.C., Inspection reliability of components with multiple critical locations, Proc. 14th Symposium on NDE, San Antonio, TX, April 19021, 1983(b).

Yang, J.N. and Trapp, W.J., Reliability analysis of fatigue-sensitive aircraft structures under random loading and periodic inspection, AFML-TR-74-2, Air Force Materials Laboratory, WPAFB, Ohio, 1974(a).

Yang, J.N. and Trapp, W.J., Reliability analysis of aircraft structures under random loading and periodic inspection, AIAA Journal, Vol. 12, No. 12, 1974(b), pp. 1623-1630.

Yang, J.N. and Trapp, W.J., Inspection frequency optimization for aircraft structures based on reliability analysis, Journal of Aircraft, AIAA, Vol. 12, No. 5, 1975, pp. 494-496.

Shinozuka, M., Development of reliability-based aircraft safety criteria: an impact analysis, AFFDL-TR-76-36, Air Force Flight Dynamics Laboratory, WPAFB, Ohio, 1976.

Yang, J.N. and Chen, S., Fatigue reliability of gas turbine engine components under scheduled inspection maintenance, Journal of Aircraft, AIAA, Vol. 22, No. 5, 1985, pp. 415-422.

Yang, J.N. and Chen, S., An exploratory study of retirement for-cause for gas turbine engine components, Journal of Propulsion and Power, AIAA, Vol. 2, No. 1, 1986, pp. 38-49.

Structural Safety & Reliability, Schuëller, Shinozuka & Yao (eds) © 1994 Balkema, Rotterdam, ISBN 90 5410 357 4

Non-periodic inspection of fatigue-sensitive structures by Bayesian approach

George Deodatis
Princeton University, N.J., USA

ABSTRACT: This paper introduces a Bayesian analysis methodology to determine appropriate non-periodic inspection intervals of fatigue-sensitive structures, so that their reliability remains above a prespecified minimum level throughout their service life. Fatigue damage is considered to initiate in a structural element when cracks develop, whether or not they are detected. The fatigue process then continues by crack propagation, resulting in strength degradation. If a fatigue crack is detected during an inspection, the cracked component is assumed to be repaired or replaced with a new one, resulting in the renewal of its residual strength and fatigue characteristics and increasing thus the overall reliability of the structure. The Bayesian approach introduced in this paper is unique and novel in that it allows one to utilize judiciously the results of earlier inspections for the purpose of determining the time of the next inspection and estimating the values of several parameters involved in the problem that can be treated as uncertain. Numerical simulations verify the above-mentioned capabilities of the Bayesian method.

1 INTRODUCTION

Fatigue is one of the most important problems of multiple-component structures such as aircraft, off-shore and marine structures, bridges, etc., subjected to random dynamic loads. Fatigue damage is considered to initiate in a structural element when cracks develop, whether or not they are detected. The fatigue process then continues by crack propagation, resulting in strength degradation. Periodic inspections of fatigue-sensitive structures are common practice in order to maintain the reliability of such structures above a desired minimum level. If a fatigue crack is detected during an inspection and if the cracked component is repaired or replaced with a new one, then both the residual strength and the fatigue characteristics of this component are renewed, increasing thus the overall reliability of the structure. In some previous studies performed by Yang and Trapp (1974), Shinozuka (1976) and Paliou and Shinozuka (1987), the effect of periodic inspections on fatigue-sensitive structures was examined using probabilistic methods. The concept of Bayesian analysis was then applied

to the inspection procedure of such structures by Itagaki et al. (1976), Itagaki and Asada (1977), Shinozuka et al. (1981), Itagaki and Yamamoto (1985), Fujimoto et al. (1989) and Deodatis et al. (1992). In the studies done by Itagaki and Yamamoto (1985), Fujimoto et al. (1989) and Deodatis et al. (1992), the effect of non-periodic inspections using Bayesian analysis was analyzed and appropriate inspection intervals were determined to maintain the reliability of the entire structure above some prespecified minimum level.

2 BASIC ASSUMPTIONS

The structure is considered to consist of a specific number of structural elements. A structural element is defined so that it possesses only one fatigue-critical location where a crack can initiate.

2.1 Inspection procedure

All structural elements are inspected immediately after initiation of service and at the

time of each scheduled inspection. If an element is found not to be intact, the following action is taken: If a crack is detected in the element, the crack is repaired and the element regains its initial strength. If the element is found to have failed, it is replaced by a new one.

The entire inspection history of each element is considered to be known at the time of the current inspection. If a crack is detected in an element, the crack length is assumed to be measured accurately. The probability of detecting element failure during an inspection is equal to unity.

2.2 Fatigue crack initiation

The time to crack initiation (TTCI), denoted by t, is assumed to be a random variable with density function following the Weibull distribution:

$$f_c(t \mid \beta) = \frac{\alpha}{\beta}\left(\frac{t}{\beta}\right)^{\alpha-1} \exp\left[-\left(\frac{t}{\beta}\right)^{\alpha}\right]$$

$$\text{for} \quad t > 0 \qquad (1)$$

Additional uncertainty is introduced in the TTCI by scale parameter β which is considered to be a random variable. The shape parameter α is assumed to be deterministic for the sake of simplicity. The distribution function of the TTCI is expressed by:

$$F_c(t \mid \beta) = 1 - \exp\left[-\left(\frac{t}{\beta}\right)^{\alpha}\right] \qquad t > 0 \quad (2)$$

2.3 Fatigue crack propagation

Fracture mechanics theory is used to determine the length of a propagating crack under random stress. For the purposes of this study, it is assumed that a crack grows according to the following law:

$$\frac{da}{dt} = C_1 \cdot (\Delta\sigma)^2 \pi a = c \cdot a \qquad (3)$$

where a is the crack length, C_1 is a material constant and $\Delta\sigma$ is the nominal stress fluctuation.

For structures with elements subjected to different stress levels, the reader is referred to Ito et al. (1992).

Integrating Eq. 3 from the initial crack length a_0 at the time of crack initiation t_c, up to the current crack length a at time t, the following expression is obtained:

$$a(t - t_c \mid c) = a_0 \exp[c(t - t_c)] \qquad (4)$$

Uncertainty in fatigue crack propagation is introduced by parameter c which is considered to be a random variable.

2.4 Probability of detection

The probability of detecting an existing crack of length a during an inspection is given by:

$$D(a \mid d) = 1 - \exp[-d(a - a_{min})] \qquad (5)$$

Uncertainty in the probability of crack detection is introduced by parameter d which is considered to be a random variable. a_{min} denotes the minimum detectable crack length.

2.5 Failure rate and element reliability

Failure of an element occurs when the random stress exceeds the strength of the element for the first time. Therefore, the problem of evaluating the failure rate is essentially that of estimating the first-passage failure probability with a constant or variable two-sided threshold depending on whether the element fails before or after crack initiation. Consequently, the failure rates before and after crack initiation at time instant t_c are given by:

Before crack initiation:

$$h(t) = \exp(r) = h_0 \qquad (6)$$

After crack initiation:

$$h(t) = \exp[q(t - t_c) + r] \qquad (7)$$

For the sake of simplicity, parameters r and q are assumed to be deterministic constants. Then, the reliability of an element before crack initiation during the service period from time instant T_l up to time instant t is denoted by $U(t - T_l)$ and given by:

$$U(t - T_l) = \exp\left\{-\int_{T_l}^{t} h(\tau)d\tau\right\}$$

$$= \exp\left\{-\int_{T_l}^{t} \exp(r)d\tau\right\} \qquad (8)$$

or

$$U(t - T_l) = \exp\left\{-(t - T_l) \cdot \exp(r)\right\}$$
$$\text{for} \quad t \leq t_c \tag{9}$$

where T_l is the time of service initiation for the element under consideration. Finally, the reliability of an element after crack initiation during the service period from the time of crack initiation t_c up to time instant t is denoted by $V(t - t_c)$ and given by:

$$V(t - t_c) = \exp\left\{-\int_{t_c}^{t} h(\tau) d\tau\right\}$$
$$= \exp\left\{-\int_{t_c}^{t} \exp[q(\tau - t_c) + r] d\tau\right\} \tag{10}$$

or

$$V(t - t_c) = \exp\left\{-\frac{1}{q}\left[\exp\left\{q(t - t_c) + r\right\}\right.\right.$$
$$\left.\left. - \exp\{r\}\right]\right\} \quad \text{for } t > t_c \tag{11}$$

3 POSSIBLE EVENTS AT TIME OF INSPECTION

At the time of the j-th inspection, performed at time T_j, of a certain element, one of the following three events may occur (knowing that this element was repaired or replaced during the l-th inspection performed at time T_l ($l < j$) or that this element initiated service at time T_l denoting the beginning of service for the structure):

1. $\{A : j, l\}$ = event that the element is found to have failed at the time of the j-th inspection T_j, or equivalently event that failure of the element occurred during the time interval $[T_{j-1}, T_j]$. This event consists of the following two mutually exclusive events:

   $E_{1,j}$ = event that the element failed before crack initiation, sometime during the time interval between the two consecutive inspections at T_{j-1} and T_j.

   $E_{2,j}$ = event that the element failed after crack initiation, sometime during the time interval between the two consecutive inspections at T_{j-1} and T_j.

2. $\{B_1(a_j) : j, l\}$ = event that the element is found not to have failed at the time of the j-th inspection T_j and a crack of length between a_j and $a_j + da_j$ is detected in the element. Event $\{B_1(a_j) : j, l\}$ is alternatively denoted by $E_{3,j}$.

3. $\{B_2 : j, l\}$ = event that the element is found not to have failed at the time of the j-th inspection T_j and no crack is detected in the element. This event consists of the following two mutually exclusive events:

   $E_{4,j}$ = event that the element did not fail in the time interval $[T_{j-1}, T_j]$ and no crack exists in the element at the time of inspection T_j.

   $E_{5,j}$ = event that the element did not fail in the time interval $[T_{j-1}, T_j]$ but a crack exists in the element which is not detected at the time of inspection T_j.

The probabilities of the above-mentioned events have been evaluated for a particular element (Deodatis et al., 1992) in terms of the probability density and distribution functions $f_c(t \mid \beta)$ and $F_c(t \mid \beta)$ of the TTCI, reliability functions $U(t)$ and $V(t)$, probability of crack detection $D(a \mid d)$ and the information that the element was repaired or replaced during the l-th inspection performed at T_l ($l < j$) or that the element initiated service at T_l denoting the beginning of service for the structure. The probabilities of events $\{A : j, l\}$, $\{B_1(a_j) : j, l\}$ and $\{B_2 : j, l\}$ are denoted by $P\{A : j, l\}$, $P\{B_1(a_j) : j, l\}$ and $P\{B_2 : j, l\}$ respectively, while $P_m\{A : j, l\}$, $P_m\{B_1(a_j) : j, l\}$ and $P_m\{B_2 : j, l\}$ are written for $P\{A : j, l\}$, $P\{B_1(a_j) : j, l\}$ and $P\{B_2 : j, l\}$ respectively, in order to identify the element number m.

4 RELIABILITY OF AN ELEMENT AFTER THE LATEST INSPECTION T_j

The reliability of two types of elements at time instant t^* after the j-th inspection (but before the $(j+1)$-th inspection, i.e. $T_j < t^* < T_{j+1}$) is examined in the following.

4.1 Elements repaired or replaced at the j-th inspection

Elements are replaced or repaired at the j-th inspection in the case of events $\{A : j, l\}$

Table 1   Values of Parameters in Numerical Example

Item	True values	Range
Parameter α in Weibull distribution	2.0	–
Parameter β in Weibull distribution	30 years	20 - 40
Parameter c in law of crack propagation	0.6 per year	–
Initial crack length a_0	10 mm	–
Minimum detectable crack length a_{min}	10 mm	–
Parameter d in probability of crack detection	0.01 mm^{-1}	0.002 - 0.018
Parameter r in failure rate	-7.5	–
Parameter q in failure rate	0.9	–

Table 2   Results for Case With Uncertain Parameters β and d

Inspect. No.	Inspect. time: T (years)	Number of failed elements	Number of detected cracks	Detected crack length (mm)
1	5.0	1(b)	1	22
2	7.9	0	1	42
3	10.1	0	4	189,189,124,23
4	12.1	0	1	1215
5	13.5	0	3	117,29,25
6	15.2	0	4	110,110,54,98
7	17.0	0	7	16,241,29,13,465,98,21
8	18.7	1(a)	6	64,178,168,104,42,117
9	20.3	0	4	68,124,241,45
10	21.9	0	3	149,140,124
11	23.5	0	4	628,189,227,158
12	24.9	0	6	168,98,64,57,42,306

(a) Failure after crack initiation   (b) Failure before crack initiation

or $\{B_1(a_j) : j, l\}$, respectively. Writing $R(t^*;$ Repair) instead of $R(t^*;$Replacement or Repair) for brevity, the reliability $R(t^*;$Repair) of an element of this type is computed as the sum of the following two probabilities:

(i) Probability that the element will survive during the time interval $[T_j, t^*]$ and no crack will initiate before t^*,

(ii) Probability that a crack will initiate in the element sometime during the time interval $[T_j, t^*]$, but the element will survive during the same time interval.

4.2 Elements not repaired at the j-th inspection

An element is neither repaired nor replaced

at the j-th inspection in the case of event $\{B_2 : j, l\}$. The reliability $R(t^*;$ No Repair) of an element of this type is computed as the sum of the following three probabilities divided by the probability of event $\{B_2 : j, l\}$:

(i) Probability that the element will survive during the time interval $[T_l, t^*]$ and no crack will initiate before t^*,

(ii) Probability that a crack will initiate during the time interval $[T_j, t^*]$, but the element will survive during the time interval $[T_l, t^*]$,

(iii) Probability that a crack initiated at some time instant t during the time interval $[T_i, T_{i+1}]$ $(i = l, \ldots, j - 1)$, this crack was not detected during all subsequent inspections (from inspection at time T_{i+1} up to inspection at time T_j inclusive) and

the element will survive during time interval $[T_l, t^*]$.

The reliabilities $R(t^*; \text{Repair})$ and $R(t^*; \text{No Repair})$ have been calculated (Deodatis et al., 1992) in terms of $f_c(t \mid \beta)$ and $F_c(t \mid \beta)$, $U(t)$ and $V(t)$, $D(a \mid d)$ and T_l.

5 BAYESIAN ANALYSIS

5.1 Uncertain parameters and their prior joint density function

As mentioned earlier, parameters β, c and d are considered as possible sources of uncertainty. Initially, it is assumed that β, c and d are jointly and uniformly distributed according to the following prior joint density function:

$$f^0(\beta, c, d) =$$

$$\frac{1}{(\beta_{max} - \beta_{min})(c_{max} - c_{min})(d_{max} - d_{min})}$$

$$= \text{constant} = f^0 \qquad (12)$$

where:

$$\beta_{min} \le \beta \le \beta_{max} \quad ; \quad c_{min} \le c \le c_{max}$$
$$d_{min} \le d \le d_{max} \qquad (13)$$

5.2 Likelihood function resulting from j-th inspection

The likelihood function LF_j for the entire structure as a result of the j-th inspection is calculated as:

$$LF_j = \prod_{m=1}^{M} LF_j^{(m)} \qquad (14)$$

where $LF_j^{(m)}$ is the likelihood function for element m resulting from the j-th inspection and M is the total number of elements in the structure.

For a specified element m, consider that replacement due to failure or repair due to a detected crack occurred at inspections T_{l_1}, T_{l_2}, ..., T_{l_r}, where r indicates the number of times the element has been repaired or replaced before the j-th inspection. Consequently:

$$l_1 < l_2 < \ldots < l_r < j \qquad (15)$$

It is pointed out that $l_1, l_2, \ldots, l_r$ are all known at the time of the j-th inspection since the entire inspection history of each element is considered to be known. It is noted that $l_1, l_2, \ldots, l_r$ as well as r take values unique to each element.

Then, the likelihood function for element m resulting from the j-th inspection is given by:

$$LF_j^{(m)} = P_m \{X : j, l_r\} \cdot \prod_{k=1}^{r} P_m \{Y : l_k, l_{k-1}\} \qquad (16)$$

In Eq. 16, X stands for either A or $B_1(a_j)$ or B_2 depending on the result of the j-th inspection for element m. Specifically, if at the time of the j-th inspection, element m is found to have failed, then X stands for A, if it is found to have a crack of length between a_j and $a_j + da_j$, then X stands for $B_1(a_j)$ and if it is found intact, then X stands for B_2. Also in Eq. 16, Y stands for either A or $B_1(a_{l_k})$ depending on the result of the l_k-th inspection for element m. Specifically, if at the time of the l_k-th inspection, element m is found to have failed, then Y stands for A and if element m is found to have a crack of length between a_{l_k} and $a_{l_k} + da_{l_k}$, then Y stands for $B_1(a_{l_k})$. Finally, for the case where element m is found intact at all inspections prior to the j-th, the product appearing in Eq. 16 is set equal to unity and Eq. 16 takes the form:

$$LF_j^{(m)} = P_m \{X : j, l_0\} \qquad (17)$$

where l_0 denotes the time of initiation of service for the structure.

5.3 Posterior joint density function of uncertain parameters

The posterior joint density function of the three uncertain parameters β, c and d, immediately after the j-th inspection, is given by:

$$f^j(\beta, c, d) =$$

$$\frac{LF_j \cdot f^0}{\int_{\beta_{min}}^{\beta_{max}} \int_{c_{min}}^{c_{max}} \int_{d_{min}}^{d_{max}} (\text{Numerator}) d\beta dc dd} \qquad (18)$$

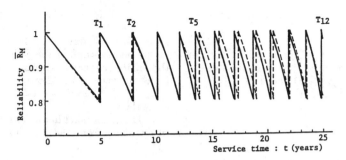

Fig. 1   Inspection schedule and structural reliability (continuous line: uncertain parameters
β and d, dotted line: true values for β and d)

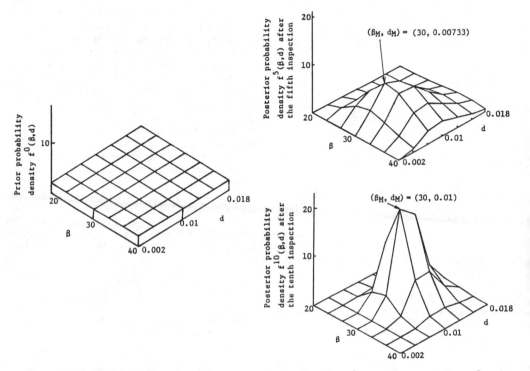

Fig. 2   Prior and posterior joint probability density functions [uncertain parameters: β and
d, true values $(\beta,d)=(30,0.01)$]

5.4   Reliability of entire structure at time instant t^* after the latest inspection T_j

The reliability of the entire structure consisting of M elements at time instant t^* after the latest inspection T_j is denoted by $\bar{R}_M(t^*)$ and calculated as:

$$\bar{R}_M(t^*) = \int_{\beta_{min}}^{\beta_{max}} \int_{c_{min}}^{c_{max}} \int_{d_{min}}^{d_{max}}$$

$$R_M(t^* \mid \beta, c, d) f^j(\beta, c, d) \, d\beta dc dd \quad (19)$$

where:

$$R_M(t^* \mid \beta, c, d) = \left[\prod_{m=1}^{M_1} R_m(t^*; \text{Repair}) \right] \cdot$$

$$\cdot \left[\prod_{m=1}^{M_2} R_m(t^*; \text{No Repair}) \right] \quad (20)$$

where M_1 = number of elements either repaired or replaced at the j-th inspection,
M_2 = number of elements found intact at the
j-th inspection and $M_1 + M_2 = M$. In Eq.
20, $R_m(t^*;\text{Repair})$ and $R_m(t^*;\text{No Repair})$ are
identical with the reliabilities $R(t^*;\text{Repair})$
and $R(t^*;\text{No Repair})$ defined earlier. The
subscript m is used to indicate that these reliabilities are associated with element m.

6 CALCULATION OF TIME T_{j+1} FOR NEXT INSPECTION

Assuming that the entire structure must
maintain its reliability above a prespecified
design level throughout its service life, the
time T_{j+1} for the next inspection after the
latest one performed at T_j, is calculated using:

$$\bar{R}_M(t^*) \geq R_{\text{design}} \qquad (21)$$

where R_{design} denotes the prespecified design
level of reliability for the entire structure.

The time T_{j+1} of the $(j+1)$-th inspection
is then estimated as the maximum value of t^*
that satisfies Eq. 21.

7 NUMERICAL EXAMPLES

The structure considered in this study is
assumed to have 100 elements (M=100).
Its service life is 25 years and the minimum reliability level for the entire structure
throughout its service life is set equal to 0.80
(R_{design}=0.80). In general, three uncertain
parameters can be considered: β, c and d. In
this paper, only the case where β and d are
uncertain has been examined because of space
limitations (refer to Deodatis et al. (1992)
for more cases). Table 1 shows the values of
all parameters (uncertain and deterministic)
involved in the problem. Note that the two
uncertain parameters β and d are given true
(deterministic) values along with their ranges
(indicating their uncertainty). The reason for
providing true values for the uncertain parameters is to use them to perform numerical simulations that will be described in the
following.

Because of the lack of actual data, numerical simulations are performed in order
to verify the validity and effectiveness of the
Bayesian analysis to determine: (a) the appropriate inspection intervals so that the reliability of the entire structure remains above
the minimum reliability level throughout its

service life, and (b) the true values of the uncertain parameters.

The numerical simulations are performed
in two stages (Deodatis et al., 1992). In the
first stage, the results of a specific inspection,
e.g. T_j, are simulated using the true values
of the two uncertain parameters β and d. In
the second stage, the simulated results of inspection T_j are used to determine the time of
the next inspection T_{j+1}. During the second
stage, parameters β and d are considered to
be uncertain. This two-stage procedure is followed at all inspections during the service life
of the structure.

At this juncture, it is very important to
note that if the Bayesian inspection methodology developed in this paper is applied to
a real structure where inspection results are
available, the first stage of the methodology
involving the simulation of inspection results
is obviously unnecessary. In such a case, the
actual (real) inspection results should be used
directly in the second stage to determine the
time of the next inspection T_{j+1}.

The results of one simulation for the case
where parameters β and d are uncertain are
displayed in Table 2. These results include
the inspection schedule, number of failed elements and number and length of detected
cracks. The corresponding structural reliability for the entire structure as a function of
time is plotted in Fig. 1. It is interesting
to note that the twelve inspections shown in
Table 2 (needed to maintain the reliability of
the entire structure above the minimum level
$R_{\text{design}} = 0.80$ throughout its 25-year service
life) are not periodic in time. The time interval between subsequent inspections becomes
shorter as a function of time. This is a logical
consequence of the fact that as the structure
gets older, more frequent inspections and repairs are needed.

Finally, the posterior joint density function of the uncertain parameters after the
fifth and tenth inspections is plotted in Fig.
2. In this figure, it can be easily seen that
the modal value of the posterior joint density
function after ten inspections coincides with
the true values of the uncertain parameters.

8 CONCLUSIONS

It has been therefore demonstrated in the numerical examples section that the proposed
Bayesian analysis methodology is capable to:
(a) determine appropriate non-periodic inspection intervals of fatigue-sensitive struc-

tures, so that their reliability remains above a prespecified minimum level throughout their service life, and (b) estimate the values of the uncertain parameters using Bayesian upgrading of their joint density function.

ACKNOWLEDGMENTS

This work was partially supported by the American Bureau of Shipping under Contract No. ABS RD-1 and by the United States Coast Guard under Contract DTCG23-86-C-20057.

REFERENCES

Deodatis, G., Fujimoto, Y., Ito, S., Spencer, J. and Itagaki, H. (1992). "Non-periodic inspection by Bayesian method I," *Probabilistic Engineering Mechanics*, Vol. 7, No. 4, pp. 191–204.

Fujimoto, Y., Itagaki, H., Itoh, S., Asada, H. and Shinozuka, M. (1989). "Bayesian Reliability Analysis of Structures With Multiple Components," *Proc. of the 5th ICOSSAR*, San Francisco, California, pp. III.2143–III.2146.

Itagaki, H. and Asada, H. (1977). "Bayesian Analysis of Inspection-Proof Loading-Regular Inspection Procedure," *Proceedings of HOPE International JSME Symposium*, pp. 481–487.

Itagaki, H. and Yamamoto, N. (1985). "Bayesian Analysis of Inspection on Ship Structural Members," *Proc. of the 4th ICOSSAR*, Kobe, Japan, pp. III.533–III.542.

Itagaki, H. et al. (1976). "On the Estimation of the Probability Distribution Function of Defects in a Structure," *Journal of the Society of Naval Architects of Japan*, Vol. 139, pp. 307–316.

Ito, S., Deodatis, G., Fujimoto, Y., Asada, H. and Shinozuka, M. (1992). "Non-periodic inspection by Bayesian method II: Structures with elements subjected to different stress levels," *Probabilistic Engineering Mechanics*, Vol. 7, No. 4 pp. 205–215.

Paliou, C. and Shinozuka, M. (1987). "Reliability and Durability of Marine Structures," *Journal of Structural Engineering*, ASCE, Vol. 113, No. 6, pp. 1297–1314.

Shinozuka, M. (1976). "Development of Reliability-Based Aircraft Safety Criteria: An Impact Analysis," AFFDL-TR-76-31, Vol. 1.

Shinozuka, M., Itagaki, H. and Asada, H. (1981). "Reliability Assessment of Structures With Latent Cracks," *Proceedings of US-Japan Cooperative Seminar "Fracture Tolerance Evaluation"*, Honolulu, USA, pp. 237–247.

Yang, J.-N. and Trapp, W.J. (1974). "Reliability Analysis of Aircraft Structures Under Random Loading and Periodic Inspection," *AIAA Journal*, Vol. 12, No. 12, pp. 1623–1630.

Structural Safety & Reliability, Schuëller, Shinozuka & Yao (eds) © 1994 Balkema, Rotterdam, ISBN 90 5410 357 4

Inspection strategy for deteriorating structures based on sequential cost minimization method Part 1: Framework of the method

Mamoru Mizutani
Tokyo, Japan

Yukio Fujimoto
Hiroshima University, Japan

ABSTRACT: A Sequential cost minimization method and its consistent formulation are presented for the inspection planning problem of fatigue deteriorating structures. The method aims to find an optimal inspection strategy so that the total cost expected in the period between the present inspection and the next be minimum. The optimization is repeatedly carried out at every inspection. Although this sequential cost minimization is not a lifetime optimization method of inspection strategy, the minimization is to be accomplished based on updated information with realistic computational effort even for a large system consisting of several kind of different member sets. Subsequently, it is found that the inspection strategy given by this method in not but quite close to the lifetime optima.

1. INTRODUCTION

Structural Reliability through the service years can be achieved with sufficient inspection and rapier maintenance, as well as by careful design. However excessive assurance of reliability is not economically accepted.

It is thought that a structure should have its optimal reliability level depending on the specification, inspection and repair expenses, risk against unexpected failure and so on. The best way to determine such reliability level is to employ decision making on the basis of lifetime cost minimization.

Several lifetime cost minimization methods have already been proposed. However, those methods are usually very complicated and need much computational effort. In addition to these, lifetime cost evaluation requires every related information at the first step; at the time of planning of structure, which induce large uncertainty in evaluation results associated with long term estimation of structural member behavior and expenses of all related items.

In this study, a sequential cost minimization method and its consistent formulation for the inspection planning of deteriorating structures are presented. The method aims to find an optimal inspection strategy so as to minimize the total cost in the period between the present inspection and the next. This is not a life time optimization method however the optimization with this method carried out based on updated information at each time of inspection and with realistic computational effort.

Here in inspection planning problems, inspection interval, inspection method, repair quality and so on are thought to be the optimization variables. For actual structures, however, the times of inspections are usually limited such as once a year or once every two years from several operational and economical reasons. Also, repair quality is determined such that the similar damage will never take place again in the member after that. On such conditions the optimization is mainly achieved by the selection of most suitable strategy from the possible combinations of inspection intervals and inspection methods allowed for the structure.

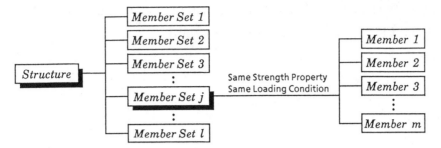

Figure 1 Hierarchy of structure, member sets and members

2. SEQUENTIAL COST MINIMIZATION METHOD

2.1 Total cost for structure during a period of inspection.

For the purpose of this study, structures can be modeled to form hierarchic configuration as shown Fig. 1 that they consist of a variety of member sets which consist of different types and numbers of members. In this study the following assumptions are made for the formulation of the problem.

All the structural members in each member set have the same strength property and are subjected to the same loading condition. Each member has a possibility of failure caused by fatigue damage. If any member fails, the service of structure is suspended urgently and the failed member is to be repaired. With certain probability, a member failure might result in a catastrophic failure; total failure of the structure. Inspections are repeatedly carried out during the service life to find defects, and the detected damages are to be repaired.

At a certain inspection time, the total cost, expected expense including risks, for the entire structure in the succeeding inspection interval is classified into two groups:
1) costs necessary in the present inspection, and,
2) risks (costs) during the period until next inspection [2].

The total cost for the whole structure of an inspection period, from time t to time t+1, denoted by CT (t, t+1), can be written as:

$$CT(t,t+1) = \sum_{j=Set1}^{Set\ l} C_j(t,t+1) + C_{sws} \qquad (1)$$

where,

$$C_j(t,t+1) = C(inspection) + C(repair) \\ + C(member\ failure) \\ + C(Catastrophic\ failure) \qquad (2)$$

$$C_{sws} = C(scheduled\ systemdown) \qquad (3)$$

In the above equations, C_j (t, t+1) represents the cost during the inspection period (t, t+1) for member set j.

2.2 Selection of optimal inspection method for a member set

If it is assumed that C_{sws} in Eq. (1) is independent on the applied inspection methods to member sets and repair qualities, then the minimization of CT(t, t+1) for the structure is achieved when the inspection method for each member set is selected so as to minimize C_j(t, t+1), (j=1 to i).

The selection is carried out at every inspection time from the following five alternative methods:
No inspection (NO), visual inspection (VI) method, mechanical (precise) inspection (MI) method, visual and conditional mechanical inspection (V&M) method, and sampling mechanical inspection (SM) method.

In the V&M method, visual inspection is carried out for all the members at first. If no defect is found then the inspection is terminated. When defects are found at least in one member, mechanical inspection is carried out for all the members.

In the SM method, first mechanical inspection is carried out for limited number of sample members. If no defect is found among

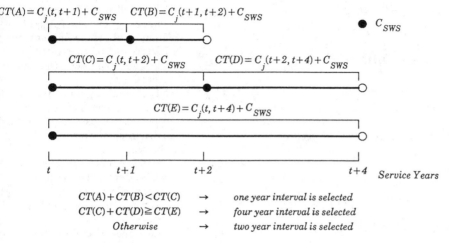

$$CT(A)=C_j(t,t+1)+C_{SWS} \qquad CT(B)=C_j(t+1,t+2)+C_{SWS} \qquad \bullet \ C_{SWS}$$

$$CT(C)=C_j(t,t+2)+C_{SWS} \qquad CT(D)=C_j(t+2,t+4)+C_{SWS}$$

$$CT(E)=C_j(t,t+4)+C_{SWS}$$

t	$t+1$	$t+2$	$t+4$ Service Years

$CT(A)+CT(B)<CT(C)$	$\rightarrow$	one year interval is selected
$CT(C)+CT(D)\geq CT(E)$	$\rightarrow$	four year interval is selected
Otherwise	$\rightarrow$	two year interval is selected

Figure 2 Selection of suitable inspection interval

the sample members, then the inspection is terminated. When defects are found at least in one of the members, mechanical inspection is applied to all the rest of the members.

2.3 Selection of suitable inspection interval

The selection of suitable inspection interval for the structure is to be achieved by comparison of total costs evaluated for all potential interval plans applying the proposed method. For simplicity to explain, let us assume a case of the choice from two inspection interval plans, once a year and once every two years.

1) First, set the inspection interval for one year (t, t+1), and select the optimal inspection methods for each member sets. Then, evaluate the total cost CT (t, t+1) using Eq. (1).
2) Repeat 1) for the next inspection interval (t+1, t+2) to evaluate $CT(t+1, t+2)$.
3) Sum up these two costs and obtain the total cost in the period from the inspection time t to time t+2 corresponding to annual inspection case:

$$CT(t,t+2)=\sum_{j=Set1}^{Set\ l}C_j(t,t+1)$$

$$+\sum_{j=Set1}^{Set\ l}C_j(t+1,t+2)+2\times C_{SWS} \qquad (4)$$

4) Secondly, change the inspection interval to two years (t, t+2), and carry out the selection of the optimal inspection method for each member set to evaluate CT (t, t+2).

$$CT(t,t+2)=\sum_{j=Set1}^{Set\ l}C_j(t,t+2)+C_{SWS} \qquad (5)$$

5) The choice of the optimal inspection interval for the structure is to be done comparing the results of above two calculations.

Fig.3 shows the selection procedure of an inspection interval for the case that three intervals of once a year, once every two years and once every four years are allowed for the structure.

3. CALCULATION OF C_j (t, t+1)

3.1 Cost evaluation equations

The total cost for a member set in an inspection period (t, t+1) is evaluated by the following equations for the five inspection methods.

No inspection (NO)

$$C_j(t,t+1|VI)=G\times P_{F1}\times C_F \qquad (6)$$

Visual inspection (VI) method

$$C_j(t,t+1|VI) = G \times \{C_{VI} + P_{DV} \times C_{RD}$$
$$+ (1-P_{DV}) \times P_{F2} \times C_F\} \qquad (7)$$

Mechanical inspection (MI) method

$$C_j(t,t+1|MI) = G \times \{C_{MI} + P_{DM} \times C_{RD}$$
$$+ (1-P_{DM}) \times P_{F3} \times C_F\} \qquad (8)$$

Visual and conditional mechanical inspection (V&M) method

$$C_j(t,t+1|V\&M) = G \times [C_{VI} + P_{DV} \times C_{RD}$$
$$+ (1-P_{DV})[(1-P_{DV})^{G-1} \times P_{F2} \times C_F$$
$$+ \{1-(1-P_{DV})^{G-1}\} \times [C_{M1} + P_{DM} \times C_{RD}$$
$$+ \{(1-P_{DV}) \times P_{F3} \times C_F\}]] \qquad (9)$$

Sampling mechanical inspection (SM) method

$$C_j(t,t+1|SM) = a \times G \times \{C_{MI} + P_{DM} \times C_{RD}$$
$$+ (1-P_{DM}) \times P_{F3} \times C_F\}$$
$$+ (1-a) \times G \times [\{1-(1-P_{DM})^{aG}\}$$
$$\times \{C_{MI} + P_{DM} \times C_{RD}(1-P_{DM}) \times P_{F3} \times C_F\}$$
$$+ (1-P_{DM})^{aG} \times P_{F1} \times C_F] \qquad (10)$$

In the above equations,

$$C_F = (1-P_{FC}) \times (C_{SWA} + C_{RF}) + P_{FC} \times C_{CF} \qquad (11)$$

$$G = m \times P_{SV} \qquad (12)$$

and

m	: Number of members in the member set.
α	: Sampling rate of members in the SM method.
P_{F1}, P_{F2}, P_{F3}	: Occurrence probabilities of member failure in succeeding inspection interval on the conditions that the NO, the VI or the MI methods are applied at the present inspection, respectively.
P_{DV}, P_{DM}	: Probabilities of detecting a defect by the VI or the MI methods respectively.
P_{FC}	: Probability that a member failure develops into a catastrophic failure of the structure.
P_{SV}	: Probability that a member has not experienced repair and failure until the present inspection.
C_{VI}, C_{MI}	: Visual and mechanical inspection expenses of a member, respectively.
C_{ZMI}	: C_{ZMI} equals C_{MI} when the detected damage needs to be inspected mechanically for sizing to determine the repair method, otherwise C_{ZMI} is zero.
C_{RD}	: Repair expense of a damaged member detected by visual or mechanical inspection.
C_{RF}	: Repair expense of a failed member.
C_{CF}	: Loss due to a catastrophic failure.
C_{SWA}	: Loss due to the service suspension caused by accidental system down.

3.2 Probability estimation using Markov Chain model

The probabilities appearing in the cost evaluation equations are calculated by the Markov Chain Model (MCM). Basically, in the simplest Bogdanoff and Kozin stationary MCM, an initial state vector A(0) and a duty cycle independent basic transition matrix P are sufficient to describe the entire probabilistic feature of fatigue Process. In this study fatigue crack initiation and propagation processes are incorporated into the transition matrix of a single MCM [1]. The probability of member failure is also considered in the absorbing term $a_m(t)$ in the state vector A(t).

Let $A_{BI}(t)$ be the state vector just before the inspection at time t.

$$A_{BI}(t) = (a_1(t), a_2(t), a_3(t), \cdots, a_{m-1}(t), a_m(t)) \qquad (13)$$

The state vector right after the visual inspection can be given by the below equation.

$$A_{AI}(t) = (a_1'(t), a_2'(t), a_3'(t), \cdots, a_{m-1}'(t), a_m(t)) \qquad (14)$$

where,

$$a_i'(t) = a_i(t) \times (1-D(i)) \quad (perfect\ repair\ model)$$

Table 1 Selected Inspection Methods for the Structure ($C_{SWS} = \text{US\$1} \times 10^6$)

Member set	Number of Members	$\overline{N}_C$ (years)	$\overline{N}_P$ (years)	P_{ID}	Selected Methods							
					4	8	10	12	14	16	18	20
A	100	20	10	0.00	M	M	M	M	M	M	M	M
B	100	30	15	0.00	-	V	M	M	M	M	M	M
C	100	50	20	0.10	M	M	M	M	M	M	M	V
D	100	50	15	0.01	M	M	M	M	M	M	M	M
E	100	100	15	0.00	-	-	-	-	-	V	M	V

- : No Inspection　　　V : Visual Inspection　　　M : Mechanical Inspection

$$a_i'(t) = a_i(t) \times (1 - D(i)) + P_{DV} \times a_i^{(0)}$$

$$(replacement \ mod\ el)$$

In which , $D(i)$ is the detection probability of a crack whose size is classified to the i-th state.

The probability P_{SV} and P_{DV} are calculated by:

$$P_{SV} = \sum_{i=1}^{m-1} a_i(t) \tag{15}$$

$$P_{DV} = \sum_{i=1}^{m-1} \left(\frac{a_i^{(t)} \times D_{VI}^{(i)}}{P_{SV}} \right) \tag{16}$$

The state vector at the next inspection with k times state transitions is calculated by the following equation.

$$A_{BI}(t+1) = A_{AI}(t|VI) \times P^k \tag{17}$$

The failure probability P_{F2} is obtained by the below equation.

$$P_{F2} = \frac{a_m^{(t+1)} - a_m^{(t)}}{P_{SV} \times (1 - P_{DV})} \tag{18}$$

in which $a_m(t+1)$ is the absorbing term of $A_{BI}(t+1)$ in Eq. (17). Similarly, if we apply the mechanical inspection and calculate $A_{AI}(t \mid MI)$, the failure probability P_{F3} can be obtained.

If the V&M method is selected as the cost minimum, then the Markov state vector right after the inspection is expressed by the combination of $A_{AI}(t \mid VI)$ and $A_{AI}(t \mid MI)$:

$$A_{AI}(t|V\&M) = (1 - P_{DV})^G \times A_{AI}(t|VI)$$
$$+ \left\{ 1 - (1 - P_{DV})^G \right\} \times A_{AI}(t|MI) \tag{19}$$

When the SM method is selected, the state vector right after the inspection is expressed as the following equation.

$$A_{AI}(t|SM) = \alpha \times A_{AI}(t|MI) + (1 - \alpha)$$
$$\times \left[(1 - P_{DM})^{\alpha G} \times A_{BI}^{(t)} \right.$$
$$\left. + \left\{ 1 - (1 - P_{DV})^{\alpha G} \right\} \times A_{AI}^{(t/MI)} \right] \tag{20}$$

4. NUMERICAL EXAMPLE

The proposed method was applied to a hypothetical structure consisting of five member sets with different fatigue properties. The number of the members was 100 for sets A, B, C and D and 500 for set E. The surface fatigue cracks initiated from the weld toe of members were treated as the deterioration damage.

The fatigue crack initiation life N_C and propagation life N_P were assumed to follow two parameter Weibull distribution with shape parameter, 2.5 and 4.0, respectively. The mean crack growth curve was described by Paris' equation with stress intensity factor range ΔK calculated by linear elastic fracture mechanics. The mean fatigue lives and the probability of existence of initial defect, P_{ID} of each member set are shown in the Table 1.

The contents of the cost items are summarized in Fig. 3, in which C_{RD} and C_{RF} are treated as functions of service time. These

Cost US $

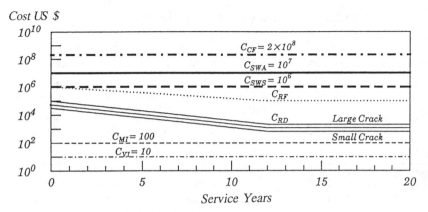

Figure 3 Cost items and their value

Cost $\times 10^6$ US$

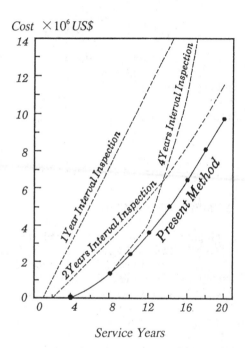

Service Years

Figure 4 Cumulative cost for the structure
(C_{SWS}=1$\times 10^6$US$)

expenses and costs were applied for all the member sets.

The following equations were assumed as the inspection capability of the VI and the MI methods, in which a is the observable crack depth at the inspection.

$$POD_{VI} = 1.0 - exp\{-0.1 \times (a - 0.5)\}$$

$$POD_{MI} = 1.0 - exp\{-0.4 \times (a - 1.0)\}$$

The scheduled system down cost, C_{SWS}, was considered for the whole structure and $C_{SWS} = 10^6$ US dollars was given. The selected inspection timing and qualities by this method are shown in Table 1.

Fig. 4 shows the superiority of the proposed method. The three dashed curve express the cumulative total costs required for the structure when the prefixed inspection intervals, one year, two years and four years, are applied throughout the life. In these cases, optimal inspection methods are selected for each member set at every inspection time. The solid curve with black circles is the cumulative cost calculated by the present method and this cost is always minimum throughout the service time. In the examples only the NO, the VI and the MI methods were selected as the optimal, and the V&M and the SM methods were never selected. This fact was also observed in the analysis of another structure [3].

The sequential cost minimization method presented herein aims to find an optimal inspection strategy so that the total cost in the period between the present inspection and the next be minimum. Therefore the method can not be regarded as an approach for the lifetime cost optimization.

The relation of the strategy obtained by the present method to the lifetime optimal was also examined. Basically, inspection strategy aiming at the lifetime cost optimization can be achieved by selecting one with the minimum lifetime cost from all of possible alternative inspection plans.

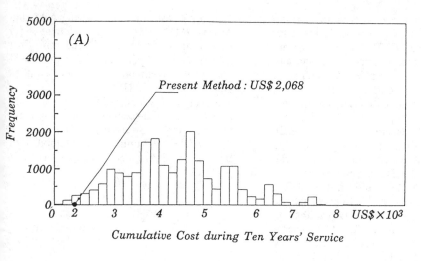

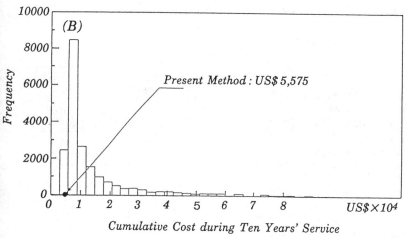

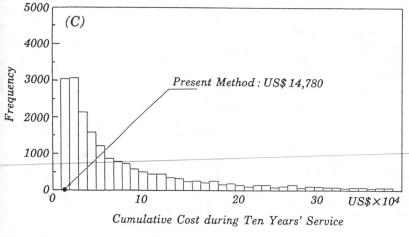

Figure 5 Histgrams of lifetime costs for all the possible plans

For simplicity, assume a structure which consists of a single member set and designed for service of ten years. The inspections are assumed to be restricted to once a year and any one of the the NO, the VI or the MI methods is applicable at each inspection. For the allowed nine inspections during the service, the number of possible inspection plans is $3^9 = 19, 683$, because three methods are applicable at each inspection. Figs. 5 shows the histograms of the lifetime costs for all the possible plans for the three cases analyzed. These three cases differ in crack initiation life and initial defect condition. It is seen from the figures that the lifetime cost distributes widely depending on the applied inspection plan. The lifetime cost calculated for the inspection plan selected by the present method is located quite near the lower bound of the distribution for all the three cases.

5. CONCLUSIONS

The framework of a sequential cost minimization method is presented for the inspection planning of fatigue deteriorating structures, with demonstration of the applicability of the method using a hypothetical structure consisting of five member sets with different fatigue properties.

The major merit of the method can be summarized:

1) the optimization can be accomplished with smaller computational effort even for a large system, and

2) the cost equation is to be calculated with updated information.

The sequential method is not based on the life time cost optimization but aims to minimize the cost in each inspection interval. However, it is revealed by several case studies that the strategies selected by the method are quite close to the lifetime optimal one in most cases.

REFERENCES

1. Fujimoto, Y., Ideguchi, A. and Iwata, M., Reliability Assessment for Deteriorating Structure by Markov Chain Model, J. of Soc. of Naval Architects of Japan, Vol.166, pp.303-314, (1989)(in Japanese).

2. Fujimoto, Y., Mizutani, M., Swilem, A.M. and Asaka, M. : Inspection Strategy for Deteriorating Structures Based on Cost Minimization Approach,, Computational Stochastic Mechanics, Elsevier Applied Science, p.421-432 (1991).

3. Fujimoto, Y., Swilem, A.M. Iwata, M. and Nagai, K. : Inspection planning for Deteriorating Structures Based on Sequential Cost Minimization Method, J. of Soc. of Naval Architects of Japan, Vol.170, p.755 (1991).

Structural Safety & Reliability, Schuëller, Shinozuka & Yao (eds) © 1994 Balkema, Rotterdam, ISBN 90 5410 357 4

Inspection strategy for deteriorating structures based on sequential cost minimization method (Part 2: Considerations of inevitable uncertainties)

Yukio Fujimoto
Hiroshima University, Japan

Mamoru Mizutani
Tokyo, Japan

ABSTRACT: In the companion paper, a sequential cost minimization method is presented for the inspection planning problem of fatigue deteriorating structures. In this paper, first of all, the applicability of this method is examined for an actual member set consisting of structural elements with round fillet weld. Then the influence of inevitable uncertainties of parameters on inspection planning are discussed though a Bayesian analysis and two types of sensitivity analyses. It is made clear that the visual and conditional mechanical inspection method and sampling mechanical inspection method become profitable when a large uncertainty exists in the prediction of fatigue life characteristics of members. It is also found that the cost optimization approach of inspection strategy itself has an effect to decrease the influence of large uncertainty in the operating cost parameters.

1. INTRODUCTION

The lifetime cost minimization is the optimal criterion of decision making for inspection and repair maintenance of structures. In the companion paper [1], the sequential cost minimization method is presented for the inspection planning problem of fatigue deteriorating structures. The method aims to find an optimal inspection strategy which minimize the total cost in the period between the present inspection and the next.

The optimization is repeatedly carried out at every inspection by selecting the optimal inspection quality for each member set included in a structure, and also by the selection of most suitable inspection interval for the structure from possible interval plans. The selection is from the following five method; the no inspection (NO), the visual inspection (VI) method, the mechanical (precise) inspection (MI) method, the visual and the conditional mechanical inspection (V&M) method and sampling mechanical inspection (SM) method.

The major advantage of this method to existing lifetime cost minimization methods is that the formulation is rather simple and the computational effort is smaller. In addition to this, the sequential method has a merit in the treatment of uncertainties involved in the cost problems. It is partly because of this method's simplicity and inexpensiveness of data crank out and partly because it is based on rather short term estimation.

In this paper, first, the applicability of the method is examined for an actual member set consisting of structural elements with round fillet weld. Then the influence of uncertainties inevitably involved in the estimation of parameters used in the inspection planning on the optimization is discussed through a Bayesian analysis and sensitivity analyses.

2. APPLICABILITY OF SEQUENTIAL COST MINIMIZATION METHOD

The sequential cost minimization method was applied to a structural member set consisting of 200 structural elements with a round fillet weld. The surface fatigue cracks initiated from the weld toe of the members were treated as the deterioration damage. The mean crack growth curve obtained by the model fatigue test [2] was employed in the analysis. The

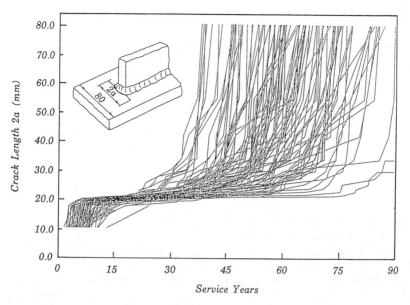

Figure 1 Sample Functions of Fatigue Process Generated by Markov Chain Model

	10 mm	20 mm	21 mm	24 mm
Case A	10 %	—	—	—
Case B	10 %	5 %	5 %	—
Case C	10 %	5 %	5 %	2 %

Figure 2 Initial Crack Conditions

fatigue crack initiation life, N_c, and propagation life, N_P, were assumed to follow two parameter Weibull distributions with shape parameter 3.0 and 5.0, respectively. The member failure was defined when the surface crack length reaches the plate width of 80mm. Fig. 1 shows the fatigue sample functions generated by Markov Chain model. Fig. 2 shows the initial crack conditions for three cases analyzed.

The detection probability of a crack of visual and mechanical inspections, POD_{VI} and POD_{MI} respectively, were assumed to be functions of surface crack length $2a$ as follows, which represent the ability of inspection method :

$$POD_{VI} = 1.0 - exp\{-0.025 \times (2a - 10.0)\}$$

$$POD_{MI} = 1.0 - exp\{-0.10 \times (2a - 10.0)\}$$

The following values were given for the cost items: Expense of the visual inspection; C_{VI} = US$10, and that of the mechanical inspection; C_{MI} = US$100. Loss due to scheduled system down; C_{SWS} = US$10^4, and accidental system down; C_{SWA} = US$10^6. Repair expense of a damaged member ; C_{RD} = US$10^3, and that of a failed member; C_{RF} = US$5×10^5. Loss due to catastrophic failure; C_{CF} = US$2×10^8. R_{FC} = 0.01 was assumed as the transition probability from member failure to a catastrophic (total) failure.

Table 1   Results of Inspection Planning

Case	Selected Inspection Years and Qualities								C_{OP} US$	P_f
A	4	8	12	16	20	24	28	32	3.2×10^5	2.06×10^{-4}
	-	-	-	-	V	V	M	M		
B	4	8	12	16	20	24	28	32	4.0×10^5	2.27×10^{-4}
	-	V	M	M	M	M	M	M		
C	4	8	12	16	20	24	28	32	4.6×10^5	2.57×10^{-4}
	M	M	V	V	M	M	M	V		

- : No Inspection   V : Visual Inspection   M : Mechanical Inspection

Table 2   Results of Inspection Planning - Fixed Interval Inspection

Case	Selected Inspection Qualities																C_{OP} US$	P_f
	2	4	6	8	10	12	14	16	18	20	22	24	26	28	30	32		
A	-	-	-	-	-	-	-	-	-	-	V	V	V	M	M	M	4.2×10^5	3.04×10^{-4}
B	-	-	-	-	V	V	V	M	V	V	V	M	V	V	M	V	5.7×10^5	4.28×10^{-4}
C	-	V	M	V	V	V	V	V	V	V	V	V	V	V	M	V	5.3×10^5	2.83×10^{-4}

- : No Inspection   V : Visual Inspection   M : Mechanical Inspection

Table 1 shows the inspection timing and qualities optimized by the sequential cost minimization method. The values of C_{OP} and P_F in the table are the accumulated total costs and the failure probabilities expected to the selected inspection strategy during 32 years' service.

$$C_{OP} = \sum_{t=0}^{Life} CT(t,t+i), \qquad i = 1\,or\,2\,or\,4 \qquad (1)$$

The results show that according to the initial crack condition, the first inspection timing changes, and that the optimal interval is selected as four years for all the cases.

Table 2 shows the result of inspection planning for the same member set when the inspection interval was fixed as two years. The VI's were often selected in these cases, and the accumulated costs, C_{OP}'s were larger than those in the Table 1.

Comparing the above two examples the effectiveness of the optimization of inspection timing is clearly understood. In both examples and the example in the companion paper, only the NO, the VI and the MI methods were selected as the optimal, and the V&M and the SM methods were never selected. The benefit of the V&M and the SM method is to be discussed in the chapter 4.

3.   CONSIDERATION OF INEVITABLE UNCERTAINTIES

Most of the probabilities and the expenses referred in the cost analysis generally contain uncertainties in the actual field.   These uncertainties may be classified into two categories: engineering uncertainty   and economical/social uncertainty. With respect to the engineering uncertainty, it is further divided into the following two groups.
(1) Reducible uncertainty: Initial uncertainty can be gradually reduced by updating the information gained though inspection and/or service   experiences.   Estimation   of deteriorating   property   of   members, inspection expense, etc. contain this type of uncertainty.
(2) Irreducible uncertainty: Initial uncertainty remains through the whole service of the

structure because of their rare occurrences. Evaluated expenses and probability of a catastrophic failure, loss due to unexpected system down, etc. have this type of uncertainty.

For reducible uncertainty, the Bayesian approach is known to be effective [3]. In this study, a Bayesian analysis is applied in conjunction with the sequential cost minimization method to a structure which has an uncertainty in the estimation of fatigue property of members.

As for irreducible uncertainty, subjective information of engineers and/or owners must be effectively utilized in the analysis. In order to support the decision making by experts, the influence of the uncertain parameters on inspection content, cumulative operating cost and the failure probability must be made clear. For this purpose, two types of sensitivity analysis were carried out.

With respect to the economical/social uncertainty which corresponds to the difficulties in forecasting the change in economical and social condition; price inflation, bank interest ratio and supply and demand balance, it is suggested that the sequential cost minimization method has a merit as it requires rather short term estimation comparing with lifetime optimization methods.

4. BAYESIAN ANALYSIS

4.1 Method of analysis

The influence of fatigue life uncertainty of members on inspection planning was discussed based on a Bayesian analysis [3]. In general fatigue nature of materials is known to some extent, however, the probabilistic nature of particular member may not be predicted precisely.

First, the uncertainty of fatigue life was expressed by a prior distribution function, which was subjectively determined with general information so as to include the true fatigue life in the distribution range. In this study a discrete uniform distribution was assumed.

The information gained in the former inspections was expressed by the likelihood function to update the fatigue life. Then, the

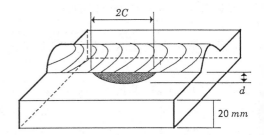

Figure 3   Surface Crack Initiated from Butt Weld Joint

posterior density function of fatigue life denoted by $f^{(n)}(N_c, N_p)$, was calculated by the Byes equation. The total cost C_j (t, t+1) was calculated by the following marginal integration using $f^{(n)}(N_c, N_p)$.

$$C_j(t,t+1) = \int \int C_j(t,t+1/\bar{N}_C, \bar{N}_P) \\ \times f^{(n)}(\bar{N}_C, \bar{N}_P) d\bar{N}_C d\bar{N}_P \qquad (2)$$

The above equation was repeatedly applied at the selections of inspection qualities and intervals.

4.2 Numerical example

Assume a single member set consisting of 200 members with welded joints as shown in the Fig.3. The surface crack initiated from the weld toe of a plate with a 20mm thickness is treated as the deterioration damage. The mean lives of crack initiation and propagation are known as N_c = 25 years and N_p =15 years respectively. The crack growth curve is described by Paris's equation with the stress intensity factor $\triangle K$ calculated by linear elastic fracture mechanics. N_c and N_p are assumed to follow two parameter Weibull distribution with the shape parameters of 3.0 and 4.0 respectively.

Inspections are periodically carried out once a year, and the inspection quality is selected from the following four methods: the No, the VI, the MI and the V&M methods. The assumed POD curves as a function of crack depth d were as follows:

$$POD_{VI} = 1.0 - exp\{-0.20 \times (d-3.0)\}$$

Table 3 Results of Bayesian Analysis

	Inspection Year	1	2	3	4	5	6	7	8	9	10	11	12	13	14	15	16	17	18	19	20
Bayesian Approach	Inspection Method	-	-	V	V&M	V&M	V&M	V&M	V&M	V	V	V	V	V	V	V	V	V	V	V	M
	Cracks Detected	0	0	0	0	0	0	0	0	0	0	0	0	0	1	0	2	1	2	6	16
Truth	Inspection Method	-	-	-	-	-	-	-	-	-	V	V	V	V	V	V	V	M	V	V	M

- : No Inspection   V : Visual Inspection   M : Mechanical Inspection

V&M : Visual and Conditional Mechanical Inspection

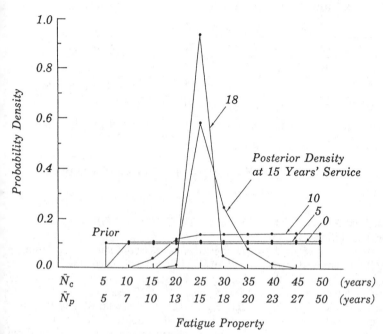

Figure 4 Posterior Density of Fatigue Life

$$POD_{MI} = 1.0 - exp\{-0.40 \times (d-1.0)\}$$

The expenses and costs assumed in the analysis are: C_{VI} = US\$10, C_{MI} = US\$100, C_{SWS} = US\$2×10^5, C_{SWA} = US\$10^6. Repair expense of damaged member, C_{RD}, and that of failed member, C_{RF}, were treated as the time dependent values; C_{RD} = from \$10^3 to \$10^5, C_{RF} = from \$10^5 to \$10^6. Loss due to catastrophic failure; C_{CF} = US\$2×10^8, and the transition probability to a catastrophic failure, P_{FC}, was assumed to be 0.005.

In the analysis, it was assumed that the uncertainty exists only in N_c and N_p, and all the other parameters were statistically determined or deterministic. The ranges of an assumed

prior distribution were (5 years $\leqq \overline{N}_c \leqq$ 50 years) and (5 years $\leqq N_p \leqq$ 30 years). The entire fatigue and inspection processes of the 200 members were simulated by the Monte Carlo method.

Tables 3 shows the results of the Bayesian analysis with the inspection qualities and the numbers of detected cracks at every inspection. The "Truth" in the table expresses the inspection strategy obtained under the condition that the true fatigue property of member set is known in advance.

In the table, the V&M method is selected until the 8th inspection, and the VI method is selected after that. Fig. 4 shows the change of the posterior density of the fatigue life as well as the prior density. The peak is not seen in

1017

Table 4　Influence of Change of P_{FC} on Inspection Strategy

P_{FC}	C_F	Inspection Years and Qualities											C_{OP} (Without Inspection) US$
		4	6	8	10	12	14	16	18	20	22	24	
0.00	1.1×10^6 to 2.0×10^6	V		M		M		V		V		M	2.11×10^5 (1.8×10^6)
0.01	3.1×10^6 to 4.0×10^6	V		M		M		M		M		M	2.66×10^5 (5.1×10^6)
0.05	1.1×10^7 to 1.2×10^7	M		M	M	V	V	V	M	M	V		3.42×10^5 (1.8×10^7)
0.10	2.1×10^7 to 2.2×10^7	M		M	M	V	V	M	V	M	M		3.91×10^5 (3.4×10^7)
0.20	4.1×10^7 to 4.2×10^7	M	M	M	M	V	M	V	M	M	M		4.06×10^5 (6.8×10^7)
0.50	1.0×10^8 to 1.0×10^8	M	M	M	M	M	V	M	M	M	M		4.82×10^5 (1.7×10^8)

$$C_F = (1.0 - P_{FC}) \times (C_{SWA} + C_{RF}) + P_{CF} \times C_{CF}$$

V : Visual Inspection　　M : Mechanical Inspection

the posterior distribution until the 10th inspection. However, the possibilities of the first two combinations of fatigue lives, (N_c = 5, N_p =5) and (N_c = 10, N_p = 7) are removed from the posterior distribution after the 10th inspection. The first crack is detected at the 14th inspection. With accumulation of inspection information, the peak of the posterior distribution becomes clearly apparent at the point of true fatigue life.

Through the similar analysis, it was found that the SM method also becomes profitable when the degree of uncertainty is large in the estimation of fatigue life of members.

When we evaluated the model in the chapter 2 of which all the parameters were statically deterministic or deterministic, each selected inspection quality as the optimal was the NO, the VI or the MI method. However, when large uncertainty exist in the fatigue life information, the V&M and SM would be the optimal method. It may be reasonable when considering that the repair quality is largely related to the member failure risks, and that if the failure risks could be well predicted complicated procedure would not be used as it costs more. The V&M and SM methods are profitable for gaining the fatigue life information.

5. SENSITIVITY ANALYSIS

A single member set consisting of 200 number with welded joints shown in the Fig.3 was here again chosen as the model. In this case, N_c = 50 years, N_p = 20 years, C_{SWS} = US2×10^4 and P_{FC} = 0.05 were assumed and all other parameters are the same as in the chapter 4.

The first sensitivity analysis was carried out by changing a individual parameter to a half or to twice of the basic condition. The parameters were ranked from the viewpoints of sensitivities to the inspection content, the cumulative operating cost and the failure probability. From this analysis, it was found that (N_c, N_p) affect most to the inspection content and the operating cost. When the inspection expenses, C_{VI} and C_{MI} became twice, the use of the VI increased. For the bad quality of POD curve, the VI method with short interval was preferred. On the other hand, the MI method with long interval was selected for a good quality of POD curve. The changes of the scheduled system down loss, C_{SWS} the initial defect condition, P_{ID}, and the transition probability to a catastrophic failure, P_{FC}, considerably influenced the inspection content and the operating cost. The changes of C_{RD}, C_{RF} and C_{SWA} were insensitive for this

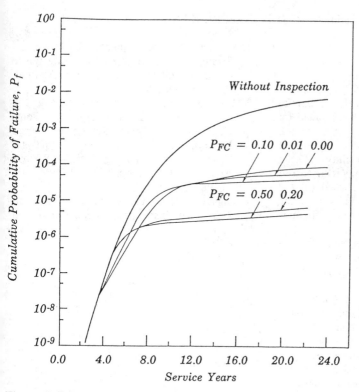

Figure 5 Cumulative Probability of Failure Respective P_{CF}

member set. The obtained sensitivities and insensitivities of each parameter coincide well with the subjective judgment of engineers.

Second, the sensitivity analysis was carried out by giving a large change to a single parameter P_{FC}, the transition probability from a member failure to a catastrophic failure. The initial uncertainty in P_{FC} is usually remain unchanged during the whole service life because of rare occurrence. We must prepare for a wide range of uncertainty of P_{FC} at the inspection planning.

The analysis was carried out changing the P_{FC} gradually from 0.0 to 0.5. $P_{FC} = 0$ means that member failure will never cause a catastrophic failure and $P_{FC} = 0.5$ correspond that member failure will be develop into a catastrophic failure with a probability of 50%.

Table 4 shows the results of inspection planning for respective P_{FC}. With the increase of P_{FC}, the frequency and the quality of inspections became higher. The table compares two C_{OP}'s accumulated total costs though entire service life. The first C_{OP} was

obtained on the condition that the member set was put into operation following the predicted inspection schedule based on the sequential cost minimization method. The second one was on the condition that the member set was put into operation without inspection all through the service life. When no inspections were carried out, the values of C_{OP}'s were influenced dramatically by the change of P_{FC}. However, following the inspection strategy obtained by the method, the effect of the change of P_{FC} to the C_{OP} was considerably small: The change of P_{FC} from 0 to 0.5 results only to twice the C_{OP}. This means that the inspection planning based on the cost minimization approach itself would have an effect to decrease the influence of uncertainty in the estimation of risk parameters.

Fig.5 shows the relationship between the cumulative failure probability, P_F, and the service life for these cases. It is seen that the P_F's are controlled to respective levels depending on the P_{FC}; the more the P_{FC}, the lower the P_F is controlled. This results in

reducing the differences of the estimated C_{OP}'s of evaluated cases.

6. CONCLUDING REMARKS

In this paper the applicability of the sequential cost minimization method is first examined for an actual member set consisting of structural elements with a round fillet weld. Then the influence of inevitable uncertainty of the parameter on inspection planning is discussed utilizing the proposed method based on the Bayesian and the sensitivity analyses. From the numerical analyses the following findings are summarized.

1) A sequential cost minimization method is applicable for the inspection planning of actual structures if the parameters used in the analysis are given.

2) The visual and conditional mechanical inspection method and the sampling mechanical inspection methods become profitable when a large uncertainty is included in the fatigue life of members.

3) The inspection planning based on the cost minimization approach itself would have an effect to decrease the influence of large uncertainty in the estimation of risk parameters.

The advantage of this sequential cost minimization approach is its rather simple formulation and the adoptability to the treatment of uncertainty.

For these decades, huge facilities have been built at many places with the progress of technology. Our lives now depend on such huge facilities in many aspects and their reliability become more and more significant to us. However, the development of construction has been so rapid and we have not have enough time to accumulate knowledge and information of performance of materials, members, components and systems. The information we rely on has large uncertainty.

Considering this state, the authors insist the need for immediate development of practical tools which can treat the uncertainty of information and offer good information to the decision makers. We hope the sequential cost minimization method presented here would contribute to this purpose.

REFERENCES

1. Mizutani, M. and Fujimoto, Y. "Inspection Strategy for Deteriorating Structures Based on Sequential Cost Minimization Method, Part I", ICOSSAR '93.

2. Kawano, H., Kawasaki, T., Sakai, D., Noda, S., Fushimi, A. and Hagiwara, K., Some Considerations on Reference stress Definition in Fatigue strength Analysis, Trans. of The West-Japan Society of Naval Architects, No.83, pp.207-214 (1992) (in Japanese)

3. Itagaki, H. and Yamamoto, N: Bayesian Analysis of Inspection on Ship Structural Members, ICOSSAR '85 Vol.3, p.533 (1985)

Structural Safety & Reliability, Schuëller, Shinozuka & Yao (eds) © 1994 Balkema, Rotterdam, ISBN 90 5410 357 4

Influence of repairs of components on the reliability and the availability of redundant structural systems

M. Fujita
Shimizu Corporation, Tokyo, Japan

R. Rackwitz
Technical University of Munich, Germany

ABSTRACT: A new calculation method of the structural system reliability and availability is proposed accounting for the effect of repair works. Due to the assumption that loads, which damage the structure concerned, rarely occur and they are statistically independent of each other, the concept of Markov's chain can be adopted in calculating the time-variant reliability of the structural system. Availability of the structural system is newly defined based on the reliability theory. From the results of the numerical analysis using Daniels-system, it is concluded that the repair of the system may not improve the reliability but substantially affect the availability.

1 INTRODUCTION

Structures with high risk potential such as nuclear power plants, LNG storage tanks or off-shore platforms must be inspected during their service-time and repaired, if necessary. If they are inspected or repaired, they frequently cannot be used. Therefore, inspection and repair strategies must satisfy not only requirements on the reliability, but also on the availability of such systems. In order to achieve efficient repair procedures it is necessary to calculate the structural system reliability and availability accounting for the effect of repair works.

Recently a new calculation method for this purpose was proposed by Fujita (1990), where it is assumed that severe damage or even loss of the facility can take place only in some rare but extreme loading conditions such as storms for off-shore platforms or operational transients for the other mentioned facilities. The loads form an independent sequence. Each loading event has relatively short duration as compared with the interval times of the loading events. In this paper the results of this research is summarized. As the model of the structural system a highly redundant Daniels-system (Daniels 1945) with ductile components is used. The concept of Markov's chain is utilized in the calculation of structural reliability updated by loadings and repairs.

Researches on system availability has been carried out especially in electrical field (e.g. Henley and Kumamoto 1981), but few researches are available in structural engineering field. So, in this study the availability of structural systems is defined based on the reliability theory and the calculation method of the availability is formulated.

2 CALCULATION MODEL

2.1 System model

Daniels-system shown in Fig.1 is used in this study.

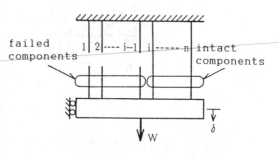

failed components 1 2 ---- i-1 i ------ n intact components

Fig.1 Daniels System

The characteristics of Daniels-system are as follows.

1) Loads are equally shared among intact components.
2) In case of the failure of component, the shared loading effect of the component is redistributed among other intact components.
3) The strengths of components have the same mean value and the same standard deviation and are statistically independent of each other.
4) Failure of all components causes the failure of the system.

Real structures can not be modelized with a single Daniels-system, but with some combinations of parallel and series systems. Calculations among strengths of components and local redistribution of stresses are also to be considered. However, in order to develop and verify new calculation methods simple calculation models like a Daniels-system are required.

In this paper a Daniels-system with n components is used as the system model. It is assumed that each component has 2 states, that is, "intact" and "failure". In case that the degree of damage of components is taken into consideration in the repair strategy, the number of the state needs to be increased.

The system state index Z is introduced and it is defined as follows.

$Z = 1$ ···· The system is intact.
$Z = i$ ···· $(i-1)$ components have failed.
$Z = n+1$ ···· The system has failed.

Z is larger as the degree of damage get to be larger.

2.2 Loading and repair models

Loading effect and repair effect are considered as the effects which change the system state. Fig.2 shows a life of a structure subjected to both loadings and repairs. Structural systems, in general, get damages due to excessive loadings. It is assumed that the events of such excessive loadings occur rarely and they are statistically independent of each other. This assumption seems reasonable in case of extreme earthquakes and waves.

On the other hand, there are problems of deterioration due to long-term loadings, such as normal waves, winds, traffic loads and so on. As to these problems, several methods to determine the optimal inspection times have been recently proposed under the condition of minimizing the maintenance cost keeping the required reliability of structure. (see Fujita, Schall and Rackwitz 1989 and Madsen, Sorensen and Olesen 1989)

The repair is defined as "the replacement of a failed component by an intact one". It is also assumed that the repair starts just after the loading has finished.

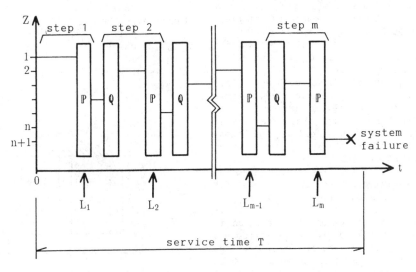

Fig.2   Transition of System State

3 CALCULATION METHOD OF RELIABILITY AND AVAILABILITY

3.1 Reliability in service-time

A pair of a repair and a following loading is called "step". (see Fig.2) The state probability vector of the system after the m-th step is defined as follows.

$$\underline{S}_m{}^T = (S_{m1}, \cdots, S_{mi}, \cdots, S_{m,n+1}) \tag{1}$$
$$0 \leq S_{mi} \leq 1 \quad \sum_{i=1}^{n+1} S_{mi} = 1$$

where S_{mi} denotes the probability that the system state is i just after the m-th step. The system is in the intact state at the beginning of service. Then, the initial state probability vector of the system $\underline{S}_0$ can be represented as follows.

$$\underline{S}_0{}^T = (1, 0, \cdots\cdots, 0) \tag{2}$$

The state transition probability matrices of loading and repair denote P and Q, respectively. By using the concept of Markov's chain, the system probability vector $\underline{S}_m$ can be calculated as follows.

$$\underline{S}_m{}^T = \underline{S}_m{}^T_{-1} Q P = \underline{S}_0{}^T P (Q P)^{m-1} \tag{3}$$

where

$$P = \begin{bmatrix} P_{1,1} & \cdots\cdots\cdots & P_{1,n+1} \\ & P_{i,i} & \cdots P_{i,j} \cdots P_{1,n+1} \\ & & P_{n,n+1} \\ \mathbf{O} & & 1 \end{bmatrix} \tag{4}$$

$$\sum_{j=i}^{n+1} P_{i,j} = 1 \qquad P_{i,j} \geqq 0$$

$$Q = \begin{bmatrix} 1 & & & \mathbf{O} \\ Q_{2,1} & & & \\ Q_{i,1} \cdots Q_{i,j} \cdots Q_{i,i} & \\ 0 \cdots & 0 \cdots & 0 \cdots & 1 \end{bmatrix} \tag{5}$$

$$\sum_{j=1}^{i} Q_{i,j} = 1 \qquad Q_{i,j} \geqq 0$$

$P_{i,j}$ is the probability that the system state of Z=i changes into that of Z=j ($i \leq j$) by a loading. This transition probability is determined by FORM/SORM methods.

As to dynamic systems, the calculation of the transition probability becomes much more difficult. Recently the calculation method

for $P_{i,j}$ at dynamic systems including dynamic effects were proposed by Fujita, Grigoriu and Rackwitz (1988a, 1988b).

While $Q_{i,j}$ is the probability that the system state of Z=i changes into that of Z=j ($i \geq j$) by a repair. Although the calculation method of $Q_{i,j}$ depends on the model of repair and the probability distribution function of repair time, $Q_{i,j}$ is basically calculated as the probability that the repair has finished until the next loading.

$$Q_{i,j} = P (T_R < T_P) \tag{6}$$

where, T_R and T_P are the repair time and the return period of loading, respectively. For example, if the probability density functions for T_R and T_P are

$$g_{TR}(t) = r_o \exp [-r_o t] \quad \text{and}$$
$$f_{TP}(t) = \lambda \exp [-\lambda t], \text{ then,}$$

$$Q_{i,j} = \frac{1}{\lambda / r_o + 1} \tag{7}$$

Strictly speaking, the equation (3) is only correct in the case that the transition of system state at each step is independent of each other. Such case corresponds to the replacement of all components at each repair. However, non-failed components are not replaced in reality and the strength of components has some correlation during several steps. In order to take this correlation of the strength of component during several steps into account, the correction factor is introduced. It is assumed that the strength of the component before repairs is the same variable as the initial strength of the component. The correction factor can be calculated as follows.

$$\alpha_i{}^r = \frac{1}{N} \sum_{k=1}^{N} (\widehat{R}^r_{ik} / \widehat{R}^0_{ik}) \quad (i = 1, \cdots, n) \tag{8}$$

where $\underline{\widehat{R}}^0$ is the order statistics of the initial strengths of components and $\widehat{R}^r$ is the order statistics of the strengths of components after the repair of r components. The correction factor for the strengths of components after the repair of r components α^r can be calculated by using numerical simulations, where N is the sample size. The correlation of the strength of component during several steps can be approximately considered by using the correction factor in the calculation of $P_{i,j}$. Then, the concept of Markov's chain can be used as shown in eq.(3).

The system failure probability after the m-th loading in the service-time of T can be calculated from

$$P_f(m, T) = P_L(m, T) S_{m, n+1} \qquad (9)$$

where $P_L(m,T)$ is the probability that the loading occurs m times in the period of T. Based on the assumption of loading in this study, the loading can be modelized as a Poisson's process.

$$P_L(m, T) = \frac{(\lambda T)^m}{m !} \exp (-\lambda T) \qquad (10)$$

Finally, the system reliability R(T) in the service-time of T is given as:

$$R(T) = 1 - P_f(T) = 1 - \sum_{m=1}^{\infty} P_f(m, T) \qquad (11)$$

3.2 Availability in service-time

For example, at nuclear power plants and off-shore platforms, the operation (productive activities) is temporarily stopped during inspections and repairs. Since long-time stop of operation yields a big loss, more efficient inspection and repair procedures are required. In order to study this problem the definition and the calculation method for the system availability are necessary.

The system availability, in the case that the m-th loading occurred at t_L and the present time is t, is defined as:

$$A_m(t) = P(\overline{V}_m(t_L) \cap \overline{RE}(t)) \qquad (12)$$

where $\overline{V}_m(t_L)$ represents the event that the system did not fail until t_L and $\overline{RE}(t)$ represents the event that the system is not under repairing at t. (see Fig.3)

The system is not available in the state of Z=n+1 (system failure) and the repair is not done in the state of Z=1 (intact system). Then, eq.(12) can be calculated as follows.

$$A_m(t) = P(\overline{V}_m(t_L, Z_1))$$
$$+ \sum_{i=2}^{n} P(\overline{V}_m(t_L, Z_i)) \cdot P(\overline{RE}(t, Z_i)) \qquad (13)$$

Since $P(\overline{V}_m(t_L, Z_i))$ is the probability that the system state is Z=i after the m-th loading, it corresponds to the following state probability.

$$P(\overline{V}_m(t_L, Z_i)) = S_{m, i} \qquad (14)$$

$P(\overline{RE}(t_L, Z_i))$ is the probability that the system state is Z=i and the repair has finished before t. So, it can be basically calculated with eq.(6). Then, the system availability at t can be calculated from the equation below.

$$A(t) = \sum_{m=1}^{\infty} P_L(m, t) A_m(t) \qquad (15)$$

4 NUMERICAL EXAMPLE

4.1 Calculation conditions and results

According to the above-mentioned formulations, the system reliability and availability in the service-time are calculated and the influence of the repair's efficiency on both the probabilities is studied.

The conditions of the calculation are as follows.

1) The system is a static Daniels-system with 10 ductile components.
2) The stress-strain relationship of each component is shown in Fig.4.

m-th loading (m+1)-th loading

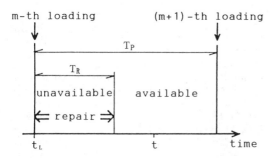

Fig.3 Definition of Availability

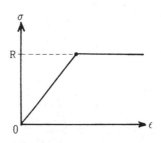

Fig.4 Stress-Strain Relationship

3) The probability distribution function of the strength of component is normal-distribution with the mean value of 0.5 and the coefficient of variation of 10% and 20%.

4) The probability distribution function of the intensity of loading is Gumbel-distribution with the mean value of 3.0 and the coefficient of variation of 20%.

5) The mean value of the return period of loading is 100 and the probability distribution function of the occurrence time of loading is exponential-distribution with the parameter of $\lambda = 0.01$.

6) Failed components are repaired at the same time.

7) The probability distribution function of the repair duration is log-normal-distribution. The repair rate r_0 is 0.0, 0.01 and 0.1. The average repair duration is $1/r_0$.

Fig.5 and Fig.6 show the calculation results for the cases that C.O.V. of the strength of component is 10% and 20%, respectively. The horizontal axis is the service-time and the vertical axis is the system reliability or availability. $R_{max}(t)$ in the figures means the system reliability in the case that failed components are repaired instantaneously. In other words the duration of repairs is equal to zero ($T_R=0$). This is also the maximum value of the system availability, because the repair rate r_0 is infinity ($r_0/\lambda = \infty$). $R_0(t)$ means the system reliability in the case that failed components are not repaired. The curves with $r_0/\lambda = 0$, 1, 10 are the system availabilities at each repair rate.

4.2 Influence of repair on system reliability

The difference between $R_{max}(t)$ and $R_0(t)$ indicates the maximum influence of repairs on the system reliability. In the case that the C.O.V. of the strength of component is 20%, it can be seen that the system reliability is improved by repairs. (see Fig.6) On the contrary, in the case of 10% repairs do not contribute to the improvement of the system reliability. (see Fig.5)

The reason why the influence of repairs on the system reliability is very small is that the damage probability of the system, i.e. the failure probabilities of components, is very small and the state probabilities of $Z=2\sim n$ are very small as compared with those of $Z=1$ or $Z=n+1$. This tendency becomes much clearer as C.O.V. of the strength of component is smaller.

It can be said from these results that the repair works are effective for the improvement of the system reliability in case of comparatively high damage probability of the system. For example, precast concrete armor blocks are usually set as wave dissipation revetment. When some blocks are washed away by waves, new blocks will be supplemented. Such repairs can improve the reliability of the revetment system considerably.

4.3 Influence of repair on system availability

It can be seen in Fig.5 and Fig.6 that the variation of the strength of component has significant influence on the system

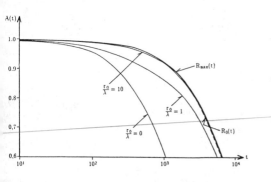

Fig.5   Reliability and Availability
(C.O.V.=   10%)

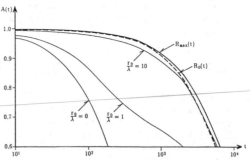

Fig.6   Reliability and Availability
(C.O.V.=   20%)

availability. Therefore, a strict quality control of the strength of the components seems to be required in order to achieve an efficient operation of the system.

It can be also seen that the difference of the repair rate has large influence on the system availability.

However, in case of $r_0/\lambda = 10$ or more the system availability does not vary very much and keeps high value. It means that efficient repairs are important for the improvement of the system availability and that in case of considerably efficient repairs the system availability is not improved very much, even if much more efficient repairs are carried out at great expense.

5 CONCLUSION

A new calculation method of the structural system reliability and availability is proposed accounting for the effect of repair works. The influence of repairs on both the probabilities is studied by using a highly redundant Daniels-system with ductile components.

From the calculation results the following conclusions can be drawn.

1) Under the conditions mentioned before repairs can hardly improve the reliability of the system.
2) The availability of the system increases with the efficiency of repairs.
3) However, in order to improve both reliability and availability strict quality control of the strength of the components appears to be more efficient than sophisticated repair procedures.

These findings which are somewhat unexpected hold only for the special loading scenarios typical for the above-mentioned technical facilities and must not be applied to other cases without appropriate modifications.

REFERENCES

Daniels,H.E. 1945. The Statistical Theory of the Strength of Bundles of Threads, Part1, Proc. of the Royal Society, Series A, Vol. 183, pp.405-435.

Fujita,M., Grigoriu,M. and Rackwitz,R. 1988a. Reliability of Daniels-System Oscillators Including Dynamic Redistribution, Proc. of the 5th ASCE Speciality Conf., pp.424-427, Blacksburg.

Fujita,M., Grigoriu,M and Rackwitz,R. 1988b. Reliability of Daniels-System with Brittle Components, Berichte zur Zuverlässigkeits-theorie der Bauwerke, Heft 84, Technische Universität München.

Fujita,M., Schall,G. and Rackwitz,R. 1989. Adaptive Reliability-based Inspection Strategies for Structures Subject to Fatigue, Proc. ICOSSAR'89, Vol.II, pp.1619-1626, San Francisco.

Fujita,M. 1990. Zur Zuverlässigkeit redundanter Tragsysteme unter Berücksichtigung der dynamischen Wirkungen bei Komponentenversagen und von Reparaturen geschädigter Komponenten, Dissertation, Technische Universität München.

Henley,E.J., Kumamoto,H. 1981. Reliability Engineering and Risk Assessment, Prentice-Hall, Englewood Cliffs.

Madsen,H.O., Sorensen,J.D. and Olesen,R. 1989. Optimal Inspection Planning for Fatigue Damage of Offshore Structures, Proc. ICOSSAR'89, Vol.III, pp.2099-2106, San Francisco.

Structural Safety & Reliability, Schuëller, Shinozuka & Yao (eds) © 1994 Balkema, Rotterdam, ISBN 90 5410 357 4

The fatigue/fracture reliability and maintainability process of structural systems: A summary of recent developments

C. Julius Kung
American Bureau of Shipping, N.Y., USA (Previously: University of Arizona, Tucson, Ariz., USA)
Paul H. Wirsching
Aerospace and Mechanical Engineering Department, University of Arizona, Tucson, Ariz., USA

ABSTRACT: Structures dominated by oscillatory stress are vulnerable to fatigue and/or fracture. But fatigue and fracture design factors are subject to considerable uncertainty. Therefore, probabilistic and statistical methods are appropriate as a tool for managing this uncertainty. The integrity of a fatigue weakened structure can be ensured by a maintenance program of periodic inspection and repair. However, the inspection process also introduces uncertainties that complicate the analysis. This paper presents a summary of recent studies at the University of Arizona analyzing the fatigue/fracture reliability of structural systems subject to a maintenance program. The strategy employed is to use (1) an efficient simulation method, and (2) an efficient structural modeling procedure to obtain accurate solutions with a minimum of computation. Performance of the application of the analysis procedures to a tension leg platform is presented.

1 INTRODUCTION

Structures exposed to random oscillatory environmental loads that produce tensile stresses are vulnerable to fatigue and fracture failure. Fatigue cracks initiate and propagate at the points of stress concentration until failure occurs. A crack weakened structural member is also subject to fracture failure resulting from the application of large load (quasi-static load). But fatigue and fracture design factors are subject to considerable uncertainty. Therefore, reliability methods are appropriate as a tool for making design decisions and risk assessments of structural systems. System reliability can be improved by a maintenance program of periodic inspection and repair. However, inspection itself also introduces uncertainty making reliability analysis difficult.

The fatigue/fracture reliability and maintainability (FRM) process has attracted considerable interest. Efforts have been made to apply the FRM process to United States Air Force airframe and gas turbine engine components, for example, Yang (1976), Manning and Yang (1987), Harris, et al. (1980), and Yang and Chen (1985). Madsen (1985) and Madsen et al. (1987, 1990) developed a numerical algorithm for analyzing the FRM process using first-order reliability method to estimate conditional reliability of the aging structures. Recent contributions to structural maintenance include those of Lotsburg and Kirkemo (1989), Hanna and Karsan (1989), and Paliou et al. (1987).

This paper summarizes recent work at the University of Arizona on reliability assessment of structural systems subject to FRM process (Torng, 1989, Torng and Wirsching, 1991, and Kung, 1991). Parts of this paper have been published (Kung and Wirsching, 1992, 1993). The basic model and suggested methods of analysis for the FRM process are presented herein. An application of the analysis to a tension leg platform (TLP) tendon system is also provided.

2 PROBLEM DESCRIPTION

2.1 Structural System and Failure Modes

The elementary structural model is a parallel/series system (Figure 1). There are M members in parallel. Each member is a series (chain) system of elements having J joints as points of stress concentration. The dynamic load process, $Q(t)$, is applied to the system over the service period. Each member carries the same load equal to Q/M, and each joint is assumed to have the same load. It is assumed that the system will fail when all of the members fail.

2.2 Applied Stresses

Three types of stresses in the members are considered: (1) random fatigue stress, (2) extreme stress, and (3) impulsive separation stress resulting from failure of one or more members. A stress modeling error is employed to account for the uncertainties associated with the assumptions made in stress analysis. The stress modeling error, B, is a random variable and is defined as,

$$B = \frac{\text{actual stress in component}}{\text{predicted stress in component}} \qquad (1)$$

The random variable B applies to the above considered stresses.

2.2.1. Random Fatigue Stress.

The applied load, Q, shown in Figure 1, is a random process. Therefore, fatigue stress range, S, in the members will be a random variable. Basic assumptions in the fatigue analysis are: (1) The stress process is assumed to be narrow band. (2) Mean stresses are not considered for fatigue. (A reasonable assumption for welded joints.) (3) Stress sequence effects are ignored. (4) The long-term distribution of stress ranges has a Weibull distribution; the shape of the stress range spectrum is characterized by the Weibull shape parameter, ξ. (5) The fatigue strength of a material is described by a characteristic S-N curve, $NS^m = A$, where S is stress range, N is cycles to failure, and m and A are the fatigue strength exponent and coefficient, respectively. (6) It is assumed that there is no stress endurance limit. (7) The Paris law is used to described the fatigue crack growth process. (8) There is no threshold level for the stress intensity factor range in the fatigue crack growth process and, therefore, the fracture mechanics model can be cast in a characteristic S-N format. (9) A geometry correction factor, $Y(a)$, can be specified so that cracks growing from a stress concentration can be described. The stress process is "characterized" by an equivalent constant-amplitude (Miner's) stress (Almar-Ness, 1985),

$$S'_e = S_0 \, (\log N_s)^{-1/\xi} \, [\Gamma(\frac{m}{\xi}) + 1]^{1/m} \qquad (2)$$

where S'_e is the best estimated fatigue stress, N_S is the total number of stress cycles in the service life, $\Gamma(\cdot)$ is the gamma function, and S_0 is the once-in-a-lifetime stress (or design stress), i.e.,

$$\Pr(S > S_0) = \frac{1}{N_s} \qquad (3)$$

The actual fatigue stress is

$$S_e = B \cdot S'_e \qquad (4)$$

2.2.2. Extreme Stress

The best estimated extreme stress, S'_E, is the largest quasi-static stress experienced by the structure during the service life. The occurrence of the extreme stress is assumed to be equally likely at any instance during the service life. The actual extreme stress is

$$S_E = B \cdot S'_E \qquad (5)$$

2.2.3. Impulsive Separation Stress

Impulsive separation stresses will develop throughout the whole system upon failure of one or more members. It is assumed that upon failure of k members $(1 \leq k \leq M)$, the

instantaneous impulsive stress will be equal in each remaining number and uniform throughout each member. The impulsive separation stress, S_I, is a random variable by virtue of the relationship to the stress modeling error,

$$S_I = B \cdot S_I'$$ (6)

where S_I' is the impulsive separation stress best estimated by the model of the analyst's choice. The impulsive separation stress is treated in two ways, depending on the mode of first member failure.

2.2.3.1. First Member Failure is Fracture

It is assumed that fracture failure occurs instantaneously, so the load is transferred to the remaining intact member of the system and is modified by three factors that relate to the mechanics of the system. Following failure, the best estimated peak impulsive stress, $S_{I,fr}'$, in the intact members can be written as

$$S_{I,fr}' = \alpha_1 \cdot \alpha_2 \cdot \alpha_3 \cdot S_E'$$ (7)

where α_1 is the system redistribution factor (SRF) accounting for the load sharing throughout the system following the failure of one or more members, α_2 is the load redistribution factor (LRF) for load transfer,

$$\alpha_2 = \frac{m}{m-k}$$ (8)

and α_3 is the dynamic load transfer (DLF) factor for any impulsive dynamic response in addition to the static response defined by α_2. Derivation of the α's is described in (Kung, 1991).

2.2.3.2. First Member Failure is Fatigue

Fatigue failure occurs where crack becomes unstable. It is assumed that a member having a fatigue crack will be more vulnerable to fatigue in the high stress level. Let F be a random variable denoting the stress at the instance prior to fatigue failure. F is assumed to have a three-parameter Weibull distribution. The three Weibull parameters can be obtained using the following conditions: (1) F will be bounded from below by static pretension, S_P, (2) the extreme stress defines the right tail, i.e.,

$$\Pr(F > S_E') = 1/N_s$$ (9)

and (3) a specified median value of F, $\tilde{F}$, i.e., $\Pr(F > \tilde{F}) = 0.5$. $\tilde{F}$ is expressed as

$$\tilde{F} = S_P + \varphi(S_E' - S_P)$$ (10)

where φ, a weighting factor to be determined by engineering judgment, is $0 < \varphi < 1$. Following failure, the best estimated peak impulsive stress is

$$S_{I,fa}' = \alpha_1 \cdot \alpha_2 \cdot \alpha_3 \cdot F$$ (11)

2.3 Failure Modes

Two failure modes are considered in this study. They are (1) a fatigue failure of any joint where there is a stress concentration and (2) a ductile fracture due to a quasi-static load exceeding the yield (ultimate) strength of an element of the member.

2.3.1. Fatigue Failure

A fracture mechanics model is used to describe fatigue in a joint. The model is calibrated to S-N data using the following: (1) S-N data are used to defined the basic parameters in the crack growth model, (2) an equivalent initial flaw size is computed so that the fracture mechanics model will predict the same life as the S-N data, and (3) the Paris coefficient, C, is modeled as a lognormally distributed random variable to account for scatter in the data. Fatigue failure occurs when the instantaneous crack depth, a, exceeds a_f, a specified failure crack depth. An equivalent initial crack depth, a_{0e}, is chosen so

that the Paris model provides the exact same life prediction as obtained from S-N data. Integration of Paris law yields

$$NS_e^n = \frac{1}{C(\sqrt{\pi})^n} \int_{a_{0e}}^{a_f} \frac{dx}{[Y(x)\sqrt{x}]^n} \qquad (12)$$

where n is the Paris exponent. Equation (12) has the same form as the characteristic S-N curve. Assume that $n=m$, it follows that the fatigue strength coefficient can be written as

$$A = \frac{1}{C(\sqrt{\pi})^n} \int_{a_{0e}}^{a_f} \frac{dx}{[Y(x)\sqrt{x}]^n} \qquad (13)$$

2.3.2. Ultimate Strength Failure

The "ultimate" strength of a member is assumed to be

$$R_u = \frac{R(A_0 - A_a)}{A_0} \qquad (14)$$

where R is assumed to be the yield strength and is lognormally distributed, A_0 is the original cross-sectional area, and A_a is the sectional area defined by the crack. It is assumed here that (1) brittle fracture is not a failure mode, and (2) a member will not experience large-scale gross yielding prior to fracture.

2.4. Maintenance Program

The structural integrity may be ensured by a maintenance program having three components, viz., inspection of joints, repair of joints, and replacement of members.

2.4.1. Inspection

Inspection performance is quantified by a POD (probability of detection) curve. The POD curve relates to crack depth, a, and is assumed to be lognormal. For an aspect ratio of 0.1, the half surface crack length would be $10 \cdot a$. Three assumptions are made: (1) no consideration is made for false positives, (2) when a crack is found, its size is accurately measured, and (3) all joints are inspected in each inspection event.

2.4.2 Repair of Joints

An a_{rep} is specified as the criterion of repair. Repair is assumed to be instantaneous. Crack is repaired if $a > a_{rep}$.

2.4.3. Replacement of Members

Replacement of members takes place on the failed members of the survived structure. It is assumed that replacement of members is done immediately.

3 RELIABILITY ANALYSIS

The FRM process is complicated and the reliability analysis of a structural system subject to the FRM process is difficult. One difficulty is that the limit state is not a continuous function of the design variables. The limit state function contains discrete random events which results from maintenance, i.e., cracks repaired and member renewal. The number of random variables resulting from the maintenance program can not be predicted in advance. Moreover, for a large scale system, the number of random variables can be large which further complicates the problem. Because of the complexity of the FRM process, a strictly analytical approach does not seem feasible. And direct Monte Carlo simulation is not generally appropriate because of large sample sizes required to estimate the small probabilities of failure anticipated.

Kung (1991) has proposed an "efficient" method to address the FRM process. The strategy employed in his method is to use (1) an efficient simulation method, and (2) an efficient structural (equivalent member) modeling procedure.

Table 1. Parameters for example used to verify the equivalent-member concept

PARAMETERS	
Service life, N_T	10^7 cycles
Extreme stress, S'_E	299.3 MPa
Design stress, S_0	137.8 MPa
Weibull shape parameter, ξ	1.5
Equivalent Miner's stress, S'_e	27.2 MPa
Geometry factor, $Y(a)$	$\lambda = 1.12, \theta = 0.0$
Paris exponent, m	3.0
Equivalent initial crack depth, a_{oe}	0.472 mm
Failure crack length, a_f	254 mm

RANDOM VARIABLES			
Variables	Distribution	Median	COV
Fatigue strength coefficient, A	lognormal	$3.27 \times 10^{12*}$	0.63
Sress modeling error, B	lognormal	1.0	0.2
Paris coeff., C	lognormal	$3.42 \times 10^{-12*}$	0.63
Yield strength, R	lognormal	551.2 MPa	0.08
P.O.D.	lognormal	5.59 mm	0.51

* in MPa units

3.1. Efficient Simulation

The random variable B is the only variable considered in the simulation. An estimate is made of the probability of failure of the system, p_f. Then, the sampling distribution for B is the right tail of the distribution. The lower bound is B_0, chosen so that the right tail area is about twenty times of p_f. The estimate of the probability of failure of the system is

$$\hat{P}_f = [1 - F_B(B_0)] \cdot \frac{N_f}{N_{sm}} \qquad (13)$$

where F_B is the distribution function of B, N_{sm} is the number of sampled B in area of $B > B_0$, and N_f is the number of observed failures. This procedure is the most crude form of importance sampling but is effective for this problem.

3.2 Equivalent Member Concept

Consider a single member having J joints. The ultimate strength of each joint is a random

Table 2. Parameters for TLP tendon system: reference case

PARAMETERS	
Tendon length	780.3 m
Length/element	9.75 m
Dimension of tendon	
diameter, D	60.9 cm
thickness, t	2.54 cm
area	486.5 cm^2
Number of member, M	4
Number of joints/tendon, J	80
Service life, N_T	10^7 cycles
Fatigue	
design stress, S_0	110.2 Mpa
Weibull shape parameter, ξ	1.45
fatigue exponent, m	3.0
Paris exponent, n	3.0
Miner's stress, S'_e	20.86 Mpa
Equivalent initial crack size, a_{oe}	0.005 mm
Geometry factor, $Y(a)\lambda_1=1., \theta_1=.125$ (penetration)	
$\lambda_1=1.0$, (through crack)	
Fracture	
tendon pretension, Sp	122.6 MPa
current, etc., Sw	43.41 Mpa
extreme stress, S'_E	221.2 Mpa
Impulse response factor,	
fatigue stress, ϕ	0.5
fracture stress, φ	1.0
Failure crcak length, a_f	25.4 cm
Crack length for repair, a_{ren}	7.6 mm

RANDOM VARIABLES			
Variables	Distribution	Median	COV
Fatigue strength coefficient, A	lognormal	$3.27 \times 10^{12*}$	0.63
Sress modeling error, B	lognormal	1.0	0.2
Paris coeff., C	lognormal	$5.8 \times 10^{-12*}$	0.63
Yield strength, R	lognormal	551.2 MPa	0.08
P.O.D.	lognormal	5.59 mm	0.51

* in MPa units

variable denoted as R_i. The fatigue strength of each joint is described by the Paris coefficient, C_i. Thus there are $2J$ random variables. It is assumed that all R_i and C_i are mutually independent. To simplify the model, an equivalent-member approach is proposed. Each

1031

Table 3. Comparison of the reliability estimates using different solution methods (3 inspections; probabilities and rates given in 10^{-4}).

	I[a]	II[b]	III[c]
Total simulation	304,153	290,870	200,000
Total sampled[d] of B_S	5,000	5,000	--
Estimated:			
Safety index, $\hat{\beta}$	3.141	3.145	3.143
90% Confidence Interval on $\hat{\beta}$			
lower limit, $\hat{\beta}_l$	3.113	3.116	3.115
upper limit, $\hat{\beta}_u$	3.172	3.176	3.174
Prob. of failure, $\hat{P}_f$	8.417	8.320	8.362
fatigue-initiated[e]	5.195	5.363	5.250
under extreme load[f]	3.222	2.957	3.112
Expected rate:			
Member replaced			
failure by fatigue	3.72	3.85	3.74
failure by extreme load	0	0	0
Cracks detected	7.46	7.49	7.55
Cracks repaired	5.33	5.12	5.62
CPU seconds[g]	13	975	39,000

[a] Strategy proposed in this paper.
[b] Importance sampling employed in Torng and Wirsching (1991).
[c] Conventional Monte Carlo simulation.
[d] Total number of samples (cases of $B_0 \leq B_S \leq \infty$).
[e] First member failure is due to fatigue.
[f] First member failure is due to extreme load.
[g] CPU seconds on a mainframe computer CONVEX C240 .

member having J joints is modeled as a single member having one joint. Thus the equivalent "ultimate strength" R_e of a member will be the minimum of a sample size of J of R. The equivalent Paris coefficient, C_e, of a member will be the maximum (lowest fatigue strength) of a sample of size J of C. For computation efficiency, R_e and C_e are modeled with Weibull and Frechet distribution, respectively. These distributions are fit to the exact extreme value distributions at two points in the lower and upper tails, respectively. The detail is described in Kung (1991).

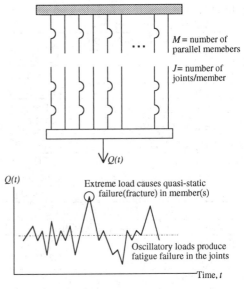

Fig. 1. A parallel series system subject to an external dynamic load, $Q(t)$.

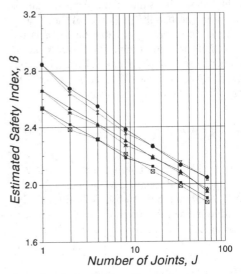

Fig. 2. Estimated safety index as a function of joints and members --- three inspections.

4 VERIFICATION AND APPLICATION

A large number of case studies are performed to verify the equivalent-member concept. The FRM simulation program is capable of analyzing the complete system, providing an estimate of the exact results (Kung and Wirsching, 1991, 1992). Results of the analyses of an example system defined in Table 1 are presented in Figure 2. Similar results obtained from simulation of a variety of examples suggest that the proposed efficient structural modeling procedure provides accurate reliability estimates to the complete structural system.

The performance of the proposed solution strategy was quantified using data for a tension leg platform (TLP) tendon system (Table 2). Comparison of the reliability estimates by this proposed approach with other methods is presented in Table 3. Simulation of 200,000 structures by direct Monte Carlo (Column III, Table 3) requires about 39,000 seconds of CPU time on a CONVEX C240. When importance sampling is employed (Column II, Table 3), the consumed CPU time is about 16 minutes. When both importance sampling and the equivalent-member concept are used (Column I, Table 3), the CPU time drops to only 13 seconds.

5 SUMMARY AND CONCLUSION

The fatigue/fracture reliability and maintainability (FRM) process of a structural system is described in this paper. The FRM process is complicated, and simulation has been suggested as the only viable approach for reliability analysis. Studies at the University of Arizona of reliability of structural systems subject to FRM process are described. These studies have proposed an efficient method that combines importance sampling with an equivalent-member concept. It has been demonstrated that the proposed method provides a dramatic improvement in efficiency without loss of accuracy of the overall reliability or maintenance statistics.

REFERENCES

Almar-Ness, A., 1985, *Fatigue Handbook*, Tapir Publishers, Trondheim, Norway.

Hanna, S. Y. and Karsan, D. I., 1989, "Fatigue Model for Reliability Based Inspection and repair of Welded Tubular Offshore Structures," *Proc. of the 8th Int. Conf. on OMAE*, ASME, New York.

Harris, J. A., Sims, D. L., and Annis, C. G., 1980, "Concept Definition: Retirement for Cause of F100 Rotor Components," AFWAL-TR-80-4118, Wright Patterson Air Force Base, Ohio.

Kung, C. J., 1991, "Fatigue and Fracture Reliability and Maintainability Process for Structural Systems," Ph. D. Dissertation, University of Arizona.

Kung, C. J. and Wirsching, P. H., 1993, "Fatigue and Fracture Reliability and Maintainability of TLP Tendons," *Journal of Offshore Mechanics and Arctic Engineering*, 115(2), pp. 137-141.

Kung, C. J. and Wirsching, P. H., 1992, "Fatigue/Fracture Reliability and Maintainability of Structural Systems: A Method of Analysis," *ASCE 6th Specialty Conference on Probabilistic Mechanics and Structural and Geotechnical Reliability*, New York.

Kung, C. J. and Wirsching, P. H., 1993, "Fatigue/Fracture Reliability and Maintainability Process for Structural Systems," to be published as a chapter in *Computational Stochastic Mechanics*.

Lotsburg, I. and Kirkemo, F., 1989, "A Systematic Method for Planning In-Service Inspection of Steel Offshore Structures," *Proc. of the 8th Int. Conf. on OMAE*, ASME, New York.

Madsen, H. O., 1985, "Random Fatigue Crack Growth and Inspection," *Proc. of the 4th Int. Conf. on Structural Safety and Reliability (ICOSSAR'85)*, Vol. 1, Elsevier, Amsterdam, pp.475-484.

Madsen, H. O., Skjong, R., Tallin, A.G., and Kirkemo, F., 1987, "Probabilistic Fatigue Crack Growth of Offshore Structures with Reliability Updating Through Inspection," *Proc. of the Marine Structural Reliability Symposium*, Vol. SY-23, pp.45-56, SNAME, Jersey City, N.J.

Madsen, H. O., Sorensen, J. D., and Olsen, R., 1990, "Optimal Inspection Planning for Fatigue Damage of Offshore Structures," *Structural Safety and Reliability*, Vol. 3, ASCE, New York.

Manning, S. D. and Yang, J. N., 1987, " USAF Durability Design Handbook: Guidelines for Analysis and Design of Durable Aircraft Structures," AFWAL-TR-86-3017, Wright Patterson Air Force Base, Ohio.

Paliou, C., Shinozuka, M., and Chen, Y. N., 1987, "Reliability and Durability of Marine Structures," *J. of Struct. Engrg.*, ASCE, Vol. 113, N0. 6, pp. 1297-1314.

Torng, T. Y., 1989, "Reliability Analysis of Maintained Structural System Vulnerable to Fatigue and Fracture," Ph. D. Dissertation, University of Arizona.

Torng, T. Y. and Wirsching, P. H., 1991, "Fatigue and Fracture Reliability and Maintainability Process," *J. of Struct. Engrg.*, ASCE, Vol. 117, No. 12, pp. 3804-3822.

Yang, J. N., 1976, "Statistical Estimation of Service Cracks and Maintenance Costs for Aircraft Structures," *J. of Aircraft*, Vol. 13, No.12, pp.929-937.

Yang, J. N. and Chen, S., 1985, "Fatigue Reliability of Gas Turbine Engine Components Under Scheduled Inspection Maintenance," *J of Aircraft*, Vol. 22, No.5, pp.415-422.

Structural Safety & Reliability, Schuëller, Shinozuka & Yao (eds) © 1994 Balkema, Rotterdam, ISBN 90 5410 357 4

Reliability-based fatigue inspection and maintenance for steel bridges

Zhengwei Zhao & Achintya Haldar
University of Arizona, Tucson, Ariz., USA

Florence L. Breen
MARTA, Atlanta, Ga., USA

ABSTRACT: Fatigue is a principal failure model for steel structures subjected to repeated loadings and exposed to environmental effects. Steel bridges are very common in the United States. Accumulated fatigue damage leading to eventual fracture is one of the major factors in deciding whether the life of such bridges can be safety extended. A reliability-based fatigue evaluation method through inspection is proposed in this paper. Information from imperfect inspection will be incorporated in the model. Uncertainties in the crack size measurement and the detectability of the equipment will be considered. This will help the engineer to maintain the overall integrity of the structures by suggesting when to inspect, when to repair and to replace, when to schedule the next inspection. The model combines high technology-based inspection results with sophisticated reliability-based analysis procedure. The method is used to evaluated fatigue reliability of full-penetration butts welds in bridges.

INTRODUCTION

Over one-half of the approximately 600,000 highway bridges in the United States are more than 30 years old (Albrecht and Yazdani 1986). Most of them are over or near the end of their design life and must be repaired, strengthened or reconstructed to insure safety considering present and future traffic needs. In a study under the sponsorship of the American Society of Civil Engineers (ASCE 1982), it was indicated that 80% to 90% of the failures in steel structures are related to fatigue and fracture. Conceptually, the problem can be addressed in two ways. The fatigue sensitive bridge components can be designed properly by careful selection of design loads, methods of analysis, quality control procedures involving material properties and fabrication, and the reliability assessment of the completed structure before it is put into service. Alternatively, since fatigue is a slow process and routine

inspections of bridges are required at regular intervals, inspection results can be used to prevent failure of bridges due to fatigue; this is the main subject of this paper. A probabilistic model is developed in such a way that it can incorporate information from inspections including the inherent uncertainty associated with the inspection outcomes.

PROBLEM DESCRIPTION

Fatigue analysis procedures applicable to the design of steel bridges can be classified into two groups: (1) the studies based on experimental data which led to the development of the S-N curves, including the American Association of State Highway and Transportation Officials' Guide Specifications for Fatigue Design of Steel Bridges; and (2) the studies based on crack propagation theory. Hereafter, the first approach will be called the AASHTO method,

and the second approach will be called the linear elastic fracture mechanics (LEFM) approach. Some of the available LEFM methods may not be directly applicable to the fatigue problem of bridges. Noting this deficiency, Albrecht and Yazdani (1986) proposed a fatigue damage evaluation model for steel bridges using the Paris Equation and the Monte Carlo simulation technique. However, the basic drawback of all AASHTO and LEFM methods, including that of Albrecht and Yazdani, is that they can not incorporate information from the non-destructive inspection (NDI). Since the underlying fatigue reliability of the AASHTO approach is accepted by the profession, any LEFM-based alternative procedure must also have at least the same reliability.

The NDI could be an essential and important tool in fatigue damage evaluation. For a particular NDI technique, several factors are expected to affect inspection results, including modeling effects, human factors and inspection factors. Various NDI techniques have been used for the purpose of detecting cracks. Some of the most common NDI techniques are visual inspection, ultrasonic inspection, liquid penetrant inspection, magnetic particle inspection, magnetic field inspection, and radiographic inspection. Ultrasonic inspection will be emphasized in the following discussion. During the lifetime of a structure, NDI could be conducted several times to insure the integrity of the structure, as required by professional guidelines. Whether any crack is detected or not, each inspection provides additional information and results in changes in the prior estimated reliability as well as in the uncertainty in the basic variables. After an inspection, decisions can be made in a systematic way regarding such issues as modification of inspection plans, changes in inspection methods, and whether it is time to repair or replace the damaged components. The development of such a plan is one of the objectives of this paper.

The problem can be simply stated as follows. A fatigue-sensitive and fracture-critical bridge component has been inspected by an NDI technique k times after the element has been subjected to N stress cycles. During the kth inspection, the results of the inspection can be classified as: (1) no fatigue crack is detected; (2) a crack is detected but its size is not known; or (3) a crack is detected with its size measured. Since the NDI is not perfect, all the inspection results should be treated as random. Now the problem is how to incorporate the information from the previous (k - 1) inspections as well as from the kth inspection using an imperfect NDI to evaluate the current state of fatigue damage, and to make a decision based on the inspection information about what to do next. Incorporation of inspection results in the fatigue damage evaluation would update the uncertainties in the basic random variables and the corresponding reliability index.

AASHTO FATIGUE DESIGN SPECIFICATIONS

The AASHTO fatigue design specifications are based on the S-N curve approach based on experimental data. The current AASHTO fatigue design curves are classified into seven categories: Category A through Category E'. With an identical slope constant of 3 for each curve, the fatigue strength can be defined by a single equation; the intercept log A is different for each category. The linear damage accumulation hypothesis, commonly known as Miner's Rule, is widely used in fatigue design. Combining the S-N curve and the mean stress effect, Miner's damage index was proposed.

FRACTURE MECHANICS APPROACH

The most commonly used crack growth model, the Paris Equation, is used in this study to overcome the major disadvantage of the AASHTO approach. Based on this crack growth model, several important fatigue damage related issues can be addressed. According to Madsen (1985), a function reflecting the damage

accumulation from crack size α_1 to α_2 can be defined as:

$$\Psi(\alpha_2, \alpha_1) = \int_{\alpha_1}^{\alpha_2} \frac{d\alpha}{[F(\alpha, Y)\sqrt{\pi\alpha}]^m} \quad (1)$$

where $F(\alpha, Y)$ is the geometry function; and α and m are the crack size and the fatigue growth exponent parameter, respectively. This damage accumulation function is related to the load accumulation by:

$$\Psi(\alpha_2, \alpha_1) = \overline{CS^m}(N_2 - N_1) \quad (2)$$

where $\overline{S^m}$ is the mean stress range; C is the fatigue growth parameter; and N is the number of stress cycles. For bridge structures, the stress range parameter S is considered to follow the Rayleigh distribution.

An actual bridge is usually subjected to a variable amplitude load process. To consider fatigue under variable amplitude stresses, two possibilities are cycle-by-cycle counting and the mean stress method. The mean stress method may be the most appropriate to study the fatigue damage accumulation in bridges and is used in this study. The critical crack length α_c is another important parameter in the model. It can be defined as the size of the crack causing failure or a design crack size beyond which the serviceability requirements can not be satisfied. Crack sizes corresponding to a specified number of stress cycles can be obtained by the crack growth law. Once α_N exceeds the critical crack size, it can be considered to cause failure.

RELIABILITY ANALYSES OF AASHTO AND LEFM METHODS

It is necessary to calculate the reliability of fatigue-sensitive bridge component design according to the AASHTO and the LEFM methods. The Advanced First-Order Second Moment (ASM) reliability method (Ayyub and Haldar 1984) is used in this study for this purpose. For a limit state function of $g(Z) = 0$, the corresponding failure probability can be calculated as:

$$P_f = P(g(\mathbf{Z}) \le 0) \approx \Phi(-\beta) \quad (3)$$

where ß is the reliability index. In a fatigue damage evaluation process, it is important to keep the reliability index above a preassigned value during the service life of the bridge.

Uncertainty Analyses of AASHTO and LEFM Design Variables

After a comprehensive literature review, the statistical characteristics of all the random variables in the AASHTO and LEFM approaches are identified. They can not be discussed here due to lack of space, but are summarized in Table 1. Two sources of uncertainty in an NDI are considered in this study. For the ultrasonic NDI, the crack detectability is considered to have a lognormal distribution with a mean of 0.254 cm (0.1 inch) and a COV of 0.50. The accuracy of the crack size measurement is modeled by a normal distribution with the measured crack size as the mean and a COV of 0.25.

Fatigue Reliability Analysis Using AASHTO and LEFM Approaches

The limit state function for the AASHTO method can be defined as:

$$\frac{N}{A}E(S^B) - \Delta = 0 \quad (4)$$

where N is the total number of applied stress cycles, A and B are the fatigue strength coefficient and exponent, respectively, S is the variable amplitude stress range parameter, $E(S^B)$ is the expected value of S^B and Δ is the fatigue damage index parameter.

For the LEFM approach, the limit state equation can be expressed as:

$$g(\mathbf{Z}) = \Psi(\alpha_c, \alpha_0) - \overline{CS^m}(N - N_0) = 0 \quad (5)$$

where α_c is the critical crack size, α_0 is the

Table 1. Summary of Statistical Characteristics of Variables

VARIABLE	TYPE	MEAN VALUE	COV
A	LOGNORMAL	3.514×10^{12}	0.450
B	CONSTANT	3.000	0.000
Δ	LOGNORMAL	1.000	0.300
N	LOGNORMAL	1.122×10^7	0.558
S_0	CONSTANT	43.673	0.000
N_s	CONSTANT	2,000,000	0.000
α_0	LOGNORMAL	0.051	0.500
α_c	CONSTANT	5.080	0.000
C	LOGNORMAL	3.94×10^{-12}	0.630
m	NORMAL	3.000	0.100
w	CONSTANT	42.000	0.000
S	RAYLEIGH	54.726	0.655

initial crack size and N_0 is the initial crack period. All other parameters are described earlier. For the welded structures under consideration, N_0 is assumed to be zero.

A geometric function needs to be considered in this case. For bridge components, Paris proposed a geometric function to treat the case of a center-notched specimen, which can be expressed as:

$$F(\xi) = \frac{1-0.5\xi+0.37\xi^2-0.044\xi^3}{\sqrt{1-\xi}} \quad (6)$$

where $\xi = 2\alpha/w$, α is the crack size and w is the width of the specimen. Since α is a random parameter in the proposed model, the geometry function is also random. As an illustration, this geometry function with its associated uncertainties is used in this study.

The limit state function for the LEFM approach represented by Eq. 5 needs to be modified considering inspection outcomes. A fracture-critical bridge component is expected to be inspected several times during its lifetime. Consider that the component is inspected k times and it has experienced N stress cycles at the time of the kth inspection. Three limit state functions can be identified as discussed below.

Case 1 - Event Without Crack Detection: using the damage accumulation function, the limit state function of the event of non-detection of a crack during the kth inspection can then be defined as (Madsen, 1985; Jiao, 1989):

$$I_k = \overline{CS^m}(N - N_0) - \Psi(\alpha_d, \alpha_0) \leq 0 \quad (7)$$

where $\Psi(.)$ and $\overline{S^m}$ are the fatigue damage accumulation function and mean stress effect, respectively.

Case 2 - Event With Crack Detection Without Size Measurement: the limit state function for this event can be obtained by reversing the sign of Eq. 7, since an event without crack detection and an event with crack detection are complementary.

Case 3 - Event With Crack Detection With Size Measurement: generally, when a crack is detected, its size is also measured. The limit state of the crack size measurement can be expressed as:

$$M_k = \Psi(A_k, A_{k-1}) - \overline{CS^m}(N_k - N_{k-1}) = 0 \quad (8)$$

where A_k and A_{k-1} are the measured crack size at the kth and $(k-1)th$ inspection and the corresponding stress cycles are N_k and N_{k-1}, respectively. If A_k is the first measured crack size in the history of inspection, then A_{k-1} will be replaced by the initial crack size α_0 in Eq. 8.

Once the limit state functions are developed for both methods and the statistical descriptions of all the random variables are available, the corresponding reliability indexes can be calculated as discussed earlier. This will be discussed further in the example section.

MODEL UPDATING THROUGH INSPECTION

Whether any crack is detected or not, each inspection provides additional information and results in changes in the estimated reliability and the uncertainties in the basic random variables. Thus, mathematical updating models must be available to consider information from all the inspections including the present one. The updated information on the distribution of basic design variables can then be used to calculate the fatigue reliability. This necessitates the updating of information after each inspection. The Bayesian method is used in this study to update information on the statistical characteristics of basic random variables, and this updated or posterior information on the basic random variables is used to calculate the posterior reliability index. This can not be given here due to lack of space but will be discussed during the presentation.

UPDATING AFTER REPAIR AND REPLACEMENT

The information on the reliability index obtained in the previous section needs to be compared with the predetermined acceptable value of the reliability index. If the updated reliability index is considerably above the targeted acceptable value, the bridge component under consideration is safe. Thus, no action of repair or replacement needs to be undertaken and the bridge can be scheduled to be inspected at the next regular inspection time. If the updated reliability index is around the recommended value, immediate remedial action may not be necessary. However, it may be recommended that the suspected component be inspected more frequently. If the updated reliability index falls far below the acceptable value, the bridge component needs to be repaired or replaced. If the decision is to repair the damaged component, then the initial crack size of the repaired component is expected to be a little larger than the original initial crack size, α_0. However, in this study, the initial crack size after the repair is considered to be α_0. It is must be noted at this stage that the component has already undergone N_{rep} stress cycles before repair.

EXAMPLES

A steel box-girder with a 21.34 meter (70 foot) span, representing a typical bridge in a public transportation system, is considered here for illustrative purposes. The simply supported span has two full penetration butt welds in its tension flange. The welded connection is very conservatively assumed to belong to Category E' of the AASHTO specifications. The bridge is subjected to train loading. A train may consist of any number of cars from one to eight. Considering the peak hour, off peak hour and idle time, the butt welds under consideration are subjected to 112 stress cycles per day. This represents approximately 2,000,000 stress cycles for a design life of 50 years. In order to insure the structural integrity of the bridge, the ultrasonic inspection is assumed to be conducted once every two years. It is assumed that all the butt welds are accessible for inspection.

Fatigue Reliability of Butt Welds According to the AASHTO and LEFM Methods

The reliability index of a typical full penetration

butt weld is evaluated using both the AASHTO and LEFM methods. The critical crack size α_c is considered to be 5.08 cm (2 inches). The reliability indexes corresponding to different numbers of stress cycles are compared using both methods. It was observed that as the number of stress cycles increases, the reliability indexes calculated by the two methods decreases, as expected. It can also be observed that for a 50-year design life (about 2 million cycles), the reliability index of each of the full penetration butt welds is expected to be about 3.32. Both approaches give almost identical reliability indexes. It is interesting to note that a similar reliability index was considered in developing design codes like the Load and Resistance Factored Design proposed by the American Institute of Steel Constructions (AISC).

(1) Updating the Fatigue Reliability Index

Suppose that a butt weld has been inspected by the ultrasonic NDI at the end of the fourth year. In this case, the weld has been subjected to about 183960 stress cycles. The measured crack size versus the updated reliability index, $\beta_{L,up}$, is plotted in Fig. 1, assuming that the COV of the measured crack size is 0.25. It can be observed that the updated reliability index decreases as the measured crack size increases. This is expected.

(2) The Influence of COV of the Measured Crack Size

The COV of the measured crack size A_k is very difficult to estimate. The two different COV values of 0.1 and 0.25 of the measured crack size are considered to study the sensitivity of the COV on the detectability and the updated reliability index. The updated reliability index versus the crack size is plotted in Fig. 1 for the two different COV values considered. It can be observed that the effect of the COV values on the updated fatigue reliability index is small, particularly when the measured crack size is less than about 0.508 cm (0.2 inches). Although the crack size of 0.508 cm (0.2 inches) seems very small, the corresponding reliability index is only about 1.7; a small value compared to the value of 3.0 recommended in the LRFD code. This implies that the bridge component was damaged seriously after four years of operation, even though the crack size is relatively small. The influence of the COV of A_k can be ignored for the butt welds considered in this study when the crack size is relatively small.

Reliability Updating After Repair

In order to insure the integrity of the bridge, a critical reliability index, β_c, should be selected. Suppose that after an inspection, it is decided that the butt weld needs to be repaired. Usually, the material properties will not be expected to change during the repair. Thus, the statistical properties of the material parameters, C_{new} and m_{new}, can be assumed to remain the same. However, defects may occur during the repair procedure as in the original fabrication process. So $\alpha_{0,new}$ is considered to be the same as α_0. Thus, after a repair, the corresponding limit state function can still be represented by Eq. 5, except that the new statistical information on the parameters needs to be used.

Suppose that during the second inspection (4 years of operation and undergoing 183960 stress cycles), a crack size of 0.1524 cm (0.06 inch) is measured in a butt weld. Based on this information, the updated fatigue reliability index, $\beta_{L,up}$, is estimated to be 2.86. If the decision is made that the weld needs repair, the updated reliability index after the repair can be estimated as 3.22, indicating a significant improvement in the reliability index after the repair.

Practical Fatigue Control Curves

To develop fatigue control curves for practical uses, the updated reliability index versus the number of stress cycles can be plotted for

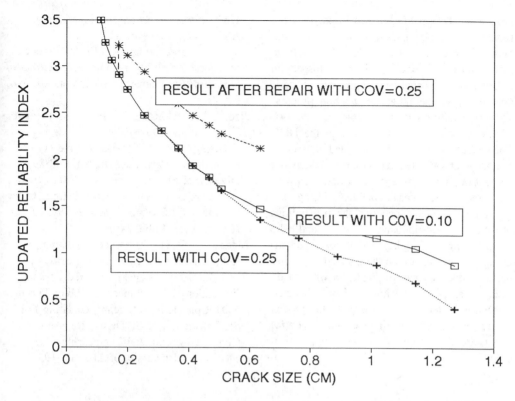

Figure 1.   The Effect of COV of Measured Crack Size on Updated Reliability Index

various crack sizes. For a given number of stress cycles, if the detected crack size is smaller, the updated reliability index is expected to be higher. This is quite logical. Also, for a given crack size, if it takes a longer time to detect the crack, the updated reliability index is expected to be larger. If it takes a longer time to develop a particular crack size, the crack growth rate is smaller and the corresponding reliability index should be higher. It can also be observed that an approximately linear relationship exists between the updated reliability index and the number of stress cycles, particularly when the measured crack size is smaller than 0.254 cm (0.1 inch). When the measured crack size is relatively small, the changes in the updated statistical characteristics of random variables are also expected to be small. From the Paris equation, it can be observed that the crack growth rate drops sharply when the propagated crack becomes

large. However, for relatively small size cracks, which are expected to be present during most of the design life of the steel structure, the linear behavior of these curves can be used for a quick check. Once a series of control curves is developed, the information on the crack size and the number of stress cycles at the time of inspection can be correlated immediately to the updated fatigue reliability index. This will help to make a decision on what to do next.

CONCLUSIONS

The fatigue reliability of bridge components designed according to the AASHTO and LEFM methods are evaluated considering several sources of uncertainties for full penetration butt welds in the tension flange of a steel box girder used in a public transportation system. The reliability indexes are almost identical, at least

around the design life. Since inspections are required at regular intervals, the results from these inspections are used for fatigue maintenance of steel bridges. Inspection information is beneficial; however, it adds uncertainty to the fatigue evaluation process. The detectability and the accuracy are the two additional sources of uncertainty in the NDI. The detection events are classified into three cases: no crack detection, crack detection but without measurement, and crack detection with the crack size measurement. Using the information from k inspections, the distributions of the basic random variables and the corresponding reliability index are updated. Using the information on the updated reliability index, a decision can be made on what to do next, in terms of reducing the next scheduled inspection interval, or whether to repair, replace, or do nothing. Fatigue damage control curves are also proposed for everyday practical use.

ACKNOWLEDGMENTS

This paper is based upon work partly supported by the National Science Foundation under Grant No. MSM-8896267. Financial support received from the Metropolitan Atlanta Rapid Transit Authority (MARTA) is also appreciated. Any opinions, findings, conclusions and recommendations expressed in this paper are those of the writers and do not necessarily reflect the views of the sponsors.

REFERENCES

Albrecht, P., and Yazdani, N. (1986) "Risk Analysis of Extending the Service Life of Steel Bridges." *FHWA/MD*, No. 84/01, Department of Civil Engineering, University of Maryland.

ASCE, (1982) Committee on Fatigue and Fracture Reliability of the Committee on Structural Safety and Reliability of the Structural Division, 1982 "Fatigue Reliability 1-4", *Journal of Structural Engineering*, ASCE, Vol. 108, No. ST1.

Ayyub, B., and Haldar, A. (1984) "Practical Structural Reliability Techniques", *Journal of Structural Engineering*, ASCE, Vol. 110, No. 8.

Jiao, G.Y., and Moan, T. (1989) "Methods for Fatigue Reliability Analysis Considering Updating Through Non-Destructive Inspections", *Report MK/R*, No. 108/89, Division of Marine Structures, NIT.

Madsen, H.O. (1985) "Random Fatigue Crack Growth and Inspection", *Proceedings of ICOSSAR'85*, Kobe, Japan.

Madsen, H.O. (1987) "Model Updating in Reliability Theory", *Proceeding of ICASP5*.

Zhao, Z. (1991) "Reliability-Based Fatigue Evaluation Under Random Loading Through NDT Considering Modeling Updating For Steel Bridges", M.S. Thesis, Department of Civil Engineering and Engineering Mechanics, University of Arizona, 1991.

Fatigue and fracture

Structural Safety & Reliability, Schuëller, Shinozuka & Yao (eds) © 1994 Balkema, Rotterdam, ISBN 90 5410 357 4

Risk analysis input for fleet maintenance planning

A. P. Berens
University of Dayton, Research Institute, Ohio, USA

J. G. Burns
Wright Laboratory, Wright-Patterson Air Force Base, Ohio, USA

ABSTRACT: This paper describes the methodology implemented in the structural risk analysis computer program, PRobability Of Fracture (PROF) and illustrates its application on structural zones of a "representative" aging fleet of transport/bomber aircraft. The effects of the timing of the structural maintenance actions on the fracture probabilities and on the expected repair costs are evaluated for an extended period of operational usage. The expected costs of alternative inspection systems are also compared.

1 INTRODUCTION

PRobability Of Fracture (PROF) is a risk analysis computer code whose objective is to stochastically assess the structural integrity of aging fleets of US Air Force aircraft [1]. The assessment is made in terms of both safety (as quantified by the probability of fracture of a population of structural details) and durability (as quantified by the expected number and sizes of the cracks requiring repair that will be detected at an inspection). The objective in writing PROF was to provide an additional tool for making decisions concerning the timing of inspection, replacement, and retirement maintenance actions. This paper briefly describes the methodology that has been implemented in PROF and presents an example application for a representative scenario.

2 PROF METHODOLOGY

Implementing structural risk analyses involves compromises between the ability to model reality and the data that is available to feed the analytical models. Because of the Aircraft Structural Integrity Program (ASIP) requirements of MIL-STD-1530A [2], the US Air Force has an extensive data base on each system for the deterministic evaluation of structural integrity. Of particular use for risk analyses are the data associated with the damage tolerance [3,4] and durability [3,5] analyses that are performed for all potential airframe cracking sites and the data

associated with the force management tasks of ASIP [6]. The risk analysis model (PROF) was developed to exploit these data systems. A detailed description of the complete PROF methodology can be found in [1].

2.1 Growing population of cracks

PROF is applicable to a population of structural elements which is defined in terms of details which experience essentially equivalent stress histories. The model predicts the growth of a distribution of cracks using the deterministic crack size versus spectrum hours curves from damage tolerance life analyses [7,8]. Spectrum hours encompass the expected stress sequences that the structural elements will encounter and provide a mechanism for correlating the tracked usages experienced by individual airframes. The distribution of crack sizes for initiating the analysis would be estimated from the best available data and could be obtained from routine inspections, teardown inspections, equivalent initial crack size distributions, or engineering judgement.

2.2 Maintenance effect on crack size distribution

At a maintenance action, the population of details are inspected and all detected cracks are repaired. The maintenance action will change the crack size dis-

tribution and the change is a function of the inspection capability and the quality of repair. Inspection capability is modeled in terms of the probability of detection as a function of crack size, $POD(a)$ [9]. Repair quality is expressed in terms of an equivalent repair crack size distribution, $f_r(a)$. If $f_{before}(a)$ and $f_{after}(a)$ represent the density function of crack sizes in the population of structural details before and after a maintenance action, then

$$
\begin{aligned}
f_{after}(a) &= P \cdot f_r(a) \\
&+ [1 - POD(a)] \cdot f_{before}(a)
\end{aligned}
\tag{1}
$$

where P is the percentage of cracks that will be detected

$$
P = \int_0^\infty POD(a) \cdot f_{before}(a) da
\tag{2}
$$

The post maintenance crack size distribution, $f_{after}(a)$, is then projected forward for the next interval of uninspected usage. The process is continued for as many inspection intervals as desired.

2.3 Probability of fracture

Safety is quantified in terms of the probability of fracture (POF) and is calculated as the probability that the maximum stress encountered in a flight will produce a stress intensity factor that exceeds the critical stress intensity factor for a structural detail. This calculation is performed in two contexts. The single flight POF is the probability of fracture in the flight given that the detail has not fractured previously. The single flight POF is the hazard rate. This number can be compared to other single event types of risks, such as the risk of death in an automobile accident in an hour of driving. The interval POF is the probability of fracture during any flight between the start of an analysis (reference time of zero or after a maintenance action) and a fixed number of spectrum hours, T. This POF is useful in predicting the expected fractures in a fleet of aircraft in an interval and is required for the expected costs associated with a maintenance schedule.

2.3.1 Single flight probability of fracture

The equation for calculating the probability of fracture at a single stress raiser in a single flight at T hours is given by:

$$
POF_E(T) = P[\sigma_{max} > \sigma_{cr}(a, K_c)]
\tag{3}
$$

$$
= \int_0^\infty \int_0^\infty f_T(a) \cdot g(K_c) \cdot \bar{H}(\sigma_{cr}(a, K_c)) dK_c da
$$

where $f_T(a)$ is the probability density function of crack sizes at T flight hours; $g(K_c)$ is the probability density function of the fracture toughness of the detail; and, $\bar{H}(\sigma_{cr}(a, K_c))$ is the (Gumbel) cumulative probability that the maximum stress in the flight exceeds the critical stress determined by a and K_c.

The single element POF, $POF_E(T)$, is interpreted as the probability that one of the elements in an airframe with T equivalent flight hours will experience a fracture due to a combination of crack size, K_c, and stress. This calculation is based on the assumption that the size of the crack in the stress raiser of the element and the critical stress intensity are independent.

To calculate the single flight probability of a fracture from any one of the k equivalent elements (stress raisers) in a single airframe at T flight hours, $POF_A(T)$, it is assumed that the fracture probabilities between elements are independent. Then

$$
POF_A(T) = 1 - [1 - POF_E(T)]^k
\tag{4}
$$

Similarly, $POF_F(T)$, the probability of a fracture in any of the N airframes in the fleet as they age through T flight hours, is calculated as

$$
POF_F(T) = 1 - [1 - POF_A(T)]^N
\tag{5}
$$

All three of these single flight POFs are calculated at 10 equally spaced increments in each usage interval. The results are printed in a summary output report.

2.3.2 Interval probability of fracture

Fracture can result during any flight in a usage period and the probability of a fracture during an entire period is required in order to estimate the expected costs of a fracture. Since the fracture toughness of an element does not change from flight to flight, single flight POFs as obtained above cannot be combined to obtain interval POF. The assumption of independence needed to make this calculation possible is not valid.

An approach to estimating interval POF which accounts for the constancy of fracture toughness over the interval was formulated as follows: a) determine

the contribution to the total POF from each possible pairing of fracture toughness and crack size at the beginning of the usage interval, say $PF(a,K)$; b) weight each contribution by the probability of the crack size-fracture toughness combination, say $f(a)da \cdot g(K)dK$; c) sum the weighted contributions over all possible combinations of crack size and fracture toughness. To calculate the contribution to the total POF from a crack size-fracture toughness pair, the total usage interval is divided into m subintervals. It is assumed that the crack size is essentially constant in a subinterval and the critical stress is calculated for the crack size of the subinterval and the fracture toughness. The distribution of maximum stresses in a subinterval is calculated from the distribution of maximum stresses in a flight and the probability of fracture in a subinterval is the probability that the maximum stress exceeds the critical stress for the subinterval. The POFs from the subintervals are combined to obtain the POF to the total usage interval for the initial crack-fracture toughness pair.

The interval POF calculation at a single stress raiser in the jth usage interval is given by the equation:

$$POF_E(I_j) = \int_0^\infty f_j(a) \int_0^\infty g(K_c) \cdot PF(a,K_c)dK_c\, da \quad (6)$$

where $f_j(a)$ = probability density function of crack sizes at the start of the jth analysis interval;

$$PF(a,K_c) = 1 - \prod_{i=1}^m H[\sigma_{cr}(a(T_i),K_c]^{\Delta T} ;$$

$H[\sigma_{cr}(a(T_i),K_c)]^{\Delta T}$ = probability that maximum stress in ΔT flights is less than the critical stress;

$\sigma_{cr}(a(T_i),K_c) = K_c/\sqrt{\pi a(T_i)} \cdot \beta(a(T_i))$;

ΔT = number of flights in a subinterval; $T_i = i \cdot \Delta T$, $i = 1,...,m$.

Interval fracture probabilities for the aircraft and for the fleet are calculated using equations analogous to equations (4) and (5).

2.4 Expected maintenance costs

Given the predicted crack size distribution at the time, T_j, of an inspect/repair maintenance action and the $POD(a)$ function, the expected number and sizes of the cracks that will be detected can be calculated. In particular, PROF calculates the cumulative proportion of cracks that will be detected as a function of crack size as

$$P(a_i) = \int_0^{a_i} POD(a) \cdot f_{before}(a)\, da \quad (7)$$

The proportion of detected cracks in the arbitrary range defined by $\Delta a_i = a_{i+1} - a_i$ is given by;

$$P(\Delta a_i) = P(a_{i+1}) - P(a_i) \quad (8)$$

Expected costs of maintenance are not currently calculated in PROF. However, PROF output can be used to estimate the expected costs of a maintenance scenario (as defined by flight hours between inspections, inspection capability, and repair quality). If the total population being modeled comprises k details in each of N airframes, then the expected number of cracks to be repaired at T_j between sizes a_i and a_{i+1} is $k \cdot N \cdot P(\Delta a_i)$. If C_i represents the cost of repairing a crack in size range i, C_F represents the cost of a fracture, and I represents the cost of inspecting each detail, then the expected costs of fracture and repairs in the usage interval are given by

$$E_j(C) = POF(T_j) - N \cdot C_F \\ + k \cdot N \cdot [I + \sum_i P(\Delta a_i) \cdot C_i] \quad (9)$$

Summing over usage intervals (maintenance periods) yields the total expected maintenance costs.

3 EXAMPLE APPLICATION

To illustrate the application of the risk analysis computer code, representative data for an aging military transport/bomber will be used to evaluate the timing of inspections and the capability of the inspection method. In particular, the objectives of the analyses were: a) to seek the most cost effective inspection intervals for a population of structural details; and, b) to determine if it is cost effective to use a better but more expensive inspection method. Assume there are 75 aircraft in the fleet which experience the same expected operational usage and that all of the aircraft will have undergone maintenance at a fixed reference number of flight hours. The risk analysis will pertain to periods of operational usage (or inspection or maintenance intervals) after this reference age, whatever it might be.

3.1 Baseline input

The assumed population of structural details com-

Table 1. Expected total fracture and maintenance costs as a percentage of total costs for 7200 hour inspection intervals.

First Inspection (Hours)	Inspection Interval (Hours)	Fracture Cost	Maintenance Cost	Total Cost
5143	5143	6.1	24.0	30.7
6000	5000	4.8	24.8	29.6
7200	4800	3.5	25.0	28.5
8400	4600	2.4	25.2	27.6
9000	5400	8.5	24.1	32.6
12000	4800	3.5	24.8	28.3
16000	4000	1.2	24.8	26.0
16000	5000	6.0	23.9	29.9
20000	4000	7.9	23.2	31.1

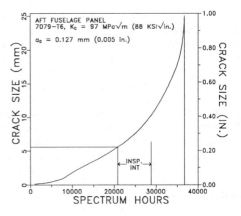

Figure 2   Crack growth for projected usage spectrum.

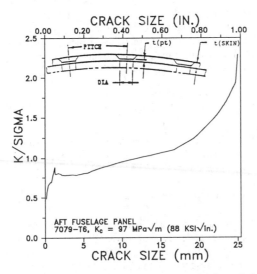

Figure 1 Stress intensity factor geometry correction for analysis region.

prises rows of fastener holes in a fail safe zone of equivalent stress experience on the upper rear fuselage. Figure 1 presents a schematic of the holes in the region and the geometry correction for crack growth calculations. Critical crack size is approximately 2.50 cm (0.986 in.). Cracks that are detected before fracture can be repaired by a patch. Assume that each airframe contains 50 separate regions such that the repair patch for any single crack in a region repairs all of the cracks in the region. However, if fracture (uncontrolled rapid crack growth) occurs, the entire panel must be replaced. The fracture toughness of the 7079-T6 aluminum alloy has an average value of 97.2 MPa $\sqrt{m}$ (88.4

KSI $\sqrt{in}$.) with a standard deviation of 4.8 MPa $\sqrt{m}$ (4.4 KSI $\sqrt{in}$.).

Figure 2 presents the projection of crack growth from a flight by flight spectrum of planned mission usage for the fleet. For the visual inspections of the region of interest, the reliably detected crack size was assumed to be 5.59 mm (0.220 in.). Under Air Force guidelines for establishing inspection intervals, subsequent inspections would be set at one half the time required for a crack of the reliably detectable size to grow to critical. For the example application, the baseline damage tolerance re-inspection interval was set at 7200 flight hours. The Gumbel fit to the maximum stress per flight of the flight by flight stress spectrum is presented in Figure 3.

At the start of the analysis (reference time of zero), it was assumed that the distribution of the largest cracks in each region was described by a Weibull distribution with a scale parameter of 0.151 mm (0.006 in.) and a shape parameter of 0.768. For this distribution, 1 in 1000 of the holes will have cracks larger than 1.905 mm (0.075 in.) and 3 in 10,000 will have cracks larger than a = 2.54 mm (0.100 in.). Cracks are repaired by patches and it is assumed that the repair quality of a patch is described by a uniform distribution of equivalent crack sizes on the interval 0 to 1.27 mm (0 to 0.050 in.). That is, a patch replaces the largest crack in the patched region with an equivalent flaw that is equally likely to be any size between 0 and 1.27 mm (0.050 in.).

For the baseline analysis, the reliably detected crack size of 5.59 mm (0.220 in.). is assumed to be the result of a close visual only inspection. This capability is interpreted as a 90 percent detection

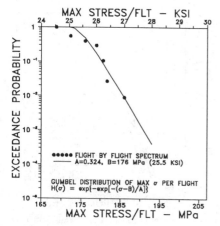

Figure 3 Gumbel distribution fit to maximum stress per flight of projected spectrum.

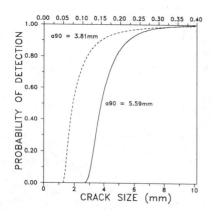

Figure 4 Crack detection probability for competing inspection methods.

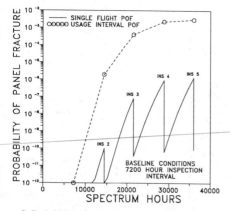

Figure 5 Probability of panel fracture in an airframe for baseline conditions.

capability at 5.59 mm (0.220 in.). Because of the fastener heads, no crack smaller than 2.54 mm (0.100 in.) could be detected, i.e., POD(a) = 0 for a ≤ 2.54 mm (0.100 in.). To complete the definition of the POD(a) function, it was also assumed that a 3.81 mm (0.150 in.) crack would be detected half of the time. The cumulative log normal POD(a) function that meets these specifications is shown in Figure 4. Also shown in Figure 4 is the POD(a) function for a potential eddy current inspection system with a smaller reliably detected crack size to be discussed in Subsection 3.3.

Because of the comparative nature of the analysis objectives, inspection and repair costs need only be specified on a relative basis. For baseline analyses, it was assumed that the cost of the visual inspection of each region is one, the cost of patching the region is 100 and the cost of replacing a fractured panel is 100,000. Expected costs for different maintenance scenarios are normalized in terms of the total expected costs for the baseline inspection interval (7200 hours) and inspection capability.

3.2 Inspection interval analysis

The probability of fracture (POF) for any one of the 50 panels on a fuselage under the baseline conditions is presented as a function of spectrum hours in Figure 5. The solid line represents the fracture probability during a single flight and the dashed line (circles) represents the probability of a fracture in any panel of an airframe at any time during the previous usage period. The large changes in single flight probability result from the removal of large cracks at the inspection/repair maintenance cycles and the growth of the population of cracks during the usage periods. PROF does not output fracture probabilities below 10^{-12}, so smaller POF values are plotted at this value. Since the structure under analysis is fail-safe and the costs are driven by the fracture probability in the entire usage period and the costs of maintenance, the single flight fracture probabilities will not be considered further.

To investigate the effect of a constant usage interval between inspections, a total analysis period of 36000 hours was assumed. Equally spaced inspection intervals were then defined to provide between three and twelve inspections in the 36000 hour period. Figure 6 presents the probability of fracture in each interval between maintenance (inspection and repair when necessary) actions for seven of the inspection intervals. The fracture probabilities display some-

what similar behavior in the early period during which the upper tail of the initial crack size distribution grows to potentially significant sizes. Following this initial period, the interval fracture probabilities tend to stabilize at distinct levels - the shorter the inspection interval the lower the equilibrium fracture probability.

Because of the equilibrium POF levels, the expected costs associated with the possibility of panel fractures at the longer inspection intervals will be greater than those of the shorter intervals. On the other hand, the costs associated with the more frequent inspections may be greater than the expected costs of panel fracture. To evaluate the trade-off, the total expected maintenance and fracture cost for each of the inspection intervals was calculated. These expected costs are presented as a function of inspection interval in Figure 7. As noted earlier, the costs are normalized by the total expected cost for the baseline inspection interval of 7200 hours. (Inspection intervals of 9000 and 12000 hours were also analyzed but the expected total costs were, respectively, 4.1 and 25.0 times greater than those of the 7200 hour increment. These intervals were not included in Figure 7 in order to provide more resolution for the shorter intervals.)

The expected total costs decrease with inspection interval down to about a 4000 hour interval and then tend to increase slightly. The decrease is due to the large decrease in the expected costs associated with panel fractures at the longer intervals. The equilibrium fracture probability for inspection intervals of 4500 hours and less produce only minor additions to the total expected costs. The costs due to the inspections and repairs increase but at a very slight rate. From a practical viewpoint, any interval less than 4500 hours would have essentially equivalent expected total costs.

To investigate the potential for reducing total costs by extending the timing of the first inspection, various combinations of initial inspection interval and equal repeat inspections thereafter were analyzed. Table 1 presents a summary of the expected normalized costs due to fracture, maintenance, and the total. As noted earlier, the expected maintenance cost were approximately equal for all scenarios considered. The expected costs due to panel fracture varied somewhat depending on the particular combination. It is interesting to note that the minimum expected total cost was achieved at a 16000 hour first inspection followed by 4000 hour intervals thereafter. The expected cost for this combination was slightly less than that of inspecting every 4000 hours.

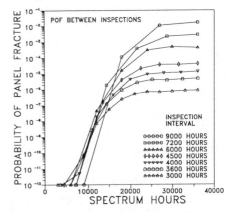

Figure 6 Probability of panel fracture in an airframe between inspections for selected inspection intervals.

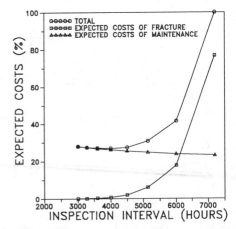

Figure 7 Normalized expected maintenance costs as a function of inspection interval.

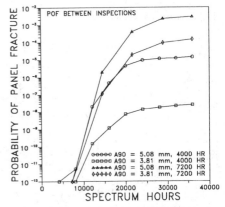

Figure 8 Probability of panel fracture in an airframe between inspections for different methods and intervals.

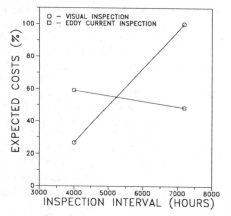

Figure 9 Expected total costs of inspections, repairs, and fractures in 36000 spectrum hours normalized by costs from 7200 hour increment.

For the assumed conditions, the above analyses imply that an inspection schedule with shorter intervals would provide a significant saving in expected fracture and maintenance costs over those determined by the deterministic damage tolerance "rule." Although a minimum was achieved under the equal interval analysis, once the inspection interval was sufficiently short, the expected costs did not change significantly. This was true regardless of the timing of the first inspection. This latitude in setting inspection intervals could be important as the actual schedule should be determined by considering the many different populations of structural details on an airframe, each of which may have different optimum schedules.

3.3 Inspection capability analysis

The inspection assumed for the baseline analysis was a close visual inspection that is inexpensive. The question might arise as to whether it would be cost effective to perform a more expensive inspection with an attendant increase in capability. Toward this end assume that an eddy current (EC) inspection could be used to inspect for cracks in the regions and that the cost of the EC inspection is 10 times that of the visual. However, the reliably detected crack size is reduced to 3.81 mm (0.150 in.). Because the eddy current probe can detect cracks under the fastener, assume the minimum detectable crack size is 1.27 mm (0.050 in.). The 50 percent detectable crack size will be assumed to be 1.90 mm (0.075 in.). The cumulative log normal POD(a) function that meets

these requirements is shown in Figure 4 along with the POD(a) of the baseline analysis.

The usage interval fracture probabilities for the two inspection capabilities at 4000 and 7200 hour inspection intervals are presented in Figure 8. The eddy current inspection significantly reduces the chances of a panel fracture in the 36000 hour period for both inspection intervals. However, when the total expected costs of inspections, repairs, and fractures are considered, the cost effectiveness of the eddy current inspection depends on the inspection interval. Figure 9 presents the normalized total expected costs in a 36,000 hour period for the two inspection systems and two inspection intervals. At the 4000 hour inspection interval, the inspection and repair costs associated with the eddy current inspection are 2.2 times greater than those of the visual inspection. At this 4000 hour inspection interval, the expected costs due to panel fracture are small (almost negligible) for both inspection methods. The better (EC) inspection system apparently requires more cracks to be repaired at each of the inspections, and these cracks are too small to be an imminent threat to the panel.

When the two inspection capabilities were analyzed at the 7200 hour inspection interval, the reverse conclusion is drawn. The chances of panel fracture at the longer usage interval is sufficiently great that the total expected costs over the 36000 hour period are significantly reduced by repairing the smaller cracks. This conclusion held for the EC inspections costing as much as 50 times those of the visual and for repair costs ranging as high as 500 times the visual inspection costs.

No clear conclusion can be drawn on the cost effectiveness of the EC inspection system as compared to the visual. When the shorter, and more cost effective intervals of this example, are used, the visual inspection capability provides the more economical choice. If the longer damage tolerance defined inspection interval is used, the additional costs associated with the eddy current inspections would be justified.

4 SUMMARY

This example application pertained to fatigue crack growth in a fail-safe panel on a transport/bomber fuselage. In this application, a splice could be applied to repair a noncritical crack at a reasonably nominal cost. If a crack grew to unstable size, the panel would fracture and the repair costs of panel replacement would be orders of magnitude greater

than splicing. Applied stresses were not close to critical before the onset of unstable crack growth. Trade-offs in inspection intervals and inspection capabilities were evaluated in terms of total expected costs of inspections, splices, and fractures over a long usage period. For the assumptions of this example application, the following conclusions could be drawn:

1. The first inspection can be delayed without a significant effect on expected maintenance costs. The delay is a function of the flight time required for a significant proportion of the initiating crack size distribution to grow close to the unstable size.

2. Expected costs using the damage tolerance "rule" for determining repeat inspection intervals were about five times greater than those of the optimum repeat inspection interval.

3. For a reasonable range of repeat inspection intervals around the optimum, expected total maintenance costs were essentially equivalent. Once the repeat inspection interval is sufficiently short, there is considerable freedom in acceptable choices.

4. A more expensive and better inspection system produced significantly higher expected costs at the optimum inspection interval. At the optimum interval, the contribution to expected costs from the chances of panel fracture is minimal, while the better inspection system finds significantly more cracks to be repaired. Thus, the "better" inspection system leads to higher repair costs with no added fracture protection.

5. The more expensive and better inspection system produced significantly lower expected maintenance costs at the longer repeat inspection intervals of the damage tolerance "rule." Postponing the inspections significantly increases the risk of fracture so that repairing the smaller cracks is cost effective for the longer inspection intervals.

These conclusions are highly dependent on the initial conditions assumed for the example application. They are presented to demonstrate the types of insights that can be obtained from risk analysis computer codes. These conclusions should be re-evaluated before being applied for a different set of conditions.

REFERENCES

1. Berens, A.P., Hovey, P.W., and Skinn, D.A., "Risk Analysis for Aging Aircraft Fleets, Volume 1 - Analysis," Wright Laboratories, WL-TR-91-3066, October, 1991.

2. Military Standard, "Aircraft Structural Integrity Program, Airplane Requirements," MIL-STD-1530A (USAF), December 1975.

3. Military Specification, "General Specification for Aircraft Structures," MIL-A-87221, February 1985.

4. Gallagher, J.P., Giessler, F.J., and Berens, A.P., "USAF Damage Tolerant Design Handbook: Guidelines for the Analysis and Design of Damage Tolerant Structures," Air Force Wright Aeronautical Laboratories, AFWAL-TR-82-3073, May 1984.

5. Manning S.D. and Yang, J.N., "USAF Durability Design Handbook: Guidelines for the Analysis and Design of Durable Aircraft Structures," Air Force Wright Aeronautical Laboratories, AFWAL-TR-83-3027, January 1984.

6. Berens, A.P., et al., "Handbook of Force Management Methods," Air Force Wright Aeronautical Laboratories, AFWAL-TR-81-3079, April 1981.

7. Yang, J.N., "Statistical Estimation of Service Cracks and Maintenance Cost for Aircraft Structures," J. of Aircraft, Vol. 13, No. 12, December 1976.

8. Lincoln, J.W., "Risk Assessment of an Aging Military Aircraft," J. of Aircraft, Vol. 22, No. 8, August, 1985.

9. Berens, A.P., "NDE Reliability Data Analysis," Metals Handbook, Volume 17, 9th Edition: Non-destructive Evaluation and Quality Control, 1988.

Structural Safety & Reliability, Schuëller, Shinozuka & Yao (eds) © 1994 Balkema, Rotterdam, ISBN 90 5410 357 4

On random fatigue crack propagation

V.V.Bolotin

Russian Academy of Sciences, Moscow, Russia

ABSTRACT: Fatigue crack propagation in media with random nonhomogeneities is considered in the framework of the theory of fatigue based on the synthesis of micro- and macromechanics of fracture. Constitutive equations of the theory include the equations of microdamage accumulation along the crack path, equations governing crack tips blunting and sharpening, and relationships between crack driving and resistance generalized forces. Results of computational experiments are discussed, in particular, from the standpoint of the comparative contribution into the fatigue life distribution. of specimen-to-specimen and within-a-specimen scatters of material properties.

1 INTRODUCTION

There are two ways to description of the stochastic fatigue crack growth. The first way consists in the randomization of parameters entering into the well-known semi-empirical equations, such as Paris or Forman equations, i.e. in the substitution of deterministic values with random variables and/or random functions. Another way is to use mathematical models suitable for irreversible processes with the interpretation of entering parameters and probabilistic measures in terms of fatigue damage. Both ways are discussed by Sobczyk and Spencer (1992). In this paper, the third approach is developed that differs from the listed above in two aspects. Firstly, a more physically substantiated model of fatigue crack propagation is used proposed primarily by Bolotin (1983) and developed in a number of publications which survey is given in the book by Bolotin (1989). Secondly, a more sound, in the author's opinion, approach to the randomization of the input data is applied compared with the used before.

The constitutive equations of the theory are too complicated to obtain analytical solutions, at least without ad hoc assumpti- ons and simpli- fications. Therefore, computational experimentation is applied to obtain both qualitative and quantitative results. Such experimentation, aside of the assessment of probabilistic measures, produces sets of samples that are of interest even without any statistical treatment. One of the aims of this research is to analyse the crack samples behaviour on various stages of fatigue damage, to study the interaction of samples, and to assess the comparative contribution of specimen-to-specimen scatter of material properties in the fatigue crack growth rate and the fatigue life distribution.

The initial point of this research is a generalized version of fracture mechanics developed by Bolotin (1989). A cracked body under loading or interacting with a loading device is considered as a mechanical system which generalized coordinates are divided into two groups: the conventional Lagrangian generalized coordinates (briefly, L-coordinates), and parameters characterizing the shape, position and dimensions of cracks. The latters are named, in honour of Griffith, Griffithian generalized coordinates or, briefly, G-coordinates. Since cracks in structural materials are usually irreversible, G-coordinates

can be chosen in such a way that only unilateral variations are admissible for them. As a result, we come to mechanical systems with unilateral (and, by the way, nonholonomic) constraints. The problem is to formulate the conditions of equilibrium and stability of cracked body in a most general form.

It the loading is quasistatic, the condition of equilibrium is given with the principle of virtual work. In the presense of unilateral constraints this condition takes the form of inequality

$$\delta A \leq 0. \tag{1}$$

The virtual work in the left-hand side is to be calculated with respect to variations of all generalized coordinates. But if, by definition, only equilibriums (in conventional sense) are considered both in unperturbed and perturbed states, and all these states are stable with respect to L-coordinates, condition (1) is to be replaced with

$$\delta_G A \leq 0. \tag{2}$$

The subscript G denotes that only G-coordinates are subjected to variation.

Depending on the sign of $\delta_G A$, several types of states of the system cracked body - loading are to be distinguished. If $\delta_G A < 0$ for all admissible variations of G-coordinates, a state is stable. Such states were named by Bolotin (1985) sub-equilibrium states. If $\delta_G A > 0$, no equilibrium exists at all, and it means instability of the state with respect to G-coordinates. At $\delta_G A = 0$ the signs of the following terms in the expansions $\Delta_G A = \delta_G A + \delta_G(\delta_G A) +$ + ... are to be considered, and an equilibrium can be stable, unstable, or neutral.

For the further purposes, reformulate the condition (2) in terms of generalized forces:

$$G_j \leq R_j, \quad j = 1, \ldots, n. \tag{3}$$

Here G_j are driving generalized forces, and R_j are corresponding resistance generalized forces. In simplest situations, say, in the framework of the linear fracture mechanics, the driving forces can be identified with the known parameters such as the strain energy release rate, J-integral, etc., and the resistance forces with the respective critical mafnitudes of these parameters. Details are discussed by Bolotin (1989).

2 THEORY OF CRACK GROWTH

To expand the theory upon fatigue crack propagation, one has to include into the analysis the dispersed damage accumulation such as microcracking, micropore nucleation, growth and coalescence, etc. Such a theory based on the synthesis of micro- and macromechanics of fracture was suggested by Bolotin (1983, 1985, 1989). The central point of theory is the account of the influence of microdamage on mechanical material's properties and, in the first line, on the resistance against macrocrack propagation. Here and later on, the label micro- is attributed to phenomena on the scale of grains, fibers and other local nonhomogeneities, and the label macro- to phenomena on the scales that are one and more orders of magnitudes larger, such as a scale of structural components. Due to microdamage, the balance between generalized driving and resistance forces is violated and, as a result, stability of the cracked body under loading changes into instability. The crack advances at a certain distance, enters into the less damaged material, and stability recovers again. Therefore, the fatigue crack growth process is interpreted as a chain of alternative jump-like transitions from one stable equilibrium state to another. Such jumps are small and local, and the growh process may be treated as continuous one. However, the crack propagation associated with striations which is typical for final stages of fatigue damage is also enters into the patterns of the theory.

To develope the analytical part of the theory, equations of microdamage accumulation are to be introduced. Let $\omega(x,N)$ is the microdamage field which may be a scalar or tensor function of the reference vecfor x, and

of the cycle or loading block number N. This field is a functional of loading and environmental actions as well as of the crack growth history:

$$\omega(x,N) = \underset{n=0}{\overset{N}{\Omega}} \ \{s(x,n),a(n)\}. \quad (4)$$

Here s(n) is a set of parameters characterizing loading, and a(n) is a set of current G-coordinates. The microdamage measures at the crack tips are of primary significance in the theory. Introduce for the set of these measures a special notation such as

$$\psi(N) = \omega[a(n),N] \quad (5)$$

where substitution x = a(N) means that the points at the tips are considered.

Since we are interested in the microdamage just at the tips, we cannot treat a crack, alike in the classical fracture mechanics, as a mathematical cut. Moreover, the fatigue crack growth is often accompaned with an alternaiting blunting and sharpening of the tips. To take this into account, introduce a set of parameters ρ(N) characterizing the effective curvature of the tips, crack tip opening displacements, etc. The general form of the governing equation with respect to ρ(n) is

$$\rho(N) = \underset{n=0}{\overset{N}{R}} \ \{s(n),a(n)\}. \quad (6)$$

Stability of cracks depends on the relationship between the generalized forces similar to (3). Both parts of this relationship depend now on s(N), a(N), ψ(N), and ρ(N). Including instability, we write instead of (3) as follows:

$$G_j[s(N),a(N),\psi(N),\rho(N)] \underset{>}{\overset{<}{}}$$

$$\underset{>}{\overset{<}{}} R_j[s(N),a(N),\psi(N),\rho(N)],$$

$$j=1,\ldots,n. \quad (7)$$

Of course, to solve a problem, one needs to know stress-strain field in a cracked (and, generally, micro--damaged) body, and that required the solution of the corresponding problem of continuum mechanics.

To make the above general formulation more transparent, consider a special case, a single-parameter fatigue crack, say, a central opening mode crack in a body of linear elastic material. Use the scalar microdamage measure ω(x,N) similar to introduced by Kachanov (1986) and Rabotnov (1979) with the magnitudes from [0,1] and the governing equation (4) in the form

$$\frac{\partial \omega}{\partial N} = \begin{cases} 0 \quad, \quad \Delta\sigma \leq \Delta\sigma_{th}, \\[2mm] \left[\dfrac{\Delta\sigma - \Delta\sigma_{th}}{\sigma_f} \right]^m, \quad \Delta\sigma > \Delta\sigma_{th} \end{cases}$$

$$(8)$$

Here $\Delta\sigma$ is the opening stress range at the crack prolongation, say, at

$|x| \geq a$, $y = 0$ (Fig. 1), σ_f, $\Delta\sigma_{th}$, and m are material parameters (microdamage resistance stress, microdamage threshold stress, and microdamage accumulation rate exponent, respectively). All these parameters, as well as many of following ones, generally, depend on the stress ratio $\sigma_{max}/\sigma_{min}$, temperature, loading frequency, etc. Equation (6) with respect to the effective tip radius ρ is specified as

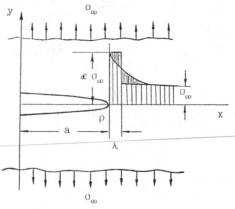

Fig.1 Stress distribution near the elliptical tip of crack and its schematization

$$\frac{d\rho}{dN} = \frac{\rho_s - \rho}{\lambda_\rho} \frac{da}{dN} + (\rho_b - \rho) \frac{d\psi}{dN} \quad (9)$$

where ρ_s is the "sharp", and ρ_b is the "blunt" tip radius, λ_ρ is a characterictic length. The first term in the right-hand side of (9) describes the sharpening of the tip during the crack propagation, and the second one - the blunting due to the microdamage accumulation at the tip including when the crack is steady.

At last, assume the following equations for the generalized forces:

$$G = K^2(1-\nu^2)/E,$$

$$R = \gamma_0(1 - \psi^\alpha). \quad (10)$$

Here E and ν are the Young modulus and Poisson ratio, respectively; γ_0 is the specific fracture work for undamaged material; α is a positive power exponent. The first of equations (10) defines the strain energy release rate in a linear elastic body, and K is the current stress intensity factor. Usually $K = Y\sigma (\pi a)^{1/2}$ with the remote stress σ and the form-factor Y (see Liebowitz, 1968). Here for simplicity it is assumed that microdamage does not effect on the material compliance.

Equations (8)-(10) remains to be too complicated to solve them analytically. However, under certain additional assumptions, they can be reduced to differential equations similar, at least in their structure, to semi-empirical equations of fatigue crack growth widely used in the engineering analysis. For example, assume $\rho = const$ and present the microdamage ahead of the crack in the form (Bolotin, 1989)

$$\omega(x,N) = \begin{cases} \psi(N) &, a \leq |x| \leq a + \lambda, \\ \omega_{ff}(N), |x| > a + \lambda. \end{cases}$$
$$(11)$$

It means the uniform stress distribution within the processing zone with the lengh λ (Fig. 1), and the consideration the path beyond this zone as the far field with the damage $\omega_{ff}(N)$. Then

$$\frac{da}{dN} = \lambda \left[\frac{\Delta K - \Delta K_{th}}{K_f} \right]$$

$$\left[\left[1 - \frac{K^2_{max}}{K^2_{1c}} \right]^{1/\alpha} - \omega_{ff}(a) \right]^{-1},$$

$$\frac{d\omega_{ff}}{dN} = \left[\frac{\Delta\sigma - \Delta\sigma_{th}}{\sigma_f} \right]^m, \quad (12)$$

where notations are introduced: $\Delta K = Y\Delta\sigma (\pi a)^{1/2}$, $K_f = (Y/Z)\sigma_f (\pi\rho)^{1/2}$, $K_{th} = (Y/Z)\sigma_{th}(\pi\rho)^{1/2}$, $K^2_{IC} = \gamma_0 E/ /(1-\nu^2)$. The stress concentration factor $1 + Z(a/\rho)^{1/2} \approx Z(a/\rho)^{1/2}$ is used here with the form-factor Z (say, Z = 2 for an elliptical hole).

3. RANDOMIZATION OF INPUT DATA

Generally, all the values entering in equations (8)-(10) and (11)-(12) are eigher random variables or random functions of the cycle number and/or of the coordinate measured along the crack path. To minimize the scale of randomization as much as possible, we consider strictly deterministic, say, stationary sinusoidal loading with the given stress range $\Delta\sigma$, and omit such a kind of randomness as the batch-to-batch scatter of commercial materials. Then the main source of randomness becomes nonhomogeneities of material properties and initial data such as the initial crack size, the initial curvature and microdamage level at the tip.

Even then a large number of random functions and random variables remain to be in consideration. The most important among them seems to be the resistance stress σ_f which is a random function of x. The threshold resistance stress $\Delta\sigma_{th}$ is to be correlate with σ_f, and, for simplicity, i can be assumed as a part of the lo-

cal σ_f, say, $\Delta\sigma_{th} = 0.05\sigma_f$. Another random function of x is certainly the specific fracture work γ_o entering in the second equation (10). But, from physical considerations, one may assume that γ_o is proportional to $\sigma_f^2\rho$ where ρ is the current tip radius which is a random function of N evaluated in the process of solution. As to remaining material parameters, we consider them as deterministic, not to multiply difficulties without an extreme necessity.

Represent the random function $\sigma_f(x)$ in the form

$$\sigma_f(x) = \sigma_f^0 + (\sigma_f^+ - \sigma_f^-)v + \sigma_{f1}u(x).$$

(13)

Here σ_f^0 is the lowest resistance stress. The difference $\sigma_f^+ - \sigma_f^-$ accounts for the specimen-to-specimen scatter, and σ_{f1} characterizes the local randomness of the resistance stress. All these parameters are considered as deterministic. Normalizing the random variable v, e.g., assuming it distributed in [-1,1], we interprete $\sigma_f^+ - \sigma_f^-$ as the range of the specimen-to-specimen scatter. Normalizing the random function u(x), we may consider σ_{f1} as a measure of local, within-a-specimen random nonhomogeneities.

Generally, u(x) is a nonstationary function of x. As a matter of fact, the stress σ_f characterizes material properties averaged upon the crack

front and, generally, is to vary when a crack propagates through the bulk of a specimen. For small cracks which are short upon the specimen's surface (Figure 2,a) the number of grains on the front is comparatively small. That results to a large variability of $\sigma_f(x)$ for small cracks (Miller and de los Rios, 1986). When the crack front widens (see Figures 2,b and c), variability of σ_f decreases staying practically stationary during the future crack propagation (Figure 3). In addition, the initial damage near the tip can also contribute into variability of the fatigue resistance stress. Thus, we come to the specific thickness effect in fatigue. It is not connected, as usually, with plastic strain fields, but is born with the averaging influence of the crack front. If a crack propagates from the beginning through a specimen of constant thickness, this effect, evidently, vanishes. In the further numerical examples we treat u(x) as a segment of a stationary ergodic random process with the given spectral density that takes into account the characteristicate size and the correlation scale of material nonhomogeneities.

4. NUMERICAL EXAMPLES

At the beginning, consider the within-a-spesimen scatter of material properties only. Then in equation (13) $\sigma_+ - \sigma_- = 0$. Let the random function u(x) in that equation is a Rayleigh's stochastic process presented by means of two independent

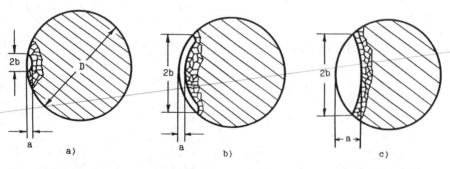

Fig.2 Crack front as an averager of material resistance against microdamage accumulation

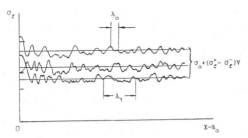

Fig.3 Material resistance against microdamage as a random function of the distance from the initial crack tip position

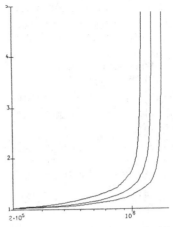

Fig.4 Numerical simulation of fatigue crack growth; the "worst", the "best", and the "average" samples among 20 samples are plotted

normalized Gaussian processes $n_1(x)$ and $n_2(x)$, i.e. $u^2(x) = n_1^2(x) + n_2^2(x)$. The spectral density for $n_{1,2}(x)$ is taken in the form

$$S(k) = \frac{2k_0 k_1}{\pi} \frac{1}{(k^2 + k_0^2)^2 + 4k_1^2 k^2}$$

(14)

where k is the wave number, k_0 and k_1 are material parameters. The length $\lambda_0 = 2\pi/k_0$ characterizes the size of nonhomogeneities, and $\lambda_1 = 2\pi/k_1$ is the spacial correlation scale. An initially edge cracked plate specimen of the width b is considered with applied (remote) tensile stress range $\Delta\sigma$ assumed constant up to the final failure.

The following numerical data were used in the numerical experimentation: $\sigma_f^0 = 9$ GPa, $\sigma_{f1} = 1$ GPa, $\Delta\sigma_{th} = 0.025\sigma_f$, $\rho_b = 0.1$ mm, $\rho_s = 0.01$ mm, $\lambda_\rho = 10$ mm, $\lambda_0 = \lambda_1 = 1$ mm, $E = 200$ GPa, $\nu = 0.3$.

Figures 4 – 7 are plotted for the case $\Delta\sigma = 100$ MPa, $b = 500$ mm. Initial conditions were taken $a_0 = 1$ mm, $\rho_0 = 0.05$ mm and ω 0 at all x a_0. As to the initial values of the random field $u_0 = u(a_0)$, they were taken equal to randomly chosen

magnitude of $u(x)$, i.e. material properties at the initial tip were identified with those randomly distributed ahead of the tip. Such a choice provides randomness of initial conditions. Generally, all the initial values a_0, ρ_0, u_0, and ψ_0, and can be treated as random variables. But we limit ourselves with the above randomization taking into account the most physically significant sources of randomness.

In Fig.4 three samples of fatigue crack growth are presented. The set of 20 sets was generated, and among them the "worst", the "best", and an average samples were chosen. Roughly speaking, they correspond to cumulative probabilities estimated as 0.95, 0.5 and 0.05, respectively. More informative are Figures 5 and 6 where the microdamage measure ψ at the tip and the effective curvature radius ρ are plotted versus the cycle number N. On the first stage, when a crack is steady, the damage is accumulating at the tip, and crack is blunting. When a crack begins to grow, the damage measure remains to be rather close to the ultimate level $\psi = 1$. Closer to the final failure the crack growth rate increases essentially, and the crack propagates at lower damage level at the tip. At the moment of final failure this level is near to that attained at the far field, i.e. to $\omega_{ff}(N)$ from equa-

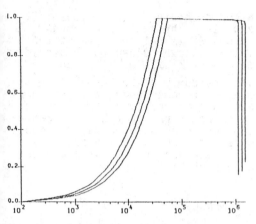

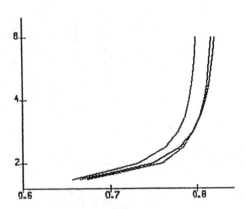

Fig.5 The same as in Fig.4 for the microdamage measure at the tip

Fig.7 Typical mutual disposition of samples of fatigue crack growth including a couple of intersecting samples

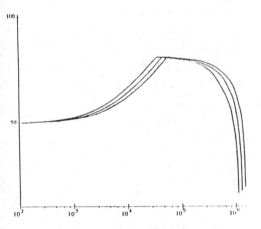

Fig.6 The same as in Fig.4 for the effective curvature radius at the tip

tions (11) and (12). A similar tendency shows the crack tip radius. Comparing Figures 5 and 6, one can see that the sharpening of the tip begins much earlier than the microdamage at the tip diminishes significantly. This apparent discrepancy may be understood considering equation (9) that takes into account both tendencies, blunting and sharpening.

It is interesting to follow the samples behaviour in particular, their mutual relationship during the crack propagation. Descrepancy of crack growth begins, as a rule, on the earlier stage of the fatigue life. The "worst" sample usually remains to be "worst" up to the final failure, and the "best" usually remains "best" one. Only a couple from 15 - 20 samples intersect during the crack growth (see Figure 7 where such a case is reproduced). That indicates to the significance of initial conditions and material properties in the vicinity of the initial crack tip. Note that only the within-a-specimen scatter of properies is taken into account there, and the deterministic intial conditions are used except initial microdamage resistance stress chosen randomly from the function $\sigma_f(x) = \sigma_f^0 + \sigma_{f1} u(x)$.

An example of crack growth diagram with account of the specimen-to-specimen scatter is shown in Figure 8. It was assumed that the random value v from equation (15) is beta-distributed, i.e. the probability density function is

$$p(v) = \frac{\Gamma(\mu + \nu) v^{\mu-1}(1 - v)^{\nu-1}}{\Gamma(\mu)\Gamma(\nu)}$$

(15)

In computation it was assumed σ_+ - σ_- = 2G Pa, and $\mu = \nu = 4$. As in Figure 7, the "best" and the "worst" samples are presented in Figure 8 taken from 20 samples. Only one pair of samples happened to be intersec-

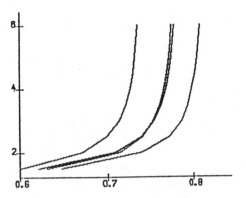

Fig.8 Typical mutual disposition of samples of fatigue with account of specimen-to-specimen scatter

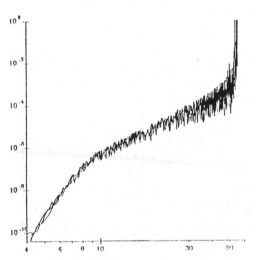

Fig.9 Three samples of the fatigue crack growth diagramme with account of specimen-to-specimen scatter

ting, and they are shown in Figure 8 too. Again the earlier history of crack propagation (including initial conditions) appears to be a decisive factor.

Experimental results in fatigue are usually presented as crack growth rate diagrams using the stress intensity factor range ΔK as a governing parameter (at the fixed stress ratio, temperature, frequency, etc). An example of such a diagram is shown in Figure 9. It is obtained with computational simulation of equations (8) - (10) rando-

mized according to equations (13) and (14). The scatter of da/dN versus ΔK is rather large at the initial stage, close to threshold $\Delta K_{th} = (Y/Z)\Delta\sigma_{th}(\pi\rho)^{1/2}$ which is also randomly distributed. Then the scatter becomes moderate, and increases significantly at the final stage of crack growth. The upper envelope of these samples may be considered as a conservative estimate of the crack growth rate with upcrossing probability near 0.05.

REFERENCES

Bolotin, V.V. 1983. Equations of fatigue crack growth. Sov. Mech. Solids (VNN) 4: 153-160 (in Russian).
Bolotin, V.V. 1985. A unified approach to damage accumulation and fatigue crack growth. Eng. Fracture Mech. 22(3): 387-398.
Bolotin, V.V. 1989. Prediction of service life for machines and structures, New York: ASME Press.
Kachanov, L.M. 1986. Introduction in continuum damage mechanics. The Hague: Martinus Nijhoff.
Liebowitz, H. (ed.) 1968. Fracture. An advanced treatise, 2, New York: Academic Press.
Miller, K.J. and de los Rios E.R. (eds.) 1986. The behaviour of short cracks. London: Mech. Eng. Publ. Rabotnov, Yu.N. 1979.
Mechanics of deformable solids. Moscow: Nauka (in Russian).
Sobzyk, K. and Spencer, B.F. 1992. Random fatigue: from data to theory, New York: Academic Press.

Structural Safety & Reliability, Schuëller, Shinozuka & Yao (eds) © 1994 Balkema, Rotterdam, ISBN 90 5410 357 4

Probabilistic model for damage analysis of Inconel 718 subjected to mechanical and thermal fatigue and creep at high temperatures

L. Boyce & V. Balaraman
The University of Texas at San Antonio, Tex., USA

C. Bast
Colorado Springs, Colo., USA

ABSTRACT: The development of methodology for a probabilistic damage analysis model is presented. It provides for quantification of uncertainty in the lifetime material strength of structural components of aerospace propulsion systems subjected to a number of diverse random effects. Presently, the model includes up to five effects that typically reduce lifetime strength: high temperature, high-cycle mechanical fatigue, creep, thermal fatigue and low-cycle mechanical fatigue. Model calibration for Inconel 718 is carried out for the five effects using statistical analysis of the data. Using a computer program, PROMISS, a sensitivity study is performed with the calibrated random model to illustrate the effects of high-cycle mechanical fatigue, creep and thermal fatigue upon lifetime strength, at $1000\,^{\circ}$F.

1 INTRODUCTION

Previously, a general material property degradation model for composite materials, subjected to a number of diverse effects was postulated to predict mechanical and thermal material properties (Chamis 1984). The resulting multifactor equation summarizes a proposed composite micromechanics theory and has been used to predict material properties for a unidirectional fiber-reinforced lamina based on the corresponding properties of the constituent materials.

Later, the equation was modified to predict the degradation of lifetime strength of a single constituent material due to diverse effects (Boyce, et al 1991). These effects could include high temperature, creep, mechanical fatigue, thermal fatigue, corrosion or even impact. Generally, strength decreases with an increase in the effect value. The general form of the modified multifactor equation forms a damage analysis model and is given by

$$\frac{S}{S_O} = \prod_{i=1}^{n} \left[\frac{A_{iU} - A_i}{A_{iU} - A_{iO}} \right]^{a_i},$$ (1)

where A_i, A_{iU} and A_{iO} are the current, ultimate and reference values, respectively, of a particular effect; a_i is the value of an empirical

material constant for the ith product term of the model and S/S_O is the lifetime strength of the material. Each term has the property that if the current value equals the ultimate value, the lifetime strength will be zero. Also, if the current value equals the reference value, the term equals one and strength is not affected by that variable. The multiplicative nature of the model assumes independent effects. An important purpose of this paper was to extend the damage analysis model to degradation of lifetime strength due to low-cycle mechanical fatigue.

Ideally, experimental data giving the relationship between all effects and strength is obtained. Thus, the deterministic model, given by equation (1) may be calibrated by least squares multiple linear regression of experimental data perhaps supplemented by expert opinion. Generally however, data for one effect is available and an approximate calibration of the model, yielding the empirical material constant for one effect, may be obtained by linear regression. The multifactor equation, given by equation (1), when written for an individual effect, results in the following single effect model (Bast 1993):

$$\frac{S}{S_O} = \left[\frac{A_U - A}{A_U - A_O} \right]^{a} = \left[\frac{A_U - A_O}{A_U - A} \right]^{-a}.$$ (2)

The probabilistic treatment of this model includes randomizing the deterministic multifactor equation, performing probabilistic analysis by simulation (Siddall 1982) and generating a probability density function (p.d.f.) estimate for lifetime strength, using the non-parametric method of maximum penalized likelihood (Scott 1976). Integration of the probability density function yields the cumulative distribution function (c.d.f.) from which probability statements regarding lifetime strength may be made. This probabilistic material strength degradation model, predicts the random lifetime strength of an aerospace propulsion component subjected to diverse random effects.

The general probabilistic damage analysis model, whose corresponding deterministic equation given by equation (1), is embodied in two FORTRAN programs, PROMISS: Probabilistic Material Strength Simulator and PROMISC: Probabilistic Material Strength Calibrator (Boyce et al 1991). PROMISS calculates the random lifetime strength of a component subjected to as many as eighteen random effects. Results are presented in the form of p.d.f's and c.d.f.'s of lifetime strength, S/S_0. PROMISC calibrates the model by calculating the values of the empirical material constants.

2 SINGLE EFFECT MODELS

Single effect models given by equation (2) require selection of ultimate and reference values. For example, for the high temperature effect, the average melting temperature of Inconel 718 is 2369 °F. Hence, it is a logical choice for the ultimate temperature value, T_U.

For the high temperature model, equation (2) is modified as follows:

$$\frac{S}{S_o} = \left[\frac{T_U - T_O}{T_U - T}\right]^{-q} = \left[\frac{2369 - 75}{2369 - T}\right]^{-q}, \qquad (3)$$

where $T_U = 2369$ °F, the ultimate temperature of the Inconel 718, $T_O = 75$ °F, the reference (room) temperature, T is the current temperature of the material and q is the empirical material constant found from a power law regression of high temperature data.

For the high-cycle mechanical fatigue model, equation (2) is written as

$$\frac{S}{S_o} = \left[\frac{N_U - N_o}{N_U - N}\right]^{-s} = \left[\frac{10^{10} - 0.25}{10^{10} - N}\right]^{-s}, \qquad (4)$$

where $N_U = 10^{10}$, the ultimate number of high-cycle mechanical fatigue cycles for which fatigue strength is very low, $N_O = 0.25$, the reference number of high-cycle mechanical fatigue cycles for which fatigue strength is very high, N is the current number of high-cycle mechanical fatigue cycles the material has experienced and s is the empirical material constant found from a power law regression of high-cycle mechanical fatigue data.

For the creep model, equation (2) has the following form:

$$\frac{S}{S_o} = \left[\frac{t_u - t_o}{t_u - t}\right]^{-v} = \left[\frac{10^5 - 0.25}{10^5 - t}\right]^{-v}, \qquad (5)$$

where $t_U = 10^5$, the ultimate number of creep hours for which rupture strength is very low, $t_o = 0.25$, a reference number of creep hours for which rupture strength is very high, t is the current number of creep hours the material has experienced and v is the empirical material constant found from a power law regression of creep data.

For the thermal fatigue model, equation (2) is written as

$$\frac{S}{S_o} = \left[\frac{N_U' - N_o'}{N_U' - N'}\right]^{-u} = \left[\frac{5 \times 10^4 - 0.25}{5 \times 10^4 - N'}\right]^{-u}, \qquad (6)$$

where $N'_U = 5 \times 10^4$, the ultimate number of thermal fatigue cycles for which fatigue strength is very low, $N'_O = 0.25$, the reference number of thermal fatigue cycles for which fatigue strength is very high, N' is the current number of thermal fatigue cycles the material has undergone and u is the empirical material constant found from a power law regression of the thermal fatigue data.

When the current value and the reference value are small compared to the ultimate value, it is necessary to transform the model such that the $\log_{10}$ of each value is used. This transformation significantly increases the sensitivity of an effect to the data used within it. Also, it usually results in better statistical fits of the data from linear regression. Hence for certain single effect models, equation (2) is better represented by

$$\frac{S}{S_o} = \left[\frac{\log A_U - \log A}{\log A_U - \log A_o}\right]^a = \left[\frac{\log A_U - \log A_o}{\log A_U - \log A}\right]^{-a} . \quad (7)$$

In addition, the ultimate and reference values in equations (2) or (7) may be adjusted to reflect the value of yield strength for Inconel 718 (Bast 93). Thus they became model parameters for the multifactor equation.

3 LOW-CYCLE MECHANICAL FATIGUE

For low-cycle mechanical fatigue the model uses stress-life (σ-N) data obtained from experimental strain-life (ε-N) data. Total strain amplitude and plastic strain amplitude data for Inconel 718 yield the strain-life curve. The plastic portion of the curve may be represented by:

$$\frac{\Delta \varepsilon_P}{2} = \varepsilon'_F (2N_F^{"})^c, \quad (8)$$

where $\Delta \varepsilon_P / 2$ is the plastic strain amplitude and $2N"_F$ are the reversals to failure. A power law regression analysis of the data yields two low-cycle mechanical fatigue properties, the fatigue ductility coefficient, ε'_F, and the fatigue ductility exponent, c. Regression statistics, such as the coefficient of determination, r^2, indicate whether or not a power law representation of the relationship between plastic strain amplitude and reversals to failure is satisfactory.

Stress amplitude, $\Delta\sigma/2$, is calculated using the modulus of elasticity, E, and the total and plastic strain amplitudes, $\Delta\varepsilon_T/2$ and $\Delta\varepsilon_P/2$, respectively, by the following equation:

$$\frac{\Delta\sigma}{2} = E\left[\frac{\Delta\varepsilon_T}{2} - \frac{\Delta\varepsilon_P}{2}\right]. \quad (9)$$

When the resulting stress amplitude is plotted against plastic strain amplitude the cyclic stress-strain plot results. Again, a power law function may be satisfactory for expressing the relationship as given by

$$\frac{\Delta\sigma}{2} = K'\left(\frac{\Delta\varepsilon_P}{2}\right)^{n'}, \quad (10)$$

where K' is the cyclic strength coefficient and n' is the cyclic strain hardening exponent.

When the stress amplitude is plotted against reversals to failure, the stress-life plot results. The following function may approximate this relationship:

$$\frac{\Delta\sigma}{2} = \sigma'_F(2N_F^{"})^b, \quad (11)$$

where σ'_F is the fatigue strength coefficient and b is the fatigue strength exponent.

The low-cycle mechanical fatigue model, depicted by equation (7), now has the form

$$\left[\frac{\log N"_U - \log N"_o}{\log N"_U - \log N"}\right]^{-r} = \left[\frac{\log 5 \times 10^4 - \log 0.5}{\log 5 \times 10^4 - \log N"}\right]^{-r},$$

where $N"_U = 5 \times 10^4$ the ultimate number of low-cycle mechanical fatigue cycles for which fatigue strength is very low, $N"_o = 0.5$, the reference number of low-cycle mechanical fatigue cycles for which fatigue strength is very high, $N"$ is the current number of low-cycle mechanical fatigue cycles the material has undergone and r is an empirical material constant found from a power law regression of the low-cycle mechanical fatigue data.

Thus, the model, as given by equation (1) may include terms representing single effects appropriate for a particular application. For example, equation (7), when modified for high-cycle mechanical fatigue, creep and thermal fatigue effects only, becomes,

$$\frac{S}{S_o} = \left[\frac{\log 10^{10} - \log 0.25}{\log 10^{10} - \log N}\right]^{-s} \left[\frac{\log 10^5 - \log 0.25}{\log 10^5 - \log t}\right]^{-v}$$
$$\left[\frac{\log 5 \times 10^4 - \log 0.25}{\log 5 \times 10^4 - \log N'}\right]^{-u} . \quad (12)$$

4 EXPERIMENTAL DATA FOR INCONEL 718

A literature search for Inconel 718, a nickel-base superalloy, was conducted to obtain data on high temperature tensile strength (INCONEL Alloy 718 1986), high-cycle mechanical fatigue strength (INCONEL Alloy 718 1986), creep rupture strength (Barker et al 1970), thermal fatigue strength (Kuwabara and Kitamura 1983) and low-cycle mechanical fatigue strength (Brinkman and Korth 1974). This data resulted from tests done on various hot worked specimens, including sheets and hot rolled bars.

The above-referenced thermal fatigue data and low-cycle fatigue data are physically distinct. Generally, low cycle fatigue data

results from both low-cycle mechanical fatigue tests, as well as, thermal fatigue tests. Low-cycle mechanical fatigue produces material damage by the cyclic application of strains that extend into the plastic range, at a constant temperature. Low-cycle thermal fatigue produces damage by the cyclic application of a changing temperature. Low-cycle fatigue failure typically occurs under 10^5 cycles.

As previously mentioned, the general model for the low-cycle mechanical fatigue effect uses stress-life data obtained from experimental strain-life data. Low cycle mechanical fatigue strain-life data for Inconel 718 is presented in Table 1 and displayed as strain-life curves in Figure 1. The tests were conducted at a constant temperature of 1000 °F for a strain rate of 4x10^{-3} sec^{-1}. They were closed-loop strain controlled tests performed in air with induction heating. Using an average value of $E = 24.5 \times 10^6$ psi for the modulus of elasticity for Inconel 718 at 1000 °F (INCONEL Alloy 718 1986), the stress amplitude can be calculated from equation (9). Hence, stress amplitude is plotted against plastic strain amplitude to produce the cyclic stress-strain curve shown in Figure 2. Using the power law regression techniques indicated in equations (8), (10), and (11), and the data from Table 1, the low-cycle mechanical fatigue properties for Inconel 718 can be calculated. These material properties are displayed in Table 2 and indicated graphically, along with their coefficient of determination, r^2, in Figures 3 - 5.

Table 1 Low-cycle mechanical fatigue data for Inconel 718 (Brinkman and Korth, 1974).

Cycles to Failure N''_F	Total Strain Amplitude, $\Delta\varepsilon_T/2$	Plastic Strain Amplitude, $\Delta\varepsilon_P/2$	Stress Amplitude, $\Delta\sigma/2$ (psi)
540	0.01475	0.00725	177,600
1500	0.01040	0.00400	156,800
3400	0.00750	0.00248	134,800
13,000	0.00615	0.00105	123,700
20,800	0.00550	0.00080	115,200

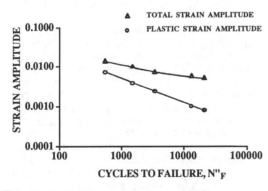

Fig. 1 Strain-life curve for Inconel 718.

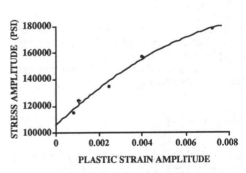

Fig. 2 Cyclic stress-strain curve for Inconel 718.

Table 2 Low-cycle mechanical fatigue material properties for Inconel 718.

Fatigue Ductility Coefficient	0.5176
Fatigue Ductility Exponent	-0.608
Cyclic Strength Coefficient	442,588 psi
Cyclic Strain Hardening Exp.	0.189
Fatigue Strength Coefficient	391,742 psi
Fatigue Strength Exponent	-0.116

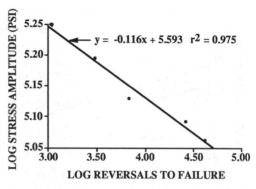

Fig. 5 Regression of equation (11) yielding fatigue strength coefficient, σ'_F, and fatigue strength exponent, b.

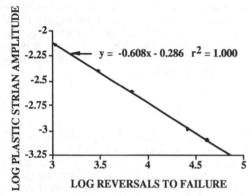

Fig. 3 Regression of equation (8) data yielding fatigue ductility coefficient, ε'_F, and fatigue ductility exponent, c.

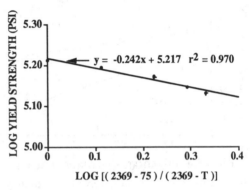

Fig. 6 Effect of temperature (°F) on yield strength for Inconel 718.

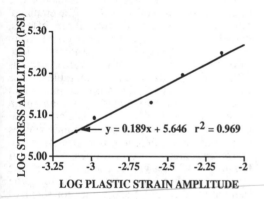

Fig. 4 Regression of equation (10) data yielding cyclic strength coefficient, K', and cyclic strain hardening exponent, n'.

To obtain the empirical material constants, for the single effect models, data were plotted in the form used in equations (2) or (7). Figure 6 shows the effect of temperature on yield strength. Figure 7 shows the effect of high-cycle mechanical fatigue (cycles) on fatigue strength for given test temperatures. Figure 8 shows the effect of creep time (hours) on rupture strength for given test temperatures. Figure 9 shows the effect of thermal fatigue (cycles) on fatigue strength for a mean thermal cycling temperature of about 900 °F. Figure 10 shows the effect of low-cycle mechanical fatigue (cycles) on fatigue strength for a constant temperature of 1000 °F and at a strain rate of 4×10^{-3} sec^{-1}. It is noted that for each effect strength decreases as temperature, cycles or time increases.

As seen in Figures 6 - 10, linear regression of the data for temperature, high-cycle mechanical

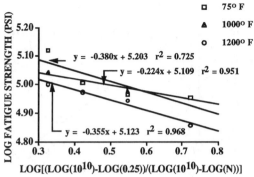

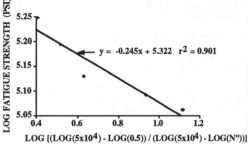

Fig. 7 Effect of high cycle mechanical fatigue (cycles) on fatigue strength for Inconel 718.

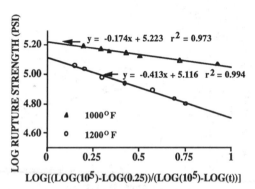

Fig. 8 Effect of creep time (hours) on rupture strength for Inconel 718.

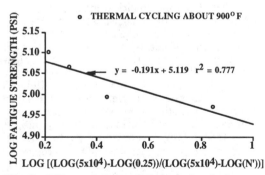

Fig. 9 Effect of thermal fatigue (cycles) on fatigue strength for Inconel 718.

Fig. 10 Effect of low-cycle mechanical fatigue (cycles) on fatigue strength for Inconel 718.

fatigue, creep, thermal fatigue and low-cycle mechanical fatigue, produces good estimates of the empirical material constants, namely, q, s, v, u and r, respectively. The estimated values of these empirical material constants are given by the slopes of the regression lines.

5 SENSITIVITY STUDY

Using the probabilistic damage analysis model embodied in PROMISS, a sensitivity study was conducted. Three effects were included in this study, high-cycle mechanical fatigue, creep and thermal fatigue. The temperature effect was not included since data for high-cycle mechanical fatigue, creep and thermal fatigue resulted from tests conducted in a high temperature environment (about 1000 °F). The multifactor equation, when applied to high-cycle mechanical fatigue, creep and thermal fatigue effects is given by equation (12). The empirical material constants for these three effects, namely s, v, and u, as determined from regression analysis of the experimental data, calibrated the model. Current values used as PROMISS input data are given in Table 4. The first three rows of the table contain bold typeface values corresponding to the mean current values of high-cycle mechanical fatigue. Likewise, the following rows contain values that indicate mean current values of creep hours and mean current values of thermal fatigue. NASA Lewis Research Center expert opinion supplied statistical distribution type (normal for this study) and values for standard deviation. The results, in the form of c.d.f.'s, are given in Figures 11 - 13, one figure for each effect. For example, Figure 11 shows the effect of high-cycle mechanical fatigue cycles on lifetime strength. Note that the c.d.f. shifts to the left, indicating a lowering of lifetime

Table 4 Mean current values used in sensitivity study of probabilistic material strength degradation model using PROMISS.

High-cycle Mechanical Fatigue (cycles)	Creep (hours)	Thermal Fatigue (cycles)
2.5 x 10^5	1000	2000
1.0 x 10^6	1000	2000
1.75 x 10^6	1000	2000
1.0 x 10^6	250	2000
1.0 x 10^6	1000	2000
1.0 x 10^6	1750	2000
1.0 x 10^6	1000	500
1.0 x 10^6	1000	2000
1.0 x 10^6	1000	3500

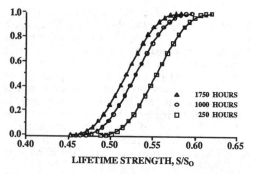

Fig. 12 c.d.f. for comparison of uncertainty of creep time (hours) on probable strength for Inconel 718: 1x10^6 high-cycle mechanical fatigue cycles, 2000 thermal fatigue cycles, at 1000 °F.

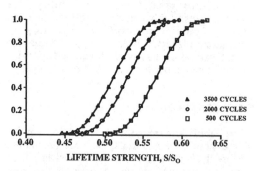

Fig. 13 c.d.f. for comparison of uncertainty of thermal fatigue (cycles) on probable strength for Inconel 718: 1x10^6 high-cycle mechanical fatigue cycles, 1000 creep hours, at 1000 °F.

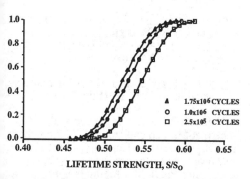

Fig. 11 c.d.f. for comparison of uncertainty of high-cycle mechanical fatigue (cycles) on probable strength for Inconel 718: 2000 thermal fatigue cycles, 1000 creep hours, at 1000 °F.

strength for an increase of high-cycle mechanical fatigue cycles. In this manner, PROMISS computes the sensitivity of lifetime strength to any effect .

6 DISCUSSION

Model calibration for the sensitivity study used the transformed effects model, equation (7) as seen by Figures 7 - 9. This resulted in a considerable increase in r^2 values obtained for each of these three effects, over a model

without the log transformation. Since r^2, a statistical measure of the goodness of fit, is relatively high, the model well-represented the actual experimental data.

Also inherent in the model given by equation (13), is the assumption that the variables are independent and there are no synergistic effects. This is not actually the case. An attempt has been made to take this into account. Creep effects are not applicable at low temperatures, therefore they are not included in models for sensitivity studies that consider temperatures below the creep threshold of 900 °F. Also, temperature effects are not explicitly included in models for sensitivity studies involving data that results from tests conducted at elevated temperatures.

Note in Figures 7 and 8 that the effect of temperature is inherent in the empirical material

constants, since their values change according to the test temperature. For example in Figure 8, the value of the empirical material constant for creep increases from a value of 0.291 at a temperature of 1000 °F to a higher value (steeper slope) of 0.650 at a temperature of 1200 °F. An increase in the material constant (slope) with an increase in temperature is expected. As Figure (7) indicates, however, the calculated value of the empirical material constant is lower at 1000 °F than at 75 °F. A plausible explanation for this phenomenon is the insufficient or bad data.

Simultaneous calibration of the model for all three effects, to build a synergistic model and better represent the interdependence of effects, may be advantageous. To do this additional experimental data would be necessary. In addition, the comparison of a synergistic model with the independent model considered here, will achieve greater model reliability (via statistical testing) when single effect data are used to predict material damage due to dependent effects. Finally, a technique for model validation has been developed, providing sufficient experimental data are available

7 CONCLUSION

A probabilistic damage analysis model, applicable to aerospace materials, has been postulated for predicting the random lifetime strength of structural components for aerospace propulsion systems subjected to diverse effects. The model takes the form of a randomized multifactor equation and contains empirical material constants. Data is available from the literature for nickel-base superalloys, especially Inconel 718, for five individual effects: high temperature, high-cycle mechanical fatigue, creep, thermal fatigue and low-cycle mechanical fatigue. Most recently, a model for low-cycle mechanical fatigue has been developed.

Thus, a general computational simulation structure is provided for describing the scatter in lifetime strength in terms of probable values for a number of diverse effects or variables. The sensitivity of random lifetime strength to each variable can be ascertained. Probability statements allow improved judgments to be made regarding the likelihood of lifetime strength and hence structural failure of aerospace propulsion system components.

8 ACKNOWLEDGMENTS

The author gratefully acknowledges the support of NASA Lewis Research Center.

9 REFERENCES

Barker, J. F., Ross, E.W. and Radavich, J. F., "Long Time Stability of INCONEL 718," Journal of Metals, Jan., 1970, Vol. 22, p. 32.

Bast, C., Probabilistic Material Strength Degradation Model for Inconel 718 Components Subjected to High Temperature, Mechanical Fatigue, Creep and Thermal Fatigue Effects, M.S. Thesis, The University of Texas at San Antonio, San Antonio, TX, April, 1993.

Boyce, L., et al, Probabilistic Lifetime Strength of Aerospace Materials Via Computational Simulation, NASA CR 187178, Aug., 1991.

Brinkman, C.R. and Korth, G.E., "Strain Fatigue and Tensile Behavior of Inconel 718 from Room Temperature to 650°C", Journal of Testing and Evaluation, JTEVA, Vol. 2, No. 4, July, 1974, pp. 249-259.

Chamis, C. ,Simplified Composite Micromechanics Equations for Strength, Fracture Toughness, Impact Resistance and Environmental Effects, NASA TM 83696, Jan., 1984.

INCONEL Alloy 718, Inco Alloys International, Inc., Huntington, WV, 1986, pp. 8-13.

Kuwabara, K., Nitta, A. and Kitamura, T., "Thermal-Mechanical Fatigue Life Prediction in High-Temperature Component Materials for Power Plant," Proceedings of the Advances in Life Prediction Methods Conference, ASME, Albany, N.Y., April, 1983, pp. 131-141.

Scott, D.W., "Nonparametric Probability Density Estimation by Optimization Theoretic Techniques," NASA CR-147763, April 1976.

Siddall, J. N., "A Comparison of Several Methods of Probabilistic Modeling," Proceedings of the Computers in Engineering Conference, ASME, Vol. 4, 1982.

Structural Safety & Reliability, Schuëller, Shinozuka & Yao (eds) © 1994 Balkema, Rotterdam, ISBN 90 5410 357 4

Prediction of machine elements reliability for changed operation conditions: The synthetized spectra method

Lech Bukowski
University of Mining & Metallurgy, Cracow, Poland

Harald Zenner
Technical University Clausthal, Germany

ABSTRACT: In the paper an application of a modified cause-effect type model for predicting reliability of machine systems at the design state is presented. Input data required by the model are obtained from a synthetized load spectrum method developed by the authors. The method has been verified in industrial practice for reliability prediction of a rolling mill shop for changed operation conditions.

1.INTRODUCTION

In most cases reliability of machine systems in a design state is assessed on the basis of datareceived from similar machine systems whichwork in similar conditions. Such estimations often lead to serious errors especialy when load variation causes fatigue damage of machine elements. Known computation methods used in fatigue reliability calculations are based on statistical models. These models describe time or number of load cycles to failure using Markov process or probability distribution e.g. Weibull. They are shown, e.g., in [4,5,7]. The application of the methods requires special experimental data. The methods are not proper if any significant load changes are expected in the future. In these cases it seems that models of cause–effect type should be useful.

In the paper an application of a modified cause–effect type model for predicting reliability of machine systems at the design state is presented. Input data required by the model are obtained from a synthetized load spectrum method developed by the authors.

2.METHOD TO DESCRIBE LOADINGS OF MACHINE SYSTEMS ELEMENTS

The first step in the process of machine elements reliability determination should be adoption of assumptions regarding loads that will occur in operation. These assumptions have to be supported by a detailed analysis of the machine system, work conditions and any possible enviromental interactions on the particular system's elements. Loads in machine elements can be specified due to their time functions as:

-steady, which means static (or more often quasistatic);

-changing with time.

In the case of static or quasistatic loads their time functions can be substituted by one value which represents loading of an element during operation or by the loading probability distribution. However in the second case that way of data reduction does not account for degradation phenomena (e.g., fatigue of material). So in such cases load time functions are transformed into cumulative chracteristics by estimation of load spectra.

Since two decades a considerable number of measurements during operation have been made all over the world.So it became possible to estimate cumulative load characteristics of many machines and plants. The characteristics can be utilised effectively in the design state, using the synthetized spectra method developed by the authors.

The following procedure is proposed to

estimate cumulative load spectra using the partial load spectra synthesis method [5]:
-determination of the total utility period T, not necessarily expressed in time units;
-defining the kinds of operational process or the states of a system as featured by clearly different loads in the form of a vector E={E };
-estimation of representative partial spectra of all particular states E in the form:

$$H_i(Y,T_m) = H(Y|E_i)|T_m \qquad (1)$$

where:
H, H_i - cumulated frequencies;
Y - load;
T_m-spectrum range expressed by measurment period.

The spectra have to be estimated for a load scatter, which should result in a form of probability distribution or at least should be estimated for special levels of happenstances probability e.g. P=50% and P=1% .
-determination of parts of particular states E_i in the whole operation process, that means assigning a value u_i to each state E_i, while $\Sigma u_i =1$;
-extrapolation of estimated partial load spectra into adequate ranges $T_i = u_i T$ that results in partial spectra in the form of probability distribution or two spectra for the above mentioned happenstances probabilities;
-superposition of extrapolated partial spectra to get cumulative load spectrum in the form of probability distribution or cumulative load spectra for the two levels of probability (50% and 1%).
Partial spectra can be described by analytical models or estimated by the method of statistic simulation. In the case of analytical models the Beta-, Normal- and Pearson- probability distributions [2] are the most often used ones. Anyway, the most covenient one in use and additionally sufficiently accurate and universal, is a model based on Weibull distribution in the form:

$$H(Y,H) = \{exp -(Y/b)^a\} H_o \qquad (2)$$

where:
Y-load, e.g.,the amplitude of normal or tangential stress;
H_o-conventional spectrum range, mostly 1.000.000 cycles;

H- cumulated frequency;
a,b- parameters of form and scale of the spectrum.

In the case that statistical characteristics of load functions for particular phases of operation are already known for a machine system in the designe state it is recomended to apply statistical simulation methods. For this purpose the so called equivalent load time functions [2] are utilized (examples are presented in fig.1 and fig.2). On the other hand if there are no data on partial spectra for the designed mechanical system, the demand arises to determine the spectra by a simulation method in a dynamic model of the system. This method has been described and illustrated by a practical application example in [3].

3. MODELS OF STRENGTH OF MACHINE SYSTEMS ELEMENTS

The strength of machine elements wich are liable to sudden-type damage is equal to ultimate strength or yield limit depending on the adopted usability criterion. The values of these strength parameters can be found for a majority of materials in technical publications along with data on their scatter (mostly mean value and variation coefficient or minimum value and variation coeffeicient).
Three different fatigue life ranges can be distinguished for elements liable to fatigue-type damage. For every range the loadability is described in a different way. In the range of unlimited life it is equal to an endurance limit z , which is described by its expected value and variation coefficient. In the range of limited life S-N curves are used. They show a dependance between life (expressed by number of load cycles) and load level (expressed in the form of stress amplitude). In the range of small number of cycles (LCF range) the strain versus cycles curves in lgϵ - lgN system are used. For typical materials they are accessible in [1].

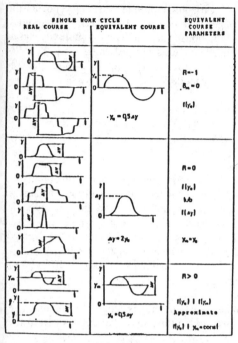

Fig.1 Equivalent loads time functions and their parameters for typical single work cycles.

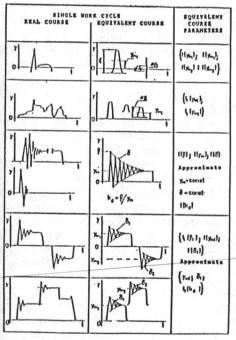

Fig.2 Equivalent loads time functions and their parameters for typical complex work cycles.

4.DETERMINATION OF MACHINE SYSTEMS ELEMENTS RELIABILITY

4.1.Reliability of elements subjected to static and quasistatic loads

The degree of reliability of machine elements subjected to static and quasistatic loads is the probability that ultimate strength W of an element is higher then its loading due to the following relation:

$$R = P(W > Y) = \int_{-\infty}^{+\infty} f(w)[\int_{-\infty}^{w} f(y)dy]dw \qquad (3)$$

where:
w,y -realizations of random variables W and Y .

The value of probability can be determined:
-analyticaly, for some typical distributions of strength and load probability;
-by numerical integration;
-through Mellin's transformation;
-by simulation, e.g,. using Monte Carlo method.

4.2.Reliability of elements subjected to time variable loads

4.2.1.Computations for the unlimited life range

If the amplitudes of variable stress do not axceed endurance limit in the whole predicted operation period, then calculations are similar to those in 4.1.using endurance limit z instead of ultimate strength W.
The following formula is used:

$$R(N) = P(N > N_G) = \int_{-\infty}^{+\infty} f(Z_G)[\int_{-\infty}^{Z_G} f(y_a)dy_a]dz_G \qquad (4)$$

where:
N-number of cycles;
N_G, z_G -number of basic cycles and endurance limit;
Y-load amplitude.

4.2.2 Computations for limited fatigue life range

In the case that loading has a sinusoidal form with constant amplitude and if this amplitude is not known the following can be determined:

1071

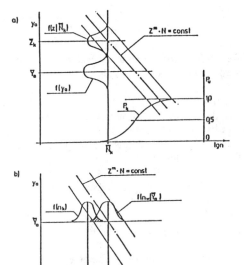

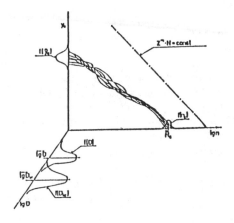

Fig.3 Determination of no-damage probability in
the range of limited fatigue life:
a) at the set number of cycles N_k
b) at the set loads level y_a.

Fig.4 Determination of reliability in the range of
limited fatigue life with application of damage
cumulation hypothesis.

-probability of no-damage for a given number of cycles;
-expected life (number of cycles) for a required no-damage probability.

Computations are realized according to formula of type (4). Graphical illustration of the both cases is shown in fig.3 .

If variable-amplitude loading occurs the damage cumulation hypothesis is applied. Computations of different complexity based on that hypothesis are presented in [2]. It is proposed to adopt the following general reliability model:

$$R(x) = P(X>x) =$$
$$= P\{D[H(Y,x)]<D_{cr}[H(Y,x_{cr})]\} ;$$
$$X \in [0,x] \qquad (5)$$

where:
$D[H(Y,x)]$- an arbitrary value of cumulated damage index determined according to damage cumulation hypothesis for the cumulative loading spectrum $H(Y,x)$ of the range x;
$D[H(Y,x_{cr})]$-critical value of cumulated damage index at which damage occurs for a cumulative loading spectrum $H(Y,x_{cr})$ of the range x_{cr} .

When that model is applied the machine elements reliability may be determined in the following way (fig.4) :
-adoption of a conventional S-N curve for the given element and type of loading;
-estimation of $f(D_{cr})$ distribution for the element and the adopted S-N curve;
-estimation of $f(D)$ distribution for the representative set of realizations of loading spectra for an assumed number of cycles N_c;
-determination of an element reliability from the formula (5) for the assumed number of cycles N_c .

The method is very accurate, but requires a huge number of experimental data. The data are often unaccessable for a designer. In these cases the simplified method developed by the authors [8] is proposed to use. It has been accepted, based on Schütz and Zenner's research [6] that index D can be described by the log-normal distribution. Two cumulative load spectra are estimated for happenstance probabilities P = 50% (mean spectrum) and P = 1% (pessimistic spectrum) respectively. Using the S-N curve appropriate for no-damage probability P = 50% (mean curve) damage indexes D(P=50%) and D(P=1%) can be determinated for both spectra. Computation results are plotted on log-normal distribution grid in the form of points S and P (fig.5) . Subsequently, a straight line is drawn through these points and the plot of D probability

1072

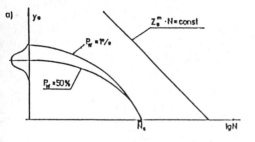

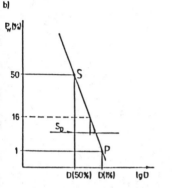

Fig. 5 Determination of distribution parameters of
damage index D:
  a) estimation of the mean (50%) and pessimi-
     stic (1%) load spectra;
  b) determination of the expected value (mean)
     and standard deviation.

much higher then previously observed. Partial
spectra were determined on the basis of short
test measurements for all seven product range
groups that were planned to roll in the future.
Then the spectra were extrapolated onto ranges
corresponding to percent share of a particular
group of the total expected manufacturing
programme and sumarized. As a result a
cumulative spectrum has been obtained, then
transformed into an amplitude spectrum.
Further, with application of the simplified
method according to fig.5 the reliability of the
weakest drive elements was determined for
changed operation conditions. Obtained results
were found satisfactory and the rolling mill
was permitted to operate in more intensive
working conditions without design changes.
Further multiyears operation of the rolling mill
has confirmed the validity of predictions and
effectivenes of the developed method.

distribution is obtained. Using this plot we
can estimate distribution parameters of D -
expected value $E(D) = D(P=50\%)$ and standard
deviation, as

$$s_D = [\log D(P=1\%) - \log D(P=50\%)]/u_o \quad (6)$$

where:

u_o - distribution quantile (for $P = 1\%$ $u_o = 2.33$)
   When distribution parameters D_{cr} are known,
the reliability can be determined from formula
(5) using known normal distribution relations.

5.SUMMARY

The method has been verified in industrial
practice for reliability prediction of a rolling
mill shop for changed operation conditions. The
changes resulted from the introduction of rolling
new sorts of tool steels. The expected loadings
of the rolling mill drive system have been

REFERENCES

1. Boller Ch., Seeger T. Materials Data for
   Cyclic Loading (Part A to D). Elsevier, 1987
2. Bukowski L. Complex Method of Metallur-
   gical Machine - Systems Modernization,
   Paying Special Attention to their Reliability.
   ZN Mechanika 24, p.1-181, AGH Krakow,
   1989
3. Bukowski L. Wykorzystanie symulacji na
   modelach dynamicznych do modernizacji
   ukladow maszynowych, na przykladzie
   przesiewacza aglomeratu. ZN Mechanika
   3\1990, AGH Krakow, p.5-17
4. Provan J.W., Bohn S.R. Stochastic Fatigue
   Crack Growth and the Reliability of
   Deteriorating Structures. Proc. of the IV.
   Int. Conference on Fatigue , 1990 in
   Honolulu-Hawaii, p.2259-2264.
5. Ranganathan R. Fatigue Reliability Analysis
   of Structures. Int. Symp. on Fatigue and
   Fracture in Steel and Concrete Strucyutres,
   1991, Madras-India, p.1415-1430.
6. Schütz W , Zenner H. Schadensakkumu-
   lationshypothesen zur Lebensdauervorhersage
   bei schwingender Beanspruchung - Ein
   kritischer Uberblück. Z.f.Werkstofftechnik,
   1973, Nr.1 - 2.
7. Wirsching P.H.,Wu Y.T. Advenced Reliability
   **Method for Fatigue Analysis, Journal of
   Eng. Mechanics. ASCE , April 1984,
   p.536-553,**

8. Zenner H., Bukowski L. Konstruierte Last und Beanspruchungskollektive unter Berücksichtigung unterschiedlicher Belastungsereignisse . Materialprüfung. 1988, Nr.7/8, p.221-224.

Structural Safety & Reliability, Schuëller, Shinozuka & Yao (eds) © 1994 Balkema, Rotterdam, ISBN 90 5410 357 4

Probability fatigue reliability

Bernard Jacob & Li Jiang

Laboratoire Central des Ponts et Chaussées, Paris, France

ABSTRACT: A full probabilistic calculation of the fatigue lifetime for steel bridge details, based on the Miner's model, is developed. The fatigue strength model comes from the Eurocode 3. Real detailed traffic records are used as an input in a simulation program, which computes accurately the stress variations and their distributions. The central limit theorem applied to the total damage gives explicit simple formulations of the reliability ß-indices in different cases. A sensitive study of ß to the model parameters and an application to three existing bridges are presented, showing that operational tools are provided for code calibration and bridge evaluation.

1 INTRODUCTION

The fatigue checking of plans for steel or composite bridges, the development of enginee-ring rules, and the problems of maintenance and evaluation of the residual safety of structures in service make use of calculations of fatigue damage of welded assemblies. Because of the randomness of traffic loads and the large uncertainties on the fatigue strengths of such assemblies, deterministic life calculations are insufficient to assess the true safety of structures in context.

For about twenty years, knowledge of traffic actions on bridges and of their probability laws has progressed considerably thanks to the acquisition of abundant data (Bruls, Jacob, Sedlacek 1989). The same is true of fatigue laws and the uncertainties of their parame-ters, thanks to very many tests (Brozzetti & al. 1989). Advanced methods in structural reliability have been introduced to estimate the probability of failure of a component or a system with respect to given limit states. They are based on the Hasofer–Lind reliability index β (Hasofer & Lind 1974). The so-called first- and second- order methods and numerical algorithms have been developed (Rackwitz & Fiessler 1978), to allow effective calcula-tions of these indices.

However, these methods apply rigorously only when the limit states are independent of time, i.e. when the model deals with random variables. Difficulties appear if the failure functions involve explicitly the time; that is the case of fatigue, where the damage is gradual.

Some applications of reliability methods to fatigue were presented in (IABSE 1990). But very simplified theoretical approaches were not really operational, because they involved basic random variables modelling parameters that were already approximate, with little data available. In other cases, the sophisti-cation of the models was an obstacle to development up to the operational stage.

An original, operational and efficient calculation of the fatigue reliability index is proposed. It combines the use of statistics of extensive fatigue tests (Brozzetti & al. 1989) and available traffic data (Bruls, Jacob, Sedlacek 1989) implemented through a numerical calculation of stress variation distributions in bridges (Eymard & Jacob 1989), with a simple probabilistic formulation of Miner's model. Application of the central limit theorem to the very large number of stress cycles, with the assumption of statio-narity, leads to an explicit calculation of β.

2 PROBABILISTIC FATIGUE MODEL

The Miner's model assumes a linear summing of the elementary damages. It is still widely used even if its theoretical grounds are debatable. The elementary damage induced by a stress cycle of amplitude S, written d(S), is given by: $d(S)=1/N(S)$, where $N(S)$ is the number of constant amplitude cycles S leading to failure. For a whole stress spectrum $(S_i, n_i)_i$, the total damage D_T is:

$$D_T = \sum_i d(S_i) = \sum_i n_i / N(S_i)$$

2.1 Strength

The number $N(S)$ is linked to S by the S-N (Woehler's) curve: $N(S)=C.S^{-m}$. Parameters C (C_m in fact) and m are deduced from experimental results by a linear regression in log-log co-ordinates. In accordance with Eurocode 3, $m=3$ then 5 is used, with a change of slope at $S=S_F$ (fatigue limit) for $N=5.10^6$, and a cut-off at $S=S_T$ for $N=10^8$. Here, only to simplify the presentation, we work some calculations for $m=3$ or 5, but constant; however general results are given, and the numerical applications are based on them. Each S-N curve is defined by the parameter C or the value S_c at 2.10^6 cycles.

The results of fatigue tests for a given geometry of detail show a large dispersion of the points around the straight regression line, due to the uncertainties on the material (in particular local faults) and the weld. To take account of these uncertainties and in accordance with the gaussian character of the variations of $\ln(N(S))$ for a given S, we use the following probabilistic model:

$$\ln(N(S)) = \ln(N_0(S)) + X \qquad (1)$$

where $X \equiv N(0,v)$ is normal, $v = s_X^2$.

$N_0(S)=C.S^{-m}$ is the deterministic number and $N(S)$ the actual random number of cycles of the S-N curve associated with S. The experimental results show that v does not depend on S, and therefore that the law of X is independent of S. Moreover, for a given specimen and a given quality of weld, the variable X represents the uncertainty on the strength of the detail. Then in formula (1) X doesn't depend on S.

s_X is linked to the standard deviation s_N of $\log(N)$, estimated from the statistics of the test results, by: $s_X = s_N \ln(10)$.

The foregoing model can also be written:

$N(S)=N_0(S)/Z$, where $Z=\exp(-X)$ is a lognormal variable with a mean and variance given by:

$$\mathbb{E}[Z] = e^{v/2} \text{ and } \mathbb{V}[Z] = e^{2v}-e^v \qquad (2)$$

2.2 Loadings

The stress variations time history in a bridge detail is the convolution of the random traffic process with the stress influence line (or area) of the detail in question. Because of the complexity of the traffic process, e.g. for a multiple traffic lanes bridge, in which both load intensities and occurrences are random, the combination of the effects of several simultaneous loads (lorry platoons or meetings), it is almost impossible to build a realistic and tractable probabilistic traffic model to calculate a theoretical law of variations of stress S. On the other hand, the many accurate traffic records available at the LCPC provide several weeks of continuous multi-lane traffic on various roads and motorways. Used as an input of the CASTOR-LCPC computer simulation program (Eymard & Jacob 1989), which makes a numerical convolution, they provide accurate stress variation time history samples. Experimental stress cycle spectra are derived by the "rain-flow" counting procedure. s_m and V_m are the mean and the variance of S^m.

Traffic analysis shows that, if the long term trends are removed, the stationarity may be assumed over long time periods ($\geq$ 1 year), and shorter periods as a week are representative. Let us note: ν the mean number (rate) of stress cycles over short representative time period τ (here τ= 1 week), linked to the traffic flow and to the shape of the area of influence, T the lifetime required of the detail and q the number of weeks in T ($T=q\tau$).

ν may be either considered as a random variable, representing the randomness variability of the rate over several short periods τ, with the mean $\bar\nu$, the variance $\mathbb{V}_\nu$ and a coefficient of variation (c.o.v.) c_ν, or as a deterministic constant (the mean rate over T). In the first case, the total number of stress cycles over T is deterministic and egal to νT. In the second case, this number becomes a random variable N' with the mean $\bar\nu T$, the variance $q\tau^2\mathbb{V}_\nu$, and a c.o.v. $c_\nu/\sqrt q$.

Parameters as ν (or $\bar\nu$, $\mathbb{V}_\nu$ and c_ν), s_m and V_m may be estimated on the short time period τ, from the traffic records and the above "rain-flow" stress cycle spectra, either for one traffic route or a set of traffic routes. In this latest case ν (or $\bar\nu$ and $\mathbb{V}_\nu$), s_m and V_m are weighted averages of the parameters deduced from each traffic route with weights egal to the mean traffic rates.

The amplitudes of successive stress cycles are represented by a series of independent identically distributed random variables S_i, for $1 \leq i \leq \nu T$. In case of a single traffic route (mainly considered hereafter) it can be easy checked that the variability (randomness) of the total number of cycles νT for long time period T (e.g. 50 years or more) is negligible. In other cases νT may be taken as a random variable.

The assumption of the independence of successive cycle amplitudes is strictly valid only for short influence lines (L≤20m, for example) having a single peak, for which each vehicle induces one cycle and two vehicles that follow one another do not interact. In case of a single vehicle induces several

cycles (action of axles on very short influence line or of vehicle on long influence line with several peaks), the foregoing model remains valid provided that the cycles are processed in groups in the calculation of the subsequent damage (Jacob & Jiang 1993). Moreover the assumption of the independence of successive cycles groups is clear in case of lorry platoons, if a group contains all the cycles between two instants where the bridge becomes empty. It was checked that the assumption of independance of individual cycles instead of the cycle groups didn't change significantly the final results (Jacob & Jiang 1993). It comes from the linearity of the Miner's model (the instant of a stress cycle doesn't change its elementary damage).

If the two-slope formulation of the S-N curve is used, let us introduce the random variables:

$$S'^3 = S^3 . \mathbb{1}_{[S_F, \infty[} \quad \text{and} \quad S'^5 = S^5 . \mathbb{1}_{[S_T, S_F[},$$

with the parameters:

$$s'_3 = E[S'^3], \quad V'_3 = V[S'^3] \tag{3}$$
$$\text{and} \quad s'_5 = E[S'^5], \quad V'_5 = V[S'^5]$$

v_1 and v_2 are the rates of cycles in $[S_F, \infty[$ and $[S_T, S_F[$ (either random or constant).

2.3 Fatigue damage

In the following, v is mainly considered as deterministic, but the results for random v are given, from (Jacob & Jiang 1993).

For a cycle of amplitude S, the elementary damage d(S) is a random variable:
d(S) = Z.S^m/C, and for a given set of cycles $(S_i)_{1 \le i \le p}$, the total damage is also a random variable: $D_T = Z(\sum_{i=1}^{p} S_i^m)/C$. If the two slopes of the S-N curve are considered, the total damage becomes:

$$D_T = Z[(\sum_{S_i > S_F} S_i^3)/C_3 + (\sum_{S_T < S_i < S_F} S_i^5)/C_5].$$

Using the independence of S and Z, the mean and variance of the damage D_T are:

$$E[D_T] = \frac{vT}{C} e^{v/2} s_m \quad \text{and}$$
$$V[D_T] = \frac{vT}{C^2}\left(e^{2v} V_m + (vT)(e^{2v} - e^v)s_m^2\right) \tag{4}$$

In many cases s_m^2 and $\overline{V_m}$ are of the same order of magnitude and vT is large (e.g.: $s_5 = 3.4 \ 10^4$, $V_5 = 3.6 \ 10^9$ and $vT \cong 10^6$ in one of our examples); then it leads to the following useful approximation :

$$V[D_T] \cong \frac{(vT)^2}{C^2}(e^{2v} - e^v) s_m^2 \tag{4a}$$

(V_m becomes negligible).

For random v, (4) becomes:
$$E[D_T] = \frac{\overline{v}T}{C} e^{v/2} s_m \quad \text{and} \tag{4b}$$
$$V[D_T] = \frac{\overline{v}T}{C^2}\left(e^{2v} V_m + (\overline{v}T)(e^{2v} - e^v)s_m^2\right) + E[D_T]^2 c_v e^v/q$$

The law of total damage D can be written:

$$D_T = \frac{Z}{C} \sum_{i=1}^{vT} S_i^m = \frac{Z\sqrt{vT}}{C}\left(\frac{\sum_{i=1}^{vT} S_i^m - vTs_m}{\sqrt{vT}}\right) + \frac{vT}{C}s_m Z$$

$$D_T \cong \frac{vT}{C}s_m Z(1+U) \tag{5}$$

by applying the central limit theorem, given the independence of the S_i and the fact that vT is large (for T of the order of several years). U is the centered normal variable having the variance: $\sigma_U^2 = V_m/(s_m^2 vT)$.

In case of the approximation (4a), $vT \cong 10^6$, $\sigma_U \cong 6.10^{-3}$, and therefore:

$$P(|U| \le 1.5 \ 10^{-2}) = 0,99; \quad \text{we may then write:}$$

$$D_T \cong \frac{vT}{C}s_m Z \quad (D_T \text{ becomes log-normal}) \tag{5a}$$

If v is random, the damage calculation is slightly modified. For a short period τ, let us note the random sum of a random number $v\tau$ of independant S_i^m: $\psi = \sum_{i=1}^{v\tau} S_i^m$.

It is easy to prove that:

$$E[\psi] = \overline{v}\tau s_m \quad \text{and} \quad V[\psi] = \overline{v}\tau V_m + \tau^2 s_m^2 V_v.$$

For the total lifetime T splited into q short periods of τ, we get a series of q (ψ_i) which are i.i.d. random variables. The damage D_T may be written: $D_T = Z(\sum_{i=1}^{q} \psi_i)/C$, and the central limit theorem still applies (q large). (5) remains valid, with v replaced by $\overline{v}$ and:

$$\sigma_U^2 = V_m/(s_m^2 \overline{v}T) + c_v^2/q.$$

With the two slope S-N curve formulation (5) and (5') are replaced by:

$$D_T \cong TZ\left(\frac{s'_3 v_1}{C_3}(1+U_1) + \frac{s'_5 v_2}{C_5}(1+U_2)\right)$$
$$\cong TZ\left(\frac{s'_3 v_1}{C_3} + \frac{s'_5 v_2}{C_5}\right) \tag{5b}$$

where U_1 and U_2 are normal, centered, and independent with the variances:

$$\sigma_{U_1}^2 = V'_3/(s'^2_3 v_1 T) \quad \text{and} \quad \sigma_{U_2}^2 = V'_5/(s'^2_5 v_2 T).$$

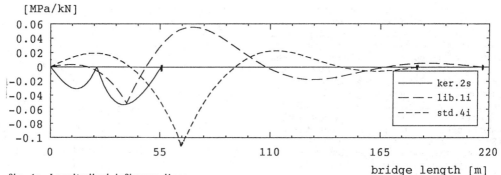

fig. 1a: Longitudinal influence lines

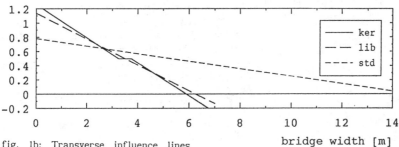

fig. 1b: Transverse influence lines

3 FATIGUE RELIABILITY INDEX

3.1 Limit state and probability of failure

The investigation of the reliability of a detail consists of evaluating its probability of failure during the specified lifetime T. The problem can be formulated in reverse, by looking for the lifetime T such that the probability of failure during T has some specified value. The fatigue limit state function of which the graph in the space of the basic variables of the problem marks the border between the domains of failure and of safety, is written:

$$g(Z,S_1,\ldots S_p,\Delta) = D_T - \Delta = 0 \qquad (6)$$

Δ is a random variable centered on 1 and having the variance δ^2, which reflects the uncertainty on the Miner's model and the criterion of failure. This variable is assumed to be log-normal (IABSE 1990). Let we introduce the normal variable $D=\ln(\Delta)$, with mean $-\frac{1}{2}\ln(1+\delta^2)$ and variance $\ln(1+\delta^2)$.

Using (5), the probability of fatigue failure becomes:

$$P_f = \mathbb{P}(D_T > \Delta) = \mathbb{P}\left[\frac{\nu T}{C} s_m Z > \frac{\Delta}{1+U}\right]$$

$$= \mathbb{P}\left[X+D < -\ln\left(\frac{C}{\nu T s_m}\right) + \ln(1+U)\right]$$

$$P_f \simeq \mathbb{P}\left[X-U+D < -\ln\left(\frac{C}{\nu T s_m}\right)\right] \qquad (7)$$

with the 2nd-order approximation: $\ln(1+U)\approx-U$.
If the 1st-order (5') approximation is used, the foregoing expression becomes:

$$P_f \simeq \mathbb{P}\left[X+D < -\ln\left(\frac{C}{\nu T s_m}\right)\right] \qquad (7a)$$

3.2 Reliability index

The Hasofer-Lind index β (Hasofer & Lind 1974) is easy to calculate from (7):

$$\beta = \ln\left(\frac{C}{\nu T s_m \sqrt{1+\delta^2}}\right)\left(\nu + \ln(1+\delta^2) + \frac{V_m}{\nu T s_m^2}\right)^{-\frac{1}{2}}$$

$$\approx \frac{\ln\left(\frac{C}{\nu T s_m \sqrt{1+\delta^2}}\right)\left(1-\frac{1}{2}\frac{V_m}{\nu T s_m^2(\nu+\ln(1+\delta^2))}\right)}{\sqrt{\nu+\ln(1+\delta^2)}} \qquad \ldots(8)$$

using the 2nd-order approximation (7) and since variables X, U, and D are normal and

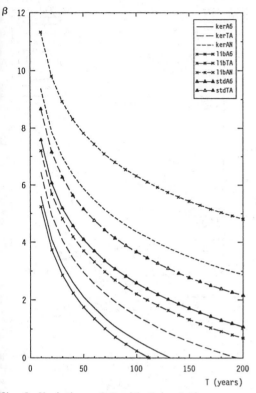

fig. 2: Variations of β with T (s_N=0.2)

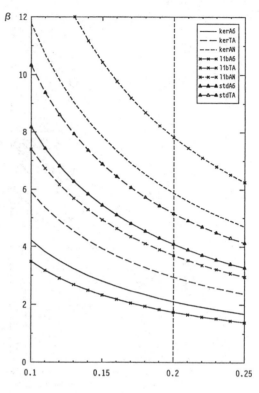

fig. 3: Variations of β with s_N (T=50 yrs)

centered and have as respective variances v, $V_m^2/\nu Ts_m^2$ and $\ln(1+\delta^2)$. We get: $p_f=\Phi(-\beta)$. With approximation (5a) and (7a) the reliability index becomes:

$$\beta \simeq \ln\left(\frac{C}{\nu Ts_m \sqrt{1+\delta^2}}\right)\left(v + \ln(1+\delta^2)\right)^{-\frac{1}{2}} \qquad (8a)$$

$$\beta \simeq v^{-\frac{1}{2}}\ln\left(\frac{C}{\nu Ts_m}\right), \quad \text{if } \Delta\equiv 1 \ (\delta=0) \qquad (8b)$$

If ν is random, the first part of formula (8) is slighty extended:

$$\beta=\ln\left(\frac{C}{\nu Ts_m \sqrt{1+\delta^2}}\right)\left(v+\ln(1+\delta^2)+\frac{V_m^2}{\nu Ts_m^2}\frac{c_\nu^2}{q}\right)^{-\frac{1}{2}} \qquad (8c)$$

But in most of the road bridge cases, as those considered in section 4, c_ν^2/q is negligible ($q\approx1000$ for $T\geq20$ years), and ν may be taken as deterministic.

In case of a two-slope S-N curve formula, (8a) becomes:

$$\beta \simeq \frac{\ln\left(\dfrac{1}{T\left(\dfrac{s_3\nu_1}{C_3}+\dfrac{s_5\nu_2}{C_5}\right)\sqrt{1+\delta^2}}\right)}{\sqrt{v+\ln(1+\delta^2)}} \qquad (8d)$$

4 APPLICATION TO COMPOSITE BRIDGES

4.1 Cases considered

We examine here the cases of three composite bridges recently built in France, at Kervitou in Britanny (ker), Libourne near Bordeaux (lib), and Saint-Denis just north of Paris (std). Kervitou is a 2-span bridge 56m long and Libourne is a 5-span bridge 216m long; both are two-main girder bridges with two traffic lanes. Saint-Denis is a 4-main girder bridge 185m long, supporting the four traffic lanes of the A86 motorway. The details studied are the junctions between the vertical stiffening web and the bottom flange of a beam, between bearings (lib & std), or the top flange, on a bearing (ker). The corresponding

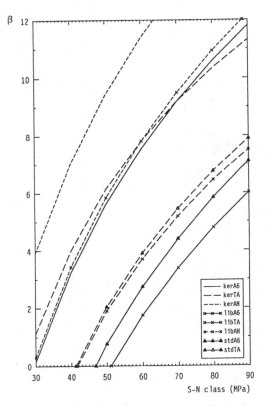

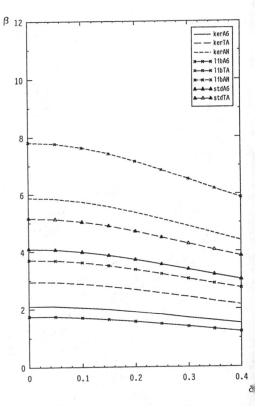

fig. 4: Variations of β with the S-N class   fig. 5: Variations of β with δ

longitudinal and transverse stress influence lines are shown in fig. 1a and 1b. Three traffic records were used to calculate the stress variations, measured continuously for one week: those of the A6 motorway (Paris-Lyon) at Auxerre (lanes 1 and 2 of the North-South direction), RN182 on the Tancarville bridge, and RN23 near Angers (slow lanes 1 and 3 of each direction for both RN), the last one used only for Kervitou and Libourne. The traffic of the A6 is one of the most aggressive in France and in Europe (Bruls Jacob & Sedlacek 1989).

4.2 Results and commentaries

The CASTOR-LCPC program was used to calculate the "rain-flow" spectra of stresses for each traffic and each detail (for a total of 8 cases). For the calculation of the annual damages, it was assumed that a year contains 50 weeks (given the lower traffic in summer). It was checked on the moments of the stress variations that the approximation of formula (5a) in fact applies here. Index β is therefore calculated from formula (8d).

Constants C_3 and C_5 are those of the mean

S-N curves of each detail, deduced from the characteristic curves (at 95%) by a translation of $2s_N$ to the right along the axis of the log(N). For this operation, s_N was taken equal to 0.20, based on the recommendations of (Brozzetti & al. 1989), for the details in question. The classes (at 95%, corrected for the effect of thickness) are 36 MPa (ker), 60 MPa (lib), and 68 MPa (std).

The default values of the various parameter for the calculations of indices β are T=50 years, s_N=0.20, δ=0. Sensitivity studies wer performed by varying, in turn, parameters (from 10 to 200 years), s_N (from 0.1 to 0.25) the fatigue class (from 30 to 90 MPa), and (from 0 to 0.4), with the default values used for the other parameters. The results ar shown in figs. 2 to 5 for the 8 cases i question.

For these cases, the most aggressive traffi is that of the A6 motorway, the least aggres sive that of RN23. In all cases for T=50 years, the index β is greater than or equal t 2.0, but in two cases the value of 3. required by the draft of Eurocode 3 fo ultimate fatigue limit states is not reache

with T exceeding 30 or 40 years. The Saint-Denis bridge has the highest β indices of the 3 bridges, but is also the only one that actually carries the traffic of a motorway. β is especially sensitive to the lifetime T and to the fatigue strength of the detail (class of fatigue), then to the uncertainty of the S-N curve, related to the dispersion of the fatigue tests. The influence of parameter δ is much smaller, which indicates that Miner's model is fairly robust and explains why generally the uncertainty of the model is not explicitly taken into account.

Additionally to the above study, we looked for the β of the Kervitou and Libourne bridges under a unknown traffic, estimated as a weighted average of the 3 traffics considered. In our cases, the estimated variances of S^m don't increase significantly in comparison with those related to each traffic. Then (5a) and (8d) remains valid and the β's stay in the range of the previous ones.

4.3 Conclusions

In conclusion, we have provided an explicit calculation of the fatigue reliability index β for Miner's model, making it possible to take account of uncertainties on the characteristics of the traffic and its effects on bridges, on the fatigue strength of the detail, and even on the criterion of failure. Applied to the case of actual structures, with accurate traffic data and a simulation computer program to obtain the parameters of the stress variation probability distribution function, it can be used to quantify the reliability in fatigue of planned or existing structures and to study its sensitivity with respect of changes in the traffic or in the strength of the welds. It is a useful tool for bridge survey and maintenance. It will also help to provide a database of realistic β-values for the calibration of the γ-safety factors of the codes, as the Eurocodes.

Such an operational approach seems to be a good compromise between very sophisticated but untractable theoretical models and too simple calculations based on rough assumptions about the traffic loads and load effects.

REFERENCES

Brozzetti, J., I. Ryan, G. Sedlacek 1989. "Background Informations on Fatigue Design Rules - Statistical Evaluation", Eurocode 3 part 1, chap 9, Doc 9.01.

Bruls, A., B. Jacob, G. Sedlacek 1989. "Traffic Data of the European Countries", Eurocode 1, Traffic Loads on Bridges, *report of WG 2.*

Eymard, R., & B. Jacob 1989. Un nouveau logiciel: le programme CASTOR-LCPC pour le Calcul des Actions et Sollicitations du Trafic dans les Ouvrages Routiers, *Bull. liaison Labo. P. et Ch.*, n°164, pp. 64–78.

Hasofer, A.M., & N. Lind 1974. "An Exact and Invariant First Order Reliability Format", *J. of Engineering Mechanics*, ASCE, Vol.100, N° EM1, pp.111-121.

Jacob, B., & L. Jiang 1993. "Indice de fiabilité et coefficients de sécurité pour la fatigue des ponts routiers", *Construction Métallique*, N° 4-93.

Rackwitz, R., & B. Fiessler 1978. "Structural Reliability under Combined Random Load Sequences", *Computers & Structures*, Pergamon Press, New-York, vol 9, pp.489-494.

IABSE 1990. "Remaining Fatigue Life of Steel Structures", International Workshop, *IABSE report vol.9*, Lausanne.

Structural Safety & Reliability, Schuëller, Shinozuka & Yao (eds) © 1994 Balkema, Rotterdam, ISBN 90 5410 357 4

Finite element fracture reliability of stochastic structures

J-C. Lee & A. H-S. Ang
University of California, Irvine, Calif., USA

ABSTRACT: This study presents a methodology for the system reliability analysis of cracked structures with random material properties, which are modeled as random fields, and crack geometry under random static loads. The finite element method provides the computational framework to obtain the stress intensity solutions, and the first-order reliability method provides the basis for modeling and analysis of uncertainties. The ultimate structural system reliability is effectively evaluated by the stable configuration approach. Numerical examples are given for the case of random fracture toughness and load.

1 INTRODUCTION

A problem of considerable and increasing importance within the fields of civil, mechanical, aeronautical, nuclear, marine and military engineering is the predominantly brittle fracture of structures. The failure process initiates with the presence of small cracks which can cause catastrophic fracture. The fracture behavior of a linear elastic structure can be inferred by comparing the applied stress intensity factor with the fracture toughness of the material. In real situations, there are usually some degrees of uncertainty associated with the flaw sizes and material properties including fracture toughness. Extraordinary loads can result in stresses significantly above the intended design level. Because of these complexities, fracture should be viewed probabilistically rather than deterministically. In general, the solution of the response field in other than very simple structures can not be obtained in closed form, but must be computed approximately. In structural reliability analysis, the finite element method is well suited for dealing with random spatial variabilities in the material properties due to the segmentation of the structure into elements, each of which can be represented by its own properties. Previous works (Besterfield et al. 1990, Der Kiureghian and Ke 1988, and Mahadevan and Haldar 1991) in structural reliability analysis are limited to the evaluation of the probability of initial damage of the component or structure using given failure criteria on deflection or strength. In this study, the system reliability of structures with multiple cracks will be determined through the finite element method employing the first-order reliability method. The uncertainties in load, crack geometry, and material properties including fracture toughness are taken into consideration. These variables are modelled as random fields on the entire domain of a structure. In order to model singularities at the crack tips, appropriate crack tip elements are employed. In the presence of multiple cracks in a structure, there are multiple failure modes, which are correlated. The system failure probability, that is, the collapse probability, is effectively computed by the stable configuration approach. The branch event corresponding to each crack is defined by appropriate fracture criteria. The system failure event is derived as the intersection of the unions of branch events by the stable configuration approach.

2 FINITE ELEMENT MODELING OF CRACK TIP SINGULARITY

The use of finite elements in fracture predictions requires two distinct considerations: (i) crack tip singularity modelling, and (ii) interpretation of the finite element results. In this study the degener-

ate isoparametric quadrilateral elements (Barsoum 1976) are utilized since these elements are simple to implement without any change in a standard finite element program and can give accurate results for mixed-mode fracture. The stress intensity factors are easily computed from the nodal displacements along the element edge of the degenerate element using the displacement matching method. For plane crack problems, the finite element mesh idealization dictates the free surfaces of the crack ($\theta = \pm 180°$) to be the most convenient choices for the evaluation of the stress intensity factors. By combining the stress intensity factor expressions along the $\theta = \pm 180°$ rays emanating from the crack tip as shown in Fig. 1, the resulting expression is obtained as follows.

$$
\left\{ \begin{array}{c} K_I \\ K_{II} \end{array} \right\} = \frac{\mu}{\kappa+1} \sqrt{\frac{2\pi}{L}} \left[\begin{array}{cc} -\sin\phi & \cos\phi \\ \cos\phi & \sin\phi \end{array} \right]
$$
$$
\cdot \left[\begin{array}{cccccccc} 4 & 0 & -1 & 0 & -4 & 0 & 1 & 0 \\ 0 & 4 & 0 & -1 & 0 & -4 & 0 & 1 \end{array} \right]
$$
$$
\cdot \{U_B \ V_B \ U_C \ V_C \ U_D \ V_D \ U_E \ V_E\}^T \quad (1)
$$

where μ is the shear modulus, and κ is $(3-\nu)/(1+\nu)$ for plane stress, $3-4\nu$ for plane strain. U_i and V_i represent the nodal displacements of the node i in the X and Y directions, respectively.

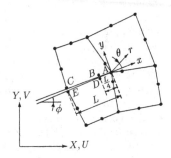

Figure 1: Degenerate isoparametric elements at the crack tip.

3 PERFORMANCE FUNCTION

3.1 Mode-I fracture

Mode I loading has the most practical importance. For mode I fracture, the fracture criterion which is commonly used states that crack propagation will occur when the stress intensity factor K_I reaches a critical value K_{Ic}, termed the fracture toughness which is a mechanical property of the material. Thus, the performance function for the mode I fracture can be expressed as

$$
g = K_{Ic} - K_I \quad (2)
$$

Accordingly, the failure state is defined as $[g < 0]$. When only mode I fracture is present, the direction of crack propagation measured from the current crack orientation, θ, is equal to 0; that is, the crack extends along a straight path.

3.2 Mixed-mode fracture

Practical structures are not only subjected to tension but may also experience shear and torsional loadings. Cracks may therefore be exposed to tension and shear, which lead to mixed.mode cracking. There are currently several fracture criteria available, which include the maximum tangential stress criterion, the maximum energy release rate criterion, the strain energy density criterion, and the elliptic rule criterion. The elliptic rule criterion (Yishu 1990) is the general criterion superceding all the mixed-mode fracture criteria. This criterion describes the loci of critical points by the fracture envelope as

$$
\left(\frac{K_I}{K_{Ic}} \right)^2 + A \left(\frac{K_{II}}{K_{Ic}} \right)^2 = 1 \quad (3)
$$

where the material constant $A = (K_{Ic}/K_{IIc})^2$. The elliptic rule criterion is employed in the present study because it can be easily formulated. The performance function for the mixed mode I-II fracture is then expressed as

$$
g = K_{Ic}^2 - \left(K_I^2 + AK_{II}^2 \right) \quad (4)
$$

Accordingly, $[g < 0]$ defines the failure state which represents that the crack extension occurs in one direction. It is noted that the fracture angles predicted by various criteria are basically in close with the measured one (Yishu 1990). Thus, the maximum tangential criterion (Erdogan and Sih 1963) is employed to calculate the fracture angle for the next configuration as follows.

$$
\theta_o = 2\tan^{-1} \left[\frac{1}{4} \frac{K_I}{K_{II}} \pm \frac{1}{4} \sqrt{\left(\frac{K_I}{K_{II}} \right)^2 + 8} \right] \quad (5)
$$

and

$$K_I(1 - 3\cos\theta_o)\cos\frac{\theta_o}{2} + K_{II}(5 + 9\cos\theta_o)\sin\frac{\theta_o}{2} < 0 \tag{6}$$

4 REPRESENTATION OF RANDOM FIELD

Material properties, structural geometry, and external loads have random spatial variabilities and are modeled by random fields rather than random variables. For finite element reliability analysis, it is necessary that such random fields be represented in terms of random variables. The midpoint method is used in this study. The field value for an element is assumed to be constant as the value at the centroid and represented by a random variable. The correlation coefficient beween any two random variables is directly defined in terms of the autocorrelation function of the random field. One important consideration in this representation is the size of the random field element. This size is controlled by the correlation measure. In the first-order reliability method, the basic random variables V are transformed into a set of statistically independent, standard normal variables

$$Y = Y(V) \tag{7}$$

Let Z_i and Z_j be a pair of standard normal variates obtained by marginal transformations of V_i and V_j. The correlation coefficient, ρ'_{ij}, between Z_i and Z_j can be expressed in terms of the correlation coefficient ρ_{ij} of V_i and V_j (Ang and Tang 1984, Der Kiureghian and Liu 1986), and the transformation to the standard normal space is then given by

$$y = \Gamma'z = \Gamma' \left\{ \begin{array}{c} \Phi^{-1}[F_{V_1}(v_1)] \\ \vdots \\ \Phi^{-1}[F_{V_n}(v_n)] \end{array} \right\} \tag{8}$$

in which $\Gamma' = (L')^{-1}$, where L' is the lower triangular matrix obtained from the Cholesky decomposition of the correlation matrix R' of Z. The preceding transformation of Eq. 8 is unique for an arbitrary number of variables with arbitrary marginal distributions and correlation coefficients and is computationally much simpler than the Rosenblatt transformation.

5 FINITE ELEMENT RELIABILITY IMPLEMENTATION

System reliability problems can be solved by replacing each individual limit-state surface by a first-order approximation surface at the corresponding minimum-distance point. Efficient solution methods for the optimization problem require the gradient vector of the limit-state function with respect to the basic variables. The gradient vector can be determined using the element partial stiffness matrices and load vectors, each of which is established analytically or numerically. By the finite element formulation, the nodal equilibrium equation for the whole structure is obtained as

$$KU = R \tag{9}$$

where K is the stiffness matrix, U the nodal displacement vector, and R the nodal load vector for the whole structure. The structural response, S, which are the stress intensity factors at the crack tip can be expressed in terms of U as

$$S = QU \tag{10}$$

where Q is the displacement-response transformation matrix which is obtained from Eq. 1. The basic random variables which are discretized from the corresponding random fields can be represented by a vector V. For convenience, the basic random variables can be divided into three groups: (1) material and geometry variables, V_M, such as Young's moduli, Poisson's ratios, coordinates of crack tips, (2) load variables, V_L, such as distributed or concentrated loads, and (3) resistance variables, V_R, such as fracture toughness. The limit-state function can be expressed as an explicit function of resistance variables V_R and response quantities S, i.e.,

$$g(V) = g(V_R, S) \tag{11}$$

in which the structural response S are functions of the basic random variables. Using the chain rule of differentiation, the gradient vector of the limit-state function with respect to the basic random variable vector V is

$$\nabla_V g = \nabla_{V_R} g \, J_{V_R,V} + \nabla_S g \, J_{S,V} \tag{12}$$

where the computation of $\nabla_S g$, $\nabla_{V_R} g$ and $J_{V_R,V}$ are easily carried out in closed form. Since the response S is a function of material and geometry variables V_M and load variables V_L only, $J_{S,V}$ can be expressed with submatrices

$$J_{S,V} = \begin{bmatrix} J_{S,V_M} & J_{S,V_L} & O \end{bmatrix}$$
$$= \begin{bmatrix} C_Q & O & O \end{bmatrix}$$
$$+ Q K^{-1} \begin{bmatrix} -C_K & J_{R,V_L} & O \end{bmatrix} \quad (13)$$

where $C_Q = [\frac{\partial Q}{\partial V_{M1}} U \ldots \frac{\partial Q}{\partial V_{Mm}} U]$ and $C_K = [\frac{\partial K}{\partial V_{M1}} U \ldots \frac{\partial K}{\partial V_{Mm}} U]$, in which m is the size of the vector V_M. The matrices K, Q, C_K, C_Q, and J_{R,V_L} are first set up for each element and then assembled in global sense.

6 PROBABILITY OF SYSTEM FAILURE

6.1 The first-order approximation

The limit-state function for a crack tip i in a given configuration j can be denoted as

$$G_{ij}(Y) = g_{ij}(V(Y)) = 0 \quad (14)$$

in which Y denotes the independent standard normal variables. The limit-state surface in the independent standard normal space can be replaced by its tangent hyperplane at the point nearest to the origin. This point is denoted by y^*. By expanding the limit-state function $G_{ij}(y)$ in a Taylor series at the point y^*, the first-order approximation of the function $G_{ij}(y)$ is as follows:

$$G_{ij}(Y) \approx \nabla G_{ij}(y^*)(Y - y^*) \quad (15)$$

where $\nabla G_{ij}(y^*)$ is the gradient of $G_{ij}(y)$ computed at y^*. The minimum distance point can be obtained by the following iteration scheme (Rackwitz and Fiessler 1978).

$$y_{k+1} = \left[\alpha_k{}^T y_k + \frac{G(y_k)}{|\nabla G(y_k)|} \right] \alpha_k \quad (16)$$

where $\alpha_k = -\nabla^T G(y_k)/|\nabla G(y_k)|$, and the gradient vector is obtained as

$$\nabla G = \nabla_V g \, J_{V,Y} = \nabla_V g \, J_{Y,V}^{-1} \quad (17)$$

6.2 Formulation of the stable configuration approach

In the stable configuration approach, the failure of the system can be defined as

$$E = \bigcap_{i=1}^{n} \bar{C}_i = \bigcap_{i=1}^{n} \left(\bigcup_{j=1}^{k_i} B_{ij} \right) \quad (18)$$

where $\bar{C}_i$ refers to cut-set i not being realized and B_{ij} is the failure of component j in cut-set i. For each B_{ij}, a performance function $G_j(Y)$ is defined such that $G_j(Y) < 0$ and $G_j(Y) > 0$ imply the failure and survival of component j, respectively. For structures exhibiting brittle behavior, the event $\bar{C}_i$ can be simplified as the event corresponding to the further damage of the configuration i (Quek and Ang 1990). For practical purposes, only a limited number of configurations can be considered. In addition to using only the stochastically dominant stable configurations, the number of essential configurations may be further reduced, because the configurations with low damage levels are more stable than those with high damage. This is particularly true for brittle structures. In mode I fracture, a crack propagates along the straight line, i.e., the crack extension angle measured from the initial crack line is 0°. The dominant stable configurations are naturally taken as the configurations which have the straight line cracks with appropriate crack extension. However, in the mixed-mode fracture the crack propagates along a curved path. The dominant stable configuration for the mixed-mode fracture can be taken as the configuration that has the crack extended from the current configuration in the direction determined by the fracture criteria using the basic random quantities corresponding to the minimum distance point of the limit-state surface in the independent normal space.

7 EXAMPLES

Three simple examples are presented to illustrate the reliability analysis of cracked structures. The fracture toughness and load are modeled as a random field and random variable, respectively. These examples can be solved using a conventional finite element program. The statistics of the variables are summarized in Table 1.

Table 1: Statistics for the example plates.

variable	unit	mean	c.o.v.	distribution
crack length, a	in.	a	0.0	-
applied load, w	$lb/in.$	$\bar{w}$	0.15	normal
modulus, E	ksi	3×10^4	0.0	-
Poisson ratio, ν	-	0.3	0.0	-
toughness, K_{Ic}	$ksi\sqrt{in.}$	43.0	0.15	normal (field)

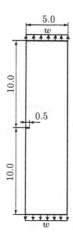

Figure 2: A stochastic plate with a single edge crack under random tensile load.

7.1 A plate with a single edge crack under tension

The example structure is a 1 $in.$ thick, 5×20 $in.^2$ rectangular plate with a 0.5 $in.$ single edge crack as shown in Fig. 2. The autocorrelation coefficient function for the fracture toughness K_{Ic} is specified as

$$\rho(\Delta x, \Delta y) = \exp\left[-\frac{\Delta x^2 + \Delta y^2}{(cL)^2}\right] \quad (19)$$

where $\Delta x^2 + \Delta y^2$ is the square of the distance between any two points on the plate and L is taken as the plate width, and c is a dimensionless measure of the correlation length. The correlation length, which is a measure of the fluctuation rate of a random field, may be defined as cL in Eq. 19. This example involves only mode I fracture where the crack extends straightly along the crack line. For the reliability analysis by the stable configuration approach, several finite element meshes for different crack length including the initial crack length are used. For the sake of simplicity, one half the plate is modeled in finite element analysis. In each mesh, two degenerate isoparametric elements are employed for the singularity at the crack tip and the remainder of the plate is modeled by regular 8-node isoparametric elements. Let Δa denote the difference between crack lengths in adjacent configurations. Thus, Δa also denotes the segment size of the fracture toughness field. For a short correlation length for the random field of fracture toughness, the rate of fluctuation is high and, thus, a small value of Δa is required in evaluation of the system reliability. To investigate the effect of Δa

for a given correlation length cL, the probabilty of system failure, i.e., the probability of collapse is computed by simulation for different values of Δa as shown in Fig. 3, where three different cases are considered. From the results in Fig. 3 it is apparent that the convergence in the probability of collapse is effectively achieved when Δa is one third of the correlation length. Thus, the appropriate value of Δa is given as

$$\Delta a = \frac{cL}{3} \quad (20)$$

When $cL = 0.475$ and $\bar{w} = 15$, the probability of collapse is obtained as 7.45×10^{-4} with seven dominant configurations, whereas the probability of fracture initiation is 1.01×10^{-3}.

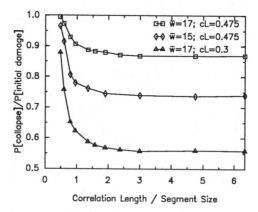

Figure 3: The effect of the segment size upon the probability of collapse.

7.2 A plate with two single edge cracks

The example structure is a 1$in.$ thick, 5×20 $in.^2$ rectangular plate with two single edge cracks as shown in Fig. 4. Each crack length is 1.0 $in.$ and the mean of the distributed load is 8.5 $lb/in.$ This example also involves only mode I fracture. However, this example, in the presence of two cracks, has many failure modes which are correlated. The system failure probability can be effectively evaluated by the stable configuration approach. Using Eq. 20, Δa is taken as 0.2 when $cL = 0.6$. By simulation, the probability of collapse is obtained as 4.7×10^{-4} with 19 configurations, whereas the probability of fracture initiation is 8.5×10^{-4}. To investigate the effect of the correlation length of the fracture toughness on the probability of collapse, three different values of the correlation length,

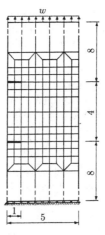

Figure 4: Finite element mesh for the initial configuration of a plate with two single edge cracks under tension.

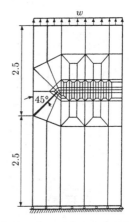

Figure 6: Finite element mesh for the initial configuration of the plate with a slant edge crack.

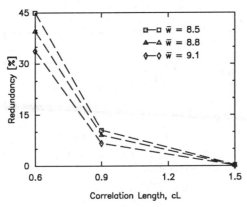

Figure 5: Redundancies for different correlation lengths.

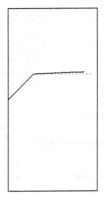

Figure 7: Crack propagation path for the plate with a slant edge crack under tension.

7.3 A plate with a slant edge crack

0.6, 0.9 and 1.5, are considered. As a measure of redundancy in the structure, the percentage of redundancy is defined as

$$R = \left[1 - \frac{P(\text{collapse})}{P(\text{initial damage})}\right] \times 100\% \quad (21)$$

Fig. 5 shows the redundancies for the different correlation lengths. When $cL = 1.5$, the redundancy is zero, which means that collapse is imminent once any initial damage occurs.

The example structure is a $1in.$ thick, $2.5 \times 5\ in.^2$ rectangular plate with a 45-degree slant edge crack as shown in Fig. 6. The crack length is 1.0 $in.$ and the mean of the distributed load is 11.0 lb/in. This example involves modes I and II with the crack extending along a curved path. Available data suggest that $K_{IIc} = 0.75K_{Ic}$. Thus, A in Eq. 3 is assumed to be 1.78. Fig. 7 shows the crack propagation path from which the possible configurations are determined. For $cL = 0.225$, the probability of collapse is obtained as 5.44×10^{-4} with seven configurations, whereas the probability of fracture initiation is 7.08×10^{-4}.

8 CONCLUSION

The ultimate structural reliability of cracked structures is evaluated on the basis of the stable configuration approach and FEM. For random fracture toughness, the appropriate segment size is found as one third of the correlation length. The results show that for the small correlation length of the fracture toughness, the probability of ultimate system failure is smaller than the corresponding probability of fracture initiation, which is due to the reserved safety margin of the structure. The collapse probability appears to depend on the correlation length of the toughness.

REFERENCES

Ang, A. H-S. and Tang, W.H. 1984. Probability Concepts in Engineering Planning and Design, 2, John Wiley & Sons, New York, N.Y.

Barsoum, R.S., 1976. "On the use of isoparametric finite elements in linear fracture mechanics," Int. J. Num. Meth. Engng., 10(1), 25-37.

Besterfield, G.H., Liu, W.K., Lawrence, M.A., and Belytschko, T.B. 1990. "Brittle fracture reliability by probabilistic finite elements," J. Engng. Mech., 116(3), 642-659, ASCE.

Der Kiureghian, A. and Liu, P-L.1986. "Structural reliability under incomplete probability information," J. Engng. Mech., 112(1), 85-104, ASCE.

Der Kiureghian, A. and Ke, J-B. 1988. "The stochastic finite element method in structural reliability," Probabilistic Engng. Mech., 3(2), 83-91.

Erdogan, F. and Sih, G.C. 1963. "On the crack extension in plates under in-plane loading and transverse shear," Trans. ASME, J. Basic Engng., 85, 519-527.

Mahadevan, S. and Haldar, A. 1991. "Practical random field discretization in stochastic finite element analysis," Structural Safety, 9, 283-304.

Quek, S-T. and Ang, A. H-S. 1990. "Reliability analysis of structural systems by stable configurations," J. Str. Engng., 116(10), 2656-2670, ASCE.

Rackwitz, R. and Fiessler, B., 1978. "Structural reliability under combined load sequences," Computers and Structures, 9, 489-494.

Yishu, Z. 1990. "Elliptic rule criterion for mixed mode crack propagation," Engng. Fracture Mech., 37(2), 283-292.

Structural Safety & Reliability, Schuëller, Shinozuka & Yao (eds) © 1994 Balkema, Rotterdam, ISBN 90 5410 357 4

Kurtosis effects on stochastic structural fatigue

Loren D. Lutes
Texas A&M University, College Station, Tex., USA

Jin Wang
Bechtel Corporation, Houston, Tex., USA

ABSTRACT: The rate of fatigue damage accumulation is investigated for structural elements subjected to non-Gaussian stress time histories. Rainflow analysis of simulated time histories is used as a basis of comparison in testing the accuracy of a set of simple analytical approximations for characterizing the non-Gaussian effect. Simulation results demonstrate that the analytical procedure can give quite good results for various stress power spectral density (PSD) functions.

INTRODUCTION

Studies of stochastic structural fatigue involve making probabilistic predictions of fatigue life for structural elements subjected to random stress loadings. It is known that fatigue can be affected by at least four factors which characterize the random stress time history. These four factors are: the standard deviation or rms, the average frequency, the shape or bandwidth of the power spectral density (PSD) function, and the probability distribution or non-Gaussianity. Analytical models for describing the non-Gaussian effects on fatigue damage have been proposed in recent studies. In particular, simple analytical expressions for a non-Gaussian correction factor can be derived based on a Hermite polynomial expansion and a narrowband assumption (Winterstein 1988, Wang and Lutes 1992).

In this study, analytical approximations are compared with simulation results using rainflow analysis for various stress PSD functions. Only symmetric probability distributions are considered and kurtosis is used as a measure of non-Gaussianity. The simulated time histories have been obtained from models used in a study of the fatigue of offshore structures subjected to random wave loads (Wang and Lutes 1992). However, some situations with exaggerated parameter values have been included in order to provide a more complete picture of the

phenomena and a more severe test of the analytical models. The simulation results demonstrate that the analytical methods work better than one might expect from theoretical considerations. In particular, even though the analytical procedures have been derived from narrowband assumptions, the simulation results demonstrate that they may also give quite good approximations for the non-Gaussian factor in situations where power spectral density is non-narrowband and/or even non-unimodal.

NON-GAUSSIAN FATIGUE FORMULATION

A commonly used technique for expressing a non-Gaussian process is through a nonlinear transformation of a Gaussian process. This nonlinear function can be approximated by a Hermite polynomial expansion. Letting $X(t)$ and $Y(t)$ denote normalized (i. e., mean zero and unit variance) Gaussian and non-Gaussian processes, respectively, one may have (Winterstein 1988),

$$Y(t) = g[X(t)] = \kappa \left[X + \sum_{n=3}^{N} c_n H_{e_{n-1}}(X) \right] \quad (1)$$

in which κ is a scaling factor ensuring that $Y(t)$ has unit variance, and $H_{e_{n-1}}(X)$ is the nth Hermite polynomial. Truncating Eq. (1) at

$N = 4$, one has

$$Y(t) = \kappa \left[X + c_3(X^2 - 1) + c_4(X^3 - 3X) \right]$$

(2)

in which $c_3 = 0$ for the symmetric situations considered here. Using Taylor series expansions for the Hermite polynomials, c_4 can be written in terms of the corresponding Hermite moment h_4, where $h_n = E\{H_{e_n}[Y(t)]\}/(n!)$ (Winterstein 1985). For the first order approximation, one has

$$c_4 = h_4 = \frac{k_y - 3}{24}$$

(3a)

$$\kappa = 1$$

(3b)

in which k_y denotes the kurtosis of the non-Gaussian process $Y(t)$. A second order approximation is (Winterstein 1988)

$$c_4 = \frac{\sqrt{1 + 36h_4} - 1}{18} = \frac{\sqrt{1 + 1.5(k_y - 3)} - 1}{18}$$

(4a)

$$\kappa = (1 + 6c_4^2)^{-1/2}$$

(4b)

Let S be a peak of the standardized Gaussian process $X(t)$. Assuming that $X(t)$ is also narrowband, the stress range (i. e., double amplitude) of the zero-mean non-Gaussian process $Y(t)$ may then be expressed as $R = g(S) - g(-S) = 2g(S)$, and S has the Rayleigh probability distribution. The usual simplified fatigue formulation then presumes that the expected rate of fatigue damage accumulation is proportional to $E(R^m)$, in which m is a parameter of the S-N curve describing constant amplitude fatigue results. The non-Gaussian effect then may be characterized by using the ratio of $E(R_y^m)$ for the non-Gaussian process $Y(t)$ to $E(R_z^m)$ for a Gaussian process $Z(t)$ having the same PSD as $Y(t)$. Using the narrowband assumption for $Z(t)$ gives $E(R_z^m) = 2^m E(S^m)$ and a non-Gaussian factor can be written as

$$f_{NG} = \frac{E(R_y^m)}{E(R_z^m)} = \frac{E[g^m(S)]}{E[S^m]}$$

(5)

It should be noted that the restriction that Y and Z have the same PSD assures that they have the same values of the key parameters of standard deviation and average frequency.

It can be shown that a very simple expression for the non-Gaussian factor may be derived based on the first order approximation (Winterstein 1985)

$$f_{NG} = 1 + m(m-1)h_4 = 1 + m(m-1)\frac{k_y - 3}{24}$$

(6)

A more accurate second order expression was also proposed by Winterstein (1988) using an approximate Weibull distribution to fit R. This gives

$$f_{NG} = \left(\frac{\sqrt{\pi}\kappa}{2p!} \right)^m \frac{(mp)!}{(m/2)!}$$

(7)

in which p is a parameter of the Weibull distribution which is related to V_R, the coefficient of variation of the stress range R:

$$V_R^2 = \frac{(2p)!}{(p!)^2} - 1$$

(8)

The value of V_R may be determined from the second order Hermite model as $V_R = (4/\pi)(1 + h_4 + c_4) - 1$, and Eq. (8) then specifies the value of p.

Winterstein (1988) used $p = V_R$ as a simple approximation for Eq. (8). This gives exact results for $p = 0$ or 1, but has errors as large as 20% for intermediate situations. For the simple Gaussian situation, $p = 0.5$, and the error in Eq. (8) at this point is about 4.5%. This causes this approach to give a f_{NG} value not exactly equal to unity (greater than 1.0) for the Gaussian case. An improved approximation for p using the exponential function may be written as $p = V_R[1 - e^{-4V_R}/3]$. This formula fits Eq. (8) within 3% for the V_R range of 0 to 1, corresponding to kurtosis values up to 10.

A consistent second order expression for the non-Gaussian factor may also be derived based on Eq. (4) by using a second order approximation of the relationship between c_4 and h_4 (Wang and Lutes 1992). This second order non-Gaussian factor is based entirely on Hermite moments and can be written as

$$f_{NG} = \kappa^m \left[1 + m(m-1)c_4 \right.$$
$$\left. + (m^2 + 5)m(m-1)c_4^2 / 2 \right] \qquad (9)$$

Note that f_{NG} is a nonlinear function of the kurtosis in the above equation, and it gives the same expression as the first order formula Eq. (6) if the second order term c_4^2 is neglected and c_4 and κ take their first order forms as given in Eq. (3). It also gives an exact f_{NG} value of unity for the Gaussian process. It should be noted, though, that going to a second order theory does not necessarily improve the accuracy for a strongly non-Gaussian process, even though it should be an improvement for k_y near 3.

SIMULATION STUDIES

In order to test the accuracy of the analytical approximations for the non-Gaussian factor, time domain simulation has been performed for a simple model of an offshore structure under Morison equation wave excitation. For simplicity, a single degree of freedom system is used. The equation of motion may be written as (Sarpkaya, T. and Isaacson, M. 1981)

$$m\ddot{Y} + c\dot{Y} + kY = K_m \dot{u} + K_d u|u| \qquad (10)$$

where Y, $\dot{Y}$, and $\ddot{Y}$ are the displacement, velocity, and acceleration of the response; c and k are the structural damping and stiffness, respectively; m is structural mass plus an added fluid mass; K_m and K_d correspond to the inertia and drag of the hydrodynamic force in the Morison equation; and u and $\dot{u}$ denote the water particle velocity and acceleration. Assuming a Gaussian sea and no current, u and $\dot{u}$ will be zero-mean Gaussian processes. However, the wave force will not be Gaussian due to the nonlinear drag. This will in turn result in a non-Gaussian response.

In this study, the Gaussian time histories of u and $\dot{u}$ were generated from a wave spectrum using a procedure for simulating a time sequence having a distribution which is almost perfectly Gaussian (Lutes and Wang, 1991). The non-Gaussian response time history $Y(t)$ was obtained by solving the equation of motion using the Newmark beta method, and the fatigue damage rate was calculated based on the rainflow analysis of the simulated stress response time history. As presented in the preceding section, finding the f_{NG} also requires knowledge of the fatigue rate for a Gaussian process with the same PSD as $Y(t)$. For some of the following results this Gaussian fatigue rate was also determined by simulation, and in other situations it was approximated based on the results of prior research.

The first situation studied is one in which the $Y(t)$ time history has exactly the same form as the hydrodynamic excitation on the right-hand side of Eq. (10). This is not a very realistic situation, but represents a pseudo-static response of the structure, as would occur for extremely large value of stiffness. The PSD of the water velocity u in this example was taken to have a rectangular shape with a bandwidth of 20% of the average frequency. For this situation rainflow analysis was performed on a simulated Gaussian stress time history which had the same PSD as the non-Gaussian time history. This was achieved by first simulating the non-Gaussian $x(t)$ time history, then finding its PSD and finally simulating a new Gaussian time history using this PSD. After performing rainflow analysis on both of these time histories, the non-Gaussian factor was obtained as the ratio. Figure 1 shows these simulated values of f_{NG} along with the results of the various analytical models. The results are plotted versus kurtosis k, and the range of k values was obtained by varying the relative magnitudes of K_m and K_d in the Morison equation excitation. The kurtosis range shown corresponds to the ratio of rms drag force to rms inertia force varying from 0.1 to 10. Note that results are given for two values of the fatigue parameter m, these values of 3 and 5 bracket the values typically encountered in practice with welded or bolted connections.

From Fig. 1, one can see that for $m = 3$, the second order model and the two Weibull

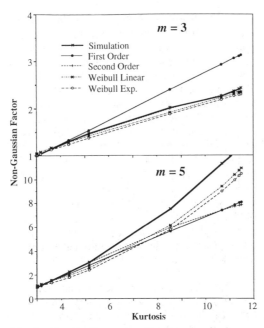

Fig. 1. Results for narrowband processes

approximations give results which are quite close to each other and to the simulation results. The first order model, though, significantly overpredicts the f_{NG} value. For $m = 5$ all the analytical models somewhat underpredict the non-Gaussian factor for larger kurtosis values. Both the first order and second order models particularly underpredict f_{NG} for the severely non-Gaussian situations with kurtosis greater than about 9. The difference between the two Weibull approximations is never large, and the primary benefit of the exponential form is that it gives $f_{NG} = 1$ for a kurtosis value of 3. Overall, both of the Weibull models work quite well for these relatively narrowband non-Gaussian situations, agreeing within about 20% with the simulation data. The second order model is also good for $k \leq 8$.

The robustness of the analytical models has been further tested for non-narrowband and non-unimodal situations. In this case the Pierson-Moskowitz (P-M) spectrum was used to represent the ocean wave elevations (Sarpkaya, T. and Isaacson, M. 1981). Results have been obtained for both compliant and

fixed offshore structures, meaning stuctures with resonant frequency below and above, respectively, the dominant frequency of the excitation. In particular, results have been obtained for $r = 0.5$ and 8 where r is the ratio of the structural frequency to the peak energy frequency of the P-M spectrum.

For realistic parameter values it has been found that there are few situations in which the structural responses are both severely bimodal and severely non-Gaussian. In order to provide a more complete picture of the non-Gaussian dynamic response and a more severe test for the analytical models,though, it was desired to to include such cases in this study. To that end, situations have been included with structural damping ratios of 5%, 15% and 30%, even though realistic damping ratios for offshore structures are usually no more than 5%. Fig. 2 shows the normalized response PSDs of the compliant and fixed structures obtained from the simulated response time histories. It can be seen that the response PSDs are non-narrowband and their shapes vary from unimodal to bimodal, depending on the damping ratio and r.

Simulated non-Gaussian factors were found from the ratio of the damage rates for the non-Gaussian time histories which yielded the PSDs in Fig. 2 to that for Gaussian time histories simulated from Fig. 2. These values, along with the results of the analytical models are shown in Fig. 3, in a form similar to that of Fig. 1. In Fig. 3, the three points with the smallest kurtosis values came from the $r = 0.5$ structure, and the others came from the $r = 8$ structure. In each structure, the kurtosis value increases with increasing damping value. It may be noted that as one moves from the smallest to the largest kurtosis value in Fig. 3, the progression is from the most narrowband unimodal PSD in Fig. 2 for $r = 0.5$ toward the most bimodal PSD in that figure, then to the most bimodal PSD for $r = 8$, and on toward the unimodal (but relatively broadband) PSD for that r value. The simulation results in Fig. 3, show that the f_{NG} values for these broadband and bimodal PSDs are lower than for narrowband processes with the same values of kurtosis. For $m = 3$ this results in errors of the analytical results which are larger than those

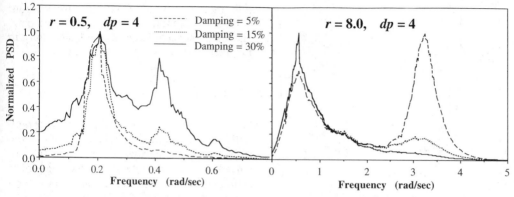

Fig. 2. Broadband and bimodal PSDs

for narrowband processes, while for $m = 5$ the analytical results are now more accurate than they were in Fig. 1. Again, the Weibull results generally agree quite well with the simulation results. In fact, the analytical approximations generally seem to be as satisfactory in these non-narrowband and non-unimodal situations situations as in the narrowband situations shown in Fig. 1.

The final set of data to be presented here

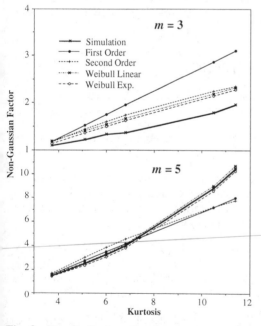

Fig. 3. Results for broadband processes

have been obtained from a fairly extensive simulation study of the non-Gaussian fatigue of offshore structures using a wide range of values of the frequency ratio r and of a drag parameter dp which gives the ratio of the rms drag force to the rms inertia force in the Morison equation hydrodynamic loading (Wang and Lutes 1992). Approximately 30 different combinations of r and dp were considered, with each parameter varying from 0.1 to 10. These results have all been obtained from the P-M spectrum for sea elevation and linear wave theory, and using a structural damping value of 5%. For these situations an empirical approximate technique was used to estimate the fatigue damage rate for the Gaussian process with the same PSD as $Y(t)$, rather that performing the additional simulation and rainflow analysis of those time histories. This approximation is called the single-moment (SM) method and was derived from rainflow analysis of simulated time histories from a variety of unimodal and bimodal PSDs (Larsen and Lutes 1991).

Figure 4 and 5 show some of the results obtained by using the SM approximation along with simulation and rainflow analysis of non-Gaussian response time histories. The only analytical method shown is the second order model of Eq. (9). It will be recalled that the Weibull models are more accurate than the second order model in some situations, but that distinction is immaterial for Figs. 4 and 5, since the kurtosis in the situations shown does not exceed a value of 8.

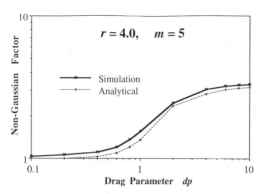

Fig. 4. Effect of drag parameter

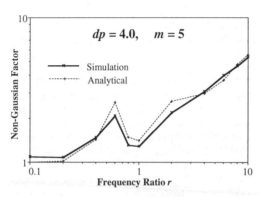

Fig. 5. Effect of frequency ratio

The results in Figs. 4 and 5 are presented as f_{NG} versus dp for $r = 4$ (a fixed structure) and as f_{NG} versus r for $dp = 4$ (a drag dominated situation). The results are plotted this way, rather than versus kurtosis, to emphasize the extent to which the approximate formula succeeds in capturing the magnitude of the non-Gaussian effect as physical parameters are varied. For these results with the fatigue parameter of $m = 5$, it is seen that the f_{NG} can vary from unity to about 5, and that the analytical approximation predicts this behavior very well.

Note that $f_{NG} = 5$ means that fatigue failure should be expected to occur in 20% of the time that would be predicted by a Gaussian fatigue analysis, so that these results may be regarded as being of significant practical importance. It is hoped that the existence of simple analytical approximations to estimate

this major change in fatigue life without extensive simulation studies may help to encourage consideration of non-Gaussianity in fatigue calculations for real structures.

CONCLUSIONS

In general the analytical results agree quite well with the simulation results even when the response is not narrowband or unimodal. In particular, the analytical models can give estimates of the non-Gaussian factor which agree with simulation with less than 20% error. This is better than one might expect from theoretical considerations, since the analytical models were derived based on narrowband assumptions. Overall it appears that the Weibull models are the most accurate, with the exponential version having some advantage for nearly Gaussian processes. The linear first order model is the least accurate of those considered.

ACKNOWLEDGEMENT

Portions of this research were sponsored by the Offshore Technology Research Center, including support from the NSF Engineering Research Centers Program Grant No. CDR-8721512.

REFERENCES

Larsen, C. E. and Lutes, L. D. 1991. "Predicting the fatigue life of offshore structures by the single-moment spectral method," *Probabilistic Engineering Mechanics*, Vol. 6, No. 2, pp. 96-108.

Lutes, L. D. and Wang, J. 1991. "Simulation of an improved Gaussian time history," *Journal of Engineering Mechanics*, ASCE, Vol. 117, No. 1, pp. 118-224.

Sarpkaya, T. and Isaacson, M. 1981. *Mechanics of wave forces on offshore structures*, Litton Education Publishing, New York, NY.

Wang, J. and Lutes, L.D. 1992. "*Effects of Morison equation nonlinearity on stochastic*

dynamics and fatigue of offshore structures, Technical Report, Offshore Technology Research Center, Texas A&M University.

Winterstein, S. R. 1985. "Non-normal responses and fatigue damage," *Journal of Engineering Mechanics*, ASCE, Vol. 111, No. 10, pp. 1291-1295.

Winterstein, S. R. 1988. "Non-linear vibration models for extremes and fatigue," *Journal of Engineering Mechanics*, ASCE, Vol. 114, No. 10, pp. 1772-1790.

Structural Safety & Reliability, Schuëller, Shinozuka & Yao (eds) © 1994 Balkema, Rotterdam, ISBN 90 5410 357 4

Approximate time variant analysis for fatigue

Mark J. Marley
Offshore Design A.S., Billingstad, Norway

Torgeir Moan
The Norwegian Institute of Technology, Trondheim, Norway

ABSTRACT: Time variant reliability calculations for fatigue degradation are demanding, in particular due to the integration of the upcrossing rate over the service life. It is observed that, when the rate of change of the conditional failure probability is large, a simple approximation which assumes a time invariant resistance equal to the strength at the end of the reference life may be applied. For many practical problems, this time invariant approximation provides highly accurate reliability estimates at greatly reduced computational cost.

1. Introduction

Fatigue and fracture behavior is an important consideration in the design of metallic structures. These structures inevitably contain defects which, under cyclic loading, may propagate as fatigue cracks. Service failures of metal structures subjected to tensile stresses generally involve either *i*) brittle fracture in the presence of fabrication defects or following the development of fatigue cracks, *ii*) ductile fracture in the presence of large fatigue cracks, or a mixed mode failure. Although it is well recognized that the structure's ultimate capacity is reduced by the presence of defects, the general current practice for reliability assessment is separate evaluation of the ultimate and the fatigue limit states. In the former, the reliability of the non-degraded structure is assessed. In the latter, fatigue failure is defined as the propagation of a fatigue crack to some predetermined critical size, e.g. through thickness, or to the fracture critical length given some characteristic extreme load. There are situations where this separate assessment is defensible, but in general this artificial separation of the limit states is inconsistent.

A methodology for coupling the ultimate and fatigue limit states is available through a time variant reliability model, where the structure's capacity is a decreasing function of time and failure defined as the first upcrossing of this threshold by

the load process. However, this approach is relatively complex and there is little experience with its application to practical engineering problems. In investigating the behavior of the time variant model it is observed that the rate of change of the conditional failure probability is high for many practical problems. In such case, a simple and conservative approach which assumes a time invariant resistance equal to the strength at the end of the service life is proposed (Marley, 1991). Failure is defined as the exceedance of this time invariant random threshold by the global maximum value of the load, and random variable methods can be applied directly.

2. Time Variant Reliability under Fatigue

The reliability may be posed as an outcrossing problem, with the resistance $\xi(t)$ decreasing due to fatigue, Figure 1. A random process model is required, however for slowly growing cracks caused by long duration load processes the resistance deterioration may be modelled deterministically in terms of a conditional mean crack growth. Then $\xi(t)$ is uniquely determined given $\mathbf{Z} = \mathbf{z}$, a vector of time invariant basic variables. Failure is defined as the first upcrossing of $\xi(t)$ by a realization of the load process $X(t)$. For independent crossings of a high threshold, the conditional probability of

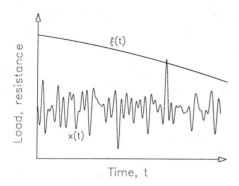

Figure 1: First-passage formulation for deteriorating resistance

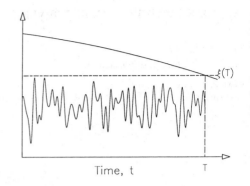

Figure 2: TI approximation of the first-passage problem

failure may be approximated (Cramer, 1967)

$$P_F(\mathbf{z}) \approx 1 - \exp\left(-\int_0^T \nu_\xi^+(\mathbf{z}, t)dt\right) \quad (1)$$

where $\nu_\xi^+(t)$ is the mean upcrossing rate. The unconditional failure probability $E_\mathbf{z}[P_F(\mathbf{z})]$, may be calculated by simulation, or alternatively by FORM/SORM with the limit state function

$$g(\mathbf{u}, u_{n+1}) = u_{n+1} - \Phi^{-1}\left\{P_F[\mathbf{T}^{-1}(\mathbf{u})]\right\} \quad (2)$$

where $\mathbf{U}$ is a vector of standard normal variables, $\mathbf{T}$ is the transform $\mathbf{U} = \mathbf{T}(\mathbf{Z})$, and U_{n+1} is an auxiliary variate (Wen and Chen, 1987). The threshold is established by a fracture mechanics model which describes crack growth and the relation between crack size and the capacity of the reduced cross section. The major difficulty is the integration of $\nu_\xi^+(t)$. Also, $P_F(\mathbf{z})$ is nearly zero or one for most points in u-space, leading to numeric problems in the evaluation of (2).

3. Approximation by Time Invariant Methods

A simple approximation is to assume a time invariant resistance equal to the strength at the end of the reference life, see Figure 2. Then a lower bound reliability estimate may be obtained using time independent methods with the limit state function

$$g(\mathbf{z}) = \xi(T; \mathbf{z}) - s_T(\mathbf{z}) \quad (3)$$

where $\xi(T)$ is the resistance at $t = T$ and s_T is a extreme value of the stress process in $[0, T]$.

I.e., crack growth from $a(0)$ to $a(T)$ and the corresponding resistance are determined, and reliability calculated for a constant threshold $\xi = \xi(T)$. The computation effort, in terms of both programming and CPU time, required with (3) is significantly reduced compared to (2). Also, convergence in the design point search is more readily achieved.

Eqs. (1),(3) are derived for a stationary load process. Extension of the models to nonstationary environmental load processes, in particular waves or wind, is discussed by Marley and Moan (1992).

Conditions for Accurate Time Invariant (TI) Estimates

The TI method is an linearization of the threshold; for the approximation eq. (3) this linearization is selected at $\tau^* = T$ for simplicity and conservatism. Clearly, there is a time $0 < \tau^* < T$ for which the time invariant threshold $\xi = \xi(\tau^*)$ would give identical results as the outcrossing rate integration. Fortunately, this linearization at T usually provides good accuracy due to three features characteristic of most practical fatigue problems:

   i)   the crossing rate is exponentially
        dependent on $\xi(t)$

   ii) the threshold level is sensitive to $\mathbf{z}$

   iii) the uncertainties in $\mathbf{z}$ are rather large

For example, for most random load processes i) holds. Fatigue is characterized by extreme sensitivity to the loading, geometry and environment; e.g., crack growth is related to the stress range to the power 3 or 4, so ii) is generally valid. Finally, the COV of the equivalent stress range often exceeds 0.25, and the COVs for the Paris crack

Table 1: Modelling of the basic variables.

Variable	Description	Dist. Type	Mean	COV	Units
a_0	initial crack size	Triangular	1.0	$1/\sqrt{6}$	mm
σ_x	std dev of $X(t)$	Normal	varies: [16, 24]	0.1	MPa
K_{IC}	fracture toughness	Lognormal	5000	0.1	Nmm$^{-3/2}$
U_4	auxiliary variate	Normal	0.0	1.0	—
μ_x	mean of $X(t)$	Fixed	100	—	MPa
C	crack growth	Fixed	1×10^{-13}	—	N, mm
m	parameters	Fixed	3.0	—	—
$\nu_0 T$	cycles	Fixed	5×10^6	—	—

growth parameter C and for the initial crack size a_0 are typically of the order 1.0, thus condition *iii*) is met.

Because of *i*) the "correct" linearization point is near T. Due to *ii*), a small change in $\mathbf{z}^*$ leads to a relatively large change in $\xi(T)$; i.e., a change in the threshold from $\xi(T) \to \xi(\tau^*)$ is similar to a minor perturbation in the linearization point in $\mathbf{z}$-space. Lastly, due to *iii*), this corresponds to a small change in $\mathbf{u}^*$ and hence in the reliability index. To define these conditions more rigorously: the TI method gives good results if the directional derivative of $P_F(\mathbf{u})$ in the direction $\boldsymbol{\alpha}^*$ is large (say, greater than about 5), and if this maximum occurs near the TI design point, $\mathbf{u}_{TI}^*$ (Marley, 1991). This directional derivative is

$$
\begin{aligned}
D_\alpha(P_F(\mathbf{u})) &= \nabla P_F(\mathbf{u}) \cdot \boldsymbol{\alpha}^* \\
&= \left(\frac{\partial P_F(\mathbf{u})}{\partial u_1}, \cdots, \frac{\partial P_F(\mathbf{u})}{\partial u_n} \right) \cdot \boldsymbol{\alpha}^*
\end{aligned} \tag{4}
$$

See Figure 7 for an illustration. Note that, consistent with *iii*), increased uncertainty in the load and strength enlarges the $\max[D_\alpha(P_F(\mathbf{u}))]$, hence improving the TI estimate.

4. Numerical Example

The example considers a propagating fatigue crack with a brittle fracture failure criterion. SORM results are obtained and compared. Crack growth is calculated by the Paris' equation, the geometry function is constant $Y(a(t)) = Y$, and the Gaussian load process $X(t)$ stationary. For $m = 3$ the brittle fracture threshold is given by:

$$
\xi(t) = \frac{K_{IC}}{Y\sqrt{\pi}} \left[\frac{1}{\sqrt{a_0}} - \frac{(Y\sigma_x\sqrt{8\pi})^3}{2} \Gamma\left(\frac{5}{2}\right) C\nu_0 t \right] \tag{5}
$$

Here the threshold is linear in t, a result of $m = 3$ and a constant Y. Table 1 describes the modelling of all variables and parameters. The mean (deterministic for each analysis) of σ_x is varied over the range 16 to 24 MPa and the COV(σ_x) is 0.1.

Time Variant (TV) Formulation

For the linearly degrading threshold, the conditional failure probability may be written:

$$
\begin{aligned}
P_F(\mathbf{z}) = {} & 1 - \exp\left\{ \frac{\sqrt{2\pi}\nu_0}{\dot{\eta}(\mathbf{z})} \right. \\
& \left. \times \left[\Phi\{\eta(0; \mathbf{z})\} - \Phi\{(\eta(T; \mathbf{z})\}\right] \right\}
\end{aligned} \tag{6}
$$

where the normalized resistance is

$$
\eta(t; \mathbf{z}) = [\xi(0; \mathbf{z}) + \dot{\xi}(\mathbf{z})t - \mu_x]/\sigma_x
$$

The limit state function is eq. (2). Although the integration of the outcrossing rate is analytic for this simple example, there remain some computational difficulties in the g-function calculation. For example, $P_F(\mathbf{z}^i)$ for $\mathbf{z}^i$ near $E[\mathbf{Z}]$ is extremely small: for case 1 with $E[\sigma_x] = 16$, $P_F(\mu_z) \approx 10^{-2760}$ (for comparison, the unconditional $P_F \approx 3 \times 10^{-6}$).

Time Invariant (TI) Formulation

Let the auxiliary random variable account for the uncertainty in the extreme load:

$$F_{S_T}(s) = \left[1 - \exp\left(-\frac{1}{2}\left(\frac{s - \mu_x}{\sigma_x}\right)^2\right)\right]^N$$

$$= F_{U_{n+1}}(u_{n+1}) = \Phi(u_{n+1}) \qquad (7)$$

For high cycle fatigue, the asymptotic relation

$$(1 - e^{-x^2/2})^N \approx \exp(-Ne^{-x^2/2})$$

may be applied, leading to the limit state function

$$g(\mathbf{z}) = \left(\frac{\xi(T) - \mu_x}{\sigma_x}\right) - \sqrt{-2\ln\left(\frac{-\ln(\Phi(u_{n+1}))}{\nu_0 T}\right)} \qquad (8)$$

Comparison of Results

Reliability indices for $E[\sigma_x]$ in the range 16–24 MPa are shown in Figure 3. For the lowest stress the reliability is high, $\beta = 4.5$, while for the largest stress failure is likely: $\beta \approx 0.3$, $P_F \approx 40\%$. In Figure 3 it is seen that the simpler TI method provides an excellent estimate over the entire range: the conservatism in P_F varies from about 4–30%. Table 2 presents importance factors; they are nearly identical by the two methods. Uncertainties in σ_x and a_0 have the dominant influence on the reliability.

The decrease from $\xi(0)$ to $\xi(T)$ is large, thus $\nu_\xi^+(t)$ increases rapidly near T, see Figure 4. $\nu_\xi^+(t)$ is essentially zero except for a short time period just prior to T (say, $t > 0.99$), and it could be expected that assuming a constant $\xi = \xi(T)$ would greatly overpredict P_F. Insight as to why this is not the case may be gained by considering the contours of the conditional probability of failure $P_F(\mathbf{u})$. Figure 5 shows plots of $P_F(\mathbf{u})$ for a low and high value. Although representing two extremes, the probability contours are close; their separation may be measured by: $P(0.000001 \leq P_F(\mathbf{u}) \leq 0.999999) = 0.00663$, which is about 30% of the unconditional probability of failure. Figure 5 clearly illustrates the rapid transition from a very low to a very high $P_F(\mathbf{u})$. Also the plotted contours bound the TI limit state function, see Figure 6. The TI g-function is a very close approximation to the TV g-function: $g(\mathbf{u}) = u_4 - \Phi^{-1}(P_F(\mathbf{u})) = 0$.

The directional derivative is calculated by eq. (4). The inset in Figure 7 illustrates the geometry, and the graph in shows a plot of $D_\alpha(P_F(\mathbf{u} = r\,\boldsymbol{\alpha}^*))$ as a function of the distance r. The maximum

Table 2: Importance factors and reliability indices.

Case 5: $E[\sigma_x] = 20$	TV	TI
$\alpha^2(a_0)$	0.2973	0.2482
$\alpha^2(\sigma_x)$	0.7193	0.7144
$\alpha^2(K_{IC})$	0.0013	0.0014
$\alpha^2(u_4)$	0.0001	0.0001
β_{FORM}	1.948	1.901
β_{SORM}	2.037	1.989

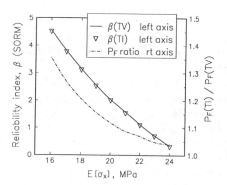

Figure 3: SORM results for fixed C, m.

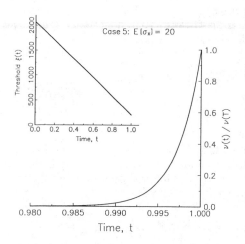

Figure 4: Threshold and normalized $\nu_\xi^+(t)$.

$D_\alpha(P_F(\mathbf{u}))$ is large, approximately 40, and the location is $r \approx 1.95$. This is very close to the first-order reliability index for the time variant analysis, $\beta = 1.948$, supporting the proposition that proximity of the $\max[D_\alpha(P_F(\mathbf{u}))]$ to β_{TI} may be used

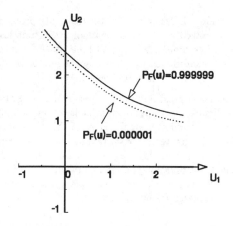

Figure 5: Level curves of $P_F(\mathbf{u})$. U_1 = transformed a_0, U_2 = transformed σ_x.

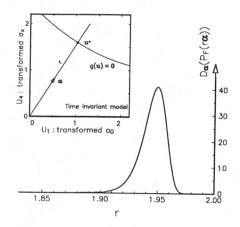

Figure 7: (inset: a) Geometric illustration of the directional derivative; (b) Plot of $D_\alpha(P_F(\mathbf{u}))$ as a function of distance along the unit vector α^*.

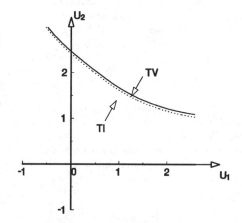

Figure 6: Time variant and time invariant limit-state surfaces.

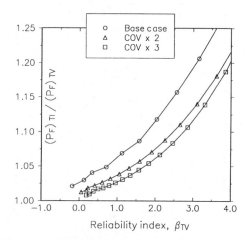

Figure 8: Effect of the increased COVs of the basic variables.

to predict the adequacy of the reliability estimate from the TI method.

Increased uncertainty in load or strength parameters enlarges the magnitude of $\max[D_\alpha(P_F(\mathbf{u}))]$ and also tends to reduce the distance from this maximum to $\mathbf{u}_{TI}^*$, hence improving the TI estimate. This effect is shown in Figure 8, where $(P_F)_{TI}/(P_F)_{TV}$ is plotted for three values of COV for the basic variables a_0, σ_x, and K_{IC}. The upper curve is for the 'base-case' (Table 1), and the middle and lower curves are for two times and three times larger COVs, respectively. For the middle curve, the COV for the load and for fracture toughness is 20%, and the COV of the initial defect size is 80%. These uncertainty levels are representa-

tive for many realistic fatigue problems, and it is promising that the TI results are accurate for this case.

Random Crack Growth Parameters

In the above, the crack growth parameters were deterministic. This assumption is now removed and LnC, m are modelled as binormal with parameters: $(-29.93, 0.5, 3.0, 0.1, -0.95)$. All other basic variables are as shown in Table 1. The threshold is:

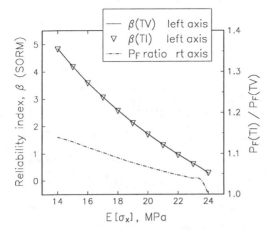

Figure 9: SORM results for random C, m.

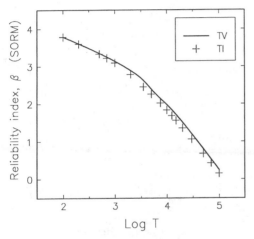

Figure 10: Example 2. β_{SORM} by TI and TV.

$$\xi(t) = \frac{K_{IC}}{Y\sqrt{\pi}} \left[(a_0)^{1-m/2} + C\nu_0 t \left(\frac{2-m}{2} \right) \times \left\{ \left(Y\sqrt{8\pi}\sigma_x \right)^m \Gamma \left(\frac{2+m}{2} \right) \right\}^{1/(m-2)} \right] \quad (9)$$

For $m \neq 3$ the threshold is nonlinear. For the time variant formulation, $P_F(z)$ must be found from eq. (1) rather than (6). An adaptive Gauss quadrature algorithm is found to be efficient and accurate for this integration. The TI limit state function is eq. (8), with $\xi(t)$ by eq. (9). Evaluation of this is significantly faster than for the time variant formulation: the TI limit state function is evaluated in less than 1% of the time needed for the TV

function call; this results in the reliability problem being solved in about an order of magnitude less computer time. Also, the TI limit state function is well-conditioned for all relevant values of $\mathbf{Z}$ and no under- or overflow problems arise.

Results are presented in Figure 9. Introducing random crack growth parameters leads to significantly lower reliability for small values of the stress variance. Of greater interest, randomizing LnC and m leads to even closer agreement between the TI and TV estimates: the maximum difference in β is only about 0.03, and the simple approximation overestimates P_F by at most 12%.

5. Example 2

This example was presented by Guers and Rackwitz (1986), who evaluated the reliability by time variant methods only. The parameters are based on experimental data reported by Yang and Lin (1987). The normalized threshold is

$$\eta(\mathbf{Z}, t) = Z_1 \left[1 - Z_2 Z_1^{-B} t \right]^{1/C} \quad (10)$$

The parameters B and C are modelled as deterministic: $B = C = 12$. The $\mathbf{Z}$ variables are lognormal: $Z_1 : LN(6, 0.1)$, $Z_2 : LN(15000, 0.1)$. The load process is stationary Gaussian and its mean zero-upcrossing frequency, ν_0, is 1. The TV g-function follows from eqs. (1,2)

$$g(z; t) = z_3 + \Phi^{-1} \left[\exp \left\{ -\nu_0 \int_0^t \exp \left[-\frac{(z_1)^2}{2} \right. \right. \right.$$
$$\left. \left. \left. \times \left\{ 1 - \frac{z_2 \tau}{(z_1)^B} \right\}^{2/C} \right] d\tau \right\} \right] \quad (11)$$

Guers and Rackwitz [2] obtained reliability estimates by Laplace asymptotic approximations to the integral in (11). However, numeric integration of the upcrossing rate is straightforward, and used for the present analysis.

The TI limit state function is

$$g(\mathbf{z}; t) = z_1 \left[1 - \frac{z_2}{z_1^B} t \right]^{1/C} - \sqrt{-2 \ln \left(1 - \Phi(z_3) \right)^{1/\nu t}} \quad (12)$$

which may be evaluated much more rapidly than the time variant formulation.

1104

Table 3: Example 2. Reliability indices and ratio of the failure probabilities by TV and TI.

t (×1000)	.1	1	10	100
β_{TV}	3.786	3.139	1.976	0.2427
β_{TI}	3.783	3.095	1.833	0.1487
P_F^{TI}/P_F^{TV}	1.014	1.161	1.382	1.091

Table 4: Importance factors for z_3.

t (×1000)	.1	.5	5	20	50	100
$(\alpha^2(z_3))_{TV}$	.68	.60	.24	.01	.00	.00
$(\alpha^2(z_3))_{TI}$	.68	.59	.11	.02	.01	.00

Comparison of Results

SORM analyses are performed for $T \in [10^2, 10^5]$; results are plotted in Figure 10. It is seen that the simplified approach provides a good approximation over the entire range, and particularly at low probabilities of failure. Some results are also presented in Table 3. The absolute difference in β is less that 0.14, and the simplified approach overestimates the probability of failure by at most about 40%. Such accuracy is generally sufficient for reliability calculations.

An interesting result is the correspondence between the auxiliary variable z_3 in the two methods, see Table 4. These variables originate for reasons which appear quite different: U_{n+1} in the time variant method (see eqs. 2) is introduced to transform the limit state function into a format suitable for solution by FORM/SORM; while U_{n+1} in the time invariant method (eq. 7) accounts for the uncertainty in the extreme load. The results in Table 4 show that the importance factors for the auxiliary variables are nearly identical by the two methods. Note also that the importance factor for Z_3 decreases with an increasing number of load cycles. This is as expected since the uncertainty in the extreme value of the load is small for large T.

This may reduce the computation time required for the g-function call by two orders of magnitude. Moreover, the TI reliability problem is well conditioned and converges quickly to the design point, whereas convergence with the TV formulation is often problematic. Examples are presented demonstrating that the approximate approach is accurate. In [3], reliability estimates by the time variant model and the time invariant approximation are compared for a variety of problems including low and high cycle fatigue and with failure in the brittle fracture, elastic-plastic, and plastic collapse regimes. For all cases analyzed, the conservatism of the TI approach is small, with the failure probability overestimated by a factor typically less than about 1.2.

References

[1] Cramér, H., and Leadbetter, M.R. (1966). *Stationary and Related Stochastic Processes*, John Wiley & Sons, New York.

[2] Guers, F. and Rackwitz, R. (1986). *Crossing Rate Based Formulation in Fatigue Reliability*, Structural Engineering Lab, Technical University of Munich, Heft 79.

[3] Marley, M.J. (1991). *Time Variant Reliability under Fatigue Degradation*, Dr. Eng. Thesis, Div. of Marine Structures, Norwegian Institute of Technology.

[4] Marley, M.J. and Moan, T. (1992). "Time Variant Formulation for Fatigue Reliability", *Proc 11th Intl Conf on Offshore Mech & Arctic Engr, OMAE'92*, Calgary.

[5] Wen, Y.K., and Chen, H.-C. (1987). "On Fast Integration for Time Variant Structural Reliability", *Prob Engr Mech*, 2(3).

[6] Yang, J.-N., and Lin, M.D. (1977). "Residual Strength Degradation Model and Theory of Periodic Proof Tests for Graphite/Epoxy Laminate", *J Composite Materials*, 11.

6. Conclusions

The time invariant model, eq. (3), is simple in concept and easily applied. It requires the evaluation of $\xi(t)$ only at $t = T$, and avoids the numeric integration needed for the time variant eq. (2).

Structural Safety & Reliability, Schuëller, Shinozuka & Yao (eds) © 1994 Balkema, Rotterdam, ISBN 90 5410 357 4

Probabilistic failure risk assessment for structural fatigue

N. R. Moore, D. H. Ebbeler & S. Sutharshana
Jet Propulsion Laboratory, California Institute of Technology, Pasadena, Calif., USA

M. Creager
Structural Integrity Engineering, Chatsworth, Calif., USA

ABSTRACT: A probabilistic methodology for evaluating failure risk, assessing service life, and establishing design parameters for structures subject to fatigue failure has been developed. In this methodology, engineering analysis is combined with experience from tests and service to quantify failure risk. The methodology is particularly valuable when information on which to base design analysis or failure prediction is sparse, uncertain, or approximate and is expensive or difficult to acquire. Sensitivity analyses conducted as a part of the probabilistic methodology can be used to evaluate alternative measures to control risk, such as design changes, testing, or inspections, thereby enabling limited program resources to be allocated more effectively. The probabilistic methodology and an example application to fatigue crack growth in a heat exchanger tube are presented.

1 INTRODUCTION

Risk management during the design, development, and service of structural systems can be improved by using a risk assessment approach that can incorporate information quantitatively from both experience and analytical modeling. In the probabilistic failure risk assessment approach presented here, experience and analytical modeling are used in a statistical structure in which uncertainties about failure prediction are quantitatively treated. Such probabilistic analysis can be performed with the information available at any time during the design, development, verification, or service of structural systems to obtain a quantitative estimate of failure risk that is warranted by what is known about a failure mode. This probabilistic method is applicable to failure modes which can be described by analytical models of the failure phenomena, even when such models are uncertain or approximate.

By conducting risk sensitivity analyses probabilistically for selected failure modes, sources of unacceptable failure risk can be identified and corrective action can be delineated. Design revision, additional characterization of loads and environments, improvement of analytical model accuracy, and improved characterization of material behavior are among the options for controlling risk that can be quantitatively evaluated by probabilistic sensitivity analyses, enabling limited financial resources to be allocated more effectively to control failure risk. Test and analysis programs focused on acquiring information about the most important risk drivers can be defined.

Probabilistic failure risk assessment can be employed in the design and development process to avoid the compounding of design conservatisms and margins that unnecessarily increase cost or weight when conventional design approaches are used. Probabilistic analysis is of particular value in the design and development of systems or components when uncertainties exist about important governing parameters or when design conservatism and redundancy used in the past must be reduced to meet more stringent cost, weight, or performance requirements.

A general approach to probabilistic failure risk assessment and an application of the approach to fatigue crack growth in a heat exchanger tube are presented in the following.

2 PROBABILISTIC FAILURE RISK ASSESSMENT

Information from experience can be combined with information from analytical modeling to estimate failure risk quantitatively using the approach shown in Figure 1. This approach is applied individually to those failure modes identified for analysis. Probabilistic failure modeling is based on available knowledge of the failure phenomenon and of such governing parameters as loads and material properties, and it provides the prior failure risk distribution of Figures 1 and 2. This prior distribution can be modified to reflect available success/failure data in a Bayesian statistical analysis. The probabilistic failure risk assessment approach shown in Figures 1 and 2 is discussed in detail by Moore, et al. (Dec., 1992; Nov., 1992; June, 1992; and 1990).

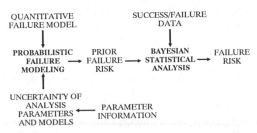

Fig. 1 Probabilistic failure risk assessment

Experience can include physical parameter information in addition to success/failure data. Information about physical parameters can be derived from measurements taken during tests or service, from analyses to bound or characterize parameter values, from applicable experience with similar systems, or from laboratory tests. Measurements of physical parameters used in analytical modeling, e.g., temperatures and loads, can be an important information source in failure risk assessment. Physical parameter information is incorporated into probabilistic failure modeling and is reflected in the prior failure risk distribution.

Success/failure data can be acquired from testing or service experience. The failure risk distribution resulting from the combination of the prior distribution and the success/failure data is the description of failure risk which is warranted by the information available. As additional information regarding governing physical parameters be-

comes available it can be incorporated into analytical modeling to obtain a revised prior failure risk distribution. Additional information in the form of success/failure data can be processed by the Bayesian statistical analysis of Figure 1 to update the prior failure risk distribution using the procedure given by Moore, et al. (June, 1992 and 1990).

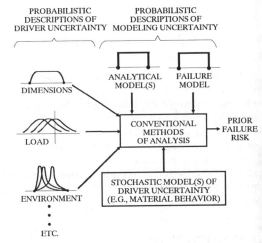

Fig. 2 Probabilistic failure modeling

The analysis procedures used in probabilistic failure modeling, shown in Figure 2, are directly derived from deterministic methods for analyses of failure modes which express failure parameters, such as burst pressure or fatigue life, as a function of governing parameters or drivers. For fatigue failure modes, the drivers include dimensions, loads, material behavior, model accuracy, and environmental parameters such as local temperatures. The accuracy of the models and procedures used in probabilistic failure modeling should be probabilistically described and treated as a driver. Probabilistic descriptions of model accuracy are based on experience in using the models and procedures, and when available, on tests conducted specifically to evaluate their accuracy.

A driver for which uncertainty is to be considered must be characterized by a probability distribution over the range of values it can assume. That distribution expresses uncertainty regarding specific driver values within the range of possible values. A driver probability distribution must represent both intrinsic variability of the driver and uncertain

knowledge or limited information on which to base the driver characterization.

Stochastic drivers are characterized by using the information that exists at the time of analysis. If driver information is sparse, the probabilistic characterization of such a driver must reflect that sparseness. If extensive experimental measurements have been performed for a driver, its nominal value and characterization of its variability can be inferred directly from empirical data. However, if little or no directly applicable empirical data is available, analysis to characterize a driver or experience with similar or related systems must be used. Driver distributions must not overstate the precision implied by the available information.

Some general guidelines for characterizing stochastic drivers have emerged from case studies conducted to date as given in Moore, et al. (Nov., 1992 and June, 1992). For drivers which have physical bounds, such as controlled dimensions or loads with physical upper limits, the Beta distribution parameterized with location, shape, and scale parameters has been successfully used. If only bounds are known, a Uniform distribution is appropriate. For a driver whose variation can be thought of as due to the combined influence of a large number of small independent effects, the Normal distribution can be used. Past experience in characterizing a particular driver such as a material property may suggest the use of a particular distribution, for example, Weibull, Normal, or Lognormal.

A hyperparametric structure for driver distributions has been found useful in describing available information about a driver. For example, to characterize inner wall temperature uncertainty for the heat exchanger tube, information from engineering analysis was used to establish upper and lower bounds for the mean temperature. In order to capture the fact that the mean value of temperature was not known with certainty, the mean value was represented by a Uniform distribution between the upper and lower bounds. This Uniform distribution is the hyperdistribution associated with the mean temperature uncertainty, and its parameters are the associated hyperparameters.

Monte Carlo simulation has been used as the principal computational method in probabilistic failure modeling because it is a general method that can be used with failure models of any complexity. Continually increasing computer power due to improving hardware and software is steadily expanding the practical application of Monte Carlo simulation. Efficient Monte Carlo techniques can be used to reduce the number of simulation trials when computational time is an issue. Certain analysis methods such as finite-element structural models, may be too computationally intensive for practical use in Monte Carlo simulation. However, the output of these models can be represented as response surfaces over the range of variation of significant parameters, see Moore, et al. (Dec., 1992). The uncertainties of response surface representations must be treated as drivers if significant.

Alternative computational methods, for example, FORM/SORM, see Madsen, et al. (1986), may fail to give accurate results for problems in which significantly nonlinear models are employed and driver uncertainty is large. Computational methods are discussed further by Moore, et al. (1990).

3 PROBABILISTIC CRACK GROWTH MODELING

In the crack growth analysis presented here, the life of a structure with initial flaws which is subjected to cyclic loading is computed probabilistically. The crack growth model used in this analysis can consider loads due to vibration, temperature gradient, and pressure. A Monte Carlo simulation procedure, shown in Figure 3, was used to calculate a life distribution.

A deterministic crack growth failure model is embedded within the simulation structure.

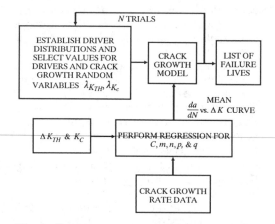

Fig. 3 Crack growth failure simulation

The failure model expresses crack growth life as a function of drivers which may be either deterministic or stochastic. The drivers consist of geometry, loads, environmental parameters, material properties, and accuracy factors which account for uncertainties in the crack growth analysis.

The generalized Forman model, NASA/JSC (1986), was chosen as the basis for the stochastic crack growth rate model. The Forman equation is

$$\frac{da}{dN} = \frac{C(1-R)^m \, \Delta K^n \, [\, \Delta K - \Delta K_{TH} \,]^p}{[\, (1-R)K_c - \Delta K \,]^q} \quad (1)$$

in which da/dN is the crack growth rate, ΔK is the stress intensity factor range, ΔK_{TH} is the threshold stress intensity factor range, K_c is the critical stress intensity factor, R is the stress ratio, and C, n, m, p, and q are the model parameters. The generalized Forman equation captures the crack growth behavior in all of the growth rate regimes, and it can be extended to a stochastic crack growth rate model.

Fatigue crack growth rate data above 10^{-6} mm/cycle and below 10^{-2} mm/cycle do not exhibit a large amount of life variation. This can be seen by examining the extensive data sets of Virkler, et al. (1979) and Ghonem, et al. (1987) in which, for the same initial crack size, the ratio between the shortest and longest life is typically much less than two. This variation in the mid-rate region is small compared to the life variation that may occur due to uncertainty in other parameters such as ΔK_{TH}, stresses, initial crack geometry, etc. Many empirical da/dN vs. ΔK plots found in the literature seem to suggest that crack growth rate data scatter is large, but the apparently large scatter is an artifact of data gathering and data reduction. By comparing the low variability in lives to the much higher scatter in growth rates derived for the same data in Virkler, et al. (1979) and Ghonem, et al. (1987) it may be seen that localized growth rate scatter is not significant. The generalized Forman model can be easily employed to model variability of crack growth rate in the mid-rate region by stochastically varying C in Equation 2, although for the reasons outlined above it was deemed unnecessary.

In contrast to the crack growth in the mid-rate region, uncertainty in the high- and low-

growth rate regions can be significant due to both intrinsic growth rate variability and lack of information in these regions. This uncertainty may be represented in terms of the values of ΔK_{TH} and K_c which are asymptotes to the crack growth rate curve at its lower and upper ends, respectively. Uncertainty about these asymptotes is readily captured by using two stochastic scale parameters $\lambda_{K_{TH}}$ and λ_{K_c}. $\lambda_{K_{TH}}$ modifies the nominal value of the lower asymptote ΔK_{TH} and λ_{K_c} shifts the upper asymptote $(1 - R)K_c$. Thus, the stochastic crack growth rate equation is given by

$$\frac{da}{dN} = \frac{C(1-R)^m \, \Delta K^n \, [\, \Delta K - \lambda_{K_{TH}} \Delta K_{TH} \,]^p}{[\, (1-R)\lambda_{K_c} K_c - \Delta K \,]^q} \quad (2)$$

The uncertainty in $\lambda_{K_{TH}}$ and λ_{K_c} may be characterized by probability distributions, or they may be treated parametrically as was done in the analyses presented here. Figure 4 shows the effect of perturbing $\lambda_{K_{TH}}$ and λ_{K_c} in the growth rate Equation 2.

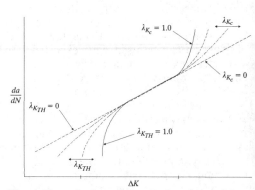

Fig. 4 Description of the stochastic crack growth equation in log-log space

As shown in Figure 3, the mean crack growth rate equation, which is an input to the crack growth model, is typically determined by performing a regression on crack growth data. The parameters C, m, n, p, and q are estimated by a least squares fit of the growth rate Equation 1. If there is uncertainty due to sparseness of data, or if the material test conditions do not closely represent the component operating environment, some of the other equation param-

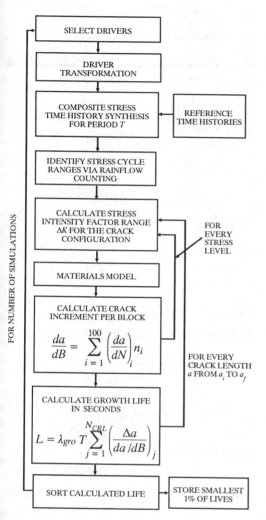

FOR NUMBER OF SIMULATIONS

Fig. 5 Flowchart for crack growth calculation

eters may also be modeled stochastically. For example, if crack growth rate data were to be only available for a single stress ratio R, the uncertainty in m could be captured by describing m stochastically, based on values observed for similar materials.

4 CRACK GROWTH CALCULATIONS

The procedure used for calculating crack growth is shown in Figure 5. In the heat exchanger tube, vibration loads are primarily responsible for crack growth which can result in structural failure. The vibration environment was represented by power spectral density (PSD) envelopes.

The analyses of loads and stresses for the heat exchanger tube and the crack growth calculations are described in detail by Moore, et al. (Dec., 1992 and June, 1992) and summarized by Sutharshana, et al. (1991). A stress history due to dynamic load sources was synthesized from the PSD envelopes. The stress cycles were obtained by performing a cycle count on the synthesized stress time history using the rainflow cycle counting method. The load interaction in growth calculations was accounted for by using the generalized Willenborg retardation model, see Gallager (1974).

Since the traditional cycle-by-cycle crack growth life calculation is computationally intensive, an extremely fast yet accurate block-by-block approach first introduced by Brussat (1974) was used. In the block approach, a block growth rate da/dB is calculated at distinct crack lengths, starting from the initial crack length a_i to the final length a_f, by summing the crack growth rates da/dN from Equation 2 that correspond to ΔK_{eff} and R_{eff} for each stress level in the load block, as follows:

$$\frac{da}{dB} = \sum_{i=1}^{100} \left(\frac{da}{dN}\right)_i n_i \qquad (3)$$

in which n_i is the number of cycles at the ith stress level. The life is computed by numerically integrating the inverted rate per block between the initial and final crack length. The life in seconds is

$$L = \lambda_{gro} \, T \int_{a_i}^{a_f} \frac{da}{da/dB} \qquad (4)$$

in which λ_{gro} is the uncertainty in the growth calculation and T is the length of a load block in seconds. This calculation is performed as a summation over unequally divided N_{CRL} crack lengths, as follows:

$$L = \lambda_{gro} \, T \sum_{j=1}^{N_{CRL}} \left(\frac{\Delta a}{da/dB}\right)_j \qquad (5)$$

The standard stress intensity factor solution for a semi-elliptic crack in a finite width plate subject to axial and bending stresses was employed to calculate ΔK for the heat exchanger tube. The temperature difference across the wall of the tube (cold inside and hot outside) induces significant thermal stresses over the thickness, whose variation across tube thickness is similar to that of bending stresses. Standard stress intensity factor solutions for cylinders with radial cracks, subjected to bending stresses over the thickness, are not available. The SIF expressions used in this analysis are given in NASA/JSC (1986).

Crack growth rate data from Rocketdyne (1989) were available for the heat exchanger tube material at stress ratios of $R = 0.16, 0.7$, and 0.9. This crack growth data set was employed to derive the parameters of the stochastic Forman model given above.

5 DESCRIPTION OF DRIVERS

From among the load, dimension, and environment parameters that appear in the crack growth analysis for the heat exchanger tube, nineteen parameters were described probabilistically. Five of these parameters account for analysis model accuracy. These parameters, i.e., drivers, and their probability distributions are given in Table 1.

The initial crack shape aspect ratio a/c was represented by a Uniform distribution with end points of 0.2 and 1.0. The crack geometry was then defined by treating initial crack length a_i parametrically. Life was simulated with the value of a_i fixed at 0.025mm, 0.063mm, 0.13mm, and 0.19mm. The crack shape distribution was based on an assessment of the crack aspect ratios that could result from the heat exchanger manufacturing process.

The heat exchanger tube wall thickness is nominally 0.312mm, which leads to the concern that "short crack" behavior may be relevant. Short crack growth rate curves have been observed by Morris, et al. (1983) not to have definite thresholds. If a threshold exists, it is a conservative assumption for the linear segment of the curve in the mid-rate region to be extrapolated down into the threshold region. Fixing $\lambda_{K_{TH}} = 0$ in the stochastic Forman equation accomplishes this, as shown in Figure 4. Analyses were performed with values of

Table 1. Description of drivers used in the heat exchanger tube analysis

DRIVER	DISTRIBUTION	RANGE
Initial crack size a_i, mm	Fixed	0.025 to 0.19
Initial crack shape a/c	Uniform	.2 to 1.0
Threshold stress intensity factor range accuracy factor $\lambda_{K_{TH}}$	Fixed	0.0 to 1.0
Fracture toughness accuracy factor λ_{K_c}	Fixed	0.0 to 1.0
Random load adjustment factor $\lambda_{D_{RANDOM}}$	$\begin{bmatrix} \text{Normal}(\mu, \sigma^2) \\ \mu = 0.77 \\ \sigma = 0.12 \end{bmatrix}$	–
Sinusoidal load adjustment factor $\lambda_{D_{SINUSOIDAL}}$	$\begin{bmatrix} \text{Normal}(\mu, \sigma^2) \\ \mu = 0.71 \\ \sigma = 0.14 \end{bmatrix}$	–
Aerodynamic load factor $\lambda_{AERO_{DYN}}$	Uniform	.5 to 1.5
Aerostatic load factor $\lambda_{AERO_{ST}}$	Uniform	.8 to 1.2
Inner wall temperature T_i (°K)	$\begin{bmatrix} \text{Normal}(\mu, \sigma^2) \\ \mu \sim \text{Uniform}(270, 370) \\ \sigma \sim \text{Uniform}(16.1, 31.4) \end{bmatrix}$	–
Outer wall temperature T_o (°K)	$\begin{bmatrix} \text{Normal}(\mu, \sigma^2) \\ \mu \sim \text{Uniform}(444, 505) \\ \sigma \sim \text{Uniform}(26.7, 27.5) \end{bmatrix}$	–
Internal pressure p_i, Mpa	$\begin{bmatrix} \text{Normal}(\mu, \sigma^2) \\ \mu \sim U(26.3, 28.8) \\ \sigma = 0.476 \end{bmatrix}$	–
Inner diameter D_i, mm	$\begin{bmatrix} \text{Beta}(\rho, \theta) \\ \rho = .5 \\ \theta \sim \text{Uniform}(.5, 20) \end{bmatrix}$	4.79 to 4.86
Wall thickness t, mm	$\begin{bmatrix} \text{Beta}(\rho, \theta) \\ \rho = .27 \\ \theta \sim \text{Uniform}(.5, 20) \end{bmatrix}$	0.29 to 0.40
Dynamic stress analysis accuracy factor $\lambda_{DYN_{str}}$	Uniform	.8 to 1.2
Static stress analysis accuracy factor $\lambda_{ST_{str}}$	Uniform	.9 to 1.1
Stress intensity factor calculation accuracy factor λ_{sif}	Uniform	.9 to 1.1
Growth calculation accuracy factor λ_{gro}	Uniform	ln 1/2 to ln 1.75
Neuber's rule accuracy factor λ_{neu}	Uniform	.6 to 1.4
Weld offset stress concentration accuracy factor λ_{OFF}	Uniform	.8 to 1.2

$\lambda_{K_{TH}}$ at 0.0, 0.1, 0.2, etc., to study the impact of the threshold location. Since growth is in the low rate region, the driver λ_{K_c} is not relevant, and its value was fixed at unity.

The stress intensity factor calculation accuracy factor λ_{sif} accounts for the error in the standard stress intensity factor solution and the uncertainty associated with employing a finite width plate solution for a crack in a cylinder. A Uniform distribution was used for λ_{sif} with a range of 0.9 to 1.1. The growth calculation accuracy factor λ_{gro} accounts for uncertainties in the block-by-block growth calculation and in transformation of a variable amplitude stress history to a constant amplitude stress vs. number of cycles table using rainflow counting. Evidence in the literature indicates that factors of two between the calculated crack growth life and tests are appropriate. Since crack propagation is the result of a number of multiplicative events, the distribution on λ_{gro} was specified in log space. A Uniform distribution was used with the lower bound set at $\ln(1/2)$. In order for the mean value of λ_{gro} to be 1.0, the upper bound was set at $\ln(1.75)$.

The Beta distributions characterizing heat exchanger tube dimensions in Table 1 are parameterized by location, scale, and range parameters which are given as ρ, θ, and the end points of the range, respectively.

6 RESULTS

Figure 6 presents the left-hand tail of the simulated failure distribution for the heat exchanger tube. The ordinate of these graphs is the failure probability. The abscissa is the life in seconds for crack growth through the thickness of the heat exchanger tube. Figure 7 illustrates the effects of the crack growth threshold and initial crack size on life at a 0.001 failure probability.

The results in Figure 6 are given for an initial crack size $a_i = 0.13$mm $\lambda_{K_{TH}} = 0$. The left curve labeled "all driver variation" is for a simulation where all the drivers were allowed to vary except a_i, $\lambda_{K_{TH}}$, and λ_{K_c}. The "nominal" value shown on the graph is for an analysis with all the drivers fixed at nominal values. Measures of the relative importance of individual drivers are given in the upper left corner in Figure 6. These were obtained by finding marginal effects of driver uncertainties using several sensitivity runs, where one driver was allowed to vary while the rest were held at nominal values. The crack shape and the growth calculation accuracy are the most important drivers with a 90% contribution to decrease in life. The right-hand curve in Figure 6 shows the shift to the left due to the variation in the crack shape and growth calculation accuracy.

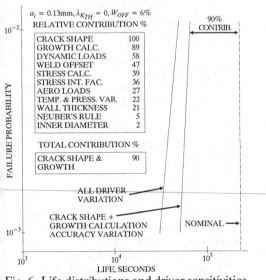

Fig. 6 Life distributions and driver sensitivities

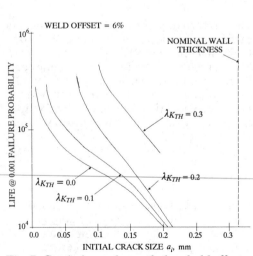

Fig. 7 Crack size and growth threshold effects

7 CONCLUSIONS

For this heat exchanger tube application, the uncertainty due to incomplete knowledge and limited information concerning the accurate characterization of analysis models and physical driver parameters have a much larger impact on failure risk than does any intrinsic parameter variability. The information available was insufficient to meaningfully characterize initial crack size and threshold stress intensity factor for "short cracks". Consequently these important drivers were treated parametrically in order to show their impact on crack growth life and to better define information that is needed to reduce failure risk. A tradeoff between knowledge of initial crack size and knowledge of short crack threshold stress intensity factor, conditioned on the uncertainties in other drivers, can be inferred from the results shown in Figure 7. For a conservative "short crack" threshold ($\lambda_{K_{TH}} = 0$) assumption, inspection techniques that can detect 0.13mm initial cracks with high reliability are required to achieve a life of about 3×10^4 seconds at 0.001 failure probability. On the other hand, if more representative crack growth data can be generated that can reliably establish a nonzero growth threshold ($\lambda_{K_{TH}} > 0$), then the requirements on the inspection may be relaxed while achieving the same life at 0.001 failure probability.

ACKNOWLEDGMENTS

The research described in this publication was carried out by the Jet Propulsion Laboratory, California Institute of Technology, under a contract with the National Aeronautics and Space Administration. The authors gratefully acknowledge the technical contributions made by engineering personnel of Rocketdyne Division, Rockwell International, Inc., Canoga Park, CA, USA.

REFERENCES

Brussat, T.R., 1974, "Rapid Calculation of Fatigue Crack Growth by Integration," Fracture Toughness and Slow Stable Cracking, ASTM STP 559, American Society for Testing and Materials, pp. 298 - 311.

Gallager, J.P., and Hughes, T.S., 1974, "Influence of the Yield Stress on the Overload Affected Fatigue Crack Growth Behavior of 4340 Steel," AFFDL-TR-74-27, Air Force Flight Dynamics Laboratory, Wright-Patterson Air Force Base, Ohio.

Ghonem, H., and Dore, S., 1987, "Experimental Study of the Constant-Probability Crack Growth Curves Under Constant Amplitude Loading," Engineering Fracture Mechanics, Vol. 27, No.1, pp. 1 - 25.

Madsen, H.O., Krenk, S., and Lind, N.C., 1986, Methods of Structural Safety, Prentice-Hall, Englewood Cliffs, New Jersey.

Moore, N., et al., Dec. 1992, An Improved Approach for Flight Readiness Certification – Probabilistic Models for Flaw Propagation and Turbine Blade Fatigue Failure, Volume I, JPL Pub. 92-32, Jet Propulsion Laboratory, California Institute of Technology, Pasadena.

Moore, N., Ebbeler, D., and Creager, M., Nov. 1992, "Probabilistic Service Life Assessment," Reliability Technology – 1992, AD-Vol. 28, American Society of Mechanical Engineers, ISBN 0-7918-1095-X.

Moore, N., et al., June 1992, An Improved Approach for Flight Readiness Certification – Methodology for Failure Risk Assessment and Application Examples, Volume I, JPL Pub. 92-15, Jet Propulsion Laboratory, California Institute of Technology, Pasadena.

Moore, N., Ebbeler, D., and Creager, M., 1990, "A Methodology for Probabilistic Prediction of Structural Failures of Launch Vehicle Propulsion Systems," Paper No. 90-1140-CP, Proceedings of the AIAA 31st Annual Structures, Structural Dynamics and Materials Conference, pp. 1092 - 1104.

Morris, W.L., and James, M.R., 1983, "Investigation of the Growth Threshold for Short Cracks," Proceedings of the International Symposium on Fatigue Crack Growth Threshold Concepts, AIME, pp. 479 - 495.

NASA/JSC 22287, 1986, Fatigue Crack Growth Computer Program, "NASA/FLAGRO" Manual, National Aeronautic and Space Administration, Johnson Space Center, Houston, TX.

Rocketdyne Division, 1989, "Fatigue Crack Growth Rate Testing of Welded 316L," Rockwell International, Canoga Park, California.

Sutharshana, S., et al., 1991, "A Probabilistic Fracture Mechanics Approach for Structural Reliability Assessment of Space Flight Systems," Advances in Fatigue Lifetime Predictive Techniques, Special Technical Publication 1122, American Society for Testing and Materials.

Virkler, D.A., Hillberry, B.M., and Goel, P.K., 1979, "Statistical Nature of Fatigue Crack Propagation," Journal of Engineering Materials and Technology, ASME, Vol. 101, April, pp. 148 - 153.

Structural Safety & Reliability, Schuëller, Shinozuka & Yao (eds) © 1994 Balkema, Rotterdam, ISBN 90 5410 357 4

Markov chain modeling of non-destructive in-service inspection

Marcelo M. Rocha & Gerhart I. Schuëller
Institute of Engineering Mechanics, University of Innsbruck, Austria

ABSTRACT: A theory for combining the effects of stochastic fatigue crack growth and non-destructive in-service inspection is presented. In this theory a Markov chain model is used to update the crack statistics, allowing an efficient computational implementation and a clear account of the involved concepts. Two kinds of action upon the structural component are distinguished, perfect and imperfect crack removal, which are idealizations defining bounds for actual situations.

1 Introduction

The presence of cracks in structural components can be statistically described by the probability density function (pdf) of the crack size and one additional parameter representing the expected number of cracks in the observed space. On one hand, this statistical description should account for a dependence on time due to the fatigue crack growth (FCG) phenomenon, which causes a decrease in the local resistance and consequently in the component reliability. On the other hand, inspection routines can be adopted to detect and eventually remove cracks with size larger than a rejection limit, resulting in the increase of reliability toward an acceptable level.

In previous investigations (see e.g. Kozin and Bogdanoff 1983, Rocha and Schuëller 1992), the Transition Probability Matrix (TPM) associated with a differential equation modeling high-cycle FCG has been used to update the crack size. This approach is based on the assumption that one-dimensional crack growth phenomena can be modeled by an *equivalent* discrete Markov process. With respect to computational aspects, the Markov chain technique requires a convenient definition of representative discrete crack sizes, denoting the system states, and the estimation of transition probabilities assembling the TPM. To update the crack size only simple vector-matrix multiplications have to be performed.

The modeling of inspection routines with TPMs results to be computationally attractive since its implementation can be easily accomplished as an extention of the system performing FCG analysis (Shimada et al 1989, Itagaki et al 1989). Further, the inspection-TPM can be easily derived from basic concepts currently used in non-destructive examination (NDE) methods. For instance, the diagonal elements (probability of no-changes in the system states) are directly associated to the probability of crack acceptance, which is related to the efficiency of the NDE-technique used.

By varying systematically the number of multiplications of the fatigue-TPM by the inspection-TPM and observing the effect in the resulting failure probabilities, conclusions can be drawn with respect to the inspection intervals and other model parameters, like the rejectable crack size. Alternatively, an upper bound for the failure probability can be specified and the required inspection intervals calculated accordingly.

2 Basic Definitions

The presence of cracks plays a central role in fracture mechanical and fatigue analysis. The size and quantity of cracks are parameters that can be — within certain limits — directly observed and measured for many types of structural components. However, measuring these indices is feasible only in a statistical sense due to random nature of crack presence, which is here described by: (1) the occurrence rate λ, repre-

senting the expected number of cracks in the observed space and (2) the probability density of the (nominal) crack size $f_A(a)$, representing the random size of individual cracks.

The number λ and the density of the size $f_A(a)$ represent here, by convention, the statistics of cracks *remaining* in the structure after having been overlooked by a first quality control. They can be combined into an auxiliary function:

$$Q(a) = \lambda f_A(a) \tag{1}$$

which represents the density for the mean number of cracks with size a that could be found in the observed space. This statistical description is mainly affected by two things: crack initiation or propagation due to fatigue, which increase λ and shift $f_A(a)$ toward larger sizes, and the performace of inspection routines followed by repair or replacement, which may decrease λ and shift $f_A(a)$ toward smaller sizes.

The Markov chain technique, applied in the following sections, requires the continuous crack size distribution to be replaced by an equivalent discrete form where probability masses p_i are associated with discrete crack sizes a_i and used to construct a *row vector* **p**, called vector of state probabilities. The probability masses can be derived from the cumulative distribution function (cdf) $F_A(a)$ by means of a finite differences scheme as:

$$p_i = \begin{cases} 0 & \text{if } i = 1, a_1 = 0 \\ F_A(a_i) - F_A(a_{i-1}) & \text{if } i > 1, a_{i-1} < a_i \end{cases} \tag{2}$$

where the number and spacing of discretization points should be chosen according to the intended accuracy, accounting for the largest possible crack sizes after fatigue growth. The arbitrary definition $a_1 = 0$ is necessary for representing a repaired crack (a *zero-size* crack), as it will be seen in Section 4. After the above discretization, the probabilities p_i should be checked for normalization in order to add up to 1.

The exact meaning of the random crack size has a decisive importance in the modeling of inspection. The possibilities considered here are:

Case I: it represents an uncertainty with respect to one single crack that could be found in a critical point of a structural component (here λ can be either 1 or defined as the probability of this single crack existing).

Case II: it represents an uncertainty with respect to one single crack that could be found in the same critical point of many similar structural components (here λ is the number of components likely to present the considered crack).

Case III: it represents an uncertainty with respect to many cracks that could be found in different critical points of the same structural component (here λ is the mean number of cracks likely to be found).

For the third case, a combined analysis of fatigue and inspection effects respectively on the crack statistics is possible only if all considered cracks have the *same geometry factor* connecting the local stress field to the *same reference load*, as required in the evaluation of stress intensities. If this is not true, cracks must be subdivided into groups fulfilling this condition and each group analysed separately.

The efficiency of an inspection routine is here modelled by considering two main uncertainties:

1. *The error in sizing e*, assumed to be a random variable with a distribution function $F_E(e)$:

$$e = a_{obs} - a \tag{3}$$

2. *The crack detection probability $P_D(a)$*, likely to present some dependence on the crack size:

$$P_D(a) = Q_{obs}(a)/Q(a) \tag{4}$$

where the subscript "obs" stands for the corresponding information as obtained directly from the inspection performance. Note that the detection probability can be alternatively defined as the number of times a crack of size a has been detected divided by the total number of trials, each trial performed by a different inspector or inspection team using the same inspection technique. However, in many cases both definitions can be considered to be equivalent.

To complete the basic definitions an additional parameter should be introduced: the *rejectable crack size a_R*. This limit size is not measured or estimated from data, but *specified* according to some criteria based both on safety and economic aspects. It can afterwards be subject to optimization. The rejectable crack size a_R (R stands also for *repair* or *replacement*) represents a limit for the essential decision of accepting or rejecting a crack after its detection. Defining a_R does not mean that all larger cracks will be rejected, since inspection bears the uncertainties already mentioned. Nevertheless, it is possible to make some predictions concerning what could occur in practice if an arbitrary a_R is specified. The rejectable crack size can be used to evaluate the following probabilities, where independence between sizing and detection is assumed:

1. *The probability of rejecting a crack with size a,* calculated as the product of the detection probability and the probability of sizing the detected crack larger than a_R:

$$P_R(a) = P_D(a)\left[1 - F_E(a_R - a)\right] \qquad (5)$$

2. *The probability of accepting a crack with size a,* calculated as the product of the detection probability and the probability of sizing the detected crack smaller than a_R, added to the non-detection probability:

$$P_A(a) = P_D(a)F_E(a_R - a) + [1 - P_D(a)] \qquad (6)$$

For a given crack size a the sum of these two probabilities is equal to unity, since a crack must always be either rejected or accepted. For a particular case where $a \geq a_R$ the function $P_R(a)$ is called, the probability of *correct rejection*, while for $a < a_R$ the function $P_A(a)$ is called, the probability of *correct acceptance*. It can be observed that eqs. (5) and (6) depend both on the efficiency of the inspection technique and on the specified rejection limit a_R.

The uncertainties in the efficiency of inspection can be included in two different stages of the analysis: correcting the estimated statistics of initial cracks and predicting the effect of future inspections on these statistics. The latter is the subject of Section 4, but first some additional remarks on the detection probability are in order.

3 Bayesian Updating of Detection Probabilities

Estimates of the crack detection probability $P_D(a)$ are of paramount importance in the meaningful assessment of inspection reliability. Efficiency in detecting, however, depends strongly on many factors like the type of crack tip, crack position and orientation, surface conditions and accessibility, among others. Many of these factors have been investigated in the well known PISC Project for ultrasonic inspection of steel components, along with other factors inherent in the inspection routine itself, like calibration of the equipment and interpretation of results (Crutzen et al 1989).

Within the scope of this work, an important experience gained from the PISC results is the possible existence of a significant uncertainty in the relation between detection probability and defect size. This uncertainty is illustrated in Fig.1a, where each point corresponds to one of 11 planar defects inspected by a maximum of 9 teams [1] performing ultrasonic inspection in the spirit of ASME 20% DAC. [2]

From eqs. (1) and (4) one can note the essential difference between the possibilities of accounting or not for a crack size dependence when modeling inspection efficiency. If a dependence does exist, both λ and $f_A(a)$ can be affected by the detection probability after an inspection routine. Otherwise any change in $f_A(a)$ will result only from the choice of the rejectable crack size; if any size is rejectable, $f_A(a)$ may remain the same and solely λ can be altered by inspection. It is not difficult to imagine a kind of inspection technique not changing the crack size distribution but the quantity of cracks. These observations are very important in the reliability assessment, implying that *both* both λ and $f_A(a)$ should be used in the evaluation of failure probabilities.

If it is decided that $P_D(a)$ must be regarded as size-dependent, subsequent uncertainties in this dependence can be accounted for by considering one or more parameters in the detection function to be random variables. If the detection function $P_D(a)$ contains an uncertain parameter α_1, it is possible to start with a (subjectively defined) prior distribution $f'_{A_1}(\alpha_1)$ and use the Bayesian approach to improve the knowledge as soon as additional information becomes available.

The choice of a likelihood function will define how the new information can be accounted for. If a fast convergence toward the expected value $E[\alpha_1]$ is aimed two types of events can be distinguished, where the likelihood function is the detection probability itself: (1) A crack of size a is detected:

$$f''_{A_1}(\alpha_1) = \frac{P_D(a, \alpha_1)}{E\left[P_D(a, \alpha_1)\right]} f'_{A_1}(\alpha_1) \qquad (7)$$

and (2) A crack of size a is not detected (but its existence is known):

$$f''_{A_1}(\alpha_1) = \frac{1 - P_D(a, \alpha_1)}{1 - E\left[P_D(a, \alpha_1)\right]} f'_{A_1}(\alpha_1) \qquad (8)$$

where

$$E\left[P_D(a, \alpha_1)\right] = \int_0^\infty P_D(a, \alpha_1) f'_{A_1}(\alpha_1)\, d\alpha_1 \qquad (9)$$

---

[1] It should be observed that due to the small number of teams a non-negligible statistical uncertainty is also responsible for the scatter. Furthermore, each of the 11 planar defects presents its own particularities and may not necessarily be considered as being described by the same probability model.

[2] Section XI of the ASME Boiler and Pressure Vessel Code, using a medium amplitude cut-off level of 20% DAC (Distance-Amplitude Correction).

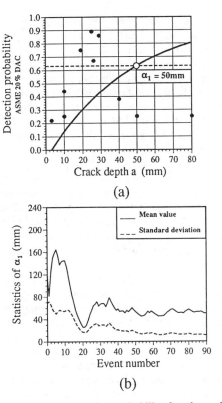

(a)

(b)

Figure 1: a) Detection probability for planar defects in Assembly 3 with ultrasonic inspection in the spirit of ASME 20% DAC (PISC II project), plotted as a function of the crack depth a (Crutzen et al 1991); b) Convergence of the random parameter α_1 with successive inclusion of events in the updating equations.

These updating equations are advantageous with respect to efficiently combining information available for *different cracks sizes*. However, they rely upon the validity of the model and will converge to the average detection function no matter how the actual detection probabilities deviate from this average. In practice, the events "detect" or "not detect" a crack of size a come from the past experience in using the same inspection technique under the same conditions, arranged, for instance, as a databank.

To illustrate these ideas an exponentially increasing detection function can be used:

$$P_D(a) = \begin{cases} 0 & \text{if } a \leq \alpha_0 \\ P_L\left[1 - \exp\left(\dfrac{\alpha_0 - a}{\alpha_1 - \alpha_0}\right)\right] & \text{if } a > \alpha_0 \end{cases} \quad (10)$$

The parameter P_L represents the *probability of de-*

tecting a very large crack, which is not necessarily equal to unity due to, for instance, gross human errors. The size α_0 represents a lower limit for cracks being detected, or the *smallest detectable crack*, while α_1 is a *reference crack size* to be detected with probability $\approx 0.63 P_L$.

The same information used to calculate the individual estimates presented in Fig.1a can now enter eqs. (7) or (8) as a sequence of related events. The convergence process is illustrated in Fig.1b, where it was assumed that $P_L = 1$ and $\alpha_0 = 3$mm. The prior distribution $f'_{A_1}(\alpha_1)$ was initially chosen as uniform between 3 and 240mm. Different priors have been also tried, without significant changes in the results obtained. A fast convergence to $\alpha_1 = 50$mm can be observed, which corresponds to the detection function plotted in Fig.1a. It is interesting to note that the sequence of events has no influence on the final distribution, even though the convergence path can be quite different.

Finally, comparing the mean detection function and individual estimates in Fig.1a one can conclude that, despite the fast convergence of $E[\alpha_1]$, the validity of the detection model cannot be ensured. The goodness of any model can be easily checked if individual estimates of $P_D(a)$ are available, like for the PISC data. If this is not the case, the crack size domain can be divided into two or more subdomains (depending on the amount of information) and corresponding data used separately in eqs. (7) to (9). If in each subdomain the random parameter α_1 converges to a value significantly different from the others, the model should be refused. This can be again illustrated with Fig.1a, where one would surely get different parameters by considering separately cracks smaller and larger than 40mm.

4 Use of Markov Chains

4.1 General

The definitions presented in the previous sections can now be used to predict what would happen to the crack statistics if one of the following actions upon a structural component is taken: (1) An inspection routine is performed and each crack detected and sized as rejectable is individually removed without the possibility of introducing a new one. This kind of action is herein called *perfect removal*. (2) An inspection routine is performed and each crack detected and sized as rejectable is individually removed with the possibility of introducing a new one, which is assumed to

follow the same statistics as for the whole initial population of cracks. This kind of action is herein called *imperfect removal*.

The word "removal" is chosen intentionally: it represent both repair and replacement actions, and must be also interpreted according to the particular case defined in Section 2. For instance, if cases I or II are considered, there could be few differences between replacing a component with a cracked weld and welding it again. Most probably, the actual results of any action should lie somewhere between the cases of perfect and imperfect removal. However, investigating the implications of both idealizations would show whether or not additional refinements are required.

4.2 Perfect removal of cracks

This modality of action is equally applicable to any of the three interpretations (cases I, II and III) of crack statistics, and no further particularizations seem to be necessary. For cases I and II, the perfect crack repair is equivalent to a component replacement only if the substitute component is surely *not* cracked in the considered location. The corresponding formulation is derived from an optional interpretation for the probability of crack acceptance:

$$P_A(a) = Q_1(a)/Q_0(a) \tag{11}$$

where $Q_0(a)$ is the density of cracks with size a before the first repairs and $Q_1(a)$ the corresponding density after the first repairs.

Performing successive inspection routines and using the last updated statistics as initial conditions for the subsequent calculations leads to the updating equation:

$$Q_n(a) = Q_0(a) [P_A(a)]^n \tag{12}$$

from which the damage indices can be recovered with:

$$\lambda_n = \int_0^{+\infty} Q_n(a)\, da \tag{13}$$

$$f_A(a,n) = \frac{Q_n(a)}{\lambda_n} \tag{14}$$

where index and exponent, denoted by n, represent the number of successive inspections performed.

Eq. (12) can be rewritten into a discrete form and the operation of updating the density function $Q_0(a)$ represented by successive multiplications with a suitable TPM. Considering the definitions in eq. (2), this

TPM can be defined as:

$$I_{ij} = \begin{cases} 1 & \text{if } j = 1, j = i \\ P_R(a_i) & \text{if } j = 1, j \neq i \\ P_A(a_i) & \text{if } j \neq 1, j = i \\ 0 & \text{if } j \neq 1, j \neq i \end{cases} \tag{15}$$

and the updating equations in matrix form are:

$$\mathbf{p}(n) = \mathbf{p}(0) \cdot \mathbf{I}^n \tag{16}$$

$$\lambda_n = [1 - p_1(n)] \lambda_0 \tag{17}$$

The inspection-TPM in the case of perfect crack removal includes only two possibilities: a crack keeps its size (no change of state) or it is brought back to zero-size (state of non-existence). In the next item, a different possibility is also considered: detected cracks resume the statistical description in the beginning of operational life.

4.3 Imperfect removal of cracks

This case is related to repairs or replacement actions bearing the possibility of reintroducing cracks into the component. Here a distinction must be made among the three cases defined in Section 2. The most straightforward formulation can be derived for case I:

$$I_{ij} = \begin{cases} 1 & \text{if } j = 1, j = i \\ 0 & \text{if } j = 1, j \neq i \\ P_A(a_i) + p_i(0) P_R(a_i) & \text{if } j \neq 1, j = i \\ p_j(0) P_R(a_i) & \text{if } j \neq 1, j \neq i \end{cases} \tag{18}$$

while the occurrence rate is not expected to change, implying that $\lambda_n = \lambda_0$.

The formulations for cases II and III, respectively, are quite similar to case I, but a new interpretation of the terms "rejection" and "acceptance" must be considered. For case II, the global crack size distribution will return to the initial state only if all components are simultaneously replaced or repaired, which depends also on the total number of components. Further, the occurrence rate is expected to change, since either a cracked component can be replaced by a non-cracked one or a repair might not introduce a new crack. For case III, the actions of repair or replacement lead to quite different formulations. If only repair is specified, then the situation is equivalent to that of case II. If replacement is specified, the recovery of initial statistics occurs as soon as any one of the λ cracks is detected and sized as rejectable. In this case the probability of replacement will be higher for a larger number of cracks. Also here the occurrence rate is not likely to change, as long as crack initiation is not considered.

It should be observed that, from the computational viewpoint, the TPMs defined in eqs. (15) and (18) must not be fully stored during calculations, since each element I_{ij} can be easily recovered from the basic parameters whenever necessary.

5 Combined Fatigue-Inspection Analysis

Due to the fatigue phenomenon, the crack statistics is likely to change between two subsequent inspection routines. If the effects of fatigue can be modeled with a TPM denoted by $\mathbf{F}$, a combined updating equation can be used:

$$\mathbf{p}(N + h\Delta N) = \mathbf{p}(N) \cdot \mathbf{F}^h \cdot \mathbf{I} \tag{19}$$

where $h\Delta N$ is the inspection interval and ΔN the number of cycles advanced after each multiplication by $\mathbf{F}$. It is important to remember that eq. (19) holds only if the considered population of cracks has a common geometry factor, relating local stresses to a same reference load, as previously mentioned.

To focus the analysis on the uncertainties related to the efficiency of inspection a deterministic high-cycle fatigue crack growth process is assumed. Generality of the formulation can be preserved by defining a "deterministic" TPM as:

$$F_{ij} = \begin{cases} 1 & \text{if } j = 1, j = i \\ 0 & \text{if } j = 1, j \neq i \\ 1 & \text{if } j \neq 1, j = i+1 \\ 0 & \text{if } j \neq 1, j \neq i+1 \end{cases} \tag{20}$$

which produces a deterministic unit-jump of each state probability, with an exception for the first state [3]. The state definitions must accordingly correspond to the deterministic solution of the crack growth equation, which can be easily accomplished with a finite differences scheme. For instance when using the Paris law:

$$\left. \begin{aligned} a_{i+1} &= a_i + \Delta a_i \\ \Delta a_i &= \Delta N \left[C (\Delta K_i)^m \right] \end{aligned} \right\} \quad \text{for all } i > 1 \tag{21}$$

where ΔK_i is the stress intensity range associated to the crack size a_i. The state a_2, following the zero-size state a_1, is the initial value for the discretization

---

[3]The probability of crack growth from zero-size state could be used to model the probability of crack initiation in the time step ΔN. This model feature ought to be mentioned, even though the crack initiation phenomenon does not belong to the scope of this work.

scheme and should be chosen according to the intended accuracy. It should be observed that any other crack growth model could be considered here as well.

After defining the system states (reference points over the crack path) the initial state probabilities can be calculated from eq. (2). For many types of structural components, however, the starting point is not a random crack size but a random time for crack initiation, which can be similarly modeled with a distribution function $F_T(t)$ (Yang 1976). Even in this case, the crack path discretization eq. (21) can be adopted to define an *equivalent* crack propagation process. If a_k is the (detectable) crack size indicating the transition from crack initiation to crack propagation, the equivalent initial state probabilities are calculated as:

$$p_{k-i}(0) = F_T \left(\frac{i\Delta N}{\nu} \right) - F_T \left[\frac{(i-1)\Delta N}{\nu} \right] \tag{22}$$

where ν is the mean number of loading cycles within the time unit. In eq. (22) the probability associated with a given time to initiation is assigned to the crack size requiring exactly this time to reach a_k. With this approach it becomes possible to avoid the complications in combining the initiation and propagation stages.

Using eqs. (19) to (21) it is now possible to evaluate the crack size distribution at any time multiple of ΔN, by multiplying successively the vector of initial state probabilities by the fatigue-TPM. For the case of deterministic growth just described, this multiplication in practice becomes unnecessary since a shift of the system states with respect to the associated probabilities produces the same result in a more efficient way. If inspection should be performed after $h\Delta N$ cycles the state probabilities by that time are obtained as follows:

$$p_i(N) \longrightarrow p_{i+h}(N + h\Delta N), \quad \text{for all } i > 1 \tag{23}$$

If the failure state is defined now as the crack reaching a critical size, also considered to be a random variable with distribution function $F_{A_{cr}}(a)$, the first passage probability can be calculated as:

$$p_f(N) = 1 - \exp \left[-\lambda_0 \sum_{i>1} p_i(N) F_{A_{cr}}(a_i) \right] \tag{24}$$

where it must be ensured that the state probabilities p_i add up to 1. The direct multiplication by λ_0 results from the assumption of independence among the growth process of different cracks.

The distribution $F_{A_{cr}}(a)$ can be evaluated by utilizing fracture mechanical considerations, where the random nature of material ductility and fracture toughness may also be accounted for.

Table 1: Summary of parameters required for the analysis and respective values used in the numerical example.

Crack Statistics			
Interpretation	Case I	Initial number of cracks (λ_0)	1
Expected value of the initial size (E [a])	6mm	Type of size distribution	Shifted exponential
Variance of the initial size (VAR [a])	(4mm)^2		
Detection probability function			
Probability of detecting a large crack (P_L)	1	Smallest detectable crack size (α_0)	3mm
Reference crack size (α_1)	Variable		
Normal additive sizing error			
Expected value of sizing error (E [e])	0mm	Variance of sizing error (VAR [e])	(5mm)^2
Fatigue crack growth process (Paris law)			
Effective loading range ($\Delta\sigma$)	120MPa	Geometry factor ($Y = \Delta K / \Delta\sigma$)	$\sqrt{\pi a}$
Paris' exponent (m)	3	Crack growth resistance (C)	2×10^{-13}
Time step (ΔN)	500 cycles		
Further parameters			
Acceptable failure probability	1×10^{-6}	Crack size defining failure (a_α)	80mm
Rejection limit (a_R)	12.5mm		

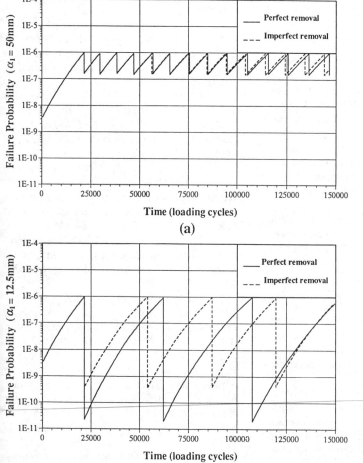

Fig.2: Failure probabilities accounting for high-cycle fatigue and periodic inspection, where an upper bound of 1×10^{-6} has been imposed: a) $\alpha_1 = 50$mm, b) $\alpha_1 = 12.5$mm.

A critical crack size may also be deterministically defined according to practical aspects like the dimensions of the component, quality standards or the efficiency of inspection itself. In the case of adopting a deterministic critical size a_c, the first passage probability can be more readily evaluated by:

$$p_f(N) = 1 - \exp\left[-\lambda_0 \sum_{i \geq c} p_i(N)\right] \qquad (25)$$

The use of the formulation presented above is illustrated by means of a numerical example. The set of used parameters is listed in Table 1. The example consists in the evaluation of the time-dependent failure probability, where an hypothetical inspection is performed each time a specified upper bound is reached. The results for two different levels of detection efficiency is shown in Fig.2.

6 Conclusions

A theory for combining the effects of stochastic fatigue crack growth and non-destructive inspection routines has been presented. In this theory a Markov chain technique is used to update the crack statistics after a non-destructive inspection. No restrictions have been made with respect to crack growth models and type of detection probability function respectively. The approach is straightforward and therefore suitable to be coupled with optimization schemes.

In the numerical example performed, it can be observed that the distinction made between perfect and imperfect crack removal lead to significantly different results only in the case of relatively high detection probabilities. This is an obvious consequence of the fact that, for low inspection efficiency, cracks are not likely to be removed anyway and the probable crack size after inspection becomes less relevant. For high inspection efficiency, however, the difference in the the resulting inspection intervals is significant and cannot be neglected. In this point, the proposed approach seems to be useful for investigating whether or not the distinction between perfect and imperfect removal is necessary at all.

Most important, however, is the observed sensibility of the resulting inspection intervals with respect to the detection parameter α_1, which has been shown to bear a significant level of uncertainty even when estimated from data obtained in carefully conducted inspection exercises. In view of the uncertainty inherent in the functional relation between crack size and detection probability, illustrated in Fig.1, the decision

which of the two time-variant failure probability estimates shown in Fig.2 is the appropriate one to define inspection intervals requires obviously further engineering judgment.

7 Acknowledgments

This research was partially supported by the Austrian Industrial Research Promotion Fund (FFF) under contract No.6/636, the Austrian Ministry of Foreign Affairs (North-South Dialogue Program, ÖAD project EH-894) and the Brasilian National Council for Science and Tecnology (CNPq) under contract 204071/89.5/EC. The authors sincerely thank Mr. *S. Crutzen* of JRC, Ispra, for providing the PISC II data as used in Fig.1.

References

CRUTZEN, S.; JEHENSON, P.; NICHOLS, R.W.; MCDONALD, N.; The major results of the PISC II RRT, *Nuclear Engineering and Design*, Vol.**115**, pp. 7-21, 1989.

CRUTZEN, S.; LOOPUYT, P.; VAN DER BERGH, R.; VINCHE, C.; *Illustration of PISC II data as available for probabilistic failure assessment procedures*, Technical Note, JRC Ispra, 1991.

ITAGAKI, H.; ITOH, S.; YAMAMOTO, N.; Bayesian reliability analysis for evaluating in-service inspection, *Current Japanese Materials Research*, Vol.**5**: Recent Studies on Structural Safety and Reliability, pp. 167-189, 1989.

KOZIN, F.; BOGDANOFF, J.L.; On the probabilistic modeling of fatigue crack growth, *Engineering Fracture Mechanics*, Vol.**18**, pp. 623-632, 1983.

ROCHA, M.M.; SCHUELLER, G.I.; Some remarks on BK-Models for fatigue crack growth, *Proceedings of the sixth specialty conference, ASCE*, Denver, Colorado, pp. 316-319, 1992.

SHIMADA, Y.; NAKAGAWA, T.; TOKUNO, H.; Application of the Markov chain to the reliability analysis of fatigue crack propagation with non-destructive inspection, *Current Japanese Materials Research*, Vol.**5**: Recent Studies on Structural Safety and Reliability, pp. 135-151, 1989.

YANG, J.-N.; Reliability analysis of structures under periodic proof tests in service, *AIAA Journal*, Vol.**14**, No.9, pp. 1225-1234, 1976.

Structural Safety & Reliability, Schuëller, Shinozuka & Yao (eds) © 1994 Balkema, Rotterdam, ISBN 90 5410 357 4

Fatigue life prediction by calculation: Facts and fantasies

W. Schütz
Industrieanlagenbetriebsgesellschaft mbH, Ottobrunn, Germany

Abstract: For any reliability analysis of the fatigue life of a structure the reliability of the basic fatigue data must be known.

Since most components and structures are stressed in service by variable load amplitudes, while most test results are available in the form of constant amplitude fatigue data, this means that the reliability of the fatigue life prediction hypothesis used must be known.

In the first section, the paper states in a general way the requirements for a useful hypothesis for fatigue life prediction; it differentiates between scientific and industrial fatigue life prediction, describes how to check the reliability of such a hypothesis and finally discusses what a "good" or "good enough" hypothesis for fatigue life prediction means, i.e. what scatter or variability of the prediction is acceptable.

The general experience with the following hypotheses is then presented, based on tests with specimens and actural components:
Miners Rule (original, modified, elementary)
The local approach
Paris
Forman
Improved Forman
Willenborg
Crack Closure hypotheses (Onera, Corpus, Loseq etc.)
Proprietary hypotheses of some aircraft manufacturers
All these hypotheses will be checked against the only real requirement - how well did they predict fatigue life under variable amplitudes.
It is proved that neither Miner's Rule nor any of the more recent local approach predictions meets the requirements, i.e. are reliable enough; some modern crack propagation models based on crack closure come near, but serious uncertainties remain, despite innumerable claims to the contrary in the literature.

1 INTRODUCTION

The service fatigue loads or stresses of practically all components, structures and vehicles consist of varying amplitudes that usually are a mixture of stochastic and deterministic components, see Fig. 1.

On the other hand, 99 % or more of the fatigue data available are of constant amplitude. If we want to calculate the fatigue life under such varying amplitudes from constant amplitude data, we need a Fatigue Life Prediction Hypothesis. This is true for the complete fatigue life to failure as well as that to crack initiation or from crack initiation to failure.

An incredible number of papers on fatigue life prediction has been piling up over the last few years. Most of these come from the aircraft field, where for obvious reasons the problem of fatigue is especially critical. In recent years, additional branches of industry beyond the aircraft community, for example the automotive or the offshore industry, have become involved in fatigue life prediction. Often, these newcomers appear to be completely oblivious of previous work and start, so to speak, from scratch.

Anyway, most of these authors claim to have more or less "solved" the problem of fatigue life prediction. Some even claim that the fatigue life of components

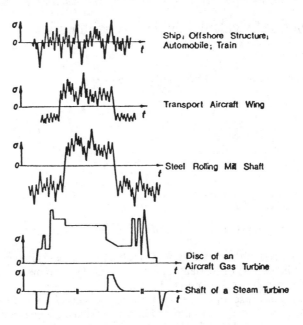

Figure 1: Typical load sequences (schematical)

	Scientific	Industrial
Spectrum	known	at best "statistically" known
Stress Sequence	known	unknown
Critical Section	known	unknown
Crack Shape	known	not known accurately or unknown
Stress Intensity Faktor	known	not known accurately or unknown
Nominal Stress	known	not known accurately
Local Stress	known	usually unknown
Materials Constants	known	not known accurately

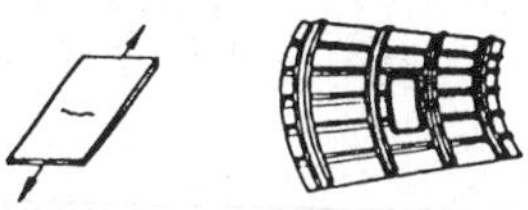

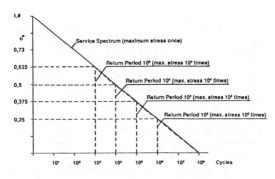

Figure 2: Fatigue life prediction

Figure 3: The spectrum shape at long lives under too short return periods

can be predicted by pushing a button, or rather a large number of buttons on a computer. Software houses sell computer packages pretending that the problem of Fatigue Life Prediction is one of clever programming.

In chronological order the development of a fatigue life prediction method usually proceeds like this:

— An engineer or scientist develops a new method, slightly different from previous ones; he then "proves" his method by a few often, unsuitable tests and presents a paper. If he is clever, he will, however, introduce restrictions. Milton Miner [1] was an early example of such cleverness:

$$D = \sum \frac{n_i}{N_i} = 1,0$$

Restrictions:

— $D \geq 1,0$
— All cycles above fatigue limit.
— Only for Al-Alloys.
— Life to crack initiation.

If strictly applied the second restriction alone - all cycles of the spectrum must be above the fatigue limit - would have made Miner's Rule useless.

— The original author and others present many more papers, using the "new" method without giving a thought to the restrictions. This happened with Miner himself [2] and with all of us who

have ever used Miner's Rule.
— Somebody else uses the method or - with the modern, complex methods - at least claims he has used it and gets very disappointing results. Again, this has happened with Miner's Rule and - to the author's knowledge - with all the more modern methods since then.
— At least one of the new authors, having obtained disappointing results, develops his own new, "improved" method avoiding some or all of the disadvantages of the previous method (or so it is claimed). The new method inevitably is more complex, or more restricted or requires more extensive test data than before. This third author then "proves" his method with another set of tests. Another scientist then comes along, tests the second, "improved" method and again gets disappointing results, etc., etc.

2 WHAT IS MEANT BY "FATIGUE LIFE PREDICTION"?

Fatigue Life Prediction (FLP for short) in the context of this paper is the prediction, by numerical means, of the fatigue life for **realistic** variable stress-time histories, neither for two step, single or multiple overload tests nor the prediction of SN-curves of notched specimens from SN-curves of unnotched specimens. As mentioned before, fatigue life in this context may be the life to complete failure, or to crack initiation or from crack initiation to failure.

We also must distinguish between two different types of fatigue life prediction, namely, scientific and industrial FLP.
— **Scientific FLP** is characterised by having all the necessary input data available, as shown in Fig. 2. The spectrum shape and stress sequence are known; because simple specimens or, at most, simple components are used, the critical section, the stress intensity factor K or the stress concentration factor K_t, whatever the case may be, are known, as well as the other parameters shown. The relevant constant amplitude data in the form of SN-curves or da/dN vs ΔK-curves are available. The prediction is also made for the same batch of material from which the SN-curves were determined. The tests under a realistic stress-time history for assessing the prediction method are also carried out with specimens from this batch of material. For all of these reasons only this type of FLP can be assessed at all. In short, these are the typical laboratory test programmes which abound in the literature and which have been the

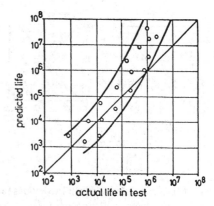

Figure 4: Effect of numbers of cycles to failure on Miner predictions

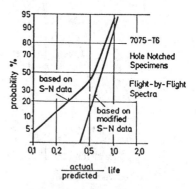

Figure 5: Fatigue life prediction according to a local strain concept [10]

topic of many Ph.D. theses in most western countries.
There is one additional characteristic of scientific FLP: Although the whole problem of fatigue life prediction centers around the problem of "damage", there actually is no damage done if the prediction fails.
— For **industrial FLP** exactly the opposite is true in every respect: All the necessary parameters are either unknown or at best known very inaccurately. The stress-time history for individual structures will always be unknown. (Strictly speaking, all FLP methods which require prior knowledge of the exact stress-time history therefore cannot be used for industrial FLP. Instead, assumptions on a kind of "most damaging" stress-time history have to be used, but what is a "most damaging" and still logical stress-time history?) The load spectrum also may vary considerably from one

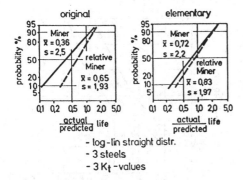

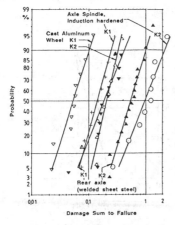

- log-lin straight distr.
- 3 steels
- 3 K$_t$-values

Figure 6: Miner and Relative Miner Predictions [22]

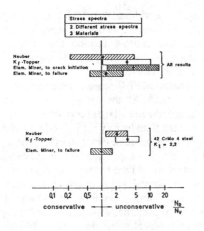

Figure 7: Miner predictions for automobile components and two different spectra

Figure 8: Comparison between fatigue lives predicted by Miner's rule and by the local approach [22]

structure to another, nominally identical one. The basic materials data in the form of SN- or da/dN vs ΔK-curves are usually available only for one type of material, i.e. sheet, and for one batch while the structures in question certainly are manufactured from many batches of material and from sheet, plate, extrusions, forgings etc.

The biggest difference to scientific FLP, however, is the consequence of a wrong prediction. Here the material and immaterial damage may be very large; consider the fatigue failure of an aircraft wing or of 1000 automobile stub axles. All FLP Hypotheses at best give the fatigue life for a probability of survival P_s = 50 %, completely useless for industrial applications. Therefore the relevant safety factors to achieve the much higher P_s-values necessary must be obtained by some other means.

It is typical for **modern** FLP methods that not only the calculating effort is increased but also the experimental effort. Simple SN-data are no longer sufficient, **variable** amplitude or overstrain tests are required, or new so-called "constants" have to be determined whenever the spectrum type changes for which the life is to be predicted.

3 ASSESSMENT OF FLP HYPOTHESES

This is only possible by comparing the predicted fatigue life or crack length with the actual life or crack length in test. That sounds trivial, but it is here that most frequently errors have been made: The prediction method must be assessed against the results of **realistic** tests. By this tests are meant with realistic stress-time histories, realistic stress levels and therefore realistic lives to failure and realistic return periods. With respect to the return period the literature is full of horrible blunders, the worst probably being the large SAE programme [3]: Because of the limited computer capacities available in the participating laboratories at that time (1970), return periods of only 1500 to 4000 cycles were used. The three stress-time histories decided upon were supposed to be typical for automobile components. Because of the short return periods the truncation levels were much too low, all stresses higher than about 40 to 45 % of the maximum to be expected in real service were left out, see Fig. 3. Had the methods developed in this large cooperative programme predicted the fatigue life correctly - none of them did - this only would have proved that the methods would work for a spectrum shape which does not occur in actual service.

Here the many standardized stress-time histories like "Gauss" [4], "Twist" [5], "Falstaff" [6], "Helix-

Felix" [7], "Wash" [8] and so on have their merit, because in every case the truncation levels were based on realistic assumptions. So the solution to this problem is at hand.

Another, and more difficult, problem is that of realistic stress levels and lives to failure. For structures which have to withstand 10^8 cycles or more during their service-life, like trucks, cars, oil rigs etc., this means extremely expensive and time-consuming variable amplitude tests.

However, the assessment of an FLP method at realistic lives and realistic stress levels is absolutely necessary, because while it might be good at 10^6 cycles, this does not prove it is necessarily good at 10^8 cycles. A number of assessments carried out, for example at the LBF and at the IABG, showed Miner's Rule and modern notch root prediction methods to be highly sensitive to the life to failure, because as the stress levels were decreased and the lives therefore increased, more and more cycles fell below the fatigue limit and therefore, at least according to Miner's Rule, do no damage. They **are,** however, damaging and the effect on the prediction is shown in Fig. 4.: For long lives, the predictions become more and more unconservative.

This can be, and has been, corrected by artificially reducing the fatigue limit, as in some modern notch root prediction methods [9, 10], or by using a SN-curve which is a straight line down to a stress amplitude of zero, the so-called elementary SN-curve. This usually results in an improvement in the predictions, see Fig. 5 [10].

Correct and reliable basic data is another requirement, if we want to assess an FLP method. Quite often these are not available and assumptions have to be used, for example on the effect of mean stress on crack propagation. Some authors even use SN-curves taken from standards, although these usually contain safety factors or are lower-bound curves. This procedure may be all right for industrial FLP, but it certainly is not the way to assess the reliability of a method.

Correct basic data are necessary. The importance of using **correct** basic data was shown by Hangartner [11]. The constant amplitude results of Australian full scale tests on F-51 wings of many years ago were plotted incorrectly by the Engineering Society Data Unit [12]. At that time this led to the statement: "Miner's Rule for the F-51 wings is unconservative". If plotted correctly, however, Hangartner states Miner's Rule to be conservative in all cases.

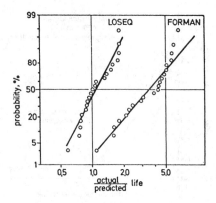

Figure 9: Crack propagation life predictions using the Loseq [23] and Forman [23] models

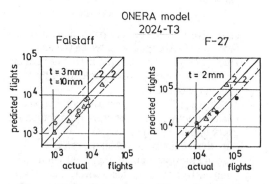

Figure 10: Crack propagation life prediction using the Onera model [25]

4 WHAT IS A GOOD FATIGUE LIFE PREDICTION?

Every author considers his method good, otherwise he would not have published it. The already mentioned SAE programme is a case in point: A number of different notch root prediction methods were presented in [3]. All authors at least gave the impression that their individual methods were "good" or "good enough". Many others, including this author, would consider none of them good enough by a large margin, irrespective of the other criticisms mentioned in chapter 3.

In the literature the following references can be found stating quantitative requirements:
— Buch [13]: All predictions between 0,5 and 2.0
— Gassner [14]: 90 % of predictions between 1,0 and 2,0
— W. Schütz [15]: 90 % of all predictions between 0,6 and 1,6 (s = 1,35)

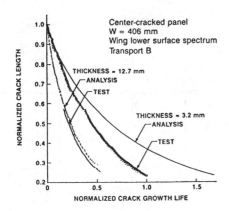

Figure 11: Crack growth comparison between prediction and test for 2324-T39, using the Boeing Model [27]

— J. Schijve [16]: Good: All predictions between 0,66 and 1,5 Acceptable: between 0,5 and 2,0
— Hück [17]: Must correctly predict the slope of the fatigue life curve (Gassner curve) over the relevant range of cycles.
— J. Schijve [18]: A crack propagation hypothesis must correctly predict the crack propagation for every individual flight.

Thus, the requirements became more severe over the years, the most severe being the recent requirement by Schijve. But even the least severe one is very difficult to attain with even the most modern FLP hypotheses. Some of the modern crack propagation hypotheses based on crack closure come near in some, but not all applications, as well as relative methods like Wheeler [19] or relative Miner [20], as it will shown below, but no local strain method, and certainly not Miner's Rule.

Some examples are shown in Fig. 6: Two Miner calculations gave standard deviations of 2,5 and 2,2, much worse than the author's requirement of 1,35. Relative Miner calculations based on the same data gave slightly better, but still not good enough results.

The accuracy of fatigue life predictions using Miner's Rule generally is very poor. In Fig. 7 results of a large programme with actual automobile components are shown [21]: Especially under the typical customer spectrum, damage sums to failure of 0,1 and less occur. In one extreme case a cast aluminum wheel failed at 2 % of the predicted life! The more severe test driver spectrum resulted in higher damage sums to failure for all three components, but still far below 1,0!

The fatigue life can also be predicted, as mentioned before, by the socalled "local approach".

Notwithstanding innumerable (and unfounded) claims in the international literature it has been shown by a large cooperative programm in Germany [22] to be even less accurate than the normal Miner approach, see Fig. 8. This may in part be due to the inaccuracy of Miner's Rule itself which, in the end, is also employed in the local approach, assuming a damage sum at crack initiation of 1,0.

Some assessments of crack propagation hypotheses are shown in Fig. 9 to 11, many more in [26]. The Loseq hypothesis [23] is based on the Dugdale yield model and crack closure; the Onera hypothesis[25], mentioned in Fig. 10, also assumes a ΔK_{eff}-related crack growth. For extensive descriptions of the basic assumptions of these hypotheses, see [26]. In Fig. 11 the recently discoverd effect of sheet thickness on crack propagation under variable amplitudes is shown and compared to a hypothesis developed by Boeing [27]: The thickness effect of 12,7 mm thick plate could be predicted by a correction factor, but not that of 3,2 mm thick sheet.

5 CONCLUSIONS

Fatigue life prediction to crack initiation by the local approach and to complete failure by Miner's Rule are still unreliable and may err by a factor of 10 or more on the unconservative side. The prediction of crack propagation life by several hypotheses based on crack closure is more reliable. However, a lot of effort still is necessary to account for the effect of material thickness or of very long lives and low stresses, in order to fulfill the requirements for a "good" prediction hypothesis.

6 REFERENCES

[1] Miner, M.A.: Cumulative Damage in Fatigue. J.Appl. Mech. 12, 159, 1945
[2] Miner, M.A.: Estimation of Fatigue life with Particular Emphasis on Cumulative Damage. In: Metal Fatigue, McGraw-Hill, New York, 1959.
[3] Wetzel, R.M.: Fatigue Under Complex Loading: Analysis and Experiments. Society of Automotive Engineers. 1977.
[4] Fischer, R., M. Hück. Köbler, H.G. Köbler and W. Schütz: Eine dem stationären Gaußprozeß verwandte Beanspruchungs-Zeit-Funktion für Betriebfestigkeitsversuche. Düsseldorf, VDI-Forschungsberichte, Reihe 5, Nr. 30, 1977.
[5] Schütz, D. H. Lowak, J.B. De Jonge and J. Schijve: Standardisierter Einzelflug-

Belastungsablauf für Schwingfestigkeits-versuche an Tragflächenbauteilen von Transport-flugzeugen. NLR Report TR 73, LBF Bericht No. FB-106, National Aerospace Laboratory, The Netherlands, 1973.

[6] Aicher, W. J. Branger, G. Van Dijk, M. Ertelt, M. Hück, J. De Jonge, H. Lowak, H. Rhomberg, D. Schütz and W. Schütz: Description of a Fighter Aircraft Loading Standard for Fatigue Evaluation "Falstaff". Common Report of F+W Emmen, LBF, NLR, IABG, March 1976.

[7] Edwards, P.R. and J. Darts: Standardized Fatigue Loading Sequence for Helicopter Rotors (Helix and Felix). RAE TR 84084, Royal Aircraft Establishment, Parts 1 and 2, Aug. 1984.

[8] Schütz, W., H. Klätschke, M. Hück and C.M. Sonsino: Standardized Load Sequence for Offshore Structures - Wash 1. Fatigue Fract. Engng. Mater. Struct. Vol. 13, No. 1, pp. 15-29, 1990.

[9] Heuler, P.: Crack Initiation Life Prediction for Variable Amplitude Loading Based on Local Strain Concepts. Rep. No. 40, Dept. Civil Eng., Techn. University Darmstadt, FRG (in German), 1983.

[10] Ekvall, J.C., L. Young and M.D. McMaster: Fatigue Life Prediction Based on Local Effective Stress. Proc. Fatigue '84, Birmingham, 987, 1984.

[11] Hangartner, R.: Correlation of Fatigue Data for Aluminium Aircraft Wing and Tail Structures. Nat. Aeron. Establ. Canada NRC-No. 14555. Aeron. Rep. No. LR-582.

[12] N.N.: Data Sheets-Fatigue. Roy. Aeron. Soc., Vol. 1; Sheet E.02.01, 1962.

[13] Buch, A.: Verification of Fatigue Crack Initiation Life Prediction Results. Techn. Israel Inst. of Techn., Haifa, TAE No. 400, 1980.

[14] Gassner, E.: U_D-Verfahren zur treffsicheren Vor-hersage von Betriebsfestigkeits-Kennwerten nach Wöhler-Versuchen. Materialprüfung 22, 155, 1980

[15] Schütz, W. Lebensdauervorhersage schwingend beanspruchter Bauteile. Proc. DVM-Colloqu. Berlin, FRG, 341, 1980.

[16] Schijve, J.: Fatigue crack growth predictions for variable amplitude and specturm loading. Proc. Fatigue 87, Charlottesville, USA, 1987.

[17] Hück, M.: Gemeinschaftsarbeit Pkw-Industrie/ IABG: Relative Minerregel; Teil 1, IABG-Report TF-2022, 1988.

[18] Siegl, J., J. Schijve and U.H. Padmadinata: Fractographic observations and predictions on fatigue crack growth in an aluminium alloy under Mini-Twist flight-simulation loading. Int. J. Fatigue 13, No. 2 (1991), pp 139-147.

[19] Wheeler, O.E.: Crack Growth under Spectrum Loading. General Dynamics, Rep. No. FZM 5602, 1970.

[20] Schütz, W.: The Fatigue Life under Three Different Spectra – Tests and Calculations. In: AGARD CP 118, 1972.

[21] Hück, M., P. Heuler und J. Bergmann: Gemein-schaftsarbeit Pkw-Industrie/IABG, Teil II, Er-gebnisse der Bauteilversuche. IABG-Report No. TF-2904, 1991.

[22] Buxbaum, O., H. Oppermann, H.-G. Köbler, D. Schütz, Chr. Boller, P. Heuler und T. Seeger: Vergleich der Lebensdauervorhersage nach dem Kerbgrundkonzept und dem Nennspannungskonzept. LBF-Bericht Nr. FB-169, 1983.

[23] Führing, H.: A Model for Nonlinear Fatigue Crack Propagation Prediction with Consideration of Load Sequence Effects (LOSEQ). LBF-Rep. No. FB-162, Darmstadt, 1982.

[24] Forman, R.G., V.E. Kearney and R.M. Engle: Numerical Analysis of Crack Propagation in Cyclic Loaded Structures J. Basic. Eng. 89, 459, 1967.

[25] Baudin, G. and M. Robert: Crack Growth Life-Time Prediction Under Aeronautical Type Loading. Proc. 5th Europ. Conf. on Fracture, Lisbon, Vol. 2, 779, 1984.

[26] Heuler, P. and W. Schütz: A Review of Fatigue Life Prediction Models for the Crack Initation and Crack Propagation Phases. Advances in Fatigue Science and Technology. M. Branco and L.Guerra-Rosa, eds., Nato ASI Series, Kluwer Academic Publishers, 1989.

[27] Miller, M., V.K. Luthra and V.G. Goransson: Fatigue Crack Growth Characterisation of Jet Transport Structures. New Materials and Fatigue Resistant Aircraft Design. D. Simpson, ed. EMAS 1987.

Structural Safety & Reliability, Schuëller, Shinozuka & Yao (eds) © 1994 Balkema, Rotterdam, ISBN 90 5410 357 4

Life distribution of semi-elliptical surface crack propagation in finite plates

Akira Tsurui & Hiroaki Tanaka
Department of Applied Mathematics & Physics, Kyoto University, Japan

Hiroshi Nogami & Koji Ueno
Kyoto University, Japan

ABSTRACT: The random propagation of a semi-elliptical surface crack in a finite plate as well as the probabilistic property of its residual life is investigated. Under some assumptions, the random crack growth law is first expressed as a system of random differential equations. With the aid of a Markov approximation method, the probability distribution of the solution processes is analytically derived, and the result is applied to calculate the residual life distribution. Finally, parameter-sensitivities on the residual life distribution are numerically evaluated.

1 INTRODUCTION

Probabilistic analysis for the random surface crack growth, due to some uncertainties associated with the material's property, the loading process and initial flaws, is desired to assess the safety of pressure vessels or piping systems from a viewpoint of the structural reliability. Especially, the probabilistic Leak-Before-Break assessment in nuclear power plants depends mainly on it [Harris 1981].

In this paper, we investigate the random growth process of semi-elliptical surface cracks by removing some restrictions assumed in the authors' previous study [Tanaka 1989] — (i) the specimen's size is infinite, (ii) the material's property is not random, and (iii) applied loads are of tensile type, — to reflect the engineering reality more precisely in the model.

We first construct a system of crack growth equations by the use of the Paris-Erdogan's law and the Newman-Raju's K-expression [Newman 1984], in which we make a basic assumption that the surface crack is always semi-elliptical until it penetrates the specimen. In Section 3, we extend the growth equations to stochastic differential equations by taking into account the randomness of the material's property as well as the loading. With the aid of a Markov approximation method [Tsurui 1986], we derive a probability distribution of the solution processes, in which we

proposed an approximation method using a mean trajectory of the crack propagation process. The result is utilized, in Section 4, to derive a residual life distribution, which plays an important role in the reliability-based safety assessment. Finally, we show some numerical examples.

2 CRACK GROWTH EQUATIONS

Let us consider a surface crack in a plate of width $2B_1$ and depth B_2, under cyclic tension or bending load, which is shown in Fig.1.

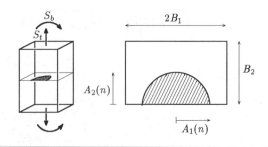

Figure 1. Semi-elliptical surface crack in a finite plate.

In this study, we assume that the crack is always semi-elliptical until it penetrates the plate, and

that its propagation can be decomposed into two directions — surface direction and depth one.

Let $A_1(n)$ and $A_2(n)$ be a half crack length and a depth after n cycles of loading respectively. Making use of the Paris-Erdogan's law as a crack growth equation for each direction, we have

$$\frac{dA_j(n)}{dn} = \varepsilon_{0j}(\Delta K_j)^{2(\lambda+1)} \qquad (j = 1, 2), \qquad (1)$$

where ΔK_j, ε_{0j}, λ represent a stress intensity factor range after n cycles at the point P_j, a crack propagation resistance at the point P_j, a material constant, respectively. Introducing a non-dimensional crack length $X_j = A_j/B_j$ $(j = 1, 2)$, and making use of the well-known Newman-Raju's expression for the stress intensity factors [Newman 1984], we can obtain the following differential equations;

$$\frac{dX_j}{dn} = \varepsilon_j Z_n^{2(\lambda+1)} g_j(X_1, X_2) \qquad (j = 1, 2) \qquad (2)$$

in which $Z_n = \Delta S_t/S_0$ or $\Delta S_b/S_0$ represents a normalized stress amplitude at the n-th cycle (S_0 : reference stress), and ε_j is defined as $\varepsilon_j = \varepsilon_{0j}\,\pi^{\lambda+1} B_2^{\lambda} S_0^{2(\lambda+1)}$. The functions $g_j(X_1, X_2)$ $(j = 1, 2)$ are defined as follows;

$$g_1(X_1, X_2) = B^{*\lambda+2} \left\{ H_1(X_1, X_2) \frac{M(X_1, X_2)}{\sqrt{Q(X_1, X_2)}} \right.$$

$$\left. \times \frac{X_2}{\sqrt{X_1}} g(X_1, X_2) f_w(X_1, X_2) \right\}^{2(\lambda+1)} \qquad (3)$$

$$g_2(X_1, X_2) = \left\{ H_2(X_1, X_2) \frac{M(X_1, X_2)}{\sqrt{Q(X_1, X_2)}} \sqrt{X_2} \right.$$

$$\left. \times f_w(X_1, X_2) \right\}^{2(\lambda+1)} \qquad (4)$$

$$Q(X_1, X_2) = 1 + 1.464 \left\{ \min\left[\frac{B^* X_2}{X_1}, \frac{X_1}{B^* X_2} \right] \right\}^{1.65} \qquad (5)$$

$$M(X_1, X_2) = M_1 + M_2 X_2^2 + M_3 X_2^4 \qquad (6)$$

for $\dfrac{B^* X_2}{X_1} < 1$

$$M_1 = 1.13 - 0.09 \frac{B^* X_2}{X_1} \qquad (7.a)$$

$$M_2 = -0.54 + \frac{0.89}{0.2 + \dfrac{B^* X_2}{X_1}} \qquad (8.a)$$

$$M_3 = 0.5 - \frac{1}{0.65 + \dfrac{B^* X_2}{X_1}} + 14\left(1 - \frac{B^* X_2}{X_1}\right)^{24}, \quad (9.a)$$

for $1 < \dfrac{B^* X_2}{X_1}$

$$M_1 = \sqrt{\frac{X_1}{B^* X_2}}\left(1 + 0.04\frac{X_1}{B^* X_2}\right) \qquad (7.b)$$

$$M_2 = 0.2\left(\frac{X_1}{B^* X_2}\right)^4 \qquad (8.b)$$

$$M_3 = -0.11\left(\frac{X_1}{B^* X_2}\right) \qquad (9.b)$$

$$g(X_1, X_2) = 1.1 + 0.35 X_2^2 \min\left[1, \frac{X_1}{B^* X_2}\right] \qquad (10)$$

$$f_w(X_1, X_2) = \sqrt{\sec\left(\frac{\pi}{2}X_1\sqrt{X_2}\right)} \qquad (11)$$

$$B^* = \frac{B_2}{B_1}, \qquad (12)$$

in which $H_j(X_1, X_2)$ is a correcting function due to bending, which takes on unity for tension (see ref.[Newman 1984] for detail). It should be noted that when $H_2(X_1, X_2)$ takes on a negative value it has to be replaced by zero.

As the finite width correction f_w given by eqn.(11) has a singularity on the curve $X_1\sqrt{X_2} = 1$, the randomization procedure stated in the next section becomes complicated. Thus, to avoid it, we slightly change f_w in the following way;

1134

$$f_w(X_1, X_2)$$

$$= \begin{cases} \sqrt{\sec\left(\dfrac{\pi}{2}X_1\sqrt{X_2}\right)} & (X_1\sqrt{X_2} < x_{fw}) \\ a_1(X_1\sqrt{X_2})^{a_2} & (x_{fw} \le X_1\sqrt{X_2}), \end{cases} \quad (13)$$

where x_{fw} is a constant satisfying $0 < x_{fw} < 1$, and the constants a_1, a_2 are selected so as to make f_w differentiable in the whole domain $\{(X_1, X_2) \,|\, 0 < X_1, 0 < X_2\}$. This correction does not bring us a serious problem, since x_{fw} can be arbitrarily near unity.

3 PROBABILITY DISTRIBUTION OF THE CRACK LENGTH

As is well known, the fatigue crack growth rate scatters because of the material's inhomogeneity even under constant amplitude loading. In this study, to take the material's inhomogeneity into consideration, we let $\varepsilon\, C_{j,n} = \varepsilon_j$ and ε be constant, where $C_{j,n}$ represents a normalized random propagation resistance at the point P_j at the n-th cycle. Without loss of generality, we can assume $E[C_{1,n}] = 1$. According to this procedure, eqn.(2) is rewritten as

$$\frac{dX_j}{dn} = \varepsilon\, C_{j,n} Z_n^{\,2(\lambda+1)} g_j(X_1, X_2) \ (j = 1, 2), \quad (14)$$

which have to be treated as a system of stochastic differential equations, since the coefficients $C_{j,n}$ and Z_n show temporally random variation.

As eqn.(14) has a kind of cross effect term, it is very difficult to treat it analytically. To avoid this difficulty, we introduce a mean trajectory of solution processes. To this end, replacing $C_{j,n}$ by its mean value $E[C_{j,n}]$ and eliminating the time variable n in eqn.(14), we obtain the following differential equation describing the solution's mean trajectory in $\mathbf{X}$ plane;

$$\frac{dX_1}{dX_2} = \frac{1}{\alpha} \cdot \frac{g_1(X_1, X_2)}{g_2(X_1, X_2)}, \quad (15)$$

in which $\alpha = E[C_{2,n}]/E[C_{1,n}]$. With the aid of the solution of eqn.(15) under the condition

$\mathbf{X}(0) = \mathbf{x}_0 = (x_{01}, x_{02})$, denoting $X_1 = q_1(X_2; \mathbf{x}_0)$ and $X_2 = q_2(X_1; \mathbf{x}_0)$, we construct the following new equations;

$$\frac{dX_j}{dn} = \varepsilon\, C_{j,n} Z_n^{\,2(\lambda+1)} \tilde{g}_j(X_j; \mathbf{x}_0) \ (j = 1, 2) \quad (16)$$

$$\tilde{g}_1(X_1; \mathbf{x}_0) = g_1\big(X_1, q_2(X_1; \mathbf{x}_0)\big),$$

$$\tilde{g}_2(X_2; \mathbf{x}_0) = g_2\big(q_1(X_2; \mathbf{x}_0), X_2\big) \quad (17)$$

which do not contain the cross-effect term. It can be considered that eqn.(16) is equivalent to eqn.(14) in the point that it gives the same trajectory as eqn.(14) if the initial condition is identical. In the following, we adopt eqn.(16) instead of eqn.(14) as a simplified model for the random crack growth.

Let $W(\mathbf{x}, n \,|\, \mathbf{x}_0)$ be a conditional probability distribution function of the solution processes, that is,

$$W(\mathbf{x}, n \,|\, \mathbf{x}_0) = \Pr[\mathbf{X}(n) \le \mathbf{x} \,|\, \mathbf{X}(0) = \mathbf{x}_0], \quad (18)$$

and $w(\mathbf{x}, n \,|\, \mathbf{x}_0)$ be its density. By the use of the Markov approximation method [Tsurui 1986], the generalized Fokker-Planck equation describing a temporal variation of the density $w(\mathbf{x}, n \,|\, \mathbf{x}_0)$ can be obtained as follows;

$$\frac{\partial w}{\partial n} = -\sum_{j=1}^{2} \beta_j(n) \frac{\partial}{\partial x_j}\big\{\tilde{g}_j(x_j)w\big\}$$

$$-\sum_{j=1}^{2} \gamma_{jj}(n) \frac{\partial}{\partial x_j}\left\{\tilde{g}_j(x_j)\frac{d\tilde{g}_j(x_j)}{dx_j}w\right\}$$

$$+\sum_{j=1}^{2}\sum_{k=1}^{2} \gamma_{jk}(n) \frac{\partial^2}{\partial x_j \partial x_k}\big\{\tilde{g}_j(x_j)\tilde{g}_k(x_k)w\big\}, \quad (19)$$

where

$$\beta_j(n) = \varepsilon\, E\big[C_{j,n} Z_n^{\,2(\lambda+1)}\big] \quad (j = 1, 2) \quad (20)$$

$$\gamma_{jk}(n) = \varepsilon^2 \int_{-\infty}^{0} \boldsymbol{K} \Big[C_{j,n} Z_n{}^{2(\lambda+1)},$$

$$C_{k,n+n'} Z_{n+n'}{}^{2(\lambda+1)} \Big] dn' \quad (j,\,k = 1,\,2) \quad (21)$$

$$\boldsymbol{K}[A, B] \equiv E[AB] - E[A]E[B]. \quad (22)$$

It should be noted that the death point has been taken into account in deriving eqn.(19). Thus, the normalizing condition for the solution of eqn.(19) has to be extended to the following form;

$$\int_0^\infty \int_0^\infty w(\boldsymbol{x}, n \mid \boldsymbol{x}_0) dx_1 dx_2 + \mathrm{P_D}(n;\, \boldsymbol{x}_0) = 1 \quad (23)$$

in which $\mathrm{P_D}(n;\, \boldsymbol{x}_0)$ represents the probability that $\boldsymbol{X}(n)$ lies in the death point D that corresponds to the state of infinite crack length.

According to ref.[Tanaka 1989], the exact solution of eqn.(19) can be analytically obtained as follows;

$$w(\boldsymbol{x}, n \mid \boldsymbol{x}_0) = \frac{1}{4\pi \tilde{g}_1(x_1) \tilde{g}_2(x_2) \sqrt{G(n)}}$$

$$\times \exp \left[-\frac{f(\boldsymbol{x}, n)}{4G(n)} \right] \quad (24)$$

$$f(\boldsymbol{x}, n) = G_{22}(n)\, \xi_1(x_1, n)^2 - \big\{ G_{12}(n) + G_{21}(n) \big\}$$

$$\times \xi_1(x_1, n)\, \xi_2(x_2, n) + G_{11}(n)\, \xi_2(x_2, n)^2 \quad (25)$$

$$G(n) = G_{11}(n) G_{22}(n) - \big\{ G_{12}(n) + G_{21}(n) \big\}^2 / 4 \quad (26)$$

$$G_{jk}(n)r = \int_0^n \gamma_{jk}(n') dn' \quad (j,\,k = 1,\,2) \quad (27)$$

$$\xi_j(x_j, n) = \int_{x_{0j}}^{x_j} \frac{dx_j{}'}{\tilde{g}_j(x_j{}')} - \int_0^n \beta_j(n') dn' \, (j = 1,\,2). \,(28)$$

To determine the uncertainty factor $G(n)$, we make the following assumptions with respect to the random processes $C_{j,n}$ and Z_n [Tanaka 1987].

(i) $C_{j,n}$ ($j = 1,\,2$) and Z_n are statistically independent of one another.

(ii) $Z_n{}^{2(\lambda+1)}$ is a stationary process whose correlation decays exponentially as the time difference increases, that is,

$$\frac{\boldsymbol{K}\Big[Z_n{}^{2(\lambda+1)}, Z_{n+n'}{}^{2(\lambda+1)} \Big]}{\boldsymbol{K}\Big[Z_n{}^{2(\lambda+1)}, Z_n{}^{2(\lambda+1)} \Big]} = \exp\left[-\frac{|n'|}{n_z} \right],$$

$$(29)$$

where n_z is a constant called the correlation time.

(iii) $C_{j,n}$ ($j = 1,\,2$) are spatially stationary with the exponentially decaying correlation, which results in the following relationship;

$$\frac{\boldsymbol{K}\big[C_{j,n}, C_{j,n+n'} \big]}{\boldsymbol{K}\big[C_{j,n}, C_{j,n} \big]}$$

$$= \exp\left[-\frac{|n'| \tilde{E}[dX_j/dn]}{\xi_{0j}} \right] \quad (j = 1,\,2), \quad (30)$$

where $\tilde{E}[dX_j/dn]$ represents a kind of typical value for the crack growth rate, and ξ_{0j} is a constant corresponding to a spatial correlation distance.

Under these assumptions, we obtain the following expressions for β_j ($j = 1,\,2$) and $G_{jk}(n)$ ($j,\,k = 1,\,2$);

$$\beta_j(n) = \varepsilon M_{cj} M_z \quad (j = 1,\,2) \quad (31)$$

$$G_{jj}(n) = \frac{\varepsilon\, M_z\, \sigma_{cj}{}^2\, \xi_{0j}}{M_{cj}} \int_0^n \frac{dn'}{\tilde{E}[dX_j/dn]}$$

$$+ \varepsilon^2 M_{cj}{}^2 \sigma_z{}^2 n_z\, n \quad (j = 1,\,2) \quad (32)$$

$$G_{12}(n) = G_{21}(n) = \varepsilon^2 M_{c1} M_{c2} \sigma_z{}^2 n_z\, n. \quad (33)$$

$$M_{cj} = E[C_{j,n}], \quad \sigma_{cj}{}^2 = Var[C_{j,n}] \quad (j = 1,\ 2) \quad (34)$$

$$M_z = E\left[Z_n{}^{2(\lambda+1)}\right], \quad \sigma_z{}^2 = Var\left[Z_n{}^{2(\lambda+1)}\right] \quad (35)$$

In deriving eqn.(33), the so-called cross effect term [Tanaka 1989] between two kinds of randomness has been neglected.

The typical growth rate $\tilde{E}[dx_j/dn]$ can be approximately evaluated as followings [Tanaka 19887, Tsurui 1991];

$$\tilde{E}\left[\frac{dx_j}{dn}\right] = \varepsilon\, M_{cj}\, M_z\, g_j\!\left(\hat{X}_1(n),\ \hat{X}_2(n)\right)(j = 1,\ 2)\ (36)$$

where $\hat{X}_j(n)$ is the solution of a kind of averaged growth equations

$$\frac{d\hat{X}_j}{dn} = \varepsilon\, M_{cj}\, M_z\, g_j(\hat{X}_1,\ \hat{X}_2) \quad (j = 1,\ 2). \quad (37)$$

4 RESIDUAL LIFE DISTRIBUTION

Suppose that the plate with a surface crack breaks when the crack growth process $\boldsymbol{X}(n)$ arrives at the region

$$\Omega_{\mathrm{F}} = \left\{(X_1, X_2)\,|\,X_2 \geq l(X_1)\right\} \quad (38)$$

for the first time. Figure 2 illustrates Ω_{F} and its boundary curve $C : X_2 = l(X_1)$. The residual life of the plate, say N, is defined as the number of cycles that the process $\boldsymbol{X}(n)$ arrives at Ω_{F} for the first time.

The residual life N is a random variable because of the random nature of $\boldsymbol{X}(n)$. We denote its probability distribution function under the condition of $\boldsymbol{X}(0) = \boldsymbol{x}_0$ as $H(n\,|\,\boldsymbol{x}_0, \Omega_{\mathrm{F}})$. It is then related to $w(\boldsymbol{x}, n\,|\,\boldsymbol{x}_0)$ as follows;

$$H(n\,|\,\boldsymbol{x}_0, \Omega_{\mathrm{F}}{}^c) = \Pr\left[\,N > n\,\right]$$

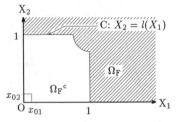

Figure 2. Limit state function and break zone.

$$\simeq 1 - \iint_{\Omega_{\mathrm{F}}{}^c} w(\boldsymbol{x}, n\,|\,\boldsymbol{x}_0)dx_1 dx_2, \quad (39)$$

since we can neglect the probability that the crack length decreases [Tsurui 1986, Tanaka 1989].

5 NUMERICAL EXAMPLES AND SENSITIVITY ANALYSES

In this section, supposing the case of constant amplitude tensile stress, we execute numerical calculations of the residual life distribution given by eqn.(39). For convenience, we introduce a new time variable $\tau = \varepsilon M_{c1} M_z n$ instead of n.

Under the assumption that the stress amplitude Z_n is constant, G_{12} and G_{21} given by eqn.(33) becomes zero. This leads to the following crack length distribution function;

$$W(\boldsymbol{x}, \tau\,|\,\boldsymbol{x}_0) = \int_0^{x_1}\!\!\int_0^{x_2} w(\boldsymbol{x}, \tau\,|\,\boldsymbol{x}_0)dx_1 dx_2$$

$$= W_1(x_1, \tau\,|\,\boldsymbol{x}_0)W_2(x_2, \tau\,|\,\boldsymbol{x}_0) \quad (40)$$

$$W_j(x_j, \tau\,|\,\boldsymbol{x}_0) = \Phi\left[\frac{\xi_j(x_j, \tau)}{\sqrt{2G_{jj}(\tau)}}\right] \quad (j = 1,\ 2), \quad (41)$$

where $\Phi(\cdot)$ is the standardized normal distribution function. Equation (40) indicates that the surface length is statistically independent of the depth if the stress amplitude is not random.

To determine the break zone Ω_{F}, we assume that the plate breaks if at least one of the following

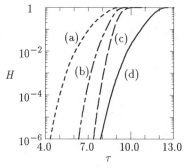
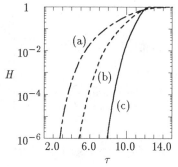

(i) Sensitivity on initial crack states
    ($\xi_{01} = 0.001$).
(a) $(x_{01}, x_{02}) = (0.10, 0.05)$,
(b) $(0.07, 0.07)$,
(c) $(0.05, 0.10)$,
(d) $(0.05, 0.05)$.

(ii) Sensitivity on correlation distances
    ($x_{01} = x_{02} = 0.05$).
(a) $\xi_{01} = 0.015$,
(b) 0.005,
(c) 0.001.

Figure 3. Residual life distributions.

three events occurs; (i) the crack penetrates the plate's width, (ii) the crack penetrates the plate's thickness, and (iii) the crack area becomes greater than a certain fraction, say S_C, of the plate's area. The event (iii) is based upon the so-called net-section stress criterion. Hence, the non-break zone $\Omega_F{}^c$ is given as

$$\Omega_F{}^c = \left\{ (X_1, X_2) \,|\, 0 < X_1 < 1, 0 < X_2 < 1 \right.$$

$$\left. \text{and } X_1 X_2 < S_C \right\}. \tag{42}$$

In the numerical calculations, we set the values of parameters as

$$x_{f_w} = 0.6, \lambda = 0.5, \alpha = 0.9^{-2(\lambda+1)},$$

$$B^* = B_2/B_1 = 0.5, \sigma_{c1} = 0.3, \sigma_{c2} = 0.3, S_C = 0.6$$

and choose (ξ_{01}, ξ_{02}) so as to satisfy $\xi_{01} = B^* \xi_{02}$ because of the material's isotropy. Figure 3 (i) examines the sensitivity on the initial state (x_{01}, x_{02}), which indicates that the residual life of a flatter initial crack is shorter in the high-reliability range compared with a deep crack of same cracked area. On the other hand, in Fig.3 (ii) the sensitivity on the correlation distance ξ_{01} is investigated. We can see that the residual life more and more scatters as ξ_{01} increases. It follows from this tendency that the residual life in the high-reliability range becomes drastically shorter.

6 CONCLUSIONS

In this paper, we have theoretically investigated the random properties associated with the surface crack propagation, and derived an analytical expression of the residual life distribution function as well as the crack length probability density function through a suitable approximation technique using a mean trajectory. The numerical results show that the reliability of the cracked component degrades as the initial crack becomes flatter.

It should be emphasized that the obtained results can be easily evaluated through a simple numerical integration, which will be of great advantage in the reliability-assessment procedure in pressure vessels of piping systems.

REFERENCES

Harris, D. O., Lim, E. Y. and Dedhia, D. D. 1981 Probability of Pipe in the Primary Coolant Loop of a PWR Plant. *NUREG/CR-2189* 5: 43-49.

Newman, Jr. J. C. and Raju, I. S. 1984.Stress-Intensity Factor Equations for Cracks in Three-Dimensional Finite Bodies Subjected to Tension And Bending Loads. *NASA Technical Memorandum* No.85793.

Tanaka, H. and Tsurui, A. 1987. Reliability Degradation of Structural Compo-

nents in The Process of Fatigue Crack Propagation Under Stationary Random Loading. *Eng. Frac. Mech.* 27: 501-516.

Tanaka, H. and Tsurui, A. 1989.Stochastic Propagation of Semi-elliptical Surface Cracks. *Proc. of ICOSSAR '89* II: 1507-1514.

Tanaka, H. and Tsurui, A. 1989.Random Propagation of A Semi-Elliptical Surface Crack as A Bivariate Stochastic Process. *Eng. Frac. Mech.* 33: 787-800.

Tsurui, A. and Ishikawa, H. 1986.Application of The Fokker-Planck Equation to A Stochastic Fatigue Crack Growth Model. *Structural Safety* 4: 15-29.

Tsurui, A., Nogami, A. and Tanaka, H. 1991.On The Life Distribution of Semi-Elliptical Surface Crack Propagation in Finite Size Specimens. *Proc. of JCOSSAR '91*: 571-574 (in Japanese).

Structural Safety & Reliability, Schuëller, Shinozuka & Yao (eds) © 1994 Balkema, Rotterdam, ISBN 90 5410 357 4

An experimental study and reliability analysis of fatigue crack propagation under random loading

W. F. Wu, C. S. Shin & J. J. Shen

Department of Mechanical Engineering, National Taiwan University, Taipei, Taiwan

ABSTRACT: Mechanical components are frequently subjected to fluctuated loads of random nature. In order to predict the fatigue life and structural reliability of such components, an analytical model is introduced in the present paper. The model focuses on the prediction of the component's fatigue crack growth behavior under different random loading histories which, however, possess the same statistical nature. Not only the average of all fatigue crack growth curves but also their statistical variability can be predicted by this model. The latter is directly or indirectly related to the structural reliability. Experimental work is carried out to verify the prediction model. It is found that by applying Elber's crack closure model and using an equivalent constant-amplitude loading concept one can obtain a fairly good and conservative prediction. It is also found that fatigue crack growth rate is influenced by the m-th statistical moment of the probability density function of the random loading although different density functions all have the same first and second statistical moments.

1 INTRODUCTION

Owing to its importance, fatigue crack growth under random loading has been investigated for years. Most researchers were interested in finding a way to predict the fatigue crack growth or fatigue life of a component under random loading (Chang & Hudson 1981, Shin & Au-Yang 1983). Experimental work has been performed to verify the applicability of the prediction models. Unfortunately, it seems that no unique conclusion with regard to the prediction model has been drawn so far. One possible reason for this result is the conceptual gap between researchers majoring in experimental fatigue work and those experienced in theoretical random signal analysis. For example, the definition of random loading may be different among different researchers. Some statistical treatments in obtaining a characteristic value to characterize a random loading history may violate the basic probability law. To overcome such difficulties, joint theoretical and experimental efforts from two sides of researchers become important in unraveling the problem.

Because of the scatter exists in fatigue data, elementary statistics has been employed to deal with the fatigue data since the sixties. Through the quick development of structural safety and reliability theories as well as the need in aerospace and nuclear industries, more sophisticated probabilistic or stochastic methods have been introduced to study the reliability issues related to fatigue and fracture of materials and structures (Yao, et al. 1986, Tanaka & Tsurui 1987, Yang & Donath 1983). However, many of the theoretical propositions still need further experimental verification, but the experimental work is usually tedious and time consuming.

In the present paper, an experimental study of the fatigue crack growth under random loading is introduced and some already obtained data is reported. A theoretical model which is primarily based on Huang and Hancock's proposition (1989) is employed to analyze the data and study the reliability related problems of the specimens. Some conclusions are drawn from the present analysis and future work is pointed out for further investigation at the end of the present paper.

2 PREDICTION MODEL

Many of the fatigue crack growth analyses are based on the following Paris law,

$$\frac{da}{dN} = c(\Delta K)^m \tag{1}$$

where a is the crack size, N is the number of stress cycles, c and m are material constants which are usually determined from experimental result, and ΔK is the stress intensity factor range depending on the stress range as well as the geometry of the specimen. For a CT specimen used in our experiment, according to ASTM's standard test method, the stress intensity factor range is

$$\Delta K = \frac{\Delta P}{B\sqrt{W}} g\left(\frac{a}{W}\right) \tag{2}$$

where ΔP is the range of the load, B and W are the thickness and width of the specimen, and $g(a/W)$ is the geometry factor defined as

$$g\left(\frac{a}{W}\right) = \left[\frac{2 + \frac{a}{W}}{\left(1 - \frac{a}{W}\right)^{3/2}}\right]$$

$$\times \left[0.886 + 4.64\left(\frac{a}{W}\right) - 13.32\left(\frac{a}{W}\right)^2\right.$$

$$\left. +14.72\left(\frac{a}{W}\right)^3 - 5.6\left(\frac{a}{W}\right)^4\right] \tag{3}$$

For constant amplitude loading, Paris law leads to accurate results for most situations. For variable amplitude loading, the use of Paris law is not very appropriate because of crack growth retardation or acceleration results from the interaction among neighboring stresses. Many propositions based on residual stresses, strain hardening, crack tip blunting or crack tip branching have been proposed to explain such phenomena. Among them, the crack closure effect proposed by Elber (1870) has been recognized by many researchers. According to Elber's idea, crack growth occurs only when the applied load is greater than a crack opening load P_{open} and, hence, Paris law should be modified to be

$$\frac{da}{dN} = c'(\Delta K_{eff})^{m'} \tag{4}$$

where c' and m' are material constants, and ΔK_{eff} is the effective stress intensity factor range. For a CT specimen,

$$\Delta K_{eff} = \frac{\Delta P_{eff}}{B\sqrt{W}} g\left(\frac{a}{W}\right) \tag{5}$$

where

$$\Delta P_{eff} = \begin{cases} P_{max} - P_{open}, & \text{if } P_{min} \leq P_{open} \\ P_{max} - P_{min}, & \text{if } P_{min} \geq P_{open} \\ 0, & \text{if } P_{max} \leq P_{open} \end{cases} \tag{6}$$

in which P_{open} is the opening load which is obtained from experimental result. This crack opening load is, in general, related to the size of the plastic zone near the crack tip.

Based on either Paris law or crack closure model, the fatigue crack growth curve under a given load history can be predicted once an initial crack size is specified. In the present study, different load histories are generated from a given type of random loading. These load histories are different but have the same statistical quantities. Fatigue crack growth curves are determined for each of the load histories, and we are interested in the mean growth curve as well as the variability behavior among these curves. The latter is directly related to the reliability estimation of the structural components.

During the fatigue crack growth, many parameters in Eqs. (1) and (4) in addition to the applied random load can be considered random from a probabilistic fracture mechanics point of view. In the present analysis, based on real physical judgments, the material constant c (or c') and the initial crack size a_0 are considered random in addition to the load amplitude ΔP. The mean values and variances of these variables are denoted by μ_c, μ_{a_0}, μ_P and V_c, V_{a_0}, V_P, respectively. They are assumed to be known in advance from either experimental result or analytical assumptions. We are now interested in predicting the statistics or probability distribution of the crack sizes after a given number of random loading cycles is applied. We are also interested in finding the statistics or probability distribution of the numbers of loading cycles when a given crack size is specified in advance. For these purposes, the following derivations are performed. Some of the derivations are referred to the work by Huang and Hancock (1989).

Consider the case of crack closure model. From Eqs. (4) and (5), the increment of crack size at the n-th cycle, Δa_n, can be derived as

$$\frac{\Delta a_n}{g^{m'}(a_{n-1}/W)} = c'_n \left(\frac{\Delta P_{eff,n}}{B\sqrt{W}}\right)^{m'} \tag{7}$$

For a given number of loading cycles, N, the crack size can be obtained by the direct summation of all increments starting from an initial size as we did before. Since the increments are small, the summation applied to the LHS of

Eq. (7) can be replaced by an integration. If we let

$$I = \int_0^a \frac{dx}{g^{m'}(x/W)} \qquad (8a)$$

and

$$I_0 = \int_0^{a_0} \frac{dx}{g^{m'}(x/W)} \qquad (8b)$$

then from Eq. (7), we obtain

$$I = \sum_{n=1}^{N} c_n'(\frac{\Delta P_{eff,n}}{B\sqrt{W}})^{m'} + I_0 \qquad (9)$$

which determines the general shape of the crack growth curves. It is interesting to note that c_n' in Eq. (9) has replaced c' in Eq. (7) to denote a random outcome of the material constant for each cycle. The other two random parameters are $\Delta P_{eff,n}$ and I_0. The latter is a function of a_0. Since N is usually very large in the fatigue crack growth study, from central limit theory, the probability density function of the summation term at the RHS of Eq. (9) approaches a normal distribution whose mean and variance equal N times the mean and variance of the original population. By using the first-order reliability method (FORM), the following mean value and variance of I can be derived

$$E[I] = N\mu_{c'}(\frac{\mu_{P_{eff}}}{B\sqrt{W}})^{m'} + \int_0^{\mu_{a_0}} \frac{dx}{g^{m'}(x/W)} \qquad (10)$$

$$V[I] = N\left\{(\frac{\mu_{P_{eff}}}{B\sqrt{W}})^{2m'} V_{c'}\right.$$
$$+ \left[m'\mu_{c'}(\frac{\mu_{P_{eff}}}{B\sqrt{W}})^{m'-1} \frac{1}{B\sqrt{W}}\right]^2 V_{P_{eff}}\right\}$$
$$+ \frac{V_{a_0}}{g^{2m'}(\mu_{a_0}/W)} \qquad (11)$$

where μ and V represent the mean and the variance of the associated quantities respectively, and ΔP_{eff} is replaced by P_{eff} for simplicity. From statistical maximum entropy theory, it is reasonable to assume that I is normal distributed. Then, from Eq. (8), the probability distribution of the crack size can be found through a simple random variable transformation from I to a.

To find the probability distribution of numbers of cycle for a crack to reach the given crack size, it can be shown from Eq. 9 that $AN = I - I_0$ where A is the average of the N quantities within the summation sign of Eq. (9), i.e.

$$A = \frac{\sum_{n=1}^{N} c_n'(\frac{\Delta P_{eff,n}}{B\sqrt{W}})^{m'}}{N} = \frac{\sum_{n=1}^{N} S_n}{N} \qquad (12)$$

In the above equation, each of N quantities within the summation sign has the same distribution as a random variable S. The mean value and standard deviation of S, i.e. μ_S and σ_S can be found by the FORM again.

From Eq. (12) and a variation of the central limit theorem, sometimes described as the Lindeberg-Lavy theorem, we know that A has a normal distribution with a mean value μ_S and a standard deviation $\sigma_S/\sqrt{N}$. If the initial crack size is given, the distribution of cycles, $p(N)$, to reach a given crack size can be obtained through a transformation of random variable from A to N according to $AN = I - I_0$. The probability density of N is then found to be

$$p(N) = \frac{I - I_0}{\sqrt{2\pi}\sigma_S N^{3/2}} \exp\left[-\frac{\mu_S^2}{2\sigma_S^2} \frac{\left(\frac{I-I_0}{\mu_S} - N\right)^2}{N}\right] \qquad (13)$$

If a_0 is also a random variable, then the distribution function of N can also be found approximately by applying the FORM and a Gaussian distribution assumption (Wu, et al. 1991).

Before a quantitative reliability analysis can be performed, the failure mode has to be defined in advance. In the probabilistic fracture mechanics analysis, failure of a component can be defined as the growth of an existing crack size to reach a pre-specified critical crack length or depth. Another failure mode which is also used frequently is the brittle fracture of the component. Since our fatigue test is carried out until fracture occurs, the latter failure mode is adopted in the present analysis. For this failure definition, the reliability of the component after n stress cycles can be written as $R_n = $ Prob. $(K_n < K_C)$ where K_C is the fracture toughness which can be considered as a random variable. The other random variable K_n is the stress intensity factor at the n-th stress cycle which is, in turn, a function of crack size and the applied maximum load. The reliability of the specimen at a specified loading cycle can then be calculated approximately.

3 EXPERIMENT

In order to verify the accuracy of the fatigue crack growth and fatigue reliability models discussed in the preceding section, experimental work is performed in the present study.

The experimental setup is illustrated in Fig. 1 which consists of a dynamic test system, a crack closure measurement system, a crack length measurement system, a spectral analysis system, and a computer control system. The procedure of a typical test begins with the digital generation of a series of random loads based on a given spectral density function or a given probability distribution function. The series of random loads is then converted to analog signal through an arbitrary waveform function generator. The controller controls the dynamic test machine and input the analog random load to the specimen. To check whether the input load is the real random load we designed, the peak and trough values of each load cycle are recorded and analyzed by a spectral analyzer. During the dynamic fatigue test process, the crack length and crack opening load are measured and recorded at a certain time interval. The compliance method is used for the measurement of the crack opening load while a direct current potential drop (DCPD) method is used for the measurement of the crack length. The calibration as well as the accuracy of both methods used in this study are recorded in a report (Wu, et al. 1991). Additional descriptions for the experimental work can also be found in the same report.

The specimens used in the present experimental study are compact tension specimens made of 4340 steel. The dimensions of the specimens are prepared according to ASTM standard. The chemical compositions and mechanical properties of the specimens are documented in Wu et al. (1991).

The original design for the digital generation of random load are based on the superposition of stationary random fluctuation components to a given mean load. The random fluctuations are simulated from a given narrow-band power spectral density function. However, owing to the limited capacity of the arbitrary waveform generator we have, a Rayleigh probability density function is assumed for the distribution of both peaks and troughs of the random fluctuation load. It is assumed that a randomly generated trough follows an independently generated peak at a constant frequency. The situation resembles a narrow-band stationary random process which has a theoretical Rayleigh distribution for the peaks and troughs, and a very narrow band-width in the frequency domain. The generation of a Rayleigh distributed

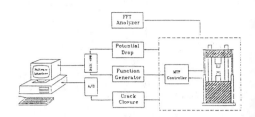

Fig. 1 Illustration of the experimental setup

random quantity can be found in a standard applied probability textbook or performed directly by the use a standard statistical software package such as IMSL.

In order to check the influence of the probability distribution of the random loading on the fatigue crack growth behavior, uniform, triangular as well as normal probability density functions are generated in addition to Rayleigh probability density function. Since the mean value and variance are used extensively in our previous derivations, the first- and second-order statistic quantities for all probability density functions are selected to be the same as those of the Rayleigh density function. The fatigue crack growth and fatigue life under differently distributed random load are then compared.

4 RESULT

Three different load ratios ($r = 0.1, 0.3$ and 0.5) are used in the constant-amplitude load test. Both crack length and crack opening load are measured. The growth rate versus ΔK curve is plotted and the result shows that load ratio has little influence on the growth rate for the material we currently use, which coincides with the conclusion made by many other researchers. Regression analysis is performed and the following values are found for the parameters needed in applying Paris law in our probabilistic crack growth analysis: $m = 3.752$, $\mu_c = 1.263 \times 10^{-9}$ and $V_c = 9.009 \times 10^{-20}$. If the crack opening load is taken into consideration, the growth rate versus ΔK_{eff} curve can then be plotted and the result gives us $m' = 3.557$, $\mu_{c'} = 6.794 \times 10^{-9}$ and $V_{c'} = 5.060 \times 10^{-18}$.

Since probability density functions are specified for the random load we generated, the statistic quantities regarding to the load can

easily be calculated analytically without applying the elementary statistics to the cycle-by-cycly counted load history as usually done by other researchers. Tremendous computer time can be saved by doing so. In the present study, it is found that the analytically obtained loading statistics are very close to those obtained from cycle-by-cycly counted real applied load.

Crack opening loads are measured from time to time during our fatigue test under random loading. However, it is found that the average crack opening load is below the mean value of the minimum load we applied. Therefore, we have $P_{eff} = P_{max} - P_{min}$ in Eq. (6). The statistic quantities related to the applied random loading can be calculated accordingly.

In the present study, nine tests are performed for Rayleigh distributed random loading, six for uniformly distributed random loading, and four for other distributed random loading. Prediction for the fatigue crack growth curves is made for each type of the random loading. It is found that Elber's closure model gives us more accurate result than that obtained from Paris law. It is also found that the prediction based on the discussed method as proposed by Huang and Hancock is not very accurate but on the conservative side. To improve the accuracy of the prediction, the RMn effective load defined as $\mu_{P_{eff}} = (E[P_{max}{}^{n}])^{1/n} - (E[P_{min}{}^{n}])^{1/n}$ is introduced to replace the originally defined effective load. In this equation, $n = 2, 3, \ldots,$ or $m(m')$ which corresponds to the RMS, RMC, ..., or RMM value used by some researchers. The result shows that RMC and RMM values give us the most accurate predictions. Typical results regarding to the variability and reliability of the fatigue crack growth behavior for the case of Rayleigh distributed random loading are shown in Figs. 2 to 4.

Comparison among the mean growth curves under random loading of different probability density functions shows that, under the same first- and second-order statistics, the crack growth faster for random loading which has higher RMM value. This conclusion coincides with the effective load concept defined previously.

5 CONCLUSIONS

The present study shows that Elber's crack

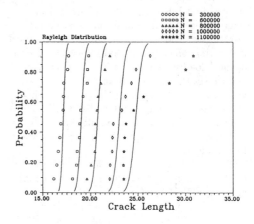

Fig. 2 Distributions of crack lengths

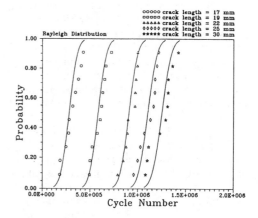

Fig. 3 Distributions of numbers of cycle

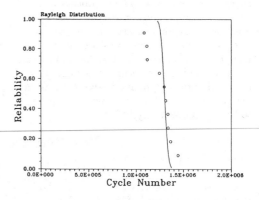

Fig. 4 Reliability against fracture

closure model gives us more accurate result than that obtained from Paris fatigue crack growth law. Elber's model emphasizes the load interaction effect by taking into account a crack opening load which results in the effective load and causes the crack to propagate. However, when and how to measure the crack opening loads experimentally and how to include the random opening load in the reliability analysis still remains problems for a component subject to random loading.

Although the effective load defined based on the mean value of the random load is compatible with FORM used in the reliability analysis, the prediction thus obtained is not accurate enough. On the contrary, the RMM and RMC values give us more accurate predictions even though they do not possess strong probability meaning. More sophisticated and probabilistically meaningful prediction models are therefore needed, at least for the present case which, in its governing fatigue crack growth equation, involves a highly nonlinear function of random crack size. How to make the compromise between a more accurate reliability analysis and an easier numerical calculation becomes a major consideration as many other problems we have faced in the study of structural safety and reliability.

The random loads employed in the present study ignore the correlation between neighboring loading cycles. Although similar approaches have been used previously in performing fatigue tests under random loading conditions, it has been pointed out that this kind of loading is unrealistic and the result thus obtained deviates from that exists in the real situation. To remedy this shortcoming, in our future study, sample functions of a random process will be generated from a given power spectral density function of the applied load. Either time series method or spectral decomposition method can be applied to achieve this purpose. How to let the generated random signal be accepted by the slowly driven dynamic testing machine in performing the fatigue tests remains a problem to be solved eventually.

ACKNOWLEDGEMENT

The research is partially supported by the National Science Council of the Republic of China under Grant NSC79-0401-E002-28. The writers are grateful for this support.

REFERENCES

Chang, J.B. & C.M. Hudson 1981. *Methods and models for predicting fatigue crack growth under random loading*. ASTM STP 748.

Elber, W. 1970. Fatigue crack closure under cyclic tension. *Engineering Fracture Mechanics* 2:37-45.

Huang, X. & J.W. Hancock 1989. A reliability analysis of fatigue crack growth under random loading. *Fatigue and Fracture of Engineering Materials and Structures* 12:247-258.

Shin, Y.S. & M.K. Au-Yang 1983. *Random fatigue life prediction*. New York: ASME Publication.

Tanaka, H. & A. Tsurui 1987. Reliability degradation of structural components in the process of fatigue crack propagation under stationary random loading. *Engineering Fracture Mechanics* 27:501-516.

Wu, W.F., C.S. Shin, W.H. Huang & J.J. Shen 1991. Reliability analysis of fatigue crack propagation under random loading. Technical Report, Department of Mechanical Engineering, National Taiwan University, Taipei, Taiwan.

Yang, J.N. & R.C. Donath 1983. Statistical fatigue crack propagation in fastener holes under spectrum loading. *AIAA Journal of Aircraft* 20:1028-1032.

Yao, J.T.P., F. Kozin, Y.K. Wen, J.N. Yang, G.I. Schuëller, & O. Ditlevsen 1986. Stochastic fatigue, fracture and damage analysis. *Structural Safety* 3:231- 267.

Structural Safety & Reliability, Schuëller, Shinozuka & Yao (eds) © 1994 Balkema, Rotterdam, ISBN 90 5410 357 4

A stochastic crack growth analysis for aircraft structures

J.N.Yang
Department of Civil Engineering, University of California, Irvine, Calif., USA

S.D.Manning & J.M.Norton
Lockheed/Fort Worth Company, Tex., USA

ABSTRACT: A stochastic crack growth analysis (SCGA) method, based on a deterministic initial flaw size, for predicting statistical crack growth accumulations in metallic aircraft structures is presented and validated. This method is an enhancement of the deterministic crack growth approach for durability and damage tolerance analyses. It can be implemented using a single initial flaw size, deterministic crack growth analysis results and an estimate of the crack growth life dispersion. A full-scale fighter wing component is used to validate the stochastic crack growth analysis (SCGA) method. The extent of damage predictions correlated reasonably well with actual tear-down inspection results for a wide range of crack sizes. It was found that the extent of damage is more sensitive to variations in stress level than to comparable variations in crack growth life dispersion.

1 INTRODUCTION

A stochastic crack growth analysis (SCGA) method developed by Yang, Manning, et al (1983, 1987, 1990) is useful for predicting the statistical crack growth accumulation in metallic aircraft structures. Based on a deterministic initial flaw size, this method has been extended and demonstrated for reliability centered maintenance analysis [Manning, Yang, Pretzer, 1992] and for the evaluation of structural maintenance, supportability and maintainability requirements in terms of risk [Manning, Yang, Welch, 1992; Yang, Manning, 1992]. The SCGA method has been validated using fractographic results for dog-bone specimens. However, it has not been validated for the durability and damage tolerance analyses of full-scale aircraft structures using tear-down inspection results.

Deterministic and probabilistic-based durability analysis tools are available for metallic airframes. Three different approaches are currently being used. Approach 1 [Gallagher, et al, 1984], based on a single initial flaw size and deterministic crack growth, is used to analytically ensure structural durability for a single detail at any service time. Approaches 2 and 3 are based on an equivalent initial flaw size distribution (EIFSD) with deterministic and stochastic crack growth, respectively. These two approaches can be used to estimate the statistical crack growth accumulation (or extent of damage due to fatigue cracking) for one or more structural details at any service time. Approaches

2 and 3 are described in the Air Force's Durability Design Handbook [Manning, Yang, 1989]. Another possible approach, referred to as Approach 4, uses a single initial flaw size with stochastic crack growth as shown in Fig. 1. This approach, an enhancement of Approach 1, is also promising for the probabilisitic durability and damage tolerance analyses of metallic airframes and it is presented in this paper.

The objectives of this paper are to: (1) present a stochastic crack growth analysis method based on the concept of a single initial flaw size for durability and damage tolerance analyses, (2) validate the stochastic durability analysis method, Approach 4, using tear down inspection results for the lower wing skins of a fighter aircraft, (3) compare the durability analysis results with those obtained based on Approach 2, and (4) investigate the sensitivity of SCGA predictions to variations in key analysis variables, such as stress level, crack growth life dispersion, etc.

2 STOCHASTIC DURABILITY ANALYSIS (APPROACH 4)

2.1 Lognormal random variable model

The simple lognormal random variable model for the stochastic crack growth rate can be written as [Yang, Manning, 1983, 1987, 1990]

$$da(t)/dt = Xg(\Delta K, R, a) \qquad (1)$$

where X is a lognormal random variable with a median value of 1.0 and a log standard deviation σ_z for $Z = \ln X$. In Eq. (1), X accounts for the statistical variability of

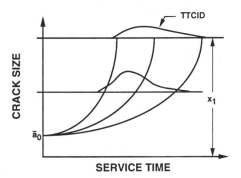

Fig. 1: Durability analysis approach 4 - single
initial flaw size with stochastic crack growth

the crack growth rate and $X = 1.0$ represents the median crack growth rate. The distribution function of X is given by

$$F_X(x) = P[X \le x] = \Phi[lnx/\sigma_z] \quad (2)$$

in which $\Phi[\ ]$ is the standard normal distribution function and σ_z is the standard deviation of $Z = ln\ X$.

A durability critical component is divided into m stress regions. Let $\bar{t}_i(x_1,\bar{a}_0)$ be the median service time to reach any given crack size x_1 from an initial flaw size $\bar{a}_0$ in the ith stress region. Similarly, let $T(x_1)$ be a random variable denoting the service time to reach any crack size x_1 from an initial crack size, $\bar{a}_0$. $T(x_1)$ can be expressed as a function of $\bar{t}_i(x_1,\bar{a}_0)$ and X by integrating Eq. (1) [Yang, Manning, 1990]

$$T(x_1) = \bar{t}_i(x_1,\bar{a}_0)/X \quad (3)$$

The cumulative distribution function of the service time, $T(x_1)$, to reach any crack size x_1 from $\bar{a}_0$, is denoted by $F_{T(x1)}(\tau) = P[T(x_1) \le \tau]$. Similarly, the probability that the crack size $a(\tau)$ at a structural detail in the ith stress region will exceed x_1 in the service time

$(0,\tau)$ is denoted by $p(i,\tau) = P[a(\tau) > x_1] = P[T(x_1) \le \tau] = F_{T(x1)}(\tau)$. $p(i,\tau)$ is also called "probability of crack exceedance." An expression for $F_{T(x1)}(\tau)$ and $p(i,\tau)$, with a constant σ_z, can be derived from Eqs. (2) and (3) as follows [Yang, Manning, 1990]

$$p(i,\tau) = F_{T(xl)}(\tau) = \Phi\left[\frac{ln\ \tau\ -\ ln\ \bar{t}_i(x_1,\bar{a}_0)}{\sigma_z}\right] \quad (4)$$

Equation (4) has been extended [Yang, Manning, 1990] to allow σ_z to vary for different crack size regions as shwon in Fig. 2. Essential features and results are summarized as follows. Suppose the crack size is divided into n regions $(a_0,a_1,a_2,...,a_n)$. The stochastic crack growth rate for the jth crack size region can be written as

$$da(t)/dt = X_j g_j(\Delta K,R,a) \ ; \ a_{j-1} \le x_1 \le a_j \quad (5)$$

where X_j is a lognormal random variable with a median value of 1.0 and a log standard deviation σ_{zj} for $Z_j = lnX_j$, and $g_j(\Delta K,R,a)$ is the crack growth rate function. In Eq. (5), X_j can be expressed in terms of the baseline random variable X defined in Eq. (2) as follows [Yang, Manning, 1990]

$$X_j = X ** h_j \ ; \ j = 1,2,...,n \quad (6)$$

where $X ** h_j$ represents X to the power h_j, and h_j is a normalization ratio

$$h_j = \sigma_{zj}\ /\ \sigma_z \ ; \ j = 1,2,...,n \quad (7)$$

In Eq. (7), σ_z is the baseline (or reference) log standard deviation for X given by Eq. (2).

The resulting expression for $p(i,\tau)$ and $F_{T(x1)}(\tau)$ for the jth crack size region $[a_{j-1} \le x_1 \le a_j]$ is obtained as [Yang, Manning, 1990] as follows

$$p(i,\tau) = F_{T_{x_1}}(\tau) = \Phi\left[-y_j(\tau)/\sigma_z\right] \ ; \ j = 1,...,n \quad (8)$$

where $y_j(\tau)$ is determined from the following equation using the Newton-Raphson numerical procedures

$$\tau = \frac{\bar{t}_i(x_1,\bar{a}_0)\ -\ \bar{t}_i(a_{j-1},\bar{a}_0)}{y_j(\tau)**h_j} + \sum_{k=1}^{j-1}\frac{\bar{t}_i(a_k,\bar{a}_0)\ -\ \bar{t}_i(a_{k-1},\bar{a}_0)}{y_j(\tau)**h_k} \quad (9)$$

In Eq. (9), τ is given and the summation term on the right hand side is zero for $j = 1$. Equation (4) or (8) will be used to predict the extent of damage for structural details in a given stress region. These results can be combined to predict the total extent of damage for a part, a component, or the entire airframe.

2.2 Statistical estimation of extent of damage

A component is divided into m stress regions. The number of structural details, $N(i,\tau)$, in the ith stress region with a crack size greater than x_1 at service time τ is a statistical variable. The mean value, $\bar{N}(i,\tau)$, and the standard deviation, $\sigma_N(i,\tau)$, are determined using the binomial distribution

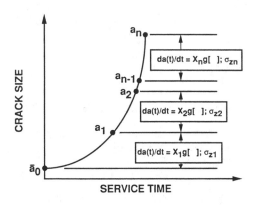

Fig. 2: Multi-segment stochastic crack growth approach

$$\bar{N}(i,\tau) = N_i\, p(i,\tau)\;;\;\sigma_N^2(i,\tau) = N_i\, p(i,\tau)[1 - p(i,\tau)] \qquad (10)$$

where N_i = total number of details in the ith stress region and $p(i,\tau)$ is defined by Eq. (4) or (8). The number of details, $L(\tau)$, with a crack size exceeding x_1 at the service time τ for m stress regions is a statistical variable following the binomial distribution. The average value, $\bar{L}(\tau)$, and the standard deviation, $\sigma_L(\tau)$, can be determined using Eq. (10)

$$\bar{L}(\tau) = \sum_{i=1}^{m} \bar{N}(i,\tau)\;;\;\sigma_L(\tau) = [\sum_{i=1}^{m} \sigma_N^2(i,\tau)]^{1/2} \qquad (11)$$

The binomial distribution for $L(\tau)$ can be approximated by the normal distribution reasonably. Therefore, the extent of damage, $L(\tau)$, can be predicted for a selected probability level.

2.3 Median crack growth curve determination

To compute the probability of crack exceedance, $p(i,\tau)$, a median crack growth curve, $\bar{t}_i(x_1,\bar{a}_0)$, is required for each stress region of the durability-critical component to be reflected in the durability analysis, Eq. (4) or (8). This curve can be determined using a deterministic crack growth analysis computer code [Roach, et al, 1987], and/or suitable fractographic results, if available. When applicable fractographic results for only one of ℓ details per specimen are used, the median crack growth curve for an equivalent single detail can be determined using the statistical scaling and transfer function methods described in the following.

2.4 Statistical scaling of fractographic results

A statistical scaling method is described in the following for determining the median crack growth curve, $\bar{t}_i(x_1,\bar{a}_0)$, for an equivalent single hole for the ith load

condition (test condition) when fractographic results are available for only 1 of ℓ holes per specimen. It was assumed that: (i) the service time to reach a given crack size x_1 follows the two-parameter Weibull distribution [Manning, Yang, 1989] and (ii) the crack growth in each hole per specimen is statistically independent. Let α and $\beta(x_1)$ be the shape parameter and scale parameter, respectively, of the Weibull distribution for the service time to reach any given crack size x_1 for a single hole. The shape parameter α, representing exclusively the statistical dispersion, is assumed to be constant for any crack size x_1. Further, let α and $\beta_\ell(x_1)$ be the corresponding shape parameter and scale parameter of the fractographic data, representing the worst case crack (fastest growing crack) in a specimen with ℓ holes. Then, it has been shown theoretically [Manning, Yang, 1989] that

$$\beta(x_1) = \beta_\ell(x_1)\,\ell^{1/\alpha} \qquad (12)$$

The scaling relationship given in Eq. (12) is schematically shown in Fig. 3(a).

Based on the lognormal stochastic crack growth model, the service time to reach any given crack size, x_1, follows the lognormal distribution as shown in Eq. (4). The median crack growth curve $\bar{t}_i(x_1,\bar{a}_0)$, starting from a single initial flaw size $\bar{a}_0$, for an equivalent single hole subjected to the ith load condition (specimen test condition) can be derived from the fractographic results, denoted by $\bar{t}_\ell(x_1)$, using the order statistics and the lognormal distribution function. $\bar{t}_\ell(x_1)$ is obtained from the fractographic data, which represents the median service time to reach any given size x_1 for the worst case crack (fastest growing crack) in a dog-bone specimen with ℓ holes. The dog-bone specimens are subjected to the ith load condition. However, using the order statistics and the lognormal distribution is cumbersome and an analytical closed-form solution is

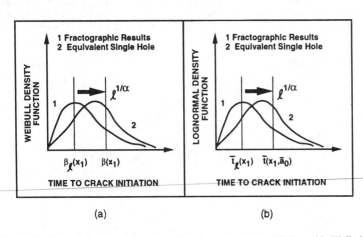

(a) (b)

Fig. 3: Scaling of fractogaphic results for time to crack initiation: (a) Weibull distribution; (b) Lognormal distribution

not available. Numerical results show that the Weibull scaling factor $\ell^{1/\alpha}$ in Eq. (12) is reasonable, i.e.,

$$\bar{t}_i(x_1,\bar{a}_0) \approx \bar{t}_\ell(x_1)\,\ell^{1/\alpha} \qquad (13)$$

For simplicity, Eq. (13), schematically shown in Fig. 3(b), is used to determine the median crack growth curve for an equivalent single hole under the ith load condition (specimen test condition).

2.5 Transfer function for median crack growth curve

The median crack growth curve $\bar{t}_i(x_1, \bar{a}_0)$ for a single hole under specimen (ith) load condition obtained from Eq. (13) does not necessarily reflect the same load condition (e.g., the same stress level, load spectra, amount of bolt load transfer, or $\sigma_{brg}/\sigma_{ten}$ ratio, etc.) as the jth stress region of the component. The median service time to reach any given crack size x_1 starting from an initial flaw size $\bar{a}_0$ for the jth load condition, denoted by $\bar{t}_j(x_1, \bar{a}_0)$, is related to that for the ith load condition, $\bar{t}_i(x_1,\bar{a}_0)$, by the following transfer function [see Fig. 4 and Manning, Yang, 1993 for derivations].

$$\bar{t}_j(x_1,\bar{a}_0) = K_{ji}\,\bar{t}_i(x_1,\bar{a}_0) \qquad (14)$$

in which K_{ji} is the transfer function for converting the median service time for the ith load condition to that for the jth load condition given by

$$K_{ji} = \bar{t}_j^*(x_1,a_0) \,/\, \bar{t}_i^*(x_1,a_0) \qquad (15)$$

In Eq. (15), $\bar{t}_j^*(x_1,a_0)$ and $\bar{t}_i^*(x_1,a_0)$ are median service times to reach any given crack size x_1 starting from a reference crack size $a_0 > > \bar{a}_0$ for the jth and the ith load conditions, respectively, using crack growth parameters commonly used in fatigue and fracture analyses (i.e., without "tuning" or "curve fitting"). $\bar{t}_i^*(x_1,a_0)$ and $\bar{t}_j^*(x_1,a_0)$ can be determined for the ith and jth load conditions using an analytical crack growth computer code [Roach, et al, 1987]. Hence, for a selected x_1 value, the median crack growth curve for the jth load condition, $\bar{t}_j(x_1,\bar{a}_0)$ can easily be determined using

Eq. (14). $\bar{t}_j(x_1,\bar{a}_0)$ and $\bar{t}_i(x_1,\bar{a}_0)$, being functions of the crack size x_1, represent the median crack growth curves as x_1 varies for the jth and the ith load conditions, respectively [see Fig. 5].

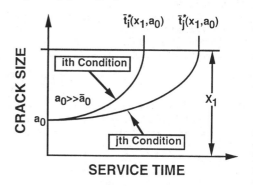

Fig. 5: Development of transfer function K_{ji} using analytical crack growth program

After obtaining the median crack growth curve, $\bar{t}_i(x_1,\bar{a}_0)$, for a single hole under the ith load condition from the fractographic results, $\bar{t}_\ell(x_1)$, using Eq. (13), the median crack growth curve, $\bar{t}_j(x_1,\bar{a}_0)$, for each of the m stress regions, j=1,2,...,m, can be computed using Eq. (14).

3 DURABILITY ANALYSIS DEMONSTRATION

Durability test results are available for a full-scale wing component that was fatigue tested using a 500-hour, two-block spectrum to 16000 flight hours (or two service lives). Tear-down inspection results were obtained for the lower wing skins by Gordon, et al [1986]. Each wing skin is made of 7475-T7351 aluminum and contains 1614 countersunk fastener holes with MS 90353-08 typical fasteners. A typical $\sigma_{brg}/\sigma_{ten}$ ratio of 1.73 (i.e., 15% load transfer through the

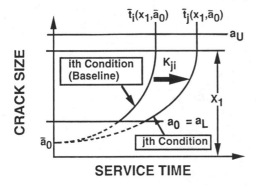

Fig. 4: Median crack growth curve for the jth loading condition based on ith loading condition and transfer function

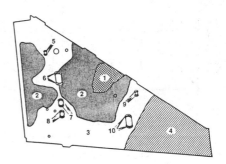

Fig. 6: Stress regions, no. of fastener holes and average stress levels for lower wing skin

Table 1: Typical durability analysis results for lower wing skin for $\tau=16000FH$ and $x_1=2.54m$ (Approach 4)

STRESS REGION (1)	σ_i (MPa) (2)	NO. HOLES (3)	$\bar{t}(x_1,\bar{a}_0)$ (4)	$p(i,\tau)$ (5)	$\bar{N}(i,\tau)$ (6)	$\sigma_N(\tau)$ (7)
1	191.0	59	41734	2.11E-02	1.25	1.10
2	177.2	320	55410	4.18E-03	1.34	1.15
3	157.9	680	78990	3.39E-04	0.23	0.48
4	115.2	469	141239	1.61E-06	0.00	0.03
5	195.9	8	37961	3.37E-02	0.27	0.51
6	201.4	30	34425	5.25E-02	1.57	1.22
7	223.5	8	24757	1.78E-01	1.43	1.08
8	180.7	8	51638	6.46E-03	0.05	0.23
9	180.7	12	51638	6.46E-03	0.08	0.28
10	177.2	20	55410	4.18E-03	0.08	0.29
		1614	$\bar{L}(\tau)$ = 6.30 ; $\sigma_L(\tau)$ = 2.43			

fastener) applies. This lower wing skin of a fighter aircraft is divided into ten stress regions as shown in Fig. 6.

The average stress level and the number of countersunk fastener holes in each stress region are shown in columns (2) and (3) of Table 1.

Fractographic results for a reversed dog-bone specimen data set, referred to as the "WAFXMR4 data set" [Gordon, 1986], has the same type of material, countersunk fasteners and similar load spectra as the lower wing skin. The test specimens are wide enough (76.2mm) to provide useful crack growth data for a wide range of crack sizes (e.g., 0.762-12.7mm). Furthermore, these specimens have a comparable $\sigma_{brg}/\sigma_{ten}$ ratio (1.73) as the lower wing skin, where σ_{brg} = fastener load/(bearing area) and σ_{ten} = bypass stress. Each specimen contains two countersunk fasteners (i.e., MS 90353-08), 25.4mm apart on the specimen centerline as shown in Fig. 7. Thirteen specimens were fatigue tested to failure using a fighter 400-hour spectrum. Each specimen contains four holes in two-matched pairs. Fractographic results for the

WAFXMR4 data set are available for the largest fatigue crack (or worst case crack) in one of four holes. The baseline stress for the WAFXMR4 data set, based on the maximum load in the 400-hour spectrum, is 234.5 MPa (gross section). Each of the four fastener holes per specimen are assumed to have the same bypass stress (234.5 MPa) and the same $\sigma_{brg}/\sigma_{ten}$ ratio (1.73). This WAFXMR4 data set serves as the baseline data set from which theoretical predictions for the extent of damage are made using Approaches 2 [Manning, Yang, 1989] and 4 (current approach).

The durability analysis procedures using the WFXMR4 fractographic data set are briefly described in the following:

(i) The median service time to reach a crack size x_1, for 1 of ℓ holes, denoted by $\bar{t}_\ell(x_1)$, is estimated using time-to-crack-initiation results for the baseline fractographic data set (WAFXMR4) and the maximum likelihood method; (ii) The baseline median crack growth curve for an equivalent single hole, denoted by $\bar{t}_i(x_1, \bar{a}_0)$, is obtained from $\bar{t}_\ell(x_1)$ using Eq. (13) in which $\ell=4$ for selected crack sizes ranging from 0.762mm to 12.7mm. This baseline curve is for σ_{ten} = 234.5 MPa, $\sigma_{brg}/\sigma_{ten}$ = 1.73 and 40-hour spectrum. The shape parameter α is obtained by fitting the service time to reach a crack size $x_1=0.762$mm, for the fractographic data, to the Weibull distribution; with the result $\alpha=1.972$; (iii) A transfer function factor K_{ji}, based on Eq. (15), is developed for each stress region using the ADAMSys crack growth analysis computer code [Roach et al, 1987]. For convenience, an initial flaw size of $a_0=0.127$mm was used. The median crack growth curve, $\bar{t}_j(x_1,\bar{a}_0)$, for a given stress region is obtained from the baseline median crack growth curve, $\bar{t}_i(x_1, \bar{a}_0)$, using the transfer function factor and Eq. (14). This procedure is repeated for j=1,2,...,10 to cover all the stress regions in the lower wing skin; (iv) Two different σ_z values were used in the durability analysis for approximation. σ_z is the log standard deviation for the service time to reach any crack size x_1,

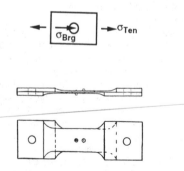

Fig. 7: Geometric configuration of test specimens

denoted by $\sigma_z(x_1)$, and it can be estimated from the fractographic data set using the maximum likelihood method. One σ_z value was obtained using the time-to-crack-initiation data at $x_1 = 0.762$mm with the result, $\sigma_z(x_1 = 0.762$mm$) = 0.497$. This value is used in the durability analysis for crack size ≤ 0.762mm. Another value was obtained using the average value of σ_z for the 0.762mm to 12.7mm crack size range with the result $\sigma_z(x_1) = 0.354$. This value is used in the durability analysis for crack size greater than 0.762mm; (v) $p(i,\tau)$ is determined for each stress region using Eq. (8). These results are used to estimate the total extent of damage for the durability-critical component using Eqs. (10) and (11).

3.1 Durability analysis predictions

Extent of damage predictions for the lower wing skin using the current approach (Approach 4) are correlated with experimental results (denoted as solid circles) in Fig. 8. Results are shown for $\tau = 16000$ flight hours and for three different probabilities: P=0.05, 0.50 and 0.95. Typical results and analysis details at $\tau = 16000$ flight hours and for $x_1 = 2.54$mm are shown in Table 1.

Extent of damage predictions for the lower wing skin were also made using Approach 2 described in [Manning, Yang 1989], based on the same baseline fractographic data set, i.e., WAFXMR4. In this appraoch the parameters for the equivalent initial flaw size distribution (EIFSD) whould be estimated. The EIFS distribution parameters were estimated using WAFXMR4 data set in the 0.762mm to 1.27mm crack size range with the results; $x_u = 0.762$mm, $\alpha = 1.972$ and $\phi = 13.6$. Predictions for Approaches 2 and 4 are

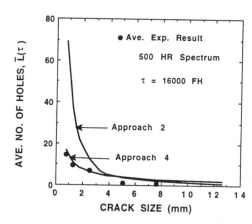

Fig. 9: Correlation of extent of damage predictions for approaches 2 and 4 for fighter lower wing skin

compared and correlated with average experimental results in Fig. 9 for P=0.50 and $\tau = 16000$ flight hours.

The following conclusions and observations are based on the durability analysis results presented in Figs. 8 and 9 and in Table 1.

1. Overall, the extent of damage predictions for the current approach (Approach 4) for the lower wing skin correlated better with the average experimental results than those based on Approach 2 (see Figs. 8 and 9). For example, in Fig. 8 extent of damage predictions based on Approach 4 for P=0.5 correlated well with average experimental results. Also, the upper and lower bound limits for the extent of damage (defined by P=0.05 and 0.95, respectively) encompass the average experimental results well (see Fig. 8).

2. More crack growth damage accumulation was predicted for the lower wing skins at $\tau = 16000$ flight hours for Approach 2 than for Approach 4 up to a crack size of approximately 5.08mm (see Fig. 9). For crack sizes larger than 5.08mm the predicted extent of damage was comparable for both durability analysis approaches as shown in Fig. 9. Hence, it is seen that greater damage is predicted for Approach 2 than for Approach 4 in the smaller crack size region; but predictions for the two approaches are very comparable for the larger crack size region. This situation is the result of the philosophical differences between the two approaches (see Fig. 10) as follows. Approach 4 uses a single initial flaw size (median). When the initial flaw size is grown forward, the crack growth life dispersion is reflected by the variation of crack growth rates. On the other hand, the crack growth life dispersion is reflected by the equivalent initial flaw size distribution for Approach 2. Both approaches, when initial flaws are grown forward, should satisfy the cumulative

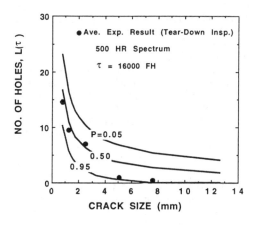

Fig. 8: Correlation of extent of damage predictions based on approach 4 for fighter lower wing skin with tear-down inspection

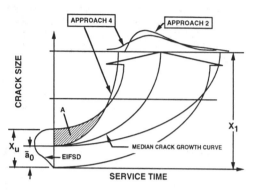

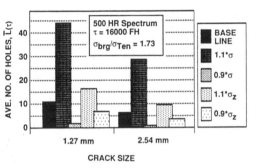

Fig. 10: Direct comparisons of approaches 2 and 4

Fig. 11: Sensitivity of extent of damage predictions based on approach 4 for fighter lower wing skin to variations in stress level and σ_z

distribution of the time to crack initiation (in particular in the lefthand tail of the distribution) for a selected crack size x_1. Since the crack growth life dispersion is treated differently by Approaches 2 and 4, greater damage would be expected for Approach 2 than for Approach 4 for crack sizes less than x_1. This situation is conceptually described by the cross-hatched area A noted in Fig. 10.

3.2 Sensitivity studies

Based on the current approach (Approach 4) the extent of damage predictions were made for + or - 10% variations in the baseline stress level for each of ten stress regions (see Table 1) while holding the baseline σ_z values constant, i.e., $\sigma_z(x_1=0.762mm) = 0.497$ and average $\sigma_z=0.354$ (0.762mm-12.7mm). Similarly, the baseline σ_z values were varied by + or - 10% while holding the baseline stress levels for each of the ten

stress regions constant. The sensitivity study reflected: $\sigma_{brg}/\sigma_{ten} = 1.73$, P=0.50 and $\tau=16000$ flight hours. The results are shown in Fig. 11 for two different x_1 values (i.e., 1.27mm and 2.54mm). These results show that both the stress level and σ_z variations (+ or - 10%) have a significant effect on extent of damage predictions and that the effect of the stress level is more significant than that of σ_z.

4 CONCLUSIONS

A stochastic crack growth analysis (SCGA) method based on the concept of a single flaw size (Approach 4) for metallic airframes has been described, evaluated and demonstrated using full-scale wing components of a fighter aircraft. The capability and potential of this approach for durability analysis have been evaluated. Statistical crack growth accumulations in countersunk fastener holes in the lower wing skin were predicted. Theoretical predictions were compared and correlated with actual tear-down inspection results. The sensitivity of the predictions to variations in key analysis variables was investigated.

Extent of damage predictions for the current approach (approach 4) and a previous approach (Approach 2) correlated well with tear down inspection results in the large crack size region (e.g., > 5.08mm). In the small crack size region, durability analysis predictions for Approach 4 correlated well with experimental results; whereas, Approach 2 was more conservative (i.e., Approach 2 predicted more damage than was observed). The extent of damage predictions were more sensitive to the variation in stress level than to the comparable variation in the crack growth life dispersion. Both approaches are useful for the durability analysis of metallic airframes for a wide range of crack sizes. The transfer function and statistical scaling concepts presented are useful and practical for durability analysis applications.

Although durability analysis was emphasized, Approach 4 presented herein also apply to damage tolerance analysis. The transfer function concept, used in conjunction with statistical scaling, is useful for determining the crack growth curve for an equivalent single detail, when fractographic results are available for only 1 of ℓ details per specimen and each detail is subjected to the same loading conditions. The statistical scaling technique can be extended to allow variable loading conditions at each of ℓ details per specimen. The authors are pursuing this research.

ACKNOWLEDGMENT

This research was supported by the "Supportable Hybrid Structures" program (Air Force contract F33615-87-C-3207), sponsored by the Wright Research and Development Center, Wright-Patterson Air Force Base, Ohio.

REFERENCES

Gallagher, J.P., Giessler, F.J., Berens, A.P. and Engle, R.M., USAF damage tolerance design handbook: guidelines for the analysis and design of damage tolerant aircraft structures, Air Force Wright Aeronautical Laboratories, WPAFB, Ohio, AFWAL-TR-82-3073, 1984.

Gordon, D.E., et al, Advanced durability analysis. volume III - fractographic data, AFWAL-TR-86-3017, Air Force Flight Dynamics Laboratory, Wright-Patterson Air Force Base, OH, August 1, 1986.

Manning, S.D. and Yang, J.N., USAF durability design handbook: guidelines for analysis and design of durable aircraft structures, Air Force Flight Dynamics Laboratory Technical Report, AFWAL-TR-83-3119, Wright Patterson Air Force Base, OH, First edition 1984, 2nd edition, 1989.

Manning, S.D., Yang, J.N., Pretzer, F.L. and Marler, J.E., Reliability-centered maintenance for metallic airframes based on a stochastic crack growth approach, Advances in Fatigue Life Predictive Techniques, ASTM, ASTM-STP-1122, 1992, pp. 422-434.

Manning, S.D., Yang, J.N., and Welch, K.M., Aircraft structural maintenance scheduling based on risk and individual aircraft tracking, Proc. Int'l. Conf. on Theoretical Concepts and Numerical Analysis of Fatigue, May 25-27, 1992, Birmingham, England (in press).

Manning, S.D. and Yang, J.N., Probabilistic durability analysis methodology for metallic airframes, paper to be presented at ICAF '93 Symposium, International Committee on Aeronautical Fatigue, Stockholm, Sweden, June 7-11, 1993.

Roach, G.R., McComb, T.H. and Chung, J.H., ADAMSys user's manual, Structural and Design Dept., General Dynamics, Fort Worth Division, July 1987.

Speaker, S.M. et al, Durability methods development. volume VIII - test and fractography data, AFFDL-TR-79-3118, Air Force Flight Dynamics Laboratory, Wright-Patterson Air Force Base, OH, 1982.

Yang, J.N., and Manning, S.D., Distribution of equivalent initial flaw size, Proc. Annual Reliability and Maintainability Symposium, 1980, pp. 112-120.

Yang, J.N. and Donath, R.C., Statistical fatigue crack propagation in fastener hole under spectrum loading, J. of Aircraft, AIAA, Vol. 20, No. 12, December 1983, pp. 1028-1032.

Yang, J.N. Hsi, W.H., Manning, S.D. and Rudd, J.L., Stochastic crack growth models for application to aircraft structures, Probabilistic Fracture Mechanics and Reliability, Chapter IV, edited by J.W. Provan, Martinus Nijhoff Publishers, 1987, pp. 171-212.

Yang, J.N. and Manning, S.D., Stochastic crack growth analysis methodologies for metallic structures, J. of Eng. Fracture Mech., Vol. 37, No. 5, December 1990, pp. 1105-1124.

Yang, J.N. and Manning, S.D., Application of probabilistic methods to aircraft structural maintenance, in Reliability Technology, edited by T.A. Cruse, ASME-AD, Vol. 28, pp. 127-142; ASME 1992 Annual Meeting, Anaheim, CA, November 8-13, 1992.

Structural Safety & Reliability, Schuëller, Shinozuka & Yao (eds) © 1994 Balkema, Rotterdam, ISBN 90 5410 357 4

Reliability analysis of pressure vessels based on proof test and multiple failure modes

J.N.Yang
Department of Civil Engineering, University of California, Irvine, Calif., USA

H.Y.Chen, C.M.Chang & H.K.Shee
Quality Assurance Center, Chung Shan Institute of Science and Technology, Lungtan, Taoyuan, Taiwan

ABSTRACT: Proof testing has been used extensively in industry for quality assurance of some types of structural components, such as pressure vessels, rocket motor cases, etc. It has also been used as an alternative means of maintenance for structural components in which the critical locations are not accessible for inspection maintenance. In this paper, the effect of proof testing on structural reliability in service is investigated. Emphasis is placed on practical applications of the reliability theory to the quality assurance of pressure vessels. Multiple failure modes, including fracture failure, plastic yielding collapse and their interactions, are considered. All the physical quantities, such as the fracture toughness, yield stress, flaw length, flaw geometry, service loads, geometric dimensions, etc., are considered as statistical variables. It is demonstrated that proof testing is most effective when (i) the statistical variability of the service load is small or (ii) the statistical dispersion of the overall strength of the structure is large. Proof testing is shown to be ineffective if the variability of the overall structural strength is small or the dispersion of the service load is large. In the latter case, the effectiveness of proof testing can be enhanced by increasing the design safety actor.

1 INTRODUCTION

Proof testing has been used extensively in practice for quality assurance of some types of structural components, such as pressure vessels, rocket motor cases, etc. [e.g., Barnett, Hermann 1965, Tiffany 1968, 1970, Shinozuka, Yang 1969, Heer, Yang 1971]. The purpose of proof testing is to screen out weak components or components containing unacceptable flaw length or defects to ensure the safety and integrity of the structure in its design service life. It is well-known, however, that the structural strengths, geometric dimensions, fracture toughness, stiffness, defect or flaw length, service loads, etc., involve statistical variabilities. As a result, the effect of proof tests can be interpreted rationally in terms of the improvement of the structural reliability [e.g., Shinozuka, Yang 1969, Heer, Yang 1971]. A structural component passing a proof test at a particular proof load level will guarantee a certain level of reliability in service. The effect of proof tests on the structural reliability in service has been investigated in the literature [e.g., Shinozuka, Yang 1969, Heer, Yang 1971, Yang 1976, 1977(a), 1977(b), 1977(c), 1980].

Proof testing has also been used as an alternative means of maintenance for certain structural components in which critical details or locations are not accessible for inspections. It is also a viable means for quality assurance and maintenance for components made of composite materials [e.g., Yang, et al 1977(b), 1977(c),

1980]. The investigation of the structural reliability under proof test maintenance as well as the determination of the optimal proof-test maintenance interval has been conducted by Yang [1976, 1977(a), 1977(b)].

In the previous works, however, either a simple failure mode or the fracture failure mode was used to evaluate the effect of proof testing on structural reliability. For a simple failure mode, structural failure is assumed to occur when the ultimate strength is exceeded by the service load. For the fracture failure mode, fracture occurs when the stress intensity factor exceeds the mode I critical stress intensity factor. The number of random variables considered was restricted to be small so that the solution can be obtained either analytically or by a direct numerical integration technique. Likewise, the previous works emphasized either the optimal structural design based on reliability and proof tests or the optimal proof test maintenance, such as the optimal proof load level and the optimal proof test maintenance interval.

In this paper, the failure mode considered is based on the ASME R-6 method [Nichols 1988, Milne 1988], in which the fracture failure mode, the plastic yielding and their interactions are taken into account. All the physical quantities are considered as statistical variables, including the fracture toughness, yield stress, radius, thickness, crack size, crack geometry, maximum operating pressure, etc. In performing proof testing, the proof load level may not be controlled precisely and it

varies slightly from test to test. Hence, the proof load level is also considered as a statistical variable. The first order reliability methods (FORM) in conjunction with the importance sampling technique are used to evaluate the structural reliability after the pressure vessel is proof-tested at a particular proof pressure level. Emphasis is placed on the practical applications of the reliability method to the quality assurance of pressure vessels.

A sensitivity study is conducted to investigate the conditions under which proof testing may or may not be effective for the quality assurance of structures. It is found that proof testing is most effective when the statistical dispersion of the service load is smaller than the statistical variability of the overall strength of the structure. If the statistical variability of the service load is equal to or larger than that of the overall structural strength, proof testing is not effective. Under this circumstance, a proof test in conjunction with a larger design safety factor will be effective for the quality assurance of the structure. In order to maintain a specified level of structural reliability in service, the required proof load level depends on many factors, such as the statistical variability of the service load, the design safety factor, the statistical dispersion of the overall structural strength, etc. Consequently, many factors should be accounted for in the determination of the proof load level for the quality assurance of structures. Numerical results are obtained to demonstrate the effect of proof testing and other variables on the reliability and quality assurance of pressure vessels.

2 RELIABILITY THEORY OF PROOF TEST

Based on the structural integrity assessment method (or R-6 method) proposed for pressure vessels, the failure criterion is expressed as [Nichols 1988, Milne 1988]

$$G(\underline{X}) = S_r \left\{ \frac{8}{\pi^2} \ln \left[sec \left(\frac{\pi}{2} S_r \right) \right] \right\}^{-1/2} - K_r = 0 \quad (1)$$

in which

$$S_r = \frac{\sigma_{eff}}{\sigma_y} = \frac{\sqrt{3} P R^*}{2 t \sigma_y} \quad ; \quad K_r = \frac{K_I}{K_{IC}} = \frac{P R^* \beta \sqrt{\pi a}}{2 t K_{IC}} \quad (2)$$

In Eq. (1), $\underline{X}$ is a vector consisting of random variables $X_1, X_2, ..., X_n$, where $P = X_1$ = maximum internal pressure in service, $R^* = X_2$ = vessel's radius, $t = X_3$ = vessel's thickness, $\sigma_y = X_4$ = yield stress, $a = X_5$ = crack length in transverse direction, $\beta = X_6$ = crack shape factor, and $K_{IC} = X_7$ = mode I critical stress intensity factor or fracture toughness. The plot of Eq. (1) as shown in Fig. 1(a) is referred to as the Failure Assessment Diagram. Failure occurs in the region where $G(\underline{X}) \leq 0$. The pressure vessels considered are

small cylindrical vessels made of two cylindrical shells welded together. The welded joint is the critical region and the flaws in the welded joint perpendicular to the longitudinal direction are most critical. Hence, the stress $PR^*/2t$ in the longitudinal direction is used in Eq. (2).

Basically, the failure criterion (or limit state function) given by Eq. (1) consists of two failure modes and their interactions; namely, elastic fracture and plastic collapse. Without interactions of failure modes, elastic fracture occurs when

$$K_r \geq 1.0 \quad (3)$$

and plastic collapse takes place when

$$S_r \geq 1.0 \quad (4)$$

All the physical quantities appearing in Eqs. (1)-(2) are considered as statistical variables. Another statistical variable considered is the proof load level (or proof pressure), denoted by $r_0 = X_8$. Random variables considered are summarized in Table 1.

To facilitate the presentation and formulation, the following random vectors $\underline{X}$ and $\underline{Y}$ are introduced for convenience,

$$\underline{X} = [X_1, X_2, X_3, ..., X_7]' \quad ; \quad \underline{Y} = [X_8, X_2, X_3, ..., X_7]' \quad (5)$$

Without being proof-tested, the probability of failure of a pressure vessel in service is given by

$$P_{f0} = Pr[G(\underline{X}) \leq 0] = \int_{G(\underline{X}) \leq 0} f_{\underline{X}}(x) dx \quad (6)$$

in which $f_{\underline{X}}(x)$ is the joint probability density function of all components of the random vector $\underline{X}$. The probability of failure of a pressure vessel under the proof pressure $r_0 = X_8$ is given by

$$P_0 = Pr[G(\underline{Y}) \leq 0] \quad (7)$$

The probabilities P_{f0} and P_0 in Eqs. (6) and (7) can easily be computed using various available techniques, such as the first order reliability method (FORM), the second order reliability method (SORM), importance sampling simulation techniques, etc. [e.g., Schueller, Stix 1987, Schueller 1989, Wu, et al 1990, Liu, Der Kiureghian 1988]. Further, various commericaly available general computer programs can also be used [e.g., Bourgund, Bucher 1986, Wu, et al 1989, Liu, Der Kiureghian 1989]. In using any available general computer program, care should be taken that the derivatives of $G(\underline{X})$, Eq. (1), at $S_r = 1$ and $K_r = 1$ do not exist and that the value of S_r should be restricted to be $S_r \leq 1$ otherwise Eq. (1) is not valid, because the argument of ln should be positive.

Pressure vessels are proof-tested at r_0 before being used in service. Proof tests are conducted until a vessel

is accepted. A pressure vessel is rejected when it either fails or leaks during the proof test. The probability of failure, P_0, under proof pressure, r_0, is given by Eq.(7), whereas the probability of leakage is given by

$$P_1 = Pr[t \le a] = P[X_3 \le X_5] = \int_0^\infty F_t(x) f_a(x) dx \quad (8)$$

in which $F_t(x)$ is the distribution function of the thickness t and $f_a(x)$ is the probability density function of the crack length a. The probability of rejecting a vessel by proof testing, denoted by $\bar{P}_0$, is given by

$$\bar{P}_0 = Pr[\{G(\underline{Y}) \le 0\} \cup \{X_3 \le X_5\}] \quad (9)$$

It is the usual practice that a pressure vessel is designed such that leakage occurs before burst. In other words, the component may not fail even in the presence of a through-the-thickness crack. As a result, the probability of rejecting a vessel by the proof test, $\bar{P}_0$, can reasonably be approximated by

$$\bar{P}_0 \approx P_0 + P_1 \quad (10)$$

in which P_0 and P_1 are given by Eqs. (7) and (8).

Components are used in service only after they are accepted by the proof test and proof tests are conducted until one component is accepted. The number, N, of vessels rejected by the proof test before one that is accepted, is a discrete random variable following the geometric distribution [Shinozuka, Yang 1969, Heer, Yang 1971], with the probability mass function

$$P_k = Pr[N = k] = \bar{P}_0^k (1 - \bar{P}_0) \quad (11)$$

The expected number of components to be rejected, before one that is accepted, denoted by $\bar{N}$, is given by

$$\bar{N} = \bar{P}_0 / (1 - \bar{P}_0) \quad (12)$$

After being proof-tested, the probability of failure

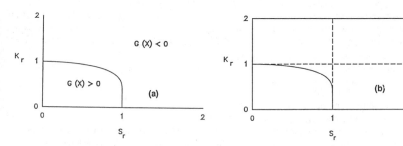

Fig. 1: Failure assessment diagram for pressure vessels; (a) exact limit state function; (b) approximate limit state function

Table 1: Physical quantities considered as random variables

Lognormal Variables	Symbols	Mean, μ_i	Standard Deviation, σ_i	Coefficient of Variation, V_i
X_1	P	1.18 kg/mm^2	0.118	10%
X_2	R*	249mm	4.98	2%
X_3	t	3mm	0.06	2%
X_4	K_{IC}	290kg/$\sqrt{mm^3}$	29	10%
X_5	a	1mm	0.15	15%
X_6	σ_y	106kg/mm^2	10.6	10%
X_7	β	1.12	0.112	10%
X_8	r_0	1.27kg/mm^2	0.0127	1%

of a pressure vessel in service is obtained as

$$P_f = Pr[G(X) \leq 0 \mid G(Y) > 0] \qquad (13)$$

$$= Pr[G(X) \leq 0, G(Y) > 0] / Pr[G(Y) > 0]$$

The probability of failure, P_f, given above can be computed using available general reliability computer programs [e.g., Bourgund, Bucher 1986, Wu et al 1989, Liu, Der Kiureghian 1989]. It should be mentioned that the computation of the probability of joint events is much more involved and less accurate. Usually, the importance sampling simulation technique is used.

3 APPROXIMATION AND SIMPLIFICATION

The formulation and solution presented in the previous section are straightforward and mathematically rigorous. However, since the solution involves many random variables, it is difficult to gain insights into the following issues which are of practical importance to a designer or managers of quality assurance programs: (i) what should the proof pressure be in order to guarantee a certain level of reliability in service?; (ii) What is the rationale, limitations and consequences associated with the proof load level specified in some design specifications?; (iii) Under what conditions proof testing is most effective and beneficial?; (iv) Under what conditions proof testing is not effective and other NDI techniques should be considered for quality assurance?; and (v) Should the level of proof pressure be dependent on the design safety factor? Better insight can be gained for the issues described above using the following approximations: (1) The dispersion of the proof pressure r_0 is negligible, i.e., the proof pressure r_0 is a deterministic value; (2) Since all the statistical variables considered represent the positive physical quantities, the distribution of all random variables is assumed to follow the lognormal distribution; and (3) Based on the limit state function given by Eq. (1) and Fig. 1(a), the failure criterion is approximated by two statistically independent failure modes; namely, plastic collapse and elastic fracture as shown in Eqs. (3) and (4). The safe and failure regions based on the separation of failure modes are shown in Fig. 1(b). A comparison between Figs. 1(a) and 1(b) indicates that the approximation is probably as good as the results obtained using the exact formulation and the importance sampling simulation technique. Based on the approximations above, the failure probability of a pressure vessel without being proof-tested, P_{f0}, given by Eq. (6) becomes

$$P_{f_0} = Pr[S_r \geq 1 \cup K_r \geq 1] \qquad (14)$$
$$\approx Pr[S_r \geq 1] + Pr[K_r \geq 1]$$

The two limit state functions in Eqs. (3) and (4), i.e., $S_r = 1$ and $K_r = 1$ can be written as

$$P/R_1 = 1 \quad and \quad P/R_2 = 1 \qquad (15)$$

in which R_1 is the resisting strength for the plastic collapse failure mode and R_2 is the resisting strength for the elastic fracture failure mode.

$$R_1 = \frac{2t\sigma_y}{\sqrt{3}R^*} \quad ; \quad R_2 = \frac{2tK_{IC}}{R^*\beta\sqrt{\pi a}} \qquad (16)$$

and failure occurs either $R_1 \leq P$ or $R_2 \leq P$.

Let μ_i and σ_i be, respectively, the mean value and standard deviation of X_i ($i=1,2,\ldots,7$), and $\bar{\mu}_i$ and $\bar{\sigma}_i$ be the mean value and standard deviation of $\ln X_i$, respectively. The relations between (μ_i, σ_i) and ($\bar{\mu}_i$, $\bar{\sigma}_i$) are well-known

$$\bar{\mu}_i = \ln\left[\mu_i / \sqrt{1 + V_i^2}\right] \; ; \; \bar{\sigma}_i = [\ln(1 + V_i^2)]^{1/2} \qquad (17)$$

in which V_i is the coefficient of variation of X_i, i.e., $V_i = \sigma_i/\mu_i$. Since the distribution functions of all random variables $X_1, X_2, \ldots, X_7$ are lognormal, it follows from Eq. (16) that the distribution functions of R_1 and R_2, denoted by $F_{R1}(x)$ and $F_{R2}(x)$, are also lognormal,

$$F_{RI}(x) = \Phi\left(\frac{\ln x - \bar{\mu}_{RI}}{\bar{\sigma}_{RI}}\right); F_{R2}(x) = \Phi\left(\frac{\ln x - \bar{\mu}_{R2}}{\bar{\sigma}_{R2}}\right) \qquad (18)$$

in which $\Phi( )$ is the standardized normal distribution function, $\bar{\mu}_{R1}$ and $\bar{\sigma}_{R1}$ are the mean value and standard deviation of $\ln R_1$, and $\bar{\mu}_{R2}$ and $\bar{\sigma}_{R2}$ are the mean value and standard deviation of $\ln R_2$. Parameters $\bar{\mu}_{R1}$, $\bar{\sigma}_{R1}$, $\bar{\mu}_{R2}$ and $\bar{\sigma}_{R2}$ can be obtained analytically from $\bar{\mu}_i$ and $\bar{\sigma}_i$ ($i=2,3,\ldots,7$) using Eq. (16) as follows

$$\bar{\mu}_{RI} = \ln(2/\sqrt{3}) + \bar{\mu}_3 + \bar{\mu}_6 - \bar{\mu}_2$$

$$\sigma_{RI} = [\bar{\sigma}_3^2 + \bar{\sigma}_6^2 + \bar{\sigma}_2^2]^{1/2} \qquad (19)$$

$$\bar{\mu}_{R2} = \ln(2/\sqrt{\pi}) + \bar{\mu}_3 + \bar{\mu}_4 - \bar{\mu}_2 - \bar{\mu}_7 - 0.5\,\bar{\mu}_5$$

$$\bar{\sigma}_{R2} = [\bar{\sigma}_3^2 + \bar{\sigma}_4^2 + \bar{\sigma}_2^2 + \bar{\sigma}_7^2 + 0.25\,\bar{\sigma}_5^2]^{1/2} \qquad (20)$$

The probability of rejecting a pressure vessel by the proof test is contributed by the probability of plastic collapse, the probability of elastic fracture, and the probability of having a through-the-thickness crack,

1158

$$\bar{P}_0 = Pr\,[t \le a] + Pr\,[r_0/R_1 > 1] + Pr\,[r_0/R_2 > 1]$$

$$= \Phi\left(\frac{\bar{\mu}_5 - \bar{\mu}_3}{\sqrt{\bar{\sigma}_5^2 + \bar{\sigma}_3^2}}\right) + \Phi\left(\frac{\ln r_0 - \bar{\mu}_{R1}}{\bar{\sigma}_{R1}}\right) + \Phi\left(\frac{\ln r_0 - \bar{\mu}_{R2}}{\bar{\sigma}_{R2}}\right) \quad (21)$$

in which Eq. (18) has been used.

After passing the proof test, the resisting strengths for both failure modes are denoted by R_{10} and R_{20}, respectively. The distribution functions of both R_{10} and R_{20} can be derived from that of R_1 and R_2 [e.g., Shinozuka, Yang 1969, Heer, Yang 1971, Yang 1977]; with the results

$$F_{R10}(x) = [1 - F_{R1}(r_0)]^{-1}[F_{R1}(x) - F_{R1}(r_0)]H(x - r_0) \quad (22)$$

$$F_{R20}(x) = [1 - F_{R2}(r_0)]^{-1}[F_{R2}(x) - F_{R2}(r_0)]H(x - r_0) \quad (23)$$

in which $H(x-r_0)$ is the unit step function starting at r_0 and the distribution functions $F_{R1}(x)$ and $F_{R2}(x)$ have been obtained analytically in Eq. (18).

The probability of failure of a pressure vessel in service, P_f, is contributed by the failure probability, P_{f1}, due to plastic collapse, and the failure probability, P_{f2}, due to elastic fracture. The results can easily be obtained as follows

$$P_f = P_{f1} + P_{f2} \quad (24)$$

in which

$$P_{f1} = \int_{r_0}^{\infty} f_p(x)\,F_{R10}(x)\,dx$$
$$\quad (25)$$
$$P_{f2} = \int_{r_0}^{\infty} f_p(x)\,F_{R20}(x)\,dx$$

where $f_p(x)$ is the probability density function of the maximum service load P. The analytical closed-form solution for Eq. (25) is not possible, with the exception of $r_0 = 0$. However, a straight-forward numerical integration can be used easily to compute P_{f1} and P_{f2}.

4 DEMONSTRATIVE EXAMPLES

The pressure vessels considered for demonstrations are made of two small cylindrical shells welded together. The welded joint is the critical location and flaws in the joint perpendicular to the longitudinal direction are considered. The mean value, μ_i, standard deviations, σ_i, and coefficient of variation, V_i, of all the physical quantities, which are assumed to be lognormal random variables, are given in Table 1. The results based on the simplified approximate solutions presented in Section 3 will be demonstrated first; namely, the proof pressure r_0 is a deterministic value and the interaction of two failure modes will be neglected, Eqs. (15) and (16).

The resulting mean values and coefficients of variation of the vessel strength for plastic collapse and elastic fracture failure modes are $\mu_{R1} = 1.475$ kg/mm^2, $V_{R1} = 10.4\%$, $\mu_{R2} = 3.59$ kg/mm^2 and $V_{R2} = 16.3\%$. From these results, it is obvious that the plastic collapse failure mode predominates. The probability of failure, P_f, of the pressure vessel in service as a function of the proof pressure r_0 is computed from Eqs. (22)-(25) and presented in Fig. 2 as Curve 1. In Fig. 2, the proof pressure r_0 in the abscissa is expressed in terms of average strength $\mu_R = \mu_{R1}$ for the plastic collapse failure mode. The average number of pressure vessels, $\bar{N}$, to be rejected by the proof test, before one is accepted, is computed from Eq. (21) and shown in Fig. 3 as Curve 1. The case considered above is referred to as Case 1. It is observed that as the proof pressure r_0 increases, both the reliability of the pressure vessel in service and the average number of rejected vessels increase.

If the coefficient of variation of the yield strength, σ_y, is reduced to 5%, i.e., $V_6 = 5\%$ as shown in Case 2 of Table 2, then μ_{R1}, μ_{R2} and V_{R2} remain the same but $V_{R1} = 5.75\%$. The probability of failure, P_f, and the average number of rejected vessels are presented in Figs. 2 and 3, respectively, as Curve 2. This case is referred to as Case 2. For Case 3, $V_6 = 10\%$ is identical to Case 1, but the dispersion of the maximum service load is reduced to 5%, i.e., $V_1 = 5\%$ (Table 2). The failure probability, P_f, is presented in Fig. 2 as Curve 3, whereas the average number of rejected vessels is the same as Curve 1 of Fig. 3. For Case 4 in which $V_1 = 5\%$ and $V_6 = 5\%$, see Table 2, the failure probability, P_f, is plotted in Fig. 2 as Curve 4, whereas $\bar{N}$ is shown in Fig. 3 as Curve 2. For Cases 5 and 6 in which $V_6 = 10\%$, the dispersion of the maximum service load is further reduced to $V_1 = 2\%$ and $V_1 = 1\%$, respectively, as shown in Table 2. The failure probabilities, P_f, are displayed as Curve 5 and Curve 6 in Fig. 2, respectively, for both cases. The corresponding $\bar{N}$ for these two cases is identical to Curve 1 of Fig. 3. The variations of the input parameters from those given by Table 1 for six cases considered above are summarized in Table 2.

Let v be the central safety factor defined by the ratio of the average vessel strength, $\mu_R = \mu_{R1}$, to the average maximum service load, μ_1, i.e.,

$$v = \mu_R/\mu_1 \quad (26)$$

Table 2: Coefficients of variation of input data

Cases	V_1	V_6	$V_R = V_{R1}$
1	10%	10%	10.4%
2	10%	5%	5.75%
3	5%	10%	10.4%
4	5%	5%	5.75%
5	2%	10%	10.4%
6	1%	10%	10.4%

For the six cases considered above, the central safety ν is identical, i.e., $\nu=1.25$. However, the effect of proof testing is significantly different. For Cases 1 and 2 in which the dispersion of the service load is $V_1=10\%$, the improvement of the failure probability is minimal and the penalty for rejecting pressure vessels by the proof test is quite heavy. For instance, at $r_0 = \mu_R$, for Case 1, there is only one order of magnitude improvement in the failure probability; however, on the average, one vessel will be rejected before a vessel is accepted, i.e., $\bar{N}=1$, as indicated in Fig. 3. Consequently, the proof test is not effective at all. On the other hand, for Cases 3, 5 and 6 in which $V_6=10\%$ and the dispersion of the service load V_1 is 5% or smaller, the improvement of the failure probability is quite drastic without much penalty in rejecting the components as shown by Figs. 2 and 3. From the results presented above, it is observed that proof testing is very effective if the dispersion of the service load is smaller than that of the resisting strength of the components. It is further observed that proof testing is effective for Case 4 only in the region where $r_0>0.95\ \mu_R$. This is because both V_6 and V_1 are 5%.

The observations obtained above can be explained in the following. Proof testing is capable of eliminating weak components or truncating the lower tail portion of the strength distribution but it has no effect on the service load. Hence, if the failure probability P_f is mainly due to the dispersion of the resisting strength, such as Case 3, proof testing is very effective. However, if the failure probability is mainly due to the dispersion of the service load, such as Case 2, proof testing is not cost effective.

The numerical results presented in Figs. 2 and 3 were computed using the approximate solutions in

Section 3. Numerical results using the exact solutions presented in Section 2 were also obtained. The difference between the approximate solution and the exact solution for the average number, $\bar{N}$, of components rejected by proof testing is negligible, and the difference for the failure probability, P_f, ranges from 2% for large failure probability to 35% for small failure probability. Likewise, the proof pressure r_0 has also been considered as a lognormal random variable. For the dispersion of r_0 smaller than 3%, numerical results indicate that $\bar{N}$ increases slightly (less than 5%) and P_f remains almost the same comparing with the case where r_0 is considered a deterministic value. Therefore, the results presented in Figs. 2 and 3 are reasonable for the present case in which the plastic collapse failure mode predominates. In particular, the approximate solutions are very convenient for sensitivity studies.

It may not be cost effective to perform proof testing when the average number, $\bar{N}$, of rejected components exceeds 0.5; namely, for every two components accepted, there is one component rejected by the proof test. A trade-off between the reduction of the failure probability, P_f, and the increase of $\bar{N}$ should be made. Curves 1 and 2 of Fig. 2 indicate that proof testing is not effective for Cases 1 and 2. To reduce the failure probability for these two cases, the design safety factor, ν, Eq. (26), should be increased. Suppose the mean value of the thickness of the vessel is increased to $\mu_3=3.7$mm and $V_3=2\%$ for Case 1, and $\mu_3=3.5$mm and $V_3=2\%$ for Case 2. These two new cases are referred to as Cases 7 and 8, respectively. The probabilities of failure, P_f, versus the average number, $\bar{N}$, of rejected components are presented in Fig. 4 as Curves 7 and 8, respectively, for Cases 7 and 8. Also displayed in Fig. 4 as Curves 1 and 2 are the

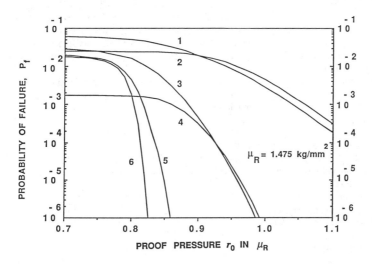

Fig. 2: Probability of failure in service vs. proof pressure for six cases

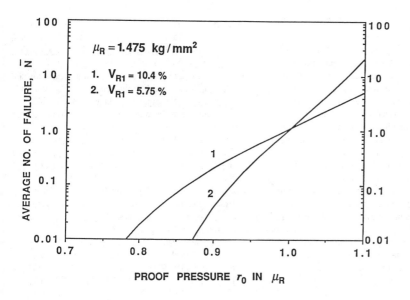

Fig. 3: Expected number of components rejected by proof testing

corresponding results for Cases 1 and 2, respectively. As observed from Fig. 4, P_f reduces faster as $\bar{N}$ increases for Cases 7 and 8 than for Cases 1 and 2. This indicates that the effectiveness of the proof testing improves as the design safety factor increases. Comparing Curves 1 and 7 in Fig. 4 for example, the failure probability is reduced from $P_f = 6.1 \times 10^{-2} (\bar{N}=0$ without proof testing) to $P_f = 1.38 \times 10^{-2} (\bar{N}=0.3)$ for Case 1, whereas P_f is reduced from $P_f = 1.32 \times 10^{-3} (\bar{N}=0)$ to $P_f = 1.7 \times 10^{-5} (\bar{N}=0.31)$ for Case 7. An improvement of two orders of failure probability is achieved for $\bar{N}=0.31$ in Case 7, whereas the improvement of the failure probability for $\bar{N}=0.3$ in Case 1 is less than one order of magnitude. Extensive numerical results indicate that the effect of proof testing is more significant when the design safety factor is higher.

In some areas of applications, there is an upper bound for the service load; namely, the distribution of the service load is limited to the right, say γ. In this case, a proof load r_0 greater than γ will guarantee a 100% reliability, provided that r_0 is a deterministic value and that the structure is not damaged by the proof test. Frequently, however, a reasonable estimation of γ may not be easy, whereas a conservative estimate of γ may not be cost-effective. As a result, the investigations of the cost-effectiveness of the proof test in conjunction with the optimal structural design is necessary [e.g., Shinozuka, Yang 1969, Heer, Yang 1971, Yang 1976].

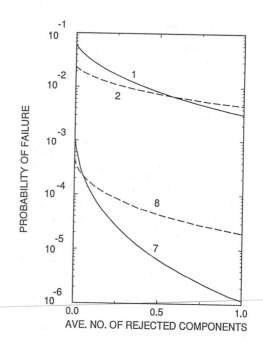

Fig. 4: Probability of failure vs. average number of rejected components

5 CONCLUSION

A reliability approach of proof test for the quality assurance of pressure vessles has been presented. Multiple failure modes have been taken into account in the reliability analysis. In addition to the general formulation, an alternative formulation, based on the concept of the structural strength associated with the failure modes, and approximate solutions have been presented. The approximate solutions are very convenient for sensitivity studies. Numerical results demonstrate that proof testing is effective in improving the structural reliability when the statistical dispersion of the structural strength is larger than that of the service load. In particular, proof testing drastically improves the structural reliability when the statistical dispersion of the service load is small. When the statistical variability of the service load is large, proof testing is not cost effective. In this case, a larger design safety factor in conjunction with proof testing can be used to guarantee a desirable level of structural reliability in service.

REFERENCES

Barnett, R.L. and Hermann, P.C., "Proof Testing in Design with Brittle Materials", Journal of Spacecraft and Rockets, Vol. 2, June 1965, pp. 956-961.

Bourgund, J. and Bucher, C.G., "Importance Sampling Procedures Using Design Point (ISPUD) - A User's Manual", Report No. 8-86, Institute of Engineering Mechanics, University of Innsbruck, Austria, Oct. 1986.

Heer, E. and Yang, J.N., Structural Optimization Based on Fracture Mechanics and Reliability Criteria", AIAA Journal, Vol. 9, No. 5, April 1971, pp. 621-628.

Liu, P.-C. and Der Kiureghian, A., "Optimization Algorithms for Structural Reliability", Computational Probabilistic Mechanics, ADM-93, ASME, 1988, pp. 185-196.

Liu, P.-L., Lin, H.-Z., and Der Kiureghian, A., "CALREL User Manual", Report No. UCB/SEMM-89/18, Dept. Civil Engineering, Univ. of Calif., Berkeley, CA, 1989.

Milne, I., et al, "Assessment of the Integrity of Structures Containing Defects", International Journal of Pressure Vessels and Piping, Vol. 32, 1988, pp. 3-104.

Nichols, R.W., et al, "The Revisions to the Structural Integrity Assessment Method CEGB/R6", International Journal of Pressure Vessels and Piping, Vol. 32, 1988.

Schueller, G.I. and Stix, R., "A Critical Appraisal of Methods to Determine Failure Probabilities", Structural Safety, Vol. 4, 1987, pp. 293-309.

Schueller, G.I., et al, "On Efficient Computational Schemes to Calculate Structural Failure Probabilities", Journal of Probabilistic Engineering Mechanics, Vol. 4, No. 1, 1989, pp. 10-18.

Shinozuka, M. and Yang, J.N., "Optimum Structural Design Based on Reliability and Proof-Load Test", Annals of Assurance Science, Proc. of the Reliability and Maintainability Conf., Vol. 8, July 1969, pp. 375-391.

Tiffany, C.F., "On the Prevention of Delayed Time Failures of Aerospace Pressure Vessels", Journal of the Franklin Institute, Vol. 290, June 1970, pp. 567-582.

Tiffany, C.F., Master, J.N. and Paul, F.A., "Some Fracture Considerations in the Design and Analysis of Spacecraft Pressure Vessels", American Society of Metals, TR-C6-2.3, 1968.

Wu, Y.-T., Millwater, H.R. and Cruse, T.A., "An Advanced Probabilistic Structural Analysis Method for Implicit Performance Functions", AIAA Journal, Vol. 28, No. 9, 1990, pp. 1663-1669.

Wu, Y.-T., et al, "Fast Probability Integration (FPI) User's Manual", Southwest Research Institute, San Antonio, Texas, 1989.

Yang, J.N., "Reliability Analysis of Structures Under Periodic Proof Test in Service", AIAA Journal, Vol. 14, No. 9, September 1976, pp. 1225-1234.

Yang. J.N., "Optimal Periodic Proof Test Based on Cost-Effective and Reliablity Criteria", AIAA Journal, Vol. 15, No. 3, March 1977(a), pp. 402-409.

Yang, J.N., "Reliability Prediction for Composites Under Periodic Proof Test in Service", Composite Materials, Testing and Design, ASTM, ASTM-STP 617, March 1977(b), pp. 272-295.

Yang, J.N. and Liu, M.D., "Residual Strength Degradation Model and Theory of Periodic Proof Test for Graphite/Epoxy Laminates", J. of Composite Materials, Vol. 11, April 1977(c), pp. 176-203.

Yang, J.N. and Sun, C.T., "Proof Test and Fatigue of Composite Laminates", J. of Composite Materials, Vol. 14, April 1980, pp. 801-817.

Structural Safety & Reliability, Schuëller, Shinozuka & Yao (eds) © 1994 Balkema, Rotterdam, ISBN 90 5410 357 4

A stochastic model for the fatigue crack propagation life considering material inhomogeneity

J.H.Yoon
Samsung Heavy Industries Co. Ltd, Korea

H.Y.Yoon
Department of Mechanical Engineering, Mokpo National University, Korea

Y.S.Yang
Department of Naval Architecture & Ocean Engineering, Seoul National University, Korea

ABSTRACT: The experimental results of fatigue crack propagation under constant amplitude loading show that the inter-specimen variability and the intra-specimen variability exist due to the material inhomogeneity. In this paper, a stochastic model, considering these variabilities in crack propagation life, is presented. To take into account the intra-specimen variability, the material resistance against crack propagation is treated as a one-dimensional spatial stochastic process (random field) varying along the propagation path. For the inter-specimen variability, C in Paris-Erdogan equation is assumed to be a random variable. Compared with experimental results reported, the present model well estimates the variation in fatigue crack propagation life. It is confirmed that the thicker the specimen thickness is, the less the variation of propagation life is.

1 INTRODUCTION

A variation of fatigue crack propagation under identical conditions is principally caused by material inhomogeneity. It has been recognized that the probabilistic approach may be reasonable for the rational analysis of crack propagation. For the probabilistic approach of fatigue crack propagation, Paris-Erdogan(PE) equation has been widely used, in which the relation between the crack growth rate da/dn and the stress intensity factor range ΔK is expressed, as follows:

$$\frac{da}{dn} = C\,(\ \Delta K\ )^m \tag{1}$$

where C and m are material coefficients. When PE equation is applied to the experimental data, it is shown that the inter-specimen (specimen-to-specimen) variability and the intra-specimen (within a specimen) variability exist. Fig.1 illustrates these variabilities. The most common approach for the inter-specimen variability is to randomize the coefficients C or m in the PE equation. The intra-specimen variability has mainly been studied under the assumption that the material resistance is a stochastic process. Ishikawa and Turui (1987), and Lin and Yang (1985) proposed stochastic models by assuming the material resistance to be a process varying with time. Ortiz and Kiremidjian (1986,1988) suggested a stochastic model by treating the material resistance as a stochastic process with spatial variation, i.e. random field.

In this paper, the inter-specimen variability is considered by taking C as a random variable. Actually, both C and m have variabilities, respectively. However, the experimental results of fatigue crack propagation show that there is a strong negative correlation between $\ln C$ and m. (Ortiz 1983, Ichikawa 1987) Consequently, it is possible that m is assumed to be constant and only C is treated as a random variable. Simultaneously, the intra-specimen variability is modeled as a stochastic process varying from point to point along the crack propagation path within a specimen.

A stochastic model for fatigue crack propagation is presented and its validity is compared with the experimental results. The effect of the specimen thickness on the variation of crack propagation life is investigated.

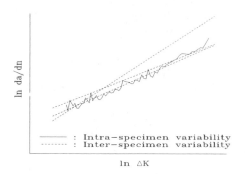

———— : Intra−specimen variability
-------- : Inter−specimen variability

ln ΔK

Fig.1 Illustration of inter- and intra-specimen variability

2 FORMULATION

In order to consider the inter- and the intra-specimen variability in fatigue crack propagation, PE equation could be randomized as follows based on Ortiz & Kiremidjian's (1986,1988) suggestion:

$$\frac{da}{dn} = \frac{Z}{X(a)}\ C\,(\ \Delta K\ )^m \tag{2}$$

If the deterministic PE equation is assumed to be the mean crack growth rate, Z is a random variable representing the inter-specimen variability with the expectation of one and $X(a)$ is a spatially varying stochastic process which represents the intra-specimen variability with the expectation of one. Fig.2 shows the stochastic process $X(a)$. By discretizing crack

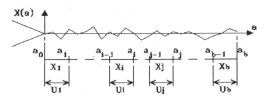

Fig.2 Discretization of one-dimensional stochastic process along crack propagation path

propagation path into b according to the crack size, as shown in Fig.2, the crack propagation life N can be calculated by the following equation:

$$N = \sum_{i=1}^{b} T_i = \sum_{i=1}^{b} \int_{a_{i-1}}^{a_i} \frac{X(a)}{Z} \frac{da}{C(\Delta K)^m} \qquad (3)$$

where T_i is the propagation life required for a crack to pass through the i-th interval, i.e. from a_{i-1} to a_i. The right-hand side of Eq.(3) involves a stochastic integral of $X(a)$. To avoid the difficulty of stochastic integral, $X(a)$ could be approximated as a series of random variables. At this time the concept of moving average is adopted. Let X_i be a moving average of $X(a)$ at the i-th interval, and T_i can be expressed as:

$$T_i = \int_{a_{i-1}}^{a_i} \frac{X(a)}{Z} \frac{da}{C(\Delta K)^m}$$
$$\approx \frac{X_i}{Z} \int_{a_{i-1}}^{a_i} \frac{da}{C(\Delta K)^m} \qquad (4)$$

In the right-hand side of Eq.(4), since the remaining part except , X_i / Z, is the mean life for a crack to propagate from a_{i-1} to a_i, it is possible to substitute as follows:

$$T_i = \frac{X_i}{Z} \overline{T_i} \qquad (5)$$

Therefore, N becomes as follows:

$$N = \sum_{i=1}^{b} T_i = \sum_{i=1}^{b} \frac{X_i}{Z} \overline{T_i} \qquad (6)$$

Since N can be expressed as a combination of random variables, i.e. Z and X_i $(i=1,2,...,b)$, the variation of N could be evaluated by the first order approximation:

$$E[N] \approx \sum_{i=1}^{b} \frac{E[X_i]}{E[Z]} \overline{T_i} = \sum_{i=1}^{b} \overline{T_i} \qquad (7)$$

$$\begin{aligned} \text{Var}[N] &\approx \sum_{i=1}^{b}\sum_{j=1}^{b} (\frac{1}{E[Z]})^2 \overline{T_i}\, \overline{T_j} \text{Cov}[X_i,X_j] \\ &+ \sum_{i=1}^{b}\sum_{j=1}^{b} (-\frac{E[X_i]}{E[Z]^2})^2 \overline{T_i}\, \overline{T_j} \text{Var}[Z] \\ &+ 2\sum_{i=1}^{b}\sum_{j=1}^{b} (\frac{1}{E[Z]})(-\frac{E[X_i]}{E[Z]^2})\overline{T_i}\, \overline{T_j}\text{Cov}[Z,X_i] \\ &= \sum_{i=1}^{b}\sum_{j=1}^{b} \overline{T_i}\, \overline{T_j} \{ \text{Cov}[X_i,X_j] + \text{Var}[Z] - 2\text{Cov}[Z,X_i] \} \end{aligned} \qquad (8)$$

Provided that the statistical characteristics of Z and X_i are given, Eq.(7) and (8) show that the mean and the variance of the crack propagation life can be calculated. X can be assumed to be a stationary process. Therefore, the mean and the variance of X are invariant to the spatial location and the covariance and the correlation function of X are governed by only the spatial distance. The moving average X_i of X along the i-th discretized propagation path U_i can be obtained by:

$$X_i = \frac{1}{U_i} \int_{u-U_i/2}^{u+U_i/2} X(u)\,du \qquad (9)$$

If $\rho_X(u_1-u_2)$ is the correlation function between two arbitrary points u_1 and u_2 within a specimen, the mean and the variance of X_i are:

$$E[X_i] = E[X] \qquad (10)$$

$$\begin{aligned} \text{Var}[X_i] &= \frac{\sigma_X^2}{U_i^2} \int_0^{U_i}\int_0^{U_i} \rho_X(u_1-u_2)\,du_1\,du_2 \\ &= \frac{2\sigma_X^2}{U_i} \int_0^{U_i} (1-\frac{u}{U_i})\rho_X(u)\,du \end{aligned} \qquad (11)$$

where σ_X^2 is the variance of X and ρ_X is the correlation function of X. Generally, as the spatial distance u between two arbitrary points goes from zero to infinity, the correlation function decreases from one to zero. Therefore, ρ_X can be assumed to be an exponential function of spatial distance u, as follows:

$$\rho_X(u) = \exp(-|u|/\theta_X) \qquad (12)$$

where θ_X means the correlation distance of the stochastic process X. Substituting Eq.(12) for Eq.(11), the variance of X_i can be expressed as:

$$\text{Var}[X_i] = \frac{2\sigma_X^2}{U_i}(\theta_X U_i + \theta_X^2 \exp(-U_i/\theta_X) - \theta_X^2) \qquad (13)$$

Also, the covariance between X_i and X_j, as shown in Fig.2, can be written as follows:

$$\begin{aligned} \text{Cov}[X_i,X_j] = \frac{1}{2U_iU_j} (&U_0^2 \text{Var}[X_0] - U_1^2 \text{Var}[X_1] \\ + &U_2^2 \text{Var}[X_2] - U_3^2 \text{Var}[X_3]) \end{aligned} \qquad (14)$$

$$U_i = a_i - a_{i-1}, \quad U_j = a_j - a_{j-1}$$

where, $\quad U_0 = a_{j-1} - a_i, \quad U_1 = a_{j-1} - a_{i-1}$

$$U_2 = a_j - a_{i-1}, \quad U_3 = a_j - a_i$$

From Eq.(13) and (14) the statistical characteristics of X_i can be estimated. If the statistical characteristics of the random variable Z and X_i are obtained from experimental data as material properties, the variation of crack propagation life can be calculated from Eq.(7) and (8).

3 ANALYSIS OF EXPERIMENTAL DATA

The statistical characteristics of Z and X can be determined by applying the standard statistical procedure to the experimental data. In this paper, the experimental data is used, in which the numbers of load cycles required for the crack to grow to known sizes are recorded at R points spaced at equal increments Δa.

First, the statistical characteristics of Z can be obtained from experimental data. With the number of specimen as M, C_k and m_k in PE equation can be determined from experimental data of the k-th specimen. Calculate the mean $\overline{m}$ of m_k ($k=1,2,...M$) and obtain a new material coefficient $\overline{C_k}$ of the k-th specimen by fixing m_k with $\overline{m}$. Provided that the mean C of $\overline{C_k}$ ($k=1,2,...M$) is calculated, Z_k can be defined as $Z_k = \overline{C_k}/C$. Subsequently, the statistical characteristics of Z can be computed by applying statistics to Z_k ($k=1,2,...M$).

Next, σ_X^2 and θ_X of the process X should be determined. For this purpose, sampled data at points Δa apart is needed. Provided that $\overline{C_k}$ and $\overline{m}$ are true values of the k-th specimen, the magnitude of X at the crack size $r\Delta a$ can be evaluated by:

$$X(r\Delta a) = \frac{(da/dn)_{\Delta a,r}}{\frac{1}{\Delta a} \int_{a_{r-1}}^{a_r} \overline{C_k}(\Delta K)^{\overline{m}} da} \quad (15)$$

where,

$$(da/dn)_{\Delta a,r} = \frac{a_r - a_{r-1}}{N_r - N_{r-1}}$$

$$= \frac{\Delta a}{N_r - N_{r-1}} \quad \text{for } r=1,2,...,R$$

$$a_r = a_0 + r\Delta a$$

a_0 ; initial crack size

R ; number of data

N_r is the number of load cycles required for a crack to propagate from the initial state to a_r in the experiment. The next step is to compute the auto-correlation function

and to determine the correlation distance θ_X of process X. The auto-correlation function can directly be evaluated by statistical calculations or alternatively by Fast Fourier Transform (FFT) (Ortiz 1986,1988).

4 EXAMPLE CALCULATIONS AND DISCUSSIONS

Yoon(1990) conducted an experiment for fatigue crack propagation under constant amplitude loading with standard compact tension (CT) specimens made from the same 2024-T3 aluminum sheet of 1.0, 3.0, 5.0 and 12.5mm thickness. The experiment was conducted under the same load conditions of stress ratio 0.14 by one person. Crack size was measured by converting the compliance, which were obtained from the load signal and the output of strain gage attached at the back-face of the specimen. Data of fatigue crack propagation were obtained from 20 identical specimens of the same thickness.

The stochastic model needs the experimental data in which the numbers of load cycles are recorded at R points spaced at equal increments Δa. Since this is not the case in Yoon's experiment, the necessary data is obtained from Yoon's data by linear interpolation with an increment of 0.2mm. From this data, the statistical characteristics of Z and X_i, in the region II of crack growth rate, are computed. To facilitate the calculation, C is treated as $\ln C$. The following stress intensity factor equation for the standard CT specimen is used:

$$K = \frac{P}{W\sqrt{B}} g(\xi) \quad (16)$$

$$g(\xi) = \frac{2+\xi}{(1-\xi)^{3/2}} (0.886 + 4.46\xi - 13.32\xi^2 + 14.72\xi^3 - 5.6\xi^4)$$

where $\xi = a/B$, P is applied tensile load, W is the plate thickness, B is the plate breadth and a is the crack size.

From Yoon's experimental data, the relationship between $\ln C$ and m is first investigated. Table 1 shows the relationship between $\ln C$ and m at each thickness. It can be known that $\ln C$ and m have strong negative correlation. Thus, the assumption that C is random and m is constant seems to be reasonable. Table 1 also shows $\ln C$ and $\overline{m}$ at each thickness and the statistical characteristics of Z. Next, for the stochastic process X, the ensemble auto-correlation functions are also shown in Fig.3 ~ 6. The correlation distances θ_X are estimated with

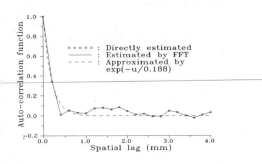

Fig.3 Ensemble auto-correlation function(thickness=1.0mm)

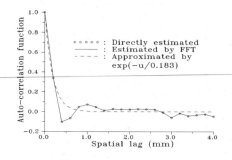

Fig.4 Ensemble auto-correlation function(thickness=3.0mm)

1165

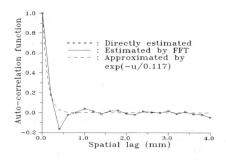
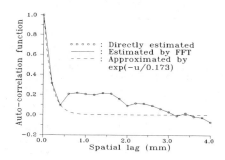

Fig.5 Ensemble auto-correlation function(thickness=5.0mm)   Fig.6 Ensemble auto-correlation function(thickness=12.5mm)

Table 1 Statistical characteristics of parameters obtained from experimental data

Specimen thickness	Parameter	Expected value	Standard deviation	Correlation distance	Remark
1.0 mm	$\ln C$	-19.2848	–	–	
	m	2.9524	–	–	$\rho_{\ln C,m}$=-0.996
	Z	1.0	0.1251	–	
3.0 mm	$\ln C$	-18.2644	–	–	
	m	2.6718	–	–	$\rho_{\ln C,m}$=-0.972
	Z	1.0	0.0703	–	
5.0 mm	$\ln C$	-18.9645	–	–	
	m	2.9081	–	–	$\rho_{\ln C,m}$=-0.993
	Z	1.0	0.0454	–	
12.5 mm	$\ln C$	-20.1546	–	–	
	m	3.2628	–	–	$\rho_{\ln C,m}$=-0.997
	Z	1.0	0.0350	–	

0.117 ~ 0.188mm, as listed in Table 1. Since these are smaller than the crack length, it can be said that the stochastic process X has little correlation between two arbitrary points. Also, σ_X of each thickness is shown in Table 1.

The results of analysis of experimental data show that the standard deviations of Z and X decrease according to the increase of the specimen thickness. Therefore, it could be assumed that the variations of Z and X are inversely proportional to the power of thickness, respectively. Let the standard deviation of Z with thickness 1.0mm be $(\sigma_Z)_{1mm}$, and then the standard deviation of Z with thickness W is well approximated by $(\sigma_Z)_{1mm}/W^{0.5}$, as shown in Fig.7. Also, Fig.8 shows that the standard deviation of X with thickness W can be approximated by $(\sigma_X)_{1mm}/W^{0.3}$, where $(\sigma_X)_{1mm}$ denotes the standard deviation of X with thickness 1.0mm. With these statistical values of Z and X, the variation of propagation life are predicted as shown in Fig.9 ~ 12. In the figures, M1 is the case considering the intra-specimen variability only and M2 is the case for the inter-specimen variability only. M3 is the case considering both variabilities under the assumption that there is no correlation between intra- and inter-specimen variability, that is between Z and X_i. On the other hand, M4 represents the case considering both variabilities under the assumption that any correlation

between intra- and inter-specimen variability exists. In the case of M1, the predicted coefficient of variation (COV) is under-estimated wholly. It is also shown that COV is small for short cracks in the case of M2. Consequently, both intra- and inter-specimen variability should be considered simultaneously. Under the assumption that PE equation describes the true crack growth behavior, the correlation between intra- and inter-specimen variability tends to zero. In other words, the covariance between Z and X_i, Cov[Z,X_i], should be zero. (M3) If comparing the case of M3 with experimental results, the predicted COV however, seems to be overestimated. This is due to the limitation of the PE equation, which is the deterministic model for the crack growth in macro viewpoint. Although there would not be any correlation between Z and X_i logically, it is thought that the possibility that some correlation physically exist cannot be neglected because the specimens are made from an identical sheet. Consequently, in the case of M4, in which the correlation coefficients between Z and X_i are considered by 0.04 ~ 0.10, it is shown that the present model (M4) is a good agreement with experimental results.

5 CONCLUSIONS

Based on the numerical calculations and discussions, the following results of this paper would be of much

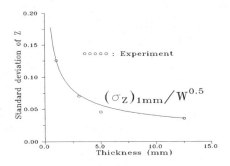

Fig.7 Thickness effect on inter-specimen variability

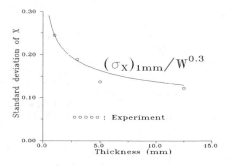

Fig.8 Thickness effect on intra-specimen variability

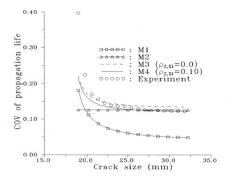

Fig.9 COV of propagation life (thickness=1.0mm)

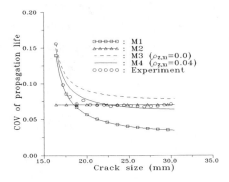

Fig.10 COV of propagation life (thickness=3.0mm)

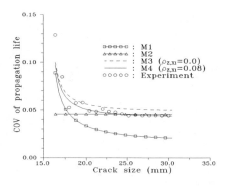

Fig.11 COV of propagation life (thickness=5.0mm)

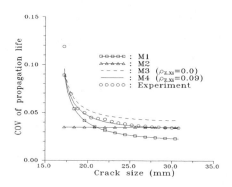

Fig.12 COV of propagation life (thickness=12.5mm)

importance. The inter- and the intra-specimen variability should be considered for the rational estimation of the variation of crack propagation life. Compared with experimental data, the stochastic crack growth model, suggested in this paper, well estimates the variation due to material inhomogeneity in crack propagation. The inter-specimen variability decreases as the specimen thickness becomes thicker. Its standard deviation would be approximated by $(\sigma_Z)_{1mm}/W^{0.5}$. The standard deviation of the intra-specimen variability could be approximated by $(\sigma_X)_{1mm}/W^{0.3}$.

REFERENCES

Benjamin, J.R. & Cornell, C.A. 1970. *Probability, Statistics and Decision for Civil Engineering.* New York: McGraw-Hill.

Ichikawa, M. 1987. Probabilistic Fracture Mechanics Investigation of Fatigue Crack Growth Rate. *Current Japanese Material Research, Vol.2.* Elsevier.

Ishikawa, H. & Tsurui, A. 1987. Stochastic Fatigue Crack Growth Model and Its Wide Applicability in Reliability-based Design. *Current Japanese Material Research, Vol.2* Elsevier.

Lin, Y.K. & Yang, J.N. 1985. A Stochastic Theory of Fatigue Crack Propagation. *AIAA, Vol.24, No.1.*

Ortiz, K. & Kiremidjian, A.S. 1986. Time Series Analysis of Fatigue Crack Growth Rate Data. *Engineering Fracture Mechanics, Vol.24, No.5.*

Ortiz, K. & Kiremidjian, A.S. 1988. Stochastic Modelling of Fatigue Crack Growth Rate Data. *Engineering Fracture Mechanics, Vol.29, No.3.*

Vanmarke, E. 1983. *Random Fields : Analysis and Synthesis.* Boston: MIT press.

Yoon, H.Y. 1990. A Study on the Probabilistic Nature of Fatigue Crack Propagation Life (II) - The Distribution of Crack Propagation Rate. *Trans. of KSME, Vol.14, No.6.* (in Korean)

Structural Safety & Reliability, Schuëller, Shinozuka & Yao (eds) © 1994 Balkema, Rotterdam, ISBN 90 5410 357 4

Fatigue crack growth of hysteretic structures under random loading

Weiqiu Zhu
Department of Mechanics, Zhejian University, Hangzhou, People's Republic of China

Ying Lei
Department of Civil Engineering, Technical University of Vienna, Austria

ABSTRACT: A probabilistic approach is proposed to analyze the fatigue crack growth, fatigue life and reliability of elastic structural components with random material resistance in hysteretic structural systems under random loading. Analytical expressions are derived for the special case of randomized Paris-Erdogan type law. Numerical example is given to illustrate the proposed procedure and the results are compared with those obtained from digital simulation.

1 INTRODUCTION

Fatigue is known to be a major cause of failure for mechanical and structural components. From a fracture mechanics point of view, fatigue damage of a component subject to dynamic load can be measured by the size of the dominant crack, and the failure occurs when this crack reaches a critical magnitude. It is widely recognized that the fatigue crack growth is fundamentally a random phenomenon. The two main reasons for the randomness in fatigue crack growth behavior are the random material resistance and the random loading. During the last decade, several stochastic models have been proposed (see Lin and Yang (1985) and etc.) for the analysis of fatigue crack growth of the components with random material resistance and under constant amplitude loading. The fatigue crack growth of linear components with constant fatigue strength under random loading has also been studied by several authors (see Sobczyk (1986) and ect.). Recently, a probabilistic analysis of fatigue crack growth, fatigue life and reliability of elastic structural component has been presented by Zhu, Lin and Lei (1992) on the basis of fracture mechanics and theory of random processes. Both the material resistance to fatigue crack growth and the time-history of the stress are assumed to be random. Analytical expressions are obtained for the special case in which the random stress is a stationary narrow-band Gaussian random process, and a randomized Paris-Erdogan law is applicable. However, it seems that no study has been made on the fatigue growth of nonlinear components with random material resistance and under random loading. In the present paper, the procedure is generalized to analyze the fatigue crack growth of elastic structural components with random material resistance in hysteretic structural systems subjected to stationary wide-band random loading.

2 GENERAL PROCEDURE

Neglecting secondary factors in a deterministic model of fatigue crack growth, the mathematical functions describing the fatigue propagating rate of elastic material under cyclic loading have the general form of

$$\frac{da}{dn} = f(\Delta k) , \qquad (1)$$

where a is crack size (half length for a though crack); n is the number of stress cycles; f is a nonnegative function; Δk is stress intensity range. For opening mode, which is the predominant mode of macroscopic fatigue, noting the dependence of Δk on Δx and a, i.e.,

$$\Delta k = \Delta x \sqrt{\pi a} \eta(a) , \qquad (2)$$

where Δx is stress range; $\eta(a)$ is a non-negative function whose functional form depends on the geometry of the crack and the component, Eq.(1) can be rewritten as

$$\frac{da}{dn} = g(a, \Delta x) . \qquad (3)$$

To account for the random material resistance to fatigue growth, Eq.(3) may be randomized to read

$$\frac{dA}{dt} = \mu \, g(A, \Delta x) \, Y(t) \qquad (4)$$

where $Y(t)$ is a slowly varying stationary random process with long correlation time describing the random material resistance. Its statistical values are obtained from experimental testing records. μ is the average number of cycles per unit time. The symbol for the crack size is capitalized to signify that it is now a random process.

For a stationary random process $X(t)$, a logical generalization of Eq.(4) is to treat the stress range as the absolute valve of the difference between a local maximum and a neighboring local minimum, i.e.,

$$\Delta X(t) = |X(t_1) - X(t_2)| \, , \, t_1 \le t \le t_2 \, , \qquad (5)$$

where t_1 and t_2 are the times at which two neighboring extrema of $X(t)$ occur. In this case, ΔX is a stationary random sequence and independent of $Y(t)$, and μ can be interpreted as the number of maxima per unit time. Thus, the fatigue crack growth of a component with random material resistance under random loading is described by the following equation

$$\frac{dA}{dt} = \mu \, g[A(t), \Delta X(t)] \, Y(t) \, . \qquad (6)$$

Assume that fatigue crack size $A(t)$ is a slowing varying random process compared with the stress process $\Delta X(t)$. This is a reasonable assumption for high cycle fatigue since the correlation time τ_c of $\Delta X(t)$ is expected to be much smaller than the fatigue life of a component. According to the Stratonovich-Khasminskii limit theorem (see Khasminskii (1966)) the crack size $A(t)$ is approximately a diffusive Markov process. Upon applying the stochastic averaging procedure (see Stratonvich (1963) and Zhu (1988)) to Eq.(6), we obtain a Fokker-Planck equation for $A(t)$ as follows

$$\frac{\partial q}{\partial t} = -\frac{\partial}{\partial a}[m(a)q] + \frac{1}{2}\frac{\partial^2}{\partial a^2}[\sigma^2(a)q] \qquad (7)$$

$$m(a) = \mu E[g] \, E[Y]$$
$$+ \mu^2 \int_{-\infty}^{0} cov(\frac{\partial g}{\partial A}\Big|_t, g_{t+\tau}) cov[Y(t), Y(t+\tau)] d\tau \qquad (8)$$

$$\sigma^2(a) = \mu^2 \int_{-\infty}^{\infty} cov(g_t, g_{t+\tau}) cov[Y(t), Y(t+\tau)] \, d\tau \, , \qquad (9)$$

where $q = q(a, t \, | \, a_0, t_0)$ is the transition probability density of $A(t)$; $g_t = g[A(t), \Delta X(t)]$; $g_{t+\tau} = g[A(t+\tau), \Delta X(t+\tau)]$; $E[\cdot]$ denotes an ensemble average,

and $cov(\cdot,\cdot)$ denotes a covariance. Eq.(7) is solved subjected to the initial condition

$$q(a, t \, | a_0, t_0) = \delta \, (a - a_0), \quad t = t_0 \qquad (10)$$

The solution will provide a complete probability description of $A(t)$ or, more precisely, a Markovin approximation of $A(t)$.

Obviously, $A(t)$ is a non-decreasing process. The reliability of the component at time t, conditional on a known initial crack size a_0 at time $t=t_0$ can be obtained from $q(a, t \, | \, a_0, t_0)$ as follows

$$R(a_{cr}, t \, | \, a_0, t_0) = \int_{a_0}^{a_{cr}} q(u, t \, | \, a_0, t_0) \, du \qquad (11)$$

where a_{cr} is critical crack size. The conditional probability density of fatigue life T is then obtained as

$$p(T) = -\frac{\partial R}{\partial t}\Big|_{t=T} \qquad (12)$$

The conditional mean and variance of fatigue life are simply

$$m(T) = \int_0^{\infty} T \, p(T) dT \qquad (13)$$

$$\sigma^2(T) = \int_0^{\infty} (T - m_T)^2 \, p(T) \, dT \qquad (14)$$

In general, precise knowledge of the initial crack is not available, but its probability distribution can be assumed. Thus the unconditional counterparts of Eqs.(11) through (14) can be obtained by averaging over the range of initial conditions.

In applying the above procedure, the mean and variance of $Y(t)$ are specified in the stochastic model for the random material resistance to fatigue crack growth, and a key ingredient is the covariance of $\Delta X(t)$ at two different times, which is needed to formulate the covariance of g_t and $g_{t+\tau}$. Of course, the covariance of $\Delta X(t)$ is obtainable if the second order probability is known. Unfortunately, to the authors' knowledge, this latter information is not available for a general stationary stress process. We have studied an important special case of a stationary narrow-band Gaussian stress process (Zhu *etal* (1992)). In what follows, we shall investigate in detail the case in which stress in the elastic structural component where crack propagates is proportional to the displacement of a hysteretic structural system under stationary wide-band random excitation.

3 HYSTERETIC STRUCTURES

Displacement of a SDOF hysteretic system is governed by the following nonlinear differential equation

$$\ddot{x} + 2\zeta\dot{x} + \alpha x + (1-\alpha)z = \xi(t) \ , \qquad (15)$$

where Z= hysteretic component of restoring force, which is modeled by the following first order nonlinear differential equation

$$\dot{z} = -\gamma|\dot{x}|\,z\,|z|^{n-1} - \beta\,|\dot{x}|\,|z|^n + A\dot{x} \ , \qquad (16)$$

where A, n, β and γ are positive constants controlling the hysteretic loop. If $\xi(t)$ is stationary wide-band random process which can be modeled as a Gaussian white noise with intensity 2D, then applying the Stochastic averaging of energy envelope (see Zhu and Lin (1991)) leads to the following Fokker-Planck equation

$$\frac{\partial p}{\partial t} = -\frac{\partial}{\partial e}[V(e)p] + \frac{1}{2}\frac{\partial^2}{\partial e^2}[U^2(e)p] \ , \qquad (17)$$

where p = p(e, t | e_0, t_0) is the transition probability density of energy envelope $E(t) = \dot{x}^2(t)/2 + G[x(t)]$, and

$$U(e) = -(2\zeta\Phi(e) + A_r)/T(e) + D \qquad (18)$$

$$V^2(e) = 2D\Phi(e)/T(e) \qquad (19)$$

$$\Phi(e) = 2\int_{-a}^{a}\sqrt{2e - 2G(x)}\ dx \qquad (20)$$

$$T(e) = 2\int_{-a}^{a}\frac{dx}{\sqrt{2e - 2G(x)}} \qquad (21)$$

For the case of A=n=1, $\beta=\gamma=0.5$ and $\dot{x} \geq 0$,

$$G(x) = \alpha x^2/2 + (1-\alpha)(x+x_0)^2/2 \ , \ -a\leq x\leq -x_0 \quad (22a)$$

$$G(x) = \alpha x^2/2 + (1-\alpha)[1-e^{-(x+x_0)}]^2/2, \ -x_0\leq x\leq a \ (22a)$$

$$A_r = (1-\alpha)[4x_0 - (x-x_0)^2] \ , \qquad (23)$$

where a and x_0 are determined by nonlinear equations which have been derived by Zhu and Lin(1991). The expressions of G(x) and A_r in the region $\dot{x} < 0$ and those for other values of A ,n, β and γ are similar (see Cai and Lin (1990)).

The stationary probability density of E(t) is obtained from solving the reduced version of Eq.(17) without the time-derivative term

$$p_s(e) = C_1\ T(e)\ \exp\left\{-\int_0^e\left[\frac{2\zeta}{D} + \frac{A_r(e')}{D\Phi(e')}\right]de'\right\} \quad (24)$$

where C_1 is a normalizing constant. Therefore,

$$p(e) = p_s(e) \qquad , \qquad (25)$$

while the transition probability density p = p(e_2, τ| e_1) of E(t) can be evaluated numerically. According to the characteristic of Markov process E(t), the second order probability density of stationary process E(t) is

$$p(e_1,e_2; \tau) = p(e_2, \tau | e_1)\ p(e_1) \ , \qquad (26)$$

where $\tau = t_2 - t_1$. Define an amplitude envelope process S(t) by

$$G[S(t)] = E(t) \qquad (27)$$

The first order and second order probability densities of stationary process S(t) are then

$$p(s) = p(e)\ de/ds = p(e)\left|G'(s)\right|_{e=G(s)} \qquad (28)$$

$$p(s_1,s_2; \tau) = p(e_1, e_2,\tau)\left|G'(s_1)\ G'(s_2)\right|_{e_i=G(s_i)}, \quad (29)$$

in which i=1,2

The average number of cycles per unit time μ can be evaluated by

$$\mu = v_0^+ = \int_0^\infty \dot{x}p(0,\dot{x})\ d\dot{x} \qquad (30)$$

where

$$p(x,\dot{x}) = p(e)/T(e)\ |_{e=\dot{x}^2/2+G(x)} \qquad (31)$$

If the displacement response of the hysteretic structure is a rather narrow band process, the displacement range can be replaced approximately by 2S(t). Assuming further the stress in the elastic component where crack propagates is in proportional to the displacement of the whole system, $\Delta X(t)$ in Eq.(6) is also proportional to 2S(t). The covariance of $\partial g/\partial A \mid_t$, $g_{t+\tau}$ and g_t in Eqs.(8) and (9) can be evaluated in the terms of Eqs.(28)and (29). Thus, the general procedures described in section 2 can be applied.

4 EXAMPLE

Consider a thin square plate 1×1 shown in Fig.1, with an initial central crack of length $2a_0$ and supporting an infinitely rigid heavy mass M at its end together with an another component. The mass M is subjected to a stationary wide-band load process $\xi(t)$ perpendicular to the crack with a one side power spectral density S_0. The plate is idealized to be elastic, massless, homogeneous, isotropic, and with light linear damping, whereas plasticity might occur in the other component under severe load. Due to the yielding of the other component, the total restoring force of the system is assumed to be hysteretic form.

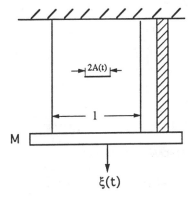

Fig.1 Hysteretic structural system under load $\xi(t)$ with initial crcak in the elastic plate

In this example, the stress in the plate is in proportion to the displacement of the mass M. We are interested in the probabilistic and statistical properties of the fatigue crack length, fatigue life and reliability as function of time and critical crack size a_{cr}. This problem has been studied for the case of linear restoring force (see Zhu *etal* (1992)).

Let $2A(t)$ and $X(t)$ be crack length and displacement. The displacement $X(t)$ satisfies the differential equation

$$M\ddot{X} + C\dot{X} + K[\alpha X + (1-\alpha)Z] = \xi(t) \quad , \quad (32)$$

where Z is described by Eq.(16). Eqs.(32) and (16) can be nondimensioned by introduction non-dimensional time $t' = \omega_0 t$ with $\omega_0 = K/M$. Assume that the following Paris-Erdogan law is applicable

$$\frac{dA}{dt'} = \mu\delta(\Delta k)^\upsilon \quad , \quad (33)$$

where δ and υ are two martial parameters,

Let h(a) be the stress intensity factor at the crack tip corresponding to a crack length 2A and unit

stress range. An approximate expression for h(a) has been provided (see Zhu (1992)) as follows:

$$h(a) = \sqrt{u} \ (0.467 - 0.514u + 0.96u^2 + \cdots) \quad (34)$$

with $u = 2a/l$. Thus

$$\Delta k = K\Delta X \ h(A) \ , \quad (35)$$

Eq.(33) can be rewritten as

$$\frac{dA}{dt'} = \lambda Q(A)S^\upsilon \quad ,, \quad (36)$$

where

$$\lambda = 2^\upsilon \mu\delta K^\upsilon \quad (37)$$

$$Q(A) = h^\upsilon(A) \ . \quad (38)$$

Eq.(36) is a special case of Eq.(6) with ΔX replaced by 2S and $Y(t)=1$.

Applying the above procedure, we obtained the following transition probability density of crack size

$$p(a,\tau'|a_0) = \frac{\exp\left[-(b_{cr} - m\tau')^2/2\sigma^2\tau'\right]}{\sqrt{2\pi\tau'} \ Q(a)\Phi(m\sqrt{\tau'}/\sigma)} \ , \quad (39)$$

where $\tau' = t_2' - t_1'$, $\Phi(\cdot)$ is the standard normal distribution function, and

$$b_{cr} = \int_{a_0}^{a_{cr}} \frac{du}{Q(u)} \quad (40)$$

$$m = \lambda \int_0^\infty s^\upsilon p(s) \ ds \quad (41)$$

$$\sigma^2 = 2\lambda^2 \int_0^\infty [\int_0^\infty \int_0^\infty s_1^\upsilon s_2^\upsilon \ p(s_1,s_1;\tau')ds_1 ds_2 - m^2]d\tau \quad (42)$$

p(s) and $p(s_1,s_2,\tau')$ are defined in Eqs.(28) and (29), respectively. The conditional reliability function is

$$R(a_{cr},\tau'|a_0) = 1 - \Phi\left[\frac{m\tau' - b_{cr}}{\sigma\sqrt{\tau'}}\right] / \Phi\left[\frac{m\sqrt{\tau'}}{\sigma}\right] \ . \quad (43)$$

The probability density of fatigue life T' is

$$p(T') = \frac{b_{cr} + mT'}{2\sigma T'\sqrt{2\pi T'}\Phi(m\sqrt{T'}/\sigma)} \exp[-\frac{(mT' - b_{cr})^2}{2\sigma^2 T'}]$$
$$- \frac{m\Phi[(mT' - b_{cr})/\sigma\sqrt{T'}] + mT'}{2\sigma T'\sqrt{2\pi T'}\Phi^2(m\sqrt{T'}/\sigma)} \exp(-\frac{m^2 T'}{2\sigma^2}) \quad (44)$$

The mean and variance of fatigue life T' can be evaluated by using Eqs.(13) and (14), respectively.

Numerical results are obtained for the following parameter values:

$l=0.254m$, $M=5.35kg$, $C=4.375kgs^{-1}$
$K=2.68\times10^3 N/m$, $\delta=0.66\times10^{-6}$, $\upsilon=2.2$:
$2a_0=0.00254m$, $2a_{cr}=0.0254m$

The mean $m_{T'}$ and stand deviation $\sigma_{T'}$ of the fatigue life T' for $\alpha=0.5$, 0.1 and a range of excitation intensity are shown in Figs.2-5. Comparison of the analytical results (denoted by solid lines) with those from digital simulation (denoted by symbol +) is made.

For moderate hysteresis, $\alpha=0.5$, it is seen that the theoretical results agree very well with those from simulation. For strong hysteresis, $\alpha=0.1$, however, there is certain discrepancy when excitation level is in the intermediate level $(0.3<\sigma_1<1.0)$. The theoretical results are on the safe side. The error is due to the fact observed from response samples that displacement response for strong hysteresis is not barrow-band. Iwan *etal* (1968) also showed that the power spectral density in this case is broad-band.

Wrischig (1982) proposed an empirical relation for the fatigue damage under not narrow-band stress process as

$$D = \lambda D_N ; \quad \lambda < 1.0 \qquad (45)$$

where D_N is the damage indicator under narrow-band stress process. Therefore, the above comparison coincides with Wrischig's conclusion.

Similar conclusion is also found about the comparison of reliability function, probability density function of fatigue life and statistics of crack size.

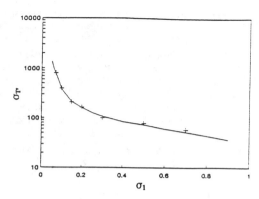

Fig. 3. Standard deviation $\sigma_{T'}$ of fatigue life vs load spectral level S_0 for $\alpha=0.5$

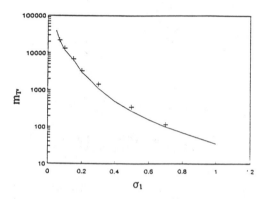

Fig. 4. Mean fatigue life $m_{T'}$ vs load spectral level S_0 for $\alpha=0.1$

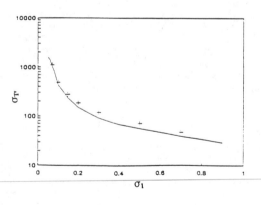

Fig. 5. Standard deviation $\sigma_{T'}$ of fatigue life vs load spectral level S_0 for $\alpha=0.1$

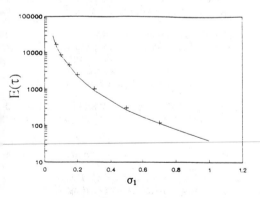

Fig. 2. Mean fatigue life $m_{T'}$ vs load spectral level S_0 for $\alpha=0.5$. $\sigma_1=\sqrt{2S_0/M^2\omega_0^3}$

Fig.6 shows the effect of hysteresis on the mean value of fatigue life. In the figure, mean value of fatigue life of hysteretic structural system with $\alpha=0.5$ is denoted by solid line while the corresponding result for elastic system, i.e., without the yielding of the other component, is denoted by dashed line. It is seen when excitation is very small ($\sigma_1 < 0.05$), hysteresis is not significant. In the interval of intermediate excitation strength, fatigue life is prolonged due to hysteresis compared to the elastic system. This effect is more apparent when υ becomes larger.

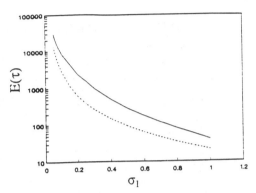

Fig.6 Mean fatigue life m_T vs load spectral level S_0 for hysteretic or elastic system

For intermediate level of excitation, the Kurtosis value defined as

$$K = E(x^4) / \sigma_x^4 \qquad (46)$$

is smaller than 3, Therefore the theoretical result agrees with the results by Lutes (1984) about the effect of non-normality on fatigue life. Moreover, it is also noted from the above comparison that the effect of nonnormality of the stress process due to nonlinear hysteretic system on the fatigue life is more significant than that of non narrow-band stress process.

5 CONCLUSIONS

A method is presented to obtain the probability density of fatigue crack size and fatigue life, and reliability of hysteretic mechanical or structural components with random material resistance under random loading. Analytical expressions are given for the important case of randomized Paris-Erdogan law. Comparison of the analytical results with those from digital simulation for an example shows that the present method yields good results for wide ranges of parameter values

ACKNOWLEDGMENT

The first author thanks the support from National Science Function of China and from The Laboratory for Strength and Vibration of Mechanical Structures.

REFERENCES

Bolotin V.V. 1989. *Prediction of Service Life for Machines and Structures*, ASME press, New York.
Cai G. Q. and Lin Y. K. 1990. On randomly excited hysteretic structures. *J. Appl. Mech.*. 57:442-448.
Hung X. and Hancock J.W. 1989. A reliability analysis of fatigue crack growth under random loading. *Fatigue Fracture of Engineering Materials and Structures*. 12: 247-258.
Iwan W.D. and Lutes L.D. 1967. Response of bilinear hysteretic system to stationary random excitation. *J. Acoust. Soc. Am.* 43: 545-552.
Khasminskii R. L. 1966. A limit theorem for the solutions of differential equation with random right-hand sides. *Theory of Probability and its Application* 11: 390-405.
Lin. Y. K. and Yang J. N. 1985. A stochastic theory of fatigue crack propagation. *AIAA Journal* 23, 257-270.
Lutes L.D., Corazao M., Hu S.L.J. and Zimmerman J. 1984. Stochastic fatigue damage accumulation. *J. Struct. Eng. ASCE* 110: 2585-2601.
Ortis K. and Kiremidjian A. S. 1988. Stochastic modelling of fatigue crack growth. *Engineering Fracture Mechanics* 29: 317-334.
Sobcyzk K., 1986. Modelling of random fatigue crack growth. *Engineering Fracture Mechanics* 24: 609-623.
Spencer Jr. B.F. and Tang J. 1989. Stochastic approach to modeling fatigue crack growth. *AIAA Journal,*, 27: 1628-1635.
Stratonovich R. L. 1963. *Topics in the theory of random noise*. Vol. 1, Gordn 6 Breach, New York.
Tsurui A. and Ishikawa H. 1986. Application of the Fokker-Planck equation to a stochastic fatigue crack growth model. *Structural Safety* 4: 15-29.
Wrischig, D.H. and Light, M.C. 1982. Fatigue under wide band random stresses. *J. Struct. Eng. ASCE* 106: 1593-1607.
Zhu W. Q. 1988.. Stochastic averaging methods in random vibration. *Appl. Mech. Rew.*, 41(5): 189-199.
Zhu W. Q. and Lin Y. K. 1988. Stochastic averaging of energy envelope. *J. Eng. Mech. Div., ASCE*, Vol. 117: 2407-2428.
Zhu W. Q., Lin Y. K. and Lei Y. 1992. On fatigue crack growth under random loading. *Engineering Fracture Mechanics* 43: 1-12.

Fatigue and fracture (ongoing research)

Structural Safety & Reliability, Schuëller, Shinozuka & Yao (eds) © 1994 Balkema, Rotterdam, ISBN 90 5410 357 4

Probabilistic fracture mechanics analysis of steam generator tubes

L. Cizelj & B. Mavko

Reactor Engineering Division, 'Jožef Stefan' Institute, Ljubljana, Slovenia

ABSTRACT: The method proposed in the paper is utilizing the probabilistic fracture mechanics techniques to assess the steam generator tube rupture probability. In particular, axially oriented stress-corrosion through wall cracks in the tubes of steam generator of a PWR nuclear power plant are studied. The effects of in-service inspection reliability, sizing accuracy and tube repair strategies are analyzed together with the crack propagation predictions in order to allow for risk-based life time optimization without sacrificing the plant safety. Numerical example considers a severely affected steam generator during hypothetical accidental conditions. Further, the efficiency of some numerical techniques employed to determine the failure probabilities is compared. Also, some comments on the applicability of different numerical techniques to determine failure probabilities are given followed by some topics for the future work.

1 INTRODUCTION

Steam generator tubing represents a substantial part of the second fission product barrier in a pressurised water reactor nuclear power plant. Various ageing processes might significantly decrease its structural reliability. In particular, the stress corrosion cracking of Inconel-600 tubes results in deep, often through wall-axial cracks in the areas with high residual stresses such as expansion transition zone at the top of the tube sheet. The number of affected tubes may easily reach up to 1/3 of the tubing in a steam generator (Hernalsteen 1991).

Maintenance procedures such as non-destructive in-service inspections and tube plugging are usually performed to control the level of the steam generator structural reliability. One of the recently developed maintenance strategies (van Vyve and Hernalsteen 1991) which allows for safe operation with cracks up to a certain crack length is investigated in this paper. The basic goal of the research work which is partially presented in this paper is to estimate the efficiency of the maintenance strategy in terms of steam generator failure probability.

A suitable probabilistic fracture mechanics model has already been proposed by Cizelj and Mavko (1992) and further elaborated by Mavko et al (1991). Some additional properties of the crack

detection reliability are addressed here by the help of a numerical example, representing a typical steam generator during hypothetical accidental conditions.

Some comments on the applicability of different numerical techniques implemented to determine failure probabilities are given and some topics for the future work are indicated.

2 STRUCTURAL RELIABILITY CALCULATIONS

To obtain the failure probability P_f, the well established procedure in statistical structural reliability theory requires the solution of the *failure integral*:

$$P_f = \int_{g(\underline{x} \leq 0)} f(\underline{x})\, d\underline{x} \qquad (1)$$

$f(\underline{x})$ being the joint probability density of basic variables $\underline{x}$ and $g(\underline{x})$ the limit state function. The value of the limit state function is by definition negative for all possible failure states and positive otherwise.

2.1 Limit state function

The limit state function may be defined following the two criteria approach (Dowling and Townley, 1975) by the limit load stress:

$$g(p, R, t, a, \sigma_Y + \sigma_M, K, \delta) = \sigma_f - m \cdot \sigma_\phi \quad (2)$$

Hoop stress σ_ϕ depends on the differential pressure p, tube mean radius R and tube wall thickness t. The flow stress σ_f is defined in the usual way (see for example (Larsson and Bernard 1978)) for the elastic-plastic material with significant degree of strain-hardening as:

$$\sigma_f = K \left(\sigma_Y + \sigma_M \right) \delta \quad (3)$$

The sum of yield and ultimate tensile stress $(\sigma_Y + \sigma_M)$, the flow stress factor K and operating temperature compensation factor δ are considered to be material properties (see for example Cochet and Flesch (1989)). Bulging factor m depends on the crack length a and tube geometry as defined by Erdogan (1976). Further details on the failure function applied in this case may be found in Mavko and Cizelj (1992).

2.2 Maintenance activities

In the operating steam generator, the vast majority of the influencing parameters and their respective probability distributions are already defined prior to the operation. We will assume here that the only change affecting the structural reliability is the stable propagation of axial cracks. Therefore, the state of the steam generator tubing under investigation is defined by the number of cracks and distribution of their lengths. Furthermore, all maintenance activities will affect only the number and distribution of the cracks and their respective lengths. This is in good agreement with maintenance strategies in field use (see for example van Vyve and Hernalsteen (1991)).

The non-destructive in-service inspection is performed to detect and size the cracks in steam generator tubing. The number and length distribution of cracks is estimated based on the in-service inspection results. The tubes containing cracks exceeding specified allowable length called *plugging limit (PL)* are then removed from service (e.g., plugged). However, a certain amount of cracks exceeding *PL* may be missed during the inspection process. This fraction is governed by the detection reliability function $(P_d(a_m))$, which is generally a function of crack length. We may summarize this behaviour in:

$$a_0 = \begin{cases} a_m + a_e, & a_m + a_e < PL \\ a_m + a_e, & a_m + a_e \geq PL \text{ and } \zeta \leq P_d(a_m) \\ 0 & \text{otherwise} \end{cases} \quad (4)$$

a_0 is a random variable representing the crack length obtained after the end of the maintenance process, while a_m and a_e represent the *as measured* crack length and measurement error, respectively. ζ stands for uniformly distributed random variable.

Eq. (4) assumes that the real crack lengths can be approximated by the measured crack lengths. Better approximations can be obtained following the procedure of Barnier et al (1992).

2.3 Detection reliability

Unfortunately, the information about the detection reliability available in the literature is limited to sentences like "All cracks in excess of 3 mm were detected" (Dobbeni 1991). To allow for the probabilistic assessment of this important parameter, an exponential distribution has been selected for the detection probability (Mavko et al 1991):

$$P_d(a) = 1 - \exp(-\lambda a) \quad (5)$$

2.4 Stable crack propagation

At the present, no suitable physical crack propagation model appears to be available. The best prediction results may be obtained by the stochastic model developed by Hernalsteen (1991) by analysing large data bases containing records of past non-destructive examinations. Essentially, this model predicts the crack propagation to follow the gamma distribution with parameters, depending on the initial crack length. Thus, the end of inspection crack length a_1 may be defined as:

$$a_1 = a_0 + a_g \quad (6)$$

Recent sensitivity analyses by Cizelj et al (1993a) showed the utmost importance of the crack propagation. Therefore, more reliable crack propagation modelling is recognized as an important topic for the future research. Equation (6) may therefore be replaced by a suitable crack propagation model (see for example Cizelj et al (1993b)).

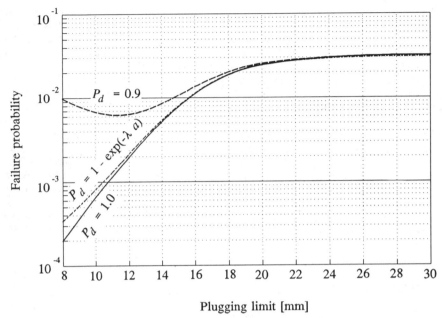

Fig. 1 Tube failure probability as a function of plugging limit

3 NUMERICAL EXAMPLE

A typical Westinghouse D-4 steam generator is considered as numerical example. The tubing with nominal dimensions $t=1.055$ mm and $R=8.525$ mm is exposed to an accidental differential pressure of 196 bar. Reader is referred to Mavko and Cizelj (1992) for more detailed input data description.

The failure probability of at least one tube in the tube bundle has been assessed for different detection reliability assumptions. While the realistic detection technique follows eq. (5) with $\lambda=0.45$, both the upper and the lower detection probability limits were considered independent of crack length and denoted as $P_d=1.0$ and $P_d=0.9$, respectively.

Results have been obtained by the Direct Monte Carlo (DMC) simulation and presented in Fig. 1 for different plugging limit (PL) values. From Fig. 1 the minimum failure probability PL can be derived in case of constant, 90% efficient detection technique ($P_f=0.9$). In remaining cases, only relatively small influence of the crack length dependent detection probability is shown, having increasing tendency when reducing the PL value. This is suggesting that very reliable detection should be used in order to apply low PL values. Such behaviour is consistent with eq. (5).

The failure probabilities presented in Fig. 1 are conditional. The hypothetical steam line break accident is assumed as an initiating event.

3.1 Applicability of numerical methods

The analysis of the perfect detection probability ($P_d=1.0$) has been also performed using First- and Second Order Reliability Methods (FORM and SORM, respectively) by Cizelj et al (1993a) and the following versions of Monte Carlo simulations: Importance Sampling MC (ISM, Brueckner 1987), Adaptive Sampling MC (ASM. Bucher 1988), and Efficient Sampling MC (ESM, Harbitz 1986) . In the PL range of practical interest (say 8 to 14 mm), the agreement of all methods implemented is quite good and favours the use of FORM and SORM because of their extremely low computing requirements. Further, in the PL range of practical interest, FORM consistently overestimates the DMC failure probability (below 10%) while SORM consistently underestimates (below 5%) it. Using both FORM and SORM together always bounded the DMC result.

4 CONCLUSIONS

A procedure to estimate the failure probability of axially cracked steam generator tubes has been proposed in the paper. Basically, a mechanical failure model is upgraded to include the maintenance strategy effects. The procedure is shown to be flexible and easy to use, especially when performing parametric studies. At the moment, First and Second order reliability methods seem to be the most cost-effective numerical method to solve for the failure probability.

The effects of the crack detection reliability are addressed in a numerical example. The influence of the real detection technique reliability increases with decreased plugging limit value. Special attention is therefore needed when applying low plugging limit values.

5 ACKNOWLEDGEMENTS

The financial support of the International Office of KFA Juelich, Germany, and the Ministry for Science and Technology of Slovenia is gratefully acknowledged by the authors.

6 REFERENCES

Barnier, M., P.Pitner and T.Riffard 1992. Estimation of Crack Size Distributions from In-Service Inspection Data for the Calculation of Failure Probabilities. *Safety and Reliability*, Copenhagen.

Brueckner, A. 1987. Numerical Methods in Probabilistic Fracture Mechanics (in Probabilistic Fracture Mechanics and Reliability, edited by J.W.Provan). Martinus Nijhoff Publishers, pp. 351-386.

Bucher, C.G. 1988. Adaptive Sampling - An Iterative Fast Monte Carlo Procedure. *Structural Safety* 5: 119-126.

Cochet, B. and B.Flesch 1989. Application of the Leak Before Break Concept to Steam Generator Tubes. *Trans. of the 9th Conf. on SMiRT*, Annaheim, Ca., USA, Vol G: 299-304.

Cizelj, L., B.Mavko and H.Riesch-Oppermann 1993a. Application of First and Second Order Reliability Methods in the Safety Assessment of Cracked Steam Generator Tubing, submitted to *Nucl Eng Des*.

Cizelj, L., B.Mavko, H.Riesch-Oppermann, A.Brueckner-Foit 1993b. Propagation of Stress Corrosion Cracks in Steam Generator Tubes. *Proc. of 12th SMiRT Conference*, Stuttgart, Germany, paper DG07/2.

Dobbeni, D. 1991. Eddy Current Inspection Methodology. *NEA-CSNI-UNIPEDE Specialist Meeting on Operating Experience with Steam Generators*, Brussels, Belgium.

Dowling, A.R. and C.H.A.Townley 1975. The Effect of Defects on Structural Failure: A Two Criteria Approach. *Int J Press Ves & Piping* 3: 77-107.

Erdogan, F. 1976. Ductile Fracture Theories for Pressurized Pipes and Containers, *Int J Press V & Piping* 4: 253-283.

Harbitz, A. 1986. An Efficient Sampling Method for Probability of Failure Calculation. *Structural Safety* 4: 109-115.

Hernalsteen, P. 1991. Prediction Models for the PWSCC degradation Process in Tube Roll Transitions. *NEA-CSNI-UNIPEDE Specialist Meeting on Operating Experience with Steam Generators*, Brussels, Belgium.

Larsson, H. and J.Bernard 1978. Fracture of Longitudinally Cracked Ductile Tubes. *Int J Press Ves & Piping* 6: 223-243.

Mavko, B., L.Cizelj and G.Roussel 1991. Steam Generator Tube Rupture Probability Estimation - Study of the Axially Cracked Tubes Case. *NEA-CSNI-UNIPEDE Specialist Meeting on Operating Experience with Steam Generators*, Brussels, Belgium.

Mavko, B., L.Cizelj 1992. Failure Probability of Axially Cracked Steam Generator Tubes: A Probabilistic Fracture Mechanics Model. *Nuclear Technology* 98: 171-177.

van Vyve, J. and P.Hernalsteen 1991. Tube Plugging Criteria for Axial and Circumferential Cracks in the Tubesheet Area. *NEA-CSNI-UNIPEDE Specialist Meeting on Operating Experience with Steam Generators*, Brussels, Belgium.

Structural Safety & Reliability, Schuëller, Shinozuka & Yao (eds) © 1994 Balkema, Rotterdam, ISBN 90 5410 357 4

Investigation and improvement of a new model for fatigue life prediction based on level crossing

M. Holmgren & T. Svensson
Swedish National Testing and Research Institute, Borås, Sweden

ABSTRACT: A new method for fatigue life prediction has been investigated experimentally and numerically. The method is developed by Holm and de Maré (1988). I the life time predictions level crossing data of the applied load are used. The results from the investigation show that the new method sometimes is superior to the traditional method for life time predictions (rain flow count in combination with the Palmgren-Miner rule). This is probably due to the fact that the new method takes small superimposed load cycles into account.

1 INTRODUCTION

The majority of failures of structural elements occur because of fatigue. This make the fatigue problem to one of the main concerns for the designer. When the expected applied load has constant amplitude, the design based on fatigue life predictions is rather straightforward, but in most situations the amplitude varies in time. In these cases it is not so obvious how to design against fatigue.

The mostly used method to predict fatigue life under variable amplitude loading is the Palmgren-Miner rule together with rain flow count (RFC). This method for fatigue life prediction has proved to give good predictions in many cases, but it is rather complicated to use and gives sometimes non-conservative results.

In 1988 Holm and de Maré presented a simple model for fatigue life using level crossing. This model has two main advantages:

- complicated cycle counts are avoided
- it is possible to include a variety of phenomenological aspects in the simple model.

From this model a method for life predictions has been developed. The method has been ex-

perimentally investigated by Svensson and Holmgren (1993) and there is ongoing research to improve the model. Some of the work done and plans for future work are summarized in this paper.

2 THEORY

The simple model for fatigue life, developed by Holm and de Maré is based upon two fundamental assumptions:

1. The linear Palmgren-Miner rule is valid
2. The damage depends only on local maxima and minima of the load and not on the waveform between them.

From these assumptions Holm and de Maré have shown that there exists a function, $G(\sigma)$, σ = stress level, which determines the fatigue life, N, of a component under variable amplitude loading. The total damage, D, can be predicted using the number of level crossings, $n(\sigma)$ and a damage density function, $g(\sigma)=dG(\sigma)/d\sigma$, according to eq.(1).

$$D = \int_{\sigma_{min}}^{\sigma_{max}} n(\sigma)g(\sigma)d\sigma \qquad (1)$$

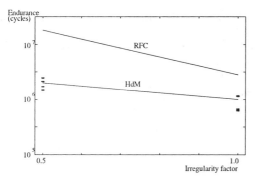

Figure 1. Test results (*) and results of life predictions (solid lines) for smooth specimen made of structural steel. Life as a function of irregularity factor.

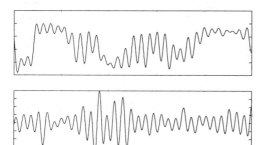

Figure 2. Parts of load sequences with irregularity factors of 0.5 and 1 respectively.

The damage density function, $g(\sigma)$, must be chosen in suitable way to satisfy for example some empirical relationship. One way is to use data from a constant amplitude test with $R = 0$, $\sigma_{max} = S$, $\sigma_{min} = 0$ and $n(s) = N$. Eq. (1) will then turn into the following form:

$$D = \int_0^S N\, g(\sigma)\, d\sigma \qquad (2)$$

By using the equation for the S-N-curve from a constant amplitude test, $N = \alpha\sigma^{-\beta}$, and the Palmgren-Miner rule, $D = 1$, the following damage density function is obtained:

$$g(S) = \frac{\beta}{\alpha} S^{\beta-1} \qquad S > 0 \qquad (3)$$

Predictions made with this model (HdM-method) show promising results. However, for some welded specimens tested with standardized blocked load sequences the new model underestimate the fatigue life. To model the special effects occurring in welded specimens, where there already exists initial defects, another damage density function has been proposed in order to take crack closure effects into account. This modified model is based on the additional assumption:

3. There exist a stress level, in this case S_{op}, below which no damage is caused. S_{op} is smaller than S.

With this new assumption the damage density function becomes

$$g(S) = \frac{b}{a(1-K)^{b-1}}\left(S - S_{op}\right)^{b-1} \quad K = \frac{S_{op}}{S_{max}} \qquad (4)$$

When the largest part of the fatigue life is used by crack propagation, this model (HdM30) predicts fatigue life better than the one with damage density function (3).

3 EXPERIMENTS, NUMERICAL SIMULATIONS AND RESULTS

To verify the new method with damage density function (3), fatigue experiments on smooth steel specimens were made. The results from the tests with variable amplitude were compared with life time predictions made with the new method. As a comparison, life time predictions for the same load sequences were made with the traditional method using rain flow count and the Palmgren-Miner law. The results are shown i figure 1.

Experimental data taken from the Gurney (1989) and Spennare, Samuelsson (1991) were used to compare predictions with tests on welded specimens. These data were analysed in the same way as the data mentioned above.

To verify the model with damage density function (4) and also to test the method on different materials a serial of tests were performed on welded structural steel and on welded stainless steel. Four different load sequences were used during the tests. The load sequences were obtained by combinations of two load-spectra

(convex and concave) and two irregularity factors (0.5 and 1). In figure 2 parts of two of the load sequences are shown and figures 3 and 4 show results.

4 DISCUSSION

The life predictions made with the HdM-method and with damage density function (3) for smooth specimens agree very well with experimental data, while life predictions made with RFC overestimate the fatigue life as much as ten times. The situation was almost the same when the method was applied on test results reported by Gurney (1989). On the other hand the RFC predictions for Spennares and Samuelssons results (1991) were much better than those made with the new method which underestimated the real life times sometimes more than ten times.

When the damage density function (4) was used for life time predictions, S_{op} was approximated to about $0.3 \times S_{max}$ according to a literature survey by Holmgren (1993). The life times predicted by the HdM-method was compared with the life times predicted by the RFC-method. With the new damage density function the HdM-method was able to predict the life time rather well in most cases, while the RFC-method overestimated it. These trends were most pronounced for the stainless steel tested under irregular load. For the welded structural steel the methods showed similar results. In figure 3 and 4 some results are shown.

The results from smooth test specimens and from stainless steel specimens indicates that the HdM-method could be superior to the traditional RFC-method in certain cases. Namely when the S-N-exponent is large (stainless steel) and when a large amount of the fatigue life is initiation (smooth specimen).

5 FUTURE WORK

The new model is promising but how useful it is in real life is not yet known. One key to find out if the model can be developed for use in practical design is to find a phenomenological explanation for its advantages and drawbacks, compared to traditional methods. Some attempts

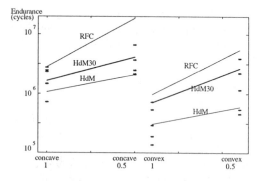

Figure 3. Test results(∗) and results of life predictions (solid lines) for welded specimens made of stainless steel under two different load-spectra. Life as a function of irregularity factor.

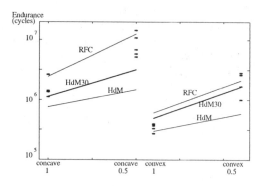

Figure 4. Test results(∗) and results of life predictions (solid lines) for welded specimens made of structural steel under two different load-spectra. Life as a function of irregularity factor.

have been made in this direction, but further research is necessary.

The new model is suitable for further improvements. As shown above it is possible to use different damage density functions. One way is to find a damage density corresponding to more than one S-N-curve, that is for different parts of the fatigue life: initiation , short crack growth or long crack growth.

The model has no memory and at the moment it is not possible to take sequence effects into account except as a constant S_{op}. The model could however be extended to contain a certain memory.

The things mentioned above are subject for presently ongoing research at the Swedish Na-

tional Testing and Research Institute in cooperation with the Department of Mathematics, Chalmers University of Technology.

6 REFERENCES

Holm, S. and de Maré, J. 1988. A Simple Model for Fatigue Life. *IEEE Transactions on Reliability* 37: 314-322.

Svensson, T. and Holmgren M. 1993. Experimental and Numerical Verification of a New Model for Fatigue Life. *Fatigue and Fracture of Engineering Materials and Structures*, Vol. 16: 481-493

Gurney, T. 1989. Cumulative damage of welded joints. Part 2-test results. *Joining and Materials,* August 1989: 390-395.

Spennare, H. and Samuelsson, J. 1991. Spectrum Fatigue Testing of a Longitudinal Non-load Carrying Fillet Weld - a Round Robin Exercise. Jernkontorets Forskning serie D nr 653, Stockholm, Sweden .

Holmgren, M. 1993. Crack Closure in Steel, a Literature Survey SP report 1993, Borås, Sweden.

Structural Safety & Reliability, Schuëller, Shinozuka & Yao (eds) © 1994 Balkema, Rotterdam, ISBN 90 5410 357 4

Inspection planning for fatigue damage using Bayesian updating

Jørgen Juncher Jensen
Department of Ocean Engineering, The Technical University of Denmark, Lyngby, Denmark

Alaa E. Mansour
Department of Naval Architecture and Ocean Engineering, University of California, Berkeley, Calif., USA

ABSTRACT: A Bayesian updating procedure suitable for handling crack inspection results for a large number of nearly identical structural details is described.

The applicability of the procedure is tested on some real inspection data for a jack-up platform.

1 INTRODUCTION

Marine structure are susceptible to fatigue damage due to the time varying hydrodynamic loads and the high stress levels. Therefore, in order to obtain a given high reliability of the structure, inspection schemes must be formulated in order to keep track of crack initiation and crack growth.

Here a Bayesian approach is taken following a procedure originally suggested by Itagaki and Shinozuka [1], [2], [3]. The statistical distributions for some of parameters describing the crack initiation, crack detection and crack propagation are updated at each inspection and used to derive better estimates for the number of severe cracks as function of time until the next inspection. Thereby it is possible to decide on an appropriate interval to the next inspection taking due account of both the cost of inspections and cost of repairs.

The procedure is based on Shinozuka's formulation [3], but extended such that any exponent in the Paris-Erdogan equation governing crack growth can be used [4]. The failure criteria is formulated in terms of an admissible maximum crack length with a given probability of occurrence and includes an empirical correction for deterioration effects due to corrosion. The crack detection strategy accounts for the possibility that only subsets of the total members are inspected.

In the present paper also different repair strategies depending on the size of the detected crack are included.

The procedure is applied to inspection data made available for three consecutive inspections of the K-joints in a jack-up rig. Based on this example, some conclusions regarding the usefulness and accuracy of the method are reached.

2 FATIGUE FAILURE

The formulation follows closely Shinozuka [3] with the extensions described in [4].

The time t_c to crack initiation is modelled by a Weibull distribution

$$F_c(t_c|\beta) = 1 - \exp\left((t_c/\beta)^\alpha\right) \; ; \; t_c > 0, \; \alpha > 0 \quad (1)$$

and an exponential distribution is used for the probability of detecting a crack of length a

$$F_D(a|d) = 1 - \exp\left(-d(a - a_0)\right) \; ; \; a \geq a_0 \quad (2)$$

The parameters α and a_0 are here given suitable deterministic values whereas β and d are taken as stochastic variables with distribution functions to be updated after each inspection.

After crack initiation at $t = t_c$ the crack growth is assumed to follow the Paris-Erdogan equation which in case of a crack-length-independent geometry function can be written ($m \neq 2$)

$$t - t_c = \frac{2}{(m-2)c}\left[1 - (a_c/a)^{m/2-1}\right] \quad (3)$$

where m is the exponent in the Paris-Erdogan equa-

tion, and a_c is the crack length at $t = t_c$. Both of these parameters are given proper deterministic values in the present procedure. The remaining parameter c in Eq (3) depends on the stress level in the detail, see e.g. [4], and is therefore taken as a stochastic variable to be updated after each inspection in the same way as β and d.

However, it should be emphasized that the procedure assumes basically identical details subjected to the same stress field. This assumption can seldom be expected to hold exactly in actual complex marine structures yielding bias factors on the value of c. Care has therefore to be taken when details are identified for inclusion in the procedure.

Member failure is defined by two failure rates valid before and after crack initiation, respectively. Details are given in [4]. Here only the resulting member reliability $V(a)$ for a member with a crack length a is stated

$$V(a) = \exp\left[-h_0/q\left\{(a/a_c)^{q/c} - 1\right\}\right] \qquad (4)$$

The parameter h_0 relates to the member reliability before crack initiation and q ($q > 0$) can be expressed in terms of an admissible maximum crack length with a given probability of occurrence. An empirical account for deterioration effects due to corrosion can be included, [4].

One notable observation from Eqs. (3)-(4) is that if $m > 2$ then

$$t \to t_c + \frac{2}{(m-2)c} \Rightarrow a \to \infty \Rightarrow V(a) \to 0 \quad (5)$$

Thus the member reliability tends to zero after a finite time $2/((m-2)c)$ after crack initiation.

3. SYSTEM RELIABILITY

The calculation of member and system reliability and the associated expected number of failed members as function of time after the last inspection is described in details in [3], [4]. Here only the modifications made to account for different repair strategies depending on the size of a detected crack will be outlined.

In the updating of the probability density functions for β, d and c after each inspection, a likelihood function is established, depending on the outcome of the inspection of each member. The possible outcomes are described by the probability $P_m(X_j : j, r)$ that

the event X_j takes place at the j-th inspection at $t = T_j$ provided the member was last repaired immediately after the r-th inspection. The events already considered in [3], [4] are X_j = 'member failure', 'detection of a crack length a_j' and 'nothing is detected'. These events cover all possible outcomes when all failures and all detected cracks are repaired immediately.

Now it is assumed that only detected cracks greater than a given size a_{min} will be repaired. Therefore events $(X_j : j, r \,|\, a_s : s)$ in which a crack length $a_s < a_{min}$ has been detected in the s-th inspection prior the j-th inspection must be included in the likelihood function. The probabilities of these events are approximated by

$$P_m \text{ (member failure} : j, r \,|\, a_s : s) = \qquad (6a)$$
$$1 - V(\bar{a}_j) / V(a_s)$$

$$P_m \text{ (detection of crack length } a_j : j, r \,|\, a_s : s) =$$
$$\begin{cases} 1 & \text{if } \bar{a}_j^- - \delta a \le a_j \le \bar{a}_j^+ + \delta a \qquad (6b) \\ \varepsilon & \text{otherwise} \end{cases}$$

$$P_m \text{ (nothing is detected} : j, r \,|\, a_s : s) = \qquad (6c)$$
$$\left(1 - F_D(\bar{a}_j \,|\, d)\right) V(\bar{a}_j) / V(a_s)$$

Here $\bar{a}_j$ is calculated from Eq. (3) using the knowledge that $(t, a) = (T_s, a_s)$ satisfies this equation. Inaccuracies in the crack length measurements have been included in Eq. (6b) by bounds calculated from Eq (3)

$$\bar{a}_j^{\pm} = a\left(t = T_j \mid a(t = T_s) = a_s \pm \delta a\right) \qquad (7)$$

with a user-defined value of the measurement accuracy δa. If the measured crack length a_j is outside this bound P_m is taken to be a small value ε, say $\varepsilon = 0.001$, as $P_m = 0$ will make the likelihood function exactly equal to zero. Similarly, if $F_D(\bar{a}_j \,|\, d) = 1$ as will be the case for large estimated cracks $\bar{a}_j$, the probability (6c) will be taken to be ε, too. The choice of ε is without significance, at least in the example to be considered.

Finally, the formulas given in [3], [4] for the member reliability $R_m(*)$ at time t^* after the last

Tabel 1

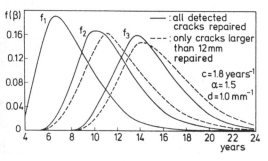

Fig. 1 Posterior function $f_j(\beta\,|\,d,c)$ for the parameter β related to crack initiation.

inspection at $t = T_j$ has to be supplemented by

$R_m(t^*$: crack of length a_s detected at $t = T_s$,
but no repair) $= V(t^*) / V(\bar{a}_j)$ (8)

As $V(t^* = T_j) = V(\bar{a}_j)$ the member reliability is 1 just after the j-th inspection in accordance with the basis assumption that all failures are detected and repaired at each inspection.

4. EXAMPLE

For a jack-up platform the results of three consecutive inspections of totally 243 hot-spot areas are given in Table 1, taken from [4].

The data were analyzed in [4] assuming that all detected cracks were repaired. Similar calculations are done here, but now only cracks larger than $a_{min} = 12$ mm are repaired, which much better models what actually was done. The measurement accuracy δa is taken to be 2 mm.

Fig. 1 shows the posterior function $f(\beta)$ conditioned on given values of d and c. As compared to the results found in [4] and included in Fig. 1, somewhat longer crack initiation periods are seen to be expected now. For the crack propagation parameter c, a comparison between Fig. 2 and the corresponding results in [4] shows that the most probable value of c now has decreased from about 1.7 years⁻¹ to about 1.5 years⁻¹. As seen from Eq. (5) this implies a slower crack growth. Both of these results are to be expected as the inspection data are the same, but fewer cracks are now assumed repaired. Clearly neither of the posterior functions $f(\beta)$ and $f(c)$ becomes independent of inspection number which they should have been if the true distributions of β and c were obtained.

However, the original uniform distributions have been improved by the Bayesian updating. As main result Fig. 3 shows the expected average number $\bar{N}_f$ of failures (or large cracks) as function of time after each inspection. The figure includes the results from [4] and it is seen that until about one year after each inspection the average numbers of failures now become slightly larger. However, later on, this difference disappears due to the low number of detected, non-repaired cracks.

Finally, it should be mentioned that even if the procedure seems to yield reasonable predictions in the present example, two contradictions are found

- The crack in hot-spot area 101 found at the first inspection is not repaired. With the present values of d and a_0 in the crack detection formula, Eq. (2), the crack should have been observed with a probability of one at the second inspection. No crack was, however, observed.
- Furthermore, this crack should have propagated to infinity (failure) in the interval between the second and third inspection, but no failure was actually observed at the third inspection.

The most probable explanation for these contradictions must be attributed to the assumption that all details are basically identical both with respect to

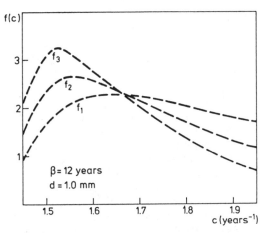

Fig. 2 Posterior function $f_i(c|\beta,d)$ for the para-
meter c related to crack growth.

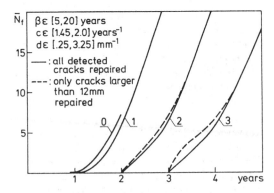

Fig. 3 Expected average number $\bar{N}_f$ of significant fatigue cracks in the 243 hot-spot
areas. 0: before inspection; 1,2,3: after
first, second and third inspection, respectively.

geometry and stress field. To overcome this problem
more elaborate methods, c.f. [5] treating the individual details separately should be used.

5. CONCLUSION

A procedure for analyzing crack inspection data has
been described and compared to real data for a jack-
up platform. Although some problems have been
identified, generally the described procedure seems to
be able to make use of the inspection data in a consistent way and thereby being a useful tool for deciding on a proper time interval to the next inspection.

6. REFERENCES

[1] Itagaki, H., Akita, Y., and Nitta, A., 1983, Appli-
cation of Subjective Reliability Analysis to the
Evaluation of Inspection Procedures on Ship
Structures, *Proc. Int. Symp. on the Role of
Design, Inspection and Redundancy in Marine
Structural Reliability*, National Academic Press,
Nov. 13-16, 1983.
[2] Itagaki, H. and Yamamoto, N., 1988, Bayesian
Reliability Analysis and Inspection of Ship Struc-
tural Members - An Application to the Fatigue
Failures of Hold Frames, *NK Tech. Bulletin and
PRADS'87*, Trondheim, June 22-26, 1987.
[3] Shinozuka, M., 1990, Relation of Inspection
Findings to Fatigue Reliability, *Ship Structure
Committee*, SSC-355.
[4] Jensen, J. Juncher and Pedersen, P. Terndrup,
1992, A Bayesian Inspection Procedure Applied
to Offshore Steel Platforms, *Proc. BOSS'92*,
London, July 17-10, 1992.
[5] Madsen, H.O., Sørensen, J.D. and Olesen R.,
1989, Optimal Inspection Planning for Fatigue
Damage of Offshore Structures, *Proc. ICOSSAR
'89*, San Francisco.

Structural Safety & Reliability, Schuëller, Shinozuka & Yao (eds) © 1994 Balkema, Rotterdam, ISBN 90 5410 357 4

Thickness effect on the statistical properties of fatigue crack propagation

T. Sasaki
Research Institute of Industrial Safety, Tokyo, Japan

S. Sakai & H. Okamura
The University of Tokyo, Japan

ABSTRACT : The statistical properties of crack growth fatigue life are often needed in a reliability-based design of fatigue-critical structures. However, recent research revealed that a decrease in specimen thickness caused an increase in the variability of crack growth fatigue life. Thus, in this paper, the dependency of the statistical properties of random crack propagation resistance on specimen thickness was investigated. Constant ΔK fatigue crack growth tests were conducted on 2024-T3 aluminum alloy in 4 cases of specimen thickness. From the experimental data, there was no negligible influence of thickness on the statistical properties of random crack propagation resistance. Using the stochastic crack growth model proposed in the previous paper, the thickness effect on the distribution of crack growth fatigue life was also investigated, and satisfactory agreement with the experiments was obtained.

1 INTRODUCTION

It is widely recognized that the fatigue crack propagation is fundamentally a random phenomenon which can be predicted only in terms of probability. Ortiz (1986) showed that the primary source of statistical variation of fatigue crack propagation is material inhomogeneity.

To explain its effect, the authors proposed a new stochastic model which treats the material's resistance against fatigue crack growth as a spatial stochastic process along the path of the crack (Sasaki *et al.* (1991a)).

In this paper, focus is centered on the effect of specimen thickness on the statistical properties of fatigue crack growth because Yoon (1988) revealed that a decrease in specimen thickness caused an increase in the variability of the distribution of crack propagation fatigue life.

2 REVIEW OF THE PROPOSED MODEL

In order to consider the random nature of fatigue crack growth, the power crack growth law is randomized using an additional non-negative non-dimensional stationary spatial stochastic process Z as

$$\frac{da}{dN} = \frac{1}{Z} C \left(\frac{\Delta K}{K_0} \right)^m \tag{1}$$

where a = crack size, N = number of load cycles, ΔK = the stress intensity factor range, C and m are constants, K_0 is a unit quantity with the same dimension as ΔK. Z can be supposed to be ergodic, and the ensemble average of Z is normalized as

$$E[Z] = 1 \tag{2}$$

where $E[\,\cdot\,]$ is expected value operator.

The spectral analysis revealed that the statistical characteristics of Z were as follows.

(1) 3-parameter Weibull distribution provides a good fit for Z.

(2) Z can be approximated by the mixture of a first-order Markov process M and a white noise process W. The two-sided power spectral density function of M and W are

$$P_M(f) = \frac{2\alpha_M \beta_M}{\alpha_M^2 + 4\pi^2 f^2} \tag{3}$$

and

$$P_W(f) = \beta_W \tag{4}$$

respectively, where f = spatial frequency, α_M, β_M and β_W are constants.

(3) The statistical properties of Z are independent of ΔK within the region of stage II fatigue crack growth.

3 EXPERIMENTAL PROCEDURE

To obtain the stochastic properties of Z, constant ΔK-controlled fatigue crack growth tests were conducted in 4 cases of specimen thickness.

Material used in this study was 2024-T3 aluminum alloy. Specimen geometry was ASTM standard compact tension type, as shown in Fig. 1, with a width $W = 50$mm and a thickness $B = 1$, 3, 5, 12.5mm.

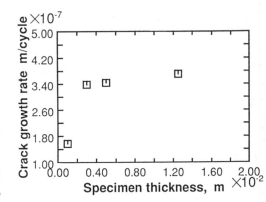

Fig. 2 Ensemble average crack growth rate vs. specimen thickness.

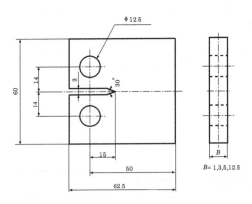

Fig. 1 Test specimen configuration.

All tests were carried out at $\Delta K = 15\text{MPa}\sqrt{\text{m}}$, stress ratio $R = 0.1$ in air at 15Hz. Number of specimen used in this study is shown in Table 1.

Table 1 Number of specimen used.

Thickness, B [mm]	1	3	5	12.5
Number of Specimen	5	4	5	4

The accumulated number of load cycles was recorded for each 0.20mm of crack growth.

4 EXPERIMENTAL RESULTS

4.1 Thickness effect on fatigue crack growth rate

The ensemble average fatigue crack growth rate

$E[\overline{da/dN}]$ is plotted vs. specimen thickness in Fig. 2.

It is observed that an increase in thickness causes an increase in da/dN. McGowan and Liu (1980) showed that this three-dimensional effect stemmed from the variation in constraint along the crack front from plane strain at the specimen center to plane stress at the specimen surface.

4.2 Distribution of Z

3-parameter Weibull distribution provides a good fit for Z as shown in Fig. 3.

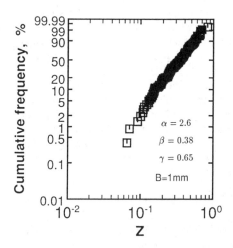

Fig. 3 Distribution of Z (Weibull Probability paper).

The distribution function of 3-parameter Weibull distribution is

$$F(Z) = 1 - \exp\left\{-\left(\frac{Z-\gamma}{\beta}\right)^{\alpha}\right\} \qquad (5)$$

where α = shape parameter, β = scale parameter and γ = location parameter.

4.3 Spectral Analysis of Z

The ensemble power spectrum density is shown in Fig. 4.

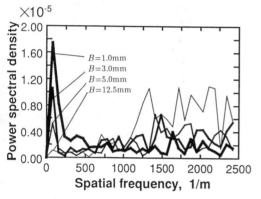

Fig. 4 Power spectral density of Z for ensemble.

This figure shows that a decrease in thickness causes an increase in the power of Z. In next section, it will be investigated whether or not the thickness effect on the distribution of crack growth fatigue life can be explained by this thickness effect on the power spectral density of Z.

5 THICKNESS EFFECT ON FATIGUE LIFE DISTRIBUTION

In order to calculate the distribution of crack growth fatigue life based on the proposed model, the parameters of eqn (3), (4) was evaluated from Fig. 4. Then the sample functions of Z were numerically generated with the aim of FFT algorithm (Yamazaki and Shinozuka (1988)). The ensemble power spectral density of simulated Z is shown in Fig. 5.

Next, using the simulated sample functions of Z, crack growth simulations were performed based

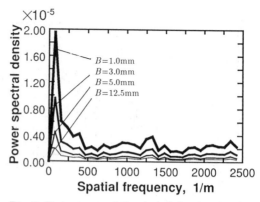

Fig. 5 Power spectral density of simulated Z for ensemble.

upon the proposed stochastic model (Sasaki et al. (1991b)). In these simulations, the same conditions as constant load amplitude fatigue crack growth tests, which were conducted by Yoon (1988) on the same 2024-T3 aluminum alloy as this study, were assumed.

Fig. 6 gives a comparison between the simulations and the experimental data. Exceptional agreement is found in every specimen thickness case.

6 CONCLUSIONS

The major conclusions of this investigation are summarized below.

(1) A decrease in specimen thickness causes an increase in the power of normalized crack propagation resistance Z.

(2) An increase in crack growth fatigue life caused by a decrease in specimen thickness can be well explained by the thickness effect on Z using proposed stochastic crack growth model.

ACKNOWLEDGEMENT

The authors wish to thank Dr. H.Y.Yoon for providing constant load amplitude fatigue crack growth data of 2024-T3.

REFERENCES

McGowan, J.J. and Liu, H.W. 1980. The role of

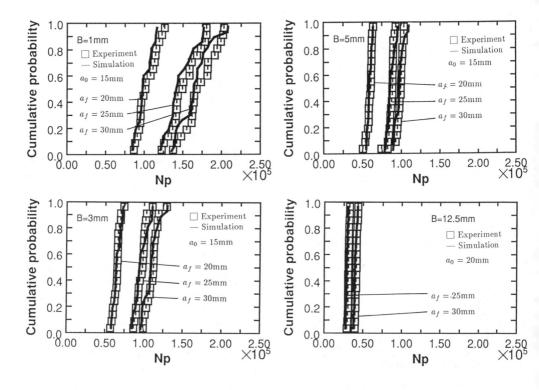

Fig. 6 Distribution of crack growth fatigue life.

three-dimensional effects in constant amplitude fatigue crack growth testing. *Transactions of the ASME, Journal of Engineering Materials Technology* 102: 341-346.

Ortiz, K. 1986. Time series analysis of fatigue crack growth rate data. *Engineering Fracture Mechanics* 24-5: 657-675.

Sasaki, T., Sakai, S. and Okamura, H. 1991a. Estimating the statistical properties of fatigue crack growth using spectral analysis technique. In *Proceedings of the 6th International Conference on the Mechanical Behaviour of Materials (ICM 6)*, Vol.1: 565-570. Kyoto: Pergamon.

Sasaki, T., Sakai, S. and Okamura, H. 1991b. Statistical evaluation of the distribution of crack propagation fatigue life by simulating the crack growth process, In *Proceedings of First International Conference on Computational Stochastic Mechanics*: 473-484. Corfu: Elsevier.

Yamazaki, F. and Shinozuka, M. 1988. Digital generation of non-Gaussian stochastic fields. *ASCE, Journal of Engineering Mechanics*, 114-7: 1183-1197.

Yoon, H.Y. 1988. A reliability engineering based study on the prediction of fatigue life of flaws (Translated by the authors). *Ph.D. thesis, The University of Tokyo*. (In Japanese)

Structural Safety & Reliability, Schuëller, Shinozuka & Yao (eds) © 1994 Balkema, Rotterdam, ISBN 90 5410 357 4

On the construction of a nonlinear damage accumulation hypothesis

H. Schäbe

Agentur für Sicherheit von Aerospace Produkten GmbH, Germany

ABSTRACT: A phenomenological method is presented to extend the linear damage accumulation theory. An additional factor is introduced taking into account varying material sensitivity. The shape of life time distribution will be changed, compared with the linear case.

1 INTRODUCTION

In many real life situations the main assumptions of linear damage accumulation are violated, as e.g. stationarity and ergodicity of the load process, the material is not free of training and relaxation effects, life length is affected by rearranging subsequences of the load sequence. Take-off and landing of an aeroplane induces seriously nonstationary load process and may serve as an example.

Attempts have been made by several authors to overcome the weaknesses of linear damage accumulation, mostly using results from crack propagation, see e.g. Xing (1990), Ben-Amoz (1990), Stallings & Frank (1991).

The aim of the present paper is to demonstrate a method to construct nonlinear damage accumulation theories. Especially a function is introduced that expresses sensitivity of the material to damage. The new damage accumulation theory is a straightforward generalization of the ansatz of Hennig [1] to linear damage accumulation.

2 NONSTATIONARY DAMAGE ACCUMULATION

2.1 Introduction of the Model

Let us introduce the following notation
S(t) - increasing stochastic process denoting the damage at time t, failure occurs as S(t) crosses level one,
T - lifetime of the specimen, identical to time when S(t) crosses level one, random variable,
V(t) - stochastic load process which is assumed to be twice differentiable in quadratic mean, V(t) is stationary and ergodic,
E - expectation,
E_V - expectation with respect to the load process V(s), s>0.
$\delta(t)$ - Dirac's delta distribution,
$\Theta(x)$ - theta function $\Theta(x)=\begin{cases} 0 & \text{if } x \leq 0 \\ 1 & \text{if } x > 0. \end{cases}$

For a cyclic load with constant amplitude v let the lifetime distribution of the specimen be of the form [5]

$$F(x|v) = F(x/g(v)), \tag{1}$$

where g(v) is a nonincreasing function and will be called the acceleration function. The distribution F(x) fulfills

$$F(0)=0, \quad \int_0^\infty xdF(x) = 1, \quad \int_0^\infty (x-1)^2 dF(x)=\sigma^2. \tag{2}$$

In the model (1) different load amplitudes v influence lifetime only via the scale, the shape of the distribution function remains unchanged. A more complex model involving also changes of the shape of the distribution caused by load can be found e.g. in [4]. In order to extend the results for loads with constant amplitudes to stochastic loads a damage accumulation

hypothesis has to be applied. The linear damage accumulation hypothesis can be formulated according to [1,3]

$$dS(t) = \frac{\ominus(V(t))\delta(V'(t)V''(t)}{g(V(t)) \, X_0}. \qquad (3)$$

Here $V(t)$ is a stochastic load process and X_0 denotes a random variable distributed according to $F(x)$. Now we will build a generalization of (3) that is the most simple. Let us introduce a function A depending on the accumulated damage $S(t)$, the load function $V(t)$ and on time t. The simplest generalization of (3) then will be to multiply the increment $d\,S(t)$ of the damage with the term in the right hand side of equation (3) so that we obtain

$$dS(t)=A(S(t),V(t),t) \, \frac{\ominus(V(t))\delta(V'(t)V''(t)}{g(V(t)) \, X_0}. \qquad (4)$$

Equation (4) can be simplified expressing A as a factor

$$A(S(t),V(t),t)=A(t)A_S(S(t))A_V(V(t)). \qquad (5)$$

It can be proved that $A_V(V(t))$ and $A_S(S(t))$ vanish if a suitable redefinition of $S(t)$ and $g(V(t)$ is provided. Hence $A(t)$ is the only substantial new term. We have

$$dS(t) = A(t) \, \frac{\ominus(V(t))\delta(V'(t))V''(t)dt}{g(V(t)) \, X_0}. \qquad (6)$$

So the simplest nonlinear damage accumulation theory shall incorporate a factor depending on time. Strictly speaking, equation (6) represents a nonstationary damage accumulation time where the material sensitivity to damage is time dependent. Such a sensibility change might reflect relaxation or training effects provided they are mainly due to time. From (6) we derive

$$Y = \frac{1}{T} \int_0^T \frac{A(t) \, (V(t))\delta(V'(t))V''(t)dt}{g(V(t)) \, X_0}.)$$

$$T = X_0 \, / \, Y, \qquad (7)$$

2.2 Derivation of the lifetime distribution under random load

Let us denote

$$M = E_V Y = \frac{1}{T} \int_0^T A(t)dt \, M_0,$$

$$M_0 = E\{ \frac{\ominus(V(t))\delta(V'(t))V''(t)}{g(V(t))} \}.$$

Result 1
Under certain regularity conditions, the distribution function of lifetime under random load is

$$F(t|V(t)) \approx F(M \int_0^t A(s)ds). \qquad (8).$$

3 PARTICULAR MODELS OF DISTRIBUTION FUNCTIONS UNDER RANDOM LOAD

In this section we study examples for distribution functions $F(t|V(t))$ given by (8).

3.1 Behaviour of hazard rate

Let

$$h(x) =f(x)/(1-F(x))$$

and $H(x)$ denote the hazard rate of $F(x)$ and its cumulative hazard function. Then we have

$$H(t|V(t)) = H(M_0 \int_0^t A(s)ds)$$

for the cumulative hazard rate of $F(t|V(t))$. The hazard rate becomes

$$h(t|V(t)) = h(M_0 \int_0^t A(s)ds) \, A(t) \, M_0.$$

Result 2:
$F(t|V(t))$ has increasing, decreasing or constant failure rate according to the function

$$\ln\{A(t)\ h(M_0 \int_0^t A(s)ds)\}$$

being increasing, decreasing or constant. We may note that the hazard rate of $F(t|V(t))$ can also be bathtub shaped.

3.2. The exponential distribution

As a particular example we will study the exponential distribution

$$F(t|v) = 1-\exp\{-t/g(v)\}.$$

Then

$$F(t|V(t))=1 - \exp\{ -M_0 \int_0^t A(s)ds\}.$$

The failure rate

$$h(t|V(t)) = M_0 A(t)$$

is increasing or decreasing depending on whether $A(t)$ is increasing or not. Since the function $A(t)$ expresses the sensitivity of the material with respect to damage a valuable choice of it could be

$$A(t) = 1- a \exp\{ -bt \},\ a \geq 0,\ b \geq 0.$$

Then we have

$$F(t|V(t)) =$$

$$1 - \exp\{ -M_0 t + (M_0 a/b)(1-\exp\{-bt\}) \}.$$

This distribution has hazard rate

$$h(t) = M_0 - M_0 a \exp\{-bt\}$$

which is increasing from level $M_0(1-a)$ at time zero to M_0 at infinity. This example illustrates that the function $A(t)$ involved in the damage accumulation theory affects the behaviour of the hazard rate.

4 THE SCALED NONSTATIONARY DAMAGE ACCUMULATION HYPOTHESIS

Equation (6) has a disadvantage. The nonstationary factor $A(t)$ depends on time t. Consequently, if the load process $V(t)$ is multiplied by a constant factor then lifetime might become smaller and the

resulting lifetime distribution under random load can not only be changed in scale but also in shape. So the effect of $A(t)$ will be quite different for load processes that are identical in shape but differ only by a constant factor. This problem can be solved introducing a new $A(t)$ that involves the scale function $g(.)$

$$dS(t) =$$

$$A(t/g(V(t)))\ \frac{\Theta(V(t))\delta(V'(t))V''(t)dt}{g(V(t))\ X_0}.\ )$$

Now the function A has a more complex dependence structure on t and V than for the simple nonstationary damage accumulation hypothesis. Using the same methods as in section two we can conclude that the lifetime distribution can be approximated by

$$F(\int_0^t a(s)\ ds),$$

$$a(s) = E_V\{ A(s/g(V(s)))\ \ \delta(V'(s))\ \ V''(s)$$

$$\Theta(V(s))/g(V(s))\ \}.$$

Let us now consider a special case. The following choices for $g(v)$ and the load process are made:
(i) The load process $V(t)$ is normally distributed with autocovariance function $R(t)$ and zero mean.

(ii) $g(v) = C\ v^{-\phi}$.

Now we can compute

$$a(t) = \frac{2^{\phi/2-1}\sqrt{R''(0)}\ R(0)^{(\phi-1)/2}}{C}$$

$$\int_0^\infty e^{-z}z^{\phi/2}A(t[2R(0)z]^{\phi/2}/C)\ dz.\ (9)$$

Equation (9) is the generalization of Miles' formula. Setting a=0 equation (9) turns out to be the Miles' formula for linear damage accumulation.
For

$A(t) = 1 - a \exp\{-bt\}, \quad \phi = 2,$

equation (9) can be evaluated explicitly and reads

$$M_0(1 - a / (1 + 2bt\, R(0) / C)^2),$$

where M_0 is defined by

$$M_0 = 2^{\phi/2-1} \sqrt{R''(0)}\; R^{(\phi-1)/2} / (\pi\, C).$$

Finally, the life time distribution is

$$F\{M_0 t(1 - 2a / (1 + 2bt R(0)/C))\}. \tag{10}$$

For the case of $F(x)$ being an exponential distribution (24) has an increasing failure rate starting at value $M_0(1-a)$ and approaching asymptotically M_0 as $t \to \infty$.

The case of an exponential distribution $F(x)$ can be used for a numerical study. For simplicity, assume $R(0)=C=M_0=1$.

Then the lifetime distribution is

$F(t) = 1-\exp\{-t(1-2a/(1+2bt)\}.$

For small a, i.e. if the sensitivity of the material to damage is slightly decreased there is

$F(t) \approx 1 - (1-2a+4abt)\exp(-t)$

The mean life time is approximately

$1 - 2a + 4ab.$

First, let b=1, i.e. the sensitivity changes are of the same order as the lifetime of the component. Then we have an increase of the lifetime by 2a 100%. The meaning of a is a reduction of the damage sensitivity of the material. In a first approximation a decrease of damage sensitivity yields an effect of lifetime prolongation that is twice.

Second, assume b=0.1, i.e. the reduced sensibility to damage occures only during a time that is approximately 10% of the lifetime. Then we have a mean lifetime of

$1 - 1.6a,$

i.e. lifetime decreases caused by changing sensibility of the material to damage.

Thus we can observe two tendencies:
- Increase of lifetime if damage sensitivity is decraesed for a sufficient long time.
- Decrease of lifetime caused by changing damage sensitivity if sensitivity is decreased only for a short time.

5 CONCLUSIONS

We have presented a phenomenological approach to derive nonlinear damage accumulation theories. The theory considered was based on a changing sensitivity of the material to damage. It has been shown that the shape of the lifetime distribution under random load is changed compared with the shape of the distribution under deterministic load with fixed amplitude. Moreover effects of increase and decrease of mean life time caused by changing

6 REFERENCES

W.Heinrich & K. Hennig 1977. *Random Vibrations of Mechanical Systems* (in German). Berlin: Akademie-Verlag.

Kallenberg, J. 1989. *Mathematical Model of Nonlinear Damage Accumulation* (in German). Dissertation, Bergakademie Freiberg.

Schäbe, H. 1990. Lifetime Computation Using Linear Damage Accumulation Hypothesis (in German), *FMC-series, Akademie d. Wiss. d. DDR, Inst. f. Mechanik Chemnitz (Karl-Marx-Stadt) No.46*: 43-66.

Viertl, R. 1988. *Statistical Methods in Accelerated Life Testing*. Göttingen: Vandenhoeck & Ruprecht.

M.Shaked, W.J. Zimmer & C.A. Ball 1979. A Nonparametric Approach to Accelerated Life Testing: *J. Amer. Statist. Assoc.* 74: 694-699.

Stallings, J.M. & K.H. Frank 1991. Cyclic Fatigue of Cables: *Eng. Fract. Mech.* 38: 341-347.

Xing, X.-S. 1990. Microscopic Statistical Theory of Damage Fracture: *Eng. Fract. Mech.* 37: 1099-1104.

Ben-Amoz, M. 1990. A Cumulative Damage Theory for Fatigue Life Prediction: *Eng. Fract. Mech.* 37:341-347.

Structural Safety & Reliability, Schuëller, Shinozuka & Yao (eds) © 1994 Balkema, Rotterdam, ISBN 90 5410 357 4

Statistical analysis of fatigue data: An overview

C.L.Shen & P.H.Wirsching
The University of Arizona, Ariz., USA

ABSTRACT: For design purposes, it is necessary to construct a statistical synthesis of fatigue test (S-N) data. Usually, a "design curve" that provides a lower bound is employed. Ongoing research at The University of Arizona is addressing the development of a tolerance-interval concept for constructing a design S-N curve.

1 INTRODUCTION

In a fatigue test of a specimen, structural system, or subsystem, typically, a constant-amplitude stress range, S_i, is chosen and the test article is subjected to this oscillatory stress until the article fails. The number of cycles to failure, N_i, is recorded. The test is repeated at different stress levels over a predetermined range of S for a sample size of n identically prepared specimens. The data from this experiment are (S_i, N_i), i = 1, n. Fatigue data thus obtained are frequently plotted on log-log paper, as shown in Figure 1.

In order to assist a designer in making decisions regarding the fatigue failure mode, it is customary to provide a statistical summary, or synthesis, of the data. Most commonly, a design curve is constructed on the lower, or safe, side of the data (Figure 1). A simple method for defining the design fatigue strength is to draw, by eye, a curve that follows the data and provides a little "white space" between the dots and the curve. A rigorous procedure following sound principles of

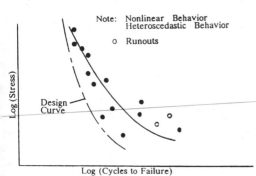

Figure 1. Illustration of fatigue data and design curve.

mathematical statistics, however, can become extremely complicated. An approach for constructing the design curve that combines mathematical statistics with engineering practicality is the subject of this paper.

There is another issue to be considered. In a probabilistic approach to design, a reliability model for fatigue strength is required. All of the model parameters are considered to be random variables. The goal of analysis would be to translate the data into random variables of the parameters.

The general goals of analysis are twofold: (1) define a design curve and (2) construct a reliability model. Only the first item is considered here.

2 ANALYSIS OF FATIGUE DATA: WHY IT IS SO DIFFICULT

Mathematical analysis of fatigue data is complicated by the following facts (many of the points are illustrated in Figure 1): (1) The fundamental relationship is nonlinear. (2) Sample sizes will be small due to the cost of testing. (3) There is large scatter in the data (life N given stress S has a large variance). (4) The data will be heteroscedastic (typically, there is broader scatter at lower stress levels). (5) There will be censored data (runouts). (6) The statistical distribution of N given S, or S given N, is unknown. (7) The existence of an endurance limit will present mathematical difficulties. (8) The statistical model should recognize that there should be good resolution in the low probability region (left tail).

3 BASIC REQUIREMENTS OF THE MODEL

The model (a design curve or reliability model) should satisfy the following basic requirements: (1) The model should be compatible with special

features of the physics of fatigue behavior and fatigue test procedures. (2) The model should have a simple form for use in design, and it should follow the trend of the data. (3) Consideration should be given to the fact that there are models in common use in the engineering community; e.g., the general strain-life model. In summary, the model should balance mathematical rigor with engineering practicality.

4 ISSUES RELATED TO THE STATISTICAL ANALYSIS OF FATIGUE DATA

Several issues related to analysis of fatigue data are addressed in this study:
1. Mathematical Models. What fundamental analytical form should be used to describe the relationship between S and N?
2. Methods of Treating Heteroscedastic Data. When the standard deviation of N (or transformed N, e.g., log N) is not a constant function of S across the range of S being considered, the data are said to be heteroscedastic. There are two methods of analyzing heteroscedastic data: (a) include a model for the standard deviation of N as a function of S and (b) make a transformation on life, U(N). A suitable "variance-stabilizing" transformation can make the scatter band of U constant.
3. Estimating the Parameters of the Model. There are two basic methods: (a) least squares analysis and (b) maximum likelihood estimators. Parameter estimates can be made by the method of least squares if the model is linear (Arnold (1981). Maximum likelihood estimators are more general and can be employed when runouts are present.
4. Methods for Treating Runouts. When a fatigue test is suspended, the information should be included in the analysis for the estimates of the model parameters. It is possible to consider runouts using both the least squares analysis (Schmee and Hahn 1979) and maximum likelihood (Nelson 1984).
5. Development of a Design Curve. There are a number of ways of determining the design curve: (a) Draw a lower-bound curve by eye. This method, of course, is subjective. (b) Draw the curve as the mean (or median) minus two or three (or any other factor) sample standard deviations of log N. This method fails to account for the fact that the parameters estimated are random variables. (c) Assume that all specimens are tested at the same stress level and use the one-sided tolerance factor K based on the normal distribution (Natrella 1963). The design curve would be the median minus Ks, where s is the sample standard deviation of log N. (d) Use Owen's tolerance limit (Owen 1968) for deriving a K-factor for a tolerance level. This ensures that the tolerance levels are satisfied at any stress level specified in the sample space of S. Unfortunately, K is a function of stress level, and this curve will have a shape different from the median. (e) An approximate Owen's curve can be constructed. This curve would follow the shape of the median curve and would, at least visually, follow the trend of the data. (f) Employ the simultaneous tolerance limit method (Miller 1981) using a K-factor to ensure that the tolerance levels are simultaneously satisfied at *all* stress levels.

5 THE OWEN LOWER TOLERANCE LIMIT

A paper by D. B. Owen (1968) provides a comprehensive review of the use of tolerance limits in engineering. His work provides the basis for the studies reported here.

Consider a set of fatigue test data, (S_i, N_i); i = 1, n. The sample size is n. Transform the data to new coordinates so as to provide a linear model, e.g.,

$$Y_i = \log N_i, \qquad X_i = \log S_i \qquad (1)$$

and the basic S-N model

$$Y = \sum_{i=0}^{M} a_i X^i \qquad (2)$$

where M = degree of polynomial chosen. It is assumed in the following that the log S-log N data are homoscedastic (constant scatter band).

At a specific stress X_0, the lower tolerance limit of Y with probability P and confidence level γ is the Y_0 defined as follows (i.e., for $100\gamma\%$ of the time, at least $100P\%$ of fatigue life is greater than Y_0):

$$Y_0 = \hat{Y}(X_0) - K(X_0; P, \gamma, n, M, \underline{x}_i)s \qquad (3)$$

where $\hat{Y}(X_0)$ is the least squares estimate of $E(Y|X_0)$, s is the sample standard deviation of $Y|X_0$,

$$s = \sqrt{\frac{1}{n-M-1} \sum_{i=1}^{n} [Y_i - \hat{Y}(X_i)]^2} \qquad (4)$$

and

$$K(X_0; P, \gamma, n, M, \underline{x}_i) = -t_\gamma \sqrt{A(X_0)} \qquad (5)$$

where

$$A(X_0) = \{X_0\}^T (X^T X)^{-1} \{X_0\} \qquad (6)$$

$$X = \begin{bmatrix} 1 & X_1 & \cdots & X_1^M \\ 1 & X_2 & \cdots & X_2^M \\ \vdots & & & \\ 1 & X_n & \cdots & X_n^M \end{bmatrix} \qquad (7)$$

$$\{X_0\} = [1 \quad X_0 \cdots X_0^M] \qquad (8)$$

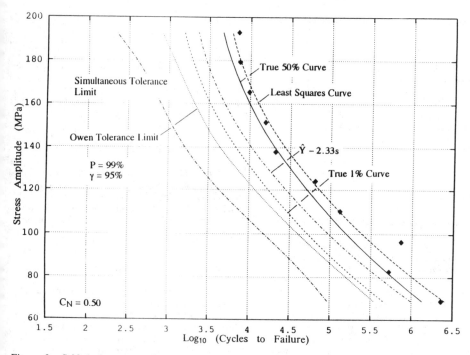

Figure 2. S-N design curve: Owen tolerance limit and simultaneous limit.

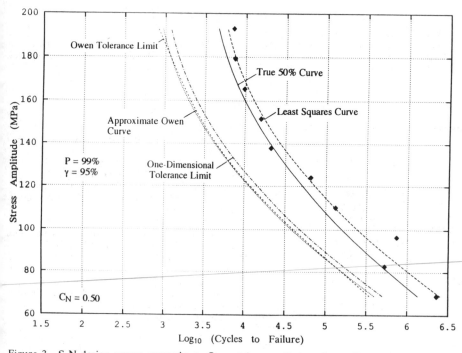

Figure 3. S-N design curve: approximate Owen tolerance limit and one-dimensional tolerance limit.

t_γ is obtained from the probability statement

$$P[T_f(\delta) > t_\gamma] = \gamma \qquad (9)$$

where $T_f(\delta)$ is the non-central Student's t variate with f degrees of freedom,

$$f = n - M - 1 \qquad (10)$$

and δ is the noncentrality parameter,

$$\delta = -K_P A^{-\frac{1}{2}} \qquad (11)$$

and

$$K_P = \Phi^{-1}(P) \qquad (12)$$

where Φ is the standard normal distribution function.

An example is shown in Figures 2 and 3. It is desired that the design curve be the lower 1% point (P = 99%). A set of data, n = 10, was sampled uniformly over a stress range. The COV of life N is 50%. The actual median curve is shown in Figures 2 and 3, along with the least squares estimate. And the actual lower 1% curve is shown in Figure 2, along with the least squares curve minus 2.33 sample standard deviations. The Owen tolerance limit for P = 99%, γ = 95%, is also shown in both Figures 2 and 3.

6 APPROXIMATE OWEN TOLERANCE LIMIT

For simplicity, it is desired to have a design curve follow the trend of the data. Therefore, an approximate Owen tolerance limit is defined that ensures that K is a constant (not a function of X_0).

Define a normalized A as

$$a_f = An \qquad (13)$$

Define a normalized stress s, the ith point, s_i,

$$s_i = \frac{S_i - S_1}{S_2} \qquad 0 < s_i < 1 \qquad (14)$$

where (S_1, S_2) represents the range of stress values of the data. For a given n, M, and test plan, it was shown that a_f will be a function of s only.

An approximate a_f is defined as the average value of a_f over the stress range,

$$a_e = \int_0^1 a_f(s)ds \qquad (15)$$

Then, let

$$A_e = a_e/n \qquad (16)$$

Using A_e in Eq. (5), it is seen that K is a constant (not a function of stress S_0).

The approximate Owen tolerance limit is illustrated in Figure 3. Also shown in Figure 3 is the one-dimensional tolerance limit.

7 SIMULTANEOUS TOLERANCE LIMIT

The Owen tolerance limit ensures a confidence limit only at a specific value of X_0. The "simultaneous limit," Y_s, derived by Miller (1981) satisfies the confidence level γ for the entire range of S. An example of the simultaneous tolerance limit is shown in Figure 2.

8 CONCLUDING REMARKS

The following issues relative to the construction of a design curve are currently being studied:

1. If the data are heteroscedastic, a variance-stabilizing transformation on N can be made.

2. When runouts are present, the maximum likelihood method is used to estimate parameters.

3. If an endurance limit is present, efficient analysis is possible by letting N and S be the independent and dependent variables, respectively.

ACKNOWLEDGMENT

The guidance, support, and encouragement of G. T. Cashman of GE Aircraft Engines, Cincinnati, Ohio, is deeply appreciated by the authors.

REFERENCES

Arnold, S. F. 1981. *The Theory of Linear Models and Multivariate Analysis*. New York: John Wiley.

Miller, R. G. 1981. *Simultaneous statistical inference*, 2nd ed. New York: Springer-Verlag.

Natrella, M. G. 1963. *Experimental statistics*. NBS Handbook 91. Washington, DC: U.S. Govt. Printing Office.

Nelson, W. B. 1984. Fitting of fatigue curves with nonconstant standard deviation in data with runouts. *J. Testing and Eval.* 12(2):69-77.

Owen, D. B. 1968. A survey of properties and applications of the noncentral t-distribution. *Technometrics* 10(3):445- 472.

Schmee, J. & G. J. Hahn 1979. A simple method for regression analysis with censored data. *Technometrics* 21(4): 417-432.

Structural Safety & Reliability, Schuëller, Shinozuka & Yao (eds) © 1994 Balkema, Rotterdam, ISBN 90 5410 357 4

Reliability analysis for the load-carrying parts of horizontal axis wind turbines

B.A.van den Horn & L.W.M.M.Rademakers
Netherlands Energy Research Foundation ECN, Petten, Netherlands

ABSTRACT: Results of a study on the possible application of probabilistic techniques for the assessment of the structural safety of mechanical components of a Horizontal Axis Wind Turbine (HAWT) are presented. As an illustration the probability of fatigue failure in the welds in the tower foot of a HAWT is determined. The wind actions as well as the structural response are considered to be of a random nature and applied to the fatigue problem by means of time domain analysis. To this end the incoming wind speed is modelled as a time series based on a wind speed spectrum. As for the long term statistics, the average incoming wind speed is assumed to be Weibull-distributed, while the principal wind direction is simply supposed to be uniform. Also external influences such as wind turbulence, surface roughness, and wind shear are included. The dynamic loads on main components of HAWT's are calculated and both the rain flow counting method and Palmgren-Miner's rule are applied to evaluate the fatigue damage in the welds. The main random variables are related to the wind speed and wind turbine properties, such as structural details in the blades, as well as to the material fatigue properties. FORM has been used in order to calculate the probability of fatigue failure. As a result insight into the importance of the various random variables with regard to failure is obtained. This insight serves as a guidance to the identification of the major gaps in knowledge and data that should be bridged for an improved safety assessment in future research.

1 INTRODUCTION

In the Netherlands wind turbines are presently being designed in accordance with deterministic design rules (Stam et al., 1991). These rules concern the design of main components, such as the rotor, the tower and the safety systems, with the purpose to minimise the likelihood of structural failure. As wind turbines increase in size and power, the rules may not be adequate to ensure a safe and well-balanced design. The rules neither facilitate the quantification of the degree of conservatism in the applied safety margins, nor do they explicitly address the reliability and availability of the wind turbine.

The project 'Probabilistic Safety Assessment for Wind Turbines' was carried out by the Netherlands Energy Research Foundation ECN to determine a methodology for the wind turbine industry that takes advantage of existing probabilistic techniques in e.g. nuclear and offshore industry (Rademakers et al., 1993, Seebregts et al., 1993). The objective is to develop a methodology applicable complementary to existing rules. This methodology should enable the wind turbine industry to make a well-balanced design and certifying bodies to assess wind turbine safety and availability, with due attention to the influence of quality assurance, inspection, maintenance procedures, and the human factor. The following project phases have been identified and have been reported (Rademakers et al., 1993):

1. Survey and comparison of methods for safety and reliability analysis
2. Case study of the Lagerwey LW 15/75 and LW 18/80 wind turbines
3. Development of a methodology including its area of application and its restrictions.

In the first project phase a distinction has been made between *system reliability* and *structural*

reliability. Although these two fields should be combined for a complete reliability analysis, this distinction is justified since there is a great difference in the required disciplines and their applicability. In this paper the applicability of structural reliability techniques has been investigated. This is significant, since the design processes of wind turbines are based on assumptions as to the scatter in loads and material behaviour which in turn are based on experience previously gained. The magnitude of uncertainties in these quantities are often not sufficiently reflected in current design procedures. Inherent variability is only accounted for by the application of material and load factors based on mean or characteristic values to obtain a acceptable margin between strength and load. However, this may lead to doubts about the integrity and safety of the structure and possibly to a non-optimal economic service life being predicted. The conservative assumptions which are presently required by the Dutch criteria to account for the uncertainties in the wind model and fatigue calculations justify a probabilistic approach to quantify the structural reliability of wind turbine components (Rademakers et al., 1993).

2 STRUCTURAL RELIABILITY ANALYSIS

Failure of the tower foot due to fatigue damage of an LW 18/80 at the coastal location Den Helder has been analysed to illustrate the use of structural reliability techniques. The LW 18/80 is a two-bladed machine with a tower height of 24 m and a rotor diameter of 17.9 m. The generator power is 80 kW. The main cause for fatigue damage is the operational mode energy production which is evaluated for a period of 20 years. Therefore, other operational modes such as starting, stopping and emergency shut-downs have not been analysed. The selection has been based on various criteria like the complexity of the component, the available material properties, and the interest of industry. Only safety aspects have been considered and the main goals are:
1. The formulation of a procedure to assess the probability of fatigue failure of a wind turbine component
2. To identify all factors affecting the failure probability and to quantify their effects
3. The identification of the major gaps in knowledge and data to fill for an improved safety assessment
4. To evaluate the sensitivity of the probability of failure to all factors that lead to failure
5. To better underpin the safety factors.

The relatively limited Mean Value Approach (Ang & Tang, 1984) has been applied to quantify the reliability of the load-carrying structure with a probability of fatigue failure P_F in the welds in the tower foot. MVA is a simple application of the First Order Reliability Methods (FORM). The reason for MVA was the enormous effort required to calculate loads with the available computer programs. Load calculations techniques are different from those used in other industries. For example in off-shore industry loads are determined in the frequency domain while in wind turbine industry calculations in the time domain are daily practice, although they are time consuming. However, the validity of the load calculations is being discussed. For instance load and fatigue data of composite materials are often inadequate.

The random variables considered are assumed to be non-correlated, mutually independent and to have a normal distribution. The reliability function Z is related to the Miner sum D. Besides $Z = 1 - D$ also $Z = -\ln D$ has been considered since the safety index β obtained with MVA is dependent on the Z formulation.

The stochastic wind field simulator SWIFT (Winkelaar, 1992) has been used to model the turbulent wind. SWIFT consists of a frequency spectrum based on ESDU (1998), a coherence function, the vertical wind profile and some turbine data such as hub height and rotor diameter. The frequency spectrum represents the turbulent wind speed at a single point in space. The coherence function represents the spatial distribution of the wind speed in the rotor plane. The wind speed variation between the cut-in and cut-out wind speed is divided into so-called wind classes with a range of 2 m/s. These wind classes are characterised by a 10 minute average wind speed U_{10} and a turbulence intensity $T = \sigma/U_{10}$. They are combined with the operational mode energy production resulting in 8 load cases. For each load case the loads on the wind turbine are determined in the time domain with the computer code PHATAS-II (Lindenburg et al., 1992) with options to accurately model flexibilities of the wind turbine. The variables incorporated are structural variables such as masses, eigenfreq-

uencies, variables describing control strategy, and aerodynamic variables such as lift and drag coefficients. With PHATAS-II the variation of the stress acting in the welds in the tower foot are determined. The method presently used for fatigue analysis for wind turbines is based on a linear Palmgren-Miner rule and the rainflow-counting method (Madsen et al., 1993). In Table 1 the random variables considered with means μ and coefficients of variation σ/μ are presented. For the determination of μ and σ, see the report of Rademakers et al. (1993c).

Table 1. Random variables.

X_i	μ_i	$(\sigma/\mu)_i$	Designation
T	19.28 %	0.052	turbulence intensity
f_t	0.97 Hz	0.050	1st natural bending frequency
C_1	1.60	0.024	lift coefficient
M	730 Nm	0.068	pitching moment
X_5	1.0	0.125	model uncertainties
t_w	10 mm	0.050	wall thickness
$logC$	12.511	0.017	constant in SN-curve
U_a	7.28 m/s	0.100	annual average wind speed

3 NUMERICAL RESULTS

The numerical results are presented in Table 2. It appears that the difference in failure probability is 3 orders of magnitude. The formulation $Z = -\ln D$ seems to be more realistic in fatigue problems, because of the lognormal distribution of the fatigue constant in the SN-curve. Importance factors for each formulation are presented in Fig. 1 and 2. The largest sources of uncertainty were the annual average wind speed

Table 2. Numerical results.

Z	β	P_F
$Z = 1 - D$	4.18	$1.4 \ 10^{-5}$
$Z = -\ln D$	2.05	$2.0 \ 10^{-2}$

and the fatigue constant. Model uncertainties appeared to be 18% to 19% which was less than expected. Also the turbulence intensity appeared to be of little influence. The importance factors are approximately the same in both formulations.

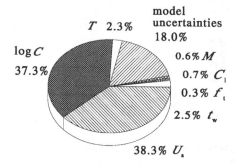

Fig.1. Importance factors for $Z = 1 - D$.

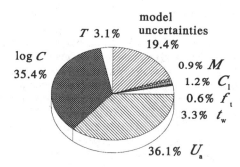

Fig.2. Importance factors for $Z = -\ln D$.

The sensitivity of β to the model uncertainties, the wall thickness, the exponent k in the SN-curve and the standard deviation of the fatigue constant has been investigated. For this end the dimensionless quantity $X\Delta\beta/\beta\Delta X$ has been calculated, (cf. Fig. 3). It appears that the difference in $X\Delta\beta/\beta\Delta X$ due to both Z formulations is approximately a factor of 2. The results appeared to be the most sensitive for the fatigue parameters.

This corresponds with the results of the work of Veers (1990) and indicates that the results obtained with MVA can only be used qualitatively. The predicted β should be handled with care. Also the siting effect has been considered. Instead of the coastal location Den Helder, the less windy location Schiphol has been investigated. It appeared that β increased from 4.18 to 4.75 for the formulation $Z = 1 - D$. This means that the reliability of the tower foot improves with a factor of 14 if a turbine is sited at a land location instead of a coastal area.

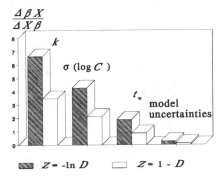

Fig.3. Sensitivity analysis.

4 CONCLUSIONS

The tower foot of the LW 18/80 has been analysed with MVA. The numerical results are only valid for the LW 18/80 and the wind characteristics considered at the coastal location Den Helder. In spite of these limitations, the analysis gives insight in how the failure probability is influenced by uncertainties in some variables. The sources of variability that have been selected in this study were based on engineering judgement and might not be exhaustive. The total variance was mainly caused by the variance in annual average wind speed and in the fatigue parameters. Model uncertainties appeared to be of less importance than might be expected and also the turbulence intensity appeared to be of little influence. It can be concluded that:
- Structural Reliability is meaningful.
- A main bottleneck in the applied procedure for the assessment of the structural integrity of wind turbine components are the time-consuming load calculations.
- It is of crucial importance to gather more data from load measurements in order to be able to

more accurately quantify the load effect on the reliability of a wind turbine.
- To enable the wind turbine industry to make a well-balanced design it is recommended to improve the probabilistic model and to make procedures on the calibration of partial safety factors with FORM in future research.

REFERENCES

Ang, A.H.S., Tang, W.H. 1984. *Probability Concepts in Engineering Planning and Design, Vol II: Decision, Risk and Reliability*, J. Wiley & Sons.

Engineering Science Data Unit (ESDU), 1988. *Wind Engineering, Vol* 1, London.

Föllings, F. et al. 1991. *Handbook Wind Data for Wind Turbine Design; Version* 3, INTRON/TNO/ECN, The Netherlands.

Lindenburg, C. et al. 1992. *PHATAS-II: Program for Horizontal Axis Wind Turbine Analysis and Simulation version II, User's Manual*, ECN-C--92-020, ECN, Petten, The Netherlands.

Madsen, P.H. et al. 1990. *Recommended Practices for Wind Turbine Testing and Evaluation No.* 3: Fatigue Characteristics, IEA.

Rademakers, L.W.M.M. et al. 1993a. Reliability Analysis in Wind Engineering, *ECWEC' 93 Conf.*, Travemünde, Germany.

Rademakers, L.W.M.M. et al. 1993b. Introduction to Probabilistic Safety Assessment in Wind Turbine Engineering, *AWEA Wind Power '93 Conf.*, San Francisco.

Rademakers, L.W.M.M. et al. 1993c. *Methodology for Probabilistic Safety Assessment for Wind Turbines; Illustrated by a Case Study of the Lagerwey LW 15/75 Design*, ECN-C--93-010, Petten, The Netherlands.

Seebregts, A.J. et al. 1993. Reliability Analysis in Wind Turbine Engineering, *to be presented at the Dutch Chapter of SRE/Dutch Reliability Society (NVvB) Symposium*, 4-6 October 1993, Arnhem, The Netherlands.

Stam, W.J. et al. 1991. *Regulations for the Type-Certification of Wind Turbines: Technical Criteria* ECN-91--01, 1991, ECN, Petten, The Netherlands.

Veers, P.S. 1990. *Fatigue Reliability of Wind Turbine Components*, Sandia National Laboratories, SAND--90-2538C, Albuquerque.

Winkelaar, D. 1992. *SWIFT: Program for Three-Dimensional Wind Simulation; Part 1: Model Description and Program Verification*, ECN-R--92-013, Petten, The Netherlands.

System reliability

Structural Safety & Reliability, Schuëller, Shinozuka & Yao (eds) © 1994 Balkema, Rotterdam, ISBN 90 5410 357 4

Probabilistic seismic reliability assessment of electric power transmission systems

A. H-S. Ang, J. Pires & R. Villaverde
University of California, Irvine, Calif., USA

ABSTRACT: A comprehensive method for evaluating the seismic reliability of electric power transmission systems is presented. The model provides probabilistic assessments of structural damage and abnormal power flow that can lead to power interruption in a transmission system under a given earthquake. Seismic capacities of electrical equipments are determined on the basis of available test data and simple modeling from which fragility functions of specific substations are developed. Earthquake ground motions are defined as stochastic processes. Probabilities of network disconnectivity and abnormal power flow are assessed through Monte Carlo simulations. The proposed model is applied to the electric power network in San Francisco and vicinity under the 1989 Loma Prieta earthquake, and the probabilities of power interruption are examined in light of the actual power failures observed during that earthquake.

1 INTRODUCTION

Electrical transmission equipment and facilities are vulnerable to seismic damage, particularly those in high-voltage transmission systems. Because electrical power is transmitted from the source (power plants) through various stages of voltage transformation to the consumers, failure of a high-voltage facility (e.g. substation) can cause major power blackout to a large area. It should be noted, however, that for a highly redundant and distributed system, damage to just a few network components may not necessarily lead to a widespread power blackout as a result of alternate paths within the system. Also as a result of its redundancy, the seismic performance and reliability of an electric power trnasmission system may be enhanced by upgrading just a few of the network components.

Although it is not economically feasible to totally prevent damage to a transmission system in the event of a major earthquake, quantitative (probabilistic) information on the likelihood of different levels of damage and extent of affected areas under different intensities of earthquake would be valuable for determining needed upgrading of an existing system, or for designing future systems. The same information for an existing system would be valuable also for emergency planning and disaster reduction preparedness. A comprehensive model for the above purposes is developed and its application to the power transmission system of the San Francisco Bay Area during the 1989 Loma Prieta earthquake is described.

2 PROBABILISTIC RELIABILITY MODEL

The proposed probability model for assessing the seismic reliability of an electrical power transmission system consists of the following components: (1) network models for transmission failure through structural damage and disconnectivity, and failure through abnormal power flow; (2) Monte Carlo simulation procedures for calculating the probabilities of disconnectivity of a given system, and of abnormal power flow in the system; (3) stochastic representation of site-specific seismic ground motions; (4) determination of the seismic capacity and fragility of each type of electrical equipment, and development of fragilities of the transmission substations.

2.1 Network connectivity model

An electrical power transmission system consists of power generating stations, substations, supervisory control and data acquisition facilities, that are interconnected by transmission lines. The system can be modeled by a network of supply and demand nodes interconnected by links. The supply nodes represent power plants or substations which feed electric power to the demand nodes. Disconnectivity of a demand node will occur when it is isolated from all the supply nodes. The graphical connectivity of a network model can be represented through the adjacency matrix $X = [x_{ij}]$ where, $x_{ij} = 1$, if node i is connected to node j and 0 otherwise. For undirected links, as in the case of electrical transmission systems, the adjacency matrix X is symmetric. The connectivity between all pairs of

nodes is then determined by the connectivity or reachability matrix $C = [r_{ij}]$, which can be obtained from the adjacency matrix by standard methods (Eaton and Cohen, 1983) as,

$$C = X + X^2 + \ldots + X^n \qquad (1)$$

where, $r_{ij} = 0$, if node i is disconnected from node j; and is $\neq 0$, if node i is connected with node j; n = number of nodes in the network. The connectivity matrix C is also symmetric for undirected networks. Computationally more efficient algorithms (Schinzinger and Peiravi, 1985) are also available for obtaining C.

Power failure at a demand node caused by disconnectivity, therefore, will occur when the demand node is totally disconnected from all the supply nodes in the network. It is possible for a power system to be split into several isolated islands. Connectivity analysis will reveal such conditions and establish the reliability of each island to continue operating autonomously.

2.2 Power flow analysis

Power failure or disruption at a demand node can occur also because of abnormal power flow conditions; e.g. caused by power imbalance or abnormal voltage in the transmission lines. Such conditions can be identified through a power flow analysis. In a transmission network, it is convenient to lump the generator power and load power of a particular bus into a net bus power S_i, defined as the difference between the generator and load powers; i.e. (Elgerd, 1982)

$$S_i^* = P_i - jQ_i = (P_{Gi} - P_{Di}) - j(Q_{Gi} - Q_{Di}) \quad (2)$$

The power flow equation for each bus is

$$S_i^* = P_i - jQ_i = V_i^* \sum_{k=1}^{n} y_{ik} V_k \qquad (3)$$

where P_i and Q_i are the real and imaginary parts of the net power at bus i, j is the unit imaginary number; V_i and V_k are the voltages at buses i and k, repectively; n is the number of buses in the network system; y_{ik} is the bus admittance matrix; and the symbol * denotes complex conjugate. If the above power flow equation is separated into the real and imaginary parts, two equations are obtained for each bus. Of the variables P_i, Q_i, magnitude of V_i, and phase of V_i at each bus any two can be specified, such as P_i and Q_i at a load bus, or P_i and magnitude of V_i at a generating bus. Thus, for an n-bus system, the 2n unknown variables are determined from the solution of the 2n independent power flow equations. These equations are nonlinear requiring iterative methods of solution, such as the Gauss-Seidel or the Newton-Raphson method. Abnormal power flow conditions are established as follows:

Power Imbalance -- In a damaged transmission network, i.e., a network in which some substations and/or power stations are damaged, the total generating power may become greater or less than the total power demand. In this study, the criterion for acceptable power balance is defined by the following tolerable limits:

$$1.05 < \frac{total\ supply}{total\ demand} < 1.1 \qquad (4)$$

If the above condition is violated, power outage will be caused by power imbalance in the system.

Abnormal Voltage -- If the power flow analysis shows the ratio of the voltage magnitude of the damaged network, V_{damage}, to that of the base case, V_{base}, at a specific node violates a specified tolerable range, a blackout condition called abnormal voltage is reached, defined by the following,

$$\left| \frac{V_{base} - V_{damage}}{V_{base}} \right| > \alpha \qquad (5)$$

where the parameter α depends on the type of transformers in the substation (node). In this study, $\alpha = 0.2$ is used.

Unstable Condition -- Cases for which a convergent solution of the power flow equations cannot be obtained are classified as unstable conditions.

Operational Power Interruption -- A bus with abnormally high or abnormally low voltage may be causing the lack of convergence. When such a bus is removed from the network convergence is obtained. The blackout at the reference bus is an operational power interruption.

2.3 Monte Carlo Simulation

Power failure can be caused by either disconnectivity or abnormal power flow. The respective probabilities at each demand node in the network are calculated through Monte Carlo simulations. The Monte Carlo procedure for this purpose consists of the following steps:

1. Define the base network model of the electric power transmission system.
2. Define the fragility function for each node and link in the network model, and the probability distribution of the site-specific ground motions.
3. For each simulation run (trial), failure of component (node or link) occurs if the sampled seismic load at the component site exceeds the sampled seismic strength of the component.
4. Remove every network component that is damaged from the initial network model (base model).
5. Perform connectivity analysis and power flow analysis on the damaged network model obtained in Step 4 based on the result of the connectivity analysis; the isolated parts are removed and the network is divided into several islands if necessary. Power balance is checked and then power flow analysis is performed for each island of the damaged network model.
6. Evaluate power failure at each demand node. The power failure modes considered at each demand are

disconnectivity from supply nodes and abnormal power flow, which are identified on the basis of the connectivity analysis and power flow analysis results of Step 5.

7. Repeat Steps 3 through 6 for a sufficient number of trials and evaluate the probability of power failure at each demand node as the number of trials with power failures divided by the total number of trials.

A computer program has been developed to perform the Monte Carlo simulations as outlined in the flow chart of Fig. 1. In this program, changes in network loading resulting from damage to the distribution system or shedding of loads by customers (Matsuda et al, 1991) are considered negligible. Also, it is assumed that the probability of failure of the network links is negligibly small. If desired, however, the model can easily be extended to consider such load changes and is capable of considering failure of the network links as well as failure of the network nodes.

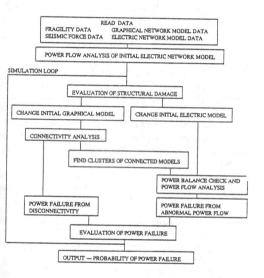

Fig. 1 Flowchart for Monte Carlo simulation procedure

3 EARTHQUAKE GROUND MOTION MODEL

Electric power transmission systems generally cover extensive areas with some generating plants at considerable distances from the major demand sites. For reliability assessment under a given scenario earthquake, e.g., the 1989 Loma Prieta earthquake, a proper characterization of the ground motions at the various sites is needed. It is obvious, given the areal coverage of a transmission system, that ground motions at various distances from the earthquake source and for a wide variety of local site conditions are likely to be required. The maximum amplitude of the ground motions (e.g. the peak ground acceleration, PGA) at a site can be determined through an appropriate attenuation equation. Corresponding to a given PGA, the ground motion time-history at a site can be modeled as a nonstationary random process. The proposed stochastic ground motion model defines the probability distribution, frequency content and duration of the ground acceleration at the site. In particular, a frequency and amplitude modulated filtered Gaussian white noise (Yeh and Wen, 1989) is used to represent the possible ground motion time histories. With this model, the ground motion process is described by the instantaneous power spectral density function, $S(\omega)$, intensity function, $I(t)$, and a frequency modulation function, $\Phi(t)$. The Clough-Penzien power spectral density function (Clough and Penzien, 1993) is used to describe the instantaneous power spectral density function. The parameters of the model, i.e., the parameters in the analytical expressions for $S(\omega)$, $I(t)$, and $\Phi(t)$ are obtained from recorded accelerograms (Yeh and Wen, 1989).

4 SEISMIC CAPACITY AND FRAGILITY FUNCTION

Critical Equipment -- Among the different equipment in an electric transmission substation, some are critical to the proper operation of the substation and thus the fragility of a substation depends on the fragilities of these pieces of equipment. The importance of a piece of equipment in the overall performance of a substation depends on the configuration of the substation. In many cases, a load is fed by two or more circuits and thus, the opening of one circuit does not necessarily mean interruption of power flow to that load. However, the uncontrolled opening of one redundant circuit may lead to an overload on the other circuits, which in turn would lead to the opening of the other circuits. Therefore, for the purposes of simplifying the proposed assessment model, it is assumed that there is effectively no redundancy in the substations and the critical pieces of equipment in these substations are simply all those that can cause the opening of a circuit. On this basis, and on the basis of the seismic equipment and structural profiles presented in (Klopfenstein, Conway and Stanton, 1976; Conway, Fong, Hawkins and Ostrom, 1978), the pieces of equipment considered most critical are: (a) potential transformers, (b) circuit breakers, (c) current transformers, (d) coupling capacitor voltage devices (CCVD), (e) switch disconnects, and (f) bus supports. Bus supports are considered critical because if a bus support fails, the bus will most likely come in contact with the ground, a short circuit will be induced, and the associated circuit breakers will open.

In addition, it is assumed that the most vulnerable components of the critical pieces of equipment are the insulating ceramic components. Past earthquakes have shown that gaskets for some of the critical pieces of equipment identified above are susceptible to leaks or rupture as a result of a combination of internal pressure and seismic loads (Matsuda et al, 1991). Such failure

Table 1. Lognormal parameters of seismic loads and fragilities

Substation or Power Plant	Load $\zeta=0.45$	Fragility				
		500 kv		230 kv		
	λ	λ	ζ	λ	ζ	
Moss Landing	-1.347	-1.356	0.1710	-1.127	0.1132	
Metcalf	-1.273	-1.370	0.1170	-0.926	0.1640	
Los Banos	-2.660	-1.351	0.1171			
Monta Vista	-1.514			-1.100	0.1350	
Jefferson	-2.226			-1.100	0.1350	
Newark	-2.230			-1.100	0.1350	
Ravenswood	-1.880			-1.100	0.1350	
San Mateo	-1.273			-1.609	0.1930	
Martin	-2.740			-1.100	0.1350	
Embarcadero	-2.740			-1.100	0.1350	

Table 2. Natural frequencies and capacities of electrical equipment at Metcalf substation

Equipment	Fundamental Natural Frequency (Hz)			Spectral Acceleration Causing Failure (g)		
	Lower Bound	Mean	Upper Bound	Lower Bound	Mean	Upper Bound
500-KV SWITCHYARD						
Westinghouse SF-6 live tank circuit breaker	1.5	2.0	2.5	1.37	1.91	2.66
Current transformer	1.5	2.0	2.5	1.65	2.29	3.21
Hitachi SF-6 dead tank circuit breaker	7.2	8.2	9.3	1.65	2.29	3.21
500/230-kV transformer bank	2.7	3.3	4.0	1.71	2.37	3.32
Disconnect switches	2.8	3.3	3.8	1.94	2.7	3.78
Bus support	2.8	3.3	3.8	1.94	2.70	3.78
230-KV SWITCHYARD						
Oil circuit breaker	8.6	9.8	11.0	1.91	2.65	3.71
GE ATB6 Live tank circuit breaker	1.5	2.0	3.0	2.02	2.81	3.93
Disconnect switch	1.6	2.1	2.8	2.35	3.27	4.58
230/110-kV Transformer bank	8.6	9.8	11.0	1.91	2.65	3.71

modes may be easily incorporated in the model once their respective fragilities are obtained. It should be mentioned, however, that such gaskets may be more easily strengthened than the ceramic insulators. Also, the leaking or failed gaskets can, in general, be more easily and quickly repaired than broken ceramic insulators, implying that it may be less difficult to restore service if the gaskets leak or fail than if the ceramic insulators fail. These two facts tend to make the ceramic insulators more critical.

4.1 Ultimate capacity of electrical equipment

The level of shaking that can cause the interruption of power flow at a given substation depends on the ultimate lateral capacity of each critical equipment. In general, the evaluation of the ultimate capacity involves many factors influencing the dynamic response and capacity of the equipment. Information or data on these factors are limited or difficult to define. For these reasons, the use of sophisticated analytical models to estimate the needed ultimate capacity of a critical equipment is not warranted.

In view of the above, a simple procedure is selected to estimate the required seismic capacities of the critical equipments. The dynamic properties of a given piece of equipment are first estimated from available test data (Fisher and Daube, 1976; Grases et al, 1989; Taylor, Smith and Klopfenstein, 1974; TEPSCO, 1988). Then, its response to a given seismic excitation is calculated by means of a simplified response spectrum approach. Additionally, simplifying assumptions include the following: (1) the equipment and its support structure are linearly elastic and respond predominantly in the fundamental mode of the combined equipment-support system; (2) the ceramic elements of the equipment are its most fragile parts and will be the first to fail under a strong earthquake; (3) failure of a ceramic component implies failure of the piece of equipment; and (4) the bushings and other ceramic elements behave as cantilever beams. With these assumptions, the ultimate lateral capacity of a given piece of equipment can be determined in terms of the dimensionless spectral acceleration, $SA(\omega,\xi)/g$, as:

$$R = \frac{Z}{WH_{cm}}(f_t + N/A_r) \qquad (6)$$

where A_r and Z are, respectively, the area and section modulus of the cross section of the ceramic element at its base; W is the weight of the equipment, H_{cm} is the distance from the base of the element to its center of mass; N is the axial force in the element induced by its own weight or a prestressing force; and f_t denotes the tensile strength of the ceramic.

The ultimate capacity of a piece of equipment, therefore, basically depends on the fundamental natural frequency and damping ratio of its equipment-support system, the geometric characteristics of its ceramic elements, and the tensile strength of ceramics. As these parameters vary widely, a range of values is estimated for each of the parameters with the assumption that the values in the range are equally likely. For example, the tensile strength of ceramic insulators varies between 400 MPa and 700 MPa (Buchanan, 1986). On this basis, uniform probability density functions are derived for the natural frequency of the equipment, ω, and for the seismic strength, R, for the ceramic components of the various pieces of equipment. The exception is the

damping ratio, ξ, which, in most of the tests reported, is found to be equal to 2 percent.

For a given ground motion intensity, $A = a$, the component fragility can be defined as the probability that the spectral acceleration response, $SA(\omega,\xi)/g$, will exceed the lateral capacity, R, of the component. Prescribing uniform probability density functions (PDF's) for both the ultimate capacity and frequency of the equipment with the respective specified ranges of values and a constant damping ratio of 2 percent, the pertinent probability of failure is

$$P_F(A=a) = \frac{1}{(\omega_u - \omega_l)(r_u - r_l)} \tag{7}$$
$$\int_{\omega_l}^{\omega_u} \int_{r_l}^{r_u} P[SA(\omega,0.02)/g > r \,|\, A=a] \, dr \, d\omega$$

where the subscripts l and u, are used to identify the lower and upper bounds of the uniform random variables that describe the seismic resistance, R, and equipment frequency, ω.

In this study, Monte Carlo simulation is used to compute the mean and standard deviation of the spectral acceleration responses for the specified input ground motions and a Type I extreme value distribution is prescribed for $SA(\omega,\xi)/g$ for a given $A=a$.

4.2 Substation fragility

The substation fragility depends on the fragility of the critical pieces of equipment in the substation which, in turn, depend on the fragilities of the ceramic elements in the component. The substation fragility is obtained assuming that the fragilities for the various ceramic elements are statistically independent. Failures of only one or a few of the critical ceramic elements in a piece of equipment have been observed (Swan, Hadjian, 1988) which lends validity to this assumption.

If $R_1, R_2,...,R_n$, denote the respective seismic capacities of the various critical equipments in a substation, the substation fragility is the cumulative probability distribution (CDF),

$$F_Y(y) = 1 - \prod_{j=1}^{n} [1 - F_{R_j}(y)] \tag{8}$$

where $F_{R_j}(y)$ is the CDF of the jth equipment. As a simplification to the simulation procedure, a lognormal distribution is fitted to $F_Y(y)$ with median λ_Y and coefficient of variation ζ_Y.

5 APPLICATION TO THE LOMA PRIETA EARTHQUAKE

The electrical power transmission system in the San Francisco Bay area was heavily impacted by the Loma Prieta earthquake of October 17, 1989. The major electrical network in the affected area covering the 500 kv and 230 kv facilities in the network and the 115 kv facilities in San Francisco is shown in Fig 2. Some of

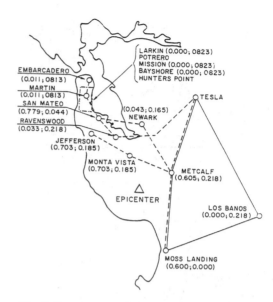

Fig. 2 Bay area transmission system with respective calculated probabilities of power failures

the power plants and major substations experienced severe damage, and caused electric power disruption to over a million customers in the area (PG&E, 1990; EQ Spectra, 1990).

During the Loma Prieta earthquake, ground motions were not recorded either at the 230kV and 500kV substations located near the rupture surface, i.e, the Metcalf, Monte Vista and Moss Landing substations, or at the San Mateo substation, a 230kV substation located far from the fault rupture but where significant damage was observed (Maley et al, 1989; Shakal et al, 1989). An attenuation equation developed on the basis of the Loma Prieta earthquake data is used to determine the median and coefficient of variation (c.o.v.) of the peak ground accelerations (PGA) at the Metcalf, Monte Vista and Moss Landing sites. The site conditions at the substations as well as their distance from the rupture surface are obtained from (Tsai, 1991). Accordingly, the median PGA's are estimated to be 0.28g, 0.22g, and 0.26g, for the Metcalf, Monte Vista and Moss Landing sites, respectively. The site conditions at the San Mateo substation (Tsai, 1991) are such that it can be classified as a soft soil site, i.e., a site where the earthquake ground accelerations are also strongly affected by the local soil conditions. In this regard, the median PGA at the San Mateo substation is taken as the PGA recorded at Foster City, at a soft woil site similar to the San Mateo site and located only 3 km (approximately 2 miles) away from it. The coefficient of variation of the PGA is 0.45 for all the sites, and the probability distribution function of the maximum ground acceleration is assumed to follow a lognormal

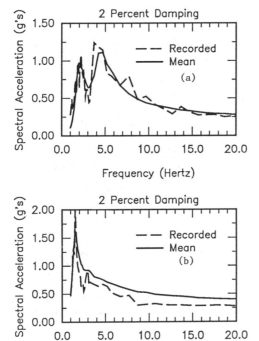

Fig. 3 Response spectra for 2 percent damping: (a) Anderson Dam record (downstream) and stochastic model (mean); (b) Foster City record and stochastic model (mean).

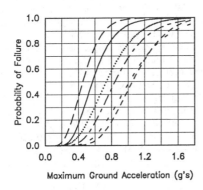

Maximum Ground Acceleration (g's)

——— Live Tank Circuit–Breaker, 500 Kv
- - Disconnect Switch, 230 Kv
— — Live Tank (ATB 7) Circuit Breaker, 230 Kv (Soft Soil)
— — · Circuit Breaker (SF 6), 230 Kv
— · Live Tank (ATB 6) Circuit Breaker, 230 Kv
· Live Tank (ATB 6) Circuit Breaker, 230 Kv (Soft Soil)

Fig. 4 Fragility curves of electric equipment

distribution with the parameters given in Table 1.

The response spectra for the 250-degree horizontal component of the Anderson Dam records (downstream), and the response spectrum of the E-W horizontal component of the Foster City records (Shakal et al, 1989) are shown in Figs. 3a and 3b, respectively. Also shown in Figs. 3a and 3b are the mean response spectra obtained using the stochastic ground motion model with parameters identified on the basis of the Anderson Dam records and the Foster City records, respectively. The parameters of the ground motion model identified on the basis of the Anderson Dam records(Ang, Pires, Schinzinger, Villaverde, Yoshida, 1992) are used to characterize the ground motions for the purposes of computing the substation fragilities at Metcalf, Monte Vista and Moss Landing, whereas the parameters obtained on the basis of the Foster City records are used to model the stochastic ground motion input for the San Mateo substation.

Using reported experimental data, the calculated natural frequencies and ultimate lateral capacities are obtained for the critical equipment in the above substations. As an example, the values obtained for Metcalf are presented in Table 2; similar results for the other substations are reported in (Ang, et al, 1992). The mean, lower and upper values for the equipment ultimate capacities are based on an average value of 6.8 ksi for the tensile strength of the ceramic reported in (Buchanan, 1986), a lower value of 4.26 ksi given in the 1976 Electrical Engineering Handbook, and an upper value of 12.626 ksi, based on NGK tests of porcelain column subassemblages reported by Fischer (1991). Fragility curves for the ceramic components of selected critical equipment at the various substations are shown in Fig. 4.

The fragility functions for the major substations in the network derived on the basis that seismic damage or failure of any piece of equipment will lead to shutdown of that facility are shown in Fig. 5. Each of those fragility functions is fitted to a lognormal distribution to be used in the Monte Carlo simulation described above. The parameters of the fitted lognormal distributions are shown in Table 1.

5.1 Disconnectivity and power flow analyses

The performance function for a substation may be defined as

$$g(X) = Y/A \qquad (9)$$

where Y = seismic capacity of the substation, and A = maximum seismic excitation at the site. $g(\underline{X}) \leq 1.0$, therefore, represents damage and/or power disruption of the substations.

Following the Monte Carlo simulation procedure outlined earlier, repeated simulations are performed for the electrical power network shown in Fig. 2, using the fragility functions and site-specific ground motions for

the respective substations. For each simulation sample, structural damage to the substation is checked using Eq. 9 and the network connectivity is checked using Eq. 1. Also, power flow analyses of the network are performed for each simulation sample to compute the loss of power at the substations resulting from abnormal power flow. The probabilities of station blackout from disconnectivity or abnormal power flow computed in this manner are summarized in Table 3 and Fig. 2.

The Monte Carlo calculations show that power loss at the stations that were subjected to high ground accelerations, such as Moss Landing, Metcalf, Monte Vista and San Mateo, were caused largely by disconnectivity -- probabilities range from 0.600 to 0.779. In contrast, for the stations in the areas of low ground accelerations, e.g., in the city of San Francisco, the probabilities of structural damage are small and the major contributor to power loss is power imbalance which was almost entirely caused by the blackout of the San Mateo substation.

5.2 Comparison with blackout data

According to post-earthquake reports (PG&E, 1990; Earthquake Spectra, 1990), the Moss Landing, Metcalf, and San Mateo substations were severely damaged during the Loma Prieta earthquake. The damage of the San Mateo substation severed transmission service to the San Francisco area, creating large generating/load imbalance (power imbalance) which caused a rapid frequency decline. Fig. 6 shows the actual line outages and station failures following the earthquake. This can be compared with the theoretical results of Fig. 2 obtained with the model. In particular, the theoretical results show very high failure probabilities caused by disconnectivity for those stations subjected to high-intensity ground motions. In the city of San Francisco, the analysis yielded high probabilities of power imbalance which are consistent also with observations following the earthquake.

6 CONCLUSIONS

Major electrical power failure during an earthquake can be caused by disconnectivity and/or abnormal power flow in a transmission network. Disconnectivity would occur primarily through structural damage of the critical equipment in the system, whereas abnormal power flow could result from power imbalance, abnormal voltage, instability, or operational power interruption. The proposed model can be used to assess the probability of either mode of failure. Implementation of the model requires stochastic definition of site-specific ground motions at the major substations, and the seismic fragility functions of the same substations. Application of the model to the San Francisco Bay area during the Loma Prieta earthquake illustrates its effectiveness.

With the proper adjustments and modifications, the

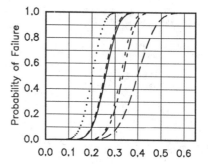

Maximum Ground Acceleration (g's)

——————— Moss Landing, 500 Kv
- - - - - Moss Landing, 230 Kv
- - - - - · Metcalf, 500 Kv
— — — Metcalf, 230 Kv
— - — - Monte Vista, 230 Kv
· San Mateo, 230 Kv

Fig. 5 Fragility curves of major substations

Table 3. Calculated probabilities of power failure (PF)

500 kv	No.	AREA	DISCONNEC-TIVITY PF	ABNORMAL POWER PF	TOTAL PF
Los Banos	6	4	0.000	0.218	0.219
Metcalf	7	2	0.605	0.218	0.823
Tesla	8	3	0.000	0.218	0.218
230 kv					
Moss Landing	9	1	0.600	0.000	0.500
Tesla	10	3	0.000	0.218	0.218
Tesla	11	3	0.000	0.218	0.218
Metcalf	12	2	0.605	0.218	0.823
Monta Vista	13	5	0.703	0.185	0.888
Jefferson	14	6	0.703	0.185	0.888
Newark	15	7	0.000	0.218	0.219
Ravenswood	16	8	0.033	0.218	0.251
San Mateo	17	9	0.779	0.044	0.823
Martin	18	10	0.011	0.813	0.823
Embarcadero	19	11	0.011	0.813	0.823
Embarcadero	20	11	0.011	0.813	0.823
115 kv					
Martin	21	10	0.011	0.813	0.823
San Mateo	22	9	0.779	0.044	0.823
Larkin	23	12	0.000	0.823	0.823
Mission	24	14	0.000	0.823	0.823
Bayshore	25	15	0.000	0.823	0.823

proposed methodology can be easily extended to include other types of equipment failure, e.g., gasket failures, as well as other types of seismic loading effects, such as surface faulting, slope failure, liquefaction, etc. Likewise, the proposed Monte Carlo simulation procedure and network analysis can be easily modified

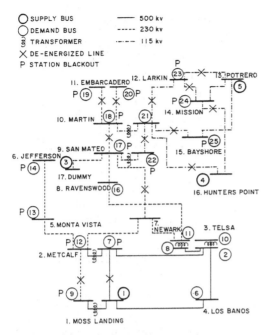

Legend:
- ○ SUPPLY BUS
- ○ DEMAND BUS
- ⊰ TRANSFORMER
- ✕ DE-ENERGIZED LINE
- P STATION BLACKOUT
- ——— 500 kv
- ----- 230 kv
- —··— 115 kv

Fig. 6 Blackout data (de-energized lines and station blackouts) following the earthquake

to incorporate different substation functional models whenever these are available.

ACKNOWLEDGEMENTS

The developments are based on a study supported primarily by the National Science Foundation under grant BCS-9011296. This support is gratefully acknowledged.

REFERENCES

Ang, A.H-S., Pires, J., Schinzinger, R., Villaverde, R. and Yoshida, I., Tech. Rept. to the National Science Foundation, C.E. Dept. Univ. of California, Irvine, 1992.

Buchanan, R.C., Ceramic materials for electronics, Marcel Dekker, Inc., 1986.

Clough, R.W. and Penzien, J. Dynamics of structures, McGraw-Hill Book Co., New York, 1975.

Conway, B.J., et al, IEEE trans. on power apparatus and systems, PAS-97/3: 703-713, 1978.

Earthquake spectra, supplement to vol. 6, 239-338, May, 1990.

Eaton, J.R. and Cohen, E., Electric power transmission systems, 2nd edition, Prentice-Hall, Inc. 1983.

Electrical engineering handbook, Siemens Aktiengesellschaft. New York, 1976.

Elgerd, O.I., Electric energy systems theory: an introduction, 2nd edition, McGraw-Hill, New York, 1982.

Fischer, E.G., Personal communication; May 29, 1991.

Fischer, E.G. and Daube, W.M., Earthquake Engineering and Structural Dynamics, Vol. 4, pp. 231-244, 1976.

Grases, J. et al, IEEE Transactions on Power Delivery, Vol. 4, No. 3, pp. 1701-1707, July, 1989.

Klopenstein, A., Conway, B.J. and Stanton, T.N., EEE transactions on power apparatus and systems, PAS-95/1: 231-242.

Maley, R.A., et al, Geological survey open-file report 89-568, US Geological Survey, 1989.

Matsuda, E.N. et al, Lifeline Earthquake Engineering, ed. ASCE, pp. 295-317, 1991.

PG&E Co., WSCC abbreviated system disturbance, Report, Power Control Department, 1990.

Schinzinger, R. and Peiravi, A. Proc. IEEE Systems, Man. & Cybernetics Soc. Nat. Mtg., 1985.

Shakal, A., et al, Report OSMS 89-06, California Department of Conservation, Division of Mines and Geology, November, 1989.

Swan, S.W. and Hadjian, A.H., EPRI Report No. NP-5607, 1988.

Taylor, G.B., Smith, C.B. and Klopfenstein, A., Nuclear Engineering and Design, Vol. 29, pp. 202-217, 1974.

Tokyo Electric Power Services Company (TEPSCO), Report to TEPSCO, May 1988.

Tsai,B., PG&E Internal Report, 1991.

Yeh, C.H. and Wen, Y.K. SRS No. 546, Univ. of Ill. at Urbana-Champaign, 1989.

Structural Safety & Reliability, Schuëller, Shinozuka & Yao (eds) © 1994 Balkema, Rotterdam, ISBN 90 5410 357 4

A Markov type model for systems with tolerable down times

G. Becker
Technical University of Berlin, Germany

L. Camarinopoulos & G. Zioutas
Aristotelian University of Thessaloniki, Greece

ABSTRACT: This paper provides a modelling framework for the use of inhomogenous Markovian techniques for components and systems, where the duration, any state is assumed, may be limited by a given value. This approach is useful e.g. to build component models with constant repair times, and to model systems or components with tolerable down times. Tolerable downtimes models are known in the field of Boolean techniques since about 10 years. A systematic Markovian approach applied to systems composed of several components allows to treat interdependencies among the components precisely, which may cause large conservativities, if Boolean modelling technique is applied.

1. INTRODUCTION

Inhomogenous Markov processes are useful for the modelling of (small) technical systems composed of components exhibiting dependencies in their failure resp. repair behavior (Barlow 1965). Examples of such dependencies are:

- System behavior depending on the sequence of component failures in time
- Failure rate of some component depending on the state of some other component(s)
- limited repair capacities

Such problems are beyond the scope of ordinary Boolean models like fault trees and reliability block diagrams, or, more precisely, they are beyond the techniques provided in conventional computer codes for fault tree analysis. An inhomogenous Markov model results, if e. g. inspections occurr at fixed times, or if the structure of the system or its success criteria vary with calendar time (phased mission problems) (Becker 1992).

A restriction for the use of Markov models is the fact, that generally, the durations, the process is in any state, have to follow an exponential distribution. This may be relieved to some extend, if semi - Markov models are applied. However, semi-Markov models impose numerical problems even in the homogenous case. Also, though duration in fact may be distributed in any way, this is in most cases not true for the life times and repair times of the components.

An semi-Markovian approach has been developed, which does not allow for deliberate distributions, but for fixed (maximum) durations of given states. It is simpler than a general semi Markov model, as it requires the solution of differential equations, rather than integral equations. It covers practical needs, because most applications of semi Markov models for reliability purposes are restricted to a mixture of exponentially distributed durations and constant maximum durations (a classical example is given in (Barlow 1965)). It is readily applicable for processes, which are inhomogenous with respect to calendar time; in this aspect, it is an extension of semi-Markovian modelling. This last property is important, because in practice, component failures can often be detected only by inspections, which usually occurr regularly at fixed calendar times. If such components exist, they tend to dominate the reliability behavior of the system. Only in special cases this can be modelled with a homogenous semi Markov model.

A typical application for this type of process

is the tolerable down time reliability problem (Camarinopoulos 1981), which is defined in the following way: There are systems, for which failure is only hazardous, if the duration of system failure exceeds some value T_{tol}, which is given by the physical properties. As an example, consider a vessel in a chemical plant, where, due to some exothermic reaction, heat is generated, which has to be removed by a cooling device. It will take some time T_{tol} (due to the thermal heat capacities involved), until temperature reaches a critical value. Should repair of the cooling device succeed, before T_{tol} elapses, the consequences of the accident may be insignificant in comparison with the case, when the critical temperature is actually reached.

Sometimes, tolerable down times occurr, which are time dependent themselves. Consider e. g. the residual heat removal in a nuclear power plant. A nuclear reactor cannot be switched off instantly. After shut down by insertion of the control rods, residual heat is produced, which (starting from some 15 % of the thermal power of the reactor) slowly decreases with time, which leads to an increasing tolerable down time.

The modelling of either type of behavior is straight forward, if an additional state can be introduced in the Markovian model, which is left after a fixed duration T_{tol}, provided it has not been left before due to a successful repair.

Certainly, the effort required is still large compared with the treatment of independent components. In practical applications, a reasonably small group of components exhibiting some inter dependencies will be a part of a large system of independent components. There is no need (and no feasibility) to model the whole system in a large Markov model. Rather, the small group should be modelled with the appropriate Markovian method and the results should be propagated to a boolean model based on fault trees or reliability block diagrams.

2. DIFFERENTIAL EQUATIONS DESCRIBING PROCESSES WITH FIXED DURATIONS

In the following sections, a system of differential equations shall be derived, which is suitable to describe a Markovian process, where some states are left after a fixed maximum duration. It should be noted at this point, that such a process does not have the strict Markovian property at all times, but it is a special variant of a (possibly inhomogenous) semi Markov process.

To formally define the problem, consider a finite state graph $G = (v, e1, e2)$. The set of vertices v shall represent the states of the process. There are two sets of directed edges e1 and e2 defined on $v \cdot v$. Start and end vertices are assumed to be different, i. e. the graph contains no slings. The elements of e1 represent ordinary transitions and are labeled with the transition rates λ_{ij} (t), which are functions of time, if the process is inhomogenous. The elements of e2 represent transitions occurring, if the maximum duration in the starting vertix is exceeded; they are labeled with according values τ_{ij}. Transitions are assumed to occurr s-independently, which implies, that in a time interval (t, t+dt), there will be at most one transition. Subsequently, members of e1 shall be refered to as λ-transitions, and members of e2 as τ-transitions. Note, that for any vertex, there may be at most one τ-transition emerging from it; only as a matter of convenience, the index τ_{ij} is used subsequently.

2.1 State probabilities and frequency densities

Let H_{rj} (t) be the expected value of the number of times, state j of the given process is reached in the interval (0, t), and H_{lj} (t) accordingly the expected value of the number of times, it is left. Then, d H_{rj} (t) = h_{rj} (t) dt and d H_{lj} (t) = h_{lj} (t) dt will define corresponding frequency density functions.

As the process has the property, that at most one transition can occurr in some interval (t, t+dt), the frequency densities may be interpreted as the probabilities, that a transition occurrs into (resp. from) state j in (t, t+dt). Formally, with

$$l_{jt} = \{\text{the event, that state j is left in (t, t+dt)}\} \tag{1}$$

$$r_{jt} = \{\text{the event, that state j is reached in (t, t+dt)}\} \tag{2}$$

the frequency densities may be written as

$$h_{lj} (t) \, dt = pr \{l_{jt}\} \tag{3}$$

$$h_{rj} (t) \, dt = pr \{r_{jt}\} \tag{4}$$

Furthermore, with

Z_{jt} = {the event, that the process is in state j at time t}   (5)

the state probability can be expressed as

$$p_j (t) = \text{pr } \{Z_{jt}\} \qquad (6)$$

<Lema>
Any finite stochastic process with independent transions and without slings obeys for all of its states (indexed with j) the equation

$$\frac{dp_j (t)}{dt} = h_{rj} (t) - h_{lj} (t) \qquad (7)$$

<Proof>
Consider $p_j (t+dt) = \text{pr } \{Z_{jt+dt}\}$. The process will be in state j at t+dt, if either, state j is reached in the interval (t, t+dt), or else, if state j is assumed already at time t and it is not left during (t, t+dt), i. e.

$$Z_{jt+dt} = r_{jt} \cup Z_{jt} \cap \neg l_{jt} \qquad (8)$$

This may be rewritten using deMorgans laws:

$$Z_{jt+dt} = r_{jt} \cup \neg \neg (Z_{jt} \cap \neg l_{jt})$$

$$= r_{jt} \cup \neg (\neg Z_{jt} \cup l_{jt}) \qquad (9)$$

All events in (9) may be treated as mutually exclusive events, as (concerning $(\neg Z_{jt} \cup l_{jt})$) a state may not be left unless it has been assumed before, and (concerning $r_{jt} \cup \neg (\neg Z_{jt} \cup l_{jt})$), a state may not be reached starting from itself by a single transition, as it has been assumed, that the state graph of the process has no slings. Thus, the resulting probability is

$$\text{pr } \{Z_{jt+dt}\} = \text{pr } \{r_{jt}\} + (1 - (1 - \text{pr } \{Z_{jt}\} + \text{pr } \{l_{jt}\})) \qquad (10)$$

which may be expressed as

$$p_j (t+dt) = h_{rj} (t) \, dt + p_j (t) - h_{lj} (t) \, dt \qquad (11)$$

Simple calculus allows to transform this to equation (7)   QED.

For an ordinary inhomogenous Markov process, the definition of the transition rates leads immediately to

$$h_{rj} (t) = \sum_{\forall i \neq j} p_i (t) \lambda_{ij} (t) \qquad (12)$$

$$h_{lj} (t) = p_j (t) \sum_{\forall j \neq k} \lambda_{jk} (t) \qquad (13)$$

which corresponds (with (7)) to the well known system of differential equation for ordinary Markov processes (Howard 1971). Note, however, that the Markovian property was not needed in any way for the proof of equation (7). Thus, (7) is qualified to be used for modelling an inhomogenous semi Markov process with given maximum durations, as described in the following section.

2.2 Frequency densities for a process with fixed durations

To be able to use the result of the last section (7), it is necessary to determine the frequency density functions $h_{rj} (t)$ and $h_{lj} (t)$. The equation for $h_{rj} (t)$ will be derived in detail, whereas for $h_{lj} (t)$, only the final result, which may be obtained in much the same way, shall be given for the sake of brevity.

To determine the frequency density, a state j is reached with, the event rjt will be split into the events

r'jt = {the event, that a λ-transition occurs into j in (t, t+dt)}

r''jt = {the event, that a τ-transition occurs into j in (t, t+dt)}

(Notation is as in sections 2 and 2.1)

As two independent transition within the same infinitesimal interval (t, t+dt) have a probability of $O (dt^2)$

$$r_{jt} = r'_{jt} \cup r''_{jt} \qquad (14)$$

$$\text{pr } \{r_{jt}\} = \text{pr } \{r'_{jt}\} + \text{pr } \{r''_{jt}\} \qquad (15)$$

From the definition of the transition rates,

$$\text{pr } \{r'_{jt}\} = \sum_{\forall i \neq j} p_i (t) \lambda_{ij} (t) \, dt \qquad (16)$$

where $\lambda_{ij} (t) = 0$, if there is no λ-transition from i to j.

With the definition

S_{it} = {the event, that state i is not left before time t} (17)

the following holds for a τ-transition r''_{jt}

$$r''_{jt} = \bigcup_{\forall i \,|\, \exists \tau_{ij}} r_{it-\tau_{ij}} \cap S_{it} \quad (18)$$

This means, there will be a τ-transition into state j in the interval (t, t+dt), if one of those states, which are linked to i via a τ-transition, has been reached in the interval $(t-\tau_{ij}, t-\tau_{ij}+dt)$, and this state has not been left before t. It should be noted here, that

$$r_{it} = \emptyset \text{ if } t < 0 \quad (19)$$

as the process is started at t=0, and there are no transitions before.

As at most one transition may occur in a given time interval, the terms of the union may be treated like mutually exclusive events, which yields

$$pr\{r''_{jt}\} = \sum_{\forall i \,|\, \exists \tau_{ij}} pr\{r_{it-\tau_{ij}}\}\, pr\{S_{it} \,|\, r_{it-\tau_{ij}}\} \quad (20)$$

or

$$pr\{r''_{jt}\} = \sum_{\forall i \neq j} \gamma_{ij}\, pr\{r_{it-\tau_{ij}}\}\, pr\{S_{it} \,|\, r_{it-\tau_{ij}}\} \quad (21)$$

where γ_{ij} is an indicator function defined as

$$\gamma_{ij} = \begin{cases} 1: \text{if there is a } \tau\text{-transition from i to j} \\ 0: \text{else} \end{cases} \quad (22)$$

The conditional probability in (21) is the inverse cumulative distribution function (or the survivor function sf) of the uninterrupted time, the process will remain in state i, given that i is reached at $t-\tau_{ij}$. This may be determined easily, as, if this duration is less than τ_{ij}, only λ-transitions have to be considered. One way to find it is to consider a simple inhomogenous Markov process consisting of two states, which are linked by a single λ-transition, which is labeled with the sum of all λ-transition rates emerging from i. With

$$\lambda_i(t) = \sum_{\forall i \neq k} \lambda_{ik}(t)$$

and

$$R_i(t-\tau_{ij}, t) = pr\{S_{it} \,|\, r_{it-\tau_{ij}}\} \quad (23)$$

it can be found that

$$R_i(t-\tau_{ij}, t) = \exp\left(-\int_{t-\tau_{ij}}^{t} \lambda_i(x)\,dx\right) \quad (24)$$

if $\lambda_i(t)$ is steady and finite.

It should be mentioned, that in practice, transition rates may exhibit discontinuities, which e. g. are used to model the effect of inspections [2]. In this case, $\lambda_i(t)$ will contain Dirac pulses with masses p_z at some times t_z. If this holds, the survivor function must be determined in a piece wise manner as

$$R_i(t-\tau ij, t) = R_i(t-\tau_{ij}, t_z)(1-p_z)$$

$$\exp\left(-\int_{t_z}^{t} \lambda_i(x)\,dx\right) \quad (25)$$

where t_z and p_z represent the last discontinuity before t.

This may be rewritten as

$$R_i(t-\tau_{ij}, t) = \left(\prod_{\forall z \,|\, t-\tau_{ij}\, \leq\, t_z\, <\, t} (1-p_z)\right)$$

$$\exp\left(-\int_{t-\tau_{ij}}^{t} \lambda_i(x)\,dx\right) \quad (26)$$

Combining these results, the frequency density hrj (t) may be given as

$$h_{rj}(t) =$$

$$\sum_{\forall i \neq j}\left(p_i(t)\,\lambda_{ij}(t) + \gamma_{ij}\,h_{ri}(t-\tau_{ij})\,R_i(t-\tau_{ij}, t)\right) \quad (27)$$

In much the same way, the frequency density $h_{lj}(t)$, with which the state j is left, can be found as

$$h_{lj}(t) = \quad (28)$$

$$\sum_{\forall k \neq j}\left(p_j(t)\,\lambda_{jk}(t) + \gamma_{jk}\,h_{rj}(t-\tau_{jk})\,R_j(t-\tau_{jk}, t)\right)$$

The equations (7), (27), and (28) together form a system of differential equations which describe

an inhomogenous Markovian process with fixed maximum durations. These latter durations occur as "dead times", showing, that the Markovian property is not completely given in the context of this type of process. Solutions have to be found numerically in most realistic cases.

2.3 An extension covering inhomogenous maximum durations

In the introduction, it has been shown, that sometimes, tolerable down times are time dependent. An obvious way to model this, is to consider maximum durations τ_{ij} (t). However, it has to be specified precisely, how this term is to be interpreted. Formally, one could allow τ_{ij} to be a function of calendar time in (27) and (28). In this case, which may be referred to as a "strict" inhomogenity, τ_{ij} would depend on the time state i is left (and state j is reached). Possibly, systems might exist, which are modeled correctly by this type of time dependency. This is, however, not true for the type of system mission, which was discussed in the introduction (heat removal from a time dependent source). In this case, and in most cases, the tolerable down time will not depend on the time, when system repair occurrs, but rather on the time, when system failure takes place. For this reason, τ_{ij} (t) is to be defined as the maximum duration, the process will stay in state i, given that it reaches i at time t.

With this definition, the evenr r''_{jt}, that that state j is reached in (t, t+dt) via a τ-transition, is different from the form given in (18). This event is

$$r''_{jt} = \bigcup_{\forall i \mid \exists \tau_{ij}} \left(\bigcup_{\forall t_x} r_{itx} \cap S_{it} \right) \qquad (29)$$

where the times t_x are all solutions of

$$t_x + \tau_{ij} (t_x) = t \qquad (30)$$

This means, state j is reached via a t-transition from state i in (t, t+dt), if state i has been reached in some interval (t_x, t_x+dt) and has not been left before t. t_x has to be chosen in such a way, that t_x and τ_{ij} (t_x) add up to t, because, in this case, the maximum duration, which has been valid at time t_x just ends at time t. In general, (30) may have more than one solution. First, consider the case, where this number of

solutions is finite. Then, as all members of the union are disjunct, the arguing of section 2.2 leads to

$$h_{rj} (t) = \qquad (31)$$

$$\sum_{\forall i \neq j} \left(p_i (t) \lambda_{ij} (t) + \gamma_{ij} \sum_{\forall t_x} h_{ri} (t_x) R_i (t_x, t) \right)$$

$$h_{lj} (t) = \qquad (32)$$

$$\sum_{\forall k \neq j} \left(p_j (t) \lambda_{jk} (t) + \gamma_{jk} \sum_{\forall t_x} h_{rj} (t_x) R_j (t_x, t) \right)$$

If the number of solutions of (30) is not finite, though, the resulting form of (31) and (32) will depend on the special form of t_{ij} (t). If this function is restricted to a positive and piece-wise steady function, the geometric location of solutions for t_x may be given as the intersections of the two functions

$$y_1 = \tau_{ij} (t) + t \qquad (33)$$

and

$$y_2 = t \qquad (34)$$

As y_2 is a horizontal line with slope 0, there will be an infinite number of intersections, if there are one or more intervals (t_s, t_e), where y_1 also has slope 0. Hence, τ_{ij} (t) must have slope -1 in such a region. If the union of all intervals with this property is denoted as G (t), this leads to

$$pr \{r''_{jt}\} = \sum_{\forall i \neq j} \gamma_{ij} \left(\sum_{\forall t_x \mid t_x \notin G (t)} h_{ri} (t_x) R_i (t_x, t) \, dt + \right.$$

$$\left. \int_{\forall t_x \mid t_x \in G (t)} h_{ri} (t_x) R_i (t_x, t) \, dt_x \right) \qquad (35)$$

For such times t, where the integral in (35) is non-zero, pr $\{r''_{jt}\}$ results to be a non-infinitesimal probability. In this case, the sum for $t_x \notin G$ (t) will disappear and the frequency density will have a probability mass given by the value of the integral at such t, which might be written using Dirac notation. This case will not be considered any further, as it will make numerical solution of the system of equations complicated. In fact, for practical calculations for inhomogenous systems, it will be sufficient to consider parameters, which are piece-wise constant functions of time. In this case, the number of solutions of (30) will always be finite.

1219

3. AN EXAMPLE

As an example, consider a storage tank, or a reaction vessel, which requires permanent heating to keep a certain temperature level. There are two identical heaters, one of which is operational, the other one is in stand-by mode and subject to scheduled inspection at some fixed time points. Should heating fail, there will be an allowable down time, which is given by the thermodynamic properties of the system, until temperature decreases to a level, which is prohibitively low. If the components behave independently, this problem could be solved with the methodological framework provided in (Camarinopoulos 1981) by calculating the frequency of the event, that both components fail and repair time exceeds the allowable down time, which may be used as a good and conservative approximation for failure probability. Now, structural restrictions make it impossible to work on both components at the same time to repair them. In this case, Markovian modelling is beneficial to take into account this dependency. In addition, it allows to find the failure probability directly and also to find the unavailability of the given event. However, it is restricted to small systems. This is not a severe restriction, though, as in practical situations, it may be applied to single minimal cut sets within the framework of conventional Boolean modelling techniques, as outlined in (Becker 1992).

3.1 Modelling of the example system

Making use of symmetries exhibited by the system, a model with five states can be derived, which are defined as follows:

1   both components intact
2   one component under repair
3   stand-by component failed (unnoticed)
4   both components defect and tolerable
    down time not exceeded
5   both components defect and tolerable
    down time exceeds repair time

The transitions are given in the state graph in fig. 1. There are three types of transition in fig. 1. Thick lines are used to denote transitions described by a constant transition rate. By these, random events, like failures of a component or repairs are described. E. g., the transition from state 1 to state 2 occurs, when the operational

component fails, hence, the transition rate is the failure rate λ. A transition from state 2 to state 1 occurs, when the defect component in state 2 is repaired, which is characterized by a transition rate ρ, which is the repair rate. The other transitions of this type are similar.

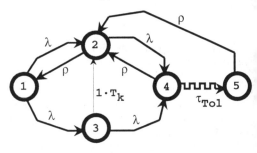

Fig. 1: State graph of the example system

The transition from state 3 to state 2 is a thin line. It has the purpose to model the inspections, which reveal failures of the stand-by component, so that repair can start. Such a transition is impossible at all times apart from some T_K, where inspections occur. Should the process be in state 3 at such T_K, however, a transition will take place with certainty, i. e. with probability 1. This corresponds to a time dependent transition rate given by a sum of Dirac pulses at all T_K. Note, that imperfect inspections could be modelled by assigning a probability less than 1 as weights to these Dirac pulses.

The transition from state 4 to state 5 is drawn by wavy line. This is to show, that the transition is characterized by a maximum duration. State 4 is left via this transition τ_{Tol} time units after it has been reached, unless it has been left before via another transition (in this case the transition from state 4 to state 2).

It should be noted, that this modelling is only correct, if the failure rates are reasonably small, i. e. the mean time to failure must be large in comparison with the time constant of the system. Otherwise, the case would have to be considered, where a second system failure occurs a short time after the first one, when the operating temperature has not been reached yet, and τ_{Tol} has a smaller value. In this case, modelling would require more states.

3.2 Example data and results

The example has been solved numerically with

the Fortran code MARK, which uses the sdriv package from the TOMS library.

The following data has been used for a simulation of 4000 hours starting with an intact system:

λ	failure rate	1.0e-05 / h
ρ	repair rate	0.05 / h
τ_{Tol}	tol. downtime	25.0 h
T_1	time of 1st inspection	730.0 h
T_2	time of 2nd inspection	2920.0 h

First, the unavailability of the system is given in fig. 2. The thick line is the time dependent unavailability, if the allowable down time is taken into consideration. Within the state graph model, this corresponds to the state probability p_5 (t). The thin line also plotted in the diagram is the unavailability, if tolerable down time is not taken into account, which corresponds to the sum p_4 (t) + p_5 (t).

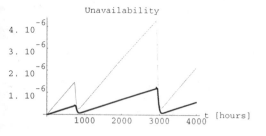

Fig. 2: Unavailability with and without tolerable down time

The unavailability is an adequate measure, if the dammage resulting from exceeding the tolerable down time is proportional to the time it is exceeded. In this case, the average unavailability is proportional to the share of time spent in such state.

Should, however, dammage occur, when the allowable down time is exceeded, independent from how long this lasts, the failure probability is the adequate measure. The failure probability can be found by making the states of interest absorbing by omitting all edges emerging from them.

Results are presented in fig. 3, where, again, the thick line represents the case with tolerable down time, and the thin line gives as a reference the case without tolerable down time.

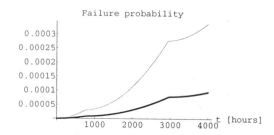

Fig. 3: Failure probability with and without tolerable down time

4. SUMMARY & CONCLUSIONS

A modelling framework for a class if inhomogenous processes, which involve maximum durations of the states, has been developed and presented.

Presently, it is being implemented in a computer code called MARK, which is be useful to model reliability problems involving tolerable down times for systems with dependencies between the basic events. The example in this paper has been calculated with a preliminary version of this code.

Experience shows, that a large number of problems from reliability and risk analysis can be solved using this modelling technique. There are, however, problems, which require the definition of a maximum allowable time not for a single state, but rather for a class of states.

Using eq. (7), the theoretical base is being extended to allow for maximum durations, which are valid for groups of vertices, and to determine failure frequency densities and renewal frequency densities. When this is given, it will be possible to determine unavailabilities and failure frequencies for systems represented by minimal cut sets, yielding a combination of conventional fault tree analysis and state graph analysis, which will have the potential to model large and complex systems.

ACKNOWLEDGEMENT

The work of G. Zioutas, University of Thessaloniki, who implemented maximum durations into the MARK code, has been a valuable aid to calculate the example given in section 3 of this paper.

REFERENCES

Barlow, R. E., Proschan, F., *Mathematical Theory of Reliability*, New York (etc): John Wiley & Sons 1965.

Becker, G., Camarinopoulos, L., Mixed discrete and continuous Markovian Models for Components and Systems with Complex Maintenance and Test Strategies, *Proc. of the European Safety and Reliability Conference* 1992 in Kopenhagen, Denmark, June 10 - 12, 1992, Elsevier: 1992.

Camarinopoulos, L., Obrowski, W., Berücksichtigung tolerierbarer Ausfallzeiten bei der Zuverlässigkeitsanalyse technischer Systeme, *Atomkernenergie*, vol 37, Lfg2 1981.

Howard, R. A., *Dynamic Probabilistic Systems*, 2 vols, New York (etc): John Wiley & Sons 1971.

Structural Safety & Reliability, Schuëller, Shinozuka & Yao (eds) © 1994 Balkema, Rotterdam, ISBN 90 5410 357 4

Problems of statistical inference in structural reliability

Karl Breitung
Munich, Germany

Yaacob Ibrahim
Department of Industrial and Systems Engineering, National University of Singapore, Singapore

ABSTRACT: An important problem in reliability is the use of suitable statistical models. The standard models, classical and Bayesian, and the usual estimation methods are often not adequate for structural reliability problems. Here empirical Bayes methods and risk based tail estimation methods are described.

1 STATISTICAL METHODS IN STRUCTURAL RELIABILITY

In probabilistic reliability models essentially two different fields of mathematics and statistics are used, probability theory and statistical inference. Probability theory deals with the mathematical properties of a given model, it is part of mathematics. Statistical inference is concerned with the relation between the data and the mathematical model. Therefore here pure mathematical reasoning is not sufficient, it is necessary to make some decision about which concepts of probability and statistics shall be used.

One of the weak points of probabilistic models in structural reliability is the neglect of statistical inference problems. The majority of papers take a mathematical model for the reality as given; then study the mathematical structure of this model and compute failure probabilities and other characteristics. Since these probabilities are normally very small, it is in general not possible to derive any predictions which these models which can be compared with actual data. This situation is different from other fields of reliability theory. If for example a model predicts a rate of 1% defective chips in a computer chip production, it is easy to check this prediction by comparing it with the actual percentage of defective in the next lot. Then, if the predicted and the observed percentage differ too much, it is an indication that something is wrong with the model. Elsewhere, if the difference is not too large, it gives some validation for the model.

In the modelling of structures by mathematical and statistical methods there is an iterative interaction between reality and models. In statistical modeling we have the following situation (for more details see West and Harrison (1989)):

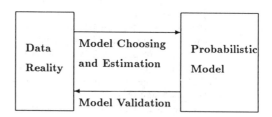

The first step usually involves making a decision about which basic concept of probability will be used, i.e. a Bayesian or frequentist interpretation. The next step involves the choice of a specific statistical and mathematical model. Here, for reliability problems usually a parametric model with a finite number of unknown parameters is chosen. Then in the third step, the parameters of the model are estimated using data and/or prior information.

The choice made in the first step is a decision which cannot be checked by mathematical or statistical methods, but for the choice of models and the parameter structure it is possible to examine the fit of the model by different methods. To check the fit of the parameters model, a number of statistical procedures are available. Statistical models are validated by comparing its predictions with data.

Almost everywhere just the standard classical or Bayesian concepts are used. But there are other concepts and modifications that could be used. The main reason for the predominance of the former concepts is that they are well suited for the bulk of statistical problems. In general, these are problems, where there

is uncertainty about some parameters of a population; then by a series of random experiments or data collection this uncertainty is reduced or removed by getting some estimates of these quantities.

The main underlying assumption is that we have a population, in general large, and by performing a series of identical random experiments, we improve our knowledge about some properties of this population.

A main difficulty in applying these models in structural reliability is that in general the assumptions above are not fulfilled. Often it is necessary to make estimates or predictions about the behavior of a single structure, where no prior information about the behavior of identical structures is available.

What makes this problem even more difficult is that in structural reliability many concepts of statistics are used without any concern about their applicability in this field. Here the principal questions are different from the usual statistical problems, where we estimate the moments and the global shape of the distribution.

In structural reliability we do not estimate means but failure probabilities. Their value depends on the tail shape of the probability distributions. And normally it is not possible to observe failures. If we check the fit of the distributions in the central part, we do not get useful results about the quality of our estimation of the failure probability.

The following problems arise if statistical inference methods are used in structural reliability.

1. The assumptions of the model are not satisfied.

2. There are no possibilities to validate the model.

3. The distribution fit is done mainly in the center part and not in the tail.

The following discussion describes some approaches that can be used to deal with these problems. Concerning the first problem, the use of more flexible models may be a solution. This means that the model assumptions are kept as general as possible. An example is the treatment of sample data given in Box and Tiao (1973), chapter 3, where the normality assumption is relaxed to allow also distributions with different skewness and curtosis. In the following section we describe yet another means of dealing with these problems - empirical Bayes models.

2 EMPIRICAL BAYES MODELS

In a number of problems in reality the assumption that we have a series of identical random experiments is not fulfilled. In the case that we have information available from similar experiments, empirical Bayes methods can be used to build more flexible statistical models. An empirical Bayes model has the following form:

1. The nonobservable parameters $\theta_1, .., \theta_n$ are n independent realizations of a random variable Θ.

2. The conditional distribution $f(x_i \mid \theta_i)$ of the observation x_i for given i is known.

3. The distribution of Θ has the form $g(\theta \mid \alpha)$ with "hyperparameter" α.

4. The likelihood of a sample $x_1, ..., x_n$ is then:

$$L(x_1, ..., x_n \mid \alpha) \propto \prod_{i=1}^{n} \int_{\Theta} f(x_i \mid \theta) g(\theta \mid \alpha) d\theta$$

(1)

5. The value of α has to be estimated from the data $x_1, ..., x_n$ by an estimator $\hat{\theta}$. Usually we take the maximum likelihood estimator $\hat{\theta}$ defined by :

$$L(x_1, ..., x_n \mid \hat{\alpha}) = \max_{\alpha} L(x_1, ..., x_n \mid \alpha) \quad (2)$$

6. Then the distribution $g(\theta \mid \hat{\alpha})$ is used as the prior distribution for further data.

Basically there is then a sequence of independent bivariate random variables (x_i, θ_i) such that only the x_i's can be observed and not the θ_i's and from the x_i's the distribution of the θ_i's has to be estimated, i.e. the value of α.

A comprehensive introduction is given in Maritz and Lwin (1989) and in chapter 13 in Martz and Waller (1982) a short introduction can be found. Hasofer and Esteva (1985) give an application in seismicity estimation.

Hierarchical Bayes models are a modification of this concept. Here it is assumed that the hyperparameter α is a random variable too. Then we do not estimate the value of α, but its distribution. This is a second stage Bayes model. Details are given in Berger (1985).

We get different posterior distributions for α, if we use a Bayesian or empirical Bayesian model. This is outlined in the following example.

We consider the case of independent normally distributed random variables X_i and compare a Bayesian and an empirical Bayes model.

1. *Bayesian* : All X_i's have the same mean μ, which has a normal prior distribution with unknown mean and known variance σ^2. For given μ the X_i's have a normal distribution with mean μ and variance 1.

2. *Empirical Bayes* : The mean μ has a normal prior distribution with unknown mean and

known variance σ^2. Each X_i has, for given μ_i, a normal distribution with mean μ_i and variance 1.

After observing n values $x_1, ..., x_n$, with sample mean $\bar{x}$, the posterior distribution of μ has the following form :

1. *Bayesian*: Normal with mean $\bar{x}$ and variance σ^2/n.

2. *Empirical Bayes*: Normal with mean $\bar{x}$ and variance σ^2 (see Maritz and Lwin (1989), p.32).

The variances are different here due to the model assumptions. In the Bayesian model it approaches zero as $n \to \infty$. In the second case it is σ^2. The reason is that in the second case there is another source of variation; the variation of the means.

3 PREDICTIVE DISTRIBUTIONS AND MODEL VALIDATION

If a Bayesian or empirical Bayesian posterior distribution has been calculated, it can be used to make predictions of further observations using the predictive distribution. Let it be given that we have a posterior distribution with p.d.f. $f_P(\theta)$. Let it also be given that the conditional distribution of an observation given a parameter value θ_0 has the p.d.f. $f(x \mid \theta_0)$. Then the predictive distribution for the next observation has the p.d.f.:

$$f_{pre}(x) \propto \int f(x \mid \theta) f_P(\theta) d\theta \qquad (3)$$

In the same way we get predictive distributions for a set of m future observations.

Geisser (1975a, 1975b and 1980) argues that the main problem of statistical inference is the prediction of further observations and not the estimation of parameters. This point of view appears to be important for statistical models in structural reliability.

Such predictive distributions can be used for model validation. By comparing a histogram of new data with the theoretical predictive distribution or by calculating the likelihood of new data under the model we can check the quality of a model.

If we look at structural reliability and the statistical models there, what is the meaning of the parameters which we estimate? If we consider a single structure, the interpretation as a mean in a fictitious population of such structures does not help. In Matheron (1989) it is outlined that these parameters are important only in the mathematical model.

In many probabilistic models in structural reliability the final results are failure probabilities for a structure; in general we have numbers between 10^{-4} to 10^{-8}. A basic problem here is the meaning of these numbers. If expert A says that the failure probability of structure X is 10^{-5} and expert B says that it is 10^{-6}; what is the correct answer? Or, does a correct or true answer exist at all?

To relate our model to reality and to show that it has any meaning here, the only way available appears to be to build models in a more predictivistic way. But since in reliability we are mainly interested in failure probabilities which are very small, it is not possible in general to compare our predictions of failure events with observed failure events. Therefore we have to proceed in an indirect way.

We make predictions about random variables which are connected with the failure probabilities and compare them with observations. This then can be used for model validation and improvement of parameter estimation in the model. For validation, we look at the likelihood of the observations in the model.

For example, if we consider the probability of failure of a component during 50 years and only an extreme value distribution is fitted, this model allows no comparison with observations during the life time of the component. We have a different situation, if we make a model in such a way that we have for each year an extreme value X_i and a probability of failure of $P(\max_{1 \le t \le 50} X_t \ge \alpha)$ with the X_i's being in the form $X_i = a_i Y_i + b_i$

Here the Y_i's are a sequence of i.i.d. random variables and the a_i's and the b_i's are some constants. Here we can use observed data to update our failure probability estimates.

Considering the last example, we can derive the predictive distributions in these cases.

1. *Bayesian*: Normal with mean $\bar{x}$ and variance $\sigma^2/n + 1$.

2. *Empirical Bayes*: Normal with mean $\bar{x}$ and variance $\sigma^2 + 1$.

To validate a model, we can calculate the likelihood of new observations for each model. Let Y and Z be independent normal random variables with zero mean and unit variance and let $X = Y + Z$. Generate n samples of Y and Z and estimate the sample mean $\bar{x} = 1/n \sum_{i=1}^n x_i$. Generate another k samples of Y and Z. The likelihood functions for both models are given below.

1. *Bayesian*:

$$(2\pi)^{-0.5k} \left[\frac{\sigma^2}{n} + 1 \right]^{-0.5k} \prod_{i=1}^{k} \exp\left[-\frac{1}{2} \left(\frac{(x_i - \bar{x})^2}{\sigma^2/n + 1} \right) \right]$$

2. *Empirical Bayes:*

$$(2\pi)^{-0.5k}(\sigma^2+1)^{-0.5k}\prod_{i=1}^{k}\exp\left[-\frac{1}{2}\left(\frac{(x_i-\bar{x})^2}{\sigma^2+1}\right)\right]$$

For $\sigma^2 = 1.0$, a prior sample of size 50 the values of the likelihood functions for both models are $0.2881E-07$ for the Bayesian and $0.5196EE-07$ for the empirical Bayes model.

4 RISK-BASED TAIL ESTIMATION

In textbooks parameter estimation is done usually by maximum likelihood or χ^2-methods. These methods consider the whole distribution and give optimal parameter estimates in the sense that the likelihood of the sample is maximal or that the squared relative approximation error is minimal.

In reliability theory the problem is different. Failures are caused normally by extreme values in one tail of the distribution, for example by small values of resistance, i.e. in the left tail, and by large values of the load, i.e. in the right tail. The failure probability is influenced mainly by the shape of the distributions in these tails and not by the distribution shape in its central part. Therefore the standard techniques of parameter estimation are not optimal for these problems since they consider mainly the central part.

A method specially meant for tail estimation has been given by Castillo (1989), p.175/6. Assume a sample of n data $x_1, ..., x_n$ from a distribution function $F(x;\theta)$ with unknown parameter θ which has to be estimated. If we are interested mainly in the right tail, a weighted estimation method should be used. Let the ordered sample be $x_{(1)} < ... < x_{(n)}$ and $p_i = i/(n+1)$. Then the parameter is chosen such that the sum

$$\sum_{i=1}^{n}\frac{1}{(1-p_i)^2}\left[p_i - F(x_{(i)};\theta)\right]^2 \qquad (4)$$

is minimal.

We consider the case of an exponential distribution with p.d.f. $f(x) = \lambda\exp(-\lambda x); x > 0$, with unknown parameter λ. The maximum likelihood estimator for λ is $\bar{x}^{-1}$. We compare two sets of data :

A : 1,1.5,2,3,4,4.5,5

B : 0.1,1,1.5,2,3,4,4.5,5,5.9

In both cases the sample mean is the same and therefore the maximum likelihood estimator of λ is the same. Using the weighted tail estimation, we get :

A : $\hat{\lambda} = 0.34796$

B : $\hat{\lambda} = 0.33483$

Here the additiional large term 5.9 leads to an increase of the estimation of large probabilities. Further eexamples are given in Maes and Breitung (1993).

5 CONCLUSIONS

The usual statistical concepts and methods are often not useful in structural reliability. Here it is necessary to modify some of them to adapt them more to the specific problems in this field.

References

[1] Berger, J.O. 1985. Statistical Decision Theory and Bayesian Analysis (second edition). New York: Springer .

[2] Box, G.F.P. and G.G. Tiao 1973. Bayesian Inference in Statistical Analysis. Reading, MA: Addison-Wesley.

[3] Castillo, E. 1989. Extreme Value Theory in Engineering. Boston: Academic Press.

[4] Geisser, S. 1975a. Bayesianism, predictive sample reuse, pseudo observations, and survival. Bulletin of the International Statistical Institute, 40(3):285-289.

[5] Geisser, S. 1975b. A new approach to the fundamental problem of applied statistics. Sankhya, 37(B4):385-397.

[6] Geisser, S. 1980. A predictivistic primer. In A. Zellner, editor, Bayesian Analysis in Econometrics and Statistics, p. 363-382, Amsterdam: North-Holland.

[7] Hasofer, A.M. and L. Esteva 1985. Empirical bayes estimation of seismicity parameters. Structural Safety, 2, pp. 199-205.

[8] Maes, M.A. and K. Breitung 1993. Risk based tail estimation. To appear in the proceedings of the IUTAM Symposium on probabilistic structural mechanics: advances in structural reliability methods, June 7-10, 1993, San Antonio, Texas, ed. Y.-T. Wu.

[9] Matheron, G. 1989. Estimating and Choosing. Berlin: Springer.

[10] Maritz, J.S. and T. Lwin 1989. Empirical Bayes Methods (second edition). London, New York: Chapman and Hall .

[11] Martz, H.F. and R.A. Waller 1982. Bayesian Reliability Analysis. New York: John Wiley.

[12] West, M. and J. Harrison 1989. Bayesian Forecasting and Dynamic Models. New York: Springer.

Structural Safety & Reliability, Schuëller, Shinozuka & Yao (eds) © 1994 Balkema, Rotterdam, ISBN 90 5410 357 4

Systems reliability: Revisited

Christian G. Bucher & Gerhart I. Schuëller
Institute of Engineering Mechanics, University of Innsbruck, Austria

ABSTRACT: The reliability analysis of structural systems frequently has to treat a complex pattern of possible failure modes involving random variables. There are two basic approaches to the problem, one based on identifying dominant failure modes, the other based on identifying important random variables. Clearly, there are certain merits and de-merits associated with each of these procedures. The paper discusses these aspects in the light of the requirements as set forth by engineering practice. Following the second approach, the use of the Response Surface Method is suggested. Numerical examples show that both computational efficiency and accuracy of mechanical modeling can be reasonably well combined leading to satisfactory results.

1 INTRODUCTION

Within the development of reliability analysis, the capability of a particular approach to take into account the most realistic structural modeling available is of utmost importance. In other words, in view of its credibility the reliability analysis should be able to utilize the state-of-the-art structural analysis, i.e. Finite Element techniques.

The failure mode approach is based on the notion that particular combinations of load and resistance quantities, which lead to structural failure can be identified. These combinations ("failure modes") are identified by specific search techniques (e.g. Branch and Bound) which are designed to find apriori the stochastically dominant modes. However, when utilizing this approach the generally very large number of failure modes - which are in many cases very hard to identify, usually in a complex structure - pose severe problems. The other major problem is the fact that failure modes usually cannot be defined in terms of analytical functions unless gross simplifications of the mechanical model are made. But, since realistic modeling is the basic requirement for acceptance of a procedure by the practicing engineering community, these methods are naturally of limited use.

The family of Response Surface Methods - based on the random variable approach - can fully take advantage of the FE-modeling since it determines failure conditions only pointwise. For this purpose it requires only deterministic FE analyses which can be carried out by any FE code available. However, its drawback is the fact that the number of calculations grows rather rapidly with an increasing number of random variables considered.

2 BRIEF REVIEW OF CURRENT METHODS

Attempts to determine structural systems reliability go back to the early sixties (see e.g. [1]). Since then a large number of procedures

have been put forward by various authors. Comparisons and classifications of these methods may be found in the literature (see e.g. [2,3,4]).

So far the requirement of a most realistic mechanical i.e. structural modeling was only met by the direct simulation procedure. It is not restricted to particular (normal) probability models for loads and resistances respectively. Moreover, all mechanical aspects, such as large deformations, PΔ effects, material nonlinearities, etc., may be directly taken into account. Despite a significant reduction of the computational effort by applying variance reduction procedures (see e.g. [5,6]), this procedure is still limited to smaller type systems. Among the methods which utilize all failure modes, bounding techniques (see e.g. [7,8]) are utilized. In order to avoid the problems caused by modal correlation the application of a probabilistic network evaluation technique (PNET) (see [9]) proved to be quite useful. By this method various groups of highly correlated modes are formed and then considered to be mutually independent. The modal failure probabilities utilized by these procedures, however, are based on severe idealizations of the structural behavior, i.e. on plastic hinge formation, no possibility of considering PΔ effects, etc.. For larger type of structures - as generally utilized in practice - attempts were made to identify directly - i.e. prior to determining all possible failure modes of a structure - the stochastically most relevant modes. In other words, before trying to identify the very large number of all possible failure modes, these modes believed to contribute most significantly to the systems failure probability are determined first.

In this respect the most effective procedure proved to be the so-called "branch and bound" method, as originally suggested by Murotsu and co-workers (see e.g. [10,11]), which is applicable to larger type structures. This method is based on the so-called "traveling salesman" problem, well known to economists. The procedure, which in fact is a "failure path" method, is applicable to truss and frame structures. By this method - which is also based on the idealization of plastic hinge formation - the sequence of plastic hinges is determined by the probability of a cross section to satisfy the yield condition. Although the algorithm - if required - is capable of finding all failure modes, modes of small probabilities may be discarded by completely cutting off failure paths of low probability. The cut off criteria can be specified by the user according to the accuracy required and the computational effort considered to be acceptable. The algorithm starts the failure path at the cross section with the largest yield probability of all possible cross sections. The structure is then modified by introducing a plastic hinge and the corresponding plastic moment capacities as fictitious loads at the respective section. Then a new structural analysis is carried out to identify the next critical section to plastify among the remaining sections. The procedure continues until a failure mechanism is found. The algorithm then returns to the preceding failure level, closes the hinge and searches for the section with the second largest failure probability, at which a new hinge is introduced. After all sections are checked at this level, the algorithm, again returns to the preceding level from where the procedure starts out as just described. Each step requires an update of the system stiffness matrix. The amount of computational effort of this branching of failure paths may be limited by appropriate bounding criteria [12] by which failure paths with small contributions to the overall systems failure probability may be neglected.

Although some efforts have been made to extend the applicability of the procedure to a wider range of material properties, besides elasto plastic behavior, it still has some severe drawbacks. So far, only very small systems can be analyzed by considering the PΔ (second order) effect [12,13]. This is an important issue, since there is a finite probability that instability occurs before the pre-determined number of hinges have developed. Furthermore, particularly for larger systems, due to the large number of mode combinations, it is not certain that the algorithm finds the stochastically most relevant modes first. Moreover, a formal proof does also not exist as yet.

3 RESPONSE SURFACE METHOD

Generally, reponse surfaces are simple, well behaved functions designed to predict the behavior of certain systems. In the context of reliability analysis, the response surface is an approximation of the limit state function separating the safe domain from the failure domain (see Fig. 1). For reasons of computational efficiency, the class of functions to be utilized for the response surface should have as few free parameters as possible [14]. Since in structural analysis the computation of each single interpolation point on the limit state surface may require an extensive nonlinear (e.g. load incrementation) procedure, the need for simple expressions to describe the response surfaces becomes apparent. It has been shown in the past that the class of second order polynomials provide such surfaces, with the additional benefit that they are also rather well-behaved functions.

$$\bar{g}(\mathbf{x}) = a + \sum_{k=1}^{n} b_k x_k + \sum_{k=1}^{n} \sum_{m=1}^{n} c_{km} x_k x_m \quad (1)$$

In this equation, x_k denote the components of the random variable vector $\mathbf{x}$, and a, b_k, and c_{km} are coefficients to be determined from interpolation.

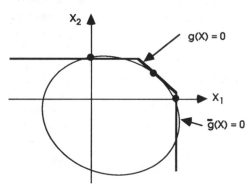

Fig.1: Schematic Sketch of Polynomial Approximation of Limit State Surface

Once such a response surface has been established, the reliability analysis can immediately proceed with the calculation of the failure probability. Here any suitable method can be used, however, it has been found that so-called "intelligent" Monte-Carlo simulation techniques like Importance Sampling [7] or Adaptive Sampling [16] can be advantageously utilized.

4 NUMERICAL EXAMPLE

In the following the numerical example is selected on the grounds to show the capabilities and advantages of the Response Surface Method (RSM) particularly in comparison to the approaches described in section 2.

A frame structure, as shown in Fig. 2, consisting of elasto-plastic elements subjected to random static loading is considered. The failure criterion is the loss of stiffness in any part of the structure due to progressive plastification. The nonlinear elements used in the analysis take

into account the spreading of plastic *zones* during plastification. So the analysis does not rely on the simplifying assumptions of plastic hinge theory.

For simplicity, first the loads are assumed to be uncorrelated Gaussian variables. Their statistical parameters are given in Table 1.

Table 1: Statistical Parameters of Loads

Variable	Mean [N]	Standard Deviation [N]
X1 - X6	15000	1500
X7	3000	1200

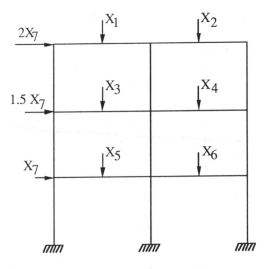

Fig.2: Two-bay three-story frame

The response surface, in the form of a second order polynomial, has been determined from interpolation of 35 points on the exact limit state. Each of these points was found by a load-incrementation procedure using each variable separately, or by using pairwise combinations.

The probability of failure using the response surface was found to be $p_f = 4.0 \cdot 10^{-7}$. For comparative purposes, a full Monte Carlo Simulation (Importance Sampling around the design point (as calculated from the response surface)) was performed using 128 simulations. This resulted in a failure estimate of $5.0 \cdot 10^{-7}$, with a statistical error of 25%. The excellent agreement indicates the good quality of a second order response surface.

Additional comparisons show the results for a response surface which has been established without pairwise combination (i.e. from only 14 load incrementation analyses). For this case, $p_f = 5.6 \cdot 10^{-7}$. Even this result is very close to the reference solution by Monte Carlo Simulation.

Also, the effect of the distribution type on the result can be very easily assessed since the mechanical behavior, and consequently the response surface, it not influenced. A re-run using lognormally distributed random variables instead of Gaussian ones yields the result $p_f = 1.2 \cdot 10^{-3}$. This underlines the importance of the appropriate modeling of the type of distribution.

5 CONCLUSIONS

The concept as suggested shows that it is possible to include state-of-the art structural modeling into reliability analysis.

The results as obtained indicate that the Response Surface Method is a computationally most efficient method, especially in situations where the reliability analysis must be repeatedly performed, each time using different sets of statistical parameters. This is for example necessary to assess the reliability under various - mutually exclusive - environmental conditions. In other methods, e.g. Branch and Bound, the resulting dominant failure modes may be influenced by the statistical parameters of the loads and resistances. Hence, in addition

to the severe simplifications necessary for the mechanical modeling, they do not necessarily allow the same degree of efficiency as the RSM.

Finally, it should be pointed out that the RSM is directly applicable to determine the collapse failure under dynamic loading, see e.g. [17,18], while other methods, such as the "branch and bound" procedure are confined to static load action.

ACKNOWLEDGEMENT

This research has been partially supported by the Austrian Industrial Research Promotion Fund under contract no. 6/636 which is gratefully acknowledged by the authors. The authors also acknowledge the assistance of Veit Bayer in context with the numerical computations.

REFERENCES

[1]   Freudenthal, A.M.: "Safety Reliability and Structural Design", *J. Structural Div.*, ASCE, Vol. 87, No. ST 3, 1961.

[2]   Grimmelt, M., Schuëller, G.I.: "Benchmark Study on Methods to Determine Collapse Failure Probabilities of Redundant Structures", *J. Structural Safety*, Vol. 1, 1982/83, pp. 93-106.

[3]   Ditlevsen, O., Bjerager, P.: "Methods of Structural Systems Reliability", *J. Structural Safety*, 3(1986), pp. 195-229.

[4]   Schuëller, G.I.: "Current Trends in Systems Reliability", Proc. 4th Intern. Conf. on Structural Safety and Reliability (ICOSSAR'85), Kobe, Japan, I. Konishi, A.H-S. Ang and M. Shinozuka (Eds.), IASSAR, New York, 1985, Volume I, pp. 139-148.

[5]   Shinozuka, M.: "Basic Analysis of Structural Safety", *J. Structural Safety*, ASCE, Vol. 109, No. 3, March 1983, pp. 721-740.

[6]   Stix, R., Schuëller, G.I.: "Problemstellung bei der Berechnung der Versagens-wahrscheinlichkeit", Internal Working Report No. 3, 1982/1986, Institut für Mechanik, University of Innsbruck, Innsbruck, Austria.

[7]   Schuëller, G.I., Stix, R.: "A Critical Appraisal of Methods to Determine Failure Probabilites", *J. Structural Safety*, 4(1987), pp. 293-309.

[8]   Ditlevsen, O.: "Narrow Reliability Bounds for Structural Systems", *J. Struct. Mech.*, 1979, pp. 453-472.

[9]   Ang, A.H-S., Ma, H.F.: "On the Reliability Analysis of Framed Structures", Proc. of the ASCE Spec. Conf. on Prob. Mech. and Struct. Rel., Tucson, 1979.

[10] Murotsu, Y., Okada, H., Niwa, K., Miwa, S.: "A New Method for Evaluating Lower and Upper Bounds of Failure Probability in Redundant Truss Structures", Bull. of the Univ. of Osaka Prefecture, Series A, Vol. 28, No. 1, 1979.

[11] Murotsu, Y., Okada, H., Grimmelt, M.J., Yonezawa, M., Taguchi, K.: "Automatic Generation of Stochastically Dominant Failure Modes of Frame Structures", *J. Structural Safety*, Vol. 2, 1984, pp.17-25.

[12] Grimmelt, M.J.: "Eine Methode zur Berechnung der Zuverlässigkeit von Tragsystemen unter kombinierten Belastungen", Diss. TU München, June 1983.

[13] Grimmelt, M.J., Schuëller, G.I.: "A Method to Determine Reliability of Structures and Combined Loading", Proc. 4th Intern. Conf. on Appl. of Statistics and Prob. in Soil and Struct. Engr. (ICASP-4), G.

Augusti et al. (Eds.), Pitagora Editrice, Bologna, 1983, pp. 261-271.

[14] Bucher, C.G., Bourgund, U.: "A fast and efficient response surface approach for structural reliability problems", *J. Structural Safety*, Vol. 7, Nr. 1, 1990, pp. 57-66.

[15] Bucher, C.G., Chen, Y-M., Schuëller, G.I.: "Time variant reliability analysis utilizing response surface approach", in: Reliability and Optimization of Structural Systems '88, P. Thoft-Christensen (Ed.), Lecture Notes in Egr., 48, Springer Verlag, Berlin, pp. 1-14, 1989.

[16] Bucher, C.G.: "Adaptive Sampling - An Iterative Fast Monte Carlo Procedure", *J. Structural Safety*, Vol. 5, No. 2, June 1988, pp. 119-126.

[17] Bucher, C.G., Pradlwarter, H.J., Schuëller, G.I.: "COSSAN - Ein Beitrag zur Software-Entwicklung für die Zuverlässigkeitsbewertung von Strukturen", VDI-Bericht Nr. 771, 1989, pp. 271-281.

[18] Schuëller, G.I., Pradlwarter, H.J., Bucher, C.G.: "Efficient Computational Procedures for Reliability Estimates of MDOF Systems", *J. Nonlinear Mechanics*, Vol. 26, No. 6, 1991, pp. 961-974.

Structural Safety & Reliability, Schuëller, Shinozuka & Yao (eds) © 1994 Balkema, Rotterdam, ISBN 90 5410 357 4

Conditional sampling for simulation-based structural reliability assessment

Chao-Yi Chia & Bilal M. Ayyub
Department of Civil Engineering, University of Maryland, College Park, Md., USA

ABSTRACT: The efficiency of simulation can largely be improved by using variance reduction techniques. In this paper, a general algorithm for conditional sampling that utilizes knowledge about the dispersions of the basic random variables was developed. Also a review and critical appraisal of variance reduction methods in simulation are provided.

1 INTRODUCTION

The performance function or safety margin that expresses the relationship between the strength and load effects of a structural member according to a specified failure mode is given by

$$M = g(X_1, X_2, ..., X_p)$$
$$= \text{Resistance - Load Effect} \quad (1)$$

in which the X_i, i = 1,..,p are the p basic random variables, with g(.) being the functional relationship between the basic random variables and failure (or survival). The performance function can be defined such that the limit state, or failure surface, is given by M = 0. The failure event is defined as the space where M < 0, and the survival event is defined as the space where M > 0. Thus, the probability of failure can be evaluated by the following integral:

$$P_f = \int\int \cdots \int f_{\underline{X}}(X_1, X_2, ..., X_p) dx_1 dx_2 \cdots dx_p \quad (2)$$

where $f_{\underline{X}}$ is the joint density function of X_1, X_2, ..., X_p, and the integration is performed over the region where M < 0. Because each of the basic random variables has a unique distribution and they interact, the integral of Eq. 2 cannot be easily evaluated. A probabilistic modeling approach of Monte Carlo computer simulation with or without Variance Reduction Techniques (VRT) can be used to estimate the probability of

failure (Ayyub and Haldar 1984; White and Ayyub 1985). Other reliability assessment methods are described by Ang and Tang (1984), and Thoft-Christensen and Baker (1983).

2 DIRECT SIMULATION

Monte Carlo simulation techniques can be used to estimate the probabilistic characteristics of the functional relationship M in Eq. 1. Monte Carlo, or direct simulation, consists of drawing samples of the basic variables according to their probabilistic characteristics and then feeding them into the performance function. It is known that failure occurs where g(.) < 0, then an estimate of the probability of failure, P_f, can be found by

$$P_f = \frac{N_f}{N} \quad (3)$$

where N_f is the number of simulation cycles where g(.) <0; and N is the total number of simulation cycles. As N approaches infinity, then P_f approaches the true probability of failure. The accuracy of Eq. 3 can be evaluated in terms of its variance. For a small probability of failure and/or a small number of simulation cycles, the variance of P_f can be quite large. Consequently, it may take a large number of simulation cycles to achieve a specified accuracy

with an unknown probability of failure. The variance of the estimated probability of failure can be computed by assuming each simulation cycle to constitute a Bernoulli trial. Therefore, the number of failures in N trials can be considered to follow a binomial distribution. Then, the variance of the estimated probability of failure can be approximately computed as

$$Var(P_f) \cong \frac{(1-P_f)P_f}{N} \qquad (4)$$

It is recommended to measure the statistical accuracy of the estimated probability of failure by computing its coefficient of variation as

$$COV(P_f) \cong \frac{\sqrt{\dfrac{(1-P_f)P_f}{N}}}{P_f} \qquad (5)$$

It is evident from Eqs. 4 and 5 that as $N \to \infty$, $Var(P_f)$ and $COV(P_f)$ approach zero.

In the direct simulation method, the probability of failure is estimated as the ratio of the number of failures to the total number of simulation cycles. Therefore, for smaller probabilities of failure, larger numbers of simulation cycles are needed to estimate the probability of failure within an acceptable level of statistical error.

3 VARIANCE REDUCTION TECHNIQUES

3.1 CONDITIONAL EXPECTATION (CE) METHOD

The performance function for a fundamental structural reliability assessment case is given by

$$M = R - L \qquad (7)$$

where R = function of structural strength or resistance; and L = function of the corresponding load effect. Therefore, the probability of failure, P_f, is given by

$$P_f = P(M < 0) = P(R < L) \qquad (8)$$

For a randomly generated value of L (or R), say l_i (or r_i), the probability of failure is given by, respectively

$$P_{f_i} = Prob\ (R < l_i) = F_R(l_i) \qquad (9a)$$

and $\qquad P_{f_i} = Prob\ (L > r_i) = 1 - F_L(r_i) \qquad (9b)$

where F_R and F_L = cumulative distribution functions of R and L, respectively. In this formulation R and L are assumed to be statistically uncorrelated random variables. Thus, for N simulation cycles, the mean value of the probability of failure is given by the following equation:

$$\overline{P}_f = \frac{\displaystyle\sum_{i=1}^{N} P_{f_i}}{N} \qquad (10)$$

The variance (Var) and the coefficient of variation (COV) of the estimated probability of failure are given by

$$Var(\overline{P}_f) = \frac{\displaystyle\sum_{i=1}^{N} \left(P_{f_i} - \overline{P}_f\right)^2}{N(N-1)} \qquad (11a)$$

$$COV(\overline{P}_f) = \frac{\sqrt{Var(\overline{P}_f)}}{\overline{P}_f} \qquad (11b)$$

For the general performance function given by Eq. 1, the conditional expectation method can be utilized by randomly generating all the basic random variables except one variable, called the control variable X_k. The randomly generated variables should be selected as the ones of least variabilities, and the resulting conditional expectation can be evaluated by some known expression, e.g., the cumulative distribution function of the control random variable that was not randomly generated. This method can be used for any performance function with any probability distributions for the random variables. The only limitation is that the control random variable, X_k, must be statistically uncorrelated to the $g_k(.)$ as shown in the following equation:

$$P_f = E_{X_i: i=1,2, ..., n\ \&\ i \neq k} [F_{X_k}(g_k(X_i: i=1,2, ..., n\ \&\ i \neq k)] \qquad (12)$$

where $g_k(X_i: i=1,2, ..., n \ \& \ i \neq k)$ is defined as follows:

$$P_f = Prob\ [\ g(X_1, X_2, ..., X_n) < 0\ ]\ \ \text{or} \qquad (13)$$

$$P_f = Prob\ [\ X_k < g_k(X_i: i=1,2, ..., n \ \& \ i \neq k)\ ] \quad (14)$$

3.2 ANTITHETIC VARIATES (AV) VRT

In this method, a negative correlation between different cycles of simulation is induced in order to decrease the variance of the estimated mean value. If U is a random number uniformly distributed in the range [0,1] and is used in a computer run to determine the probability of failure $P_{fi}(1)$, the 1-U can be used in another run to determine the probability of failure $P_{fi}(2)$. Therefore, the probability of failure in the ith simulation cycle is given by

$$P_{f_i} = \frac{P_{f_i}(1) + P_{f_i}(2)}{2} \qquad (15)$$

Then, the mean value of probability of failure can be calculated by Eq. 10. In the random generation process for each simulation cycle, say the ith cycle, a set of uncorrelated random numbers based on the uniform random variable U is used in the first stage of the ith cycle to determine the probability of failure $P_{fi}(1)$. In the second stage of the ith cycle, a complementary set of uncorrelated random numbers based on the random variable (1-U) is used to determine the probability of failure $P_{fi}(2)$. Therefore, the probability of failure in the ith simulation cycle is given by Eq. 14. This method has been used in conjunction with importance sampling technique by Schüeller (1989). The antithetic variates VRT is described in detail by Ayyub and Haldar (1984), Law and Kelton (1982).

3.3 GENERALIZED CONDITIONAL EXPECTATION (GCE) METHOD

The conditional expectation method can be generalized by allowing the number of the control variables to be larger than one. Equation 12 can be generalized as follows:

$$P_f = E_{X_i: i=1,2, ..., n \ \& \ i \notin k}\ [\ P_f(\underline{X_k})\ ] \qquad (16)$$

where $\underline{X_k}$ = a vector of control random variables, $X_{k1}, X_{k2}, ..., X_{km}$; and $P_f(\underline{X_k})$ = the probability of failure evaluated in the dimensions of $X_{k1}, X_{k2}, ..., X_{km}$. This probability can be evaluated using any method, e.g., moment methods, importance sampling, or conditional expectation.

The suggested computational steps according to this generalized approach are summarized in the following (Ayyub and Chia 1992):

1. The performance function should be defined according to Eq. 1.

2. The control random variables, $\underline{X_k}$ = $(X_{k1}, X_{k2}, ..., X_{km})$, are selected on the basis of reducing the dimensionality of the problem to the space of the control random variables. All other random variables, $X_i: i \notin \underline{k}$, are considered the conditional random variables, and they are generated randomly according to this method.

3. Therefore, the reliability assessment problem is reduced to N evaluations of the probability term (Prob) in the following expression:

$$P_f = E_{X_i: i=1,2, ..., n \ \& \ i \notin \underline{k}}\ \{Prob[g(X_1, X_2, ..., X_{k1}, X_{k2}, ..., X_{km}, ..., X_n) < 0]\} \qquad (17)$$

4. In the non-generalized conditional expectation method, the expression "$Prob[g(x_1, x_2, ..., X_{k1}, X_{k2}, ..., X_{km}, ..., x_n) < 0]$" with m=1, is evaluated using the CDF of X_k as given in Eqs. 12 and 13. It should be noted that the lower cases for $x_1, x_2, ..., x_n$ in this expression indicate generated values of the random variables $X_1, X_2, ..., X_n$. In the generalized approach, the value of m is larger than one. Therefore, the probability expression can be evaluated using any suitable (or convenient) method. For example, the first-order second-moment (FOSM) method, advanced second-moment (ASM) method (Hasofer and Lind 1974), importance sampling (Madsen et al 1986; Melchers 1987), conditional expectation (CE) method, or other structural reliability method can be used for this purpose. The choice of the m random variables of $\underline{X_k}$ should be based on

the intended method for the evaluation of the probability expression. The simplification can be, for example, in the form of (1) reducing a nonlinear performance function into a linear function that is used in the probability expression, (2) using a closed-form expression for evaluating the probability expression, and (3) removing random variables with non-normal probability distributions from the expression. These concepts are best explained using the following examples.

3.4 IMPORTANCE SAMPLING (IS) METHOD

The probability of failure of a structure according to the performance function of Eq. 1 is provided by the integral of Eq. 2. In evaluating this integral with direct simulation, the efficiency of the simulation process depends on the magnitude of the probability of failure, i.e., the location of the most likely failure point or design point (Schuëller and Stix 1987). The deeper the location of the design point in the failure domain, the larger the needed simulation effort to obtain failures. This deficiency can be addressed by using importance sampling. According to IS, the basic random variables are generated according to some carefully selected distributions with mean values that are closer to the design point than their original probability distributions. Therefore, failures are obtained more frequently and the simulation efficiency is increased. To compensate for the change in the distributions, the results of the simulation cycles should be corrected. The fundamental equation for this method is given by

$$P_f = \frac{1}{N} \sum_{i=1}^{N} I_f \frac{f_{\underline{X}}(x_{1i}, x_{2i}, ..., x_{pi})}{h_{\underline{X}}(x_{1i}, x_{2i}, ..., x_{pi})} \quad (18)$$

where N = number of simulation cycles, $f_{\underline{X}}(x_{1i}, x_{2i}, ..., x_{pi})$ = the original joint density function of the basic random variables evaluated at the ith generated values of the basic random variables, $h_{\underline{X}}(x_{1i}, x_{2i}, ..., x_{pi})$ = the selected joint density function of the basic random variables evaluated at the ith generated values of the basic random variables, and I_f = failure indicator function that takes values of either 1 for failure and 0 for survival.

3.5 STRATIFIED SAMPLING (SS) METHOD

This method is based on the theorem of total probability. The integration domain of the integral in Eq. 2 is divided into several, say k, regions ($R_1, R_2, .., R_k$). The probability of failure is then estimated as

$$P_f = \sum_{j=1}^{k} P(R_j) \frac{1}{N_j} \sum_{i=1}^{N_j} I_{fi} \quad (19)$$

where $P(R_j)$ = the probability of region R_j, N_j = number of simulation cycles performed in region j, I_{fi} = the indicator function as defined in Eq. 18 evaluated at the ith simulation cycle. Additional information about this method is provided by Law and Kelton (1982), and Schuëller, et al (1989).

3.6 LATIN HYPERCUBE SAMPLING METHOD

The importance sampling method, or the stratified sampling method require the analyst to know in advance the important regions or important variables of a problem. In many practical problems, this requirement cannot be met. The Latin hypercube sampling method (Iman and Canover 1980, and Iman and Shortencarier 1984) offers an advantage for these problems, as well as problems where the important regions or variables are time-variant. The LHS method provides a constraint sampling scheme instead of random sampling according to direct simulation. In LHS method, the region between 0 and 1 is uniformly divided into N-nonoverlapping intervals for each random variable, where N is the number of random numbers that need to be generated for each random variable, i.e., number of simulation cycles. Therefore, the corresponding N-nonoverlapping intervals for each random variable have the same probability of occurrence. Then, N different values are randomly selected from the N intervals, one value per interval. A generated random value in the ith interval can be computed as

$$u_i = \frac{u}{N} + \frac{i-1}{N} \quad (20)$$

where u = a random number in the range [0,1], and u_i (i=1,2,..., N) = the random value for the *ith* interval. Once the u_i (i=1,2,..., N) values are obtained, then inverse transformation can be used to obtained values for the generated random variables. The last step in the sampling procedure is to group the generated values for all variables, so that they can be used in the simulation-based estimation of the probability of failure. A random permutation of N integers corresponding to the N simulation cycles is used for each random variable. Then grouping is performed by associating the different random permutations for the random variables for use in the simulation cycles. The resulting permutation-based grouping can be used in any simulation-based structural reliability assessment method to estimate the failure probability. Ayyub and Lai (1989 and 1991) provide illustrative examples of this method.

3.7 ADAPTIVE SAMPLING (AS) METHOD

The adaptive nature of this method comes from updating the importance function used in the importance sampling method (Eq. 18), i.e., $h_{\underline{X}}(x_1,x_2, ..., x_p)$ during simulation. Therefore, this iterative simulation method (Bucher 1988) utilizes the results obtained from importance sampling to change the assumed sampling density function. The starting density function for $h_{\underline{X}}(x_1,x_2, ..., x_p)$ is one of the main difficulties in using this method. However, with a properly selected function, this method provides a fast convergence to the probability of failure. Bucher (1988) and Schuëller, et al (1989) provide additional information about this method with computational examples.

3.8 RESPONSE SURFACE (RS) METHOD

In some engineering problems, the performance function of Eq. 1 has a non-closed form expression, e.g., a computer code that possibly involves finite element analysis. In these problems, the actual performance function can be approximated by using curve fitting of a polynomial to selected evaluations of the performance function. Then, the approximate polynomial expression of the performance

function can be used in estimating the probability of failure. Bucher and Bourgund (1987), and Schuëller, et al (1989) provide additional information about this method with computational examples.

3.9 CONDITIONAL SAMPLING (CS) METHOD

According to this method, the sampling effort for each basic random variable is made proportional to its variability level, for example, proportional to its coefficient of variation (COV). Therefore, the simulation algorithm according to any of the previously discussed method is divided into several levels of simulation cycles, similar to impeded "Do Loops" in FORTRAN programming. The basic random variables can be grouped into these levels based on their COV values. Each level has its own number of simulation cycles which is proportional to the grouped random variables. Random variables with relatively large COV values are assigned to the inner levels, whereas random variables with relatively small COV values are assigned to the outer levels. As a result, a generated value at an outer level is used many times in conjunction with the generated values of the basic random variables at the inner levels. The number of times the value is used is equal to the product of the number of simulation cycles assigned to the inner levels. The sampling efforts according to this algorithm in combination with some simulation methods are expected to be reduced in comparison with the direct use of these simulation methods. The conditional sampling algorithm can be implemented in the Monte Carlo simulation methods with the requirements of improved accuracy and assured validity of the methods. The conditional sampling algorithm can be applied to the direct simulation method, importance sampling, conditional expectation, and to other simulation method. Also, it can be used in combination of the moment methods. A general recursive programming technique was developed for the implementation of the conditional sampling method (Chia 1992).

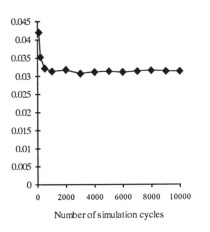

Number of simulation cycles

Figure 1a. Failure Probability -Example 1

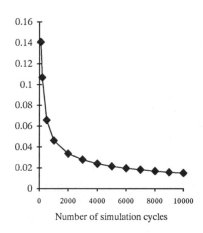

Number of simulation cycles

Figure 1b. COV(P_f) - Example 1

Table 1. Random variables in Example 1

Random variable	Mean value	COV	Distribution type
Y	38 ksi	0.05	Normal
S_e	100 in^2	0.05	Normal
w	0.3 kips/in	0.25	Normal
L_n	180 in	0.05	Normal

Table 2. Random variables for Example 2

Variable	Mean Value	COV	Case 1	Case 2
Y	275.52 MPa (40.00 ksi)	0.125	Normal	Log-normal
S_e	8.19x10^{-4} m^3 (50.00 in^3)	0.050	Normal	Log-normal
M_e	1.13x10^5 N-m (1,000.0 kip-in)	0.200	Normal	Type I - largest

4. EXAMPLES

4.1 EXAMPLE 1

The performance function that describes the flexural behavior of a simply supported beam of a span length L_n supporting a uniform load w is

$$M = Y S_e - \frac{w L_n^2}{4}$$

where Y = yield stress of the material of the beam, and S_e = elastic section modulus. In this example, failure is defined as yielding at the extreme material fibers of the cross-section of the beam. The mean values and standard deviations of the variables are given in Table 1. Using conditional expectation with w as the control variable. Therefore, Y, S_e and L_n were randomly generated, and the cumulative distribution function of w was used to compute the probability of failure at each simulation cycle. Also, the COV(P_f) was computed. The results are shown in Figure 1.

4.2 EXAMPLE 2

Consider the first-yield failure mode of a structural steel section subjected to a bending moment loading. The performance function is

$$M = Y S_e - M_e$$

where Y = yield stress of material, S_e = elastic section modulus and M_e = moment effect due to applied loading. The statistical characteristics of the variables are shown in Table 2 (Ang and Tang 1984). These variables are assumed to be statistically uncorrelated.

The probability of failure of the structural component according to the first-yield failure mode can expressed as

$$P_f = \text{Prob} \, (M < 0) = \text{Prob} \, (M_e > Y S_e)$$

The control random variable, in this case, is selected as the random variable M_e, because it has the largest coefficient of variation (COV). Therefore, conditioning on Y and S, the cumulative distribution function of M_e was used to evaluate P_f in each simulation cycle. The variables Y and S were randomly generated using the inverse transformation method. Then, CE variance reduction techniques were used to estimate the probability of failure. To improve the performance of the method, antithetic variates (AV) variance reduction technique was combined with CE to estimate the probability of failure.

In this example, two cases were considered, normal random variables (Case 1) and non-normal random variables (Case 2), as shown in Table 2. For the non-normal case, the probability of failure according to the specified performance function and the distribution types for the ith simulation cycle is given by

$$P_{fi} = 1 - F_{M_e} (y_i \, s_i) = 1 - \exp[-\exp[-\alpha(y_i \, s_i - \gamma]]$$

in which α and γ are the parameters of Type I-largest extreme value distribution for the assumed probabilistic characteristics of M_e, and F_{M_e} is the cumulative distribution function of M_e. The sample mean and coefficient of

variation (COV) of the failure probability were then determined, and are shown in Table 3 for the normal and non-normal probability distributions.

For the purpose of comparison, P_f was re-calculated using the advanced-second moment (ASM) method (Ang and Tang 1984; Ayyub and Haldar 1984). The results are 1.1×10^{-3} and 3×10^{-3} for the normal and non-normal probability distributions, respectively. Using the GCE method, M_e and Y were selected as the control variables, and S was randomly generated. For the ith simulation cycle, the probability expression is given by

$$P_{fi} = \text{Prob} \, [Y \, s_{ei} - M_e < 0]$$

where s_i = randomly generated value of S. The probability expression was then evaluated, for the normal probability distributions (Case 1), as follows:

$$P_{f_i} = 1 - \Phi \left[\frac{\mu_Y s_{ei} - \mu_{M_e}}{\sqrt{s_{ei}^2 \, \sigma_Y^2 + \sigma_{M_e}^2}} \right]$$

where μ = mean value; and σ = standard deviation. For the non-normal probability distributions (Case 2), the advanced second moment (ASM) method was used to determine P_{fi}. Then, the mean value and COV of the failure probability were determined for N simulation cycles. The resulting statistical characteristics of P_f are shown in Table 3 for the normal and non-normal probability distributions, respectively. For all cases, the number of simulation cycles was increased in increments and the resulting probability of failure at the end of each increment was recorded in order to study the convergence and effectiveness of the combined methods in estimating the probability of failure.

It is evident from Example 2 that the assessment of failure probability based on non-linear performance function can be reduced to averaging N evaluations of the probability of structural failure according to a linear expression of the performance function. This transformation can be achieved by carefully selecting the control random variables. The probabilistic evaluation of the linear expression

Table 3. Results of Example 2

Simulation Method	Number of Cycles	$\overline{P}_f$ (10^{-2})	COV $(\overline{P}_f)$
Case 1. Normal			
Direct Monte Carlo	200,000	0.128	0.0625
Conditional Expectation (CE)	40,000	0.118	0.0460
Generalized CE	500	0.118	0.0380
Case 2. Non-normal			
Direct Monte Carlo	100,000	0.325	0.0560
Conditional Expectation (CE)	2,000	0.319	0.0460
Generalized CE	500	0.300	0.0240

was performed, in the two examples, using the advanced second moment method. Other methods could have been used to achieve to this objective. The choice of the ASM method was for the purpose of illustrating merging moment reliability methods with conditional expectation in Monte Carlo simulation. This concept can be greatly utilized (or manifested) in complex performance functions to transform them into computationally manageable formats. Although the two examples have explicit performance functions, the proposed method can be used to solve problems with non-explicit performance functions.

Based on the results of the examples, the proposed GCE approach offers the advantages of expediting convergence and increased solution stability in solving structural reliability problems (Ayyub and Chia 1992).

REFERENCES

Ang, A. H-S. and Tang, W. 1984. Probability concepts in engineering planning and design, volume II, John Wiley and Sons, N.Y.

Ayyub, B.M., and Lai, K-L. 1989. "Structural reliability assessment using Latin hypercube sampling," *ICOSSAR*, Volume 2, 1171-1184.

Ayyub, B.M., and Lai, K.-L. 1991 "Selective sampling in simulation-based reliability assessment," Int. J. of Pressure Vessel and Piping, 46, 2, 1991, 229-249.

Ayyub, B.M., and Chia, C.-Y. 1992. Generalized conditional expectation for structural reliability assessment," *Structural Safety*, 11, 2, 1992.

Ayyub, B.M. and Haldar, A. 1984. "Practical structural reliability techniques." *J. of Structural Engineering*, ASCE, 110(8), 1984, 1707-1724.

Bjerager, P. 1987. "Probability integration by directional simulation." *J. of Engineering Mechanics*, ASCE, 114(8), 1285-1302.

Bourgund, U. and Bucher, C.G. 1986. "Importance sampling procedures using design points (ISPUD)- a user manual." *Report No. 8-86*, University of Innsbruck, Austria.

Bucher, C.G. 1988. "Adaptive sampling - an iterative fast Monte Carlo procedure" *Structural Safety*, 5, 119-126.

Bucher, C.G., and Bourgund, U. 1987. Efficient use of response surface methods, Report 9-87, University of Innsbruck.

Chia, C.-Y., 1992. *Simulation-based structural reliability assessment*, Ph.D. dissertation, University of Maryland, College Park, MD.

Ditlevsen, O., and Bjerager, P. 1987. "Plastic reliability analysis by directional simulation." *DCAMM Report 353*, The Tech. University of Denmark.

Grigoriu, M. 1982, "Methods for approximate reliability analysis." *Structural Safety*, 1, 155-165.

Harbitz, A. 1986. "An efficient sampling method for probability of failure calculation." *Structural Safety*, 3(2), 100-115.

Hasofer, A.M. and Lind N.C. 1974. "Exact and invariant second-moment code format." *J. of Engineering Mechanics Division*, ASCE, 100(EM1), 111-121.

Iman, R.L., and Canover, W.J., 1980. "Small sample sensitivity analysis techniques for computer models with an application to risk assessment." *Communications in Statistics, Theory and Methods*, A9(17), 1749-1842.

Iman, R.L., and Shortencarier, M.J., 1984. A FORTRAN 77 program and user's guide for the generation of Latin hypercube and random samples for use with computer models. *NUREG/CR-3624, SAND83-2365.*

Law A.M. and Kelton W.D. 1982. *Simulation Modeling and Analysis*. McGraw Hill, NY.

Madsen, H.O., Krenk, S. and Lind, N.C. 1986. *Methods of Structural Safety*. Prentice-Hall.

Melchers, R.E. 1987. *Structural reliability analysis and prediction*. Ellis Horwood.

Rubinstein, R.Y. 1981. *Simulation and Monte Carlo method*. John Wiley and Sons, NY.

Schuëller, G.I., and Stix, R. 1987. "A critical appraisal of methods to determine failure probabilities," Structural Safety, 4, 293-309.

Schuëller, G.I., Bucher, C.G., Bourgund, U., and Ouypornprasert, W. 1989. "An efficient computational schemes to calculate structural failure probabilities." *Probabilistic Engineering Mechanics,* 4(1), 10-18.

Shinozuka, M., 1983. "Basic analysis of structural safety," *J. of Structural Engineering*, ASCE, 109(3), 721-740.

White, G.J. and Ayyub, B.M. 1985. Reliability methods for ship structures, *Naval Engineers Journal*, ASNE, 97(4), 86-96.

Structural Safety & Reliability, Schuëller, Shinozuka & Yao (eds) © 1994 Balkema, Rotterdam, ISBN 90 5410 357 4

Distribution arbitrariness in structural reliability

Ove Ditlevsen
Department of Structural Engineering, Technical University of Denmark, Lyngby, Denmark

ABSTRACT: The author points at the urgent need for code standardizations of distribution types for practical design applications of structural reliability methods. In support of the author's opinion that distribution type standardizations are necessary in order to avoid reliability comparisons on the basis of incommensurable reliability measures, the paper demonstrates the fundamental difficulty of choosing the probability distributions for example for annual extreme loads such as wind and snow loads for which only limited data series are available. Moreover, it is argued that even if several distribution types may pass a statistical test not all of these are reasonable candidates for standardization. The class of standardized distributions in a code of practice for reliability analysis should be looked upon as an internally harmonized entity chosen on the basis of consequence calculations applied to the class of structures for which the code is intended. Otherwise even the imposed ordering with respect to reliability becomes dubious.

JCSS Probabilistic code text example

The distribution tail sensitivity is a well known unavoidable property of structural reliability analysis of highly reliable structures. This fact causes the computed failure probabilities to be of limited informational value except for reliability comparisons made on the basis of the same set of probability distribution types, that is, within the same model universe of probability distributions. Therefore, for the advancement of the use of modern probabilistic reliability analysis to aid rational structural engineering decisions in practice there is an indispensable need for an agreement among competing engineers and the general public on using a standardized distribution model universe as a common reference. In other words, there is an indispensable need for a code of practice for structural reliability analysis.

An attempt to formulate such a code has four years ago been published by the Joint Committee on Structural Safety (JCSS) as a text with the title: Proposal for a Code for the Direct Use of Reliability Methods in Structural Design, Ditlevsen and Madsen (1989). Without achieving general consensus the details of the code text example has been discussed within the committee and the text is published in the form of a Working Document. JCSS is supported by the international associations CEB, CIB, ECCS, FIP, IABSE, IASS, and RILEM.

An important point of this code proposal is that the distribution types to be used in the reliability analysis are standardized. The way in which these standardizations are introduced is best illustrated by direct quotation from the proposal. It is not intended to discuss whether the actual specifications of the code text example are reasonable or not or why the text at some places is ambiguous. Of course, such a text will always be controversial and subject to discussions. In the Working Document's section 9 on reliability models there is the following code type text:

If no specific distribution type is given as standard in the action and material codes this code for the purpose of reliability evaluations standardizes the clipped (or, alternatively, the zero−truncated) normal distribution type for basic load pulse amplitudes. Furthermore, the logarithmic normal distribution type is standardized for the basic strength variables.

Deviations from specific geometrical measures of physical dimensions as length are standardized to have normal distributions if they act at the adverse state in the same way as load variables (increase of value implies decrease of reliability) and to have logarithmic normal distribution if they contribute to the adverse state in the same way as resistance variables (decrease of value implies decrease of reliability).

Further:

In special situations other than the code standard-ized distribution types can be relevant for the reliability evaluation. Such code deviating as-sumptions must be well documented on the basis of a plausible model that by its elements generates the claimed probability distribution type. Asymp-totic distributions generated from the model are allowed to be applied only if it can be shown that they by application on a suitable representative example structure lead to approximately the same generalized reliability indices as obtained by application of the exact distribution generated by the model.

Experimental verification without any other type of verification of a distributional assumption that deviates strongly from the standard is only sufficient if very large representative samples of data are available.

Distributional assumptions that deviate from those of the code must in any case be tested on a suitable representative example structure. By calibration against results obtained on the basis of the standardizations of the code it must be gua-ranteed that the real (the absolute) safety level is not changed significantly relative to the require-ments of the code.

When arguing within a specific model universe of distributions it is important to ensure that in particular near zero probability value results are used for comparisons only within the model itself. Carrying the results to the outside world and attaching the usual probability inter-pretation of relative frequency of occurrence in the real world of the considered adverse event will generally be highly misleading even though the model has been carefully calibrated to real world data. This insight is not new. In fact, it has been repeatedly discussed at least during the last quarter of a century of structural reliability theory development. However, it has now be-come urgent to focus on this point due to the recent maturing of practicable reliability analysis methods as resulting from the almost explosive development of available computational power.

Example: Snow load statistics

The problem of the choice of distribution type is well illustrated in the case of historical data series that by their very nature increase in sample size as slowly as by only one new ob-servation each year. For example, this is the case for yearly extreme ground snow loads at a given location.

Ellingwood and Redfield (1983) report that the lognormal distribution fits several data sets of 28 yearly extreme ground snow loads given in terms of water equivalents, each load data set measured at a different weather station in USA. It is reported that for most weather stations the lognormal distribution fits better than a Gumbel

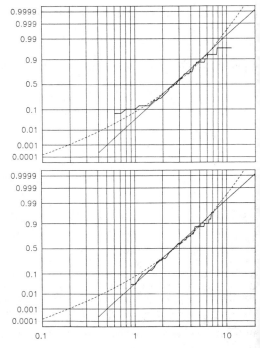

Fig. 1. Lognormal distribution function (straight line) and gamma distribution function (dotted line) with the same mean and coefficient of variation both compared to an empirical distribu-tion function obtained by simulation of two samples of size 30 from the gamma distribution. The gamma distribution gives the best fit to the empirical distribution function in the top diagram while the lognormal distribution gives the best fit to the empirical distribution function in the bottom diagram.

distribution (Fisher–Tippet Type I extreme value distribution).

Luy and Rackwitz (1978) report that samples of 30 yearly extreme snow depth observations in Germany are for the larger part better fitted by the gamma distribution than by the lognormal or the Gumbel distribution.

It is a well–known consequence of the nature of statistical uncertainty that several distribu-tion types seem to be reasonable candidates as models for the population from which the actual and only known sample of moderate size is drawn. This is easily demonstrated by a simulation experiment. Let a random variable X be distributed according to a gamma density, that is, a density proportional to $x^{k-1} e^{-ax}$, x $\in \mathbb{R}_+$, with parameters k = 3, a = 1. Then X has the mean and coefficient of variation $E[X] = k/a$ = 3, $V_X = 1/\sqrt{k} \simeq 0.577$ respectively. These

parameters are close to be representative for the ground snow load populations reported by Ellingwood and Redfield (1983). The graph of the corresponding distribution function is shown in dotted line in Fig. 1. The abscissa scale is logarithmic while the ordinate scale corresponds to fractiles of the normal distribution. Thus any lognormal distribution function appears as a straight line in this diagram. The straight line shown in Fig. 1 is the graph of the lognormal distribution function $F_{log}(x)$ corresponding to the mean 3 and the coefficient of variation $1/\sqrt{3}$. Fig. 1 also shows the graph of the empirical distribution function $S_n(x)$ = relative number of observations at most equal to x in the simulated sample $x_1,...,x_n$ of size n = 30 from the considered gamma distribution.

From the visual impression of the graphs it is difficult to draw any conclusions about whether or not $F_{gam}(x)$ fits the empirical distribution function $S_n(x)$ better than $F_{log}(x)$. Therefore it is necessary to compute some test statistics in order to try to be objective about the matter. Three different standard measures (test statistics) of the goodness of the fit are considered, Kendall and Stuart (1961,Vol.2, 450–452):

Kolmogorov–Smirnov:

$$A = \sqrt{n} \sup |S_n(x) - F(x)|$$

$$= \sqrt{n} \max_{r=0}^{n} \left\{ \left| y_r - \frac{r}{n} \right| \right\} \tag{1}$$

Cramér–von Mises:

$$B = n \int_{-\infty}^{\infty} [S_n(x) - F(x)]^2 dF(x)$$

$$= \sum_{r=1}^{n} \left[y_r - \frac{2r-1}{2n} \right]^2 + \frac{1}{12n} \tag{2}$$

Anderson–Darling:

$$C = n \int_{-\infty}^{\infty} \frac{[S_n(x)-F(x)]^2}{F(x)[1-F(x)]} dF(x)$$

$$= n \sum_{r=1}^{n} \left[\left[\frac{r-n}{n} \right]^2 \log \left[\frac{1-y_r}{1-y_{r+1}} \right] + \left[\frac{r}{n} \right]^2 \log \left[\frac{y_{r+1}}{y_r} \right] \right]$$

$$+ n \left[-1 - \log(1-y_1) - \log y_n \right] \tag{3}$$

in which $F(x)$ is the tested distribution function and $y_1 = F(x_1) \leq y_2 = F(x_2) \leq ... \leq y_n = F(x_n)$

correspond to the ordered sample $x_1 \leq x_2 \leq ... \leq x_n$. With $F = F_{gam}$ or $F = F_{log}$ we denote A as A_{gam} or A_{log} respectively, and we consider the indicator random variable I_A where $I_A = 1$ if $A_{log} < A_{gam}$, $I_A = 0$ otherwise. The indicator random variables I_B and I_C are defined correspondingly. Using the joint information from these statistics a reasonable decision rule might be to choose the gamma distribution if at least two of the three variables I_A, I_B, I_C take the value zero. Otherwise the lognormal distribution is chosen.

Even though it is known here that the data are drawn from the gamma distribution, it is an event of considerable probability to choose the lognormal distribution . The empirical distribution function in the top diagram in Fig. 1 leads to the choice of the gamma distribution while the empirical distribution function in the bottom diagram in Fig. 1 leads to the choice of the lognormal distribution. Getting a sample as in the bottom diagram of Fig. 1 may therefore erroneously lead to the adoption of a wrong distribution model for X. Of course, this is a trivial fact in the theory of mathematical statistics, but in the field of structural reliability this quite probable event of choosing the wrong distribution model can have severe consequences. For example, the 0.999 fractile in the gamma distribution of Fig. 1 is about 11.3 while the 0.999 fractile in the lognormal distribution is about 13.8 which is about 22% larger than the first. For the 0.9999 fractile the numbers are about 13.8 and 18.9 respectively with the last being about 37% larger than the first.

In order to evaluate the probability of making the wrong choice, the probability distribution of (I_A, I_B, I_C) has been calculated by repeated simulations of independent samples of X of size n = 30.

Table 1 shows the 8 probabilities obtained in the case where the exact values of E[X] and V_X are used, that is, when no parameters are estimated from the data (simple hypothesis), as well as the 8 probabilities obtained when E[X] and V_X are estimated sample by sample from the simulated data by the method of moments (composite hypothesis). These estimates are used to define the two alternative distribution functions from which the ordered samples $y_1 \leq y_2 \leq ... \leq y_n$ are calculated. Usually it is the last situation with estimated parameters that is relevant in case of samples of data related to natural phenomena.

For the simple hypothesis it is seen that C gives the smallest error probability 0.20 while A gives the largest error probability 0.32 ,

Table 1. Probability distributions of (I_A, I_B, I_C) for a sample of size $n = 30$ from the gamma distribution corresponding to the parameter values $E[X] = 3$, $V_X = 1/\sqrt{3}$.

$(I_A, I_B, I_C) =$	(0,0,0)	(1,0,0)	(0,1,0)	(0,0,1)	(1,1,1)	(0,1,1)	(1,0,1)	(1,1,0)
prob.1)	0.61	0.08	0.02	0.03	0.14	0.02	0.01	0.09
	P(choice of gamma) = 0.74				P(choice of lognormal) = 0.26			
	P(I_A=1) = 0.32 , P(I_B=1) = 0.27, P(I_C=1) = 0.20							
prob.2)	0.55	0.07	0.01	0.02	0.25	0.07	0.01	0.02
	P(choice of gamma) = 0.65				P(choice of lognormal) = 0.35			
	P(I_A=1) = 0.35 , P(I_B=1) = 0.35, P(I_C=1) = 0.35							

1) exact parameters, 2) estimated parameters (method of moments)

results that seem to fit with the intuition when observing that C puts much more weight to the deviations in the tail regions than A does. However, in the usual situation of a composite hypothesis the three test statistics surprisingly give approximately the same error probability of 0.35 .

All in all it can be concluded that in more than 3 out of 10 cases the wrong distribution model will be chosen. In case of a modest sample of data of a natural phenomenon as for example the maximal yearly ground snow load we therefore run a considerable risk of choosing the wrong distribution model when the choice is made solely on the basis of a best fit criterion.

It is also interesting to note that if by bad luck the sample is such that the best fit criterion points at the wrong distribution model, then it takes a considerable increase of sample size before it becomes evident that the model is wrong. This is illustrated in Fig. 2 which shows the simulation estimate of the conditional probability of choosing the wrong distribution as a function of the additional sample size N given that the wrong distribution was chosen for the sample size n = 30. For N = 30 the probability of maintaining the wrong model is about 50% and it is still as high as about 20% for N = 150.

The Ellingwood–Redfield distribution investigation on snow loads

Ellingwood and Redfield (1983) make similar simulation investigations in order to evaluate the probability of choosing the wrong distribution model among the two alternatives they consider. Their choice among the two alternative distributions are based on the maximum probability plot correlation coefficient criterion, Filliben (1975). When samples of size n = 28 are generated from a suitably representative lognormal distribution, about 21% of the samples show a better fit to the Gumbel distribution. Alternatively, simulation from a Gumbel distribution gives that about 25% of the samples are better fitted by the lognormal distribution.

Thus the same difficulty of choosing among the two alternatives shows up. However, these investigators have another seemingly strong argument. They have samples not only from one weather station but from 76 (38) weather stations. Among these samples about 66% (76%) are fitted best by the lognormal distribution and the remaining 34% (24%) best by the Gumbel distribution. The numbers in parenthesis correspond to the subset of weather stations that all have 28 years record of observations with non–zero water equivalent in each year. The observed similarity between the results of simulation from the lognormal distribution and the actual findings among the weather stations seems to point at the lognormal distribution as being closer to the "true" distribution model of ground snow loads than the Gumbel distribution is. Indeed, if the data from the different weather stations can be considered as statistically independent, then it is an easy probability computation to see that the Gumbel distribution should be rejected as a tenable candidate.

This decisive independence assumption is the disputable point in Ellingwood's and Redfield's argumentation. One could as well argue that the mutual dependence is very strong. Cold winters and mild winters are usually not experienced as local phenomena (relatively) but are common to larger geographical regions.

For illustration of the effect of dependency between samples consider the random vector $Z = (X+Y_1, X+Y_2, ..., X+Y_m)$ where $X, Y_1, ..., Y_m$ are mutually independent, X has the gamma density with parameters $k = k_x$, $a = 1$ while $Y_1, ..., Y_m$ all have the gamma density with parameters $k = k_y = 3-k_x$, $a = 1$. Then the elements in Z all have the gamma density with parameters $k = 3$, $a = 1$. The vector Z could be envisaged to represent the simultaneous measurements at m stations. A sample of size

n of **Z** gives m empirical distribution functions $S_{1n}(x),...,S_{mn}(x)$. Each of these empirical distribution functions defines a value of the indicator random variable $I = I_A$, say. Thus we get the m–vector $(i_1,...,i_m)$ of zeros and ones and we have that $p = (i_1+...+i_m)/m$ is the fraction of the m measuring stations for which the lognormal distribution is chosen on the basis of the test statistic A . By repeated simulations a large sample of outcomes of p is finally generated and from this sample a distribution function of p is estimated.

A simulation investigation can also be made in the case where the gamma distribution assumption is replaced by a Gumbel distribution assumption allowing another type of dependence model than above. The observations at the m stations are now most conveniently represented by the vector $\mathbf{Z} = (Z_1,...,Z_m)$ where $Z_i = [\max\{X,Y_i\}|\max\{X,Y_i\} > 0]$, i = 1,...,m , has the truncated Gumbel distribution $[F(x)-F(0)]/[1-F(0)]$ with $F(x) = \exp\{-\exp[-\alpha(x-\beta)]\}$, x ∈ ℝ . This distribution is obtained by letting X and Y_i be mutually independent random variables that have Gumbel distributions with parameters α, β_x for X and α, β_y for Y_i such that $\exp[\alpha\beta] = \exp[\alpha\beta_x] + \exp[\alpha\beta_y]$. Giving α and β those values that make the mean and variance of Z_i the same as in the gamma distribution case, and also giving β_x such a value that $E[X] = k_x$ we get a reasonable case for comparisons of the two different dependence models.

Under the independence assumption (i.e. if X ≡ 0) only a small influence of the simulation distribution type should be expected. Any difference between the two distribution functions for p is an effect of the estimation uncertainty of the moments. If the exact moments are used (simple hypothesis) the results are identical because then the values $y_1 \leq ... \leq y_n$ in both cases are observations of an ordered sample from the uniform distribution.

As in Ellingwood and Redfield (1983) the simulation experiments are run for n = 28 and m = 38 . Using the moment estimates of the parameters of the two alternative distribution functions gamma or lognormal (Gumbel or lognormal), the distribution functions of p corresponding to $k_x = 0$ (X ≡ 0, i.e. independence) and $k_x = 2.5$ are shown in Fig. 3. It is interesting that the two models also in the dependence case give almost the same distribution functions for p .

For the Ellingwood and Redfield snow data

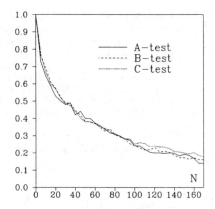

Fig. 2. *Simulated estimation of the conditional probability that the lognormal distribution fits better a sample of size 30+N than the gamma distribution given that the lognormal distribution fits better the subsample consisting of the first 30 observations. All sample values are generated from the gamma distribution.*

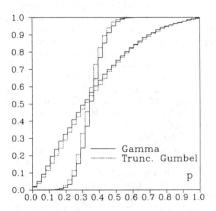

Fig. 3. *Distribution functions for the fraction p of m = 38 measuring stations at which the lognormal distribution is chosen instead of the correct gamma distribution (or the correct Gumbel distribution). The distribution choice is made on the basis of a sample of size n = 28 using the Cramer–Von Mises test statistic with estimated parameters of the two alternative distributions. The steep curves correspond to independence between measuring stations while the other curves correspond to a specific degree of dependence (in the gamma case the equicorrelation coefficient is $2.5/3 \simeq 0.83$).*

the hypothesis that the data population is lognormal implies that the fraction of cases for which the Gumbel distribution gives the best fit should be distributed approximately as in Fig. 3. It is seen that 0.24 is about the 0.05–fractile in

case of independence and about the 0.40–fractile in the considered case of dependence. This observation increases the doubt about the validity of the independence assumption. Clearly, the opposite hypothesis of having a Gumbel population is totally ruled out under the independence assumption by observing that 0.76 is far out in the upper tail of the distribution of the fraction of wrong distribution choices. However, if the independence assumption is removed the situation is different. For the considered case of dependence it is seen that 0.76 is about the 0.92–fractile.

Choice of snow load distribution type for structural reliability analysis

Without further considerations Ellingwood's and Redfield's investigation makes it reasonable to choose the lognormal distribution as the standard distribution for ground snow load in the considered region. However, there is yet another consideration to be made. As a formal load distribution model for reliability analysis the fat upper tail of the lognormal distribution may cause that the corresponding loads always dominate over loads that have been assigned distribution types with less fat upper tails. This strong and more or less arbitrary weight on some of the combining load types may be judged to be unreasonable from an engineering point of view. Besides neutralizing the tail sensitivity problem the purpose of the standardization is also to ensure that no single load type arbitrarily gets an overweight of influence on the calculated structural reliability.

In order to justify that there is an empirical evidence of a fat upper tail extending far beyond the range of the available sample of measured values, more than just the limited sample is needed. At least support should be obtained from some reasonable model of a mechanism that generates the fat tail. In this respect the lognormal distribution has a weak position in relation to most physical processes. The lognormal distribution family is closed with respect to multiplication of the random variables but not with respect to addition.

In a continental climate it is reasonable to consider the ground snow load as a result of an additive accumulation process. Taking the skewness of the observed empirical distribution into account the additive mechanism suggests the gamma distribution family as a reasonable candidate. This complies with the findings of Luy and Rackwitz (1978).

Also the Gumbel distribution type may be mechanistically defended. In particular this is the case for marine climates with isolated not accumulating snowfalls. If the snowfalls occur as the pulses of a homogeneous Poisson process with an exponentially distributed maximal snow load at each pulse, the extreme snow load during any period will have a Gumbel distribution function on the positive axis.

Independent of the JCSS there is a CIB Commission W 81: "Actions on Structures" with the task of writing reports on stochastic models for actions which are mutually consistent and which can be used both in probabilistic design and analysis, and as a basis for deterministic models of actions. Until now three reports have been published, CIB, W 81 (1989a, 1989b, 1991) on "self–weight loads", "live loads in buildings", and "snow loads" respectively. On the basis of the arguments given here it is in the snow loads report suggested to standardize the gamma distribution type (or mixtures of gamma distributions) rather than the lognormal distribution type for ground snow loads to be used in structural reliability analyses.

Conclusions

It follows from the simulation demonstrations herein that a best fit criterion is not sufficient as the basis for choosing distribution models for reliability analysis. In particular it is important to be aware of this fact if the reliability analysis is used as the basis for design decisions in competing consulting engineering companies. If there is a free choice of distribution types, competition about material savings, say, can easily end up being based on conflicting "false" information from arbitrary and practically nonverifiable modeling, that is, modeling that carries no empirical evidence. The practical answer to this dilemma is the use of codified internally harmonized standardizations of distribution types imposed on all competitors by some authorized code committee.

Acknowledgement

The computer programming has been made by Johannes M. Johannesen at the author's department. The work has been financially supported by the Danish Technical Research Council.

References

CIB, W 81 (1989a). "Actions on structures. Selfweight loads". CIB Report, Publication 115.

CIB, W 81 (1989b). "Actions on structures. Live loads in buildings". CIB Report, Publication 116.

CIB, W 81 (1991). "Actions on structures. Snow loads". CIB, W81 (1991), Publication 141.

Ditlevsen, O. and Madsen, H.O. (1989). "Proposal for a code for the direct use of reliability methods in structural design". Working document, Joint Committee on Structural Safety, ed. IABSE–AIPC–IVBH, ETH–Hönggerberg, Zürich, Switzerland. Reprinted in slightly revised form in Ditlevsen, O. and Madsen, H.O. (1994). "Structural Reliability Methods" (translation of SBI–rapport 211: Bærende konstruktioners sikkerhed (1990)).

Ellingwood, B. and Redfield, R. (1983). "Ground snow loads for structural design". *J. Struct. Engrg.*, ASCE, 109 (4), 950 – 964.

Filliben, J.K. (1975). "The probability plot correlation coefficient test for normality". *Technometrics*, 17(1), 111–117.

Kendall, M.G. and Stuart, A. (1961). *The advanced theory of statistics*, Vol. 2: Inference and relationship, Charles Griffin, London.

Luy, H. and Rackwitz, R. (1978). "Darstellung und Auswertung von Schneehöhenmessungen in der Bundesrepublik Deutschland". *Berichte zur Zuverlässigkeitstheorie der Bauwerke, Heft 31*, Technische Universität München.

Structural Safety & Reliability, Schuëller, Shinozuka & Yao (eds) © 1994 Balkema, Rotterdam, ISBN 90 5410 357 4

Reliability analysis of reinforced concrete columns

F. Duprat, M. Pinglot & M. Lorrain
Institut National des Sciences Appliquées, Toulouse, France

ABSTRACT : the reliability of reinforced concrete columns is evaluated by application of the Monte Carlo simulation techniques, in order to quantify the probability of overtaking the ultimate limit state of buckling. The mechanic model takes into account the second order effects and the material non-linearity. The probability of failure gives an estimation of the design safety and allows to show its sensitivity to changes in the deterministic parameters as steel ratio, slenderness, connection types, and in distributions of the random variables as geometrical dimensions, mechanical properties of materials and load parameters.

1 INTRODUCTION

Design codes for concrete structures take into account uncertainties on various specified values, as geometrical data, mechanical properties of materials and load parameters, that are partly covered by using safety factors and characteristic values.

However, in the case of the hardly non linear behaviour of reinforced concrete columns, design codes calculation rules could likely lead to not quiete satisfactory results, because of a non constant degree of reliability. Thus, it appears necessary to justify, by using an appropriate probabilistic approach, the choice of partial factors, geometrical tolerances and characteristic values involved in these rules, and traditionnaly selected largely on the basis of intuition and experience.

To take into account different connection types between beams and columns within a frame, this present study includes:
- a mechanical model of the behaviour of reinforced concrete columns, taking into account stiff or flexible joints at its ends,

- a probabilistic approach, in order to evaluate the probability of failure, by application of the Monte Carlo simulations method.

Various cases of steel ratio, geometrical slenderness and connection types, with several distributions of the random variables, have been tested and discussed.

2 MECHANICAL MODEL

The "general method" [1], developed on the basis of the finite elements method, appears to be suitable to take into account the two types of non linearity by which is complicated the prediction of the behaviour of reinforced concrete columns: second order effects and material non linearity. Owing to the variation of bending moment and therefore also of curvature along the column it is necessary to divide the structure into several straight elements. In order to determine the load bearing capacity, the external load is increased gradually, computing the corresponding equilibrium state with each increase of load, until the load entailing the instability

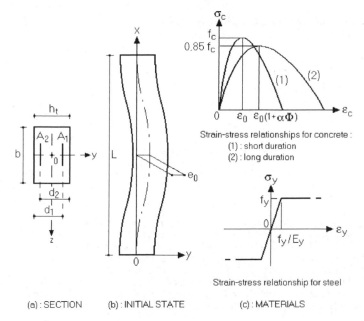

Strain-stress relationships for concrete :
(1) : short duration
(2) : long duration

Strain-stress relationship for steel

(a) : SECTION   (b) : INITIAL STATE   (c) : MATERIALS

Fig.1 : studied element

by equilibrium divergence is reached.

The main definition and assumptions related to the structure under consideration are presented in Figure 1.

The accuracy and flexibility of the "general method" allows to consider different connection types at the extremities of the column, which are estimated by the ratio r_c/r_b, where r_c is the rigidity of the column and r_b is the rigidity of the beams connected to it. Thus, the following cases have been considered :
- clamped ends $(r_c/r_b=0)$,
- elastically clamped ends $(r_c/r_b=0,1; r_c/r_b=1; r_c/r_b=10)$,
- hinged ends $(r_c/r_b=\infty)$.

3 RELIABILITY ASSESSMENT

The probability of failure P_f is the probability for the load S applied on the column to be greater than its resistance R [2]; P_f is stated by

$$P_f = \text{Prob}(S \leq R) = \int_0^\infty (1-F_S(r)) f_R(r) dr \quad (1)$$

where F_S is the cumulative probability distribution of S and f_R is the density of probability of R.

The Monte Carlo Simulations Technique allows to obtain a sample of N_{sim} columns and their computed bearing capacities $r_1, r_2, \ldots, r_{Nsim}$, which are independant generations of the variable R.

Provided to calculate numerically the distribution F_S, an equivalent expression of P_f is [3]

$$P_f = \frac{1}{N_{sim}} \sum_{k=1}^{N_{sim}} (1-F_S(r_k)) \quad (2)$$

4 LOAD EFFECT MODELLING

Load effect modelling has been achieved according to commonly accepted assumptions [4,5] (see Table 1). The ultimate limit state load is obtained by $S_u=1,35G_k+1,5Q_k$, and the yielding reinforcement is determinated for S_u by applying recommendations of section A-4-4 of BAEL rules [4], wich are close to the CEB recommendations [1,5].

Table 1 : loads characteristics

Load type	Law	Mean value	Standard deviation	Characteristic value
G	normal	G_m	$s_G=0,10G_m$	$G_k=G_m$
Q	E1max	Q_m	$s_Q=0,35Q_m$	$Q_k=Q_m+1,28s_Q$

Table 2 : material properties modelling

	Mean value	Coefficient of variation	Characteristic value	Design value
f_c	f_{cm}	C_{vc} (0,15)	$f_{ck}=f_{cm}(1-1,64C_{vc})$ (25 Mpa)	$f_{cd}=0,85f_{ck}/1,5$
f_y	f_{ym}	C_{vy} (0,08)	$f_{yd}=f_{ym}(1-1,64C_{vy})$ (400 Mpa)	$f_{yd}=f_{yk}/1,15$

Table 3 : geometrical parameters characteristics

Parameter	Tolerance	Standard deviation
b	0,05b	0,025b
h_t	$0,05h_t$	$0,025h_t$
d_1,d_2	0,05d+5mm	0,025d+2,5mm
e_0	L/250	L/250

5 RANDOM VARIABLES MODELLING

Compression concrete strength f_c, steel yielding strength f_y, total depth h_t and width b of cross-section, effective depthes d_1 and d_2, and initial eccentricity e_0 are the mechanical and geometrical parameters considered as random variables.

The distributions of the random variables, assumed to be normal, have been defined so as to respond to the design recommendations [4,5] particulary with respect to characteristic values and tolerances (see Table 2 and Table 3).

Standard deviation of a geometrical parameter is assumed to be equal to its half-tolerance [6], except for the initial eccentricity [7] : in this case the mean is zero and the standard deviation is assumed to be equal to the tolerance.

6 RESULTS

The results are presented according to the various deterministic parameters, as geometrical slenderness L/h_t, steel ratio $w=(A_1+A_2)/bh_t$, with $A_1=A_2$, load factor $\alpha=G_k/(G_k+Q_k)$ and connection types, estimated by the ratio r_c/r_b.

Figure 2, Figure 3 and Figure 4 show the global influence of these parameters on the design safety.

The probability of failure P_f strongly increases with the ratio r_c/r_b, particulary for important slenderness. While the variability of reliability may be neglected if $r_c/r_b \leq 1$, more flexible connections lead to increase the failure risk. Thus, the design safety does not appear homogeneous when connection types and slenderness vary. This anomaly may be explained by the scattering of resistances, which occurs when second order effects

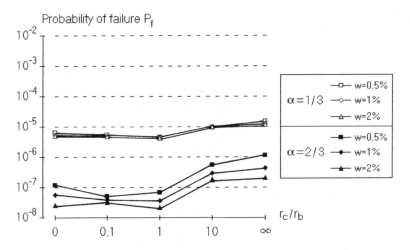

Fig.2 : influence of α and r_c/r_b for $L/h_t=16,7$

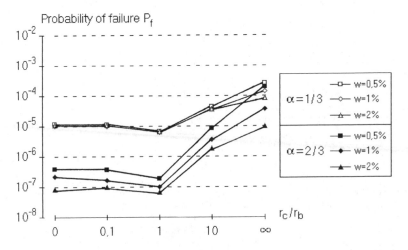

Fig.3 : influence of α and r_c/r_b for $L/h_t=23,4$

are amplified (deflection of co-lumns is very sensitive to slen-derness and connection type), hence worse consequences of geometrical and mechanical uncertainties on the design safety.

The influence of the load factor α is likewise emphasized : the fai-lure risk is more important when the variable load Q is preponderant ($\alpha=1/3$) than when the permanent load G is preponderant ($\alpha=2/3$). This deviation is likely due to the scattering of Q (coefficient of variation 0,35) in comparison with the one of G (coefficient of

variation 0,10). It would seem necessary to harmonize the safety partial factors related to loads, with the defining of their characteristic values, in order to achieve a better coherence in design codes rules.

The influence of the steel ratio w may not be neglected; in particu-lar for $\alpha=2/3$ and $r_c/r_b=10$, the probability P_f increases when w decreases. The application of de-sign codes rules leads to better safety for columns bearing more important load.

The effect on the safety of the

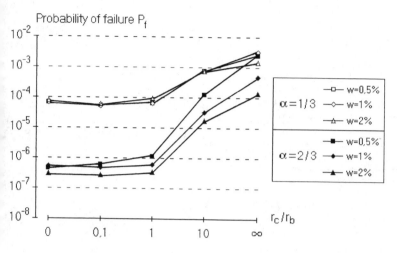

Fig.4 : influence of α and r_c/r_b for $L/h_t=30$

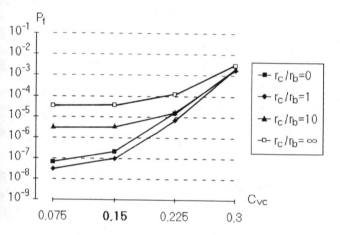

Fig.5 : influence of f_c for $w=1\%$, normal distribution

distributions of each random variable has been studied by varying its mean or/and its standard deviation. However, only results concerning variables wich the reliability is really sensitive to are reported here (for $L/h_t=23,4$).

Thus, in the case of the concrete strength Figure 5 and Figure 6 show the influence of the combination of the mean f_{cm} and the standard deviation, giving the same characteristic value f_{ck} (see Table 1), on the design safety; various values of the coefficient of variation c_{vc} are hence considered.

According to the assumption of

normal distribution, the probability P_f increases in a large manner when stronger values of c_{vc} to be taken (see Figure 5), while its variability practically vanishes if the assumption under consideration is the lognormal distribution (see Figure 6). This likely arises because of the lowertail displaying of the normal distribution, which is greater in comparison with the one of the lognormal distribution. The relationship between the characteristic value, the safety partial and the assumed distribution should be more precisely defined.

Proceeding in the same way, with

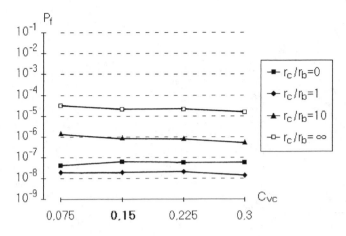

Fig.6 : influence of f_c for w=1%, lognormal distribution

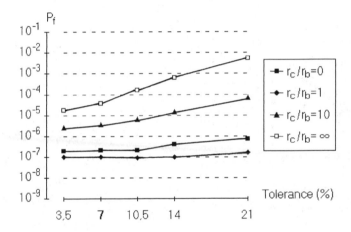

Fig.7 : influence of d, for w=1%

respect to the steel yielding
strength, one notes that reliability remains homogeneous.

The influence of tolerances related to geometrical parameters has
been emphasized by making each
standard deviation increase sepa-
ratly.

The sensitivity of the design safety appears relatively weak to
changes in tolerances related to
the cross-section dimensions, for
the usual range of these tolerances.

Contrariwise, the failure risk
strongly increases with tolerances
concerning the effective depth (see
Figure 7) and the initial eccentricity (see Figure 8). For the last

one in particular, on notes that
stiffer connections have no effect
on the sensitivity.

7 CONCLUSION

The reliability assessment of
reinforced concrete columns by application of the Monte Carlo simulation technique, allows to evaluate the sensitivity of the design
safety to changes in deterministic
parameters as geometrical
slenderness, load factor and connection type, and to changes in the
distributions of random variables
(geometrical values and mechanical
properties of materials).

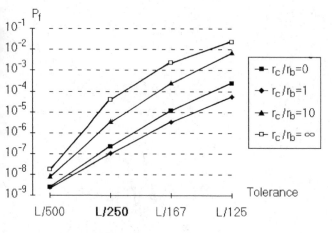

Fig.8 : influence of e_0, for $w=1\%$

In every cases tested, design codes rules have been used to determinate the design load bearing capacity of columns (section A-4-4 of BAEL rules) according to design values.

Assumptions on distributions of random variables have of course to be kept in mind when investigating results. These show that :

- taking into account stiffer connections always leads to increase safety, particulary when variable load is preponderant (load factor $\alpha=2/3$),

- slenderness is a source of unsafety, increasing the sensitivity to load eccentricity,

- various combinations of the mean and standard deviation of concrete strength f_c, giving the same characteristic value f_{ck}, do not lead to homogeneous reliability, wich is hardly decreased when high values of the coefficient of variation to be taken, provided to accept the assumption of normal distribution for f_c. The sensitivity of design safety practically vanishes when assumption under consideration is the lognormal distribution,

- the influence of steel yielding strength f_y and of the cross-section dimensions b and h_t is not important,

- tolerances on effective depth d and particulary on load eccentricity e_0 have to be strictly defined, because of their unfavourable effect on design safety.

A more exhaustive study could allow to give more accuracy in defining the means for achieving better homogeneity of reliability of similar structures. Slender columns could support "design penalty", particulary if connections are flexible.

REFERENCES

[1] Comité Euro-international du Béton : "Buckling and instability", Bulletin d'Information n°123, december 1977.

[2] R.E. Melchers : "Structural reliability. Analysis and prediction", Ellis Horwood Limited, Chichester (UK), 1987.

[3] A. Mébarki, M. Pinglot, M. Lorrain : "Fiabilité des poutres isostatiques en béton armé: sensibilité aux paramètres géométriques, mécaniques et de chargement", Annales ITBTP, n°443, march-april 1986.

[4] BAEL 91 : règles techniques de conception et de calcul des ouvrages et construction en béton armé suivant la méthode des états limites, mars 1992.

[5] Comité Euro-international du Béton : "Code Modèle CEB-FIP pour les structures en béton", Bulletin d'Information n°196, september 1990.

[7] L. Östlund : "An estimation of γ-values. An application of a probabilistic method", to be published in the Bulletins d'Information du CEB.

[8] F. Casciati, I. Negri, R. Rackwitz : "Geometrical variability in structural members and systems", JCSS (publication of IABSE-AIPC-IVBH), Working Document, january 1991.

Structural Safety & Reliability, Schuëller, Shinozuka & Yao (eds) © 1994 Balkema, Rotterdam, ISBN 90 5410 357 4

Adaptive response surface techniques in reliability estimation

I. Enevoldsen, M. H. Faber & J. D. Sørensen
Department of Building Technology and Structural Engineering, University of Aalborg, Denmark

Abstract

Problems in connection with estimation of the reliability of a component modelled by a limit state function including noise or first order discontinuities are considered. A gradient free adaptive response surface algorithm is developed. The algorithm applies second order polynomial surfaces determined from central composite designs. In a two phase algorithm the second order surface is adjusted to the domain of the most likely failure point and both FORM and SORM estimates are obtained. The algorithm is implemented as a safeguard algorithm so non-converged solutions are avoided. Furthermore, a number of checks on the solutions are suggested and illustrative examples are shown.

1. Introduction

Response surface methods (RSM) have been used in a wide range of applications within the last 20-30 years. As an example it can be mentioned that in connection with planning of experiments in the chemical industry response surface methods have been widely used. Also within the nuclear industry application of RSM has been widespread. A general description of RSM can be found in Myers (1971) and Box & Draper (1987). The main idea is that the response which consists of a complex function of some input variables and some model and measurement uncertainties is approximated by a known 'simple' function of the input variables and an error term modelled by a stochastic variable.

In structural reliability theory applications of RSM can be found in e.g. Bucher & Bourgund (1990), Faravelli (1989), Rackwitz (1982) and Engelund & Rackwitz (1992). In structural reliability RSM can directly be used in the formulation of a limit state function to estimate the reliability when the response knowledge is based on experiments. RSM can also be used to approximate a known but very complex relation between the input and output, e.g. in stochastic finite element theory, see Faravelli (1989). In this case the error term models the fitting error, i.e. the modelling uncertainty.

Mainly two different types of response surface are used in practical examples, namely polynomial surfaces (linear or quadratic), see e.g. Myers (1971), Box & Draper (1987) and Rackwitz & Engelund (1992) or response surfaces obtained by application of spline functions in combination with interpolation functions, see e.g. Schall, Scharrer, Östergaard & Rackwitz (1991).

In this paper a single component is considered. The objective is to describe an adaptive response surface technique which can be used to estimate the reliability of components with difficult limit state functions. Examples of such limit state functions are noisy limit state functions, limit state functions with a number of local minima and limit state functions which are continuous but have discontinuous first derivatives with respect to the input variables, see figure 1.

Further it is assumed that the main failure area is located in a single domain of the state space. The goal is that the adaptive response surface technique should be able to give a satisfactory estimate of the reliability of the component, i.e. it is not required that the β-point determined using the response surface has to coincide with the point on the real limit state function closest to the origin in the normalized space. Further it is required that the technique has to give an indication to the user

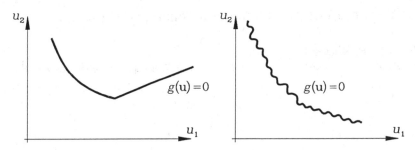

Figure 1. Limit state functions.

when the limit state function is too difficult to be modelled satisfactorily by a response surface.

The adaptive response surface technique which is presented in this paper is based on an application of second order polynomial response surfaces where the experimental plan specifying the nodal points is adjusted by an adaptive process. This implies that a SORM estimate of the reliability of the approximated limit state function can easily be calculated. An algorithm is proposed and tested in a number of examples.

2. Response Surface Methods

The problem of representing functionals in terms of approximating functional relationships is three-fold. First, the region and the points of the definition space at which the functional relationship is to be approximated to the original functional should be selected. Then the choice of type of approximating functional relationship should be selected and finally this functional relationship should be fitted to the original functional in the selected region of the definition space using an appropriate fitting method. It is clear that the first two problems are intimately related since a change of the size of the selected definition domain corresponds to a scaling of the rate of fluctuations of the original functional, i.e. a functional will have an increasingly satisfying approximate representation if the extension of the selected definition domain is decreased. Even for relatively low dimensional functionals with moderate rates of fluctuation and moderate amplitudes of fluctuation the task of fitting approximating functionals becomes impracticable for many purposes due to the necessary associated numerical efforts, unless a certain fitting error is allowed for. This fitting error is related to the difference in rate and amplitude of fluctuations between the original functional and the

approximating functional relationship and to the positioning of the points where the original function is approximated, as will be seen below.

If the original functional is reproducable any number of times at any point in its definition domain the fitting error is deterministic. However, if the original functional is only fitted at a limited number of points, as is the case in most practicable applications, the knowledge about the error is incomplete and the error is therefore an uncertain variable. If the original functional is not reproducible any number of times at any point in the definition space then the knowledge about the original functional itself even at the points where it is approximated are incomplete due to pure inherent uncertainty.

The vast literature on response surface modelling is therefore interested mainly in three tasks:
1) Selection of the most suitable approximating functional relationships.
2) Selection of the corresponding best set of points for the approximation.
3) Selection of an appropriate fitting methodology.

Appropriate fitting methodologies are understood methodologies giving means for access and analysis of the errors of the approximating functional relationship.

In the present paper the matter of interest is the use of response surface models in the application of FORM/SORM techniques in reliability estimation problems. To this end the limit state function $h(\mathbf{z})$ is defined separating the so-called safe domain $h(\mathbf{z}) > 0$ from the failure domain $h(\mathbf{z}) \leq 0$ where $\mathbf{z}$ is a realization of the vector $\mathbf{Z}$ of random variables. The integration of the failure probability using FORM/SORM is performed in a space spanned by zero mean unit variance Gaussian variables by introducing a transformation $\mathbf{u} = T(\mathbf{z})$. The limit

Table 1: Errors in response surfaces.

Sum of squares	Source	Degree of freedom
$c^T X^T y$	model	p
$\sum_{i=1}^{m} r_i(\tilde{y}_i - \bar{y}_i)^2$	lack of fit	$m - p$
$\sum_{i=1}^{m} \sum_{u=1}^{r_i}(y_{iu} - \bar{y}_i)^2$	pure error	$N - m$
$\sum_{i=1}^{m} \sum_{u=1}^{r_i} y_{ui}^2$	total	N

state function in the space spanned by u is denoted $g(\mathbf{u})$. It should be noted that the uncertainty in the problem of consideration may not entirely be captured in $\mathbf{u}$. Such cases result in limit state functions subject to random fluctuations.

Following the ideas behind sequential quadratic programming it is here assumed that the overall large-scale behaviour of the original functions (noisy or discontinuous) can be captured by a sequence of approximating second order polynomials such that the region of the probability space contributing the most to the failure probability can be identified.

Having identified this region, different approximations to the failure probability readily follow. First of all, the FORM result is available corresponding to the linear approximation of the failure domain at the most likely failure point. Secondly, if curvature information concerning the limit state function is available the SORM modification can be introduced. Alternatively or supplementary, the knowledge about the important domain in the failure domain can also simply be utilized by using the most likely failure point as the central point in a crude importance sampling.

Typically, approximating functional relationships of the polynomial form are suggested in the literature, see e.g. Myers (1971) and Box & Draper (1987). Here, for the above-mentioned reasons, second order response surfaces are considered in the u space.

For complete second order approximating functional relationships the orthogonal central composite design is especially efficient in terms of number of points where it is necessary to evaluate the original functional. This design consists of a 2^n-factorial design with each factor at the two-level $-d_{max}$ and $+d_{max}$ (see the next chapter) augmented by n" center points and completed by $2n$ points placed at the coordinates $-\phi d_{max}$ and $+\phi d_{max}$ on all axes. See e.g. Myers (1971) for a more detailed description. Even though several other designs may also be suitable this design is considered in the following.

A complete second order approximation of the limit state function in two dimensions u_1, u_2 is given as

$$f(\mathbf{u}) = y = c_0 + c_1 u_1 + c_2 u_2 + c_{11} u_1^2 + c_{22} u_2^2 + c_{12} u_1 u_2 + \epsilon \quad (1)$$

In equation (1) the coefficients c_i are so-called polynomial coefficients to be determined by fitting and ϵ is the uncertain error term. If the limit state function values $\mathbf{y}$ are known in the N points the design matrix $\mathbf{X}$ is given as

$$\mathbf{X} = \begin{bmatrix} 1 & u_{11} & u_{21} & u_{11}^2 & u_{21}^2 & u_{11}u_{21} \\ 1 & u_{12} & u_{22} & u_{12}^2 & u_{22}^2 & u_{12}u_{22} \\ \cdots & \cdots & \cdots & \cdots & \cdots & \cdots \\ 1 & u_{1N} & u_{2N} & u_{1N}^2 & u_{2N}^2 & u_{1N}u_{2N} \end{bmatrix} \quad (2)$$

then, see e.g Myers (1971) and Box & Draper (1987), the polynomial coefficients can be determined by linear regression analysis as

$$\mathbf{c} = (\mathbf{X}^T \mathbf{X})^{-1} \mathbf{X}^T \mathbf{y} \quad (3)$$

which corresponds to a maximum likelihood estimator for $\mathbf{c}$ if it can be assumed that ϵ is normal distributed. The accuracy of the approximating functional depends on the type of polynomial chosen. For a complete second order polynomial the errors in the approximating function can be evaluated as shown in table 1.

In table 1, p is the number of fitted parameters, m is the number of distinct points, N is the total number of points, r_i is the number of repetitions at the same point, $\bar{y}_i$ is the true mean value of the repeated experiments at the same sample point and $\tilde{y}_i$ is the predicted value at the point i.

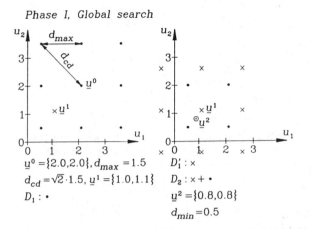

Figure 2. The principle of ARERSA in the 2-dimensional case. First, a two-step global search is performed to find the domain where the β-point is located. Next, a local search is performed in order to obtain a more precise estimation of the β-point.

3. Reliability Estimation by Adaptive Experimental Planning

In the following an adaptive scheme called the Adaptive Reliability Estimation Response Surface Algorithm (ARERSA) is presented. ARERSA uses central composite designs and second order response surface methods, as described in the previous sections, together with a two phase adaptive scheme. The algorithm is formulated in the standard normal space and illustrated in figure 2.

ALGORITHM: ARERSA

Phase I (Global Search for Domain)

First, an orthogonal central composite design D_1 is defined in the standard normal space from 1) input of a starting point u^0 as the best guess of the final β-point u^*. (u^0 is used as the center point of D_1) 2) input of d_{max} (e.g. 1.0-2.0) defined as half the side length in the design, i.e. if the dimension of the design is n the distance from the central starting point to a corner in the design D_1 will be $d_{cd} = \sqrt{n}d_{max}$.

Next, for the design D_1 a) calculate the limit state function values at all the design points, b) find the coefficients in a quadratic response surface using the mean least squares technique and c) find the β-point u^1 with the response surface as limit state function by use of a standard non-linear optimization algorithm.

The point u^1 can either be inside (defined as $|u^1 - u^0| \le d_{cd}$) or outside (defined as $|u^1 - u^0| > d_{cd}$) the initially defined domain. A solution where u^1 is outside is considered as an error state and a restart with a new guess of u^0 and/or d_{max} should be performed. A solution inside the domain is a success and phase I can progress.

With the point u^1 as center and d_{max} defining the side lengths an orthogonal central composite design D_1' is generated.

A new design D_2 is then set up by including D_1' and all the points in D_1 which are placed at a distance less than d_{cd} from u_1 and greater than d_{min} (an input parameter, e.g. 0.2-0.5) from the nearest point in the D_1'-domain. All remaining points in D_1 are neglected.

From the design D_2 the new β-point estimate u^2 is obtained in the same way as u^1 was obtained. If 1) $|u^2 - u^1| < \epsilon_c$ (e.g. $1 \cdot 10^{-3}$) convergence is obtained and ARERSA has obtained a solution. If 2) $|u^2 - u^1| \le d_{cd}$ the algorithm can progress with phase II. If 3) $|u^2 - u^1| > d_{cd}$ a restart with a new guess of u^0 and/or d_{max} should be performed.

Phase II (Local Domain Search)

Define a local domain design D_3 including the point in an orthogonal central composite design with u^2 as center and d_{min} as half the side length and the points in D_2 which are placed at a distance less

Table 2: Test of d_{min} influence.
*) Break down (d_{min} is to small compared to the noise)

d_{min}	N_g	β_F	β_S
0.1	50	*)	-
0.3	76	4.00	4.16
0.5	78	4.01	3.98
1.0	83	4.01	4.00
2.0	75	4.03	4.03

than $d_{cd} = \sqrt{n}d_{min}$ from $\mathbf{u}^2$. With this local domain a phase II iteration, $k = 2, 3, \cdots, k_{max}$, is performed until $|\mathbf{u}^k - \mathbf{u}^{k+1}| < \epsilon_c$. The design is only extended with the new β-point estimates $\mathbf{u}^{k+1}, k = 2, 3, \cdots, k_{max}$. Again, extrapolation is not allowed, i.e. $|\mathbf{u}^{k+1} - \mathbf{u}^2| > d_{cd}, k = 2, 3, \cdots, k_{max}$ is considered as an error state and a restart must be performed with another d_{min}.

After convergence, i.e. $|\mathbf{u}^k - \mathbf{u}^{k-1}| < \epsilon_c$, the estimates of the FORM-$\beta_F = |\mathbf{u}^k|$ and the SORM-β_S are calculated using the Hessian matrix of the last quadratic response surface.

Explanation and Comments on the Algorithm

The idea in phase I is to locate the domain in which the β-point is expected to be placed and then in phase II to refine the response surface fit locally for a more precise determination of the β-point. The idea behind the setup of D_2 is to avoid clustering of the points and still use as much old information as possible. D_3 is set up with the opposite idea, because clustering in general is needed for a good local approximation.

One of the major concerns applying an approximate technique is whether the technique can produce false solutions. Therefore, ARERSA is implemented as a safeguard algorithm, i.e., if the limit state function cannot be approximated by a quadratic surface with the β-point in the suggested domain (e.g. $u_i^0 \pm d_{max}, i = 1, 2, \cdots, n$) the algorithm will stop and a new suggestion of $\mathbf{u}^0$ and or d_{max} must be performed. It is therefore important that the starting point and the value of d_{max} are considered carefully. $\mathbf{u}^*$ must be in the domain and d_{max} must be selected taking into consideration the non-linearities in the limit state function. The more higher order non-linearities the smaller d_{max} should be selected. d_{min} is also selected depending on the actual problem. A small d_{min} gives a good approximation but also the possibility of extrapolation in phase II. Furthermore,

d_{min} should be selected so large that a smoothing for a noisy limit state function can be performed.

The number of point in a central composite design is $(2^n + 2n + 1)$, i.e the number of limit state calls in ARERSA will lie between $2(2^n + 2n + 1)$ and $3(2^n + 2n + 1) + 20$, with $k_{max} = 20$.

Check on the Reliability Estimates

ARERSA is recommended in the two cases mentioned in the introduction: 1) Limit state functions including first order discontinuities and 2) limit state functions including noise. In the first case the primary goal is to check whether it is a β-point which has been obtained, i.e. whether $|\mathbf{u}^* \times \nabla_u g| / |\nabla_u g| < \epsilon_1$, where $\nabla_u g$ is the gradient of the limit state function at $\mathbf{u}^*$ with respect to $\mathbf{u}$ and ϵ_1 an acceptance limit (e.g. $1 \cdot 10^{-3}$). A second check is to detect whether a first order discontinuity is so close to the assumed single β-point that it influences the probability of failure. This is performed by comparing the estimated β_S from ARERSA with $\beta_{S'}$ estimated from a quadratic surface fitted in the most important domain, which is defined as the domain which mainly contributes to the multi-normal integration in the standard normal space. The size d of the domain is here approximated from a linear limit state function as the solution $\exp(-(\beta_F + d)^2/2) = 0.05 \exp(-\beta_F^2/2)$, i.e. it is required that the value of the normal density function at the boundary of the domain is decreased to 5 % of the value at the most likely point. The half-side length in the central composite design is then found as $d_I = (\sqrt{\beta_F^2 + 6.0} - \beta^F)/\sqrt{n}$.

If $\beta_{S'}$ and β_S differ significantly the SORM-β_S cannot be expected to give a good estimate of the failure probability because the local surface obtained in ARERSA does not approximate accurately enough the surface in the interesting domain due to higher order non-linearities or maybe more likely a first order discontinuity too close to the estimated β-point.

1261

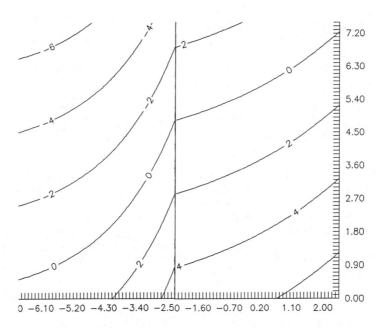

		7.20
		6.30
		5.40
		4.50
		3.60
		2.70
		1.80
		0.90
		0.00

0 −6.10 −5.20 −4.30 −3.40 −2.50 −1.60 −0.70 0.20 1.10 2.00

Figure 3. Limit state function in the standard normal space.

Table 3: Test of $\mathbf{u}^0$ influence.

$\mathbf{u}^0$	N_g	β_F	β_S
-2.0 ; 2.0	28	4.41	4.46
-3.0 ; 3.0	28	4.41	4.47
-4.0 ; 4.0	28	4.41	4.47
-2.0 ; 4.0	28	4.41	4.46

A third check is to examine how good the introduced response surface in the most important d_I-domain fits the underlying d_I-design points. This is performed by modelling ϵ in (1) as a zero mean stochastic variable Z_ϵ with a variance calculated from table 1 with contributions from both lack of fit and pure error, if pure error is present. This model uncertainty can then be taken into account as a correction in the estimation of β_F.

In the problems including noise the user must assess whether the amplitude of the noise is so large close to the estimated β-point at the response surface that it will influence the reliability significantly, i.e. whether the introduced smoothing cannot be accepted. If this is the case a simulation must be performed.

4. Examples

ARERSA has been tested in approximately 15 ex-

amples of which the first was without noise or first order discontinuities present. In the most cases ARERSA works in a stable way under the conditions given in the presentation of the algorithm with $d_{max} = 2.0$ and $d_{min} = 0.5$. In very non-linear cases ARERSA may break down. However, the selection of a smaller d_{max} may reduce the problem, but due to the safeguard implementation false solutions are not obtained.

In Karamchandani (1990) a limit state function with noise is considered: $g = 8 - a - U_1 - U_2 - U_3 - U_4 - a\cos(8\pi U_1)$ where $a = 0.125$ is the amplitude of the noise, and $\mathbf{U}$ is a 4-dimensional vector of standard normal variables. For $a = 0.125$ the noise is of a significant order and would probably not be accepted as e.g. noise from a well-scaled numerical model.

ARERSA is first tested for dependence on the guess of the initial point $\mathbf{u}^0$ with $d_{max} = 3.0$, $d_{min} = 0.5$ and $\epsilon_c = 1 \cdot 10^{-3}$. For 10 different initial points

a solution was found with $\beta_F = \beta_S = 4.03$

in all cases with only $N_g = 50$ limit state function calls, i.e. the solutions are obtained already in phase I. This introduces a test where d_{min} is varied and d_{max} is selected so small that phase II iterations must be performed.

With $d_{max} = 2.0, u^0 = \{1.5, 1.5, 1.5, 1.5\}$ and $\epsilon_c = 1 \cdot 10^{-3}$ d_{min} is varied and the results in table 2 are obtained.

From table 2 and the previous test it is seen that the ability of ARERSA to obtain a solution is clearly connected to the smoothing of the limit state function and not the initial point in this underlying linear problem. The value of the reliability index must be compared to 4.06 which is obtained by simulation in Karamchandani (1987). Whether a solution using smoothing of the noise can be accepted or a simulation must be performed is as mentioned in section 3 dependent on the magnitude of the noise and is the decision of the user.

In example no. 2 the following limit state function is considered: $g = Z_1^2 - Z_2$, if $Z_1 \leq 2.2$ and $g = 2.2^2 + 0.5(Z_1 - 2.2) - Z_2$, if $Z_1 > 2.2$. Z_1 is lognormally distributed $LN(4,1)$ and Z_2 is a standard normal $N(0,1)$. The limit state function is shown in figure 3.

The stability of ARERSA is examined by use of 4 different initial points and $d_{max} = 2.0, d_{min} = 0.5$ and $\epsilon_c = 1 \cdot 10^{-3}$. The results are shown in table 3.

The β-point is in all cases found as $u^* = (-3.51, 2.67)$ and it is seen that ARERSA is insensitive to the initial point in this example as long as the solution is in the d_{max}-domain. It must be noted that standard non-linear optimization algorithms will generally find the solution $u = (-1.61; 5.04)$ if the initial value of $u_1^0 > -2.3$ (corresponding to $z_1^0 > 2.2$) because both case functions are convex.

The obtained solution is checked as explained in the previous section. At the solution point u^* the Hessian matrix is calculated and the SORM index is obtained as $\beta_S = 4.46$. The size of the check design is for $\beta_F = 4.41$ found as $d_I = 0.45$. By use of the Hessian matrix of the quadratic surface fitted in the interesting domain $\beta_{S'} = 4.46$ is obtained.

In this case it is seen that the discontinuity in the first order derivatives has no effect on the estimated reliability. ARERSA has been used in similar cases where the presented check rejected the obtained estimate of the reliability due to first order discontinuities too close to the obtained solution.

The check of the quadratic fit in the most important domain d_I by introduction of ϵ in (1) as a stochastic variable Z_ϵ does not introduce extra uncertainty because the variance in this case has shown to be approximately zero.

5. Conclusions

In this paper the problem of estimating the reliability of a component modelled by a limit state function including either noise or discontinuities in the first order derivatives is treated. The various sources of uncertainty among other problems in approximating a limit state function in a response surface are outlined.

The reliability problem is solved by formulation of a gradient free adaptive response surface algorithm called ARERSA. ARERSA applies second order polynomial response surfaces obtained from central composite designs. First, the domain at which the most likely failure point is located is determined in a global search. Next a more precise response surface is determined in the local domain around the most likely failure point from a local search in a second phase. Hereafter, both a FORM and a SORM estimate of the reliability are obtained.

The algorithm is implemented as a safeguard algorithm to avoid false solutions and a number of checks are suggested to control the quality of the obtained estimates. Finally, the algorithm is tested in some examples and has under the safeguard conditions been working stable.

6. Acknowledgements

Part of this paper is supported by the research project "Risk Analysis and Economic Decision Theory for Structural Systems" sponsored by the Danish Technical Research Council which is greatly acknowledged.

References

Box, G.E. & N.R. Draper (1987) Emperical Model-Building and Response Surfaces. John Wiley & Sons.

Bucher, C. G. & U. Bourgund (1990) A Fast and Efficient Response Surface Approach for Structural Reliability Problems. Structural Safety, 7, pp. 57-66.

Engelund, S. & R. Rackwitz (1992) Experiences with Experimental Design Schemes for Failure Surface Estimation and Reliability. Proc. ASCE Spec. Conf. Denver, USA, July 1992, pp. 252-255.

Faravelli, L. (1989) A Response Surface Method for Reliability Analysis. J. Eng. Mech., ASCE, Vol. 115.

Karamchandani, A. (1990) New Methods in Systems Reliability Ph.D-Thesis. Report No. RMS-7, Dept. of Civil Eng. Stanford University.

Myers, R.H. (1971) Response Surface Methodology. Allyn and Bacon, Boston.

Rackwitz, R. (1982) Response Surfaces in Structural Reliability. Berichte zur Zuverlässig- keitstheorie der Bauwerke, Heft 67, Technische Universit at München.

Schall, G., M. Scharrer, C. Östergaard & R. Rackwitz (1991) Fatigue Reliability Investigation for Marine Structures using a Response Surface Me thod. OMAE 1991, Vol. II, Stavanger, Norway, pp. 247-254.

Structural Safety & Reliability, Schuëller, Shinozuka & Yao (eds) © 1994 Balkema, Rotterdam, ISBN 90 5410 357 4

Structural reliability applications in aerospace engineering

S.Gollwitzer & A.Zverev
RCP GmbH, Munich, Germany

R.Cuntze & M.Grimmelt
MAN Technologie AG, Munich, Germany

ABSTRACT: The concepts for structural reliability calculations with FORM/SORM connected with a Finite Element code are presented. The corresponding computer code is described. Some applications illustrate the efficiency and generality of the approach. Effort has been spent to help interpreting the results by suitable graphical representations.

1 INTRODUCTION

Modern probabilistic concepts and numerical techniques for reliability analysis have reached a high level of applicability and are frequently used for example in the aerospace industry and in the off-shore industry. The range of applications spans from probabilistic design over reassessment of existing structures to optimal maintenance planning.

The estimation of the performance of an engineering structure in a probabilistic sense requires that the structural response can be assessed as a function of the uncertain variables entering the problem. Not in all cases this response can be expressed by an analytical function (the so called "State Function") which, until very recently, was the requirement of commercially available computer codes for reliability analysis. The Finite Element Method (FEM) is one of the most important numerical tools for computation of the response of engineering structures. It is therefore attractive to have a direct link between the reliability analysis modules and FEM analysis modules.

In the stochastic model for the uncertain variables of an engineering structure it is useful to distinguish between load variables on the one side and variables describing material properties and geometrical quantities on the other side. The structural stiffness properties which depend on material properties and geometric quantities can be deterministic or random and structural responses under time-varying loads can be treated (quasi-) static or dynamic. Whereas many theoretical results on the various computational tasks are available now practicable computer implementations are still very rare and mostly devoted to some special tasks under rather idealizing circumstances.

Componental reliability analysis together with a specialized post processor to display the results is one attempt to a general solution of the problem described. A computer code which combines a reliability analysis module with a FEM program is used to demonstrate newly developed facilities. So far, they are restricted to linear systems and systems whose failure is identical to the failure of any and, in particular, the weakest component. Yet, these capabilities may cover the majority of tasks in practical applications.

2 CONCEPTS AND DEFINITIONS

2.1 *Reliability formulation*

In order to introduce the most important notions consider the structure as given in section 4.1 (see figure 1). The structure is loaded by a vector of time-invariant loads (internal pressure, ring loads and mass force in the figure). The load effects $Z_s = (Z_{s_1}, Z_{s_2}, ...)$ are certain functions of the loads filtered by the structure. Point i in the structure is said to have failed if a criterion of the form $V_i = \{ g_i(z_r, z_s) \leq 0 \}$ is fulfilled. An example is a yield failure mode $V_{yi} = \{ R_{po.2} - \sigma_{eq,i} \leq 0 \}$ with $R_{po.2}$ the yield strength and $\sigma_{eq,i}$ the v. Mises stress at point i in the structure, i.e. a nodal point in the FEM mesh in the figure. Here the load effects z_s are the cartesian stresses entering the v. Mises formula. The load effects are compared with resistances comprised in the vector z_r. These can be e.g. yield strength $R_{po.2}$ as in the example, ultimate strength, fracture toughness but also allowable deformations. The limit state is reached for $g_i(z_r, z_s) = 0$. It is also denoted as the failure surface. Consequently the structural point is said to be in a safe state if $g_i(z_r, z_s) > 0$. Hence, the failure

event F_i is defined if $Z \in V_i$ with $V_i = \{ g_i(z_r, z_s) \le 0 \}$ the so-called failure domain.

For the numerical analysis it is useful to introduce as basic uncertain variables $(X_1, X_2, ...)$ those quantities for which stochastic models can be assessed from test data or from literature. Thus the vector X comprising all basic variables includes random loads and all relevant uncertain structure variables. X has joint probability distribution function $F_X(x) = \mathbb{P}(\cap_{i=1}^{n} \{ X_j \le x_j \})$. In general, the distribution function is assumed to be continuous. In some cases the parameters of some distributions must also be considered as uncertain, for example due to statistical uncertainties. Formally they need not be distinguished from other uncertain basic variables.

The quantities Z_j define functions of partial vectors of the basic variables. They will be denoted by state variables. The transformation $Z = \mathbb{Z}(X)$ is performed by the FEM module. While the dimension of X can be large the dimensions of Z_s and Z_r related to a specific failure criterion usually are relatively small. This is the reason why it is useful to differentiate between basic variables X and state variables Z.

The failure probability under these circumstances then can be written as

$$P_f = \mathbb{P}(F) = \int_V dF_X(x) = \int_V f_X(x) \, dx \qquad (1)$$

where $f_X(x)$ is the probability density of X. The object fails at first loading with probability P_f or never. Alternatively, the reliability $\mathcal{R} = 1 - P_f$ may be computed. It is convenient to express P_f in terms of the reliability index $\beta = \Phi^{-1}(\mathcal{R}) = -\Phi^{-1}(P_f)$ with $\Phi^{-1}(.)$ the inverse of the standard normal integral. In general, eq. (1) has to be evaluated by suitable numerical procedures in the reliability module, applying e.g. advanced level II methods, see Madsen et al. (1986). These procedures can be augmented by importance sampling procedures as given in Fujita and Rackwitz (1988).

Also for the time variant case a reliability formulation can be given. Both time varying loads (load processes) and time variant, non-stationary resistance properties (deterioration, fatigue) can be considered. The difficulties in determining time-variant reliabilities arise from the fact that the probability distribution of time to failure is not known nor can it be determined statistically from observation with sufficient accuracy. It must be derived from stochastic and physical context. This can be done by constructing an auxiliary random point process of excursions (exits) of the state function into the physical failure domain. See STRUREL manual by RCP (1992) for a description of the theory and Grimmelt et al. (1988) for an example. It is worth mentioning here that exactly the same interface from the reliability module to the FEM module together with the sensitivities (as outlined in section 3) can be used also for time variant case. Thus it is possible to derive failure rates also for structures.

2.2 Global and local variables, locations

In addition to the differentiation between the vector of State Variables (Z) and Basic Variables (X) a partition of the X-vector is of great advantage for a reliability analysis where the transformation $Z = \mathbb{Z}(X)$ is performed by a FEM module.

One group of basic variables is denoted as global X-variables. All variables affecting the overall ("global") mechanical model are global variables. These are load-set multipliers, mechanical properties (E, G, ν) and geometrical quantities. Any change in a global variable triggers recomputations in the FEM module, i.e. the stiffness matrix has to be modified or even rebuilt for any new realization of these variables. If only load set multipliers are uncertain, the FEM module has to compute just a number of different right hand sides.

The other group of basic variables are the local X-variables. The main difference to the global variables is that these variables do neither affect the right hand side (loads) nor the stiffness matrix. The name local variable was chosen as these variables locally enter the state function without affecting the overall mechanical model. Thus they are independent of the FEM model. Uncertain material resistance properties $(R_{P0.2}$ in the example above) are typical local variables.

Usually global variables do not refer to a single element or a single nodal point but refer to a group of elements or points. In this way one can have a different parameterization in the FEM mesh and in the stochastic model which is of great importance in applications. For the local variables one has complete freedom in assigning them to single elements/points up to having e.g. the same uncertain resistance throughout the structure.

Taking into account the problem of different parameterizations it is mandatory for pre- and post processing to store the assignments of global and local variables to the locations in the FEM mesh.

Next the stochastic model of all basic uncertain variables X must be set up and stored in the reliability module. Finally, a data structure for the results of the reliability analysis, which is performed at various points or elements in the structure, has to be set up. This allows visualization of the results by color plots, contour lines or the like using standard post processing techniques from the FEM module.

3 STRUCTURAL RESPONSE AND ITS SENSITIVITIES FOR RELIABILITY ANALYSIS

The most important case in applications and also the simplest in data handling and computational effort is when only loads on the structure and local resistance properties are uncertain. Otherwise the system has deterministic properties. As outlined in section 2.1 let the failure criterion be formulated in the space of state variables $Z = (Z_1, Z_2, ... Z_m)$ some of which are related to the load effects and

the other correspond to local resistance properties. The vector of basic uncertain variables is $\mathbf{X}$.

For a deterministic stiffness matrix the equilibrium in a structure is given by

$$\mathbf{K}\, \mathbf{V}(\mathbf{X}) - \mathbf{p}(\mathbf{X}) = 0 \qquad (2)$$

where $\mathbf{V}(\mathbf{X})$ are the unknown displacements, $\mathbf{p}(\mathbf{X})$ are the nodal loads and $\mathbf{K}$ is the (deterministic) stiffness matrix. The vector of state variables in a given state function can be given by

$$\mathbf{Z}(\mathbf{X}) = \mathbf{A}(\mathbf{X})\, \mathbf{V}(\mathbf{X}) + \mathbf{Z}^o(\mathbf{X}) =$$
$$\mathbf{A}(\mathbf{X})\, \mathbf{K}^{-1}\, \mathbf{p}(\mathbf{X}) + \mathbf{Z}^o(\mathbf{X}) \qquad (3)$$

$\mathbf{A}$ is a constant matrix whose elements depend on the type of state variables, e.g. displacements, sectional forces of stresses and $\mathbf{Z}^o(\mathbf{X})$ a certain vector. It is assumed that the matrices and vectors are expanded suitably so that the matrix multiplications become consistent.

For each selected location in the structure the reliability analysis is done using FORM/SORM methods augmented by certain importance sampling schemes. The reliability module performing this task internally works in the space of standard normal independent variables (U-space). In this space the basic numerical task now is to search for the β-point, that is to solve

$$\beta = \|\mathbf{u}^*\| = \min\{\|\mathbf{u}\|\} \quad \text{for}$$
$$g(\mathbf{z}) = g(\mathbb{Z}(\mathbf{x})) = g(\mathbb{Z}(\mathbb{I}(\mathbf{u}))) \le 0 \qquad (4)$$

where $\mathbb{I}(.)$ is the probability transformation to be performed by the module for reliability analysis and $\mathbb{Z}(.)$ the mechanical transformation to be performed by the FEM module. One exact and efficient way of the transformation $\mathbf{x} = \mathbb{I}(\mathbf{u})$ is known as the Rosenblatt- transformation, see Rosenblatt (1952). An efficient search for the β-point requires at least the gradients.

They can be represented as

$$\nabla_u g^T = \nabla_x g^T\, \mathbf{J}_{x,u} = \nabla_z g^T\, \mathbf{J}_{z,x}\, \mathbf{J}_{x,u} \qquad (5)$$

where $\nabla_u g^T$ is the gradient of the state function with respect to $\mathbf{U}$, $\nabla_x g^T$ the gradient of the state function with respect to $\mathbf{X}$, $\nabla_z g^T$ the gradient of the state function with respect to $\mathbf{Z}$ and

$$\mathbf{J}_{z,x} = \left[\frac{\partial z_i}{\partial x_j}\right]_{(m,n)}$$

is the Jacobian of the mechanical transformation

$$\mathbf{J}_{x,u} = \left[\frac{\partial x_i}{\partial u_j}\right]_{(n,n)}$$

is the Jacobian of the Rosenblatt-transformation
n is the dimension of the vectors $\mathbf{X}$ and $\mathbf{U}$ and m is the dimension of the vector $\mathbf{Z}$. The vector $\nabla_z g$, whose dimension is usually small ($m < < n$), can be computed either analytically or numerically. The

Jacobian of the Rosenblatt-transformation can be computed by the reliability module in terms of the distributions and density functions of the X-variables. Provided that $\nabla_x g$ can be computed according to eq. (5) also sensitivities of the reliability index β with respect to distribution parameters and moments (mean, standard deviation) of the basic variables can be computed without extra numerical effort in the FEM module. See Abdo and Rackwitz (1990) for further details on the evaluation of $\mathbf{J}_{z,x}$ and $\mathbf{J}_{x,u}$.

When computing the Jacobian $\mathbf{J}_{z,x}$ a further simplification can be achieved if the matrix $\mathbf{A}$ in eq. (3) is assumed to be deterministic. Even if this is not so the error involved by such an assumption is usually small. Hence what has to be provided by the FEM module is the output vector $\mathbf{z}(\mathbf{x})$ and the Jacobian $\mathbf{J}_{z,x}$ where the partial vector $\mathbf{x}$ only comprises the loads on the structure being the global variables as defined in section 2.4. In a linear deterministic system this Jacobian has to be computed only once (for unit loads). An arbitrary number of local variables can be assigned to each selected point in the structure. With this information a reliability analysis can be performed for the selected points in the structure. The same concepts hold also for time-variant analyses.

If system properties are uncertain too (stochastic stiffness matrix) but time invariant no essential conceptual difficulties are met but the computational effort can increase considerably. In this case eq. (2) is written as

$$\mathbf{K}(\mathbf{X})\, \mathbf{V}(\mathbf{X}) - \mathbf{p}(\mathbf{X}) = 0 \qquad (6)$$

where now the stiffness matrix $\mathbf{K}(\mathbf{X})$ is also a function of a partial vector of $\mathbf{X}$. In analogy to eq. (3) the vector of state variables in a given state function can be given by

$$\mathbf{Z}(\mathbf{X}) = \mathbf{A}(\mathbf{X})\, \mathbf{V}(\mathbf{X}) + \mathbf{Z}^o(\mathbf{X}) =$$
$$\mathbf{A}(\mathbf{X})\, \mathbf{K}^{-1}(\mathbf{X})\, \mathbf{p}(\mathbf{X}) + \mathbf{Z}^o(\mathbf{X}) \qquad (7)$$

The dependency of $\mathbf{K}$ on the vector $\mathbf{X}$ makes reliability analysis time consuming. In a sampling scheme for each new sampled vector $\mathbf{X}$ the state variables $\mathbf{Z}(\mathbf{X})$ must be recomputed from an updated or even completely rebuilt stiffness matrix. For gradient based methods $\mathbf{Z}(\mathbf{X})$ and the Jacobian $\mathbf{J}_{z,x}$ have to be recomputed for each iteration step in the search for the β-point. Also one must note that the x-dimension of $\mathbf{J}_{z,x}$ now can be very high.

4 EXAMPLES AND DISCUSSION

The sensitivity of a basic variable X_j is the sensitivity of the corresponding standard normal variable:

$$\alpha_{uj} = (1/\|\nabla_u g^T\|)\, \partial g(\mathbb{Z}(\mathbb{I}(\mathbf{u})))/\partial u_j \approx \partial \beta/\partial u_j.$$

Table 1: Stochastic model for steel booster case

Variable X_j	Distr.F.	Mean	C.o.V.	Comment
bipn	Gumbel	7.0[MPa]	3%	Internal pressure
bdzf	Gumbel	7.0[MPa]	3%	Intern. press. on igniter
bdfk	Normal	5.0[MPa]	10%	Ring load on top ring
bdrk	Normal	3.0[MPa]	10%	Ring load on bottom ring
bcnm	Normal	62.5[MPa]	5%	Mass force (acceleration)
$R_{p0.2}$	Lognorm.	1474[MPa]	3%	Yield strength of steel
R_m	Lognorm.	1705[MPa]	3%	Ultim. strength of steel
K_{Ic}	Weibull	120[MPa√m]	8%	Fracture toughness
$R_{p0.2}$-w	Lognorm.	1327[MPa]	3.3%	Yield str., weld region
R_m-w	Lognorm.	1535[MPa]	3.3%	Ultim.str., weld region
K_{Ic}-w	Weibull	108[MPa√m]	8.9%	Fract. tough.,weld region
a/c	Fixed	1.0[-]	-	Crack aspect ratio
a_0	Rayleigh	1.0[mm]	52.3%	Crack depth
t_{wall}	Normal	8.5[mm]	1.5%	Wall thickness

These sensitivities can no more be interpreted if the X-variables are dependent (correlated) whereas the elasticities of means and standard deviations of basic variables are always meaningful. If α_{mj} denotes the sensitivity of β with respect to the mean of X_j then the elasticity is $e_{mj} = \alpha_{mj} (m_j/\beta)$.

4.1 Yield failure, ultimate failure and brittle failure of a steel booster case

As an example a steel case for a solid propellant booster is presented. The principal dimensions (only one cylindrical segment considered) are given in figure 1 together with a plot of the deformed mesh and the loading. The load sets considered are internal pressure, two ring loads and inertial force to simulate a flight load case. The basic uncertain variables (X_j) with their stochastic models are comprised in table 1. In general, the data reflect realistic estimates of the authors but not an actual and optimized design. Also the nozzle cone area is a simplified model outside the scope of this investigation. The FE model of the structure uses rotational symmetric solid elements (2088 d.o.f.).

The selected failure criteria in the form $g(z) \leq 0$ are

1: Yield $R_{p0.2} - \sigma_{eq} \leq 0$ with σ_{eq} the von Mises stress

2: Ultimate $R_m - \sigma_e \leq 0$ with $\sigma_e = 1.03\,\sigma_{eq}$ (2-dimensional ductile rupture criterion

3: Brittle $f(L_r) - K_I(\sigma,Y)/K_{Ic} \leq 0$ with $f(L_r)$ according to the R6, Rev.3 (1988), Y a suitable geometry function.

Some notes to table 1 apply:

The entries in the first column refer to the mnemonic character descriptors as used by the system.

The Gumbel distribution is selected to represent the maximum internal pressure during flight.

The structural loads are applied to the ring areas and are assumed to be negatively correlated ($\rho = -0.80$).

The inertial force due to acceleration of the booster mass is concentrated at the indicated location.

Yield strength and Ultimate strength are positively correlated ($\rho = 0.75$).

Material properties in assumed areas of 100 mm of welded joints (which are only 15 mm) below the top ring and above the bottom ring have 10% smaller means as compared to the undisturbed regions.

The aspect ratio a/c of the crack depth over its half length is assumed deterministic and

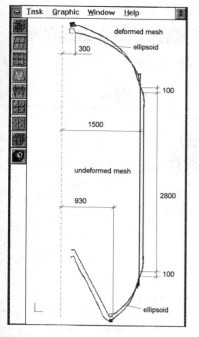

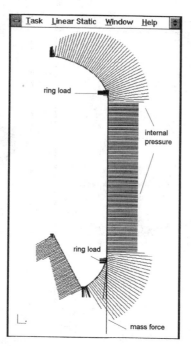

Figure 1: Geometry, deformed mesh and applied loads of steel booster case (Deformation amplification is 7.5. Internal pressure is displayed as equivalent nodal loads in outward direction to avoid confusion in the domes and the nozzle cone.)

determined by the manufacturing process.

See Gollwitzer et al. (1990) for the derivation of the stochastic model of a_0. The most unfavorable crack orientation is taken. Also it is (conservatively) assumed that a crack with depth a_0 is present in a critical region with probability = 1.

To model a local uncertainty due to manufacturing, t_{wall} is a chosen to be a random quantity. It has equal coefficient of variation for domes and cylinder.

In figure 2a the v.Mises stresses (at mean loads) in the front dome near the igniter flange are shown as a intensity plot (originally colored but set to grey shades here). Figure 2b shows the distribution of the reliability index β for the yield criterion. The qualitative distribution of the stress invariant is very similar to the distribution of β. The smallest, most critical β is 1.77 and is located in the region of shell bending. The smallest β for the ultimate, brittle failure mode is 4.23 at the same location. These β's are considered to be insufficient, therefore a redesign of the igniter flange is necessary. Another critical part is the weld region below the front dome which also has significant bending (see also deformed mesh) and in addition reduced material properties due to welding at the point indicated ("yield"-β = 2.48). In the welded region near the rear dome also influenced by shell bending the smallest yield β is 3.95 which is less critical.

Figure 3 presents some β's of the brittle failure mode together with sensitivities (pie chart) and elasticities (bar charts). The critical area is the front weld region (1st line of β-values). The next line is already in the region with undisturbed material properties and therefore shows larger β's. Both sections are in the area of shell bending. The value indicated below (β = 3.20) is valid for pure membrane behavior in the cylinder. It is smaller compared to the central value in the bending region (β = 3.58) because of larger hoop stresses.

The sensitivities and elasticities are the same for the 3 β-values in the critical region. The uncertainties of $K_{Ic}w$ and a_0 dominate as can be seen from the sensitivities and elasticities of the standard deviations. From the elasticities of the means it is obvious that a systematical variation of the internal pressure has the largest influence on β followed by $K_{Ic}w$ and a/c. As expected the e_m of a/c indicates a strong influence of the crack shape on the results which cannot be seen from the α_u's because a/c is "fixed" to its mean value. It is also interesting to compare the α_u, e_m and e_s of $R_{po.2}w$ and $R_m w$. To improve the situation proof testing and/or enlarging the mean value of $K_{Ic}w$ would be effective strategies.

From the failure modes with their β-values discussed so far one can conclude that (colored) intensity plots facilitate finding critical region but

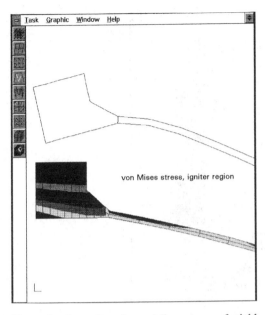

Figure 2a: Intensity of von Mises stress of yield failure mode in igniter region, front dome

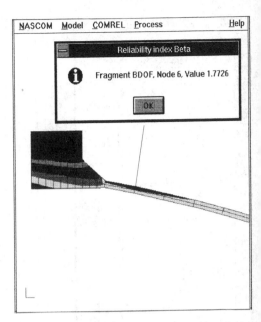

Figure 2b: Intensity of reliability index of yield failure mode in igniter region, front dome

Table 2: Stochastic Model for FRP Pressure Vessel

X_j	Distr.F.	Mean	C.o.V.
E^t_1	Normal	170000	4%
E^t_2	Normal	6500	6%
G_{12}	Normal	5000	3%
ν_{21}	Normal	0.26	5%
R^t_1	Weibull	2000	4%
R^c_1	Normal	1400	8%
R^t_2	Weibull	40	10%
R^c_2	Normal	140	9%
R_{12}	Weibull	90	4%
p_i	Normal	17	10%

Subscripts $_{1/2}$ denote parallel/orthogonal to fiber direction, $_{12}$ denotes shear, ^{t} denotes tension, ^{c} denotes compression, p_i is the internal pressure; all values are [MPa], $\nu_{21}[-]$

have to be interpreted with care. For example the minimal "yield β" indicates local yielding but not yielding of the total cross section. The "brittle β's" on the outer surface in the bending region are conservative estimates as they were derived from the (local) surface hoop stresses.

4.2 Serviceability, ultimate failure and optimal proof load of FRP pressure vessel

As a second example the cylindrical part of a pressure vessel made of fiber reinforced plastic (FRP) is taken. As material fiber T800 is selected with epoxy matrix LY556/HT976 and fiber-volume fraction $\varphi = 0.60$. The properties together with the stochastic model are given in table 4.2. The internal diameter of the pressure vessel is 120 mm, total wall thickness (Σt_i) is 3.60 mm and the stacking sequence [90°, 25°, -25° / symmetrical] with t_i [0.90, 0.45, 0.45 / symm.] (90° is hoop, 0° is axial). Internal pressure has a mean of 170 bar.

Due to the rotational-symmetric structure and a symmetric sequence of the layers, the load effects and therefore also the reliability indices are the same for layers $\{1, 6\}$ and $\{2, 3, 4, 5\}$. For (ultimate) fiber failure $\beta_{1/6} = 8.90$ and $\beta_{2-4} = 10.5$ indicate sufficient reliability. As expected R^t_1 is the dominating variable with $\alpha_u = 0.95$ followed by p_i with $\alpha_u = -0.30$. All other basic variables have vanishing influence. For (serviceability) inter fiber failure $\beta_{1/6} = 4.18$ and $\beta_{2-4} = 3.36$, the latter corresponding to P_f(serviceability) $\approx 4 \cdot 10^{-4}$ which

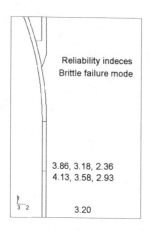

Reliability indeces
Brittle failure mode

3.86, 3.18, 2.36
4.13, 3.58, 2.93

3.20

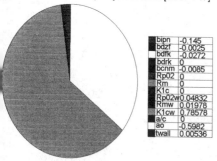

Alphas of R-variables - [bust] BUST.BCD [BCYF->17]

bipn	-0.145
bdzf	-0.0025
bdfk	-0.0272
bdrk	0
bcnm	-0.0085
Rp02	0
Rm	0
K1c	0
Rp02w	0.04832
Rmw	0.01978
K1cw	0.78578
a/c	0
ao	-0.5982
twall	0.00536

Figure 3a: Reliability indices and sensitivities of brittle failure mode at point (17) in front cylinder

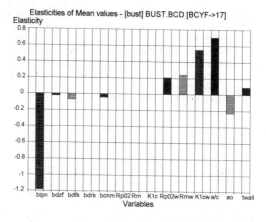

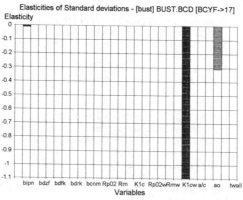

Figure 3b: Elasticities of brittle failure mode at point (BZYF: 17)

is considered to be insufficient. Rt_2 dominates with $\alpha_u = 0.92$, p_i again has a $\alpha_u = -0.30$.

Proof loading is a possible strategy to improve P_f(serviceability) as presented in figure 4. The failure probability after a proof load test is a conditional probability computed from

$$P_f[g(\mathbb{Z}(\mathbf{X})) \leq 0 \mid p_i \geq p_{proof}] =$$

$$\frac{\mathbb{P}(\{g(\mathbb{Z}(\mathbf{X}))\} \leq 0 \cap \{p_i \geq p_{proof}\})}{\mathbb{P}(p_i \geq p_{proof})}$$

and the corresponding β-values form the upper line in figure 4. The condition $p_i \geq p_{proof}$ denotes a survival event and is introduced with deterministic $p_i = p_{proof}$. The β-values depicted in the lower line in figure 4 correspond to the failure events under proof load. For $p_{proof} = 22.1$[MPa] (mean of $p_i + 3$ standard deviations) a conditional P_f(serviceability) of $3 \cdot 10^{-6}$ corresponding to $\beta = 4.53$ is reached.

The β corresponding to the failure events under

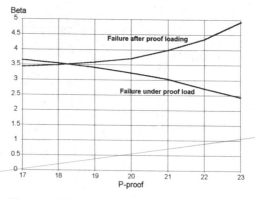

Figure 4: Influence of proof load

$p_{proof} = 22.1[MPa]$ is 2.46 or $\mathbb{P}(p_i \leq p_{proof}) = 7 \cdot 10^{-3}$ (about 7 of 1000 test specimens would "fail" during the test). It can be seen that proof loads only slightly larger than the expected value adversely affect the reliability, because they introduce some damage in the laminate without a gain in information. The results of such a study allow an optimization of the proof load magnitude.

REFERENCES

Abdo, T., Rackwitz, R., 1990
    Reliability of Uncertain Structural Systems, Proc. Finite Elements in Engineering Applications, Strasbourg, 26./27.4.1990, INTES GmbH., S.161-176
Fujita, M., Rackwitz, R., 1988
    Updating First- and Second-Order Reliability Estimates by Importance Sampling, Structural Eng./Earthquake Eng. Vol.5, No.1, pp 53-59, Japan Society of Civil Engineers (Proc. of JSCE No. 392/I-9)
Gollwitzer, S., Grimmelt, M., Rackwitz, R., 1990
    Brittle Fracture and Proof Loading of Metallic Pressure Vessels, Proc. 7th Int. Conf. on Reliability and Maintainability, Brest
Grimmelt, M., Gollwitzer, S., Rackwitz
    R., Ariane V Booster: Reliability of Unequipped Booster Case and Intersegment Connections, Proc. 6th Int. Conf. on Reliability and Maintainability, Strasbourg, pp. 140-145
Madsen, H.O., Krenk, S., Lind, N.C., 1986
    Methods of Structural Safety, Prentice-Hall, Englewood-Cliffs
R6, Rev.3, 1988
    Assessment of the Integrity of Structures Containing Defects, Central Electricity Generating Board, Rev.3
Rosenblatt, M., 1952
    Remarks on a Multivariate Transformation, Ann. Math. Statistics, Vol. 23, pp. 470-472
STRUREL, 1992
    Theoretical Manual, RCP-GmbH, 1988

Structural Safety & Reliability, Schuëller, Shinozuka & Yao (eds) © 1994 Balkema, Rotterdam, ISBN 90 5410 357 4

Reliability analysis of elastic-plastic structures based on the deformation theory of plasticity

Satoshi Katsuki
University of Colorado, Boulder, Colo., USA (Formerly: National Defense Academy, Yokosuka, Kanagawa, Japan)

Dan M. Frangopol
University of Colorado, Boulder, Colo., USA

Nobutaka Ishikawa
National Defense Academy, Yokosuka, Kanagawa, Japan

ABSTRACT: A reliability analysis method of truss structures limiting the elastoplastic displacements based on the deformation theory of plasticity is developed. This approach computes the reliability of holonomic elastic-perfectly plastic structures by using both the elastoplastic mode generation method and the sequential linear programming method. The proposed approach is applied to the reliability analysis of an elastic-perfectly plastic load-path-independent steel truss.

1 INTRODUCTION

Several recent reliability-based code developments have included extreme-event limit states for buildings, highway bridges, dams and offshore structures [e.g., Development of Comprehensive Bridge Specifications and Commentary (1992), Design Manual for Sediment Control Steel Dam (1987), Liu and Moses (1991), Moses and Liu (1992)]. These limits states are included in modern structural codes to ensure the survival of a structural system during an extreme-event by the development of significant inelastic deformations prior to failure.

Along these lines, the objective of this paper is to propose a method for reliability analysis of elastic-perfectly plastic structures based on the deformation theory of plasticity. The classical deterministic holonomic (i.e., load-path-independent) elastoplastic analysis formulation proposed by De Donato (1977) is modified for the purpose of reliability analysis. The derived formulation is used in conjunction with the elastoplastic mode generation method and the sequential linear programming method to solve the reliability analysis problem of holonomic elastic-perfectly plastic truss systems under random loads. Both excessive elastoplastic displacements and plastic collapse are considered as system limit states.

In order to illustrate and validate the proposed methodology computations are carried out on an elastic-perfectly plastic holonomic steel truss system.

2 FORMULATION OF LOAD-PATH-INDEPENDENT ELASTIC-PERFECTLY PLASTIC RELIABILITY ANALYSIS

A procedure is suggested herein for the reliability analysis of elastic-perfectly plastic structures that is based on the deformation theory of plasticity. According to De Donato (1977), this theory implies reversible material behavior (i.e., no permanent deformation after complete unloading). Therefore, no load-path dependency (i.e., the final loading situation alone is of concern) effects have to be considered.

This paper deals with pin-jointed plane holonomic truss systems made of elastic-perfectly plastic members under random independent loads. In this section, the reliability analysis formulation of these systems against excessive elastoplastic displacements presented by Katsuki et al. (1993) is summarized.

Using mathematical programming, the formulation associated with an elastoplastic mode (say, k) is as follows:

$$Given \quad : \bar{f}, , \sigma_f, C, K, K_{ep}, K_{epi}^{-1}, K_n,$$
$$\bar{N}, \tilde{N}, \bar{r}, \tilde{r}, c_{pe}, \theta, u_{ia} \dots\dots\dots\dots(1)$$

$$Find \quad : f, s \dots\dots\dots\dots\dots\dots\dots\dots(2)$$

$$Minimize : D_{kd} + c_{pe}|s|$$
$$= \sqrt{\sum_{j=1}^{m}[(f_j - \bar{f}_j)/\sigma_{fj}]^2} + c_{pe}|s| \dots(3)$$

Subject to :

$$[\cos\theta, \sin\theta] K_{epi}^{-1} f + s$$
$$= u_{ia} + [\cos\theta, \sin\theta] K_{epi}^{-1}$$
$$\times C^T K \bar{N} K_n^{-1} \bar{r} \quad\quad\quad\quad (4)$$

$$\tilde{N}^T K [C - \bar{N} K_n^{-1} \bar{N}^T K C] K_{ep}^{-1} f$$
$$< \tilde{N}^T K [C - \bar{N} K_n^{-1} \bar{N}^T K C] K_{ep}^{-1}$$
$$\times C^T K \bar{N} K_n^{-1} \bar{r}$$
$$- \tilde{N}^T K \bar{N} K_n^{-1} \bar{r} + \tilde{r} \quad\quad\quad (5)$$

$$K_n^{-1} \bar{N}^T K C K_{ep}^{-1} f$$
$$\geq K_n^{-1} \bar{N}^T K C K_{ep}^{-1}$$
$$\times C^T K \bar{N} K_n^{-1} \bar{r} + K_n^{-1} \bar{r} \quad (6)$$

In this formulation the following matrices, vectors and scalars are used: $\bar{f}$= the mean load vector, σ_f= the standard deviation load vector, C= the equilibrium matrix of the structure, K= the block-diagonal stiffness matrix consisting of member stiffness matrices, K_{ep}= the elastoplastic stiffness matrix associated with the k elastoplastic mode , K_{epi}^{-1}= the flexibility matrix associated with the i displacement vector of the flexibility matrix K_{ep}^{-1}, $K_n = \bar{N}^T K \bar{N}$, $\bar{N}$ = the unit outward normal matrix on the active yield conditions of the k elastoplastic mode (i.e., $\bar{\phi}_k = 0$, $\bar{\lambda}_k \geq 0$, where $\bar{\phi}_k$ = the active part of the yield function vector ϕ_k, and $\bar{\lambda}_k$= the plastic multiplier vector associated with active yield conditions $\bar{\phi}_k$), $\tilde{N}$= the unit outward normal matrix on the inactive yield condition of the k elastoplastic mode (i.e., $\tilde{\phi}_k < 0$, $\tilde{\lambda}_k = 0$, where $\tilde{\phi}_k$ = the inactive part of the yield function vector ϕ_k, and $\tilde{\lambda}_k$= the plastic multiplier vector associated with the inactive yield conditions $\tilde{\phi}_k$), $\bar{r}$= the plastic capacity vector associated with the active yield conditions of the k elastoplastic mode, $\tilde{r}$ = the plastic capacity vector associated with the inactive yield conditions of the k elastoplastic mode, c_{pe}= a penalty positive coefficient, θ= the angle between the x-axis and the direction of allowable displacement, u_{ia} = the allowable displacement, f= the random load vector, and s= the penalty variable. It is worth noting that the elastoplastic modes are generated by using a technique recently proposed by Katsuki *et al.* (1993), in which active and inactive set conditions are separated.

Eq. (3) defines the objective to be minimized as

the sum between D_{kd} (i.e., the distance between the origin of the standard normal load space , also called space of standardized variables, and the failure surface associated with the elastoplastic mode k) and the product of the absolute value of the penalty variable $|s|$ and the penalty coefficient c_{pe}. Eq. (4) defines the limit state condition $u_i = u_{ia}$ when $s = 0$. Eqs. (5) and (6) are constraints on the load space domain associated with inactive (i.e., $\tilde{\phi}_k < 0$) and active yield conditions (i.e., $\bar{\lambda}_k \geq 0$), respectively. If the solution of the optimization problem defined by Eqs. (1) to (6) corresponds to $s = 0$ then the minimum distance D_{kd} associated with the displacement limit function in the k elastoplastic mode is obtained.

When the result of the optimization problem (1) to (6) is represented by a positive value of the penalty variable (i.e., $s > 0$) combined with a collapse condition for mode k (i.e., the vector solution f is associated with a plastic collapse mode), the structure is in unrestricted plastic flow regime (i.e., $u_{id} >> u_{ia}$). In this case, the minimum distance D_{kc} associated with the above plastic collapse situation is calculated by modifying Eqs. (1) to (6) as follows:

$$Given \quad : \bar{f}, \sigma_f, C, K, K_{ep}, K_n,$$
$$\bar{N}, \tilde{N}, \tilde{n}_l^T, \bar{r}, \tilde{r}, \tilde{r}_l, \quad\quad\quad (7)$$
$$Find \quad : f \quad\quad\quad\quad\quad\quad\quad (8)$$
$$Minimize : D_{kc} = \sqrt{\sum_{j=1}^{m} [(f_j - \bar{f}_j)/\sigma_{fj}]^2} \quad (9)$$

Subject to :

$$\tilde{n}_l^T K [C - \bar{N} K_n^{-1} \bar{N}^T K C] K_{ep}^{-1} f$$
$$= \tilde{n}_l^T K [C - \bar{N} K_n^{-1} \bar{N}^T K C] K_{ep}^{-1}$$
$$\times C^T K \bar{N} K_n^{-1} \bar{r}$$
$$- \tilde{n}_l^T K \bar{N} K_n^{-1} \bar{r} + \tilde{r}_l \quad\quad (10)$$

$$\tilde{N}^T K [C - \bar{N} K_n^{-1} \bar{N}^T K C] K_{ep}^{-1} f$$
$$< \tilde{N}^T K [C - \bar{N} K_n^{-1} \bar{N}^T K C] K_{ep}^{-1}$$
$$\times C^T K \bar{N} K_n^{-1} \bar{r}$$
$$- \tilde{N}^T K \bar{N} K_n^{-1} \bar{r} + \tilde{r} \quad\quad (11)$$

$$K_n^{-1} \bar{N}^T K C K_{ep}^{-1} f$$
$$\geq K_n^{-1} \bar{N}^T K C K_{ep}^{-1}$$
$$\times C^T K \bar{N} K_n^{-1} \bar{r} + K_n^{-1} \bar{r} \quad (12)$$

where $\tilde{n}_l^T$ and $\tilde{r}_l$ are the l-th row vector of the outward unit matrix $\tilde{N}^T$ and the l-th element

Table 1 Input Data for Five-Bar Truss Example

(a) Member Properties

Bar	Young's Modulus (kgf/cm^2)	Section Area (cm^2)	Axial Force at Yielding	
			Tension (kgf)	Compression (kgf)
①, ⑤	2.1×10^6	3.0	7200	2300
②, ④	2.1×10^6	3.0	7200	3400
③	2.1×10^6	3.0	7200	4600
Note: 1kgf=9.81N				

(b) Load Statistics

No.	Mean Value (kgf)	Standard Dev. (kgf)	Coefficient of Variation
f1	3200	960	0.30
f2	3200	960	0.30
Note:1kgf=9.81N			

of the plastic capacity vector $\tilde{r}$ corresponding to the additional active yield condition l for collapse, respectively.

The sequential linear programming method [see, e.g. Kirsch (1981)] was used to solve the nonlinear programming problems represented by Eqs. (1) to (6) and Eqs. (7) to (12). The reliability index β_s with respect to excessive elastoplastic displacement is given by the minimum of all distances between the origin of the standard normal load space and the failure surfaces associated with both elastoplastic and collapse modes:

$$\beta_s = min \ (minD_{kd}, \ minD_{kc})$$
$$k = 1, \ .... \ ,m \quad ....(13)$$

where m=the total number of elastoplastic modes of the structural system.

3 NUMERICAL EXAMPLE: FIVE-BAR TRUSS

The problem definition for this example is summarized in Table 1 and Fig. 1. The space of independent Gaussian loads f_1, f_2 shown in Fig. 2 is used to represent both the elastic and the collapse boundaries of the truss system. The elastic region is enclosed in the domain OABCDE and the collapse boundary is represented by the four

segments LN, NQ, QS, and ST. Load combinations associated with points in the domain between the boundaries ABCDE and LNQST produce an elastoplastic behavior of the truss. The yield function vector of the truss shown in Fig. 1 is

$$\phi = [\phi_1^+, \phi_1^-, \phi_2^+, \phi_2^-, \phi_3^+, \phi_3^-, \phi_4^+, \phi_4^-, \phi_5^+, \phi_5^-]^T ..(14)$$

where the superscripts + and − denote tension and compression in member i, respectively. This vector has one component active on the elastic boundary ABCDE and two components active at the vertices B, C, and D. The active yield conditions ($\phi_i^{+,-} = 0$) and non-negative plastic multipliers ($\lambda_i^{+,-} \geq 0$) corresponding to all segments and regions of the load space are also shown in Fig. 2. The failure of the holonomic truss in Fig. 1 is defined by the displacement criterion. Contours associated with seven different values of the displacement u_{id} along the 45 degree direction in Fig. 1 are shown in Fig. 3. It is interesting to note that the contour $u_{id} = 0.05cm$ is associated with elastic behavior and the others correspond to elastoplastic or collapse states. Using the standard normal load space and the computational procedure outlined in Section 2, the problem is to find the minimum distance β_s (see Eq. 13) between the origin of this space and the failure surfaces shown in this figure. Fig. 4

1275

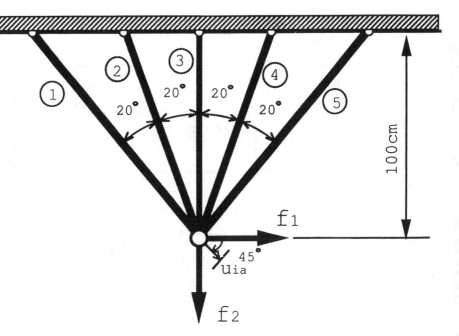

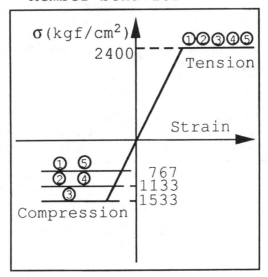

Member behavior

Fig. 1 Five-Bar Truss

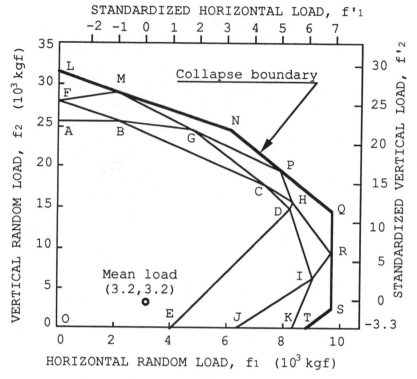

STANDARDIZED HORIZONTAL LOAD, f'_1

Collapse boundary

Mean load
(3.2,3.2)

VERTICAL RANDOM LOAD, f_2 (10^3 kgf)

STANDARDIZED VERTICAL LOAD, f'_2

HORIZONTAL RANDOM LOAD, f_1 (10^3 kgf)

AB:$\phi_3^+=0$, BC:$\phi_2^+=0$, CD:$\phi_1^+=0$, DE:$\phi_5^-=0$

FB,BG:$\phi_2^+=0,\phi_3^+=0$, GC,CH:$\phi_1^+=0,\phi_2^+=0$,

HD,DI:$\phi_1^+=0,\phi_5^-=0$, IJ:$\phi_4^-=0,\phi_5^-=0$,

FM:$\phi_2^+=0,\phi_3^+=0,\phi_4^+=0$, MG,GP:$\phi_1^+=0,\phi_2^+=0,\phi_3^+=0$,

PH,HR:$\phi_1^+=0,\phi_2^+=0,\phi_5^-=0$, RI,IK:$\phi_1^+=0,\phi_4^-=0,\phi_5^-=0$,

LM,MN:$\phi_1^+=0,\phi_2^+=0,\phi_3^+=0,\phi_4^+=0$,

NP,PQ:$\phi_1^+=0,\phi_2^+=0,\phi_3^+=0,\phi_5^-=0$,

QR,RS:$\phi_1^+=0,\phi_2^+=0,\phi_4^-=0,\phi_5^-=0$,

ST:$\phi_1^+=0,\phi_3^-=0,\phi_4^-=0,\phi_5^-=0$,

ABCDEO:Elastic Region,

ABF:$\lambda_3^+\geq0$, BCG:$\lambda_2^+\geq0$, CDH:$\lambda_1^+\geq0$, DEJI:$\lambda_5^-\geq0$,

FBGM:$\lambda_2^+\geq0,\lambda_3^+\geq0$, GCHP:$\lambda_1^+\geq0,\lambda_2^+\geq0$,

DIRH:$\lambda_1^+\geq0,\lambda_5^-\geq0$, IJK:$\lambda_4^-\geq0,\lambda_5^-\geq0$,

LFM:$\lambda_2^+\geq0,\lambda_3^+\geq0,\lambda_4^+\geq0$, MGPN:$\lambda_1^+\geq0,\lambda_2^+\geq0,\lambda_3^+\geq0$,

PHRQ:$\lambda_1^+\geq0,\lambda_2^+\geq0,\lambda_5^-\geq0$, RIKTS:$\lambda_1^+\geq0,\lambda_4^-\geq0,\lambda_5^-\geq0$

Fig. 2 Load Space with Elastic, Elastoplastic and Plastic Boundaries

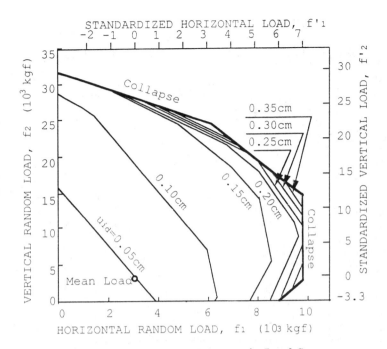

Fig. 3 Isodisplacement Contours in Load Space

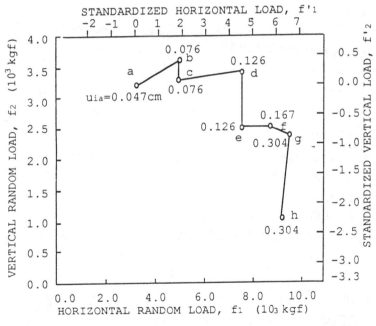

Fig. 4 Sensitivity of Design Point to Changes in Allowable Displacement

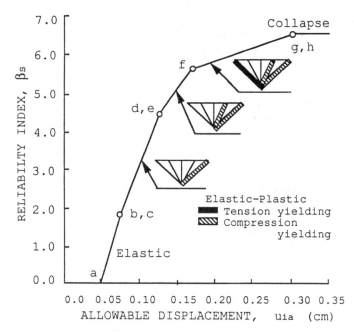

Fig. 5 Reliability Index versus Allowable Dispalcement

shows the sensitivity of the most probable failure point, also called design point, in both the original (f_1, f_2) and the standard normal (f'_1, f'_2) load spaces to changes in the allowable displacement u_{ia}. Finally, Fig. 5 shows the relationship between the reliability index and the allowable displacement corresponding to points a to h in Fig. 4.

4 CONCLUSIONS

A procedure has been proposed for computing the reliability of holonomic elastic-perfectly plastic structures based on the deformation theory of plasticity. This procedure combines the elasto-plastic mode generation method and the sequential linear programming method. It is important, however, to note that although this procedure appears to be adequate for the case of elastic-perfectly plastic holonomic (path-independent) systems under random loads, much effort is needed to deal with structures made of non-holonomic (path-dependent) materials which require knowledge of the overall load history for reliability assessment.

ACKNOWLEDGMENTS

The support from the National Defense Academy of Japan is gratefully acknowledged. It made it possible for the first writer to pursue his research work at the Department of Civil Engineering, University of Colorado, Boulder. Partial support from the U. S. National Science Foundation Grant MSM-9013017 to the University of Colorado at Boulder is also gratefully acknowledged.

REFERENCES

De Donato, O. 1977. Fundamentals of elastic-plastic analysis," Chapter 13 in *Engineering plasticity by mathematical programming*, M. Z. Cohn and G. Maier, eds., Pergamon Press. New York: 325-349.

Design Manual for Sediment Control Steel Dam 1987. Report of Sediment Control and Landslide Technical Center. Tokyo, Japan (in Japanese).

Development of Comprehensive Bridge Specifications and Commentary 1992. National Cooperative Highway Research Program. Project 12-33. Third Draft. April.

Katsuki, S., Frangopol, D. M., and Ishikawa, N. 1993. Holonomic elastoplastic reliability analysis of truss systems. I: Theory, *J. Struct. Engrg.* ASCE. Paper No. 3503. 119(6). June.

Kirsch, U. 1981. *Optimum structural design.* McGraw-Hill. New York.

Liu, Y. and Moses, F. 1991. Bridge design with reserve and residual reliability constraints, *Struct. Safety.* 11: 29-42.

Moses, F. and Liu, Y. 1992. Methods of redundancy analysis for offshore platforms. *Proceedings 11th Intl. Conf. on Offshore Mechanics and Arctic. Engrg.*, ASME, Calgary, Canada, June.

Structural Safety & Reliability, Schuëller, Shinozuka & Yao (eds) © 1994 Balkema, Rotterdam, ISBN 90 5410 357 4

Probabilistic modelling of concrete strength in heating process

Z. Keršner, D. Novák & B. Teplý
Technical University of Brno, Czech Republic

ABSTRACT: The prediction of the increase of concrete strength with time due to a heating process is studied. Two computational models are used: the modification of the time parameter according to temperature changes and the maturity model recommended by CEB. Uncertainties of deterministic models and definition of the space of basic random variables are presented. Reliability analysis is performed for the estimation of the probability distribution function of concrete strength at different points in time of the production process. A sensitivity analysis is also given.

1 INTRODUCTION

The strength and quality of concrete can be considered probably as one of the greatest sources of random variability in concrete structures. It is well known that continued hydration of cement paste causes a time increase in concrete strength. This process can be accelerated by higher temperatures; therefore heating is commonly used in the mass production of precast concrete elements.

A realistic prediction of concrete strength varying with time during a heat treatment based on a probabilistic approach is the objective of this paper. Two computational models for the prediction of concrete strength are used. The first model is based on the modification of the time parameter according to temperature changes (TM model); the second one is the maturity model recommended by CEB (CEB model, CEB Bulletin 1989).

The process causing the strength increase depends on several physical quantities. These quantities are random variables; their significance, scatter and statistical correlation are considered in order to provide a reasonable preliminary input to reliability calculations. The temperature history of heat treatment is also described by random variables.

A reliability concept is based on the limit state functions at different points in time of the process (8, 11, 24 hours and 3, 28 days). These functions represent the strength of the concrete which has been obtained by evaluating models mentioned above, compared with a certain required strength. As a measure of risk probabilities that the

required strength is not reached are calculated; we may call them the failure probabilities. The set of these probabilities represents the numerical estimation of the distribution function of the random variable "strength of concrete".

In the following two deterministic computational models are briefly described and a reliability concept is introduced.

2 DETERMINISTIC COMPUTATIONAL MODELS FOR CONCRETE STRENGTH

The basis of time modification model (TM model) is the function of strength increse of concrete in stable conditions (SC) according to Czechoslovak design codes (1988). This function is a regression relationship between the strength of concrete $R_{b,SC}$ [MPa] and time τ [$days$] at the temperature t_{SC} [$°C$]:

$$R_{b,SC} = b_{SC} \exp\left(\frac{-2\,a_{SC}}{\tau}\right) \qquad (1)$$

where b_{SC} is a limit of concrete strength [MPa] and a_{SC} is time of maximum of strength increase [$days$]. In order to express the strength during heating process, the time parameter τ in (1) is modified according to the temperature history:

$$\tau = \int_0^{\tau_i} \frac{\tau_{TF}(t_{SC})}{\tau_{TF}(t)}\,d\tau \qquad (2)$$

where $\tau_{TF}(.) = (const\ t^{-n})$ is a temperature function of the cement paste, t is the temperature

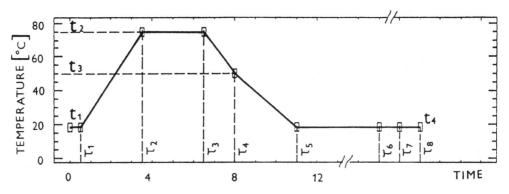

Fig. 1 Time vs. temperature

Table 1. Basic variables

Basic variable	X_i	Unit	Model	Point in time 1 2 3 4 5	Mean value	Standard deviation	Probability distribution
a_{SC}	X_1	$days$	TM	+ + + + +	1.0	0.08	lognormal
b_{SC}	X_2	MPa	TM	+ + + + +	32.7	2.3	gamma
n	X_3	-	TM	+ + + + +	1.6	0.03	normal
t_{SC}	X_4	$°C$	TM	+ + + + +	20.0	3.0	normal
t_1	X_5	$°C$	TM, CEB	+ + + + +	18.0	3.0	normal
t_2	X_6	$°C$	TM, CEB	+ + + + +	75.0	10.0	normal
t_3	X_7	$°C$	TM, CEB	+ + + + +	50.0	8.0	normal
t_4	X_8	$°C$	TM, CEB	+ + + +	18.0	3.0	normal
τ_1	X_9	h	TM, CEB	+ + + + +	0.5	0.1	lognormal
τ_2	X_{10}	h	TM, CEB	+ + + + +	3.5	0.1	lognormal
τ_3	X_{11}	h	TM, CEB	+ + + + +	6.5	0.1	lognormal
τ_4	X_{12}	h	TM, CEB	+ + + +	8.0	0.1	lognormal
τ_5	X_{13}	h	TM, CEB	+ + +	11.0	0.5	lognormal
τ_4	-	h	TM, CEB	+	8.0	0	-
τ_5	-	h	TM, CEB	+	11.0	0	-
τ_6	-	h	TM, CEB	+	24.0	0	-
τ_7	-	h	TM, CEB	+	72.0	0	-
τ_8	-	h	TM, CEB	+	672.0	0	-
τ_e	X_{14}	h	CEB	+ + + + +	50.0	5.0	lognormal
$R_{b,lim}$	X_{15}	MPa	CEB	+ + + + +	32.7	2.3	gamma
α	X_{16}	-	CEB	+ + + + +	1.0	0.08	normal

$[°C]$, n is a characteristic cement constant [dimensionless]; the integration is performed over the time-temperature diagram, see Fig. 1. Five important points in time are considered in the analysis: $8, 11, 24$ hours and $3, 28$ days ($i = 1, 2, \ldots, 5$). Strength of concrete according to (1) is:

$$R_{b,i}^{TM} = b_{SC} \exp\left(\frac{-48\, a_{SC}\, t_{SC}^n}{I_i}\right)$$

where

$$I_1 = \tau_1 t_1^n + (\tau_2 - \tau_1)\frac{t_2^{n+1} - t_1^{n+1}}{(t_2 - t_1)(n+1)}$$
$$+(\tau_3 - \tau_2)t_2^n + (\tau_4 - \tau_3)\frac{t_3^{n+1} - t_2^{n+1}}{(t_3 - t_2)(n+1)}$$
$$I_2 = I_1 + (\tau_5 - \tau_4)\frac{t_4^{n+1} - t_3^{n+1}}{(t_4 - t_3)(n+1)}$$
$$I_3 = I_2 + (\tau_6 - \tau_5) t_4^n$$
$$I_4 = I_2 + (\tau_7 - \tau_5) t_4^n$$
$$I_5 = I_2 + (\tau_8 - \tau_5) t_4^n \qquad (3)$$

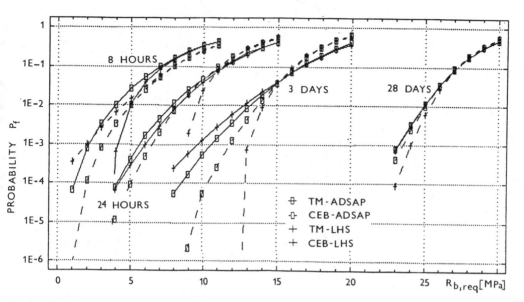

Fig. 2 Cumulative distribution functions of concrete strength

A different approach is recommended by CEB (CEB Bulletin 1989). Strength of concrete $R_{b,i}^{CEB}$ [MPa] for each i is described by:

$$R_{b,i}^{CEB} = R_{b,lim} \exp \left(\frac{-\tau_e}{M_i} \right)^{\alpha} \qquad (4)$$

where $R_{b,lim}$ is the potential final strength [MPa], τ_e is a characteristic time constant [$hours$], α is a curve parameter [$dimensionless$] and M_i is the maturity of concrete, defined by:

$$M_i = \int_0^{\tau_i} H(t) \, d\tau \qquad (5)$$

where t is the temperature [$°C$] and H is a temperature function:

$$H(t) = \exp \left(\left(\frac{1}{293} - \frac{1}{273+t} \right) \frac{E}{R} \right) \qquad (6)$$

where R is the gas constant ($R = 8.314 J/mole°C$) and E is a characteristic activation energy [$J/mole$]. For portland cement:

$$E = 33500 \qquad for \ t \geq 20°C$$

$$E = 33500 + 1470(20 - t) \\ for \ t \leq 20°C \qquad (7)$$

3 RELIABILITY CONCEPT

For a reliability analysis of concrete strength, we define a function $g(\mathbf{X})$ for each point in time i:

$$g_i(\mathbf{X}) = R_{b,i}(\mathbf{X}) - R_{b,req} \qquad (8)$$

where $\mathbf{X} = X_1, X_2, \ldots, X_m$ is a vector of m basic random variables involved in calculation of the strength according to the formula (3) or (4). $R_{b,req}$ is the required strength of concrete. We can now calculate the theoretical probability:

$$p_{f,i} = p(g_i(\mathbf{X}) \leq 0) \qquad (9)$$

which represents the probability that the required strength $R_{b,req}$ will not be reached at time i. Calculating these failure probabilities for a set of $R_{b,req}$ represents the calculation of the cumulative distribution function of concrete strength:

$$F_i(R_{b,i}) = p(R_{b,i}(\mathbf{X}) \leq R_{b,req}) \qquad (10)$$

Several well-known methods of structural reliability theory may be used for solving (9) and (10).

The vector of basic random variables $\mathbf{X}$ consists of physical quantities that describe the time-temperature history of the production process. The description and statistical parameters of these variables are presented in the following.

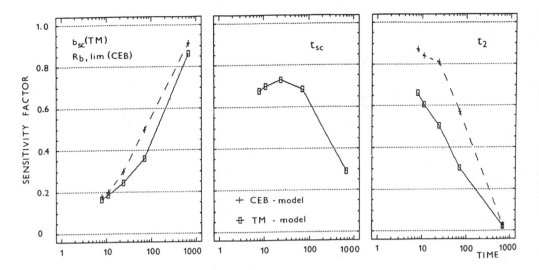

Fig. 3 Sensitivity factors vs. time

Table 2. Estimations of statistical parameters of concrete strength

Model	Point i	Mean [MPa]	St. dev. [MPa]	Coeff. of variation	Probability distribution
TM	1	11.9	3.7	0.31	Weibull
CEB		12.2	3.3	0.27	normal
TM	2	13.6	3.6	0.26	Pearson III
CEB		13.1	3.2	0.24	Pearson III
TM	3	15.8	3.4	0.22	Pearson III
CEB		14.7	2.7	0.19	Weibull
TM	4	20.7	3.1	0.15	Pearson III
CEB		19.2	2.4	0.12	Weibull
TM	5	30.1	2.3	0.08	lognormal
CEB		29.9	2.2	0.07	lognormal

4 NUMERICAL EXAMPLE

Basic variables considered are listed in Table 1. Random variables $(X_1, X_2, \ldots, X_{16})$ are statistically described by the mean, standard deviation and the type of probability distribution. These parameters were established from measurements or from the literature. Some of them had to be estimated intuitively by engineering judgment. Statistical correlation between X_5 and X_8 ($r = 0.9$ for $i = 2, 3, 4$ and 5) and between X_7 and X_{13} ($r = 0.7$ for $i = 3, 4$ and 5) is considered (where r is the correlation coefficient). The assumed relationship between time and temperature is shown in Fig. 1.

Two methods were used for the reliability assessment utilizing both the TM and CEB models.

Firstly, numerical calculations of the probabili-

ty of failure were performed by the multipurpose software package *ISPUD* (Bourgund and Bucher 1986). The mode of the program *"Adaptive Sampling"* (ADSAP), (Bucher et all. 1989) appeared as the most efficient one for the limit state functions in this example. 5000 simulations were used, which was enough to reduce the statistical error of estimation of the probabilities bellow 5%. The sensitivity analysis was also performed using sensitivity factors provided by *ISPUD* - the mode *"Importance Sampling Procedure and Design Point Calculation"*, with $R_{b,req}$ equal to 5% fractile.

Secondly, an alternative approach in which theoretical models of distribution functions for concrete strength are selected based on numerical simulation by *"Latin Hypercube Sampling"* (LHS) and a curve fitting approach (Novak and Kijawatworawet 1990, Novak and Teply 1991) was also used. For the selection of the most suitable model, the theory of comparison tests of distribution functions was applied. In this second approach 50 simulations were performed for each i by using software package *SAMPLE* (Teply and Novak 1989).

Some results are shown and compared in Figs. 2 and 3. Estimations of statistical parameters with the most suitable theoretical models of the concrete strength obtained by LHS are arranged in Table 2.

5 CONCLUSIONS

- Probabilistic models of concrete strength affected by heating process can be evalua-

ted by standard reliability methods.

- The probability features of both the TM and CEB models are of similar nature; it was confirmed that these two models provide similar results also in reliability analysis.

- The CEB model shows lower standard deviation of concrete strength in comparison with TM model at all points in time - see Table 2.

- The scatter of concrete strength assessed by reliability methods decreases with time for both models.

- The LHS approach with curve fitting has a decreasing accuracy for small probabilities (Novak and Kijawatworawet 1990). The tail of distribution function with small probabilities is more sensitive to the reliability method and to the deterministic model used. This fact was confirmed also in this example - see Fig. 2. However, in estimating statistical parameters and theoretical model of probability distribution the LHS seems to be satisfactory and only 50 simulations are sufficient.

- The basic random variables with dominating influence on p_f expressed by sensitivity factors are: t_2 - the maximal temperature in the heating process, t_{SC} - the temperature for strength increase of concrete in stable conditions (TM), b_{SC} (TM) and $R_{b,lim}$ (CEB) - the potential final strengths.

- The sensitivity factors of t_2 and t_{SC} decrease with time; factors of b_{SC} and $R_{b,lim}$ obviously increase - see Fig. 3.

The authors believe that the presented approach can serve for numerical prediction of concrete strength as a basis for optimization of energy consumption in heating process.

ACKNOWLEDGEMENT

The authors would like to express the most sincere thanks to Prof. G. I. Schuëller, the head of Institute of Engineering Mechanics, Innsbruck University, Austria, for granting the possibility to use software package *ISPUD* for numerical calculations presented. This research was partially supported by internal grants of Technical University of Brno, No. 6 and No. 54-55/93-D, which is gratefully thanked.

REFERENCES

Bourgund, U. and Bucher, C. G. 1986. *A code for importance sampling procedure using design points - ISPUD - A user's manual.* Institute of Engineering Mechanics, University of Innsbruck, Austria, Rep. No. 8/86.

Bucher, C. G., Nienstedt, J. and Ouypornprasert, W. 1989. *Adaptive strategies in ISPUD v. 3.0.* Institute of Engineering Mechanics, University of Innsbruck, Austria, Rep. No. 25/89.

CEB Bulletin D'Information - Durable Concrete Structures, CEB Design Guide 1989. No. 182, App. 1.

Czechoslovak Code ČSN 73 1311 1987. (in Czech).

Czechoslovak Code ČSN 73 2400b 1988. (in Czech).

Madsen, H. O., Krenk, S. and Lind, N. C. 1986. *Methods of structural safety.* New Yersey, USA: Prentice - Hall, Inc., Englewood Cliffs.

Novak, D. and Kijawatworawet, W. 1990. *A comparison of accurate advanced simulation methods and Latin Hypercube Sampling method with approximate curve fitting to solve reliability problems.* Institute of Engineering Mechanics, University of Innsbruck, Austria, Rep. No. 34/90.

Novak, D. and Teply, B. 1991. Estimation of Structural Failure Probability. *Proc. Europ. Conf. on New Advances in Comput. Struct. Mech.:* 237-244. Giens, France.

Teply, B. and Novak, D. 1989. Consequence of uncertainty of input data on engineering software reliability. *Software for Engineering Workstations 5:* 33-34.

Structural Safety & Reliability, Schuëller, Shinozuka & Yao (eds) © 1994 Balkema, Rotterdam, ISBN 90 5410 357 4

System reliability computation under progressive damage

S. Mahadevan & T. A. Cruse
Vanderbilt University, Nashville, Tenn., USA

ABSTRACT: A computational technique to model the effect of progressive damage in system reliability analysis is reported in this paper. The proposed method is useful for continuum structures characterized by interactive and synergistic failure modes, some of which are progressive in nature and grow in the continuum (e.g., crack growth, yielding etc.). The system critical limit states are approximated through a sequence of linear segments corresponding to various levels of progressive damage, leading to overall system reliability computation. The computational strategies to accurately and efficiently construct such segments are described and illustrated by application to a rotor system.

1 INTRODUCTION

This paper presents a method developed for system reliability and risk assessment of propulsion structures, as part of the PSAM (Probabilistic Structural Analysis Methods) project sponsored by National Aeronautics and Space Administration (NASA) Lewis Research Center. In this project, probabilistic methods have been developed for the analysis and design of space shuttle propulsion systems that operate in severe and highly uncertain environments. Perturbation-based probabilistic finite element analysis and fast probability integration have been combined to develop second-order reliability estimates and closed-form approximations to the limit states, and have already been validated for reliability estimation of individual failure modes. The recently developed system-level analysis methodology addresses the following concerns:

• The reliability of each structural component under its expected probabilistic loading and usage conditions considering all failure modes and their interactions;

• The effect of the unreliability of one component on the structural response and reliability of another component;

• The effect of the unreliability of one component on the performance, cost, and availability of the system; and,

• Development of efficient system certification strategies through integration of reliability analysis and test data.

Traditionally, system reliability computation involves modeling system failure as a network of series and parallel combinations of individual failure mode probabilities — an approach that is best suited for systems with discrete components. For mechanical and structural components used in propulsion systems, the aforementioned computations need to incorporate the progressive and synergistic nature of various damage and degradation processes, which are distributed over the continuum. Therefore, this paper proposes a new method to systematically account for progressive damage in system reliability.

2 PROGRESSIVE DAMAGE — COMPUTATIONAL ISSUES

Structural system reliability analysis techniques have largely focussed on structures with discrete components, such as trusses, frames, or composite laminates. The computational strategies

pursued in these methods have been (i) use of pre-defined plastic collapse mechanisms (in the case of frames), or (ii) load redistribution and reanalysis of the structure after each member failure to identify dominant failure sequences. The mechanism method is uniquely suited to framed structures; therefore, the method of reanalysis is pursued in this paper for continuum structures. In the reanalysis of framed structures, the partial failure condition is modeled either by removing the failed member from the structural system (for brittle failure), or by imposing a plastic hinge and a load or moment equal to the full capacity of the member (for ductile failure) (e.g., Karamchandani 1987, Thoft-Christensen and Murotsu 1986). Semi-brittle failure has also been modeled.

However, in the case of continuum structures which are discretized into a number of two- and three-dimensional finite elements, it is not possible to remove a finite element on failure or impose a plastic hinge and an artificial load/moment for reanalysis as in framed structures. Therefore a different strategy is necessary to simulate partial failure during the application of reanalysis to continuum structures.

In gas turbine engine structures, some of the failure modes of interest are related to burst, low-cycle and high-cycle fatigue and fracture, creep rupture, buckling etc. Many of these modes are progressive in nature and are distributed within the continuum. Also, certain modes (such as creep and LCF) may become synergistic at certain response levels. Therefore, system reliability analysis of such structures has to correctly account for both the physical and statistical correlations between interacting failure modes, the effect of strength degradation due to progressive damage, and the growth of damage in the continuum.

3 PROPOSED METHOD

The proposed method separates the failure modes into system failure (critical) and progressive damage (non-critical) modes. Examples of system failure modes can be burst, instability, etc. When these modes occur, the entire structural system fails catastrophically. The limit states for these modes can be formulated in a capacity vs. demand format as in most structural reliability problems. Examples of progressive damage modes may be yielding, crack growth etc. In these cases, the failure is not catastrophic in the initial stages, and it progresses through the continuum, degrading the load carrying capacity of the structure. This effect on the probability of a system critical event is computed as below.

Consider Figure 1, which shows a system failure limit state and several levels of damage corresponding to a progressive damage mode. If the progressive damage mode has any effect on the system failure mode, then the different damage level limit states should intersect the system limit state. This implies that the system failure could occur at different levels of damage, and the intersections indicate the values of the random variables that correspond to the specified levels of damage and the occurrence of that system limit state. In practical structures with a large number of random variables, neither the system limit state nor the progressive damage limit states are known as closed-form expressions. The system critical limit state can only be constructed through reanalysis of the structure after each level of damage, such that the modification in structural behavior due to progressive damage is correctly taken into account.

Since the accurate construction of the non-linear critical limit state is obviously time-consuming, an efficient strategy is to construct a set of linear approximations at significant levels of progressive damage. With such a polyhedral envelope, the failure probability is easily computed through the union of failure regions defined by the different linear segments. These linear segments are the tangents of the system limit state at its intersections with various levels of progressive damage, as shown in Figure 1. Therefore the intersections have to be determined. For multiple random variables, the intersection is a surface — many combinations of random variables correspond to the imposed level of progressive damage and system failure. The minimum distance point on the intersection surface is therefore used to obtain the system sensitivity to progressive damage. This point may be referred to as the joint MPP of the two limit states. The joint MPP estimation requires a nonlinear programming algorithm in which the objective is to minimize the distance subject

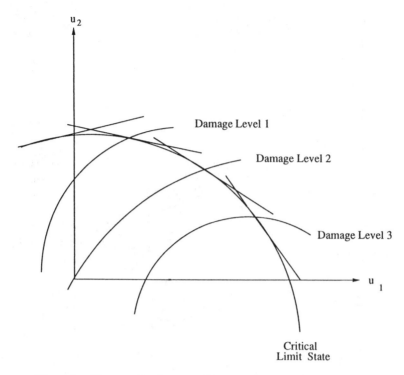

Figure 1. Progressive damage effect on system critical limit state.

to two limit state constraints. The Rackwitz-Fiessler algorithm is extended to the case of two limit states to achieve this objective, as illustrated in Figure 2.

The proposed method uses the first-order intersection of two limit states in the space of standard normal variables as the initial guess for the joint MPP. The two limit states are then evaluated at this point. If the two limit states are indeed linear, then they would be exactly equal to zero at this point; otherwise, there will be an error, since the intersection surface is different. Then, based on the slopes of the linear approximations in the first step, the two linear limit states are moved such that their values again become zero, and the new intersection surface and its minimum distance point are determined. The above procedure is repeated until convergence, i.e., until the limit states at the intersection point become close to zero. This results in a very close approximation to the actual joint failure region, without the need to find second-order derivatives. The resulting computational

benefit in the case of large structures is obvious, where the computation of second-order derivatives can be very expensive.

The overall system failure probability is then computed through the union of all the system failure events. Second-order bounds are used on the probability of the union. The above procedure also computes the sensitivity of system failure probability to the individual random variables, and to the different levels of damage (Cruse et al., 1992).

4 NUMERICAL EXAMPLE

The proposed method is illustrated with the help of a numerical example, considering a rotor system with two disks, an interference-fit spacer ring, and dummy blade masses (Figure 3). Yielding of the ring is used as a progressive non-critical damage mode, for the sake of illustration. The critical failure events are: bursting of a disk, LCF fracture at the bore, LCF fracture

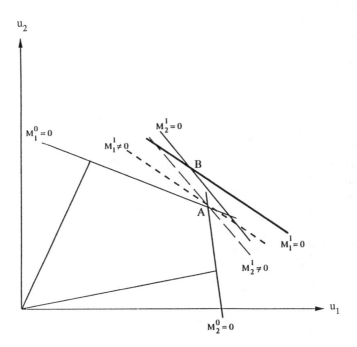

Figure 2.   Determination of joint MPP.

at the rim, and HCF of a disk caused by excessive vibration when the spacer ring disengages from a disk.

The rotor operating environment is modeled with a random rotational velocity and random temperatures. The interference fit between the rotor and the spacer is taken to be random; positive interference is taken to be needed for modal vibration of the assembly, rather than the disk alone. The parameters of the random variables are taken to be those given in Table 1.

The burst condition is assumed to occur when the average tangential stress exceeds a shape factor times the material ultimate strength. Crack initiation life (N_{LCF}) is taken as a simple power law model, given as

$$\Delta N_{LCF} = A \, \Delta \varepsilon_{-b} \tag{1}$$

The parameters A, b are defined in Table 1, where it is seen that b is taken to be deterministic. Crack propagation N_{FM} is taken to follow a simple Paris law relation

$$\Delta N_{FM} = C \, \Delta K_n \tag{2}$$

where C, n are also defined in Table 1, and where n is taken to be deterministic. Total fracture mechanics life is computed for two locations differently. In the bore region of the disk, the fracture mechanics life is the life from an initial defect size a_i, while in the rim region, the crack is initiated to a size of 0.762 mm. according to the crack initiation model given in Equation 1. Finally, the deterministic target life for both fracture mechanics limits is taken to be 10,000 cycles.

The high cycle fatigue mode in the example is taken to occur when any system natural frequency falls within a specified range of a fixed cyclic frequency. This very simple model takes the structure to have negligible damping, and is not a true engine-order resonance condition. The purpose here is to illustrate the type of event that follows a binary state for the structure — an engaged spacer which acts to stiffen the system, and a disengaged spacer.

The system reliability analysis is performed by using the algorithm formulated above. A coarse finite element mesh with 23 axisymmetric elements. In addition, 8 elements with very small stiffness are used to represent the mass of the blades. Each critical limit state is

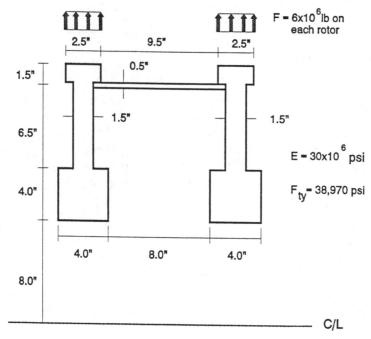

Figure 3. Simplified rotor system model.

approximated by a set of linear segments corresponding to different levels of non-critical damage (yielding of the ring, in this case). The resulting reliability estimates may be referred to as second-level estimates as compared to first-level estimates which are obtained at step 0 of the above algorithm without considering progressive damage effects. The probabilities of system critical failure events are:

[Rotor Burst:] First-level 0.00089; Second-level 0.00262.

[LCF Fracture at bore:] First-level 0.00165; Second-level 0.00729.

[LCF Fracture at rim:] First-level 0.0000017; Second-level 0.00677.

[HCF of rotor:] First-level 0.0; Second-level 0.00747.

The second-order bounds on the union of the above events are:

$$0.016992 \leq p_f \leq 0.017039 \tag{3}$$

It is seen that the second-order bounds are very narrow, implying that the joint probabilities of three or more events are negligible for this problem. Also, the search algorithm for expansion point was found to give very close linear approximations to the critical limit states at the first intersection itself, for this particular problem. The number of iterations will obviously depend on the amount of nonlinearity caused by progressive damage.

5 CONCLUSION

An iterative method for sequential probabilistic reanalysis and accurate determination of the failure probability of a structural system has been proposed in this paper. The system performance violations are separated into critical system failure events and noncritical system degradation (progressive damage) events. A system critical limit state is approximated by a series of linear segments corresponding to different levels of any one progressive damage mode. The failure probability corresponding to that system critical limit state is computed through the union of different failure regions defined by the linear segments. This procedure is repeated to construct the limit state for each critical system failure event for different modes

Table 1. Random variable definitions.

Random variable definitions

Variable	Mean	Std Dev	Distribution
Rotor Young's modulus	20 msi	0.40 msi	Lognormal
Ring Young's modulus	17 msi	0.34 msi	Lognormal
Rotor unit weight	0.0008 lb/cu.in.	0.000016 lb/cu.in.	Lognormal
Ring unit weight	0.0008 lb/cu.in.	0.000016 lb/cu.in.	Lognormal
Ultimate tensile strength	96 ksi	9.6 ksi	Lognormal
Yield strength	140 ksi	9.8 ksi	Lognormal
Fatigue life coefficient A	2.2E10	2.2E9	Lognormal
Fatigue life exponent b	4	0.0 in	Deterministic
Stress intensity factor K	50 ksi $\sqrt{in}$	5 ksi $\sqrt{in}$	Lognormal
Initial crack size a_i	0.002	0.0002	Weibull
Crack growth rate coefficient C	1.45E-9	1.45E-10	Lognormal
Crack growth exponent n	2.61	0.0	Deterministic
Interference fit δ_{INT}	0.0	0.05 in	Normal
Temperature difference ΔT	0° F	53° F	Type I
Rotor speed Ω	628 rad/sec	49 rad/sec	Type I

Note: For conversion to S.I. units, use 1 kN = 0.22 kip, 1 m = 3.24 ft.

of progressive damage, to calculate the overall system failure probability. The proposed method allows the engineer to impose any level of damage corresponding to a system degradation mode, and calculate its effect on a system critical limit state. Such sensitivity information identifies criteria for risk management and certification of large systems.

6 ACKNOWLEDGEMENT

This research was performed partly under NASA Lewis Research Center contract NAS3-24389 (Program Manager: Dr. C. C. Chamis) through subcontract to Southwest Research Institute (Project Manager: Dr. Y.T. Wu).

REFERENCES

Cruse, T.A., Burnside, O.H., Wu, Y.-T., Polch, E.Z., and Dias, J.B. 1988. Probabilistic structural analysis methods for select space propulsion system structural components (PSAM). *Computers and Structures*, 29:891-901.

Cruse, T.A., Huang, Q., Mehta, S., and Mahadevan, S. 1992. System reliability and risk assessment. *Proc. 33rd AIAA/ASME/ASCE/AHS/ASC Conference on Structures, Structural Dynamics and Materials:* 424-431. Dallas, Texas.

Karamchandani, A. 1987. Structural system reliability analysis methods. *Report No. 83.* John A. Blume Earthquake Engineering Center, Stanford University.

Mahadevan, S., Cruse, T.A., Huang, Q., and Mehta, S. 1992. Structural reanalysis for system reliability computation. *Proc. ASME*

Winter Annual Meeting: (AD-)28;169-187. (Reliability Technology 1992), Anaheim, California.

Thoft-Christensen, P., and Murotsu, Y. 1986. Application of structural systems reliability theory. Springer Verlag, Berlin.

Structural Safety & Reliability, Schuëller, Shinozuka & Yao (eds) © 1994 Balkema, Rotterdam, ISBN 90 5410 357 4

Approximations and derivatives of probabilities in structural reliability and design

K. Marti

Universität der Bundeswehr München, Neubiberg, Germany

ABSTRACT: (Approximative) Differentiation formulas for probability functions can be found by applying one of the following two methods: I) Transform the multiple integral defining the probability function such that the transformed domain of integration becomes independent of the argument x; II) if (I) can not be applied directly, insert appropriate stochastic completion terms in the response function up to Method (I) can be applied. Based on the given differentiation formulas for probability functions, error estimations for aporoximating probabilities by approximating (e.g. linearizing) the underlying response function may be derived.

1 INTRODUCTION

A very important tool in reliability based structural design and optimization are, see [1],[6],[9], probability functions of the type

$$P(x) := P(y_{1i}<(\le)y_i(a(\omega),x)<(\le)y_{2i},$$
$$i=1,\ldots,m), \quad x\in \mathbb{R}^n \qquad (1)$$

$$P_f(x) := P(\min_{1\le i\le m} g_i(a(\omega),x)<0), \quad x\in \mathbb{R}^n. \qquad (2)$$

Here, $x=(x_1,\ldots,x_n)'$ denotes the vector of design variables x_k, e.g. structural dimensions, degree of refinement of the material and/or manufacturing process, $a=a(\omega)$ is the ν-vector of random parameters, e.g. material coefficients (elastic moduli,...), tolerances, load parameters etc.

In (1), $y=y(a,x)=(y_1(a,x),\ldots,y_m(a,x))'$ is the vector of basic response quantities, e.g. certain displacement and/or stress variables, where y must fulfill the basic behavioral constraints

$$y_{1i}<(\le)y_i<(\le)y_{2i}, \quad i=1,\ldots,m \qquad (1.1)$$

with given upper and lower bounds $y_{1i}<y_{2i}, i=1,\ldots,m$. In (2), $g_i=g_i(a,x)$, $i=1,\ldots,m$, are the limit state functions of certain elements, at certain points of the structure, where the failure domain is described by

$$\bigcup_{i=1}^{m} \{a\in \mathbb{R}^\nu: g_i(a,x)<0\}. \qquad (2.1)$$

Since the failure probabilities P_f can be described by means of probability functions of the type (1), in the following we concentrate on these functions (1). In structural safety and reliability, in structural sensitivity and in the optimal structural design, the following mathematical problems occur:

A) Find sufficiently accurate approximations $\hat{P}(x)$ of the probability $P(x)$; determine estimates of the approximation error

$$\epsilon=\epsilon(x) := \hat{P}(x)-P(x).$$

While there are already several very efficient techniques for the approximative calculation of $P(x)$, see [1-4], the mathematical estimation of $\epsilon(x)$ is a still open problem, cf.[5].

B) Find efficient formulas for the (approximative) calculation of the first and higher order partial derivatives $\nabla P(x),\nabla^k P(x)$ of the probability function $P(x)$. Looking for first order differentiation formulas for probability functions in the literature, see [10],[11], one observes that their mathematical derivation is incomplete, and - as a major disadvantage - the numerical evaluation in structural reliability and design is very difficult.

Since we also need derivatives of P(x) in the estimation of the approximation error $\epsilon(x)$, based on [6], we start with the presentation of differentiation formulas for probability functions of the type (1).

2 DIFFERENTIATION FORMULAS FOR PROBABILITY FUNCTIONS

As can be seen from a FE-analysis of the underlying mechanical structure, the basic response variables $y_i = y_i(a,x)$, $i=1,\ldots,m$, of the structure have the following functional form:

$$y_i(a,x) = \eta_i(b_o, Q^{(1)}(x)b_1, \ldots, Q^{(r)}(x)b_r). \quad (3)$$

Here,

$$b_j = (a_k)_{k \in I_j} \text{ with } I_j \subset (1,\ldots,\nu),$$

$$j=0,1,\ldots,r, \quad (3.1)$$

denote certain $|I_j|$-subvectors of a, where

$$I_0 \cap I_j = \emptyset, \quad j=1,\ldots,r, \quad (3.2)$$

but $I_j, I_\ell, j, \ell \geq 1, j \neq \ell$, are not necessarily disjoint,

$Q^{(j)} = Q^{(j)}(x)$, $j=1,\ldots,r$, are positive definite $|I_j| \times |I_j|$ matrix functions on a

subset $D \subset \mathbb{R}^n$, $\quad (3.3)$

where $Q^j(x)$ is a very simple (e.g. a linear) function of the design vector x for most practical applications. Furthermore,

$$\eta_i = \eta_i(b_o, q^{(1)}, \ldots, q^{(r)}),$$

$$i=1,\ldots,m, \quad (3.4)$$

are given functions on $\mathbb{R}^{|I_0|} \times \mathbb{R}^{|I_1|} \times \ldots \times \mathbb{R}^{|I_r|}$ having important analytical properties such as continuity, differentiability.

Considering e.g. a planar truss, the vector of displacements u reads [8]

$$u = u(a,x) = \tilde{K}(a_{II},x)^{-1} p(a_I), \quad (4)$$

where a_I, a_{II} is a certain partition of a, and $p=p(a_I)$ denotes the load vector depending on random parameters $a_I = a_I(\omega)$. Moreover, the stiffness matrix $\tilde{K}$ can be represented [8] by

$$\tilde{K} = \tilde{K}(a_{II},x) = \sum_{k=1}^{\kappa} (q_k^R \tilde{K}_R^{(k)} + q_k^B \tilde{K}_B^{(k)}), \quad (4.1)$$

where $\tilde{K}_R^{(k)}, \tilde{K}_B^{(k)}$ are fixed matrices corresponding to the behavior of the k-th element of the structure as a rod, beam, resp., and

therefore

$$q_k^R = \frac{E_k A_k(x)}{L_k}, \quad q_k^B = \frac{E_k I_k(x)}{L_k},$$

$$k=1,\ldots,\kappa. \quad (4.2)$$

Obviously, $E_k, A_k = A_k(x)$, $I_k = I_k(x)$, L_k denotes the elastic modulus, the cross-sectional area, the second moment of area, the length, resp., of the k-te element of the truss. Consequently, u can be represented by

$$u = U(a_I, q^R, q^B) \quad (4.3)$$

where

$$q^R = Q^R(x)a_{II}, \quad q^B = Q^B(x)a_{II} \quad (4.3.1)$$

$$a_{II} := (E_1(\omega), \ldots, E_\kappa(\omega))' \quad (4.3.2)$$

$$Q^R(x) := (\frac{A_k(\omega)}{L_k} \delta_{\ell k}), \quad Q^B(x) := (\frac{I_k(x)}{L_k} \delta_{\ell k}). \quad (4.3.3)$$

If only pin-jointed trusses are considered, then (4.3) is reduced to

$$u = U(a_I, q^R), \quad q^k = Q^R(x)a_{II}. \quad (4.3)'$$

Note that this important case can be represented by (3) by setting r:=1, i.e.

$$y_i(a,x) = \eta_i(a_I, Q^{(1)}(x)a_{II}),$$
$$i=1,\ldots,m. \quad (5)$$

2.1 The case r=1

If formula (5) holds, and the random vector $a=(\begin{smallmatrix} a_I \\ a_{II} \end{smallmatrix})$ has the probability density

$$f(a) = f(a_I, a_{II}) = \psi(a_{II}|a_I)\varphi(a_I),$$

then by integral transformation $a_{II} := Q^{(1)}(x)^{-1}q^{(1)}$ we find

$$P(x) = P(y_{1i} < \eta_i(a_I(\omega),$$

$$Q^{(1)}(x)a_{II}(\omega)) < y_{2i}, \quad i=1,\ldots,m)$$

$$= \int \int_{y_{1i} < \eta_1(a_1, q^{(1)}) < y_{2i}, \quad i=1.\ldots,m}$$

$$\psi(Q^{(1)}(x)^{-1}q^{(1)}|a_I)$$

$$\times \varphi(a_I) \frac{dq^{(1)}}{detQ^{(1)}(x)} da_I. \quad (6)$$

Since the integration domain in (6) is now independent of the design vector x, by differentiation under the integrals we obtain - after the back-transformation $q^{(1)} := Q^{(1)}(x)a_{II}$ - the first **differentiation formula**

$$\frac{\partial P}{\partial x_t}(x) = - \int\int_{B(x)} (<\frac{\partial}{\partial x_t} Q^{(1)}(x), Q^{(1)}(x)^{-1}>$$

$$+ \left(\frac{\partial}{\partial x_t} Q^{(1)}(x) Q^{(1)}(x)^{-1} \frac{\nabla\psi(a_{II}|a_I)}{\psi(a_{II}|a_I)}\right)' a_{II})$$

$$\times \psi(a_{II}|a_I)\varphi(a_I)da_{II}da_I, \qquad (7)$$

where $<A,C> = \sum_{i,j=1}^{p} a_{ij} \cdot c_{ij}$ denotes the scalar product of $p \times p$ matrices $A=(a_{ij})$, $C=(c_{ij})$, $\nabla\psi(a_{II}|a_I)$ is the gradient of $\psi(a_{II}|a_I)$ with respect to a_{II}, x_t is the t-th component of x, and

$$B(x) = \{(a_I,a_{II}):$$
$$y_{1i} < \eta_i(a_I, Q^{(i)}(x)a_{II}) < y_{2i},$$
$$i=1,\ldots,m). \qquad (7.1)$$

2.2 The case r>1

The above transformation-based method can be generalized immediately to the case r>1, provided that, cf. (3.1),(3.2),

$$I_j \cap I_\ell = \emptyset \text{ for all } j \neq \ell, \ j,\ell=1,\ldots,r, \qquad (8)$$

which simply means that $b_1,\ldots,b_r$ are subvectors of $(a_k)_{k \notin I_0}$ containing **different** components of a.

If (8) does not hold, then approximating differentiation formulas for (1) can be obtained by the following generalized transformation-based method which was suggested first in [6]:

Step 1: The probability function $P(x)$ defined by (1) and (3) is approximated by

$$\tilde{P}(x) := P(y_{1i} < \eta_i(b_o(\omega),$$
$$Q^{(1)}(x)(b_1(\omega)+\delta_1(\omega)),\ldots,$$
$$Q^{(r)}(x)(b_r(\omega)+\delta_r(\omega))) < y_{2i},$$
$$1 \le i \le m) \qquad (9.1)$$

where the random $\nu_j(:=|I_j|)$-vectors $\delta_j(\omega)$, $j=1,\ldots,r$, e.g. zero mean normal distributed random vectors, are selected such that the derivative $\frac{\partial\tilde{P}}{\partial x_t}(x)$ can be obtained by using a similar integral transformation as described in Section 2.1 and in the beginning of this section.

Step 2: The derivative $\frac{\partial P}{\partial x_t}$ is obtained then by the limit process

$$\frac{\partial P}{\partial x_t}(x) := \lim_{\delta_j(\cdot) \to 0} \frac{\partial\tilde{P}}{\partial x_t}(x). \qquad (9.2)$$
$$1 \le j \le r$$

For simplicity of notation, this method (Step 1 and Step 2) is worked out here for

the important special case, see (4),

$$y_i(a,x) = \eta_i(a_I, Q^{(1)}(x)a_{II}, Q^{(2)}(x)a_{II}),$$
$$i=1,\ldots,m. \qquad (10)$$

According to (9.1) we replace now $P(x)$ by

$$\tilde{P}(x) := P(y_{1i} < \eta_i(a_I(\omega), Q^{(1)}(x)a_{II}(\omega),$$

$$Q^{(2)}(x)(a_{II}(\omega)+\delta(\omega))) < y_{2i},$$
$$1 \le i \le m), \qquad (11)$$

where $\delta=\delta(\omega)$ is a random ν_2-vector having stochastically independent $N(0,\sigma_k^2)$-distributed components $\delta_k(\omega)$, $k=1,\ldots,\nu_2$, which are also independent of $a(\omega)=(a_I(\omega), a_{II}(\omega))$. Defining $q^{(1)}:=Q^{(1)}(x)a_{II}$, $q^{(2)}:=(x)(a_{II}+\delta)$, hence, using the integral transformation

$$a_{II} := Q^{(1)}(x)^{-1}q^{(1)},$$

$$\delta := -Q^{(1)-1}q^{(1)}+Q^{(2)-1}q^{(2)}, \qquad (11.1)$$

corresponding to (6) we get

$$\tilde{P}(x) =$$

$$\int \int \int_{y_{1i} < \eta_i(a_I, q^{(1)}, q^{(2)}) < y_{2i}} \psi(Q^{(1)}(x)^{-1}q^{(1)}|a_I)$$
$$1 \le i \le m$$

$$\times \gamma(-Q^{(1)}(x)^{-1}q^{(1)}+Q^{(2)}(x)^{-1}q^{(2)})$$

$$\times \frac{dq^{(1)}}{\det Q^{(1)}(x)} \frac{dq^{(2)}}{\det Q^{(2)}(x)} da_I, \qquad (11.2)$$

where $\gamma=\gamma(\delta)$ denotes the density of $\delta(\omega)$. Since the transformed domain of integration is independent of x, by differentiation under the integrals and back-transformation, corresponding to (7) we have the following first result:

Theorem 2.1. Approximative differentiation formula. If the above assumptions hold, then

$$\frac{\partial\tilde{P}}{\partial x_t}(x) = -\int\int\int_{\tilde{B}(x)} \left\{ <\frac{\partial Q^{(1)}(x)}{\partial x_t}, Q^{(1)}(x)^{-1}> \right.$$

$$+ <\frac{\partial Q^{(2)}(x)}{\partial x_t}, Q^{(2)}(x)^{-1}> \qquad (12)$$

$$+ \frac{\nabla\gamma(\delta)'}{\gamma(\delta)}(-Q^{(1)}(x)^{-1}\frac{\partial Q^{(1)}}{\partial x_t}(x)a_{II}$$

$$+ Q^{(2)}(x)^{-1}\frac{\partial Q^{(2)}}{\partial x_t}(x)(a_{II}+\delta))$$

$$+ \frac{\nabla \psi(a_{II}|a_I)'}{\psi(a_{II}|a_I)} Q^{(1)}(x) - 1 \frac{\partial Q^{(1)}}{\partial x_t}(x) a_{II} \Big\}$$

$$\times \psi(a_{II}|a_I) \gamma(\delta) \varphi(a_I) da_{II} d\delta da_I,$$

where

$$\bar{B}(x) := \{ (a_I, a_{II}, \delta) : y_{1i} < \eta_i(a_I, Q^{(1)}(x) a_{II},$$

$$Q^{(2)}(x)(a_{II} + \delta)) < y_{21}, \quad i = 1, \ldots, m \}.$$

$$(12.1)$$

Remark 2.1

i) $\nabla \gamma(\delta) = -\gamma(\delta) \Sigma \delta$, where Σ denotes the diagonal matrix having the diagonal elements $1/\sigma_k^2$, $k = 1, \ldots, \nu_2$.

ii) Stochastic quasigradients, i.e. approximations of stochastic gradients, can be easily obtained from (12).

iii) The exact differentiation formula for $\frac{\partial P}{\partial x_t}(x)$ is obtained from (12) by the limit process $\sigma_k \longrightarrow 0$ for all $k = 1, \ldots, \nu_2$.

iv) Note that the **higher order partial derivatives** of $\bar{P}(x)$ can be obtained by the same procedure.

v) In the important special case

$$Q^{(1)}(x) := (q_{kk}^{(1)}(x_k) \delta_{\ell k}), \quad Q^{(2)}(x) :=$$

$$(q_{kk}^{(2)}(x_k) \delta_{\ell k}), \quad (13)$$

cf. (4.3.3), we find

$$\frac{\partial \bar{P}}{\partial x_t}(x) = - \iiint_{\bar{B}(x)} \Big\{ \frac{\dot{q}_{tt}^{(1)}(x_t)}{q_{tt}^{(1)}(x_t)} + \frac{\dot{q}_{tt}^{(2)}(x_t)}{q_{tt}^{(2)}(x_t)} + $$

$$(13.1)$$

$$+ \frac{\dot{\gamma}_t(\delta_t)}{\gamma_t(\delta_t)} (- \frac{\dot{q}_{tt}^{(1)}(x_t)}{q_{tt}^{(1)}(x_t)} a_{\nu_1 + t}$$

$$+ \frac{\dot{q}_{tt}^{(2)}(x_t)}{q_{tt}^{(2)}(x_t)} (a_{\nu_1 + t} + \delta_t))$$

$$+ \frac{\frac{\partial \psi}{\partial a_{\nu_1 + t}}(a_{II}|a_I)}{\psi(a_{II}|a_I)} \frac{\dot{q}_{tt}^{(1)}(x_t)}{q_{tt}^{(1)}(x_t)} a_{\nu_1 + t} \Big\}$$

$$\times \psi(a_{II}|a_I) \varphi(a_I) \gamma(\delta) da_{II} d\delta da_I.$$

2.2.1. The limit $\sigma_k \longrightarrow 0$, $k = 1, \ldots, \nu_2$

For sake of simplicity we consider here

only the case (13). Defining $B(x)$ by

$$B(x) := \{ (a_I, a_{II}) : y_{1i} < \eta_i(a_I, Q^{(1)}(x) a_{II},$$

$$Q^{(2)}(x) a_{II}) < y_{2i}, \quad i = 1, \ldots, m \},$$

we have the following differentiation formula:

Theorem 2.2. If $m = 1$ and $B(x)$ is given by (14.1), then

$$\frac{\partial P}{\partial x_t}(x) = - \iint_{B(x)} \frac{\partial}{\partial a_{\nu_1 + t}} \Big(\psi(a_{II}|a_I) \right.$$

$$\times \frac{\frac{\partial \eta}{\partial x_t}(a_I, Q^{(1)}(x) a_{II}, Q^2(x) a_{II})}{\frac{\partial \eta}{\partial a_{\nu_1 + t}}(a_I, Q^{(1)}(x) a_{II}, Q^{(2)}(x) a_{II})} \Big)$$

$$\times \varphi(a_I) da_{II} da_I. \quad (14)$$

Using the Gaussian divergence theorem for volume intgerals in $\mathbb{R}^\nu$, we may write $\frac{\partial P}{\partial x_t}$ also in the following form:

Theorem 2.3. If $m = 1$, then

$$\frac{\partial P}{\partial x_t}(x) = - \int_{B(x)} \text{div}_a (f(a) \frac{\partial \eta}{\partial x_t} \cdot \frac{\nabla_a \eta}{||\nabla_a \eta||^2}) da.$$

$$(15)$$

Remark 2.2
Using a further smoothing process, cf. (9), the case $m > 1$ can be reduced to the case $m = 1$.

3 ERROR ESTIMATION

In structural reliability theory approximations $\hat{P}(x)$ of

$$P(x) = P(y_1 < y(a(\omega), x) < y_2),$$

where $y(a, x) := (y_1(a, x), \ldots, y_m(a, x))'$,

$y_1 := (y_{11}, \ldots, y_{1m})', y_2 := (y_{21}, \ldots, y_{2m})'$, see (1), are defined [1-4] mostly by

$$\hat{P}(x) := P(y_1 < y(a(\omega), x) < y_2), \quad (16)$$

where $\hat{y} = \hat{y}(a, x)$ is an appropriate approximation, e.g. a linearization, of $y(a, x)$ with respect to a.

For a given, **fixed** design vector x, we suppose that $y(a, x)$ can be represented by the parametric model

$$y(a, x) = Y(\theta_0, a) := \hat{y}(a, x) + R(\theta_0, a),$$

$$\theta \in \Theta (\subset \mathbb{R}^q, 1 \leq q \leq \infty), \quad (17)$$

with the remainder

$$R(\theta,a) = R(\theta,a;x) :=$$

$$\sum_{j=1}^{q} \theta_j R_j(a), R_j(a)=R_j(a;x), \quad (17.1)$$

where $R_j=R_j(a;x)$, $j=1,\ldots,q$, are given functions, e.g. certain polynominals with respect to the variables $a_1,\ldots,a_\nu$, and $\theta_j=\theta_j(x)$, $j=1,\ldots,q$, are certain real parameters, e.g. partial derivatives of $y(a,x)$ with respect to $a_1,\ldots,a_\nu$ at $\bar{a}:=Ea(\omega)$, see [7]. Defining now for a given, fixed vector x

$$W(\theta) := P(y_1<Y(\theta,a(\omega))<y_2), \quad (17.2)$$

the error ϵ can be represented by

$$\epsilon = W(\theta_o)-W(0) = \nabla W(\lambda\theta_o)\cdot\theta_o \quad \text{(with } 0<\lambda<1)$$

$$= \nabla W(0)\cdot\theta_o+\ldots \ .$$

By the differentiation formula (15), we find - replacing there x_t by θ_j, and using that $R(\theta,a)$ is linear in θ - the following error representation:

Theorem 3.1. If $m=1$ and the above assumptions hold, then

$$\epsilon = - \int_{y_1<Y(\lambda\theta_o,a)<y_2} \mathrm{div}_a(f(a)R(\theta_o,a)$$

$$\times \frac{\nabla_a Y(\lambda\theta_o,a)}{||\nabla_a Y(\lambda\theta_o,a)||^2})da$$

$$= - \int_{y_1<\hat{y}(a,x)<y_2} \mathrm{div}_a(f(a)R(\theta_o,a)$$

$$\times \frac{\nabla_a\hat{y}(a,x)}{||\nabla_a\hat{y}(a,x)||^2})da +\ldots \ . \quad (18)$$

Note. Upper bounds for $|\epsilon|$ can be obtained now from (18).

REFERENCES

[1] Abdo, T., Rackwitz, R.: Reliability of Uncertain Structural Systems. Finite Elements in Engineering Applications, p. 161-176. Stuttgart, INTES GmbH 1990

[2] Bjerager, P.: On Computation Methods for Structural Reliability Analysis. Structural Safety 9, 79-96 (1990)

[3] Breitung, K.: Asymptotic Approximation for Multinomial Integrals. ASCE J. of the Eng. Mechanics Division 110, 357-367 (1984)

[4] Breitung, K.: Asymptotische Approximation für Wahrscheinlichkeitsintegrale. Habilitationsschrift, Fakultät für Philosophie, Wissenschaftstheorie und Statistik der Universität München, 1990

[5] Breitung, K.: Parameter Sensitivity of Failure Probabilities. In: A. Der Kiureghian, P. Thoft-Christensen (eds.): Reliability and Optimization of Structural Systems '90. Lecture Notes in Engineering, Vol. 61, 43-51 (1990)

[6] Marti, K.: Stochastic Optimization Methods in Structural Mechanics. ZAMM 70, T742-T745 (1990)

[7] Marti, K.: Approximations and Derivatives in Structural Design. ZAMM 72, T575-T578 (1992)

[8] McGuire, W., Gallagher, R.H.: Matrix Structural Analysis. New York-London, Wiley 1979

[9] Schueller, G.I.: A critical appraisal of methods to determine failure probabilities. J. Structural Safety 4, No. 4, 293-309 (1987)

[10] Streeter, V.L., Wylie, E.B.: Fluid Mechanics. New York, McGraw Hill 1979

[11] Uryas'ev. St.: A differentiation formula for integrals over sets given by inclusion. Numer. Funct. Anal. and Optimiz. 10 (7.u.8), 827-841 (1989)

Structural Safety & Reliability, Schuëller, Shinozuka & Yao (eds) © 1994 Balkema, Rotterdam, ISBN 90 5410 357 4

System reliability of redundant structures using response functions

Fred Moses
Department of Civil Engineering, University of Pittsburgh, Pa., USA

Nikhil C. Khedekar
Department of Civil Engineering, Case Western Reserve University, Cleveland, Ohio, USA

Michel Ghosn
Department of Civil Engineering, The City College of New York, N.Y., USA

ABSTRACT: Considerable interest exists in defining and calibrating the levels of redundancy for both new and existing designs. Such applications represent a culmination of recent theoretical studies aimed at analyzing system reliability for bridges, buildings, offshore platforms and other structures. The authors are implementing system reliability measures as part of the development of new highway bridge design and evaluation codes in the United States. The paper presents the formulation of the "Response Function Approach" to find the System Safety Index (β_{sys}) for highway bridge superstructures. The system reliability indices are formulated from the output of nonlinear finite element models through the use of response functions. Practical redundancy implementation allows checking both serviceability and ultimate capacity limit states for either overloads or accident scenarios. Two examples illustrating the procedure are presented.

1 INTRODUCTION

Redundancy in structural engineering can be defined as the capability of the structure to carry additional load after the failure of one or more of its components. A new generation of highway bridge codes is making an effort to consider redundancy in bridge superstructures. With this in mind, the authors are implementing system reliability measures as a part of highway bridge design and evaluation codes in the United States.

In all redundant structures, load is transmitted in more than one path and there can also be multiple system failure modes. If discrete-ductile or discrete-brittle member behavior is assumed and the functional relationship between input and response quantities is known in a closed form, then the expression for failure modes can be easily identified (Moses 1982, Moses & Rashedi 1983). But this may not be possible in the case

of complex structures, especially with nonlinear (material and/or geometry) behavior which may involve the unloading of some components.

One common approach to this problem is *numerical simulation* (Verma, Fu & Moses 1989). Each experiment is a solution of a deterministic problem, in which structural properties and loading history are realizations of random variables. Well known methods of numerical simulation are 1) Monte-Carlo simulation, 2) Importance Sampling and 3) Response Surface technique. Because high reliability levels are required in structural engineering, very low values of probability of failure have to be investigated. This requires careful investigation of the tails of the PDF's of random variables. With *crude* Monte-Carlo simulation, such tails are defined with good accuracy if the number of trials is on the order of 10^6 to 10^9. Although importance sampling may require fewer trials compared to Monte-

Carlo simulation $(10^2 - 10^3)$, this number is still very high when we take into consideration the computation time involved. One way to circumvent this problem is to introduce experiments to represent responses likely to be obtained by simulation (Augusti,Baratta & Casciati 1984). These values are then used as interpolation points for the construction of the "response surface". This approach is known as "Response Function Method". The set of responses is then investigated to obtain the required probability of failure or safety index.

The aim of the present study is to calculate the System Safety Index (β_{sys}) for highway bridge superstructures using the Response Function Approach. Two limit states are considered, they are : 1) Serviceability limit state and, 2) Ultimate capacity (or collapse) limit state. Behavior of the structure at these two limit states is checked with emphasis on two primary events : 1) *Overload* analysis during the design and evaluation process and, 2) Accident evaluation to determine the available redundancy of structures should there be a sudden *failure* (due to an accident or fatigue failure) of one of its members. The component behavior is nonlinear and may include unloading following the attainment of a member's strength limit state condition. The nonlinear behavior is incorporated into a finite element analysis using experimental moment - rotation curves for positive as well as negative (compact and non-compact) moment sections (Schilling 1989).

This paper analyzes first a one girder system to get some insight into the behavior of bridge superstructures. A description of the Response Function Method is then given. The complete procedure is finally presented with the aid of two examples of multigirder bridges.

2 OVERVIEW OF ONE GIRDER SYSTEM

The behavior of simple span girders is characterized by their experimental moment -

rotation curve (fig.1A). The nominal load applied is one wheel line of AASHTO's HS20 truck model and is placed to maximize the moment in the girder. The load magnitude is gradually increased until the girder reaches its ultimate moment capacity (M_p). Fig.2 shows the simple span girder loading curve in terms of the load factor that multiplies the applied HS20 loading, versus the maximum displacement. Dead load deflection is not included. The curve (fig.2) is straight until the maximum moment in the girder reaches the yield value. The girder stiffness is then modified and the analysis is carried out with increased load until the ultimate moment capacity is reached. The girder cannot carry any *additional* load and the load - deflection curve is horizontal. The dotted vertical line on the curve indicates that a *predefined* (span length / 300) serviceability limit is reached.

The behavior of a continuous span girder is characterized by the positive moment - rotation curve (fig.1A) for the midspan region and negative moment - rotation curve for either a compact or a non-compact section (fig.1 B&C) for the region near interior support. The maximum load PDF depends on the load pattern, hence, load applied to one as well as two spans is checked. For a *non compact* girder, unloading starts when the maximum negative moment reaches the yield moment (M_y) and instability can occur (fig.3 B&D). In the case of a *compact* girder, when the maximum negative moment reaches the ultimate moment (M_p), due to sufficient rotation capacity of the negative moment section, load redistribution occurs. This type of failure is ductile and occurs with the formation of a mechanism when the positive moment section also reaches its ultimate moment capacity (M_p).

In a multigirder bridge superstructure, in addition to the longitudinal redistribution, the load is also redistributed transversely through the deck slab and transverse members. The additional load carrying capacity (or

redundancy) of the structure comes from this transverse redistribution.

3 RESPONSE FUNCTION APPROACH

This section formulates system response functions for both serviceability and ultimate capacity limit states. Bridge response depends on several parameters such as member ductility, bridge geometry, and loading pattern. The system behavior is nonlinear and the expression for failure function g( ) is found using the Response Function Approach.

The first step is to identify the basic random variables X_i forming the failure function $g(X_i)$. These are usually member resistances (in form of yield stress or moment capacity) and loading (dead load, live load etc.). Select some trial points x_i (realizations of X_i). These x_i's are deterministic. Use these trial points to get the output point y which is the load response of the structure conditional upon the realization x_i. Each set of x_i's produces one output y. The process is repeated to cover the range of values likely to be taken by X_i. A multivariable regression fit is then made to these points to get a response surface that gives the relationship between the load output y and the input variables X_i. This can be represented by the equation:

$$y = f(X_i) \tag{1}$$

If we define g( ) as,

$$g( ) = f(X_i) - Y \tag{2}$$

where, $f(X_i)$ is the resistance term and Y is the load effect term, then, g( ) is the failure function such that if $g( ) \le 0$, failure occurs. Y depends on the load pattern and exposure period. In our problem, response functions for both serviceability and ultimate capacity limit states are constructed from a nonlinear finite element analysis of the bridge. The basic variables are member resistances and dead load. The output is the load factors needed to reach both limit states. Several different load patterns as well as both overload and accident scenario to bridge member are considered. The probability distribution of the load is based on recent field measurements of truck loadings on highway bridges (Nowak & Hong 1991). After applying a multiple linear regression fit, eq.(2) is a failure function needed in conventional structural reliability analysis. The safety index (ß) is then calculated using the Hasofer-Lind transformation and the statistical properties of the basic variables (nominal value, bias, COV, distribution type). Alternate limit state and load pattern cases lead to *different* failure functions and a system safety index ($ß_{sys}$) is calculated for each response function.

4 ILLUSTRATIVE EXAMPLES

One simple span bridge and one continuous span bridge is considered. The design is taken from US Steel Highway Structures Design Handbook, 1983. Fig.4 shows the typical cross section of the bridge. The basic variables are the resistance of each girder (Ri), dead load (D) and live load factor (S). Girder resistance is expressed in terms of ultimate moment capacity (M_p) and the dead load is expressed as load per unit length (kips/ft.) for each girder. The live load is expressed in terms of load factor i.e. the multiplier of HS20 loading. The loading patterns are shown in fig.4 where, L1 is the lateral position for maximum single lane loading in exterior girder and, L1 and L2 together show the lateral position for maximum two lane loading in interior girder. In longitudinal direction, the loads are placed to produce the maximum moment (positive or negative) in the girder. Live load statistics is used for both exposure periods of 2 years corresponding to the mandated inspection

interval and 50 years corresponding to the expected bridge life span. Table 1 shows the statistical properties of the basic variables. All basic variables have log normal distribution.

Table 1. Statistical properties of the basic variables in examples 1 and 2.

Variable	Nominal value	Bias	COV (V %)
Girder Resistance (Ri)	Example 1. 6167 kip-ft. (M_p)	1.1	12.0
	Example 2.* a) 4383 kip-ft. (M_p) b) 3267 kip-ft. (M_y)		
Dead Load (D)	1.065 kip/ft.	1.0	9.0
Live Load (S)**	1 lane 2 lane	1.0	21.4
1 span	2 yrs 3.67 1.93 50 yrs 4.00 2.30		
2 span	2 yrs 2.14 0.70 50 yrs 2.33 0.80		

* a) Positive moment section and
  b) Negative moment section

** Live load factor (S) = Total live load / Total HS20 load effect.

4.1 Example 1

The example considered is an 80 ft. simple span steel bridge with four plate girders at 8.33 ft. center to center. Resistance of each girder and dead load is varied one at a time i.e. when R1 is varied, R2,R3,R4 and dead load are held at their nominal value. When R1,R2,R3,R4,D are at their nominal value, the nonlinear FE model shows load factor (s) = 3.91 at serviceability limit state and 14.83 at ultimate capacity limit state. When a trial 0.9R1,R2,R3,R4,D is used, the load factor changes to 3.84 at serviceability limit state and 14.4 at ultimate capacity limit state, etc. A multiple linear regression is carried out to give the g( ) functions for both limit states,

bridge condition (intact or damaged) and load pattern.

i.e. $s = f(R1,R2,R3,R4,D);$

and $g(\ ) = f(R1,R2,R3,R4,D) - S;$

where S also depends on load pattern and exposure period (Table 1). The damaged girder case is simulated by giving zero resistance to one of the girders (internal or external) and performing the FE analysis. The g( ) functions and system safety indices (β_{sys}) are shown below for the case where an internal girder is damaged, i.e. R2=0.

g( ) functions for example 1 :

1. Serviceability Limit State :

a) Intact :
One lane,
$$g1 = 3.46 + 1.09 \times 10^{-4}(R1) + 0.6 \times 10^{-4}(R2) + 0.2 \times 10^{-4}(R3) - 3.312 \times 10^{-7}(R4) - 0.667(D) - S$$
Two lane,
$$g2 = 2.36 + 0.65 \times 10^{-4}(R1) + 0.45 \times 10^{-4}(R2) + 0.16 \times 10^{-4}(R3) + 1.1 \times 10^{-7}(R4) - 0.463(D) - S$$

b) Interior girder damaged :
One lane,
$$g3 = 2.66 + 0.986 \times 10^{-4}(R1) + 0.273 \times 10^{-4}(R3) + 0.038 \times 10^{-4}(R4) - 0.652(D) - S$$
Two lane,
$$g4 = 1.59 + 0.403 \times 10^{-4}(R1) + 0.284 \times 10^{-4}(R3) + 0.136 \times 10^{-4}(R4) - 0.415(D) - S$$

2. Ultimate Capacity Limit State :

a) Intact :
One lane,
$$g5 = 0.019 + 6.973 \times 10^{-4}(R1 + R2 + R3 + R4) - 2.243(D) - S$$
Two lane,
$$g6 = -0.006 + 3.504 \times 10^{-4}$$

$$(R1+R2+R3+R4)-1.146(D)-S$$

b) Interior girder damaged :
One lane,
$$g7=0.209+6.811\times10^{-4}(R1+R3+R4)-2.15(D)-S$$
Two lane,
$$g8=0.086+3.438\times10^{-4}(R1+R3+R4)-1.112(D)-S$$

Table 2. System safety index (β_{sys}) for example 1.*

Limit state	Load-ing pattern	Expo-sure period (years)	β_{sys} (intact)	β_{sys} (dam-aged)
Span/300	1 lane	2	0.53	-1.08
		50	0.13	-1.48
	2 lane	2	1.72	-0.48
		50	0.90	-1.30
Ultimate capacity	1 lane	2	6.82	5.10
		50	6.43	4.73
	2 lane	2	6.59	4.89
		50	5.80	4.12

* No member strength correlation.

Table 3. System safety index (β_{sys}) for example 1.*

Limit state	Load-ing pattern	Expo-sure period (years)	β_{sys} (intact)	β_{sys} (dam-aged)
Span/300	1 lane	2	0.53	-1.08
		50	0.13	-1.47
	2 lane	2	1.70	-0.47
		50	0.89	-1.29
Ultimate capacity	1 lane	2	5.95	4.48
		50	5.62	4.16
	2 lane	2	5.75	4.30
		50	5.07	3.63

* Complete member strength correlation.

For all the cases, system safety index (β_{sys}) is calculated with no member strength correlation and with full member strength correlation (Tables 2 & 3).

4.2 Example 2

The example considered is a continuous steel bridge with four plate girders at 8.33 ft. center to center with two 100 ft. spans. Cases with non-compact and compact girder sections at the interior support are considered. The basic variables are the resistance of the girder at midspan (R1,R2,R3,R4), at interior support (R5,R6,R7,R8) and the dead load (D). *Intact* bridge condition and ultimate capacity limit state is considered. Table 4 shows the system safety indices (β_{sys}) for example 2.

Table 4. System safety index (β_{sys}) for example 2.*

Limit state	Load-ing pattern	Expo-sure period (years)	β_{sys} (one span)	β_{sys} (two span)
a. Ultimate capacity	1 lane	2	4.46	4.91
		50	4.08	4.56
	2 lane	2	4.33	7.38
		50	3.56	6.86
b. Ultimate capacity	1 lane	2	5.44	7.65
		50	5.05	7.28
	2 lane	2	5.19	10.0
		50	4.40	9.41

*a. Non compact girder section at interior support.
 b. Compact girder section at interior support.

5 DISCUSSION

Looking at the system safety indices it is possible that either one or two lane loading

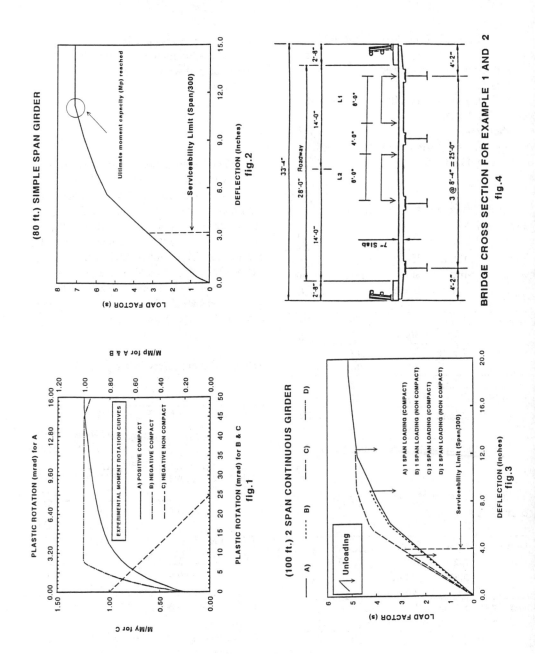

(80 ft.) SIMPLE SPAN GIRDER

fig.2

BRIDGE CROSS SECTION FOR EXAMPLE 1 AND 2

fig.4

EXPERIMENTAL MOMENT ROTATION CURVES

A) POSITIVE COMPACT
B) NEGATIVE COMPACT
C) NEGATIVE NON COMPACT

fig.1

(100 ft.) 2 SPAN CONTINUOUS GIRDER

A) ———   B) -----   C) -·-·-   D) ------

Unloading

A) 1 SPAN LOADING (COMPACT)
B) 1 SPAN LOADING (NON COMPACT)
C) 2 SPAN LOADING (COMPACT)
D) 2 SPAN LOADING (NON COMPACT)

Serviceability Limit (Span/300)

fig.3

may give the *lower* safety index depending upon load probability distribution. For simple span bridges, the ultimate capacity limit state is reached when all members reach their ultimate moment capacity (M_p) (fig.2). This is shown by the equations g5 to g8 where the coefficient for all members is same. Member strength correlation has a very little effect on β_{sys} in case of serviceability limit state but it *significantly* reduces system safety in case of ultimate capacity limit state. A damaged member has a *pronounced* effect on system safety for both limit states. The same load parameters were used in the example, so the actual risk for the damaged case must also include the probability of the damage condition occurring.

In continuous span bridge, for the same load parameters, compact girder case gives higher safety index than non-compact girder case (Table 4). Again, as in case of simple span bridge, the governing parameter for critical system safety index is dependent on the load probability distribution. Simple and/or higher order bounds then can be used to find a combined system safety index.

6 CONCLUSIONS

The examples show that system reliability indices can be formulated through the Response Function Method. This approach is used for *nonlinear* problems which involve load sharing and member interaction, particularly when there is member unloading. Practical implementation allows both serviceability and ultimate capacity limit states as well as both overload and accident scenarios. The formulation of response surface is a function of the particular structure studied and the limit state. Accuracy of response surface depends on the choice of basic variables and the range of values considered. This needs engineering judgment and understanding of the structural behavior on the part of the evaluating engineer.

ACKNOWLEDGMENT

This work was sponsored by the American Association of State Highway and Transportation Officials, in co-operation with the Federal Highway Administration, and was conducted in the National Cooperative Highway Research Program which is administered by the Transportation Research Board of the National Research Council.

REFERENCES

Augusti, G.; Baratta, A. and Casciati, F. 1984. *Probabilistic Methods in Structural Engineering*. Chapman and Hall, New York.

Melchers, R.E. 1987. *Structural Reliability, Analysis and Prediction*. John Wiley and Sons.

Moses, F. 1982. System reliability development in structural engineering. Structural Safety, 1.

Moses, F. and Rashedi, M.R. 1983. The application of system reliability to structural safety. 4th International Conference on Applications of Statistics and Probability in Soil and Structural Engineering. Universita di Firenze, Italy.

Nowak, A. and Hong, Y.K. 1991. Bridge live load models. ASCE Journal of Structural Engineering. Vol. 117.

Schilling, C.G. 1989. Unified autostress method. Report 51. American Iron & Steel Institute.

Verma, D.; Fu, G. and Moses, F. 1989. Efficient structural system reliability assessment by Monte-Carlo methods. Proceedings of ICOSSAR 89, p.895-901.

Structural Safety & Reliability, Schuëller, Shinozuka & Yao (eds) © 1994 Balkema, Rotterdam, ISBN 90 5410 357 4

Triparametric distribution and design or characteristic values

A. Mrázik, M. Križma & Z. Sadovský
Institute of Construction and Architecture of the Slovak Academy of Sciences, Bratislava, Slovakia

ABSTRACT: Strength quantities of constructional steels have as a rule asymmetric distributions with positive or negative coefficients of asymmetry. This fact is to be taken into account when determinating characteristic or design values. The triparametric gamma and log-normal distributions are suitable for this purpose. A general formula for p-fractile of the bi- and triparametric distributions is presented. The formula is formed as a product of the normalized p-fractile of the normal distribution and of a correction function, thus, design values can be expressed by the safety index defined for normal distribution. The differences resulting from aplication of biparametric or triparametric distributions are demonstrated on examples of calculations of characteristic and design values of resistance.

INTRODUCTION

Codes or standardization documents, e.g. / 1-7 /, based upon the structural reliability theory, utilize biparametric distributions for the determination of design and characteristic values of actions and resistances. For example in ISO 2394 / 1 / the normal distribution for the actions F and biparametric log-normal distribution for the resistance R has been used. It is well known that the symmetric normal distribution as well as the asymmetric log-normal distribution having zero as the infimum do not satisfactorily fit various practical cases. For example, yield and ultimate strength of steels are distributed asymmetrically with positive or negative coefficients of asymmetry (skewnes), the values of which are in some cases greater than +1 or less than -1.

Our opinions, based on the long term investigations / 9 /, show that for mild steels there prevails a tendency to positive values of coefficient of asymmetry whereas for higher strength steels there dominates the occurence of negative ones. Typical examples

of the yield strength histograms with approximating triparametric gamma distributions curves, taken from the Part 4 of the Catalogue / 9 /, are shown in Fig. 1. On studying the quality of concrete, a similar situation is encountered. Figure 2 shows column diagrams and corresponding curves of theoretical distributions of strength of concrete B-170 and B-600 according to / 10 /.

The aim of this paper is to show that despite the explicit appearance of coefficient of asymmetry the aplications of the triparametric gamma and log-normal distributions can be simplified to the format acceptable in the standardization practice retaining the use of the safety index. The suggested approximate formulas for p-fractiles of bi- and triparametric distributions are employed for alternative calculations of characteristic and design values of resistance of steel members cross-sections.

THE P-FRACTILE FORMULAS

Allowing for a posibility of applying the triparametric gamma and

Table 1 Values of quantities u_p and $b_{i,k}$

Distri-bution	p	u_p	$b_{1.3}$ / $b_{1.5}$	$b_{2.3}$ / $b_{2.5}$	$b_{3.3}$ / $b_{3.5}$	$b_{4.5}$	$b_{5.5}$
gamma	10^{-6}	-4.75	-0.754 / -0.756	0.1202 / 0.1285	0.0468 / 0.0538	-0.00349	-0.00510
	10^{-5}	-4.26	-0.671 / -0.673	0.0885 / 0.0930	0.0370 / 0.0428	-0.00107	-0.00355
	10^{-4}	-3.72	-0.573 / -0.575	0.0551 / 0.0563	0.0263 / 0.0303	0.00096	-0.00190
	10^{-3}	-3.09	-0.463 / -0.464	0.0231 / 0.0221	0.0187 / 0.0201	0.00192	-0.00087
	10^{-2}	-2.33	-0.319 / -0.319	-0.0101 / -0.0126	0.0109 / 0.0102	0.00230	-0.00020
	5.10^{-2}	-1.645	-0.173 / -0.173	-0.0290 / -0.0306	0.0032 / 0.0027	0.00108	0.00020
	10^{-1}	-1,28	-0.085 / -0.084	-0.0351 / -0.0382	0.0016 / 0.0008	0.00054	0.00008
log.-normal	10^{-6}	-4.75	-0.763 / -0.764	0.3121 / 0.3246	-0.0378 / -0.0402	-0.01440	0.00338
	10^{-5}	-4.26	-0.675 / -0.676	0.2298 / 0.2405	-0.0112 / -0.0113	-0.01314	0.00148
	10^{-4}	-3.72	-0.577 / -0.578	0.1530 / 0.1612	0.0085 / 0.0100	-0.01043	-0.00017
	10^{-3}	-3.09	-0.461 / -0.462	0.0801 / 0.0848	0.0206 / 0.0234	-0.00645	-0.00140
	10^{-2}	-2.33	-0.315 / -0.316	0.0133 / 0.0145	0.0231 / 0.0260	-0.00165	-0.00185
	2.10^{-2}	-1.645	-0.172 / -0.172	-0.0293 / -0.0349	0.0165 / 0.0185	0.00202	-0.00146
	10^{-1}	-1.28	-0.082 / -0.082	-0.0458 / -0.0484	0.0093 / 0.0102	0.00361	-0.00078

log-normal distributions, calculations of p-fractiles of random variable X can proceed by formula

$$X_p = \mu_x + u_{p,\alpha x}\,\sigma_x \qquad (1)$$

where μ_x, σ_x are the mean value, the standard deviation and $u_{p,\alpha x}$ the normalized p-fractile depending also upon the coefficient of asymmetry α_x.
We suggest the normalized p-fractile in the form

$$u_{p,\alpha x} = u_p.f(\log p,\ \alpha_x) \qquad (2)$$

where u_p is the normalized p-fractile of the normal distribution and f is a function of the arguments p and α_x. Thus, the term $f(\log p,\ \alpha_x)$ represents a corrective multiple of u_p due to the non-zero value of α_x. It will be shown in two steps that sufficiently precise approximation of f, given as finite power series in α_x and $\log p$, can be found.
Firstly, assuming

$$u_{p,\alpha x} \simeq u_p.(1 + \sum_{i=1}^{n} b_{i,n}.\alpha_x^i) \qquad (3)$$

where $b_{i,n} = b_{i,n}(\log p)$ is a function of $\log p$, and performing the least squares calculations for several practically interesting p values from 10^{-6} to 10^{-1}, it has been shown that the necessary number of terms retained in the series depends upon the range of values α_x. For the interval $-1.2 < \alpha_x < +1.2$ there suffices a cubic approximation ($n = 3$), whereas for $-2 < \alpha_x < +2$ an approximation of the 5th order ($n = 5$) is needed. The calculated values of $b_{i,3}$ and $b_{i,5}$ are shown in Table 1. Let us note that in / 8 / a bilinear approximation of $u_{p,\alpha x}$ in α_x ranging from -0.3 to $+0.8$ was applied.
In order to avoid an unconvenient interpolation of values $b_{i,n}$ from Table 1 in p the second approximation step is desirable. For any of $b_{i,n}$ functions the formula

$$b_{i,n} \simeq \sum_{j=0}^{3} a_{ij}.(\log p)^j \qquad (4)$$

has been substantiated.
The values of constants a_{ij} are given in Table 2.
Obviously, the same formula (3) can be used to obtain the biparametric gamma and log-normal p-fractiles. In a special case of zero value of lower limit of these distributions it should be taken

$$\alpha_x = 2\,\omega_x \qquad (5)$$

for the gamma distribution and

$$\alpha_x = 3\,\omega_x + \omega_x^3 \qquad (6)$$

for the log-normal distribution, where ω_x denotes the coefficient of variation of X.

CHARACTERISTIC AND DESIGN VALUES

As an example we choose to treat the resistance of compact steel cross-sections subjected to simple loadings. The resistance function is taken in the stress form

$$R = f_s.(A/A_n).(R_{exp}/R_{th}) =$$
$$= f_s.\varphi_a.\varphi_{et}, \qquad (7)$$

where f_s denotes the strength of steel (f_y - yield strength, f_u - ultimate tensile strength), A, A_n, φ_a - the geometrical quantity its real, nominal values and their ratio, respectively, φ_{et} is the ratio of experimental and theoretical resistances.
Following several standards the characteristic value R_k is calculated with $\varphi_a = \varphi_{et} = 1$, thus reducing its evaluation to that one of material strength. In accordance with commonly accepted probability $p_R = 0.05$ ($u_{0.05} = -1.645$) it is

$$R_k = \mu_f + u_{0.05,\alpha f}.\sigma_f \qquad (8)$$

Taking $n = 3$ the suggested approximation of $u_{0.05,\alpha f}$ is

$$u_{0.05,\alpha f} = -1.645.(1 - 0.173\alpha_f -$$
$$-0.029\alpha_f^2 + 0.0032\alpha_f^3) \qquad (9)$$

for the gamma distribution and

$$u_{0.05,\alpha f} = -1.645.(1 - 0.172\alpha_f -$$
$$-0.0293\alpha_f^2 + 0.0165\alpha_f^3) \qquad (10)$$

for the log-normal distribution.
The design value R_d of the resistance is according to the semiprobabilistic limit state verification defined as a p_{Rd}-fractile corresponding to the product

$$\alpha_{R(S)}.\beta_d \qquad (11)$$

Table 2   Values of constans a_j

Calculated quantity $b_{i.k}$	Gamma distribution				Log-normal distribution			
	a_0	$10\ a_1$	$10^2\ a_2$	$10^3\ a_3$	a_0	$10\ a_1$	$10^2\ a_2$	$10^3\ a_3$
$b_{1.3}$	0.2276	3.741	5.977	4.125	0.2254	3.676	5.747	3.952
$b_{2.3}$	-0.0551	-0.150	0.476	0.399	-0.0976	-0.472	0.451	0.166
$b_{3.3}$	-0.0091	-0.111	-0.098	-0.113	-0.0195	-0.391	-1.001	-0.497
$b_{1.5}$	0.2292	3.749	5.964	4.099	0.2272	3.699	5.804	3.996
$b_{2.5}$	-0.0568	-0.136	0.535	0.414	-0.1062	-0.512	0.481	0.231
$b_{3.5}$	-0.0074	-0.073	0.074	0.042	-0.0226	-0.446	-1.145	-0.588
$b_{4.5}$	-0.0031	-0.047	-0.119	-0.066	0.0091	0.054	-0.010	-0.058
$b_{5.5}$	-0.0003	-0.007	-0.035	-0.016	0.0018	0.037	0.108	0.070

Table 3   Data and results of examples (indices: g – gama, ln – log.-normal distribution; 3,2 three-, twoparametric distribution)

No.	Steel, Product	f_{yn} (MPa) t (mm)	Range (MPa) and number of data	μ_y μ_R (MPa)	σ_y σ_R (MPa)	ω_y ω_R	α_y α_R	$R_{kg.3}$ $R_{dg.3}$ (MPa)	$R_{kg.2}$ $R_{dg.2}$ (MPa)	$R_{kln.3}$ $R_{dln.3}$ (MPa)	$R_{kln.2}$ $R_{dln.2}$ (MPa)
1	11 369 plates	225 ≤20	205-405 2301	295.0 295.0	23.9 25.4	0.081 0.086	-0.023 +0.036	255.5 217.8	256.8 222.7	255.5 217.8	257.4 225.4
2	11 373 shapes	235 ≤13	215-415 4708	288.6 288.6	27.9 30.3	0.097 0.105	0.488 0.464	246.9 214.5	244.3 204.0	246.8 213.2	245.1 207.8
3	11 375 plates	216 60-80	185-425 3004	285.0 285.0	30.6 31.7	0.107 0.111	0.732 (0.556) 0.699 (0.541)	241.8 (239.9) 217.0 (210.6)	236.6 197.0	241.5 (239.8) 214.1 (208.9)	237.6 201.2
4	11 484 plates	355 25-60	285-515 689	415.2 415.2	33.3 35.5	0.080 0.085	-0.491 -0.348	356.2 287.6	362.0 314.1	356.3 286.9	362.8 317.8

STEEL 11 375   PLATES 60 - 80 mm   STEEL 11 484   PLATES 25-60 mm

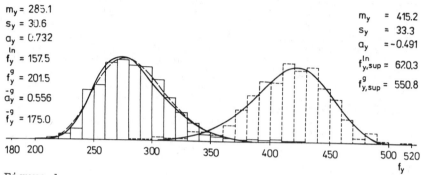

m_y = 285.1
s_y = 30.6
a_y = 0.732
f_y^{ln} = 157.5
f_y^g = 201.5
$\bar{a}_y^g$ = 0.556
$\bar{f}_y^g$ = 175.0

m_y = 415.2
s_y = 33.3
a_y = -0.491
$f_{y,sup}^{ln}$ = 620.3
$f_{y,sup}^g$ = 550.8

180 200   250   300   350   400   450   500 520
                                                                        f_y

Figure 1

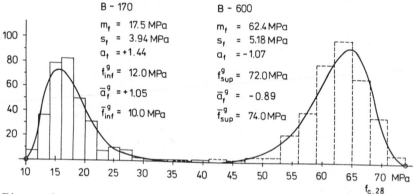

B - 170

m_f = 17.5 MPa
s_f = 3.94 MPa
a_f = +1.44
f_{inf}^g = 12.0 MPa
$\bar{a}_f^g$ = +1.05
$\bar{f}_{inf}^g$ = 10.0 MPa

B - 600

m_f = 62.4 MPa
s_f = 5.18 MPa
a_f = -1.07
f_{sup}^g = 72.0 MPa
$\bar{a}_f^g$ = -0.89
$\bar{f}_{sup}^g$ = 74.0 MPa

10   15   20   25   30   35   40   45   50   55   60   65   70 MPa
                                                          $f_{c,28}$

Figure 2

where β_d is the design value of the safety index and $\alpha_{R(S)}$ the sensitivity (weighting) factor. Alike the definition of probability of failure p_f by the standardized normal distribution: $p_f = \Phi(-\beta_d)$, the probability p_{Rd} is

$$p_R = \Phi(-\alpha_{R(S)} \cdot \beta_d) \qquad (12)$$

and consequently requiring the same design probability for the non-normal distribution it is

$$R_d = \mu_R + u_{pRd,\alpha R} \cdot \sigma_R =$$

$$= \mu_R - \alpha_{R(S)} \beta_d f(\log p_{Rd}, \alpha_R) \sigma_R \qquad (13)$$

For the normal safety class some standard documents require $\alpha_{R(S)} \beta_d$ = 0.8 x 3.8 = 3.04, others 0.75 x x 4.25 = 3.18 or 0.725 x 4.25 = 3.08 i. e. approximately $p_{Rd} = 10^{-3}$.

Taking numerical value 3.09, we find from (3), (4) for n = 3 that

$$u_{0.001,\alpha R} = -3.09 \cdot (1 - 0.463\alpha_R +$$
$$+ 0.0231\alpha_R^2 + 0.0187\alpha_R^3) \qquad (14)$$

for the gamma distribution and

$$u_{0.001,\alpha R} = -3.09 \cdot (1 - 0.461\alpha_R +$$
$$+ 0.0801\alpha_R^2 + 0.0206\alpha_R^3) \qquad (15)$$

for the log-normal distribution.

NUMERICAL EXAMPLES

The statistical data of material strength for numerical evaluations are taken from 4-th part of the catalogue / 9 /. Three examples are

for steel grade Fe 360, the fourth for steel grade 510, see Tab.3. The choise comprises (i) nearly symmetrical distribution, (ii) current positive asymmetry, (iii) elevated positive and (iv) current negative asymetry. The distributions of the yield strength f_y are plotted in Fig. 1.

The characteristic (8) and (13) values are calculated for the triparametric gamma and log-normal distributions applying the suggested approximations (9) and (10), respectively, in the former case and (14) and (15), respectively, in the latter case. For the sake of comparison the corresponding biparametric distributions given by (5) and (6) are taken into account, too.

Assuming the resistance (7) as a product of two random variables $f_s \cdot \varphi_a$ (i.e. $\varphi_{et} \simeq 1$) the statistical characteristics are obtained according to / 12 / or / 8 / by the formulas

$$\mu_R = \mu_f \cdot \mu_a = \mu_f \, ,$$
$$\omega_R \simeq (\omega_f^2 + \omega_a^2)^{1/2} \tag{16}$$
$$\alpha_R = (\omega_f^3 \, \alpha_f + 6 \, \omega_f^2 \, \omega_a^2)/\omega_R^3$$

where $\mu_a = 1$ and $\omega_a = 0.03$ for plates and $\omega_a = 0.04$ for shapes.

The results derived from the yield strength are presented in Table 3.

CONCLUSIONS

The contribution points out that the distribution of probability is for the strength of steel and concrete as well as for the resistance of structures generally asymmetric. Suggesting the triparametric gamma and log-normal distributions as suitable to reflect this fact, it presents a simple unified formula for their p-fractiles covering also the biparametric cases. On chosen examples of calculated characteristic and design values of the resistance of steel member it is ilustrated that:
(a) the characteristic values obtained by the triparametric and biparametric distributions are not substantially different,
(b) the design values as found by the biparametric distributions are either conservative (the coefficient of asymmetry α_x greater or close to the coresponding value (5) or (6) , or non-reliable (small positive and negative α_x),
(c) the design values as found by the triparametric gamma and log-normal distributions are negligibly different.

Since the triparametric distributions of probability give realistic results and their applications in standardization are desirable.

ACKNOWLEDGMENT

The autors are grateful to the Slovak Grant agency for science (grants No. 2/999213 and 2/999216) for partial supporting of this work.

REFERENCES

/ 1 / ISO 2394 General Principles on Reliability for Structures. 2nd edition, 1986-10-15.
/ 2 / CAN 3-S6-M78 Design of Highway Bridges, Suplement No. 1-1980. Rexdale, Ontario CSA 1980.
/ 3 / Load and Resistance Factor Design Specification for Structural Steel Buildings. Chicago, Ill., AISC 1986.
/ 4 / EUROCODE No. 3 Design of Steel Structures, Part 1 General Rules and Rules for Buildings. Background Documentation, Chapter 2 - Document 2.01, Chapter 7 - Document 7.01. CEC 1989.
/ 5 / DIN Grundlagen zur Festlegung von Sicherheitsanforderungen für bauliche Anlagen. Berlin - Köln, Beuth Verlag 1981.
/ 6 / RAVINDRA, M. K., GALAMBOS, T. V.: Load and Resistance Factor Design for Steel. Journal of Structural Division ASCE 104, 1978, No.ST9, pp. 1337-1353.
/ 7 / GALAMBOS, T. V., RAVINDRA, M. K.: Properties of Steel for Use in LRFD. Journal of the Structural Division ASCE 104, 1978, No. ST9, pp. 1459-1468.
/ 8 / MRÁZIK, A.: Theory of Reliability of Steel Structures (in Slovak). Bratislava, VEDA 1987.
/ 9 / MRÁZIK, A., SADOVSKÝ, Z., KRIVÁČEK, J.: Catalogue of the Statistical Data about the Yield Strength, Ultimate Tensile Strength and Elongation of Steels. Part 1: 1956-59, Part 2: 1966-69, Part 3: 1971-75, Part 4: 1983-88 (in Slovak)

/ Research report /. Bratislava,
ÚSTARCH SAV 1990-92.
/ 10 / ČOROVIČ, J., KOVÁČ, J.: Quality of Concrete and its Component
in Slovakia according to the Control Tests. In: HAVELKA, K.(ed.):
Theory of Calculation of Structures
according to the Limit States (in
Slovak). Bratislava, Slovak Academy
of Sciences 1964.
/ 11 / SALVOSA, L. R.: Tables of
Pearson's Type III Function. The
Annals of Mathematical Statistic 1,
1930, No. 3.
/ 12 / VORLÍČEK, M.: Statistical
Quantities for Functional Relations
in Constructional Research (in
Czech). Staveb. Čas. 9, 1961, No. 8,
pp. 485-515.

Structural Safety & Reliability, Schuëller, Shinozuka & Yao (eds) © 1994 Balkema, Rotterdam, ISBN 90 5410 357 4

Application of reliability theory to design of fibrous composite structures

Yoshisada Murotsu, Mitsunori Miki & Shaowen Shao
University of Osaka Prefecture, Aerospace Engineering Department, Sakai, Japan

ABSTRACT: This paper is concerned with the optimum design of fibrous composite structures under probabilistic conditions. Reliability evaluation is performed by applying the advanced first-order second-moment method. Numerical examples are provided for multiaxially laminated composites, and discussions are focused on the comparison between the probabilistic and deterministic designs.

1 INTRODUCTION

The strength and stiffness of fibrous composite structures have remarkable dependency on the kinds, volume contents, and orientations of their reinforcing fibers and stacking sequences of laminates. Therefore, optimum material design can be performed under a given loading condition. From this standpoint, optimum material design methods have been developed under criteria on maximum bending stiffness[1], maximum in-plane strength[2-4], maximum bending strength[5], maximum buckling strength[6] and maximum fundamental frequency[7]. It is found from these studies that a maximum performance can be obtained with an optimum laminate configuration, but sometimes the optimum fiber orientations change discontinuously when the loading condition changes and some performances have high sensitivities to design variables and loading conditions. Therefore, the best configurations studied so far can be optimum only under deterministic conditions, but many problems remain to be solved under probabilistic conditions.

The reliability of the strength of unidirectional and laminated composites using a macroscopic failure criterion and fundamental data on the variations of the strengths along the principal directions has been analyzed,[8,9] while the reliability analysis of fibrous laminated composites with advanced reliability techniques has been carried out.[10,11] The optimum configuration, however, has not been investigated from the view point of reliability so far.

In this paper, a method for determining the optimum fiber orientations of laminated composites is proposed by applying the advanced first-order second-moment method, and the results of the optimum design under probabilistic conditions are discussed through comparison with those of the deterministic design.

2 FAILURE CRITERION FOR COMPOSITES

There have been several criteria proposed for the failure of unidirectional fibrous composites with respect to their principal axes. These are 1) the maximum stress, 2) maximum strain, 3) Hill,[12] 4) Hoffman,[13] and 5) Tsai-Wu[14] theories. Among them, the Tsai-Wu failure criterion has been used by many researchers, and is recognized as the most general criterion for unidirectional composites. The Tsai-Wu failure criterion has the form

$$F_{xx}S_x^2 + 2F_{xy}S_xS_y + F_{yy}S_y^2 + F_{ss}S_s^2 + F_xS_x + F_yS_y = 1 , \tag{1}$$

$$
\begin{aligned}
&F_{xx} = 1/R_xR_x' , &&F_x = 1/R_x - 1/R_x' , \\
&F_{yy} = 1/R_yR_y' , &&F_y = 1/R_y - 1/R_y' , \\
&F_{ss} = 1/R_s^2 , &&F_{xy} = F_{xy}^* \sqrt{F_{xx}F_{yy}}
\end{aligned}
$$

where subscripts x and y denote the fiber and its orthogonal directions, respectively, subscript s denotes shear, and the prime denotes compressive strength. Factor F_{xy}^* is assumed to be -1/2.[15] In this paper, the Tsai-Wu criterion is used, although

Table 1 Material constants used.[15] (unit: MPa)

Material	Type	R_x	R_x'	R_y	R_y'	R_s
T300/5208	Graphite/epoxy	1500	1500	40	246	68

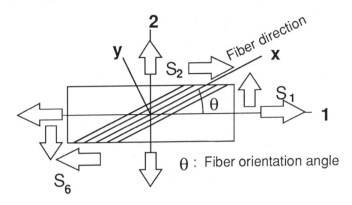

Fig. 1.

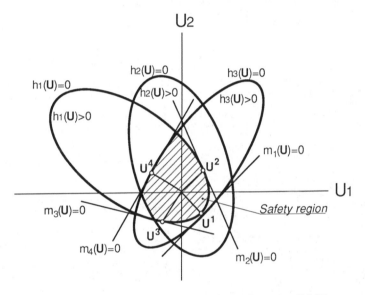

Fig. 2. Multiple-check-point method for evaluating reliability.

any other criterion can also be used to evaluate the reliability by the proposed method.

For failure under any plane stress condition, off-axis (not along the material principal axes) stresses are transformed into on-axis (along the principal axes) stresses to use on-axis failure criteria. The coordinate systems are shown in Fig. 1, where 1 and 2 represent the reference axes and θ is the angle between the 1 axis and x axis. Typical material constants for the composite used in the calculations are shown in Table 1.

When the Tsai-Wu criterion is used, a mathematical expression for the failure of unidirectional composites is given as follows:

$$M = 1 - \left[\sum_{i=x}^{y} \left\{ \left(\frac{1}{R_i} - \frac{1}{R_i'} \right) S_i + \frac{S_i^2}{R_i R_i'} \right\} + \frac{S_s^2}{R_s^2} + \frac{2F_{xy}{}^* S_x S_y}{\sqrt{R_x R_x' R_y R_y'}} \right]$$

(2)

3 RELIABILITY ANALYSIS OF MULTIAXIALLY LAMINATED COMPOSITES

3.1 Limit State Function for a Multiaxial Laminate

A symmetrical laminate composed of N plies is considered. The plate is subjected to in-plane stresses S_1, S_2 and S_6, where subscripts 1, 2 and 6 represent the major plate axis, the axis perpendicular to the 1 axis, and shear with respect to the 1-2 axes, respectively.

The stresses in the laminate system are calculated by using laminated plate theory,[15] and a limit state function M_i of each ply is evaluated by using the Tsai-Wu criterion Eq. (2). The system failure probability or system reliability for a multiaxial laminate is discussed in the next section.

1318

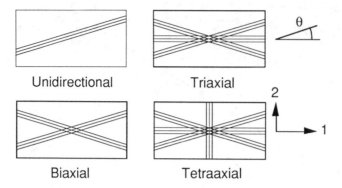

Unidirectional Triaxial

Biaxial Tetraaxial

Fig. 3. Multiaxial laminates.

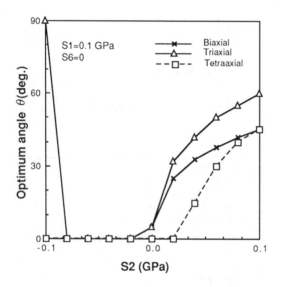

Fig. 4. Optimum fiber orientation angles for multiaxial laminates determined by the strength ratio.

3.2 System Reliability

The applied stresses (S_1, S_2 and S_6) and the strengths (R_x, R_x', R_y, R_y' and R_s) of each ply in a laminate system are treated as the basic random variables.   They are transformed into the independent standard normal vector U[23, 24].

The first-ply-failure criterion is adopted for a laminate system.   Consequently, the failure event of a laminate system becomes the union of failure events for all the plies. Then, failure probability P_f for the laminate system is given by:

$$P_f = P\left[\bigcup_{i=1}^{N} h_i(U) \leq 0\right] = \Phi(-\beta) \quad (3)$$

where $h_i(U)$ is the limit state function of the i-th ply

derived from Eq. (2). As illustrated in Fig. 2, the boundary between the safety region and the failure region for a laminate system consists of multiple nonlinear limit state functions, $h_i(U)=0$ (i=1, 2, ···, N).   Consequently, the standard AFOSM method[16,18,20] using only one linearization point is not suitable, and the multiple-check-point method[25] and the multiple-point importance sampling method[23] need to be applied to evaluate the system reliability of the laminated composite.

4 OPTIMUM FIBER ORIENTATION UNDER A DETERMINISTIC LOAD

Biaxial laminates $[+\theta, -\theta]_S$, triaxial laminates $[0°, +\theta, -\theta]_S$ and tetraaxial laminates $[0°, +\theta, -\theta, 90°]_S$ are

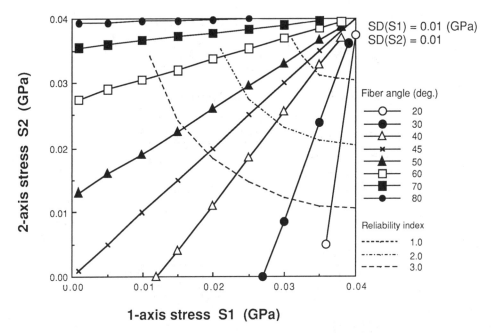

SD(S1) = 0.01 (GPa)
SD(S2) = 0.01

Fiber angle (deg.)

○ 20
● 30
△ 40
✕ 45
▲ 50
□ 60
■ 70
● 80

Reliability index

- - - - - - - 1.0
- - · - · - · 2.0
- - - - - 3.0

Fig. 5. Optimum fiber orientation angles of a unidirectional composite obtained by the AFOSM method. The standard deviations of the applied stress are 10 MPa each dotted lines indicate contours for the constant reliability indices.

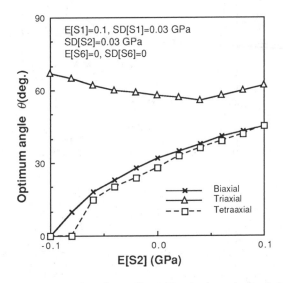

E[S1]=0.1, SD[S1]=0.03 GPa
SD[S2]=0.03 GPa
E[S6]=0, SD[S6]=0

✕ Biaxial
△ Triaxial
□ Tetraaxial

Fig. 6. Optimum orientation angles of a multiaxial laminate under probabilistic conditions.

considered, as shown in Fig. 3, where orientation angle θ is the design variable. Plies with the same fiber angle are assumed to fail simultaneously, and the volume content of each ply group with the same fiber orientation is assumed to be the same.

Tsai has proposed a method for determining the

optimum fiber angle in terms of a strength ratio.[15] This strength ratio R is defined by

$$S_{ia} = R\,S_i \quad (i=1,2,6) \tag{4}$$

The design method based on the strength ratio

implicitly assumes that the direction of the applied load is kept constant.

The strength ratio of a multiaxially laminated composite is defined as the strength ratio of the first-failure ply. The optimum fiber angles are shown in Fig. 4, in which S_1 is 0.1 GPa, $S_6=0$ and S_2 varies. The optimum orientation is 0°, that is, [±0]$_S$, [0, ±0]$_S$ and [0, ±0, 90]$_S$ are optimum for the respective types when S_2 is negative, except for the cases of a triaxial laminate and when $S_2= - S_1$. For positive S_2, the optimum orientation increases with increasing S_2, and θ=45° is optimum for biaxial and tetraaxial laminates, while θ=60° is optimum for triaxial laminates when $S_1=S_2$.

5 OPTIMUM FIBER ORIENTATION UNDER PROBABILISTIC CONDITIONS

5.1 Optimization Problem

The optimum design method mentioned above is valid as long as the applied loading is proportional and the failure envelope is deterministic. It does not, however, give the optimum solution for probabilistic conditions where the applied stresses are subject to uncertainty and the strengths have some variations.

The reliability-based optimum design problem for multiaxial laminates is to find the optimum fiber orientation angles which give the maximum system reliability for the laminate, i.e.

Find θ_i $(i = 1, 2, \cdots N)$

such that β $\rightarrow$ maximum   (5)

5.2 Optimum Angle for a Unidirectional Composite

The optimum fiber orientation angle of a unidirectional composite plate under probabilistic conditions can be obtained as the special case of $N=1$. The results are shown in Fig. 5, in which the strengths have no variation and the values for the standard deviation of the applied stresses are 10 MPa each. Contours of equal β value are also drawn in the figure.

The following points can be observed from the figure:
1) Optimum fiber orientations under probabilistic conditions are completely different from those under deterministic conditions.
2) Optimum orientations under probabilistic conditions depend on the mean stress level.
3) Optimum orientations under probabilistic conditions are not necessarily 0° or 90°, even if the mean values of the stresses are uniaxial.
4) Optimum orientations under probabilistic conditions tend to be identical to the deterministic results as the mean values of the applied stresses increase.

5.3 Optimum Angle for a Multiaxial Laminate

The optimum fiber orientation angles for the three types of multiaxial laminates are shown in Fig. 6. By comparing this result with Fig. 4, the optimum fiber angles are very different between the deterministic and probabilistic conditions These optimum values generally increase when the applied stresses have some uncertainties. This means that the optimum laminate should approach a quasi-isotropic plate under probabilistic conditions. This can be clearly seen from the result for the triaxial laminate, whose optimum fiber orientation angles are around θ = 60° under probabilistic loadings. Since there are 0° plies in a triaxial laminate, ±60° plies are needed to approach a quasi-isotropic plate which is suitable for any combination of applied stresses.

6 CONCLUSIONS

The method is proposed for the optimum design of fibrous composite laminates by applying the advanced reliability analysis method. The optimum fiber orientations for maximum reliability change remarkably between probabilistic and deterministic conditions. The optimum laminate configuration for a multiaxial laminate approaches the quasi-isotropic configuration when the variation in loadings increases.

REFERENCES

1) T.R. Tauchert and S. Adibhatla, "Design of Laminated Plates for Maximum Stiffness," J. Compos. Mater., Vol. 18, 1984, pp. 58-69.
2) W.J. Park, "An Optimal Design of Simple Symmetric Laminates Under the First Ply Failure Criterion," J. Compos. Mater., Vol. 16, 1982, pp. 341-355.
3) R.S. Sandhu,"Parametric Study of Tsai's Strength Criteria for Filamentary Composites," AFF-TR-68-168, 1969.
4) H.E. Brandmaier,"Optimum Filament Orientation Criteria," J. Compos. Mater., Vol. 4, 1970, pp. 422-425.
5) T.R. Tauchert and S. Adibhatla,"Design of Laminated Plates for Maximum Bending Strength," Engineering Optimization, Vol. 8, 1985, pp. 253-263.
6) M. Miki,"Optimum Design of Fibrous Laminated Composite Plates Subject to Axial Compression," Composites '86: Recent Advances in Japan and the United States, ed. by K. Kawata et al., Japan Society for Composite Materials, 1986, pp. 673-680.
7) R. Reiss and S. Ramachandran,"Maximum Frequency Design of Symmetric Angle-Ply Laminates," Composite Structures 4, Elsevier Applied Science, 1987, pp. 1.476-1.487.
8) Z. Maekawa and T. Fujii, "Probabilistic Design on Strength of Fiber Reinforced Composite Laminates," Progress in Science and Engineering of Composites, Proceedings of the 4th ICCM, ed. by T. Hayashi et al., 1982, pp. 537-544.
9) R.C. Wetherhold, "Reliability Calculations for Strength of a Fibrous Composite Under

Multiaxial Loading," Journal of Composite Materials, Vol. 15, 1981, pp. 240-248.

10)G. Cederbaum, I. Elishakoff, and L. Librescu, "Reliability of Laminated Plates via the First-Order Second Moment Method," Composite Structures, Vol. 15, 1990, pp. 161-167.

11)H. Nakayasu and Z. Maekawa, "Stochastic Material Design of Composite Materials," Transactions of Japan Society of Mechanical Engineers, Vol. 57, 1991, pp. 2042-2049.

12)R. Hill,"A Theory of Yielding and Plastic Flow of Anisotropic Metals," Proceedings of the Royal Society, Series A, Vol. 193, 1948, pp. 281-297.

13)O. Hoffman,"The Brittle Strength of Orthotropic Materials," Journal of Composite Materials, Vol. 1, 1967, pp. 200-206.

14)S.W. Tsai and E.M. Wu, "A General Theory of Strength for Anisotropic Materials," Journal of Composite Materials, Vol. 5, 1971, pp. 58-??

15)S.W. Tsai and H.T. Hahn., Introduction to Composite Materials, Technomic, 1980, p. 280.

16)A.M. Hasofer and N.C. Lind,"Exact and Invariant Second Moment Code Format," Journal of the Engineering Mechanics Division, ASCE, Vol. 100, 1974, pp. 111-121.

17)R. Rackwitz and B. Fiessler,"Structural Reliability under Combined Random Sequences," Computers and Structures, Vol. 9, 1978, pp. 489-494.

18)P. Thoft-Christensen and Y. Murotsu, Application of Structural Systems Reliability Theory, Springer Verlag, 1986, pp. 15-24.

19)M. Shinozuka, Basic Analysis of Structural Safety, Journal of the Struct. Division, ASCE, Vol. 109, No. 3, 1983, pp. 721-740.

20)H.O. Madsen, S. Krenk, and N.C. Lind, Methods of Structural Safety, Prentice-Hall, 1986, p. 44.

21)Y. Murotsu, M. Yonezawa, H. Okada, S. Matsuzaki, and T. Matsumoto, "A Study on First-Order-Second-Moment Method in Structural Reliability," Bulletin of University of Osaka Pref., Ser. A, Vol. 33, No. 1, 1984, pp. 23-36.

22)M. Miki, Y. Murotsu, T. Tanaka, and S. Shao, "Reliability of the Strength of Unidirectional Fibrous Composites," AIAA Journal., Vol. 28, 1990, pp. 1980-1986.

23)Y. Murotsu, S. Shao, N. Murayama, and M. Yonezawa, "Importance Sampling Method for Reliability Assessment of Structures with Multi-Modal Limit States," Proc. of ISUMA '90, The First International Symposium on Uncertainty Modeling and Analysis, Univ. of Maryland, USA, Dec. 3-5, 1990, pp. 49-54.

24)A. Der Kiureghian, and D-L. Liu, "Structural Reliability under Imcomplete Probability Information," Journal of Engineering Mechanics, ASCE, Vol. 102, January 1986, pp. 85-104.

25)S. Shao, M. Miki, and Y. Murotsu, "Optimum Fiber Orientation Angle of Multiaxially Laminated Composites Based on Reliability," AIAA Journal, Vol. 31, No. 5, May 1993, pp. 919-920; AIAA Paper 91-1032.

Structural Safety & Reliability, Schuëller, Shinozuka & Yao (eds) © 1994 Balkema, Rotterdam, ISBN 90 5410 357 4

Strain energy as a criterion of structural safety and reliability of RC slabs

T. Nürnbergerová & J. Hájek
Institute of Construction and Architecture, Bratislava, Slovakia

ABSTRACT: A possibility of using strain energy as a quantity enabling to characterize reliability and safety of a concrete structure is presented. The theoretical approach is supported by the test results on one-way slabs. An attempt is made to apply the strain energy as quantity making possible to take into account the time factor, too.

1 INTRODUCTION

As a rule, the verification of reliability and safety of reinforced and prestressed concrete structural elements is based on several characteristics decisive from the viewpoint of serviceability, as well as of ultimate limit states. However, the occurrence of a limit state can be interpreted as exhausting of certain part of strain energy what is made use of at some particular stress states or at stability problems. The aim of our contribution is to present the generalizing character of strain energy in structural reliability theory on the example of reinforced concrete slabs. In our opinion, the strain energy could be used as a quantity determining the limit states instead of separate stress or deformation characteristics. We are aware of the fact that the present contribution cannot take into account all aspects from the viewpoint of safety and reliability.

2 EXPERIMENTAL PROGRAMME

2.1 Test specimens and material properties

As test specimens the reinforced concrete slabs with nominal size 3.6x1.2x0.12 m on a span of 3.54 m were used. The welded deformed wire fabrics with longitudinal and transversal wires of 8 mm and 6 mm in diameter, respectively were used as reinforcement. The corresponding wire spacements were 100x200 mm. In Fig. 1 the cross section of slabs is illustrated.

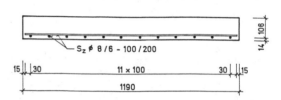

Fig. 1. Cross section of tested slabs (dimensions in mm).

Eight slabs were cast, a half of them made of concrete with aggregates of the river Danube alluvia, the second half with crushed andesite aggregates. The average mechanical properties of concrete obtained by tests at age of 28 days were: cube strength f_c = 40.6 (45.2) MPa, tensile strength f_{ct} = 3.0 (3.2) MPa, modulus of elasticity E_c = 32.9 (31.4) GPa. The properties of concrete with crushed aggregates are given in parentheses.

The average mechanical characteristics of reinforcement obtained from tests were: yield stress f_{sy}=677 MPa, strength f_{st}=719 MPa, and modulus of elasticity E_s=208 GPa.

2.2 Testing procedure

The slabs were tested in vertical position (i.e. with horizontal loading) by means of a special device (Hájek 1986). The uniformly distributed load on the whole surface of the tested slab was produced pneumatically using a loading pillow made of softened PVC. The pillow was inserted into the slot between the

tested slab and a rigid supporting plate that belonged to the loading device. Load intensity was measured by means of a hydrostatic manometer.

The unloading was applied on different load levels. The tests were carried out in a so-called mixed loading procedure under what the strain rate procedure at the first ascending branch and the stress rate procedure at the other branches of loading paths were meant. Three variations of loading procedure were chosen: 1. monotonic loading up to the failure (2 slabs), 2. loading with 4 unloading cycles at 4 different deformations levels (midspan deflections, 4 slabs), 3. three unloadings at the same deformation levels (2 slabs). The final part of the procedure with previous unloadings contained a loading up to the failure limit followed by unloading.

The loading procedures consisted of gradual increase of deflection (on the first ascending branch) with a deflection rate of 0.5 mm/min or that of load (on the following descending and ascending branches). During the tests the deflections in tenths of span, the elongations on compressed and the tensioned surfaces and the process of crack formation were registered.

2.3 Tests results

The measurements of deflections and those of elongations on the compressed and tensioned surfaces carried out independently each from other enabled the evaluation the work both of external forces and of bending moments.

The external work consumed for achieving a given deflection or a given load level was calculated by numerical integration according to formula

$$W_m = 1/2 \; b. \sum_{j=2}^{n} \sum_{k=2}^{m} (q_{k-1} + q_k).\Delta w_j, \qquad (1)$$

where n is the total number of measured deflections, m the total number of load (or deflection) levels, Δw_j the increment of the deflection w_j, q_k the intensity of uniformly distributed load at the level k, b the width of cross section.

The work of bending moments was evaluated from the moment versus curvature relationship. The measurements of elongations on the compressed and tensioned surfaces enabled to evaluate the variations of curvatures along the span of the slab. Then, the work of bending

moments at a given loading level was determined numerically in such a way that first the work of the cross section was calculated according to formula

$$wp_j = 1/2 \sum_{k=2}^{m} (M_{j,k-1} + M_{j,k}).\Delta(1/r_j), \qquad (2)$$

where $\Delta(1/r_j)$ is the increment of curvature and $M_{j,k}$ the corresponding bending moment in the cross section j at the load level k. Then, the total work W_m at a load level m was obtained from the formula

$$W_m = \Delta l/2 \sum_{j=2}^{n} (wp_{j-1} + wp_j), \qquad (3)$$

Δl denoting the length of the measurement basis (it was the same along the whole span of slab), n number of cross sections including those on the both ends of the slab.

In Fig. 2 the comparison of the values of external work obtained from measured deflections (line 1) and those calculated on the basis of curvatures (line 2) is shown. As it can be seen, the external work is a little larger than the work of pure bending moments. The obtained difference between the coordinates of the lines 1 and 2 can be supposed to be caused by neglecting of rest of work components, especially of shearing forces.

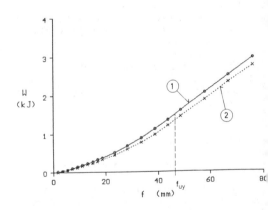

Fig. 2. Work of load versus deflection at the midspan cross section of one of slabs tested using monotonic loading up to the failure: 1 - external work of load; 2 - work of bending moments; f_{uy} - deflection at the yield limit of reinforcement.

3 THEORETICAL CONSIDERATIONS

3.1 Deterministic model

The deterministic model used in the presented theoretical considerations is based on the moment versus curvature relationship, not taking into account the rest of the strain energy components. Stress-strain diagrams of concrete and steel at monotonic increasing deformation were used to the determination of above mentioned relationship. The proposed analytical expression of the stress-strain diagram of concrete besides the initial tangent modulus E_c is determined by 12 constants $[a_{i,j}]$, $i = 1,\ldots6$, $j = 1, 2$, where $j = 1$ ($j = 2$) belong to the tension (compression), $i = 1, 2, 3$ (4, 5, 6) to the ascending (descending) branch of the diagram. In Table 1 the matrix [a] of the material properties is given. First three lines correspond to the deformation at the end of the elastic range, to the deformation at the peak of the diagram, and finally to the peak stress. The lines $i = 4, 5, 6$ are the characteristics of descending branches (ultimate deformation, stress and tangent modulus at the ultimate deformation). The elements of the matrix [a] are given in Table 1.

For the deformation $|\varepsilon_c| \le |a_{1,j}|$ the elastic behaviour of concrete is assumed, i.e. $\sigma_c = E_c \varepsilon_c$. For the deformation range $|a_{1,j}| < |\varepsilon_c| \le |a_{4,j}|$ the following formula is chosen:

$$\sigma_c = (2-k)E_c \varepsilon_c + (k-1)a_{3,j} +$$

$$+(a_{5+2k,j} + a_{6+2k,j}\lambda_c)\lambda_c^2, \tag{4}$$

where

$$\lambda_c = (\varepsilon_c - a_{k,j})/(a_{2k,j} - a_{k,j}) \tag{5}$$

$$a_{5+2k,j} = 3[a_{2k+1,j} - (k-1)a_{2k-1,j}] -$$

$$- (2-k)E_c(2a_{2,j} + a_{1,j}) -$$

$$- (k-1)a_{6,j}(a_{4,j} - a_{2,j}) \tag{6}$$

$$a_{6+2k,j} = (2-k)(a_{2,j} + a_{1,j}) +$$

$$+ (k-1)a_{6,j}(a_{4,j} - a_{2,j}) -$$

$$- 2[a_{2k+1,j} - (k-1)a_{2k-1,j}]. \tag{7}$$

Table 1. Matrix [a] of mechanical properties of concrete

i	j	
	1	2
1	0.000044	0.
2	0.000105	-0.002046
3	2.497	-37.3
4	0.0003	-0.004127
5	1.498	-23.29
6	-1498.0	-10101.0

$E_c = 33903$ MPa

By coefficients $a_{5+2k,j}$, $a_{6+2k,j}$ the matrix $[a_{i,j}]$ is enlarged to a type of (10, 2). This matrix can be then used for integration of stresses at a given curvature of cross section. It should be added that for $|\varepsilon_c| > |a_{4,j}|$ $\sigma_c = 0$. However the elements of the completed matrix [a] were not supposed to be independent from each other even in the deterministic model. Some of them were assumed to be dependent on the peak stresses of diagrams given by element $a_{3,j}$ of above mentioned matrix.

As the stress-strain diagram of reinforcement the following bilinear function was taken:

$$\sigma_s = E_s \varepsilon_s \quad \text{for} \quad \varepsilon_s \in (-\varepsilon_{sy}, \varepsilon_{sy})$$

$$\sigma_s = f_{sy} \quad \text{for} \quad |\varepsilon_{sy}| \le \varepsilon_s \le |0.01|. \tag{8}$$

In such way, the bending moment M for a given curvature $1/r_{cr}$ can be calculated from Eq. (4)-(8). In Fig. 3 the results obtained from test of one of the slabs are presented by crosses. The coordinates of the curve 2 were calculated using the average values of concrete and steel properties supposing that crack had occured in the cross section. Tension stiffening effect was taken into account by the correction coefficient $\kappa_{cm} = r_{cr}/r_m$, where $1/r_m$ is the average curvature respecting the tension stiffening (curve 3), $1/r_{cr}$ is the curvature of the cracked cross section. Before the cracks occur (M < M_{cr}) this coefficient is $\kappa_{cm} = 1$. Over the cracking limit (M > M_{cr}) the ratio κ_{cm} was expressed by the following formula derived from the test results of all slabs.

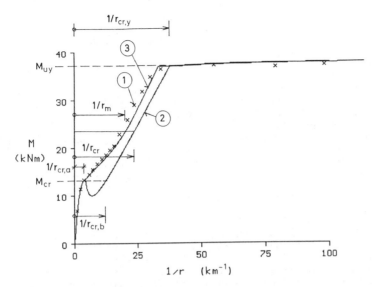

Fig. 3. Moment versus curvature relationship used in deterministic model: 1 - test results; 2 - moment calculated at cracked cross section for monotonically increasing curvature; 3 - moment calculated taking into account the tension stiffening using Eq. (9)

$$\kappa_{cm} = \kappa_{cy} + (r_{cr,a}/r_{cr,b} -$$

$$- \kappa_{cy})\exp[-0.36(1/r_{cr}-1/r_{cr,b})] \qquad (9)$$

Here, the coefficient κ_{cy} denotes a multiple of the curvature $1/r_{cr,y}$, i.e. at the beginning of yielding of steel, $r_{cr,a}$, $r_{cr,b}$ are the curvatures belonging to the moment M_{cr} with or without respect to the tension stifening, respectively. The values of the curvatures $1/r_{cr,a}$, $1/r_{cr,b}$ result from the calculation of cracked cross section (line 2) while the coefficient κ_{cy} was determined from tests (an average value of $\kappa_{cy} = 0.03/r_{cr,y}$ was obtained). In addition, it is supposed that $\kappa_{cy} \geq 0$ and $r_{cr,a}/r_{cr,b} > \kappa_{cy}$. It follows from the Eq. (9) that only two

numerical values (0.03 in κ_{cy} and 0.36 in exponential function) are used for the determination of tension stiffening effect. Obviously they may vary in certain limits. Using coefficient κ_{cm} the average curvature $1/r_{cm}$ belonging to the given moment M can be calculated. This relationship is plotted in Fig. 3, line 3. Integration of the work of cross section along the span of the slabs gives the value of work for the whole slab. Then, the work versus deflection relationship can be plotted similar to that given in Fig. 2, line 2. It should be added that only the cases where the failures were caused by deformation of reinforcement in tension were taken into consideration. The consequence of the obtained relationship represented by the line 3 in Fig. 3 is

Table 2. Statistical characteristics of randomly permuted individual factors.

Characteristic			Mean	Standard deviation
Concrete compressive strength	σ_{ce}	(MPa)	35.0	4.7
Coefficient for tensile strength	κ_{ct}		0.23	0.04
Coefficient for modulus of elast.	κ_{ec}		10000.0	300.0
Yielding point of reinforcement	f_{sy}	(MPa)	626.0	38.8
Modulus of elasticity of reinf.	E_s	(MPa)	210000.0	6300.0
Diameter of wires	ϕ	(mm)	8.0	0.12
Width of slab	b	(m)	1.19	0.0065
Depth of slab	h	(m)	0.12	0.003
Wire's distance from bottom edge	a	(m)	0.015	0.003

that the line 2 (and analogically also the line 1) in Fig. 2 is a straight line after the deflection fu corresponding to the ultimate moment Mu is reached.

3.2 Stochastic modelling

The method of random permutation (so-called Latine hypercube sampling method, see e.g. Bažant 1985) was applied for stochastic modelling. At that, the distinction between the strain energy at short-time loading and that one at sustained loading was made. A monotonic loading procedure up to the failure was supposed for short-time loading. The statistical characteristics introduced into calculation assuming the normal distribution are given in Table 2.

There were 9 factors in all. The distribution curves of individual material characteristic were divided into eight parts according to equal gradation of probability. It should be noted that the relations between tangential modulus of elasticity and compressive strength of concrete on the one hand and between tensile and compressive strength of concrete on the other hand were expressed as follows:

$$E_c = \kappa_{ec} (|\sigma_{ce}|)^{1/3} \tag{10}$$

$$\sigma_{ct} = \kappa_{ct} (|\sigma_{ce}|)^{2/3} \tag{11}$$

whereby statistical parameters of coefficients κ_{ec}, κ_{ct} are given in Table 2.

As far as long-term load is concerned, time development of mean value and standard deviation of coefficient β on the basis of our previous tests (Hájek 1974) was made. The coefficient β expresses the relative increment of deflection at the time t with regard to initial deflection. The result obtained from experiments that the variation coefficient of the characteristic β is practically constant (according to our measurements it is equal to 0.12) was used as a basis. This again enabled to generate a system of 9 factors corresponding to individual values of time presented in Table 3 and then by means of random permutations to determine also the time development of the strain energy. As the increase of the strain energy at sustained load is caused by the increase of deflections of the slabs, the result of the theoretical study can be accepted, namely, that the mean value of the strain energy corresponding to the deflection at

Table 3. Statistical characteristics of coefficient β expressing time development of deflection of slabs.

Time after applying long term load (days)	Mean	Standard deviation
1	0.0520	0.0062
3	0.0879	0.0105
7	0.1301	0.0156
28	0.2383	0.0286
60	0.3222	0.0387
90	0.3735	0.0448
365	0.5677	0.0681
1825	0.7311	0.0877
20000	0.7697	0.0924

the time t is approximately equal to the mean value of the strain energy corresponding to the short-time load at the same deflection. However, the differences in the development of the dispersion arise. As it can be deduced from the comparison of Fig. 4 and Fig. 5 the dispersion of the values of the strain energy increases in comparison with that at short-time load. For example, at the mean value of deflection f = 57.3 mm due to the short-time load the standard deviation is σ_{fst} = 0.130 kJ, while at the same deflection due to the sustained load the standard deviation is σ_{flt} = 0.153 kJ. Obviously, it is due to the increase of the standard deviation of the coefficient β as it is given in Table 3.

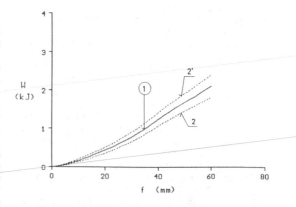

Fig. 4. Work of short time load uniformly distributed on surface of slabs versus midspan deflection relationships computed using stochastic modelling: 1 - mean values, 2, 2' - lower and upper limits for the given probability level 0.05.

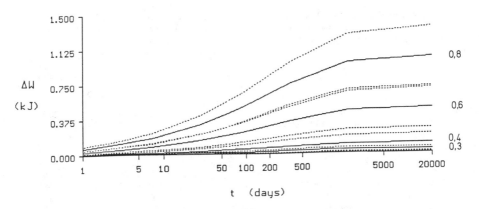

Fig. 5. Time development of the strain energy corresponding to the sustained load on the levels 0.3; 0.4; 0.6 and 0.8 multiple of the limit M_{uy}; dashed lines indicate the ranges of the toleration margins.

3.3 Reliability and safety criteria

On the basis of the results of this study the opinion was attained that the strain energy could be a unifying characteristic at the determination of reliability, not only of a cross-section and of a structural element but even the entire structure. The reliability condition in more general form could be expressed as follows

$$P(W_{ss} \leq W_{ls}) \geq p_{rel}, \tag{12}$$

where P(.) is the probability that the limit state characterised by the strain energy W_{ls} does not occur, W_{ss} means the work executed by given load and p_{rel} indicates the probability expressing the criterion of the given limit state. Obviously, the criteria for limit states of serviceability would differ from those for ultimate limit states. It should be noted that the work W_{ss}, as well as the strain energy W_{ls} are determined with regard to the dissipation of energy. Thus the process of damage due to short-time or sustained load can be taken into account. In the region of service load also the knowledge about complementary energy could be used. The special attention is to be paid to the determination of the criteria p_{rel}.

4 CONCLUSIONS

The main purpose of our contribution was to indicate the possibility of using the strain energy as a quantity enabling to check reliability and safety of concrete structures in general. The application on one-way slabs was chosen because of possibility to support the theoretical approach by test results enabling the evaluation of work of bending moments. In the case of slabs under sustained load with different loading levels an attempt was made to apply the strain energy as a quantity which makes possible to express the influence of time factor, too.

REFERENCES

Hájek, J., Fecko, L., Nürnbergerová, T.: Short term tests of concrete slabs supported on three sides. Staveb. Čas., 34, 1986, 5, 359-380 (in Slovak).

Bažant, Z.,P., Kwang Liang Liu: Random Creep and Shrinkage in Structures: Sampling. ASCE Journal of Structural Engineering, vol.111, 5, May 1985, 1113-1134.

Hájek, J., Fecko, L., Nürnbergerová, T.: Long-term deflection of precast reinforced concrete floor slabs. Comission CEB-IV.b, february 1974.

Structural Safety & Reliability, Schuëller, Shinozuka & Yao (eds) © 1994 Balkema, Rotterdam, ISBN 90 5410 357 4

A comparative study of probabilistic methods in structural safety analysis

A. Sellier
Laboratoire de Mécanique et Technologie (LMT),
ENS /CNRS/Université Paris 6, Cachan, France

M. Lorrain
Laboratoire de Mécanique des Structures, Toulouse,
France

M. Pinglot
Laboratoire de Mécanique des Structures, INSA/UPS
Toulouse, France

A. Mébarki
LMT Cachan, France

ABSTRACT: When dealing with the reliability analysis of civil constructions, several probabilistic methods known as level-2 or level-3 techniques may be used. The authors consider the case of RC columns and RC beams. The reliability, regarding ultimate limit states, is then described through safety indices denoted β, and probabilities of failure.

Classical Monte Carlo simulations and importance sampling are ran. It is shown that simulations, outside the hyper sphere having β-radius, require a reduced number of simulations and a short time calculations in presence of large β-values, i.e. small values of the probability of failure, or a small number of involved random variables. The results are compared to those collected from level-2 methods and hypercone method. They allow the identification of importance zones, in the random operating space, that have a restricted area but a major contribution in the final value of the probability of failure.

1 INTRODUCTION

When dealing with the assessment of structural reliability, one may use either simplified procedures that assume idealised forms of the failure domain Df, i.e. level-2 methods [19], or more precise techniques which consider the whole geometry of Df, i.e. level-3 methods. Monte Carlo simulations are commonly used as level-3 techniques. They are easy to run but they have to face the problem of convergence[20,21,22] in terms of both the precision and the requested calculation time duration.

The three level-3 methods that are presented in this paper are based on Monte Carlo simulations. Their main purpose is to afford precise Pf values (Probability of failure) while requiring small calculation duration. The first one separates the output random variables into two groups and simulates separately the resistance parameters and the loads. The second one simulates the random variables outside an hyper sphere having a β-radius when the basic RV (Random Variables) are transformed into gaussian RV, β being the Lind-Hasofer reliability index. The third method combines these two previous techniques to take advantage of their mutual performances. The proposed methods are tested, first, in the case of a simple structure in order to compare the obtained results and analytical calculations. Their respective efficiencies are compared.

They are, therefore, applied in the case of more realistic structures: RC columns and RC beams. The results collected for the beam are also compared to the values deduced from level-2 methods which assume a linear shape of the limit state surface. They confirm the existence, for each limit state, of importance zones, around

the design point P*, which have a great contribution on the final value of Pf.

2- STRUCTURES UNDER STUDY AND MECHANICAL MODELS

For comparative purposes, we have adopted several structures. Actually, the first structure, see **fig.1**, is a short RC column. The probability of failure can then be calculated analytically. These precise values are then compared to the values obtained by running the procedures we are proposing in this paper. The second structure, see **fig.2**, is a normally reinforced column. Non linear behaviour of the constitutive materials and geometrical imperfections are considered. The third structure, see **fig.3**, is a normally reinforced beam. Non linear behaviour of the materials is also adopted.

2.1- *Structure 1: short RC column under compression*
Figure1: Short RC column

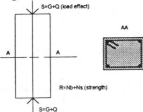

The load S is defined as: S=G+Q (1) where G=dead load and Q=live load.
The carrying capacity R is defined as: R=Nb+Ns (2), where Nb=internal effort in the concrete under compression and Ns=internal effort in the reinforcement, so limit state function is:
E=R-S   (3).

The limit state function is, for the present structure, defined as: $E=(X1+X2)-(X3+X4)$ (4).
The output random variables are: $Y1=R$ and $Y2=S$ (5)

2.2- *Structure 2: Reinforced column*
Figure2: RC column

The column is subdivided into a set of macro-elements. The second order effects can then be studied. The reinforcement area takes a constant value: $A= 6,28cm^2$. The concrete behaviour, under compression, is defined according to CEB-FIP recommendations, [18]:

$\sigma = (\mu. fc).(k\eta - \eta^2)/(1+(k-2).\eta)$

$\eta = \varepsilon/\varepsilon_0, k = \varepsilon_0' Ec/(\mu. fc), \varepsilon_0' = \varepsilon_0(1+\phi)$ (6)

$\phi = (1.35Gk)/(1.35Gk+1.5Qk)$

with $\mu=0.85$ (Rüsch effect), fc= concrete strength under compression, Ec= initial elasticity modulus, σ= stress corresponding to a strain value ε, Gk and Qk are the respective characteristical values of G and Q and $\varepsilon_0=0.002$. The steel behaviour is defined as: $\sigma = min(E.\varepsilon, fs)$ (7), with ε = strain, σ=stress, $E=2.1x10^5$MPa, and fs= reinforcement strength.

Table 1- Distribution of the involved RV (structure 1)

RV	description	Distribution	unit	mean value	standard deviation
X1	Ns	Gaussian	MN	240	10
X2	Nb	Gaussian	MN	70	20
X3	G	Gaussian	MN	90	10
X4	Q	Gaussian	MN	70	20

Table 1 shows the distributions adopted for the basic random variables (*structure 1*).

Table 2 shows the distributions adopted for the basic random variables (*structure 2*).

Table 2- Distributions of the involved RV (structure 2)

RV	Description	Distribution	Unit	Mean value	standard deviation
X1	Ec	gaussian	Mpa	μ_{Ec}	$0.07 \cdot \mu_{Ec}$
X2	fc	gaussian	Mpa	26.7	3.2
X3	fs	gaussian	MPa	475	2.85
X4	y1	gaussian	cm	11	1
X5	y2	gaussian	cm	11	1
X6	eo (eccentricity)	gaussian	cm	2	1
X7	b	gaussian	cm	40	1
X8	ht	gaussian	cm	30	0.75
X9	G	gaussian	daN/m2	250	25
X10	Q	Gumbel (E1max)	daN/m2	309	108

The limit state function is, for the present structure, defined as: E=R-S .... (8), where R = R(fc,fs,y1,y2,ht,b,Ec) = carrying capacity of the column, S=S(G,Q,eo)= loads effect involving second order effects.

2.3- Structure 3:Reinforced beam

b= width of the section and h= depth of the section. Table 3 shows the distributions that are adopted for the basic random variables.

The structure is studied under pure bending due to the applied loads. Its carrying capacity is then equal to the strength of the mid-span cross area. The limit state is defined as: E=R-S

Table 3- Distributions of the involved RV

RV	description	random distribution	unit	mean value	Coeff of variation
X1	b	experimental	cm	25.1	0.02
X2	h	experimental	cm	50.1	0.01
X3	ys	experimental	cm	17.75	0.06
X4	fs	gaussian	Mpa	470	0.11
X5	fc	experimental	Mpa	27.5	0.23
X6	G	gaussian	kN/m2	Gm (with Gk=2.5)	0.10
X7	Q	E1max	kN/m2	Qm (with Qk=4.5)	0.35

Figure3: Reinforced beams

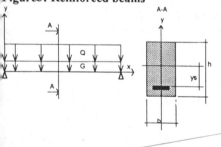

The materials have the same behaviour than for structure n°2, i.e. the normally reinforced column. The geometrical reinforcement ratio takes a constant value: ω=As/(b.h)=1% where As=reinforcement area,

2.4- Definition of the failure domain

For the three cases, the failure domain is defined as :E=R-S≤0 and denoted Ω (usually denoted as Df); its complement in the operating space is the safety domain denoted Ω (usually denoted as Ds). The probability of failure is defined as:

$$Pf = \int_{\Omega} f_i(\bar{X})d\bar{X} \qquad (9)$$

An analytical solution exists in the case of the short RC column (structure n° 1):

$$\left. \begin{array}{l} \beta(\text{Lind} - \text{Hasofer}) = \mu_e / \sigma_e \\ Pf = Pf_L = \phi(-\beta) \end{array} \right\} \qquad (10)$$

where μ_e and σ_e are, respectively, mean and standard deviation values of E.

In the other cases (RC columns and RC beams with non linear behaviour of the material), it is not possible to find the exact solution by an analytical investigation.

3- ASSESSMENT OF THE PROBABILITY OF FAILURE

3.1- Method 1: Dissociation of the output RV into R and S [2,7,11,16]

When the output random variables R (resistance) and S (loads effect) are independent, the joint density function can be defined as:

$$f_{RS}(R,S) = f_R(R).f_S(S) \quad (11)$$

then:

$$Pf = \int_\Omega f_{RS}(R,S).dR.dS = \int_\Omega f_R(R).f_S(S).dR.dS$$

$$= \int_{Smin}^{Smax} fs(S).F_R(S).dS$$

(12)

F_R (= Cumulative Distribution Function, CDF of R) is estimated by Monte Carlo simulations; a sample of basic random variable Xi is artificially generated. For each simulation, the deterministic mechanical model is ran to evaluate R. A cumulated histogram of R can then be drawn. The probability density function (PDF), fs, of S can also be estimated by Monte Carlo simulations. However, in general, simulating S requires much less time than generating R. It is then obvious that it is more interesting to ran more simulations on S than on R, for a given precision on R histogram, resulting in improvement of fs value, as stated in Eq.13. The probability of failure is, in this case estimated, by $\hat{P}f$ defined as:

$$\hat{P}f = \sum_{smin}^{smax} f_s(S)\Delta S.F_R(S) \quad (13) \text{ as shown in fig4.}$$

Figure 4- Use of R and S histograms for Pf estimation

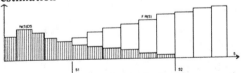

Fig 4 shows that only S values ranging between S1 and S2 have a non negligible contribution in Pf value. It seems then interesting to use a conditioning procedure in order to minimise the number of simulations outside the interval [S1,S2].

3.2- Method 2: Conditioning procedure [12]

In a first step, let us consider a domain Bn(β*), see Fig.5.

Figure 5- Hypersphere and failure domain

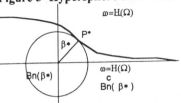

It is defined such as none simulation in this domain causes failure. Such a domain is suggested in [12]; it is defined, in the standard gaussian space, as an hypersphere having β*-radius, where β* is the Lind-Hasofer index. This hypersphere is tangent to the failure domain ω at the design point P*, see fig 5, with ω=H(Ω) and H= the Rosenblatt transformation [12]. β* may be calculated by any classical optimisation method [5,7,14,16]

$$\text{then:} \begin{cases} p(\bar{X} \in \underline{\Omega}) = p(\bar{u}(H(\bar{X})) \in \underline{\omega}) \\ = p((\bar{u} \in \underline{\omega}) \cap (\bar{u} \in B_n^c(\beta^*))) \\ \underbrace{p(\bar{u} \in \underline{\omega}/(\bar{u} \in B_n^c(\beta^*)))}_{P1} \times \underbrace{p(\bar{u} \in B_n^c(\beta^*))}_{P2} \end{cases} \quad (14)$$

$p2 = 1 - \chi_n^2(\beta^{*2}) = H_n(\beta^*)$ (15) is easily computed, [12]

$p1 = p(\bar{u} \in \underline{\omega}/(\bar{u} \in B_n^c(\beta^*)))$ (16), is estimated by the use of Monte Carlo simulations by generating $(\bar{u} \in \underline{\omega}/(\bar{u} \in B_n^c(\beta^*)))$ (17). Then the probability P1 is estimated from:

$$\hat{p}1 = \frac{1}{nsim}\sum_{nsim} 1_{\underline{\omega}}(\bar{\bar{u}})$$

with: $\bar{\bar{u}} = (\bar{u} \in B_n^c(\beta^*))$ (18)

with: $1_{\underline{\omega}}(\bar{\bar{u}}) = 1. if. E = (R-S)_{(\bar{X}=H^{-1}(\bar{\bar{u}}))} \leq 0$

else: $1_{\underline{\omega}}(\bar{\bar{u}}) = 0$

So, only P1 is estimated by Monte Carlo sampling. As (P1=Pf/P2)>Pf (19), the efficiency of this method increases with 1/P2=1/Hn(β*) (20). Table 4 gives some Hn(β*) values. The more β* is large or the

smaller is n, the more efficient is the conditioning sampling.

Table 4- Values of Hn(β*), [12]

Hn(β*)	β*=3	β*=4	β*=5
n=2	0.011	3.35e-4	3.73e-6
n=4	0.061	3.02e-3	5.03e-6
n=6	0.174	1.38e-2	3.42e-4
n=7	0.342	4.24e-2	1.56e-3
n=8	0.532	9.96e-2	5.33e-3

3.3- Method 3: Conditioning and Dissociation Coupling (CDC) [16]

3.4- Results

3.4,1- Dissociation, Conditioning, Conditioning and Dissociation Coupling methods

We have compared the convergence rates, of each of the three methods described above, in the case of the short column. Actually, Pf has been successively calculated with the dissociation technique, the conditioning sampling and coupling both of them. The obtained results are given in Table 5.

Table 5- Comparison of the convergence rates; Nsim=number of simulations performed

Nsim	Dissociation	Conditioning	CDC	Analytical results
50	0	3e-6	1.45e-6	β*=4.74
150	0.5	2e-6	1.32e-6	$Pf = \phi(-\beta^*) = 1.05e-6$
300	5.64e-6	1.08e-6	1.21e-6	
600	2.93e-6	1.62e-6	1.16e-6	
1200	1.54e-6	1.48e-6	1.15e-6	

We have seen that either dissociation of output random variables or conditioning sampling reduces the number of required simulations. To take advantage of their respective efficiencies, we have coupled dissociation technique and the conditioning Monte Carlo method. This coupling consists in using dissociation method to estimate P1 from Eq.(16). We assume then that Hasofer Lind index β* is already calculated by a classical optimization procedure, and that the output random variables R and S are mutually independent. The probability density functions of R and S are modified by conditioning sampling. We have then:

$$\tilde{R} = R(\tilde{X} = H^{-1}(\tilde{u} \in B_n^c(\beta^*)))$$
$$\tilde{S} = S(\tilde{X} = H^{-1}(\tilde{u} \in B_n^c(\beta^*)))$$
(21)

and P1 is estimated through :

$$\hat{P}1 = \sum_{i_{min}}^{i_{max}} f\tilde{s}(\tilde{S})\Delta\tilde{S}.F_{\tilde{R}}(\tilde{S})$$
(22)

As discussed for Eq.(13), Eq.(22) will be evaluated by running more simulations on $\tilde{S}$ than on $\tilde{R}$. This technique based on coupling both dissociation principle and conditioning procedures allows the computation of a term P1>Pf with a reduced number of simulations.

Figure 6: Relationship between Pf and the total number of simulations, Nsim

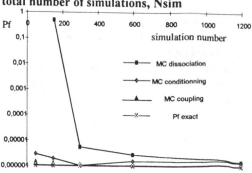

Fig.6 shows a rapid convergence, at Nsim=50 simulations, of the method wich couples the dissociation and the conditioning techniques compared to the convergence rates of the two others methods. An acceptable convergence with the dissociation method requires more than 300 simulations while the conditioning technique affords a precise Pf value for a total of 100 simulations.

We have also compared the convergence rates, of each of the three methods, in the case of the normally reinforced column. The obtained results are given in Table 6.

Table 6- Comparison of the convergence rates

Nsim	Dissociation	CDC: $\beta=\beta*$	CDC: $(\beta=3.95)<\beta*$	level II
20	0.00E+00	2.76E-05	5.80E-05	
50	6.00E-01	3.41E-05	4.10E-05	$\beta*=4.2$
100	3.00E-04	3.21E-05	3.40E-05	n=10 RV
200	3.50E-05	3.41E-05	3.42E-05	Pfl=1.34E-05

Figure 7: Relationship between Pf and the total number of simulations, Nsim.

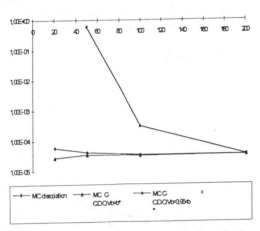

(Conditioning Dissociation Coupling). The dissociation technique requires about 200 simulations to reach a good precision. It can be noticed that CDC method with $\beta<\beta*$ shows a slower convergence rate than with $\beta=\beta*$.

3.4,2- Hypercone, Conditioning- Dissociation Coupling, and level 2 methods

We have here above shown that CDC technique has more efficiency than the dissociation or conditioning methods taken separately. We have then compared CDC technique with two other kinds of methods that are the hypercone method [14], which is basically used in case of single limit state, and a level-2 method which assumes a linear shape of the limit state surface. Pf values obtained are:

* Pf=7.25E-05 from CDC method corresponding to Nsim=200 simulations
* Pfl=$\phi(-\beta*)=\phi(-3.64)=7.54E-05$ from level-2 method while the limit state surface is supposed to be linear.

The hypercone method affords upper-bound, Pfc_{min}, and lower-bound, Pfc_{max}, values of Pf obtained respectively by substituting to the actual failure domain, Df, sets of inscribed (fig 8) or circumscribed (fig 9) hypercone sections, [7,14]. These calculations are based on the following relation, corresponding to the probability of failure for an elementary hypercone section:

$$Pfc = Vc(\beta_i, \beta_j, \theta_{ij}, n)$$

$$= (2\pi)^{-1/2} \int_{\beta_i}^{\beta_j} \exp(-\frac{t^2}{2}) \cdot \chi^2_{n-1}(t^2 \cdot tg^2(\theta_{ij})) \cdot dt \quad (24)$$

The values deduced from the hypercone method are, for n=7: Pfc_{min}=4,64E-05 and Pfc_{max}=11,3E-05.

Fig 8- Inscribed hypercone sections in Df

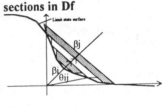

Pf value deduced from CDC method ranges within the interval obtained from the hypercone approximation

Pfl calculated **Fig 9- Circumscribing** while the limit **hypercone sections to Df**

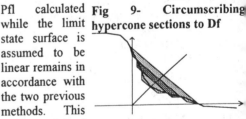

state surface is assumed to be linear remains in accordance with the two previous methods. This can be explained by the fact that the region located between the hyperplane, tangent at the design point P* to the limit state surface, and the external hypercones (circumscribing Df) or Df itself has a negligible influence in terms of "probabilistic weight", see fig.10. It means that the region having the major influence on Pf is located near the design point P* as illustrated in fig.11 which represents the

numerical values of Pfc, from Eq.24, when θ varies. We have drawn, in the case of the RC beam, the limit state surface when the operating space is constituted by the three most important random variables, ie fs=steel strength,

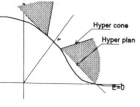

Fig 10- Importance zone around P*

fc=concrete strength and S=G+Q=load effect. The actual form of this surface, represented on

fig 11- Vc(θ) values for n=7 and β=3.64

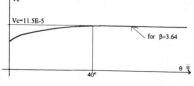

Fig.12, is far from being linear, however Pfl remains very close to Pf. It means that the real form of the surface, far from P*, has no influence on Pf since the importance zone is located in a restricted region near P*.

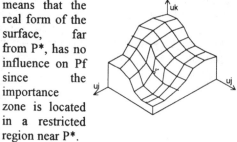

Fig12-Limit state surface

4- CONCLUSIONS

An efficient level 3 method based on Monte Carlo simulations has been introduced. This CDC method (Conditioning and Dissociation Coupling) combines both dissociation and conditioning principles to improve the convergence rate of the simulations. Actually, for the structures tested herein, a restricted number of simulations is required to obtain precise values of Pf: Nsim=20 to 50 while Pf is almost equal to 10^{-5} for short RC columns and normally reinforced column. This convergence rate is improved in the case of large β* values, whereβ*=safety index, or small

n values, where n=dimension of the operating space.

The comparison between the values of Pf obtained from the safety index (level-2 method) and those collected from CDC method gives relevant information about the accurracy of level-2 methods. From the results collected in this study, it can be drawn that level-2 method, which deduces Pf from the safety index while the limit state surface is assumed to be linear, remains acceptable since it gives results in accordance with those calculated from CDC method. Furthermore, its results range within the interval deduced from the hypercone method, ie lower and upper bound values of Pf, though the limit state surface was not linear as demonstrated in the case of RC beams. This can be explained by the fact that only a restricted region of the failure domain Df, located near the design point P* in the operating space, is important. This importance zone has the major contribution on the final value of Pf. So importance sampling should give a larger weight at this zone [23]. Furthermore, assuming that the limit state surface is an hyperplane affords acceptable values of Pf.

5- REFERENCES

(1)- M Lorrain, M Pinglot & A Mébarki; "Introduction au problème des tolérances géométriques dans les ouvrages en béton"; INSA de Toulouse, CEB réunion de la commission 1-Bruxelles 16 Novembre 1989.

(2)- M Pinglot, M. Lorrain & A. Mébarki; "Columns reliability: comparison of two mechanical models"; Proc. of First International Conference on Multipurpose High Rise Towers; Bangalore (Inde), 26-28 Octobre 1988.

(3)- F Duprat; "Instabilité des poteaux bi-encastrés en béton armé"; Thèse de doctorat; INSA de Toulouse 1993.

(4)- M Lorrain & M Pinglot; " La sécurité à l'effondrement des bâtiments. Influence des défauts d'exécution sur la probabilité de ruine des poteaux de bâtiment"; Annales de l'ITBTP n° 425; Paris, Nov. 1989.

(5)- E Leporati; "The assessment of structural safety. A comparative statistical study of the evolution and use of level 3, level 2, and level 1 methods"; Research Study Press volume 1 , Letchworth (UK), 1977.

(6)- H Mathieu; "Manuel de sécurité des structures"; Bulletin d'information n°127 et 128 du CEB; $2^°$ édition; Décembre 1979-Janvier 1980.

(7)- A Mébarki; "Evaluation de la probabilité de ruine des poutres isostatiques en béton armé. Influence des paramètres géométriques et de chargement."; Thèse de Docteur-Ingénieur; INSA de Toulouse; Juillet 1984.

(8)- M Fogli, M Lemaire & M Saint André; "L'approche de Monte Carlo dans les problèmes de sécurité: Application à l'estimation probabiliste du risque de ruine des poutres hyperstatiques en béton armé soumises à des actions aléatoires"; Annales de l'ITBTP n°403; Paris, Mars-Avril 1982.

(9)- M Fogli; "L'approche de Monte Carlo dans les problèmes de sécurité"; Thèse de Docteur-Ingénieur, INSA de Lyon, 1980.

(10)- J Bass; "Eléments de calcul des probabilités"; $3^°$ éd. Masson; Paris, 1974.

(11)- H Venstel; "Théorie des probabilités"; Edition Mir Moscou, 1973.

(12)- P Bernard & M Fogli; "Une méthode de Monte Carlo performante pour le calcul de la probabilité de ruine"; Revue du CTICM n°4, Paris, 1987.

(13)- Bulletin d'information du CEB n°124/125, Avril 1978.

(14)- A Mébarki; "Sur l'approche probabiliste de la fiabilité des structures de Génie Civil. La méthode de l'Hypercône et ses algorithmes d'application"; Thèse d'état ès-sciences, UPS de Toulouse, Décembre 1990.

(15)- A Mébarki & M Lorrain; "Algorithmes pour l'analyse de la fiabilité par la méthode de l'hypercône et le calcul de l'index de sécurité", Annales des Ponts et Chaussées, n°47, $3^°$ trimestre, 1988.

(16)- A Sellier; "Fiabilité des structures, utilisation et optimisation des méthodes de calcul", mémoire de DEA, INSA de Toulouse, 1990.

(17)- A Mébarki "Fiabilité des matériaux et des structures. Méthodes et applications", Proc. des Journées du GAMI des 10 &11 Novembre 1991, Paris.

(18)- Comité Euro-International du Béton CEIB. "Système international de réglementation technique unifiée des structures"; vol. 1 & 2, Bul. 124 & 125, 1979.

(19)- TJ Sweting and A.F Finn "A Monte Carlo method based on first and second order reliability approximation"; Structural Safety 11 , 1992, 20- 3-212 Elsevier.

(20)- A.E Wessel,E.B Hall, G.L Wise "Importance sampling via a simulacrum"; Journal of the Franklin Institute, 1990, vol 327, n°=5 pp771-783.

(21)- R.H Geist, M.K Smotherman "Ultrahigh reliability estimates through simulation"; Annual reliability an Maintainability symposium, IEEE, 1989, n°= 315-316, pp244-250.

(22)- A.Harbitz " Efficient and accurate probability of failure caculation by use of the importance sampling technique"; fourth international conference on application of statistics and probability in soil and structure engineering, 1983, Universita di Firenze (Italy), Eds Pitagioria, pp 825-836.

(23) A.Sellier A.Mebarki "Fiabilité des structures, tirage d'importance et optimisation statistique"; Annales des Ponts & Chaussées (pending).

Structural Safety & Reliability, Schuëller, Shinozuka & Yao (eds) © 1994 Balkema, Rotterdam, ISBN 90 5410 357 4

Modelling of concrete strength records by stable distribution functions

L. Taerwe

Department of Structural Engineering, Magnel Laboratory for Reinforced Concrete, University of Ghent, Belgium

ABSTRACT : In this paper it is shown that differenced concrete strength records can be modelled by means of the class of stable distribution functions. These distributions are not used frequently in statistical analysis because they are defined in terms of the characteristic function. However, it was shown previously (Mandelbrot 1960, 1963, 1967) that they proved to be very useful to represent economic time series. Due to certain very general properties they can serve as probabilistic model for a wide class of phenomena that show more complicated variation patterns than usually assumed in classical probabilistic modelling.

1 ORIGINAL FIELD OF APPLICATION

The so-called Pareto-Lévy or stable non-Gaussian distribution functions were proposed as an extension of the classical Pareto-law for the distribution of income (Mandelbrot 1960). The field of application was further extended to stock market values and the prices of certain goods such as cotton and wheat (Mandelbrot 1960, 1963, 1967). For the prices of these latter products, exceptionally long records exist. For these cases an analysis was made of the process $X(t + T) - X(t)$ where $X(t)$ denotes the value of the price at the instant t, and T is generally taken equal to 1. It was observed that the distribution of these differences is generally unimodal but that it shows a sharper peak than the normal distribution function. Moreover, it contains more extreme values and hence shows higher tails. Generally, this problem was circumvented by assuming that these extreme values were generated by another mechanism than the one valid for the majority of the observations. By discarding a non-negligible number of "outliers" the empirical information is distorted and finally also the model derived from it. A further typical feature of the process considered is that the empirical moment of second order does not tend to a limiting value as the observation period increases. Before Mandelbrot proposed his new model, it was assumed that price differences were normally distributed or that

the prices themselves followed a "random walk" process. According to the new model, abrupt changes are more likely to occur than according to the Gaussian model. This means that a Pareto-Lévy stock market is more risky for speculants than a market that obeys the classical model. The validity of the stable distributions for economic variables was also confirmed by other authors (Fama 1965 ; Fielitz and Smith 1972 ; Officer 1972).

2 DEFINITION OF STABLE DISTRIBUTION FUNCTIONS

Consider the variables $y_j = a_j x_j + b_j$ with $a_j > 0$ and be $\Phi(t)$ the characteristic function of the variables x_j. In classical probability theory it is shown that

$$\Phi_{y_j}(t) = \Phi(a_j t) \cdot \exp(it b_j) \qquad (1)$$

where i is the imaginary unit. The characteristic function of the sum of two variables y_j and y_k is equal to the product of the characteristic functions, or

$$\begin{aligned}\Phi_{y_j+y_k}(t) &= \Phi_{y_j}(t) \cdot \Phi_{y_k}(t) \\ &= \Phi(a_j t) \cdot \Phi(a_k t) \cdot \exp[it(b_j + b_k)]\end{aligned} \qquad (2)$$

If (2) has the same structure as (1), then the

characteristic function and the density function are called "stable". This terminology is related to the fact that under summation of several independent stable variables the resulting distribution function remains invariant or stable. Examples of this class of distribution functions are the normal distribu- tion and the Cauchy distribution. Among the stable distributions, the normal one is the only one with a finite moment of second order. It can be shown that all stable distributions are continuous and unimodal. Lévy has shown that the logarithm of the characteristic function of a stable distribution has as general expression

$$\ln \Phi(t) = \delta \, it - \ldots$$
$$\ldots - \gamma \, |t|^\alpha \left[1 + i\beta \, \frac{t}{|t|} \, w(|t|,\alpha) \right] \quad (3)$$

where

$$w(|t|,\alpha) = \begin{cases} \tan\left(\dfrac{\pi \alpha}{2}\right) & \text{for } \alpha \neq 1 \\ \dfrac{2}{\pi} \ell n \, |t| & \text{for } \alpha = 1 \end{cases}$$

The meaning of the four parameters is discussed in the next section. In case $\beta = 0$, (3) reduces to

$$\ln \Phi(t) = i\delta t - \gamma \, |t|^\alpha \quad (4)$$

Introducing the reduced variable

$$Y = (X - \delta) \cdot \gamma^{-1/\alpha} \quad (5)$$

it is found that

$$\ln \Phi_y(t) = - \, |t|^\alpha \quad (6)$$

The density function itself is not available in closed form, except for the normal and the Cauchy distribution and the distribution with $\alpha = 0.5$, $\beta = 1$, $\delta = 0$ and $\gamma = 1$. As the variance does not exist, this class of distribution functions is not very attractive for current use in statistical data analysis. Series expansions for the density function and the cumulative distribution function have been elaborated (Fama and Roll 1968). A method for simulating stable random variables is outlined in (Chambers, Mallows and Stuck 1976).

3 MEANING OF THE PARAMETERS

In general, the four parameters have the following meaning :

- The parameter α is a measure of the kurtosis and is limited to the interval $0 < \alpha \leq 2$. The parameter α determines the height of the tails. Noting that for the normal distribution $\alpha = 2$, the other stable distributions all have higher tails. If $\alpha = 1$ than $\beta = 0$ which applies to the Cauchy distribution.
- The parameter β is a measure for the skewness and is limited between -1 and +1. For $\beta = 0$ the distribution is symmetric.
- The parameter δ is a location parameter. If $1 < \alpha < 2$ then E[X] is finite and $\delta = E[X]$. The parameter δ has no direct meaning in case $0 < \alpha < 1$ and $\beta = 0$.
- A stable variable is standardized if $\delta = 0$ and $\gamma = 1$. If U is a standardized stable variable, then the distribution of sU is also stable with the same α and β, $\delta = 0$ and $\gamma = s^a$. This means that γ is equal to a scaling factor to the power α.
- Let U' and U" be two independent standardized stable variables with the same α and β. It can be shown that the distribution of s'U' + s"U" is the same as that of sU whereby

$$s^\alpha = s'^\alpha + s''^\alpha$$

or $\gamma = \gamma' + \gamma"$. The well known additive property for the variance of normal variables is now valid for the parameter γ.

The sum of n stable variables is also stable with parameters α, β, $n\gamma$ and $n\delta$. This property of stability or invariance under summation shows the fundamental character of these distributions. It also follows that stable distributions are the only possible limit distributions for the sum of independent identically distributed random variables.

4 PARAMETER ESTIMATION

Although graphical methods and the use of specific fractiles have been proposed (Press 1972), the most efficient method for parameter estimation appears to be based on the characteristic function (Paulson, Holcomb and Leitch 1975). The empirical characteristic function of a sample $x_1, \ldots, x_n$ can be calculated as

$$\hat{\Phi}(t) = \frac{1}{n} \sum_{k=1}^{n} \exp{(itx_k)} \quad (7)$$

According to the procedure outlined in (Fama and Roll 1968), the quadratic difference between the empirical and the theoretical characteristic

Table 1. General characteristics of strength series

Series	n	$\bar{x}$	s	$\sqrt{b_1}$	b_2	min	max
A	1787	46.5	6.56	0.147	3.124	26.3	76.3
B	945	36.8	5.30	0.386	3.099	23.0	56.6
CI	222	45.8	6.44	0.121	2.702	30.6	62.0
CII	312	39.3	5.21	-0.015	2.739	24.4	53.8
D	1468	59.9	3.68	0.828	3.199	45.6	73.6
EI	580	68.5	4.17	0.184	3.724	56.0	88.8
EII	578	76.1	4.98	-0.199	2.818	62.4	91.6
ZA	550	42.9	5.18	0.545	3.188	28.8	59.0
ZB	550	37.0	3.69	0.566	3.044	28.6	49.4
ZC	550	40.1	3.43	-0.026	2.831	28.9	48.9
ZD	550	41.5	3.60	0.160	3.121	29.6	54.0
CEM1	556	51.4	3.51	0.090	2.774	43.0	62.0
CEM2	437	56.7	3.08	0.043	2.808	47.8	66.7
CEM3	536	52.0	2.64	0.145	2.365	46.2	58.9
CEM4	402	47.4	3.53	0.599	3.383	40.2	59.6
CEM5	402	28.4	3.49	0.051	2.694	19.0	37.2

functions is weighted and integrated over the range of t values, yielding

$$I = \int_{-\infty}^{+\infty} |\hat{\Phi}(t) - \Phi(t)|^2 . e^{-t^2} \, dt \qquad (8)$$

The function $\exp(-t^2)$ is used as weight function. The integral I is a function of the observed values and of the parameters to be estimated. These latter values are determined in such a way that I becomes minimal. In case $\beta = \delta = 0$ and according to (4), the previous expression reduces to

$$I = \int_{-\infty}^{+\infty} \left[\frac{1}{n} \sum_{k=1}^{n} \cos(tx_k) - \ldots \right.$$
$$\left. \ldots - \exp(-\gamma |t|^\alpha) \right]^2 e^{-t^2} \, dt \qquad (9)$$

Now, $\hat{\alpha}$ and $\hat{\gamma}$ have to be determined such as to minimize I. It is noted that the summation in the first factor of the integrand is independent of $\hat{\alpha}$ and $\hat{\gamma}$ and only needs to be calculated once for each value of t for which the integrand is evaluated. The most efficient procedure consists in standardizing the x_k-values with respect to an arbitrary parameter g :

$$x'_k = x_k/g \qquad (10)$$

First an iteration on α and γ is performed until I becomes minimal. Subsequently g is adjusted until this minimum occurs for $\gamma = 1$ which yields $\hat{\alpha}$. The parameter $\hat{\gamma}$ is calculated from

$$g = \hat{\gamma}^{1/\hat{\alpha}} \qquad (11)$$

5 APPLICATION TO EMPIRICAL RECORDS

The following empirical strength records are considered (table 1) :
- The concrete strength series A, B, C I, C II, D, E I, E II. The observed values were obtained on large building sites.
- The concrete strength series ZA, ZB, ZC, ZD consisting of control specimens from concrete armour units, used as top layer on breakwaters.
- The cement strength series CEM 1, ..., CEM 5.

General characteristics of the series are summarized in table 1. All strength values are given in MPa. More detailed information is given by (Taerwe 1992).

The differenced series with values z_i are obtained from the original series with values x_i, by applying the difference operator ∇ as follows

$$z_i = \nabla x_i = x_i - x_{i-1} \qquad (12)$$

Table 2. General characteristics of differenced strength series

Series	n	s	$\sqrt{b_1}$	b_2	min	max	min/s	max/s
AD	1786	7.29	0.049	3.561	-27.2	28.0	-3.73	3.84
BD	944	5.41	0.038	3.906	-19.6	20.8	-3.62	3.84
CID	221	6.75	0.115	3.694	-20.8	22.0	-3.08	3.26
CIID	311	5.51	-0.064	3.872	-17.0	18.4	-3.09	3.34
DD	1467	3.64	0.045	3.822	-14.4	14.0	-3.96	3.85
EID	579	4.80	0.123	4.160	-20.0	18.4	-4.17	3.83
EIID	577	5.78	0.166	3.525	-18.4	19.6	-3.18	3.39
ZAD	549	3.77	0.354	6.935	-18.4	23.2	-3.93	6.15
ZBD	549	2.82	0.270	6.266	-13.0	13.4	-4.61	4.75
ZCD	549	2.92	0.380	6.150	-10.6	17.1	-3.63	5.86
ZDD	549	2.87	-0.227	4.883	-12.6	9.3	-4.39	3.24
CEM1D	555	2.77	0.280	3.468	- 8.0	11.0	-2.89	3.97
CEM2D	436	3.42	-0.125	2.931	- 8.8	10.2	-2.57	2.98
CEM3D	535	2.12	0.014	3.047	- 6.4	7.0	-3.02	3.30
CEM4D	401	3.74	0.089	3.052	- 8.6	12.5	-2.30	3.34
CEM5D	401	3.66	0.091	3.627	-11.5	13.5	-3.14	3.69

Table 3. Estimated parameters

Series	$\hat{\alpha}$	$\hat{\gamma}$	$\gamma^{1/\hat{\alpha}}$	$\gamma^{1/\hat{\alpha}}/s$
AD	1.83	17.65	4.80	0.66
BD	1.81	9.66	3.50	0.65
CID	1.66	10.70	4.17	0.62
CIID	1.57	6.36	3.25	0.59
DD	1.77	4.47	2.33	0.64
EID	1.73	6.69	3.00	0.63
EIID	1.84	11.78	3.82	0.66
ZAD	1.53	2.91	2.01	0.55
ZBD	1.60	2.04	1.56	0.55
ZCD	1.58	2.16	1.63	0.56
ZDD	1.65	2.40	1.70	0.59
CEM1D	1.83	3.05	1.84	0.66
CEM2D	1.97	5.61	2.40	0.70
CEM3D	1.97	2.16	1.48	0.70
CEM4D	1.95	6.44	2.60	0.70
CEM5D	1.75	4.46	2.35	0.64

The characteristics of these series are given in table 2. The designation is the same as the original series but with an additonal "D". In this case $\bar{x} = 0$. With the exception of series CEM 2D, CEM 3D and CEM 4D the kurtosis is fairly high, in particular for series ZAD, ZBD and ZCD. In table 2, the minimum and maximum values are mentioned as related to the standard deviation. However, it are not only these two values that make the kurtosis high. A more detailed analysis shows that several values contribute to the high kurtosis and not only the two extremes.

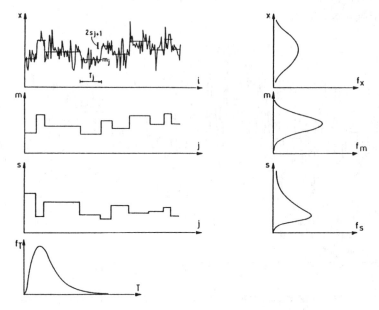

Fig. 1 - Renewal pulse process for concrete strength

The parameter estimation proceeds as outlined in section 4, whereby it is assumed that $\beta = \delta = 0$. The values of $\hat{\alpha}$ and $\hat{\gamma}$ are given in table 3. Also mentioned is the standardization factor $\hat{\gamma}^{1/\hat{\alpha}}$ and the ratio of this quantity to the empirical standard deviation s. This ratio varies between

0.59 and 0.66 for series AD to ED
0.55 and 0.59 for series ZAD to ZDD
0.64 and 0.70 for series CEM 1D to CEM 5D.

As for most differenced series $\hat{\alpha}$ is significantly lower than 2, it can be concluded that the differenced concrete strength series belong to the general class of stable distribution functions. For the cement strength series $\hat{\alpha}$ is rather close to 2 which corresponds to the normal distribution. This is quite logical since the cement production process is expected to yield more uniform results than a concrete production plant.

6 DISCUSSION

In the papers on economic time-series referred to in the first section, it is stated that significant price variations can be attributed to
a) the diffusion of new information on the market
b) the renewed evaluation of available information.

Regarding concrete strength, similar processes occur which cause jumps in the strength level :
a) the "new information" referred to above, can be considered as similar to new values of the basic variables which determine concrete strength. These can be caused by a new supply of cement or aggregates, changes in moisture content of sand, seasonal temperature variations influencing the water content and hence the workability of fresh concrete, etc.
b) The second aspect cited above can be considered similar to the feed-back from strength tests or measurements of the moisture content of sand, resulting in slight modifica-tions of the mix design of the concrete considered.
The occurrence of high tails in the differenced strength series is related to the stochastic model developed by (Taerwe 1985, 1987a). The strength values are assumed to be generated by a renewal pulse process (fig. 1), whereby the mean strength level m is subject to random jumps, resulting in segments with a length T which can be modelled by means of a truncated Gamma distribution. Also the standard deviation in the segments s is subject to random jumps. The model has been verified both for concrete and cement strength records. Fig. 2 shows a record of 200 strength values with the segmentation estimated on the basis of significant jumps in the mean strength level. When differencing the series, it becomes

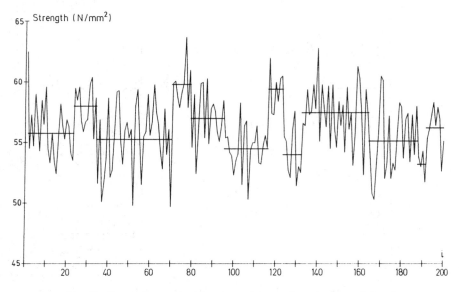

Fig. 2 - Strength record with estimated jumps in mean strength level

clear that generally at each jump, an extraordinary high positive or negative value results. Hence, the occurrence of high tails in the differenced concrete strength series can be linked to the previously established stochastic model.

It was found that in extensive concrete strength records significant autocorrelation exists. Hence, the following time series model (autoregressive series of order 2) appears to be appropriate for a broad range of practical situations

$$x_i = \phi_1 \, x_{i-1} + \phi_2 \, x_{i-1} + \varepsilon_i \qquad (13)$$

In this formula the indexed variable x_i represents a strength value from a record arranged in chronological order and ε_i is a discrete white noise series, consisting of independent contributions with a Gaussian marginal distribution $N(\mu, \sigma)$. A statistical analysis of extensive concrete strength records from different sources reveals that $\phi_1 = 0.4$ and $\phi_2 = 0.2$ are fairly representative values. The influence of autocorrelation on OC-lines for conformity criteria can be rather important and needs to be taken into account if one wants to obtain realistic probabilities of acceptance (Taerwe 1987b). The correlation structure in time is transformed into a spatial correlation of strengths in a concrete structure.

From the previous analysis it is clear that the real variation pattern of consecutive concrete strength values is much more complicated than

generally accepted. For most practical applications however, the normal or lognormal distribution approach is an appropriate tool.

7 CONCLUSIONS

- The paper gives a survey of the main properties of the class of stable distribution functions, a parameter estimation procedure and indications for its practical use. The density functions are characterized by long tails which yield a high kurtosis value. The main property of these distributions is its stability or invariance under summation.
- It is shown that long concrete strength records belong to this general class of distribution functions. Stable distribution functions were originally applied for modelling economic time-series.
- The occurrence of extreme values in the long tails is linked to jumps occurring in the mean strength level of strength records.
- Stable distribution functions can be used for simulating "realistic" concrete strength records.

REFERENCES

CHAMBERS J., MALLOWS C., STUCK B. 1976. A method for simulating stable random variables, Journal of the American Statistical

Association, Vol. 71, No. 354 : 340-344.

FAMA E. 1965. The behaviour of stock-market prices, Journal of Business, Vol. 38 : 34-105.

FAMA E., ROLL R. 1968. Some properties of symmetric stable distributions, Journal of the American Statistical Association, Vol. 63, No. 323 : 817-836.

FIELITZ B., SMITH E. 1972. Asymmetric stable distributions of stock price changes, Journal of the American Statistical Association, Vol. 67, No. 840 : 813-814.

MANDELBROT B. 1960. The Pareto-Lévy law and the distribution of income, International Economic Review, Vol. 1, No. 2 : 79-107.

MANDELBROT B. 1963. The variation of certain speculative prices, Journal of Business, Vol. 36 : 394-419.

MANDELBROT B. 1967. The variation of some other speculative prices, Journal of Business, Vol. 40 : 393-413.

OFFICER R. 1972. The distribution of stock returns, Journal of the American Statistical Association, Vol. 67, No. 340 : 807-812.

PAULSON A., HOLCOMB E., LEITCH R. 1975. The estimation of the parameters of the stable laws, Biometrica, Vol. 62, No. 1 : 163-170.

PRESS S. 1972. Estimation in univariate and multivariate stable distributions, Journal of the American Statistical Association, Vol. 67, No. 340 : 842-846.

TAERWE L. 1985. Aspects of the stochastic nature of concrete strength including compliance control, Ph.D. Thesis, University of Ghent, Belgium.

TAERWE L. 1987a. Detection of inherent hetero geneities in cement strength records by means of segmentation, Uniformity of Cement Strength, ASTM STP 961, E. Farkas and P. Klieger, Eds., American Society for Testing and Materials, Philadelphia : 42-65.

TAERWE L. 1987b. Influence of autocorrelation on OC-lines of compliance criteria for Concrete Strength, Materials and Structures, Vol. 20 : 418-427.

TAERWE L. 1992. Extension of the statistical basis and procedures for quality control of concrete (in Dutch), Thesis for the degree of "Ge-aggregeerde voor het Hoger Onderwijs" (Habilitation), University of Ghent, Belgium.

Structural Safety & Reliability, Schuëller, Shinozuka & Yao (eds) © 1994 Balkema, Rotterdam, ISBN 90 5410 357 4

Structural reliability analysis of nonlinear systems

Goran Turk, Martin R. Ramirez & Ross B. Corotis
The Johns Hopkins University, Baltimore, Md., USA

ABSTRACT: In addition to traditional reliability methods such as Monte Carlo simulation and first-order second-moment, a relatively new response surface approach has been recently introduced in reliability analysis. The main advantage of the response surface method is its independence of the method used to solve the mechanics of the problem, since the reliability analysis is performed on the fitted response surface. This enables use of the method in a wide variety of problems. Difficulties may arise, however, when using this approach. Some of them and associated solutions are introduced in this paper. The use of derivatives of the response with respect to input variables in the approximation procedure is given. A simple numerical example shows that this additional information about response may improve the estimation of the probability of failure. A discussion on the choice of response variable for the ultimate limit state is also presented.

1 INTRODUCTION

The behavior of a structure can be described by a response variable y, which is typically stress or strain, or displacement at some point of interest. The response variable is a function of input variables $\underline{x}$ (geometry, material properties, loads), all of which may be random. Performance of the structure is measured by comparison to a limiting value, which may represent the ultimate limit state (collapse of the structure), damage limit state or serviceability limit state. The limit state is thus defined by:

$$g(\underline{x}) = y_{limit} - y(\underline{x}) = 0 \qquad (1)$$

The reliability is the probability that the response exceeds the limit value, which can be evaluated by the following equation,

$$P_f = P[g(\underline{x}) \leq 0] = \int_{g(\underline{x}) \leq 0} f_X(\underline{x}) d\underline{x} \qquad (2)$$

where $f_X(\underline{x})$ is the joint probability density function of all input variables.

Equation 2 is usually impossible to integrate analytically because the integrand and the integration region are highly nonlinear, and the function $g(\underline{x})$ which defines the integration limits is often not known explicitly. This is especially true in structural systems with nonlinear material and geometry. Various approximate and iterative methods, such as the first-order second-moment method, second-order second-moment method and Monte Carlo simulation, are summarized in reliability analysis books (Melchers, 1987, Thoft-Christiansen and Baker, 1982). Haldar and Zhou (1993) and Liu and Der Kiureghian (1991) have shown that first- and second-order reliability method may be effectively used in reliability analysis of nonlinear structures.

Recently, response surface techniques have been adopted by several researchers (Faravelli, 1989, Bucher and Bourgund, 1990, Corsanego and Lagomarsino, 1990, Thacker and Wu, 1992, Böhm, Reh and Brückner-Foit). The method consists of three steps:

- The response variable is calculated for different sets of value of input variables. This part can be solved by a finite element code or some other numerical procedure.
- The response is approximated by an explicit function, usually fit by the least squares method.
- The reliability of the structure is then determined by Monte Carlo simulation, using the explicit function. This improves the efficiency of the method since the mechanics of the problem is solved fewer times than in traditional Monte Carlo simulation.

2 RESPONSE SURFACE METHOD

The response surface method was introduced in the early 1950's and used extensively in physical, biological, clinical, social and food science (Myers, Khuri and Carter, 1989). The procedure is described in many books (Petersen, 1985, Khuri and Cornell, 1987, Box and Draper, 1987).

In structural reliability analysis of nonlinear systems, the relationship between response, y, and input variables, $\underline{x}$, is usually not known explicitly. In this case, the response is determined for a discrete set of values of $\underline{x}$, based on a number of experiments. The design of these experiments determines the accuracy of the reliability estimation.

2.1 Design of experiments

A 2^k factorial design is the basic design used for first-order models, i.e., the approximation function is a linear function of the input variables. In the 2^k factorial design each input variable is measured at two levels, a low and high level. The number of experiments is therefore 2^k, where k is the number of input variables. Other methods used for first-order models include the simplex design and the Placket-Burman designs.

A 3^k factorial design, a fractional 3^k factorial design, the Box-Behnken designs and the central composite designs are the most commonly used designs for second-order models. Central composite design consists of a complete 2^k factorial design, n_0 center points and two axial points for each input variable at a distance α from the design center. The set of center points is used in the analysis of variance for estimation of pure error - the error due to uncertainty in the experiment. By repeating the experiment for the same values of input variables and assuming that pure error exists, the lack of fit error may be estimated (Petersen, 1985, Khuri and Cornell, 1987, Box and Draper, 1987). In structural reliability, the experiments are performed numerically and the pure error does not exist if it is not introduced artificially (Böhm and Brückner-Foit).

2.2 The least squares method

The second part of the response surface method consists of approximating the real response, y, by the function $\hat{g}(\underline{x})$.

$$y = \hat{g}(\underline{x}) + \varepsilon = \hat{y} + \varepsilon \tag{3}$$

where $\underline{x}$ is set of input variables, $\hat{y}$ is the approx-

imated response and ε is the total error (lack of fit error and pure experimental error). A function $\hat{g}(\underline{x})$ that is linear in parameters b_i (but not in $\underline{x}$) is usually selected (see Equation 4). The advantage of a linear model is that it leads to a set of linear algebraic equations to be solved for the unknown parameters. A nonlinear model leads to a set of equations which are usually solvable only by numerical methods. It is convenient to choose the function g such that it reflects the general relation between the response and input variables. For instance, if the response variable is displacement at some point, and one of the input variables $\underline{x}$ is Young's modulus, it is reasonable to include $\frac{1}{x}$ as one of the $z_i(\underline{x})$ in Equation 4.

$$\hat{y} = \sum_{i=1}^{p} b_i \, z_i(\underline{x}) \tag{4}$$

where b_i are parameters of the model to be determined by the least squares method, $z_i(\underline{x})$ are some functions of input variables $\underline{x}$, and p is the number of parameters of the model.

For each outcome of the experiment an approximation by the chosen model can be made

$$\hat{y}_j = \sum_{i=1}^{p} b_i \, z_{ij} = y_j - \varepsilon_j \tag{5}$$

in which z_{ij} is the function $z_i(\underline{x})$ evaluated for the set of values of the input variables $\underline{x}$ of the jth experiment. The summed squared residuals for n experiments are

$$S(\underline{b}) = \sum_{j=1}^{n} \varepsilon_j^2 = \sum_{j=1}^{n} [y_j - \sum_{i=1}^{p} b_i z_{ij}]^2 \tag{6}$$

The p normal equations are derived from Equation 6 by differentiating with respect to b_i

$$\sum_{i=1}^{p} \sum_{j=1}^{n} z_{ij} z_{kj} \, b_i = \sum_{j=1}^{n} z_{kj} \, y_j \tag{7}$$

The normal equations method is usually used to solve the least-squares problem. In many cases the system of normal equations is singular, or close to singular, and the results can't be obtained or are erroneous. Alternatively, QR decomposition and singular value decomposition (SVD) methods have been used to solve the least squares problem, although not in the context of response surfaces (Golub and Van Loan, 1991, Press, Teukolsky, Vetterling and Flannery, 1992).

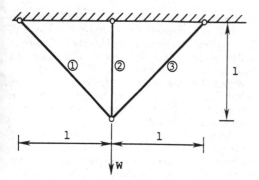

Figure 1: The three bar truss

Table 1: Probability of failure for different values of load

Load factor	P_f "exact"	P_f response surface
1.0	0.69825	0.69530
0.9	0.14675	0.15152
0.8	0.00515	0.01271

2.3 Singular value decomposition

Equation 6 may be rewritten in matrix form

$$S(\underline{b}) = \|\underline{y} - \underline{Z} \cdot \underline{b}\|_2 \tag{8}$$

The matrix $Z_{n \times p}$ can be decomposed by singular value decomposition into three matrices

$$\underline{Z} = \underline{U} \cdot \underline{W} \cdot \underline{V}^T \tag{9}$$

where $\underline{U}_{n \times p}$ and $\underline{V}_{p \times p}$ are column-orthogonal matrices, and $\underline{W}_{p \times p}$ is a diagonal matrix with non-negative elements - singular values w_i. If the system is singular, or close to singular, one or more singular values is zero or close to zero. If singular value w_i is zero or close to zero this suggests that the equation $z_i(\underline{x})$ does not contribute to the approximation of actual response, and the corresponding parameter b_i should be zero. This is obtained by setting $1/w_i$ to zero. The solution of the least squares problem is

$$\underline{b} = \underline{V} \cdot \underline{W}^{-1} \cdot \underline{U}^T \tag{10}$$

The main advantage of the SVD method is that it theoretically cannot fail, whereas the normal equation method is numerically very unstable. SVD also reduces round-off error. Disadvantages of SVD are that it requires larger storage space and is slower than the normal equation method. In general the SVD method is recommended instead of normal equations.

2.4 Sensitivity information

As mentioned previously, the response of a non-linear structural system will generally require the use of a numerical procedure, such as the finite element method. Because a complete finite element analysis for a particular set of input variable is very time-intensive, it is desirable to obtain as much information as possible from each analysis. Some researchers have recently studied techniques to determine the sensitivity of the response in terms of the input variables (Arora and Cardoso, 1989, Zhang and Der Khiureghian, 1991) by variational or analytical methods. Although advances are still needed to develop efficient methods for nonlinear systems, this approach appears to be very promising. If the derivatives of the response variable with respect to input variables are known, the least squares procedure needs to be modified. In addition to Equations 5, a set of equations which represents the derivatives of the response function with respect to the input variable is

$$\frac{\partial \hat{y}_j}{\partial x_l} = \sum_{i=1}^{p} b_i \frac{\partial z_{ij}}{\partial x_l} = \frac{\partial y_j}{\partial x_l} - \varepsilon'_j \tag{11}$$

The sum of squares becomes

$$S(\underline{b}) = \sum_{j=1}^{n+n_d} \varepsilon_j^2 = \sum_{j=1}^{n} [y_j - \sum_{i=1}^{p} b_i z_{ij}]^2 +$$
$$\rho \sum_{j=1}^{n_d} \left[\frac{\partial y_j}{\partial x_l} - \sum_{i=1}^{p} b_i \frac{\partial z_{ij}}{\partial x_l} \right]^2 \tag{12}$$

in which different values of the weight, ρ may be used. If the differentiation is achieved numerically the derivatives are less accurate than the values of the function g, and therefore values of ρ smaller than unity should generally be used. n_d is the number of additional terms given by derivatives of the response function (for instance, if derivatives with respect to three input variables are used at all points where $z_i(\underline{x})$ is computed, $n_d = 3n$).

3 NUMERICAL EXAMPLES

Some simple numerical examples are presented here to show the applicability of the response surface method and illustrate problems that may arise in using this method.

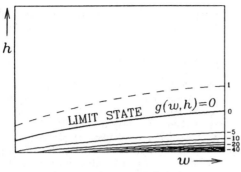

a) Original response function
$$g(w, h) = a - b\frac{w}{h^3}$$

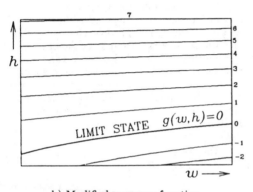

b) Modified response function
$$g(w, h) = a\,h^2 - b\frac{w}{h}$$

Figure 2: Response variable

3.1 Three bar truss

The three bar truss shown in Figure 1, with deterministic bar lengths, is loaded by a deterministic force $W = Load\ factor\ (1 + \sqrt{2})\ \mu_{Ny}$; Young's modulus is equal for all bars and deterministic; and the material is elastic perfectly-plastic. The yield forces in each bar are normally-distributed, independent random variables, with a mean of $\mu_{Ny} = 1$ and standard deviation $\sigma_{Ny} = 0.1$.

The collapse of the truss is considered as a limit state, i.e., the yield force is reached in bar 2 and one of the side bars. For small displacements, the limit state can be represented by two piecewise linear equations. The problem was analyzed by Augusti, Baratta and Casciati (1984).

$$g = W - N_{y2} - \sqrt{2}\,N_{y1} = 0 \quad \dots \quad N_{y1} < N_{y3}$$
$$g = W - N_{y2} - \sqrt{2}\,N_{y3} = 0 \quad \dots \quad N_{y3} < N_{y1} \quad (13)$$

The 3^k factorial design has been used and a quadratic approximation for $g(N_{y1}, N_{y2}, N_{y3})$ was found

by the least squares method. The Monte-Carlo method with a sample size of 100,000 was used to find the "exact" value of probability of failure as well as to find the probability of failure from the approximate response surface.

Results for three different load levels are shown in the Table 1. Agreement with exact results is very good for a higher load factor, when the probability of failure is high, but is very poor for a lower load level. This is because the computed reliability is more sensitive to the accuracy of the fitted response surface over a specific sub-range of input variables when the probability of failure is small.

3.2 Cantilever beam

A cantilever beam is subjected to uniformly distributed load with a random magnitude, w, which is normally distributed with a mean $\mu_w = 1000\frac{N}{m^2}$ and coefficient of variation $V_w = 0.2$. Material is elastic with a deterministic Young's modulus of $E = 2.6\ 10^{10}\frac{N}{m^2}$. Length of the beam, l, is 6.0 m. Height of the beam, h, is a normally distributed random variable with $\mu_h = 0.25m$ and $V_h = 0.15$. The deflection serviceability limit state is

$$g = \frac{l}{325} - \frac{12wl^4}{8Eh^3} =$$

$$= 18.46 - 7.48\ 10^{-5}\frac{w}{h^3} = 0 \quad (14)$$

The function g (Equation 14) is highly nonlinear. The values of $\frac{\partial g}{\partial h}$ change very quickly around the limit state (see Figure 2a) and a quadratic approximation does not represent the response function adequately. The response function may be changed so that the function g is smoother around the limit state. This can be achieved by multiplying Equation 14 by h^2 and then fitting a quadratic function. This approach has been used to solve this numerical problem (see Figure 2b).

A 3^k factorial design and quadratic approximation were used. The approximation was performed by the least-squares method, and the values of derivatives $\frac{\partial g}{\partial w}$ and $\frac{\partial g}{\partial h}$ were used to obtain a better approximation. The Monte-Carlo simulation with 100,000 samples was used to estimate the probability of exceeding the serviceability limit with both the "exact" and response surface approach. The results are shown in the Table 2.

The result was not accurate when the quadratic response surface approach was applied to the original response function (Equation 14). Inclusion of the mixed term, $w\,h$, in the quadratic approximation made a significant difference. Much better results were found for the changed response function. From Table 2 it can be seen that the use of

Table 2: Probability of exceeding the serviceability limit

Weight ρ	Original response		Modified response	
(see Eqn. 12)	No mixed term	Mixed term	No mixed term	Mixed term
0.0	0.00423	0.00626	0.00828	0.00904
0.1	0.00473	0.00679	0.00836	0.00911
1.0	0.00688	0.00907	0.00860	0.00944
10.0	0.01116	0.01369	0.00910	0.00996
100.0	0.01256	0.01529	0.00929	0.01012
1000.0	0.01276	0.01551	0.00931	0.01013
"exact"	0.00953			

derivatives in the approximation (with moderate values of ρ) improves the estimation of probability of exceeding the serviceability limit state.

3.3 Thick-walled cylinder

A thick-wall cylinder under internal pressure was investigated by Millwater, Wu and Fossum (1990), who used an augmented first-order second-moment method (Wu, Millwater, Cruse, 1989). The cylinder (see Figure 3) and the load are axisymmetric. The material is elastic perfectly-plastic with elastic modulus, E, 2.1×10^{11} Pa, Poisson's ratio 0.3. It is assumed that the internal pressure, p_0, and yield stress, σ_Y, are random. Internal pressure is taken to be normally distributed with the mean of 1.3×10^8 Pa and coefficient of variation 0.05. Yield stress is distributed by a Weibull distribution and has a mean of 3.1×10^8 Pa and coefficient of variation 0.1. The response variable was chosen to be the radial stress, σ_r, at r = 3 cm.

The finite element method is used to solve for the structural response. Axisymmetrical 4-node elements are used to form a FEM mesh (see Fig. 4). The Monte Carlo simulation with 100,000 samples was run to determine the "exact" cumulative distribution function (CDF) of σ_r. In this case, the quadratic function was chosen as a response surface. There were six unknown coefficients to be determined by solving the least squares problem. Two different experimental designs were used: 3^k factorial and 2^k factorial design. In the case of 2^k factorial design the derivatives with respect to input variables were found. Without this additional information it would be impossible to determine the unknown coefficients, as there are only four points in 2-D 2^k factorial design.

The cumulative distribution function of radial stress at r = 3 cm obtained by Monte Carlo simulation and the response surface approach is shown on Figure 5. The CDF obtained by the response surface approach is a good approximation of "ex-

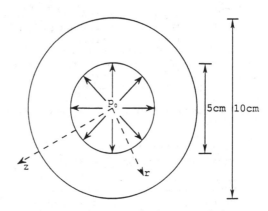

Figure 3: Thick-walled cylinder

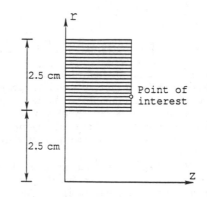

Figure 4: FEM mesh

act" results. It should be noted that the response surface method is much more efficient than Monte Carlo simulation. The FEM code was called only 9 times in the case of 3^k factorial design, 4 times (plus sensitivity analysis) in the case of 2^k factorial design, and 100,000 times in Monte Carlo simulation.

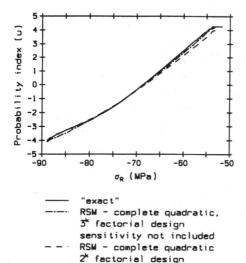

- "exact"

-·-·· RSM - complete quadratic,
      3^k factorial design
      sensitivity not included

- - · RSM - complete quadratic
      2^k factorial design
      sensitivity included

Figure 5: Cumulative distribution function of σ_r

Table 3: Means and coefficients of variation of input variables

Input variable	Mean	Coefficient of variation
EI	3.5×10^6 N m^2	0.1
L	3 m	0.05
P	4.4×10^5 N	0.2

4 CHOICE OF RESPONSE QUANTITY

In the three examples presented, simple response quantities presented themselves readily. In general, the choice of response quantity is fairly straightforward for the cases of serviceability or damage. For ultimate performance, however, the appropriate quantity is less obvious, especially for a structural system.

Since the collapse of a structure is associated with singularity of the stiffness matrix, the determinant is one such response measure. One drawback, however, is that an accurate value of the determinant may be expensive to obtain, particularly for large matrices. The inverse of the condition number, which is a non-dimensional measure of the singularity of the stiffness matrix, is a much better choice. If the matrix is singular or ill-conditioned the condition number is infinitely large. The condition number is defined as

$$cond(\underline{K}) = \|\underline{K}\| \, \|\underline{K}^{-1}\| \tag{15}$$

where $\|\cdot\|$ represents the norm of the matrix. From Equation 15 it may be seen that the inverse of the stiffness matrix is needed. Because the inverse of the matrix is numerically very expensive, an approximation of condition number has to be used. There are different ways to approximately estimate the condition number

$$cond(\underline{K})_* = \frac{\lambda_{max}}{\lambda_{min}} \tag{16}$$

$$cond(\underline{K})_{**} = \frac{w_{max}}{w_{min}} \tag{17}$$

where λ_{max} and λ_{min} are the largest and the smallest eigenvalues of the matrix $\underline{K}$, and w_{max} and w_{min} are the largest and the smallest singular values of the matrix $\underline{K}$ (Press, Teukolsky, Vetterling and Flannery, 1992).

4.1 Example

The ultimate limit state for elastic stability of a cantilever column is examined to show how the determinant and condition number of the tangential stiffness matrix may be used to perform the reliability analysis of the problem. The input random variables, stiffness EI, height L and axial force P, are chosen to be independent and normally distributed. The means and coefficients of variation are shown Table 3. The Monte Carlo simulations with a sample size of 10^6 were performed to estimate the "exact" probability of failure, $P_f = 0.00034$. 3^k factorial design (27 experimental points) was chosen. The central point of the design was at mean values of EI, L and P. Other points were $\delta \cdot \sigma_{EI}$, $\delta \cdot \sigma_P$ and $\delta \cdot \sigma_L$ from the mean, where $\sigma_.$ is standard deviation of the input random variable and δ is a parameter that varied from 1 to 3.

Before the response surface was approximated, the values of input variables were transformed

$$x_i = \frac{X_i - \mu_{Xi}}{\sigma_{Xi}} \tag{18}$$

where x_i are transformed values of input variables EI, L and P, respectively. Without the transformation the results were unstable because the absolute values of EI, L and P are so different that the system becomes ill-conditioned. The determinant of $\underline{K}$ and the inverse of the condition number of $\underline{K}$ were approximated by a complete quadratic polynomial.

The results (see Figure 6) were very inaccurate in the case of det($\underline{K}$) and were relatively better in

the case of $1/\mathrm{cond}(\underline{K})$. The terms of the tangential stiffness matrix include terms such as $\frac{1}{L}$ and $\frac{1}{L^2}$, therefore it is assumed that the following response surface function with the same number of parameters would be better than a quadratic polynomial

$$\hat{g} = b_1 + b_2 x_1 + b_3 \frac{1}{A + x_2} + b_4 x_3 +$$

$$+ b_5 x_1^2 + b_6 \frac{1}{(A + x_2)^2} + b_7 x_3^2 +$$

$$+ b_8 \frac{x_1}{A + x_2} + b_9 x_1 x_3 + b_{10} \frac{x_3}{A + x_2} \qquad (19)$$

where b_i are coefficients to be determined by least squares method and A is a constant. The value of A was first chosen to be $1/V_L = 20.0$ (g_A). A lower value of $A = 6.0$ was also used (g_B). The results improved in the case of $\det(\underline{K})$, but did not change much in the case of $1/\mathrm{cond}(\underline{K})$.

The estimated probability of failure depends on δ, the coordinates of experimental points in transformed variable space. In this case the most consistent results were obtained when δ was between 2.0 and 2.5. The function g_B gives relatively good results but should not be considered the best final solution. Other functions with higher order terms and more than 10 parameters may prove to be even better.

5 CONCLUSIONS

The response surface approach was presented as a tool for reliability analysis. This method has the advantage that it is relatively independent of how the actual response is calculated or measured; therefore different finite element or other codes may be used in evaluating the mechanics component of the problem. However, some problems may arise when the response surface method is used.

The response surface has to approximate the actual response in a way that the estimation of the probability of failure is accurate. As is shown in a simple problem, if the response is not smooth, the approximation of the response with a quadratic function may not be adequate.

If the derivatives of the response with respect to input variables are known or easy to evaluate, they may be considered in the approximation process. It is shown that the estimation of the probability of failure can be improved if derivatives are used in the approximation. Because of accuracy and efficiency considerations, analytical derivatives are favored over those calculated by finite differences.

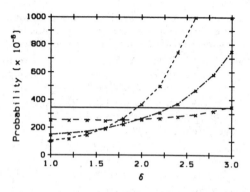

a) Response variable: $\det(\underline{K})$

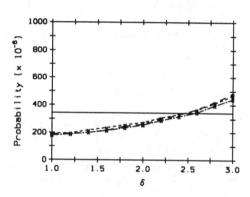

b) Response variable: $1/\mathrm{cond}(\underline{K})$

——— Exact value
--×-- Response surface (quadratic)
—×-- Response surface (function g_A)
- ×· Response surface (function g_B)

Figure 6: Probability of failure of cantilever column

Acknowledgment

This research is part of a project on reliability of nonlinear structural systems. The support of the National Science Foundation, Grant MSS-9016018 is gratefully acknowledged.

REFERENCES

Augusti, G., A. Baratta, & F. Casciati 1984. *Probabilistic methods in structural engineering*. Chapman and Hall, London.

Arora, J.S. & J.E.B. Cardoso 1989. A design sensitivity analysis principle and its implementation into ADINA. *Computers & Structures*, 32: 691-705.

Böhm, F. & A. Brückner-Foit. On a criterion for accepting a response surface model. *Probabilistic Engineering Mechanics Journal*, (submitted).

Böhm, F., S. Reh & A. Brückner-Foit. A review of response surface techniques - Their Strengths and Limitations. unpublished.

Box, G.E.P. & N.R. Draper 1987. *Empirical model-building and response surfaces*. John Willey & Sons, New York, 1987.

Bucher, C.G. & U. Bourgund 1990. A fast and efficient response surface approach for structural reliability problems. *Structural Safety* 7: 57-66.

Corsanego, A. & S. Lagomarsino 1990. A response surface technique for fuzzy-random evaluations of local seismic hazard. *Structural Safety* 8: 241-254.

Faravelli, L. 1989. Response surface approach for reliability analysis. *Journal of Engineering Mechanics* 115: 2763-2781.

Golub, G.H. & C.F. Van Loan, *Matrix computation*, The Johns Hopkins University Press, Baltimore.

Haldar A. & Y. Zhou 1992. Reliability of geometrically nonlinear PR frames. *Journal of Engineering Mechanics* 118: 2148-2155.

Khuri, A.I. & J.A. Cornell 1987. *Response surfaces, designs and analysis*. Marcel Dekker, Inc., New York.

Liu, P.-L. & A. Der Kiureghian 1991. Finite element reliability of geometrically nonlinear uncertain structures. *Journal of Engineering Mechanics* 117: 1806-1825.

Melchers, R.E. 1987. *Structural reliability, analysis and prediction*. John Wiley & Sons, New York.

Millwater, H., Y. Wu & A. Fossum 1990. Probabilistic analysis of a materially nonlinear structure. *AIAA* 1099: 1048-1053.

Myers, R.H., A.I. Khuri & W.H. Carter Jr. 1989. Response surface methodology: 1966-1988. *Technometrics* 31: 137-157.

Petersen, R.G. 1985. *Design and analysis of experiments*. Marcel Dekker, Inc., New York.

Press, W.H., S.A. Teukolsky, W.T. Vetterling & B.P. Flannery 1992. *Numerical recipes*. Cambridge University Press, Cambridge.

Thacker, B. & Y.T. Wu 1992. A new response surface approach for structural reliability analysis. AIAA 2408: 586-593.

Thoft-Christiansen, P. & M.J. Baker 1982. *Structural reliability theory and its applications*. Springer-Verlag, Berlin.

Wu, Y., H. Millwater & T. Cruse 1989. An advanced probabilistic structural analysis method for implicit performance functions. AIAA 1371: 1852-1859.

Zhang Y. & A. der Khiureghian 1991. Dynamic response sensitivity of inelastic structures. Dept. of Civil Engineering, University of California at Berkeley, Report no. UCB/ SEMM-91/06.

Structural Safety & Reliability, Schuëller, Shinozuka & Yao (eds) © 1994 Balkema, Rotterdam, ISBN 90 5410 357 4

Reliability evaluation of offshore platforms against hurricanes

Y. K. Wen
University of Illinois at Urbana-Champaign, Ill., USA

H. Banon
Exxon Production Research Corporation, Houston, Tex., USA

ABSTRACT: Offshore platforms in the Gulf of Mexico may be subjected to environmental loads due to hurricanes. Forces on platforms due to winds, waves, and currents produced by hurricanes are random, time varying and correlated. The performance of platforms under hurricanes has been an important consideration in developing reliability-based design criteria. A method has been developed for the evaluation of reliability of fixed-bottom and tension leg platforms subjected to hurricane conditions. The hurricane wind, wave, and current fields; variabilities of storm parameters and the dynamic nature of response of platforms are properly considered in this method. The methodology and risk results were used to develop reliability-based environmental design criteria.

1 INTRODUCTION

A primary concern in analysis and design of offshore platforms in the Gulf of Mexico (GOM) is the extreme environmental loads, i.e. forces due to winds, waves, and currents produced by hurricanes. These loads are time varying and stochastic in nature and because they are associated with a storm system they are often correlated in magnitude, direction and phase. Therefore the variability in the loadings and reliability of platforms under such loadings are important design considerations. The previous design practice for platforms in the GOM was to use a wave height that corresponds to a 1 % annual probability of exceedance, i.e. a 100-year return period. The currents and winds likely to coexist with the design wave have been suggested as the design values. Therefore similar to other codes and standards (e.g., for seismic loads), although the selection of the design environment was based on probability of occurrence, the reliability of the platforms designed according to the procedures was undefined and unknown. In other words, the design code avoided specifying target reliabilities and environmental event combination criteria. In order to develop a reliability-based design procedure, a methodology is needed that considers the variabilities and uncertainties in the loadings and platform resistance and accounts for the joint probability of the winds, waves, and currents. In the following, a recently developed method is outlined. The results were used to specify the design combination criteria for environmental events. Details can be found in Wen (1988, 1990, 1992).

2 RISK MODEL FOR HURRICANE LOAD AND LOAD EFFECT

In the GOM, significant data for winds, waves, and currents based on actual recordings and hindcast results have been accumulated since the turn of the century which allow prediction of site specific statistics. There is still a need, however, for developing a code procedure which can be used by those who do not have access to the data or the resources to perform a detailed statistical analysis. For this purpose a risk model for predicting probability of environmental events and their combination at a given site is needed. A model is developed in this study based on consideration of:

1 statistics and hindcast results for hurricanes in the GOM since the turn of the century
2 wind, wave, and current fields during passage of a hurricane
3 variability in hurricane parameters and

Table 1 Hurricane Parameter Statistics

Parameter	Δp (mb)	R (nm)	V_f (kts)
Sample size	69	69	69
Distribution	lognormal	lognormal	lognormal
Mean	46.38	24.92	9.75
Standard Deviation	31.64	10.55	3.74
Coeff. of Variation	0.68	0.42	0.38
Sample K-S Statistics	0.135	0.076	0.106
significance Level P[K-S Statistics > Sample Value]	16 %	82 %	41 %

$$\rho_{\Delta p,R} = -0.351, \ \rho_{\Delta p, Vf} = 0.260, \ \rho_{R,Vf} = -0.108$$

$$\rho = \text{correlation coefficient}$$

Table 2 Platform Annual Safety Index Against Overturning Moment

water depth / RSR	300 ft		600 ft		1000 ft	
	β	P_f (in 10^{-3})	β	P_f (in 10^{-3})	β	P_f (in 10^{-3})
1.8	2.79	2.63	2.89	1.92	2.84	2.25
2.0	2.91	1.80	3.01	1.30	2.95	1.58
2.2	3.01	1.30	3.11	0.99	3.05	1.14
2.5	3.20	0.82	3.25	0.57	3.19	0.71

Table 3 Comparison of Annual Limit State Probabilities of Fixed-Bottom Platforms
(Effect of Hurricane Load Directionality and Platform Resistance Asymmetry)

RSR	Platform Asymmetry		Worse Case
	1.5	1.2	
1.8	0.50×10^{-3}	1.10×10^{-3}	2.0×10^{-3}
2.0	0.34×10^{-3}	0.74×10^{-3}	1.4×10^{-3}
2.5	0.15×10^{-3}	0.34×10^{-3}	0.65×10^{-3}

uncertainty in hindcasting its environmental effects at a site

4 hurricane track geometry and platform orientation

The output of the model is the probability of exceedance for an environmental event, forces and load effects at a given site for a given period of time. Both fixed-bottom and tension leg platforms are considered.

2.1 Hurricane Parameter Distribution

The parameters of interest are the pressure difference (Δp), radius of maximum wind speed (R), and storm translation speed (V_f). The statistics of maximum values observed before storms crossed the shelf are used. The statistics of these parameters indicate that they can be modeled as jointly lognormal random variables. The fitted model parameters and the K-S statistics and corresponding significance levels are given in Table 1. It is seen that the sample evidence does not contradict the models with a significance level of at least 16%. Notice the large coefficients of variation of all the parameters and negative correlation coefficients between some storm parameters. The storm track statistics indicate that the track crossing points along the depth contour follow approximately a uniform distribution and the track direction relative to the contour line follows a beta distribution. The storm occurrence over time can be approximately modeled as a simple Poisson process (Haring and Heideman, 1978).

2.2 Environmental Events Hindcast Model

The hurricane is assumed to travel along a straight line with a given translation speed and direction (Fig.1). It is governed primarily by Δp, V_f, R and the Coriolis force. The changes in the storm parameters over the life of the storm are not considered. The wind and wave predictions are given by parametric expressions fitted to more detailed numerical hindcast results. These expressions give the wind velocity, significant wave height, and wind and wave directions as functions of the hurricane parameters and the site position relative to the storm center. The current prediction is based on a one-dimensional numerical model (Cooper,1988). Snapshots of the wave and current

fields around the center of hurricane Camille (1969) according to the models in this study are shown in Fig.2. Comparison with storm data indicated that the errors of the simplified hindcast model are small and their effect on the risk prediction is not important (Wen, 1988).

2.3 Conversion From Load to Load Effect

For fixed-bottom platforms, the global overturning moment (OTM) is due to the combined action of wind, wave and current. An empirical parametric model for OTM is used in this study;

$$OTM = C_1 (H_{max} + C_2 V)^{C_3} + C_4 W^2 \quad (1)$$

in which H_{max} is the maximum wave height, V is the component of surface current aligned with the wave direction, and W is the wind speed. C's are empirical constants. Comparisons with calculated values indicates that the r.m.s. error in this parametric equation is small (Wen, 1988).

For tension leg platforms (TLPs, which are in deeper waters), both the environmental forces and structural response are significantly different. For example, the contribution of wind to TLP response is higher due to the very low frequency content of the wind spectrum and the extremely long periods of TLP surge motions. Also, currents play a comparatively less important role because of relatively small tethers. Since the platform is compliant, responses are time -variant and involve combinations of various time dependent and time-independent components. Additionally, there are more limit states to be considered such as maximum and minimum tether tension, platform offset and air gap (i.e., the possibility of wave impact). The global response calculations were performed by a computer program, aTLPcx, developed by Offshore System and Analysis Corporation (OSAC, 1991). This program is basically a frequency domain solver which is based on first-order transfer functions of response. The global responses of interest in this study are the extreme values for a given sea state over a given duration.

During the passage of a storm, the platform global responses build up and decay; the global behavior of a TLP or a fixed-bottom platform depends on the hurricane parameters and the proximity of the platform to the storm track. The above models

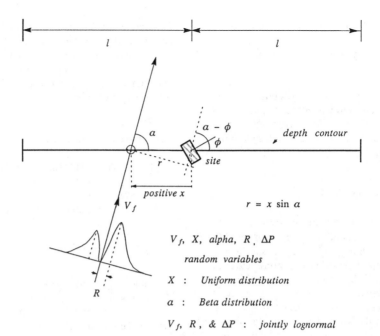

$$r = x \sin \alpha$$

$V_f,\ X,\ alpha,\ R\ ,\ \Delta P$

random variables

$X\ :\quad$ *Uniform distribution*

$\alpha\ :\quad$ *Beta distribution*

$V_f,\ R,\ \&\ \Delta P\ :\quad$ *jointly lognormal*

Figure 1 Geometry of a Random Hurricane Path Relative to a Given Site

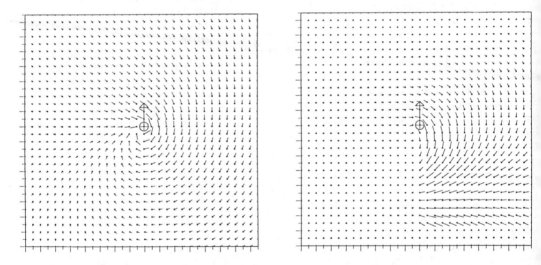

Figure 2 Hurricane Wave Field (left) and Hurricane Current Field (right)

provide the tool for conversion from force to load effect as the storm passes by the platform site.

2.4 Envelope Function and Response Surface Analysis

The risk analysis is facilitated by the use of an envelope function. This is defined as the most severe (maximum or minimum) response during the passage of a hurricane as a function of the storm parameters and the distance from the platform site to the storm track. As the storm parameters considered are Δp, R and V_f, the envelope function can be conveniently expressed as a product of two functions as follows;

$$G = S (\Delta p, R, V_f) F(r/R) \qquad (2)$$

where S is a scale function which attains its maximum (i.e., the maximum response that the platform will experience) at distance of $r = R$ to the storm track (see Fig. 1). $r = X \sin(\alpha)$, in which X is the distance of the crossing point from the site, a uniformly distributed random variable and α is the storm direction. F is a shape function that describes the decrease in response as the site is moved away from the storm track. The envelope function is therefore a parametric equation of the extreme response taking into account the dynamics of the platform, hurricane wind, wave and current fields and the geometry of the storm track relative to the site.

To determine the scale functions, a response surface method (RSM) in conjunction with a central composite design is used; this function is fit by second-order polynomials of the three storm parameters. For this purpose, the hurricane field models and the fixed platform parametric equation or the program aTLPcx are used to calculate the extreme responses of interest for a set of combinations (grid points) of the storm parameters. This method gives accurate fit of the scale function with a small number of points in the range of the storm parameters considered. The spacing between the grid points should be wide enough to cover the range of platform response under consideration. Comparisons of results of this RSM prediction of the response with data indicate that the fits are excellent with a maximum error of about 3 %. An

example is shown in Fig.3 for maximum offset of a TLP.

To determine the shape function, maximum responses are calculated at various distances from the storm track for the 40 most intense storms in GOM for the period 1900 to 1983 and normalized by the value at $r = R$. These shape functions are fitted by exponential decay functions via a nonlinear regression procedure.

2.5 Risk Analysis

The probability of a response exceeding or falling below a given threshold level is calculated using the envelope function defined in the foregoing. This method provides a means for fast evaluation of the probability and sensitivity analysis of the risk to system parameters. Referring to Fig. 1, the probability that the loading S exceeding a threshold level, s , given that a hurricane has occurred in the reference frame can be evaluated with the aid of the envelope function as follows;

$$P(S > s | H) = \frac{1}{2\ell} \int dx \int f_\alpha d\alpha \iiint_{G > s} f_{\Delta p, R, V_f} d\Delta p \cdot dR \cdot dV_f$$

$$(3)$$

in which H indicates the occurrence of hurricane in the reference frame of length 2 l, and f is the probability density function or joint probability density function of the storm parameters. The integration is over the domain G > s where G is the envelope function. The probability that the threshold level is exceeded in a time period of T years is given by

$$P_f = 1 - \exp [-v_h P (S > s / H) T] \qquad (4)$$

in which v_h is occurrence rate of hurricanes per year in the reference frame. The integral given in Equation 3 can be evaluated numerically or by a first-order reliability method. Because of its large dimensions, this integral is evaluated by the Monte-Carlo method in this study.

The probabilities are also evaluated from direct computer simulations of the problem, i. e., without using the envelope functions. The number of hurricanes that occur within the reference frame

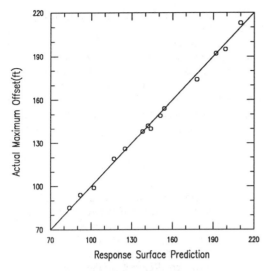

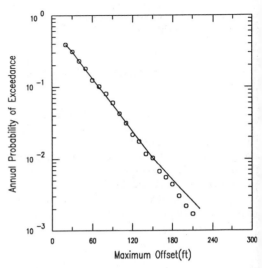

Figure 3 Comparison of Response
Surface Prediction of Maximum
Offset with Data (TLP)

Figure 5 Probability of Exceedance
of Annual Maximum Offset(TLP)

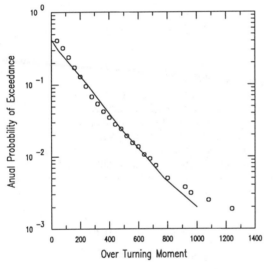

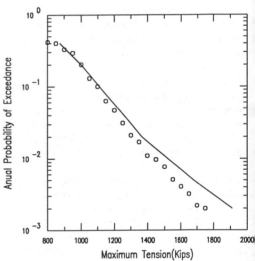

Figure 4 Probability of Exceedance
of Annual Maximum Overturning
Moment(fixed–bottom Platform)

Figure 6 Probability of Exceedance
of Annual Maximum Tension(TLP)

(Fig. 1) for the time period considered (e. g., one year) is first generated according to a Poisson distribution. For each hurricane, a set of the hurricane parameters X, α, ΔP, R, and V_f are generated according to the distributions. The time histories of the wind, wave, current and platform responses at the site as the storm passes are calculated according to the geometry of the problem. The process is repeated for a large number of years to determine the relative frequencies of various threshold levels of the environmental events and platform responses being exceeded.

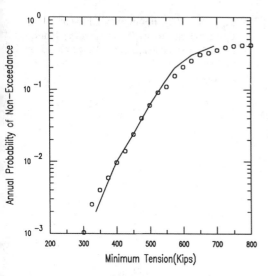

Figure 7 Probability of Non–Exceedance of Minimum Tension (TLP)

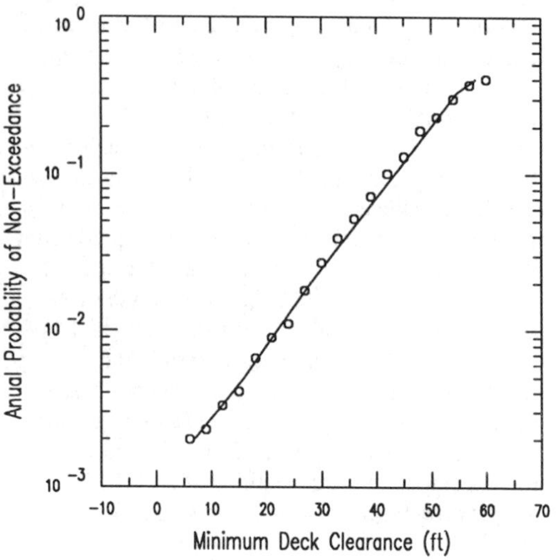

Figure 8 Probability of Non–Exceedance of Minimum Clearance (TLP)

3 PLATFORM RESISTANCE UNCERTAINTY

For a newly designed platform, the actual resistance may exceed that of the nominal design value because of the inherent conservatism in the design procedure. For example, factors of safety are typically used in the design code; the actual material strengths are higher than the nominal values; working stress is used and not the ultimate stress; and finally system redundancy generally gives a higher structural resistance than the design value. In this study, for fixed-bottom platforms the global factor of safety against global failure in a design seastate referred to as reserve strength ratio (RSR) (Lloyd and Clawson, 1983) is used. The platform RSR is defined as the ratio of mean ultimate resistance to the design load. The variability for a given design is assumed to follow a lognormal distribution and the coefficient of variation for the resistance is estimated to be 20% based on past experience. Therefore, the resistance uncertainty is completely specified for a design that has a given RSR value. The coefficients of variation of the annual load effects, e.g.,OTM, are found to be of the order of 60%, therefore, the reliability of the platform is not sensitive to the resistance variability.

4 RESULTS

Figs. 4 to 8 , for example, show the estimated probabilities of annual maximum hurricane load effects for both fixed-bottom and TLP platforms at a generic site in the GOM. Results based on detailed computer simulations are also shown for comparison. Table 2 shows the calculated safety index for a generic fixed-bottom platform in different water depths designed according to the criteria of 100-year wave, 100-year wind, and 66% of 100-year current. Results for different RSR values are shown reflecting the range of conservatism in design.

The effect of directionality of the hurricane load and platform resistance was also investigated (Wen, 1990). Table 3 shows the comparison of annual limit state probabilities of fixed-bottom platforms with resistance asymmetry with those based on the "worst case scenario" assumption that the direction of the lowest resistance always coincides with that of the highest force. The asymmetry is described by the ratio of platform resistances along the major and minor axes with the latter held constant. It is seen that consideration of the asymmetry in the problem could lower the risks by a factor of 2 to 4 compared with the those based on the worst case scenario assumption.

5 RELIABILITY-BASED DESIGN

The above methodology has been used in developing environmental design combination

criteria, i.e., selecting a combination of wind, wave, and current for design. A target annual safety index of 3.0 is chosen based on consideration of past performance of platforms in the GOM. The criterion is that the platform population (in different water depths and with different RSRs) collectively should meet the target safety level. A nonlinear programming technique is used in combination with the reliability analysis to the calculate the optimal design wind and current to be used with the 100-year wave. It is found that 100-year wind and 54% of 100-year current can be used to satisfy the reliability requirement. Recommendations based on the findings were submitted to API Committee on Standardization of Offshore Structures and were implemented in the 20 th edition of API RP2A. The effort is currently being extended to TLPs.

6 ACKNOWLEDGMENT

This paper is based on results of research projects supported by the American Petroleum Institute. The contributions from the members of the API Technical Advisory Committee (TAC) for this project are gratefully acknowledged.
Computational help from T. Yao and R. H. Cherng is also appreciated.

REFERENCES

Cooper, C. 1988. "Parametric Model of Hurricane-Generated Winds, Waves, and Currents in Deep Water," OTC 5738, Offshore Technology Conference, Houston, Texas.

Haring, R. E. and Heideman, J.C. 1978. " Gulf of Mexico Rare Return Periods", OTC 3230, 10 Offshore Technology Conference, Houston, Texas.

Lloyd, J. R. and Clawson, W.C. 1983. " Reserve and Residual Strength of pile Founded Offshore Platforms," Proc. Symposium on the Role of Design, Inspection, and Redundancy in Marine Structural Reliability, National Academic Press.

Offshore System and Analysis Corporation (OSAC), Houston,Texas, 1991. aTLPcx-TLP Response Analysis.

Wen, Y. K. 1988. " Environmental Event combination Criteria, Phase I: Risk Analysis", API Research Report for Project PRAC-87-20.

Wen, Y. K. 1990. " Environmental Event Combination Criteria, Phase II: Design Calibration and Directionality Effect", API Research Report for Project PRAC-88-20.

Wen, Y. K. 1992. "Environmental Event Combination Criteria for Tension Leg Platforms, Phase I: Risk Analysis", API Research Report for Project API 91-20A.

System reliability (ongoing research)

Structural Safety & Reliability, Schuëller, Shinozuka & Yao (eds) © 1994 Balkema, Rotterdam, ISBN 90 5410 357 4

Probabilistic estimation of the collapse risk under bending of statically indeterminate RC beams

M. Aitali, M. Pinglot & M. Lorrain
INSA de Toulouse, France

ABSTRACT: The estimation of the probability P_f that a given structure reaches a failure limit state requires the working out of a model taking into account first the mechanical behaviour of the structure itself with a great accuracy and secondly the random character of the basic data of the problem. The authors present in this paper the principle and the development of such a model in the case of a R.C. beam with fixed ends submitted to static random loads. The mechanical deterministic model is first developed, based on a unique stress–strain diagram for concrete in compression, which allows a precise non linear analysis of the structure. Then the problem is expressed into probabilistic terms and the reliability is estimated using the Monte Carlo simulations. An example of application is given showing the influence on the reliability level of parameters such as the total steel ratio and the choice of the design moments at fixed ends of the beam.

1 INTRODUCTION

The estimation of the reliability of statically indeterminate beams is complicated by the non linear behaviour of materials, which makes necessary a non linear structural analysis. Such an analysis is developed hereafter in order to calculate the bearing capacity of reinforced concrete beams with fixed ends. The reliability of these beams is then estimated using Monte Carlo simulations, the different parameters being considered as random variables.

2 STRUCTURAL ANALYSIS

2.1 Choice of materials behaviour relationships

2.1.1 Concrete

The proposed stress–strain relation for concrete in compression by Aitali Pinglot and Lorrain (1993) is :

$$\sigma_c = E_c \varepsilon_c [1 - (\varepsilon_c / \varepsilon_{cr})^{v+1}] \qquad (1)$$

The parameters E_c, ε_{cr}, v are chosen according to C.E.B. (1990) and Sargin (1971) recommendations (see Figure 1).

2.1.2 Reinforcing steel

The stress–strain diagram of reinforcing

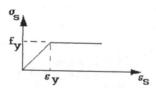

Fig.1 Stress–strain diagram for concrete

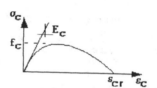

Fig.2 Stress–stain diagram for steel

steel in tension is defined by the relations :

$$\sigma_s = \begin{cases} E_s \varepsilon_s & \text{if } \varepsilon_s \leqslant \varepsilon_y \\ \\ f_y & \text{if } \varepsilon_s > \varepsilon_y \end{cases} \qquad (2)$$

where f_y and ε_y are respectively the yielding stress and strain of the reinforcing steel (see Figure 2).

2.2 Calculation of the internal forces

For a rectangular cross-section submitted to a bending moment, the compression force developed in the concrete can be expressed, using relation (1), as :

$$N_C(\varepsilon_{CM},x)=bE_C\varepsilon_{CM}\{0.5 - \frac{(\varepsilon_{CM}/\varepsilon_{cr})^{v+1}}{(v+3)}\}x, \qquad (3)$$

Where ε_{CM} is the maximum concrete strain, x the depth of the neutral axis and b the width of the cross-section.

The corresponding internal moment can be expressed by :

$$M_{int}(\varepsilon_{CM},x)=bE_C\varepsilon_{CM}\{[(\frac{-1}{6} + \frac{(\varepsilon_{CM}/\varepsilon_{cr})^{v+1}}{(v+3)(v+4)}]x^2 +$$

$$d[0.5 - \frac{(\varepsilon_{CM}/\varepsilon_{cr})^{v+1}}{(v+3)}]x \}, \qquad (4)$$

where d is the effective depth of the cross-section.

The steel force $N_S(\varepsilon_S,x)$ is calculated using (2).

2.3 Moment/curvature relationship

The moment/curvature diagram is determined by successive points calculated as follow :

- for a given value of ε_{CM}, x is deduced from the condition

$$N_C(\varepsilon_{CM},x)=N_S(\varepsilon_S,x); \qquad (5)$$

- the bending moment corresponding to a curvature ε_{CM}/x is then calculated using (4).

Owing to the choice of the stress-strain relationship for concrete, equation (5) can be solved analytically. Therefore the computing time is extremely reduced in comparaison with usual calculations Aitali (1992).

permanent load : g
variable load : q

p=g+q

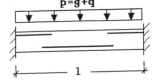

Fig.3   Studied beam

2.4 Ultimate bearing capacity

The studied beam is submitted to uniform loading and has fixed ends (see Figure 3)

The moment/curvature relationship permits to determine the equilibrium states corresponding to increasing load p, satisfying all compatibility conditions and taking into account the rotation capacity of the critical sections. The maximum possible value of p is the bearing capacity p_u.

3 PROBABILISTIC STUDY

3.1 Random variables

Table 1 give an exemple of the assumed distribution of the random variables.

The reinforcement steel areas are determined using the B.A.E.L. 1991 (1992) recommendations, after a choice of the bending moment at ends and at the midspan

3.2 Estimation of the failure probability P_f

P_f can be expressed by the well known relation :

$$P_f = \int_0^{+\infty} (1-F_S(r))f_R(r)dr, \qquad (6)$$

where $F_S(.)$ is the cumulative function of the total load s=g+q and $f_R(.)$ is the density funtion of the resistance r (in

Table 1.   Probabilistic data

*Random variables	Assumed distribution	Coefficient of variation	Caracteristic values
l	Gauss-normal	6 %	$l_k = l_m$
b			$b_k = b_m$
d			$d_k = d_m$
f_c	Gauss-normal	20 %	$f_{ck}=f_{cm}(1-1.64c_{fc})$
f_y		10 %	$f_{yk}=f_{ym}(1-1.64\,c_{fy})$
ε	Gauss-normal	10 %	$\varepsilon_k = \varepsilon_m(1+1.64c_\varepsilon)$
q	El Max (Gumbel)	35 %	$q_k = q_m(1+1.28c_q)$

*$(X_m, X_k$ and c_X denotes the mean value, the caracteristic value and the coefficient of variation of the variable X)

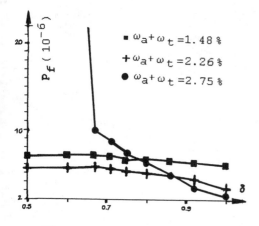

Fig.4 Faillure Probability/δ

this case r is the bearing capacity p_u of the beam).

When using Monte Carlo simulations, the derived expression :

$$P_f \approx \sum_{k=1}^{nsim} (1-F_S(p_u{}^k)),$$ (7)

give an estimation of the failure probability. In this formula nsim denotes the number of simulation and $p_u{}^k$ is the bearing capacity obtained for the "number k" simulation (k=1,...,nsim).

Using expression (7), only 500 simulations are required to get a suffisant precision

4 Exemple of result

4.1 Choice of design moments of the beam

A δ coefficient is defined as the ratio:

$$\delta = M_d/M_e.$$ (8)

where M_d is the design moment at end supports and M_e the moment calculated with a linear analysis ($M_e=pl^2/12$ in our case).

When the δ value is chosen, the design moment M_d is computed using (8). The design moment at midspan is determined using equilibrium conditions

4.2 Influence of the δ coefficient on the failure probability

Figure 4 shows the influence of the coefficient δ on the value of the failure probability, for different total steel ratios (w_a+w_t), where w_a and w_t are the steel ratios, respectively at end supports and at midspan.

As can be seen, the failure probability remains constant for δ varying from 0.5 to 1, where the total steel ratio is less or equal to about 2.26 %. For a total ratio equal for instance to 2.75 % a minimum value of 0.7 is required for δ to get approximately the same reliability level.

CONCLUSION

The chosen stress–strain relation for concrete permits to perform a non linear analysis of a fixed ends beam with a very short calculation time.
It is therefore possible to estimate the failure probability of this beam using Monte Carlo simulations. When the probabilistic distribution of the applied loads is known, only 500 simulations are required to get a suitable precision.
A possible application is the choice of the coefficient δ defining the design moments at the supports of a fixed end beam, in order to get a given level of reliability.

REFERENCES

AITALI, M., PINGLOT, M. et LORRAIN, M. 1993. Modélisation du comportement en flexion statique des sections rectangulaires en béton armé. RILEM. Materials and Structures, 26, 207–213.
Aitali, M. 1992. Contribution à l'étude de la fiabilité des poutres hyperstatiques en béton armé. Thèse de Doctorat de l'Institut National des Sciences Appliquées de Toulouse, France.
B.A.E.L. 91. 1992. Règles techniques de conception et de calcul des ouvrages et constructions en béton armé suivant la méthodes des états limites. Cahiers du C.S.T.B., France.
C.E.B. (Comité Euro-International du Béton). 1990. Code-modèle CEB-FIP 1990. Bulletin d'information n° 190a
Sargin, M. 1971. Stress–strain relationships for concrete and the analysis of structural concrete sections. SM study n°4, Solid Mechanics Division, University of Waterloo, Ontario, Canada.

Structural Safety & Reliability, Schuëller, Shinozuka & Yao (eds) © 1994 Balkema, Rotterdam, ISBN 90 5410 357 4

Reliability analysis of prestressed concrete beams

Ali S. Al-Harthy
Department of Civil Engineering, College of Engineering, Sultan Qaboos University, Al-Khod, Muscat, Oman

Dan M. Frangopol
Department of Civil Engineering, University of Colorado, Boulder, Colo., USA

ABSTRACT: First and second order methods of structural reliability analysis are used to evaluate the reliability with respect to several limit states for seventy three beams that are extensively used in the prestressed concrete industry in the United States. These beams are designed according to the ACI 318-89 Code requirements. Reliability methods are used to assess the safety levels associated with various limit states.

1 INTRODUCTION

In this study, beam sections and spans are selected to reflect the types of structures that are extensively used in the prestressed concrete industry in the United States. These include rectangular, inverted tee, double tee, and single tee sections with spans ranging from 25 ft to 100 ft (Fig. 1). These sections are designed following the ACI Building Code Requirements for Reinforced Concrete (1989) to satisfy permissible stresses at both transfer and serviceability, ultimate strength, sufficient strength to avoid premature cracking and deflection limitations. Limit state functions with respect to ultimate strength, cracking, and permissible stresses at both the initial and final stages are formulated. Appropriate statistical parameters and distributions are assumed.

Reliability methods are used to assess the safety levels associated with various limit states. The effects of different assumptions for the statistical parameters of various random variables and the effects of correlation are also investigated and reported. The range of variation of reliability index β for each limit state is given. Detailed information can be found in Al-Harthy (1992) and Al-Harthy and Frangopol (1993a, 1993b).

2 RELIABILITY INDEX

Reliability indices are computed according to the Rackwitz-Fiessler (1978) algorithm. Original variables are transformed into uncorrelated standard normal variables. For non-normal distributions, equivalent normal means and standard deviations are found (Ang and Tang 1984). Initial failure points are assumed and successive failure points are calculated according to an iteration scheme. The reliability index β is evaluated as the minimum distance from the failure point on the limit state surface to the origin in the reduced coordinate system. The iteration is repeated until β converges. In order to guarantee global minimum, different initial failure points are assumed and the reliability index is:

$$\beta = \min (\beta_1, \beta_2, ..., \beta_n) \tag{1}$$

where n denotes the number of initial failure points considered.

3 LIMIT STATES

The different limit states of a prestressed concrete beam would have different performance functions.

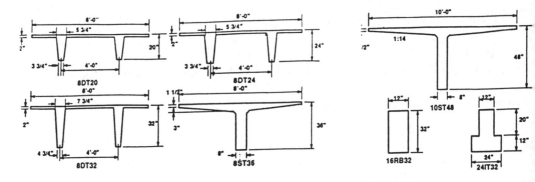

Fig. 1   Cross-Sections of Prestressed Concrete Beams (1" = 25.4mm)

Table 1   Statistical Properties of Random Variables and Associated References

Variable	Distribution Type	Mean	Coefficient of Variation	References
Allowable Tension At Transfer (psi): ($f'_{ci} = 3500.0psi$)	Normal	196.0	0.18	Al-Harthy (1992)
Allowable Compression At Transfer (psi): ($f'_{ci} = 3500.0psi$)	Normal	2440.0	0.15	Al-Harthy (1992)
Allowable Tension At Service (psi): ($f'_c = 5000.0psi$)	Normal	444.0	0.18	Al-Harthy (1992)
Allowable Compression At Service (psi): ($f'_c = 5000.0psi$)	Normal	2370.0	0.15	Al-Harthy (1992)
Compressive Strength f'_c (psi) $f'_c = 5000.0psi$	Normal	4475.0	0.15	Mirza et al. (1979)
Tensile Strength (psi) $f'_c = 5000.0psi$	Normal	555.0	0.18	Mirza et al. (1979)
Prestressing Stress (ksi)				Mirza et al. (1980)
At initial stage	Normal	189.0	0.015	
At final stage	Normal	162.5	0.04	
At ultimate capacity f_{pu}	Normal	281.0	0.025	
Concrete Density (lb/ft^3)	Normal	150.0	0.10	Ellingwood et al. (1980)
Dead Load	Normal	nominal	0.10	Ellingwood et al. (1980)
Live Load				Ellingwood and Culver (1977)
nominal= 50.0psf	Type-I	44.7	0.25	
nominal= 100.0psf	Type-I	89.4	0.25	
Area of strands A_{ps} (in^2)	Normal	0.1548	0.0125	Naaman and Siriaksorn (1982)
Model Coefficient α				
α (Ultimate)	Normal	1.01	0.03	
α (Permissible Stresses)	Normal	0.945	0.03	
			Standard Deviation	
Beam Dimensions (in)				Mirza and MacGregor (1979)
Beam width b	Normal	$b_n + \frac{5}{32}$	$\frac{1}{4}$	
Web width b_w	Normal	b_{wn}	$\frac{1}{16}$	
Flange Thick. h_f	Normal	h_{fn}	$\frac{5}{32}$	
Overall Depth h	Normal	h_n	$\frac{1}{11}$	
Depth to reinf. d_p	Normal	$d_{pn} + \frac{1}{8}$	$\frac{11}{32}$	
Span	Fixed	l_n		

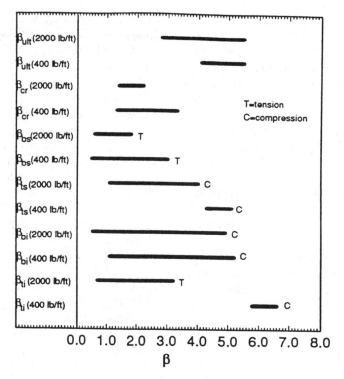

Fig. 2 Reliability Ranges for Seventy-Three Prestressed Concrete Beams Associated with Different Limit States and Live Loads (1 lb/ft = 14.59 N/m)

Such a performance function may be represented as

$$g(R,S) = R - S \qquad (2)$$

where R is the resistance and S is the load effect. At initial and final stages, R and S represent, respectively, permissible stresses and extreme fiber stresses due to both loading and prestressing. For the cases of cracking and ultimate behavior, R represents the ultimate moment capacity and the beam resistance to cracking, respectively, and S is the applied moment. Expressions for R and S can be found in Naaman (1982). In addition, model coefficients that account for the imperfection in the equations predicting the behavior of prestressed concrete beams have been assumed (Al-Harthy 1992).

4 STATISTICAL PROPERTIES OF THE VARIABLES

Statistical properties for the variables entering in the limit states equations were collected from several previous studies. The properties for the selected variables along with the associated references are shown in Table 1.

5 RESULTS AND CONCLUSIONS

The reliability ranges for all seventy three prestressed concrete beams considered in this study are plotted in Fig. 2. All of these beams were designed according to the ACI Building Code Requirements for Reinforced Concrete (1989). Based on both the results in Fig. 2 and the detailed information presented in Al-Harthy (1992) and Al-Harthy and Frangopol (1993a, 1993b), it can be concluded that: (a) the reliabilities with respect

to allowable tension (β_{ti} or β_{bs}) are small and in many cases these reliabilities are the most critical, (b) reliabilities with respect to allowable compression (β_{bi} or β_{ts}) are high and decrease more rapidly with span than do reliabilities with respect to other limit states, (c) as expected, the reliability with respect to flexural cracking (β_{cr}) is higher than the reliability with respect to permissible tension (β_{bs}), (d) the reliability with respect to ultimate limit state (β_{ult}) is high and its variation with load and span is moderate compared with the variation of β associated with other limit states, and (e) in general, all reliabilities decrease with increasing spans and load levels.

It has also been observed (Al-Harthy 1992) that reliability results using second order approximation of the limit state function β_{SORM} (PROBAN 1987) are slightly smaller than first order results β_{FORM}.

ACKNOWLEDGEMENTS

Results presented in this paper are part of a thesis prepared by the first writer under the direction of the second writer in partial fulfillment of the requirements for a Ph.D. degree in the Department of Civil, Environmental, and Architectural Engineering, University of Colorado, Boulder, Colorado. The financial support from the Government of Oman through Sultan Qaboos University is gratefully acknowledged.

REFERENCES

Al-Harthy, A.S. 1992. *Reliability Analysis and Reliability Based Design of Prestressed Concrete Structures*, Ph.D. Thesis, Department of Civil Engineering, University of Colorado, Boulder, Colorado.

Al-Harthy, A.S. and Frangopol, D.M. 1993a. Reliability Assessment of Prestressed Concrete Beams, submitted for publication.

Al-Harthy, A.S. and Frangopol, D.M. 1993b. Reliability-Based Design of Prestressed Concrete Beams, submitted for publication.

Ang, A. H-S. and Tang, W.H. 1984. *Probability Concepts in Engineering Planning and Design. Vol. II: Decision, Risk, and Reliability,* John Wiley & Sons, New York.

Building Code Requirements for Reinforced Concrete. 1989. ACI Standard 318-89, ACI Committee 318, American Concrete Institute, Detroit, Michigan.

Ellingwood, B.R. and Culver, C. 1977. "Analysis of Live Load in Office Buildings," *Journal of the Structural Division,* ASCE, 103(8): 1551-1560.

Ellingwood, B.R., Galambos, T.V., MacGregor, J.G. and Cornell, C.A. 1980. "Development of a Probability-Based Load Criterion for American National Standard A58," *NBS Special Publication 577,* Washington, D.C.

Mirza, S.A., Hatzinikolas, M., and MacGregor, J.G. 1979. "Statistical Descriptions of Strength of Concrete," *Journal of the Structural Division,* ASCE, 105(6): 1021-1037.

Mirza, S.A., Kikuchi, D.K., and MacGregor, J.G. 1980. "Flexural Strength Reduction Factor for Bonded Prestressed Concrete Beams," *Journal of ACI,* 237-246.

Mirza, S.A., MacGregor, J.G. 1979. "Variations in Dimensions of Reinforced Concrete Members," *Journal of the Structural Division,* ASCE, 105(4): 751-765.

Naaman, A.E. 1982. *Prestressed Concrete Analysis and Design,* McGraw-Hill Co., New York.

Naaman, A.E. and Siriaksorn, A. 1982. "Reliability of Partially Prestressed Concrete Beams at Serviceability Limit States," *PCI Journal,* 27(6):66-85.

PROBAN, 1987. *A Computer Program for Probabilistic Analysis,* Det norske Veritas, Norway.

Rackwitz, R. and Fiessler, B. 1987. "Structural Reliability Under Combined Random Load Sequences," *Computers and Structures,* 9:484-494.

Structural Safety & Reliability, Schuëller, Shinozuka & Yao (eds) © 1994 Balkema, Rotterdam, ISBN 90 5410 357 4

Application of a reliability approach in the verification of the design margins of the structural and mechanical system of the Ariane 5 launcher

J. P. Baudet
Mirespace, Orléans, France

S. Bianchi
European Space Agency, Paris, France

ABSTRACT: The Ariane 5 development programme is currently in the design definition phase. The result of this design effort, together with the development test results, will permit the definition of the final Ariane 5 design which shall be submitted to qualification tests. In this phase of the programme, a strong reliability effort is needed to support design engineers in gathering and analyzing test data. Through such data, the accuracy of design margins and compliance with reliability objectives of the launcher will be assessed. In this framework, a valid and easy method to help solving specific design problems has been developed. This method, based on the application of a time dependent stress/strength reliability model, and the software tool to be used for calculations, currently under development, are the subject of this paper.

1. INTRODUCTION

Classical techniques for reliability assessment of mechanical systems are not very effective when applied to launcher development where high reliability objectives correspond with severe economical constraints that limit the number of tests. The method described in this paper addresses specific concerns of reliability activities, in particular: 1) how to introduce a reliability assessment if only a small amount of data is available, 2) how to gather the maximum amount of information concerning design margins from a limited number of tests, and 3) identifying problem areas and corrective action.

This method is based on a step-by-step process and a stress/strength time dependent model of the failure mode and it is applicable to any structural or mechanical component.

Strength and stress of a component depend on different influence variables. Strength depends on material properties, manufacturing, geometry; stress on mission loads, environmental loads such as temperature, or internal loads such as vibrations. From the probability density functions of stress and strength, which are directly related to the influence variables, the reliability of a component is determined through analytical means for any mission time.

The time dependency is introduced to take into account stress and strength "time history" for the defined mission.

A graphical representation of the evolution of the probability of failure at various mission time is shown in fig.1.

The method described permits to establish this evolution with the support of a calculation tool. Different hypotheses can be made on the stress and strength distributions to account for the different influence variables, for their uncertainty and for their time evolution. This, in order to understand which of the variables involved, and which parameter of the variable distribution, are critical with respect to reliability.

stress g(y,t)   strength f(x,t)

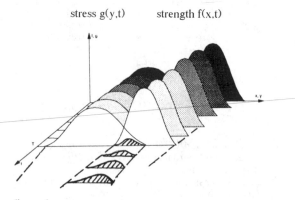

figure 1

2. THEORY AND METHOD

The classical reliability evaluation method usually applied to structures is based on the description of stress and strength distributions. From this model, a general expression of reliability may be obtained.

Considering maximum stress and minimum strength (i.e. acceptable stress) as corresponding to the most unfavourable area, and the associated distributions,

K   minimum strength
C   maximum stress,
f(x)   strength probability distribution
g(y)   stress probability distribution

the probability of failure is expressed as

$$P\{K<C\} = \int_{x<y} f(x)\,g(y)dx\,dy \qquad (1)$$

This approach represents a time independent condition in a defined area of a component. It is used in the dimensioning process, considering the maximum and minimum strength occurring during mission. The design margins used account for uncertainties in the minimum and maximum stress and strength values. Defining stress as any factor that may cause failure, and strength as the ability of the item to resist failure (then as acceptable stress) any mode of mechanical failure may be expressed as a physical process that occur with a probability defined by (1).

2.1 Time dependency

In order to consider mission time (T), a stress/strength time dependent model is introduced, where distributions at time t for strength and stress are f(x,t) g(y,t).

As shown in figure 1, time dependent f(x,t), g(y,t) represent the evolution, during a defined mission time (T), of the failure governing stress and strength distributions.

The time related variation of the probability of failure $\delta F(t)$, between t and t+δt, may be expressed as:

$$\delta F(t) = \iint_{x<y} \left\{ \frac{\partial}{\partial t}[f(x,t).g(y,t)]dx.dy \right\}.\delta t \qquad (2)$$

From equation (2) and figure 1 we may consider 3 different situations governing the probability of failure during mission:

1) in case the failure is not dependent upon usage time: (f.g)/ t = 0 at any t, the item reliability can be ascertained from initial (t=0) stress and strength distributions:

$$1-R(T)=F(0)=\iint_{x<y} f(x,0)g(y,0)dxdy \qquad (3)$$

2) for typical mission conditions, stress and strength vary with time, due to the dynamic and environmental load distributions. Variations in δF as given by (2) may then be >0 or <0 .

From combined probability considerations, between t and t+dt, it can be shown that variation in reliability is:

dR = -(dF/dt).dt,   if (dF/dt)>0,   or
dR = 0,   if (dF/dt)<0

These considerations may be extended to define the reliability function: From (2), by integrating dR for all dt intervals, reliability may be expressed in terms of stress and strength distributions, as a monotonous decreasing function of time,

$$R(T)=1-Sup\left|\iint_{x<y} f(x,t)g(y,t)dxdy\right|_0^T \qquad (4)$$

The model applies then to any failure mode, either time dependent or not.

2.2 Stress and strength distributions

Any factor that may produce failure is described as stress. In a mechanical system, the combination of different factors, - such as temperature or load - in a functional relationship results in the failure governing stress function, defined as the stress that at a certain location governs a certain failure mode.

In the proposed model, the time dependent distribution of the failure governing stress function is defined step by step during the development from the distribution of elementary stress factors $y_1,...,y_i$ From design analysis and tests results, elementary stress factors as well as their evolution during mission [y(t)] are identified. In a second phase the relationship between the stress factors and the failure governing criteria can be theoretically established to determine the failure governing stress function and its distribution.

Likewise for strength, the distribution of any parameter $x_1,..., x_n$ contributing to the ability to resist failure, may be combined through a functional relationship $F(x_1,..., x_n)$ to finally provide the principal strength distribution f(F,t) that defines the failure governing strength for each time (t) of the planned mission.

Any identified failure mode is thus des-

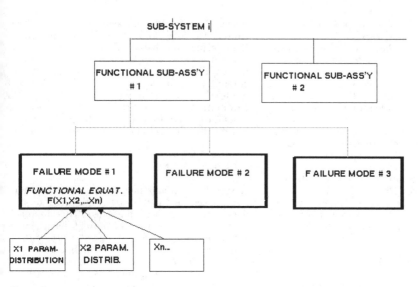

figure 2

cribed (figure 2).
Different hypotheses on:
-stress and strength distribution shapes
-functional relationships between stress or strength and elementary factors
-x_n and y_i variables distribution shapes
-time dependency
may be introduced into the calculation of R(T), using the supporting software tool.

Such direct relation, between the calculated reliability and the distribution parameters of the elementary factors contributing to stress and strength, permits a wide field of application. Some examples of these applications are presented in the next paragraphs.

3. SENSITIVITY TO DEFECTS

The proposed stress/strength model may be used to assess the risk of structural failures caused by a crack-type defect:

1) In case of "no defect" conditions, strength-stress distributions are defined as f_1, g_1, at time t_0. They correspond to the "as designed" conditions, with 100% effective inspection and maintenance.

2) If a crack-type defect is introduced, this would result in changes in f_1 and g_1 distributions. These changes can be calculated (e.g. through Finite Elements), at any given t. The resulting new stress and strength distributions f_2, g_2, take into account all significant parameters concerning defects (defect size, defect distribution and growth). To f_2, g_2, corresponds a new reliability value which can be compared to the R(T) goal. The

sensitivity to specific defects can thus be obtained, as well as the acceptable limits for defect size and distribution. Furthermore, the influence of inspection and maintenance operations on reliability may be quantified introducing in the calculation different probabilities of defect detection corresponding to different inspection tools and defect size.

4. TEST OPTIMISATION

Test objectives include verifying design, and taking into account reliability objectives. Cost limitations enhance the optimal use of tests information. For this purpose a timely identification of critical parameters is mandatory. The model may be used to quantify the influence on reliability of the stress parameters $y_1 \ldots y_i$ and strength parameters $x_1, \ldots, x_n$. Preliminary theoretical assumptions, based on design analysis, define the relationship between stress and strength distributions and the distributions of their basic variables, x_n for strength, and y_i for stress. Through variations applied to mean and/or deviation parameters of any distribution of these x_n and y_i variables, the criticality of any single parameter may be calculated. From this sensitivity analysis, potential problem area may be identified (for example, high sensitivity to the uncertainty of a certain variable can direct further analysis or test and enhance quality control). In addition, as soon as test data becomes available, a calibration of the strength and stress distributions can be made by comparing

initial theoretical assumptions with the results obtained from tests.

An application of the sensitivity analysis described relates to the development of the launcher engine and concerns the so called "functioning domain" of the engine. In reliability terms any point in the functioning domain must verify the condition: $R_c > R_o$
where:

R_c is the reliability calculated introducing variations in the x_n or y_i distributions and

R_o is the reliability objective for the mission.

Considering a variation $\delta\theta_n$ of the dispersion of the x_n strength contributor distribution and τ_n the sensitivity factor, which measures the impact of the variation $\delta\theta_n$ on resulting strength, the effect of the dispersions variations for strength is:

$$(\delta R_s/R_o) = -\Sigma_n(\tau_n.\delta\theta_n) \qquad (5)$$

in the same way the effect of a variation in dispersion $\delta\phi_i$ for the y_i stress contributor, being ξ_i its sensitivity factor, is:

$$(\delta R_s/R_o) = \Sigma_i(\xi_i.\delta\phi_i) \qquad (6)$$

To comply with the reliability objective R_o, dispersion variations must verify the following:

$$(\delta R_c/R_o)= \Sigma_i(\xi_i.\delta\phi_i) - \Sigma_n(\tau_n.\delta\theta_n) \geq 0 \qquad (7)$$

This establish a limitation in the dispersion variations. In other words, to stay within the functional domain, any dispersion variation which decreases reliability shall be adjusted by a dispersion variation that increase reliability.

The same calculation can be applied to variations in the average values of the x_n and y_i parameters.

From this analysis of marginal effect of single variations on reliability immediate and useful conclusions can be drawn. Furthermore, effect of combined variations of both average stress and strength values and of their dispersions can also be computed directly for more complete simulation.

5. CALCULATION OF ACCEPTABLE LIMITS

Test conditions above nominal values provide important information on the functioning limits of the tested system. The results of such hardened tests are analyzed introducing a generalized Weibull Model (ref.1) defined as:

$$R(T) = e^{-\left(\frac{\beta}{n^\beta}\right).\left[\int_0^1 g[C(t)]dt - t_0\right]^\beta} \qquad (8)$$

where β is the Weibull shape factor, θ is characteristic life and $g[C(T)]$ is the factor that defines the hardened conditions. C is the percent increase, with reference to the nominal conditions, of the physical parameter considered as load. Applying the above Weibull model, the stress/strength model is used to calculate β ; then the results of hardened tests are used to calculate the P{R} distribution of the reliability function.

6. CONCLUSIONS

A step by step approach to mechanical reliability, based on a stress/strength time dependent model, has been presented. This approach is considered as the theoretical basis for a software tool, currently under development, which shall support the calculations.

In the framework of the Ariane 5 launcher development programme, this method will be applied to improve the process of verifying design margins and reliability objectives. Several application have been presented such as sensitivity to defects for a structural component, identification of critical tests parameters.

An important outcome of this method is that any failure mode may be represented by the physical variables involved in the failure process. Reliability and design margins can be calculated for any mission time and for different hypotheses concerning the variables involved. The comparison of the results with the reliability objective permits to identify problem area and to assess corrective actions.

REFERENCES

Baudet,J.P., Bianchi,S. 1991. MARTA. ESA SP 316 293-296.

Structural Safety & Reliability, Schuëller, Shinozuka & Yao (eds) © 1994 Balkema, Rotterdam, ISBN 90 5410 357 4

Extended Rackwitz-Fiessler (ERF) method and its application to nonlinear load combination problem

Ying-Jun Chen & Lee Wang
Civil Engineering Department, Northern Jiaotong University, Beijing, People's Republic of China

ABSTRACT: The Extended Rackwitz–Fiessler(ERF) method is proposed on the basis of system reliability theory in this paper. The idea here is to approximate the nonlinear failure domain by plane polygon. The ERF method can be used to find the hyperplanes which compose the polygon. In the nonlinear time variant case, it is pointed out that the sensitivity factor is in the same important position as the reliability index. The method of finding sensitivity factors is proposed on the basis of polygon approximation and equivalent linear safety margin method.

1. INTRODUCTION

The limit state surface is generally a nonlinear function of the design variables. The purpose of this study is to find an efficient tool to handle this situation both in the time invariant case and in the time variant case.

2. EXTENDED RACKWITZ–FIESSLER (ERF) METHOD AND POLYGON APPROXIMATION

Generally, an approximation of the failure surface by one tangent hyperplane at the design point is satisfactory. But if the radius of the curvature is not large compared to Hasofer–Lind reliability index, this approximatoin may be too crude. In this case, we can approximate the nonlinear domain by plane polygon. The hyperplanes which compose the polygon can be obtained by ERF method according to several designated directions. The hyperplane is the tangent plane of the original nonlinear surface, and the contact point (linearization point) is in one of the designated directions. One direction corresponds to one tangent hyperplane. After this treatment, the nonlinear failure domain is substituted by the plane polygon. By system reliability method, the reliability of this plane polygon can be calculated. The Extended Rackwitz–Fiessler method can be used to find the tangent plane which the contact point (linearization point) is in arbitrary designated direction. The procedure and explanation are expressed in the follows. For simplicity this iteration method is shown in normalized space, and the basic variables are independent.

The limit state surface is defined by the equation

$$g(x_1, x_2, ..., x_n) = 0 \qquad (1)$$

in the n–space of basic variables. Introduce the reducing variates

$$u_i = \frac{x_i - \mu_{x_i}}{\sigma_{x_i}} \qquad (2)$$

Then the limit state equation becomes

$$g_1(u_1, u_2, ..., u_n) = 0$$

Designating the direction $(\theta_{u_1}, \theta_{u_2}, ..., \theta_{u_n})$ and supposing that the point on the failure surface in this direction is $(u_1^*, u_2^*, ..., u_n^*)$, the direction of the normal line is $(\theta_{P_1^*}, \theta_{P_2^*}, ..., \theta_{P_n^*})$, the

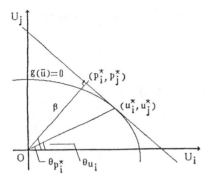

Fig.1   Extended RF method

foot is $(P_1^*, P_2^*, ..., P_u^*)$. See Fig.1. Then we have

$$u_i^{\bullet} = \frac{\beta \cos\theta_{u_i}}{\cos(\theta_{p_i^{\bullet}} - \theta_{u_i})} \quad (3)$$

in which $\cos\theta_{p_i^{\bullet}}$ are the direction cosines of the tangent plane

$$\cos\theta_{p_i^{\bullet}} = \frac{-\dfrac{\partial g_1}{\partial u_i}\bigg|_{u^{\bullet}}}{\sqrt{\sum_{i=1}^{n}\left(\dfrac{\partial g_1}{\partial u_i}\bigg|_{u_i^{\bullet}}\right)^2}} \quad (4)$$

Where the derivatives are evaluated at $(u_1^{\bullet}, u_2^{\bullet}, ..., u_n^{\bullet})$, then

$$x_i^{\bullet} = \mu_{x_i} + \frac{\sigma_{x_i}\beta\cos\theta_{u_i}}{\cos(\theta_{p_i^{\bullet}} - \theta_{u_i})} \quad (5)$$

The solution of the limit state equation

$$g(x_1^{\bullet}, x_2^{\bullet}, ..., x_n^{\bullet}) = 0 \quad (6)$$

then yields β.

The result summarized above would suggest the following:

(1)Giving direction $(\theta_{u_1}, \theta_{u_2} ..., \theta_{u_n})$;

(2)Assume initial values x_i^*, $i = 1,2,...,n$ and obtain u_i^*;

(3)Evaluate $\left(\dfrac{\partial g_1}{\partial u_i}\bigg|_{u_i^{\bullet}}\right)$ and $\cos\theta_{p_i^{\bullet}}$ of the

tangent plane at u_i^*;

(4)Form x_i^* by equation (5);

(5)Substitute above x_i^* in equation (6). and solve for β;

(6)Use the β obtained in step 5, reevaluate x_i^* by equation(5);

(7)Repeat steps 2 through 6 until convergence is obtained.

Example: In the standard normalized 2-dimensional space, Fig.2(a), x_1, x_2 are independent variables. The limit state surface is

$$\frac{x_1^2}{2} + \frac{x_2^2}{8} = 1$$

Designating: $\theta_{x_1} = \pi/3$, $\theta_{x_2} = -\pi/6$

For the first iteration, assuming: $x_1 = x_2 = 2$

Using the ERF method, we obtain:

the tangent plane is

$$0.918x_1 + 0.397x_2 = 1.716$$

the linearization point is

$$x_1 = 1.07, \quad x_2 = 1.85$$

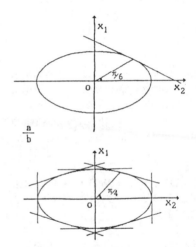

Fig.2   Examples of ERF method and polygon approximation

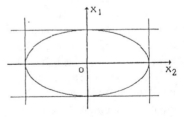

Fig.3   Example of polygon approximation in time variant case

1376

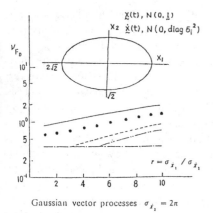

Gaussian vector processes $\sigma_{\dot{x}_1} = 2\pi$

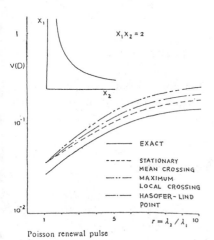

Poisson renewal pulse

EXACT

STATIONARY
MEAN CROSSING

MAXIMUM
LOCAL CROSSING

HASOFER-LIND
POINT

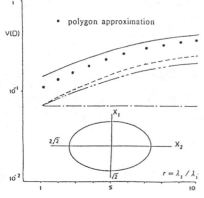

• polygon approximation

Fig.4 Comparison of the mean crossing rates of various methods

the reliability index is $\beta = 1.716$.

In order to improve the accuracy, we can use a set of tangent planes to approximate the nonlinear failure surface. The reliability of the plane polygon can be calculated by system reliability method. The procedure is:

1. Designate a set of directions, and one direction will determine one tangent hyperplane;
2. Get each hyperplane according to the ERF method proposed in this section and form the polygon;
3. Calculate the reliability of the polygon by system reliability method.

Fig.2(b) is the 8-plane approximation of the above ellipse surface. The failure probability of the polygon is $P_f = 0.1864$, The exact result of the ellipse is $P_f = 0.1956$.

3. RESULTS FOR MEAN CROSSING RATE

In the standard normal space, the mean crossing rates of a hyperplane for Poisson renewal pulse vector processes can be obtained by

$$v(D_H) = \sum_{i=1}^{n} \lambda_i [\Phi(\beta) - \Phi_2(\beta;\beta;1 - \alpha_i^2)] \qquad (7)$$

in which, λ_i, α_i are the renewal intensity and the direction cosine of the ith variable. β is the distance of the origin from the plane. and Φ_2 is the bivariate standard normal distribution.

When the process is a stationary continuously differentiable n-dimensional Gaussian process X(t), consider the standard case

$$E[X(t)] = [0], \qquad COV[X(t)] = [I]$$
$$E[\dot{X}(t)] = [0], \qquad COV[\dot{X}(t)] = [\sigma_i^2]$$

The mean crossing rate out of a hyperplane is

$$v(S_H) = \frac{1}{2\pi} \left(\sum_{i=1}^{n} \alpha_i^2 \sigma_i^2 \right)^{\frac{1}{2}} exp\left(-\frac{1}{2}\beta^2 \right) \qquad (8)$$

From formulae (7)(8) we can find that the sensitivity factors as well as the reliability index have influence on the results of mean crossing rate. Therefore, the sensitivity factors of the failure domain are very important in the nonlinear load combination problem. In order

to obtain the reliability index and sensitivity factors of the original nonlinear surface, the equivalent linear safety margin method can be used after the substitution of polygon for nonlinear domain. After this treatment, the equivalent linear function of the nonlinear failure domain is gotten. With mean crossing rate formulae (7)(8), we have the next procedure to deal with the nonlinear load combination problem:

1. Reducing a nonlinear safe domain to a plane polygon by Extended Rackwitz–Fiessler method;

2. Reducing the plane polygon to a linear surface domain by the equivalent linear safety margin method;

3. Calculating the mean crossing rate by the formulae in this section.

Example:

Consider the examples which have been mentioned in several other papers for pulse processes and continuous Gaussian processes, Fig.4.

For ellipse safe domain, when designating four directions $(\pi/2,0)(\pi,\pi/2)(3\pi/2,\pi)$ $(0,-\pi/2)$, we can obtain four planes to approximate the ellipse bound, Fig.3. Carrying the procedure in this section(in step 2, placed $\varepsilon=0.05$), we get the equivalent linear safety margin

$$0.9874x_1 + 0.158x_2 = 0.9892$$

The mean crossing rates by formulae (7)(8) are plotted in Fig.4.

Actually, the linearization in Hasofer–Lind point can be considered as the first order approximation of the polygon method proposed in this paper. The reason why this first order approximation is good for hyperbola surface and crude for ellipse surface is that the tangent plane on Hasofer–Lind point in the first situation has the same sensitivity factors with the original hyperbola failure surface, but the tangent plane in the later situation dosn't. This is the shortcoming of the one point linearization. When we approximate the nonlinear safe domain by one point linearization technique, we cann't guarantee that the direction cosines of the tengent hyperplane are the same with the sensitivity

factors of the original nonlinear domain. Fortunately, the method of polygon approximation can overcome this difficulty.

So the sensitivity factors,i.e. the direction cosines in the equivalent linear surface are very important in time variant reliability analysis. Not only should we pay our attentions to how to approximate the reliability index of the original nonlinear domain, but also should we pay our attentions to how to approximate its sensitivity factors. This is the main difference between the time variant case and the time invariant case. We can get more accurate results by increasing the number of tangent planes.

4. CONCLUSION

The polygon approximation proposed in this paper is an efficient tool to treat nonlinear reliability problem.

5. ACKNOWLEDGEMENTS

This study was supported by the National Science Foundation of China. The author would like to express his sincere gratitudes to Professor M.Shinozuka for many helpful guidances.

REFERENCE

Pearce, H.T.,and Wen, Y.K., On linearization points for nonlinear combination of stochastic load processes, Struct. Safety, 2, 1985, pp.169–176

Structural Safety & Reliability, Schuëller, Shinozuka & Yao (eds) © 1994 Balkema, Rotterdam, ISBN 90 5410 357 4

Fuzzy optimization of structural reliability

M. Holický

Klokner Institute, Czech Technical University, Prague, Czech Republic

ABSTRACT: The ability of a structure to comply with given functional requirements is described by a membership function defining a fuzzy set of damaged structures. If the malfunction cost of a structure is proportional to its damage level, the expected total cost can be expressed in terms of a fuzzy probability of failure. The optimization technique, based here on the demand for minimum total cost, is illustrated by an example of serviceability limit state analysis.

1 INTRODUCTION

The reliability of building structures and other civil engineering works can often be characterized by single reliability indicator u (e.g. force, bending moment, deflection) and may be verified using the well-known reliability requirement

$$S \leq R, \qquad (1)$$

where S denotes load effect and R structural resistance, both quantities expressed in terms of indicator u.

When describing S and R two types of essentially different uncertainties can be identified (Munro & Brown 1983). Firstly randomness, handled by the classical probability theory, and secondly imprecision or vagueness in definitions of structural resistance, handled by methods of fuzzy set theory (Blockley 1980, Brown & Yao 1983). While the first type of uncertainty is commonly recognized, the second type is often ignored or assumed to be somehow included in randomness.

It becomes, however, more and more evident (Yao 1980, Brown & Yao 1983, Shiraishi and Furuta 1983, Holický 1988), that these two types of uncertainties are different in nature, and should be carefully distinguished in order to reach an authentic and sufficiently accurate interpretation of information concerning structural reliability. Particularly in certain serviceability limit states, vagueness and imprecision in their definition appear to be the more significant uncertainty than the randomness (Holický 1988, Holický & Östlund 1993).

In the following analysis, the load effect S is considered as a classical random variable described by a probability distribution with the density function $\varphi_S(u)$, the mean μ_S, standard deviation σ_S and coefficient of skewness α_S. Three-parametric lognormal distribution, used in the following analysis, seems to be a sufficiently general and convenient model. However, the structural resistance R is assumed to involve both types of uncertainties: randomness as well as fuzziness.

2 FUZZINESS OF RESISTANCE R

In the traditional approach to reliability analysis, structural resistance R is described using the two-value logic. A structure is either acceptable and belongs to the crisp set of acceptable structures A, or damaged and belongs to the complementary set of damaged structures $D = non\ A$. In this case the membership function $\mu_D(u)$, indicating support for a structure to be a member of the set of damaged or failed structures D, is a step-wise function equal to 0 or 1 with an abrupt change at the failure point r_0. The failure point r_0 is then considered as a classical random variable described by suitable probabilistic model.

However, experimental and practical experiences clearly indicate that the transition of a structure from its fully acceptable or satisfactory state to the fully damaged or failed state is gradual rather than abrupt. Generally, with increasing values of the reliability indicator u, the ability of a given structure to comply with the specified requirements gradually decreases and its damage level increases. Evidently, the sets A and D are not crisp but fuzzy, and more complicated type of membership function $\mu_D(u)$ should be considered. A power function

$$\mu_D(u) = 0, \qquad \text{if } u < r_1 ;$$
$$\mu_D(u) = \frac{(u - r_1)^n}{(r_2 - r_1)^n}, \qquad \text{if } r_1 \le u \le r_2 ; \qquad (2)$$
$$\mu_D(u) = 1, \qquad \text{if } u \ge r_2 ,$$

having the exponent n is accepted. Here r_1 and r_2 denote the lower and upper bounds of the transition region, where a structure is gradually losing its capacity. Obviously r_1, r_2 and the exponent n characterize the fuzziness (not randomness) of the set D, and should be determined using appropriate experimental data.

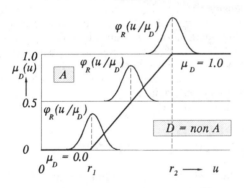

Fig.1. Membership function $\mu_D(u)$.

As indicated in Fig.1 for the linear membership function when $n=1$, the bounds r_1 and r_2, can be random variable as in the case of the previously discussed quantity r_0. In fact for any damage level μ_D, the corresponding resistance R may have a certain probability density function $\varphi_R(u/\mu_D)$ and the membership function is then actually a random function having the mean $\mu_D(u)$, given by Eq. (2). The normal distribution

with the probability density function $\varphi_R(u/\mu_D)$ having a standard deviation σ_R independent of u is considered in the following.

The above theoretical model of structural resistance R is thus characterized by fuzziness characteristics r_1, r_2, and n, and by the randomness characteristic σ_R. Four extreme combinations of both concepts may obviously be recognized:

(a) deterministic case, $r_1 = r_2 = r_0$, $\sigma_R = 0$;
(b) pure fuzziness, $r_1 \ne r_2$, $\sigma_R = 0$;
(c) pure randomness, $r_1 = r_2 = r_0$, $\sigma_R \ne 0$;
(d) fuzzy-random case, $r_1 \ne r_2$, $\sigma_R \ne 0$.

There is, however, considerable lack of appropriate experimental data and observations. Consequently, engineering judgement will have to be relied on until more relevant information becomes available. The proposed theoretical concepts describing structural resistance should, however, be considered whenever a new experimental investigation is to be conducted.

3 PROBABILISTIC ANALYSIS

3.1 Damage function

The probability of the actual resistance of a structure at a given damage level μ_D being lower than a given value of the reliability indicator u, is given by the distribution function $\Phi_R(u/\mu_D)$. The expected damage $\Psi_R(u)$ of structures, called the damage function, is defined as the weighted average of all probabilities $\Phi_R(u/\mu_D)$ with respect to damage levels μ_D taken over the entire range of μ_D (Holický 1988)

$$\Psi_R(u) = \frac{1}{N} \int_0^1 \mu_D \, \Phi_R(u/\mu_D) d\mu_D, \qquad (3)$$

where $N = 1/(n+1)$ denotes the normalizing factor reducing the damage function to $<0,1>$. Previous analyses (Holický 1988, Holický & Östlund 1993) show that in some serviceability limit states, the damage function can be effectively used to determine design values of the resistance R. Thus, Eq. (3) defines an important reliability measure, which may be considered as a generalization of the classical concept of probability.

3.2 Fuzzy probability of failure

Using the traditional approach of structural reliability, the probability $p_f(\mu_D)$ of the reliability requirement (1) being violated at a given damage level μ_D, can be expressed as

$$p_f(\mu_D) = \int_{-\infty}^{\infty} \varphi_S(u)\, \Phi_R(u/\mu_D)\, du. \qquad (4)$$

The uncertainty of the structural resistance R is expressed here by the distribution function $\Phi_R(u/\mu_D)$ related to the damage level μ_D.

The total fuzzy probability of failure π_f is then defined as the weighted average of all failure probabilities $p_f(\mu_D)$ with respect to damage levels μ_D taken again, as in the case of damage function, over the entire range of μ_D (Holický 1988, Holický & Östlund 1993):

$$\pi_f = \frac{1}{N} \int_0^1 \mu_D\, p_f(\mu_D)\, d\mu_D$$

$$= \int_{-\infty}^{\infty} \varphi_S(u)\, \Psi_R(u)\, du , \qquad (5)$$

where Eq. (4) is used for the probability $p_f(\mu_D)$.

4 RELIABILITY OPTIMIZATION

4.1 Total cost

It is assumed that the malfunction cost of a structure is proportional to its damage level μ_D. If the malfunction cost due to the full damage (when $\mu_D = 1$) is denoted as C_D, then the malfunction cost $c_f(\mu_D)$ at μ_D. is

$$c_f(\mu_D) = C_D \cdot \mu_D \cdot p_f(\mu_D) . \qquad (6)$$

Then the expected malfunction cost C_f can be expressed as the weighted average of malfunction costs $c_f(\mu_D)$ with respect to all possible damage levels μ_D by

$$C_f = \frac{C_D}{N} \int_0^1 \mu_D p_f(\mu_D)\, d\mu_D = \pi_f C_D . \qquad (7)$$

Finally the total cost C, defined as the sum of the initial cost C_0 and the expected malfunction cost C_f, is

$$C = C_0 + \pi_f \cdot C_D , \qquad (8)$$

where π_f is given by Eq.(4).

4.2 Optimization conditions

The necessary conditions for minimum total cost follow from partial derivatives of the objective function with respect to decision parameters. If both the initial cost C_0 and the fuzzy probability of failure π_f are functions of a single decision parameter x only, and the malfunction cost C_D is independent of x, then, in case of compound function $p_f(\mu_S(x), \sigma_S(x))$, the necessary condition for the minimum total cost is

$$\frac{\partial C_0}{\partial x} + \left(\frac{\partial \pi_f}{\partial \mu_S} \frac{\partial \mu_S}{\partial x} + \frac{\partial \pi_f}{\partial \sigma_S} \frac{\partial \sigma_S}{\partial x} \right) C_D = 0. \qquad (9)$$

Assume that the load effect S is expressed by

$$S = K\, x^{-m} , \qquad (10)$$

where K denotes a constant and m an appropriate exponent given by the actual loading conditions. This relationship may be used in the case of horizontal structural members exposed to vertical loads, when the decision parameter x denotes the cross-section depth and the reliability indicator u denotes the stress or deflection.

If the load effect S is given by Eq.(10), the condition (9) may be simplified to

$$-\frac{C_D}{C_0} = m\mu_S \frac{\partial \pi_f}{\partial \mu_S} + (m+1)\, \sigma_S \frac{\partial \pi_f}{\partial \sigma_S}. \qquad (11)$$

To apply the necessary condition (11), a computer program has been developed assuming general three-parametric lognormal distribution of the load effect S. Preliminary results show however, that the coefficient of skewness α_S has a negligible effect on the optimum characteristics, and therefore the normal distribution of S will be considered in the following example only.

Table 1. Optimum mean relative sag μ_s /L.

σ_s /μ_s	C_D /C_0		
	1	10	100
0.00	1/159	1/282	1/391
0.10	1/220	1/532	1/855
0.20	1/313	1/847	1/1351

4.3 Example

The example is based on experimental data (Mayer and Rüsch 1967) concerning the serviceability limit state of visual disturbance. Sags of 49 reinforced concrete floors and beams were recorded when annoying deformations were perceived. In this case, the load effect S is the sag and R represents the serviceability limit.

The observed relative disturbing sags r/L, where L denotes the effective span, were within a broad range from 0.003 to 0.018. Using this data, the membership function $\mu_D(u)$ may be well approximated by the tri-linear mean function $(n=1)$ with and the following characteristics.

$$\frac{r_1}{L} = 0.003, \quad \frac{r_2}{L} = 0.014, \quad \frac{\sigma_R}{r_2-r_1} = 0.05. \quad (12)$$

Applied experimental data does do not, however, include all the relevant information, and some additional assumptions were required (Holický 1991) to figure out these characteristics.

Assuming the normal distribution of the load effect $S=Kx^{-3}$, where x denotes the decision parameter (cross-section depth), the optimum mean sags μ_s, determined for selected ratios C_D /C_0 and coefficients of variation σ_s /μ_s using Eq.(11) are given in Table 1.

It appears, that the most frequently applied limiting values ranging from $L/360$ to $L/200$ correspond to relatively low cost of full malfunction C_D (C_D/C_0 from 1 to 5) and a high fuzzy probability of failure π_f (from 0.01 to 0.05), which can well be estimated as

$$\pi_f \approx 0.05 \, C_0 \, / \, C_D . \quad (13)$$

Thus, we may conclude that the commonly accepted serviceability constraints may frequently be uneconomical.

5 CONCLUSIONS

(1) When investigating structural reliability two kinds of uncertainties are to be generally distinguished: randomness and vagueness.

(2) Vagueness in the definition of structural resistance may be handled by methods of the developing fuzzy set theory.

(3) Proposed reliability measures enable rational formulation of design criteria and application of optimization techniques.

(4) The optimization of serviceability limit state due to visual disturbance shows that the commonly used constraints for sag of horizontal components can often be uneconomical.

(5) Further research should be concentrated on

- experimental investigations of more accurate theoretical models for vagueness in definition of structural resistance,
- fuzzy concepts of multidimensional problems.

REFERENCES

Blockley, D.I. 1980. *The nature of structural design and safety*. Ellis Horwood, Chichester.

Brown, C.B. & Yao, J.T.P. 1983. Fuzzy sets and structural engineering: *Journal of the Structural Engineering*, Vol. 109, No. 5, May: 1211-1225.

Holický, M. 1988. Fuzzy concept of serviceability limit states: *Proc. CIB Symposium/Workshop on Serviceability of Buildings:*19-31. Ottawa, Canada.

Holický, M. 1991. Optimization of Structural Serviceability, *Stavebnícky časopis,* Vol. 39, No. 9-10: 473-486.

Holický, M. & L. Östlund 1993. Design concept. *Proc. IABSE/CIB International Colloquium. Structural Serviceability of building*: 91-98, Göteborg.

Mayer, H. & Rüsch, H. 1967. *Bauschäden als Folge der Durchbiegung von Stahlbeton-Bauteilen*. Deutscher Ausschuss für Stahlbeton, Heft 193, Berlin.

Munro, J. & B.C. Brown 1983. The safety of structures in the face of uncertainty and imprecision. *Proc. ICASP 4:* 695-711, Bologna: Pitagora Editrice.

Shiraishi, N. & H. Furuta 1983. Structural design process as fuzzy decision model. *Proc. ICASP 4:* 741-752, Bologna: Pitagora Editrice.

Structural Safety & Reliability, Schuëller, Shinozuka & Yao (eds) © 1994 Balkema, Rotterdam, ISBN 90 5410 357 4

A study of reliability-based seismic design criteria

H. Hwang & H. M. Hsu
Memphis State University, Tenn., USA

ABSTRACT: This paper presents a study to develop reliability-based criteria for the seismic design of reinforced concrete intermediate moment-resisting frame buildings. The seismic criteria in the LRFD format are established on the basis of the collapse of a frame system rather than the failure of a member. The seismic LRFD criteria developed in this study are applicable to three categories of buildings (ordinary, high-risk, and essential) in various seismic zones in the United States.

1 INTRODUCTION

Structural design of earthquake-resistant buildings is complicated by large uncertainty in predicting spatial and temporal characteristics of future earthquakes. Uncertainties are also caused by the idealization of analytical models for evaluating response of structures. In addition, structural capacity cannot be evaluated precisely because of variation in material strengths, workmanship, etc. Therefore, earthquake loading, structural response, and structural capacity are probabilistic in nature. Thus, reliability-based design criteria are needed to include all the uncertainties into building codes. This paper presents a study to develop reliability-based seismic design criteria for reinforced concrete intermediate moment-resisting (IMR) frame buildings.

2 SEISMIC CODE DESIGN PHILOSOPHY

The earthquake-resistant design philosophy implied in model building codes is that a building designed according to seismic provisions will (1) resist a moderate earthquake without structural damage, and (2) resist a large earthquake without collapse. This is a qualitative statement. In the following, moderate and large earthquakes, limit states, and acceptable risk levels are established to express earthquake-resistant design philosophy in a more quantitative way.

2.1 Moderate and large earthquakes

Algermissen and Perkins (1976) of the U.S. Geological Survey (USGS) evaluated seismic hazards for the contiguous 48 states and produced generic seismic hazard curves corresponding to four levels of the design earthquake E_D ranging from 0.1g to 0.4g ("NEHRP" 1988). Since the seismic hazard map specified in model building codes is based on the USGS study, these generic seismic hazard curves are used in this study.

In model building codes, the design earthquake is defined as an earthquake with a 10% probability of exceedance in 50 years. It is denoted as a 475-year earthquake, since the mean recurrence interval of such an earthquake is 475 years. This design earthquake is neither a moderate earthquake nor a large earthquake mentioned in the design philosophy. In this study, the upper bound of a moderate earthquake is set as a 100-year earthquake, while

a 2000-year earthquake is used as the upper bound of a large earthquake.

2.2 Limit states

A limit state represents a state of undesirable structural behavior. In general, a building should be designed by considering all possible limit states such as ultimate strength, instability, drift limit, and serviceability. At the beginning of this study, two limit states, first yielding and collapse of a structure, are considered. For a moment-resisting frame structure, the first yielding is defined as the formation of the first plastic hinge anywhere in the structure. The collapse of a structure is defined as the formation of a failure mechanism. As shown in Hwang and Hsu (1991), the results of this study indicate that the collapse of a structure controls the seismic performance of the frame structures in the event of an earthquake. Thus, the discussion hereafter will focus on the collapse limit state.

2.3 Acceptable risk levels

Most model building codes do not explicitly specify the acceptable risk level. The acceptable risk level shall be established by an authority on the basis of the usage of a structure, characteristics of a limit state, and consequences upon reaching that limit state. Hence, the acceptable risk level (target limit-state probability) may not necessarily be the same for different limit states. In this study, the acceptable risk levels are set for three categories of buildings, i.e., essential, high-risk, and ordinary buildings. Essential buildings are defined as structures housing critical facilities such as hospitals and fire stations that are required to remain functional during and after an earthquake. High-risk buildings are those structures used for assembly of a large number of people, for example, schools. All structures not covered by the above two categories are ordinary buildings. For ordinary ,high-risk,

and essential buildings, the acceptable collapse probabilities are set as 1/1000, 1/2000, and 1/5000 per year, respectively (Hwang and Hsu 1991).

3 PROCEDURE FOR ESTABLISHING SEISMIC DESIGN CRITERIA

The procedure for establishing reliability-based seismic design criteria is: (1) select a load combination format, (2) establish representative frame structures, (3) design these structures according to the proposed seismic design criteria, (4) determine annual limit-state probability of structures, and (5) establish load and resistance factors by optimization with respect to the specified acceptable risk levels.

3.1 Load combination format

The load combination format used in this study is the load and resistance factor design (LRFD) criteria (Ravindra and Galambos 1978). The design of a building requires that the structural resistance should be larger or equal to the design load effects. For a structure subject to three types of loads, dead load D, live load L, and seismic load E, this requirement in the LRFD format is expressed as follows:

$$\phi R \geq 1.2 D + 0.5 L \pm \gamma_E E \qquad (1)$$

$$\phi R \geq 0.9 D \pm \gamma_E E \qquad (2)$$

where γ_E is the seismic load factor. The dead load and live load factors are preset according to ASCE 7-88 ("Minimum" 1988) to simplify optimization. The nominal resistance R and the resistance factor ϕ are those specified in ACI code 318-89 ("Building" 1989).

3.2 Proposed design criteria

In this study, six representative IMR frames are established by means of the Latin hypercube sampling technique. The combination of these six frames and four design

earthquakes yields a total of 24 samples (Hwang and Hsu 1991). Each representative frame is designed according to the proposed seismic design criteria with the design base shear V expressed as

$$V = I \ (\frac{ZC}{R\mu}) \ W \qquad (3)$$

where I is the importance factor; Z is the seismic zone factor; C is the spectral acceleration coefficient; R_μ is the elastic-to-inelastic response factor and W is the total seismic dead load. The importance factor is assigned as 1.0 for ordinary buildings. For high-risk and essential buildings, the importance factors will be determined later in this study. Structures are expected to behave in a nonlinear manner in the event of a large earthquake. In this study, the R_μ factor is used to reduce the base shear from the elastic level to the collapse level. On the basis of studies conducted by Riddel and Newmark (1979), and Hawkins (1986), the R_μ factor is taken as 2.5 for the IMR frame structures.

3.3 Seismic performance

A reliability analysis method has been developed for evaluating the collapse limit-state probability of moment-resisting frame structures (Hwang and Hsu 1991). The seismic load factors can be determined so that the computed limit-state probabilities of all the representative structures are sufficiently close to the specified acceptable risk level (target limit-state probability) by using the following objective function $\Omega(\gamma_E)$.

$$\Omega(\gamma_E) = \sum_{j=1}^{N} \{\frac{\log(PF_{C,j}) - \log(PF_{C,T})}{\log(PF_{C,T})}\}^2$$
$$\text{min} \qquad \qquad \qquad \qquad (4)$$

where $PF_{C,T}$ is the target collapse limit-state probability; $PF_{C,j}$ is the collapse limit-state probability computed for the j-th representative structure; N is the total number of representative structures.

4 SEISMIC LOAD FACTOR AND IMPORTANCE FACTOR

For ordinary buildings, the acceptable collapse probability is set as 1/1000 per year. For the case that the design earthquake E_D = 0.4g, six representative frames are first designed according to the proposed design criteria with the trial seismic load factor γ_E equal to 1.1, 1.2, 1.3, 1.4, and 1.5, respectively. For each trial value of γ_E, the computed annual collapse limit-state probabilities are substituted into Eq. 4 to compute the value of $\Omega(\gamma_E)$. The optimum γ_E is determined as 1.3 by fitting a curve through the data of γ_E versus $\Omega(\gamma_E)$.
Similarly, the optimum γ_E is determined as 1.15 and 1.0, respectively, for the case that the Z factor is equal to 0.3 and 0.2.

For high-risk and essential buildings, the acceptable collapse probabilities are set as 1/2000 and 1/5000 per year, respectively. Since the seismic load factors determined for ordinary buildings are also used for high-risk and essential buildings, the importance factor is employed for increasing the strength and stiffness to meet the more stringent target probabilities. For high-risk buildings ($PF_{C,T}$ = 1/2000 per year), the optimum I values are determined as 1.2, 1.2 and 1.1 for E_D = 0.4, 0.3 and 0.2g, respectively, by using the optimization technique similar to Eq. 4. Since these values are very close to each other, the I factor is set as 1.2 for all levels of the design earthquakes. Similarly, for essential buildings ($PF_{C,T}$ = 1/5000 per year), the optimum I value is determined as 1.5.

5 CONCLUSIONS

In this study, reliability-based seismic design criteria for reinforced concrete intermediate moment-resisting frame buildings have been developed. The proposed seismic design criteria are applicable to three categories of buildings (ordinary, high-risk, and

essential buildings) in various seismic zones in the United States. The reliability-based design criteria established in this study will produce risk-consistent structures under various design conditions, because the seismic load factor and the importance factor are determined by means of optimization with respect to the specified acceptable risk levels.

ACKNOWLEDGMENTS

This paper is prepared under the support by the National Center for Earthquake Engineering Research (NCEER) under contract No. 92-4001A (NSF Grant No. BCS-9025010). Any opinions, findings, and conclusions expressed in the paper are those of the authors and do not necessarily reflect the views of NCEER or NSF of the United States.

REFERENCES

Algermissen, S.T. & D.M. Perkins 1976. A probabilistic estimate of maximum acceleration in rock in the contiguous United States. USGS Open File Report 76-416. Reston: U.S. Geological Survey.

Building code requirement for reinforced concrete 1989. ACI 318-89. Detroit: American Concrete Institute.

Hawkins, N. M. 1986. Seismic evaluation procedure for existing structures. Seismic Design Concerns for Existing Structures, SCM-14 (86), Section II: 1-27. Detroit: American Concrete Institute.

Hwang, H., and H.M. Hsu 1991. A study of reliability-based criteria for seismic design of reinforced concrete frame buildings. Technical Report NCEER-91-0023, Buffalo: National Center for Earthquake Engineering Research, State University of New York.

Minimum Design Loads for Buildings and Other Structures 1988. Standard ASCE 7-88. New York: ASCE.

NEHRP recommended provisions for the development of seismic regulations for new buildings 1988. Washington, D.C.: Federal Emergency Management Agency.

Ravindra, M.K., and T.V. Galambos 1978. Load and resistance factor design for steel. J. Struct. Engrg., 104(ST9): 1337-1353. New York: ASCE.

Riddell, R., and N. M. Newmark 1979. Statistical analysis of the response of nonlinear systems subjected to earthquakes. Structural Research Series No. 468, Urbana: Dept. of Civil Engineering, University of Illinois.

Structural Safety & Reliability, Schuëller, Shinozuka & Yao (eds) © 1994 Balkema, Rotterdam, ISBN 90 5410 357 4

A probabilistic model of acceptance control of concrete used in the Russian standard GOST 18105-86

M.B. Krakovski

Research Institute for Concrete and Reinforced Concrete, Moscow, Russia (Presently: University of Ulsan, Republic of Korea)

ABSTRACT: A new probabilistic model for acceptance control of concrete is presented. The model reflects many peculiarities of real field control and makes it possible to calculate minimum required concrete strength in a batch in such a way as to find a compromise between the risks of customers and suppliers. Numerical results obtained by Monte Carlo simulation and used in the Russian standard GOST 18105-86 are given.

1 INTRODUCTION

Concrete is usually produced in batches. A batch is accepted or rejected if the mean strength of concrete in this batch is respectfully more or less than a specified required mean strength. The main task in the development of any field control of concrete is to fix the required mean strength of concrete in a batch. In so doing two tendencies should be taken into account:

1) If the required mean strength is high, then the safety is also high: most of the imperfect batches are rejected (the risk of customers is low). But such control can appear to be unattractive economically: together with imperfect many perfect batches can be rejected as well (the risk of suppliers is high).

2) On the other hand, when the required mean strength of concrete in a batch is low, such control is more attractive economically: consumption of cement decreases and most of the batches are accepted (the risk of suppliers is low). But the safety also decreases: together with perfect many imperfect batches can be accepted (the risk of customers is high).

Considerable difficulties have been experienced in attempts to find a compromise between these contradictory tendencies and to specify values of required mean strength of concrete in a batch in such a way as to obtain a reasonable level of risks for customers and suppliers. The analysis of the current rules for acceptance control of concrete used in different countries (USA, Great Britain, Japan among others) has shown that this problem is not completely solved.

Below a new probabilistic model for acceptance control of concrete is introduced. The model makes it possible to solve the above problem as well as some other problems of concrete strength control and to obtain a predetermined level of structural reliability.

2 DESCRIPTION OF THE MODEL

The suggested model is shown in Fig. 1. Concrete is accepted or rejected in batches. The mean values R^0_m of concrete strength in all batches produced are assumed to be normally distributed (curve 1, the mean value and coefficient of variation of this distribution are respectively R^0_{mb} and v^0_{mb}). If a

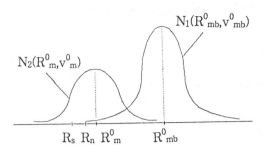

$N_2(R^0_m, v^0_m)$ $N_1(R^0_{mb}, v^0_{mb})$

R_s R_n R^0_m R^0_{mb}

Fig. 1

particular realization R^0_m is considered, then the concrete strength in this particular concrete batch is assumed to be also normally distributed (curve 2, the mean value and coefficient of variation of this distribution are respectfully R^0_m and v^0_m). The estimates R_m, v_m, v_{mb} of quantities respectfully R^0_m, v^0_m, v^0_{mb} are also used. The estimates are obtained in sequential tests of concrete specimens.

Required mean strength of concrete in a batch is denoted by R_s and the quantity R_n is defined as

$$R_n = \max\ [B_n/(1 - 1.64v^0_m);\ B_n/(\gamma_c(1 - 3v^0_m))] \qquad (1)$$

where B_n is characteristic concrete strength and γ_c is the partial safety factor for concrete strength ($\gamma_c = 1.3$ for heavy weight concrete).

The quantity R_n is the minimum mean value required to obtain the probabilities not less than 0.95 and 0.9986 that quantities B_n and B_n/γ_c respectively will be exceeded by concrete strength in sequentially tested specimens.

All batches produced can be either perfect ($R^0_m \geq R_n$) or imperfect ($R^0_m < R_n$). Probabilities of these events are denoted by a and b:

$$a = P(R^0_m \geq R_n)$$

$$b = P(R^0_m < R_n) \qquad (2)$$

$$a + b = 1$$

Imperfect batches can be correctly rejected or erroneously accepted. They are correctly rejected when two

conditions are satisfied: the mean strength of concrete R^0_m in a batch is less than R_n and the estimate R_m is less than the required mean strength R_s. Let φ be the probability of this event:

$$\varphi = P(R^0_m < R_n \cap R_m < R_s) \qquad (3)$$

Similarly the probabilities β of erroneous acceptance of an imperfect batch, δ of correct acceptance of a perfect batch and α of erroneous rejection of a perfect batch can be written as

$$\beta = P(R^0_m < R_n \cap R_m \geq R_s) \qquad (4)$$

$$\delta = P(R^0_m \geq R_n \cap R_m \geq R_s) \qquad (5)$$

$$\alpha = P(R^0_m \geq R_n \cap R_m < R_s) \qquad (6)$$

Probabilities β and α represent the risks of customers and suppliers respectively. It is clear that

$$\varphi + \beta = b$$
$$\qquad (7)$$
$$\alpha + \delta = a$$

3 ALGORITHM USED IN CALCULATIONS

The algorithm based on Monte Carlo simulation reproduced the procedure of concrete acceptance control adopted in the Russian Standard GOST 18105-86. The following input data were used: v^0_m, v^0_{mb}, R_n, R_s, R^0_{mb}, n, where n is a number of series of specimens in a batch tested for acceptance control.

The algorithm includes the following operations:

1. Knowing R^0_{mb}, v^0_{mb} (curve N_1) fix a random realization R^0_m.

2. Compare R^0_m and R_n. If $R^0_m \geq R_n$ assume that the batch with mean value R^0_m is perfect, oterwise it is imperfect.

3. Knowing R^0_m and v^0_m (curve 2) fix n random realizations R_i (i = 1,...,n) of concrete strength representing the specimens test results.

4. Find the estimate R_m of the mean value R^0_m:

$$R_m = \Sigma\, R_i/n.$$

5. Compare R_m and R_s. If $R_m \geq R_s$ then the batch with mean value R^0_m is accepted, otherwise it is rejected.

6. Carry out calculations M times in accordance with operations 1–5.

7. Calculate δ, φ, α, β (e.g. $\delta = T/M$ where T is a number of correctly accepted batches).

The algorithm was implemented in the form of a computer program.

4 THE RESULTS OF CALCULATIONS

The main aim of calculations was to fix such values of R_s for GOST 18105-86 which assure the acceptable levels of risks for customers and suppliers. With this object in view different values of R_s were fixed in calculations. The final results are presented in Tables 1, 2, 3.

According to GOST 18105-86 the value of v^0_m is determined from the tests which have been carried out during a specified design period. Then v^0_m is fixed and remains constant during a specified control period. Three values of v^0_m = 8, 12, 16% were considered. It was assumed that $v^0_{mb} = 0.7v^0_m$ and $R_n = 100$ for $v^0_m = 13.5$ ($B_n = 78$, see(1)). The values of R^0_{mb} were fixed in accordance with GOST 18105-86 as a function of R_s taking into account that R_s changes with n.

The safety of structure is reduced mainly by erroneously accepted batches. Therefore in Table 3 these bathes are divided into several groups according to the values of s_1 and s_2 where s_1 and s_2 are the probabilities that random realizations of concrete strength will exceed its design and characteristic values respectfully. The values s_1^1, s_1^2, s_1^3 denote the probabilities 0.9986 − 0.9980, 0.9980 − 0.9975, 0.9975 and less respectfully; the values s_2^1, s_2^2, s_2^3 denote the probabilities 0.95 − 0.90, 0.90 − 0.80, 0.80 and less respectfully.

Table 1. Initial data for calculations.

Case No	v^0_m %	v^0_{mb} %	R_n	R_s	R^0_{mb}	n
1				86.0	96.0	2
2	8	5.6	89.7	85.0	94.5	4
3				85.0	94.5	8
4				94.5	107.0	2
5	12	8.4	97.1	93.5	106.0	4
6				92.0	104.5	8
7				108.0	130.0	2
8	16	11.2	115.3	102.0	128.5	4
9				105.0	126.5	8

Table 2. Results of calculations: δ, φ, α.

Case No	δ %	φ %	α %
1	83.5	6.2	4.4
2	79.9	6.6	1.2
3	80.9	5.9	0.3
4	76.9	8.3	10.2
5	80.5	9.2	4.7
6	86.3	10.2	1.4
7	77.3	8.1	8.1
8	78.3	8.5	3.5
9	76.5	8.4	0.9

It is seen from Table 2 that the maximum value of the risk of suppliers $\alpha = 10.2\%$. But for the most part the risk of suppliers is significantly below this value. As is clear from Table 3 maximum risks of customers are $\beta = 14.2\%$ (for $v^0_m = 16\%$, $n = 8$) and $\beta = 12.9\%$ (for $v^0_m = 8\%$, $n = 8$). But in the majority of batches the probability that the characteristic and design strengths of

Table 3. Results of calculations: β.

Case No	In all	Batches with $s_1=$			Batches with $s_2=$		
		s_1^1	s_1^2	s_1^3	s_2^1	s_2^2	s_2^3
1	5.9	0	0	0	4.8	0.9	0.2
2	7.3	0	0	0	4.9	2.2	0.2
3	12.9	0	0	0	10.6	2.3	0
4	4.6	0.8	0.2	0.3	4.0	0.6	0
5	5.6	1.3	0.2	0.3	4.9	0.7	0
6	12.1	10.8	0.4	0.4	9.6	1.2	0
7	6.5	3.0	1.5	2.0	0.6	0	0
8	9.7	5.5	1.5	2.7	0.9	0	0
9	14.2	8.6	1.8	3.8	1.2	0	0

concrete will be exceeded by random realizations of concrete strength tends to diminish only slightly. As an example, in regard to the design strength for 8.6% batches out of 14.2% the above probability is in the range from 0.9986 to 0.9980, and in regard to the characteristic strength for 10.6% batches out of 12.9% the above probability is in the range from 0.95 to 0.90.

It is interesting to note that, as may be inferred from Table 3, risk of suppliers (the total percentage of erroneously accepted imperfect batches) increases with n. This is due to the fact that for strength control of concrete the condition $R_s < R_n$ was imposed. The risk of suppliers is defined by Eqn. (4). As n increases the estimates R_m of the mean values of concrete strength in a batch R_m^0 improve, i.e. the difference $|R_m^0 - R_m|$ decreases. In the event that R_m^0 satisfies the condition $R_s < R_m^0 < R_n$, the number of estimates R_m satisfying the same condition increases with n. Therefore the number of erroneously accepted imperfect batches increases as well.

However the risk of suppliers α and the total percentage of rejected batches $\alpha + \varphi$ decrease as n increases. In addition, the accuracy of estimates v_m of the coefficient of variation of concrete strength in batches v_m^0 improves. As a result the value of required mean strength of concrete in a batch R_s dependent on v_m is specified more precisely. As indicated above an increase in β does not involve significant deterioration in quality of accepted batches. Therefore in all cases one should use as many series of specimens n as possible.

From the preceding it is seen that the adopted values of R_s assure rather low risks for customers and suppliers.

5 CONCLUSION

The main results obtained in the present investigation are as follows:

1. A new probabilistic model for acceptance control of concrete is suggested.

2. It is shown that the model reflects many peculiarities of real field control and makes it possible to develop procedures which assure acceptable levels of risks for customers and suppliers.

3. An algorithm and a computer program based on the above model and Monte Carlo simulation are developed.

4. Numerical calculations are carried out with the aim of obtaining required mean values of concrete strength in a batch so as to find a compromise between the risks of customers and suppliers and to assure a predetermined level of structural reliability.

5. The results of calculations are used in the Russian State Standard GOST 18105-86.

Structural Safety & Reliability, Schuëller, Shinozuka & Yao (eds) © 1994 Balkema, Rotterdam, ISBN 90 5410 357 4

Method of reliability analysis of structural system with random parametric excitation

Shigeyuki Matsuho & Wataru Shiraki
Tottori University, Japan

ABSTRACT: In this study, a method of reliability analysis for dynamic stability of parametrically excited structural system is proposed. It is uncertain whether the randomly excited structural system will become unstable state or not in the problem of dynamic stability. Therefore, stochastic response of the structural system in stable state may change into unstable state with a certain probability. In this study, this probability is defined as probability of instability, corresponding to the concept of failure probability in reliability analysis. And calculation method of this probability is developed. Last, the effectiveness of proposed method of reliability analysis is demonstrated through numerical examples.

1. INTRODUCTION

In this study, a method of reliability analysis for dynamic stability of parametrically excited structural system is proposed.

Stochastic stability provides an important criterion in studying the behavior of dynamic and control systems under random parametric excitation. In the traditional analysis of such problem, the stability boundaries of the system are expressed in terms of the statistical properties of the excitation and the system physical parameters. However, it is uncertain whether the randomly excited structural system will become unstable state or not in the problem of dynamic stability. Therefore, stochastic response of the structural system in stable state may change into unstable state with a certain probability. In this study, this probability is defined as probability of instability, corresponding to the concept of failure probability in reliability analysis. And calculation method of this probability is developed. Last, the effectiveness of proposed method of reliability analysis is demonstrated through numerical examples. As numerical examples, problems of uniform strut with hinged ends subjected to the action of the axial compressive force and flutter of suspension bridge are considered.

2. METHOD OF RELIABILITY ANALYSIS FOR DYNAMIC STABILITY

In this chapter, definition of unstable states of a random process and concept concerning method of reliability analysis for dynamic stability are provided.

2.1 Definition of unstable states of random process

When a certain system in the equilibrium state is exited by initial small vibration, behavior of the system is roughly classified into three types, namely, stable state, statically unstable state and dynamically unstable state. In the last case, displacement of system increases with vibration as time goes by. In analysis concerning such dynamic stability of structural systems with random excitation, we must define the dynamic stability of random response process. There are many modes of definition (Ibrahim 1985) about unstable states of a random process, corresponding to the various forms of convergence in theory of probability. In this study, the definition of stability in the mean–square sense, to which Lyapunov's definition is extended, is used. Namely, if a random process satisfies the following Eq.(1), the random process is defined to be stable in the mean–square (2nd moment) sense:

$$E[\sup \| x(t) \|^2] < \varepsilon , \quad \| x(t) \|^2 = \Sigma_i |x_i|^2 \qquad (1)$$

where $E[\cdot]$ and $\sup(\cdot)$ indicate expected $[\cdot]$ and supremum of $(\cdot)$ during the period under consideration, respectively. This definition represents real phenomena very well, and stability in 1st moment can be also defined.

2.2 Method for calculating probability of instability

First step of reliability analysis for dynamic stability is to express the stability boundaries of the system in terms of the statistical properties of the excitation and the system physical parameters. Stability of the system is analyzed according to the definition equation (1). For example, Fig.1 shows boundaries between stable and unstable states by the relation between expected value P_0 and variance D_P of randomly varying excitation $P(t)$ as time t goes by. Namely, values on the boundary curve in Fig.1 correspond to the critical values P_0^* and D_P^* of P_0 and D_P. Therefore, assuming that the random excitation is according to normal distribution $N(P_0, D_P)$, the random excitation in the above critical state can be thought to be according to distribution $N(P_0^*, D_P^*)$. And the structural system subjected to the excitation of

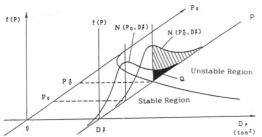

Fig.1 Concept of probability of instability

$N(P_0^*, D_P^*)$ can be interpreted to have a fifty-fifty chance to become unstable state from stable state. Fig.1 also illustrates such situation. Vertical axis f(x) indicates p.d.f. (probability density function) of random excitation P(t). Then, area of the marked part by slash lines in the figure gives probability of instability Q as follows:

$$Q = \frac{1}{\sqrt{2\pi D_P^*}} \int_{P_0^*}^{\infty} \exp\left\{-\frac{(P-P_0^*)^2}{2D_P^*}\right\} dP = 0.5 \qquad (2)$$

Namely, probability Q can be interpreted by the probability of upward excursion from the critical value P_0^*. To extend this concept to the general case that the excitation is characterized by mean P_0 and variance D_P^*, probability Q can be calculated as follows:

$$Q = \frac{1}{\sqrt{2\pi D_P^*}} \int_{P_0^*}^{\infty} \exp\left\{-\frac{(P-P_0)^2}{2D_P^*}\right\} dP \qquad (3)$$

which is provided by size of black part in the figure.

3. NUMERICAL EXAMPLE

In this chapter, how to calculate the probability of instability Q is illustrated by simple examples. As numerical examples, problems of uniform strut subjected to action of axial compressive force and flutter of suspension bridge are considered.

3.1 Problem of uniform strut

In this problem, the external force is modeled by $P(t)=P_0+A(t)$, in which P_0 is mean of force and A(t) is

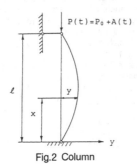

Fig.2 Column

stationary narrow-band process with zero mean. Assuming that deflection of strut with hinged ends (see Fig.2) is $\tilde{y}(x,t) = \tilde{q}(t)\sin(\pi x/l)$ in which t is time, stochastic differential equation concerning $\tilde{q}(t)$ is provided as follows:

$$\frac{d^2\tilde{q}}{dt^2} + \frac{1}{m}\left(\frac{\pi}{l}\right)^2\left\{EI\left(\frac{\pi}{l}\right)^2 - P_0\right\}\tilde{q} = \frac{1}{m}\left(\frac{\pi}{l}\right)^2 A(t)\tilde{q} \qquad (4)$$

where EI and m are flexural rigidity and weight per unit length of the strut, respectively. Condition such that the system of strut is in stable state in mean-square sense was given by various papers, e.g. Samuels et al. (Soog 1973), as follows:

$$\frac{\pi}{l} > \sqrt{\frac{\sqrt{D_P} + P_0}{EI}} \qquad (5)$$

where variance of P(t), D_P, can be calculated by variance of A(t), D_A. Such conditions for strut with various supporting conditions can be also considered.

In calculation, I-type steel strut of JIS (Japan Industrial Standard) with size of cross section 200×100mm, l=2.5m, EI=29.82tonf·m² is considered. Fig.3 is results in the case of strut with hinged ends, which correspond to the critical buckling stress σ (=P_0/S[kgf·cm²], S is cross-sectional area) versus slenderness ratio λ diagram of Euler strut in the static problem. Probability Q can be calculated by using method of section 2.2. In Fig.3, it is assumed that σ is smaller than the yield stress of steel material σ_y=2800kgf/cm² (SS41) and Q is 10⁻². λ corresponds to natural period of strut. Fig.3 shows that σ decreases as D_P increases under same λ. Critical line or curve in the case of D_P=0 agrees with the result in the static problem.

3.2 Problem of suspension bridge

Lin and Ariaratnam (Lin et al. 1980) considered the effect of turbulent components $u\xi(t)$ (u=mean wind velocity) in the wind velocity $U=u\{1+\xi(t)\}$ on the stability boundary of the torsional motion of suspension bridges (see Fig.4). They assumed that the turbulent components are stationary white noise processes for which the bridge response can be approximated as a

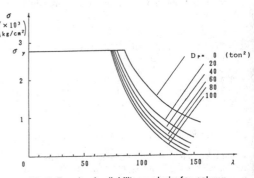

Fig.3 Result of reliability analysis for column

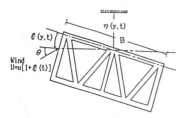

Fig.4 Cross section of suspension bridge

Markov process. And they derived the Ito–type stochastic differential equations which characterized the bridge torsional motion based on various assumptions as follows:

$$\dot{x}_1 = x_2, \quad \dot{x}_2 = -\omega_0^2 x_1 - 2\omega_0 \zeta_0 x_2 + A_6(x_3 + x_4) + 2(A_6 x_1 + \omega_0 A_7 x_2)\xi(t)$$

$$\dot{x}_3 = c_1\{1 + 2\xi(t)\}x_2 - R_1 x_3, \quad \dot{x}_4 = c_3\{1 + 2\xi(t)\}x_2 - R_3 x_4 \qquad (6)$$

$$A_6 = \frac{\rho u^2 B^2}{I} \cdot \frac{dCM}{d\alpha}, \quad A_7 = \frac{A_6 X_a}{\omega_0}, \quad \omega_0^2 = \omega_a^2 - A_6, \quad \zeta_0 = \frac{2\omega_a \zeta_a - A_6 X_a}{2\omega_0}$$

in which x_1 and x_2 are the modal torsional angle (function of time t) and its derivative concerning t, respectively. And I is torsional moment of inertia per unit span length, ω_a is the natural frequency of torsional motion, ζ_a is the damping ratio, ρ is air density, B is width of the bridge. X_a and $dCM/d\alpha$ are experimentally determined aerodynamic constants. Moreover, it was postulated that the changes in the aerodynamic indicial function $X_{M\alpha}$ due to velocity variation can be neglected and can be expressed by

$$X_{M\alpha}(t) = c_1 \cdot \exp(-R_1 t) + c_3 \cdot \exp(-R_3 t) \qquad (7)$$

where c_1, c_3, R_1 and R_3 are constants. Although $\xi(t)$ is a white noise process in Eq.(6), actual turbulent components $u\xi(t)$ can be modeled by non-white noise process, e.g. the process whose one-sided spectrum density S_u is provided by

$$S_u(\lambda) = 0.4751 \frac{\sigma_u^2}{\beta}\{1 + (\frac{\lambda}{2\pi\beta})^2\}^{-5/6}$$
$$\qquad (8)$$

$$\beta = 0.001169 \frac{U_{10}\alpha}{K_r^{1/2}}(\frac{z}{10})^{2m\alpha-1}, \quad \sigma_u^2 = 6K_r U_{10}^2$$

in which λ is circular frequency, σ_u^2 is variance of actual wind speed, z is height, K_r is coefficient of surface roughness, U_{10} is reference wind speed at z=10m, and α is power number in the case that vertical distribution of wind speed is expressed in the form of the power law. Eq.(8) is called Hino–equation and can be approximated by both–sided spectrum density $S_w(\lambda)$ of Eq.(10) (Matsuho 1989) which is yielded by response process w(t) of first–order linear stochastic differential equation (9)

$$\dot{w} + bw = \xi(t) \qquad (9), \qquad S_w(\lambda) = S_0/(\lambda^2 + b^2) \qquad (10)$$

where constant S_0 is spectrum density of $\xi(t)$, and b is constant. Relation of the following equation (11) can be derived by Eqs.(8) and (10).

$$S_0 = 0.9502\pi^2 k\sigma_u^2\beta, \qquad b = 2\pi\beta \qquad (11)$$

where k=1.340 is coefficient to modify Eq.(10) so that variance due to Eq.(10) is equal to variance of actual wind speed σ_u^2. The accuracy of calculation results by this model was checked through analysis for the buffeting phenomenon of the suspension bridge (Matsuho 1989). Using the state vector $x=\{x_1, \cdots, x_5\}^T$ in which $\{\cdot\}^T$ indicates transposed matrix of $\{\cdot\}$ and $x_5=w$, dynamic system given by Eqs.(6) and (9) can be derived in the form of stochastic moment equation as follows:

$$\dot{M}_2 = AM_2,$$

$$M_2 = \{m_{20000}, m_{02000}, m_{00200}, m_{00020}, m_{00002}, m_{11000}, m_{10100}, \qquad (12)$$
$$m_{10010}, m_{10001}, m_{01100}, m_{01010}, m_{01001}, m_{00110}, m_{00101}, m_{00011}\}^T$$

where $m_{11000}=E[x_1 x_2]$, for example. Eq.(12) is result developed by introducing the Wong–Zakai correction and Ito–Dynkin's formula. In the case that all the real parts of the eigenvalues of the matrix A in Eq.(12) are non-positive values, the dynamic system is in stable state in the mean-square sense.

Lastly, reliability analysis of the Narragansett Bay Bridge of span length l=490m is performed by the use of method in section 2.2. Parameter values (Beliveau 1977) in calculation are B=16.4m, I=8.98×10⁵kgf•m²/m, ζ_a =0.01, ω_a=0.74π rad/sec, ρ=1.226kgf/m³, α=1/7, K_r=0.003, z=50m, X_a=1.52, dCM/dα=0.93, c_1=1.64, c_3=−51.61, R_1=0.38u/B, R_3=19.74u/B and θ=0°. Fig.5

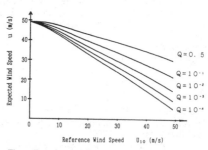

Fig.5 Result of reliability analysis for bridge

shows the boundaries on the expected wind speed – reference wind speed plane. The results in Fig.5 seem to be in contradiction with the observed fact that wind turbulence sometimes has a stabilizing effect. Lin and Ariaratnam (Lin et al. 1980) examined causes of this discrepancy. Moreover, Bucher and Lin (Bucher and Lin 1987) showed the stabilizing effect of the turbulence by means of a two–degree–of–freedom (2DOF) system model of a suspension bridge. In the appendix of this paper, reliability analysis is also performed by making use of this model.

4. CONCLUDING REMARKS

In this study, method of reliability analysis for dynamic stability problem was proposed. Effectiveness was demonstrated through several numerical examples. This study was performed by a Grant-in-Aid for Scientific Research provided by the Japanese Ministry of Education in 1990 (Grant No.02855116).

1393

APPENDIX – ANALYSIS OF 2DOF BRIDGE MODEL

In this appendix, reliability analysis of a suspension bridge is carried out by using the model of Bucher and Lin (Bucher and Lin 1987).

This structural model is a two-degree-of-freedom (2DOF) system representing the first torsional and vertical bending modes of vibration of the bridge. These modes are assumed to be uncoupled structurally but coupled aerodynamically. The self-excited loads are expressed in terms of convolution integrals with impulse-response-function type kernels.

Table A1 Analytical results of two-degree-of-freedom bridge model

Reference Speed of Wind U_{10} m/sec	Variance of Wind Speed σ_u^2 m²/s²	Mean Wind Speed m/s	
		Q=0.5	Q=0.1
29.65	15.83	53.25	48.15
34.32	21.20	54.13	48.22
38.18	26.23	55.00	48.43
50.16	45.29	57.50	48.87
59.46	63.65	60.00	49.77
70.71	90.01	62.00	49.84
76.46	105.24	60.00	46.85

Probability Q of Eq.(3) can be calculated by application of wind velocity model of Eqs.(9) and (10) to the equations of motion derived based on the above condition. Calculation results of B type bridge model (Bucher and Lin 1987) are provided in Table A1. This table shows that turbulence of the wind velocity can have a stabilizing effect.

REFERENCES

Beliveau, J.G. et al. 1977. Motion of Suspension Bridge Subject to Wind Loads: *Journal of Struct. Div. ASCE.* Vol.103(ST6): pp.1189–1205.

Bucher, C.G. and Lin, Y.K. 1987. Stochastic Stability of Bridges Considering Coupled Modes. *Report of Florida Atlantic University.* Report CAS 87–7.

Ibrahim, R.A. 1985. *Parametric Random Vibration:* Research Studies Press.

Lin, Y.K. et al. 1980. Stability of Bridge Motion in Turbulent Wind: *J. Struct. Mech.* Vol.8(1): pp.1–15.

Matsuho, S. et al. 1989. Analytical Method of Dynamic Stability for Suspension Bridge⋯. *Proc.of 44th annual meeting of JSCE.* (in Japanese)

Soong, T.T. 1973. *Random Differential Equations in Science and Engineering:* Academic Press.

Structural Safety & Reliability, Schuëller, Shinozuka & Yao (eds) © 1994 Balkema, Rotterdam, ISBN 90 5410 357 4

Probabilistic reliability analysis using general purpose commercial computer programs

I. R. Orisamolu, Q. Liu & M. W. Chernuka
Computational Mechanics Group, Martec Limited, Halifax, N.S., Canada

ABSTRACT: The essential features of a probabilistic reliability analysis computer program is summarized. A survey of general purpose computer codes that are currently available for probabilistic analysis is presented and guidelines for their application are recommended. The role of finite element modelling in the reliability analysis of complex structures is emphasized. Attention is drawn to deficiencies of existing tools and methodologies, and challenges of the future with regard to practical applications.

KEYWORDS: Reliability, probabilistic analysis, computer programs, stochastic modelling, software, finite elements.

1. INTRODUCTION

Over the years, the fields of probabilistic mechanics and computational mechanics have proceeded vigorously to what may be safely referred to as a mature stage of development. Progress in the two fields, however, has proceeded rather independently for two major reasons. First, until recently, and except for a few exceptions, research in each of the above fields have been conducted by separate groups that are essentially mutually exclusive. Secondly, the application of stochastic methods to engineering structures generally requires considerable computational resources and caused concerns regarding feasibility of application to complex models. The advent of powerful computer hardware and the development of novel numerical techniques are making it possible to address problems on a more realistic basis. In order to provide a communication link between researchers in the area of computational stochastic mechanics and to make these useful tools beneficial to practising engineers, it is desirable to have access to computer programs that could be routinely utilized by designers and analysts for uncertainty modelling. Such programs should be versatile enough to handle the wide variety of stochastic engineering problems. The programs should also have user-friendly interfaces so that they can be efficiently and effectively used by engineers and scientists who may not necessarily be experts in probabilistic mechanics.

The availability of general purpose commercial finite element analysis (FEA) packages has promoted its application over the last two decades, and significantly contributed to its wide acceptability in the engineering community. From this experience, one could infer that the availability of probabilistic analysis programs should serve to promote the merit of the stochastic approach to uncertainty modelling and thereby significantly enhance its state-of-practice. It is encouraging to note that a number of programs are currently available and more are emerging, to fulfil this need.

The present study has two main objectives. First, to provide information on the accessibility and characteristics of computational tools for reliability analysis that are available in the public domain. Secondly, to provide guidance for the selection and applications of these computer programs in engineering practice.

2. BASIC FEATURES OF RELIABILITY ANALYSIS PROGRAMS

The general reliability problem is usually formulated in terms of a finite set of basic random variables $X = (X_1, X_2, ..., X_n)$ and a limit state function $g = g(X)$ in which g is the failure or performance function. Failure is

defined by the event $\{g(X) \leq 0\}$, while $\{g(X) > 0\}$ identifies a safe state. The probability of failure, P_f, is defined as:

$$P_f = \text{Prob}\{g(X) \leq 0\} = \int_{g(X) \leq 0} f_x(x)dx \ , \quad (1)$$

where $f_x(x)$ is the multivariate joint probability density function of X.

The integral in (1) is in general very difficult to evaluate and approximate procedures have evolved as practical tools for efficiently calculating the integral. Prominent examples in this connection include the first-order reliability methods (FORM) and the second-order reliability methods (SORM) which are well discussed in classical monographs on structural reliability. These procedures are based on the evaluation of a reliability index (β) from which the failure probability can then be computed. Efficient simulation schemes are also available to evaluate P_f via "intelligent" sampling of points in the sample space of the events defined above. These computational algorithms are applicable to component reliability analysis (in which only a single limit state condition is involved) as well as systems reliability analysis in which multi-components or multiple failure modes are of interest. The primary objective of probabilistic reliability analysis computer programs, therefore, is the efficient computation of β and/or P_f. Also very important is the computation of parametric sensitivity and importance factors which are very useful in probabilistic modelling. They provide guidance in the assessment of the validity of reliability estimates and in the definition of the roles of the random variables in subsequent analysis.

3. COMPUTER PROGRAMS FOR RELIABILITY ANALYSIS

Some of the well-known software packages available in the public domain for reliability analysis include PROBAN® (1989), NESSUS® (1989), STRUREL® (1992), ISPUD (1986) and CALREL (1989). In addition to the basic FORM/ SORM and simulation capabilities, these programs have additional features to enhance computational efficiency. These features include intelligent simulation schemes such as directional simulation, axis-orthogonal simulation, Latin-Hypercube sampling and adaptive importance sampling. Some of the programs also have capabilities for computing parametric

sensitivity and importance factors for components and systems. Information on the program developers, prices, and unique features are summarized in Table 1. The programs are all

TABLE 1: Summary of Unique Features

Program/ Developer/ Vendor/Price (US $)	Unique Features
PROBAN® DnV/Veirtas Sesam Systems, Oslo, Norway $20,000*	Random distribution parameters. Special purpose modules. (e.g. reliability updating, inspection, maintenance) Extensive systems reliability modelling and analysis capabilities Excellent documentation. Graphical user interface.
NESSUS® Southwest Research Institute, San Antonio, Texas, USA $20,000*	AMV+ CDF analysis. Fully integrated PFEM module. Random field discretization. Interfaced with other FEA packages (i.e. MSC/ NASTRAN, ANSYS).
STRUREL® RCP Gmbh, Munich, Germany $10,000+	Time-variant reliability analysis. Integrated with a FEA package. Graphical user interface. Statistical data analysis.
ISPUD Institute of Engineering Mechanics, University of Innsbruck, Austria $2,200+	Adaptive numerical integration using importance sampling. Not permitted for military applications.
CALREL Dept. of Civil Eng., University of California, Berkeley, CA, USA $1,100+	Interfaced with a finite element reliability package CALREL-FEAP. Restart capability.
COMPASS Martec Limited Halifax, NS Canada $2,500	Interfaced with a stochastic finite element package STOVAST. Probabilistic data characterization. Built-in limit state functions for graphical user interface failure modes.

* Annual license fee
+ Purchase price

coded in the FORTRAN 77 language and are available for a wide variety of hardware platforms. More detailed discussions are available in an internal report published by Orisamolu et al. (1992).

In Orisamolu et al. (1992), other less known reliability analysis programs such as COMPASS (recently developed by the authors), PRADSS, and RELAN are described. Special purpose codes for probabilistic fatigue/fracture mechanics such as PRAISE, COVASTOL, BOPPER, SRRA, VISA-II, and SAPOS are also described.

4. GUIDELINES FOR APPLICATION

Software selection is the first step in the application of reliability analysis methods, especially for the nonexpert, and also for the expert who has no time to devote to computer program development. The programs listed in the preceding section possess the essential ingredients of a reliability analysis package. Like any other software, however, the choice of an analyst must be guided by certain basic considerations. Among these are: user support; user base; user interface/friendliness; features, capabilities, robustness; licensing restrictions; particular application of interest; and price. Once a decision is made, the next step is to gain familiarity with the program by solving some simple problems with known or verifiable solutions. These problems should not be limited to those recommended by the vendor or described in the example manual. Problems and solutions that have been published, or those that are similar to the user's intended application should be considered.

The formal reliability analysis starts with a description of the probabilistic model. In this crucial step, probability distribution functions and the associated statistical parameters are assigned, preferably via a statistical data characterization. Where, as often is the case, sufficient data is not available, probabilistic characteristics that are representative of the modelling parameters (for example as acquired from experience or conventional practice) may be utilized. The limit state conditions or failure modes are also defined. This usually requires the user to develop computer subroutines (for the description of these limit states) which must be compiled and linked with the rest of the program. Equally important is an accurate description of the structural or mechanical

model. Next is the choice of analysis technique and computation options. This aspect usually requires a good appreciation of the various methodologies for reliability assessment. The user has the final responsibility of analyzing the results and assessing their validity. Although probabilistic analysis results can always be checked by Monte Carlo simulation schemes, this mode of confirmation is computationally expensive and defeats the purpose of efficient algorithms available in these computer programs.

5. RECOMMENDATIONS

It now appears that the probabilistic modelling of complex systems can be accomplished without sacrificing the sophistication of well-established and broadly accepted mechanical/structural modelling procedures such as the finite element method. Existing and emerging computational tools such as COMPASS: Orisamolu et al. (1992), seem to recognize the need for accurate, efficient, and integrated stochastic and mechanical modelling in structural analysis. Stochastic Finite Element Methods (SFEMs) are developing as one of the frameworks for achieving this end. Response surface methods (RSMs) have also been recommended by Schuëller et al. (1989) as efficient schemes for the stochastic analysis of complex structures.

The main deficiency of most existing computer programs consists of the capability to handle time-dependent reliability problems and systems reliability analysis of continuum structures. Time dependence is particularly crucial to the analysis of structures subjected to stochastic dynamic loads or those with degrading system properties. In the case of system reliability, additional research is warranted into the modelling of large scale continuum structures, including those involving progressive damage processes.

With the availability of computer programs cited herein, we believe the time is now ripe for the formulation of a plan of action for the development of probabilistic analysis standards. The objective of the proposed standards should be similar to those pertaining to the National Agency for Finite Element Methods (NAFEMS) in respect of FEA. We recommend that the International Association for Structural Safety and Reliability (IASSAR) should set up a committee to seriously examine this proposition.

6. ACKNOWLEDGEMENTS

The authors express gratitude to the staff of the program developers for providing valuable information about their software packages. Special thanks are due to Dr. Rolf Skjong and Mr. Magne Mathisen for making PROBAN® available to us for evaluation. The partial support of this work given by the National Research Council of Canada through the Industrial Research Assistance Program is gratefully acknowledged.

REFERENCES

Bourgund, U. and Bucher, C.G., "Importance Sampling Procedure Using Design Points (ISPUD) - A User's Manual", Report No. 8-86, Institute of Engineering Mechanics, University of Innsbruk, Austria (1986).

Liu L., Lin, H.-Z. and Der Kiureghian, A., "CALREL User Manual", Report No. UCB/SEMM-89/18, Department of Civil Engineering, University of California, Berkeley, California, USA, August (1989).

Millwater, H.R., Wu, Y.-T., Dias, J.B., McClung, R.C., Raveendra, S.T., and Thacker, B.H., "The NESSUS Software System for Probabilistic Structural Analysis", Proceedings of the 5th International Conference on Structural Safety and Reliability (ICOSSAR '89), A.H.-S. Ang, M. Shinozuka, and G.I. Schuëller, Eds., Vol. III, pp. 2283-2290, ASCE, New York, USA (1989).

Orisamolu, I.R., Liu Q., and Chernuka, M.W., "Probabilistic Reliability Analysis Using General Purpose Commercial Computer Programs", Martec Technical Report No. TR-92-5, Martec Limited, Halifax, Nova Scotia, Canada, May (1992).

RCP, "STRUREL: Computer Program Package for Probabilistic Reliability Analysis", RCP GMBH, Munich, Germany (1992).

Schuëller, G.I., Bucher, C.G., Bourgund, U., and Ouypornprasert, W., "On Efficient Computational Schemes to Calculate Structural Failure Probabilities", Probabilistic Engineering Mechanics, Vol. 4, No. 1, pp. 10-18 (1989).

Tvedt L. and Bjerager, P., "PROBAN Version 2 Theory-Manual", A.S. Veritas Research Report No. 89-2023, Oslo, Norway, June (1989).

Structural Safety & Reliability, Schuëller, Shinozuka & Yao (eds) © 1994 Balkema, Rotterdam, ISBN 90 5410 357 4

Curvature employing algorithm for determination of the design point

Z. Sadovský

Institute for Construction and Architecture of the Slovak Academy of Sciences, Bratislava, Slovakia

ABSTRACT: A new simple algorithm for determination of the design point is proposed. In the plane of one iteration step, the curvature of the failure function contour, calcula-ted at a starting point, is taken into account. An observed remarkable feature of the algorithm enables to perform also an efficient SORM analysis by obtaining the principal curvatures without calculating the Hessian of the failure surface at the design point. The algorithm may provide an interesting alternative to the already established calcula-tion tools in structural reliability. Several reproduceable numerical examples are presented.

1 INTRODUCTION

For several important computational me-thods, developed in order to assess the failure probability of structure, the key role plays a determination of the design point. Mathematically, it means to find a point u^* of the shortest distance from the origin to the failure surface $g(u)=0$ in the n-dimensional space of independent standard normal random variables U.

   In order to determine the suitability for application in structural reliability Liu and Der Kiureghian (1991) evaluated five optimization algorithms. Based on limited number of examples, they selected as effi-cient the sophisticated sequential quadra-tic programming method (the scheme propo-sed by K. Schittkowski) and a modification of the widely used Hasofer and Lind method (Madsen, Krenk and Lind, 1986). The first method possesses a superlinear local convergence, but, it requires the storage of the approximate Hessian matrix of the Lagrangian associated with the constrained optimization problem; the second one re-quires the least amount of storage and computation in one iteration step and for most situations also converges rapidly (Liu and Der Kiureghian 1991).

   The performance of the algorithm propo-sed in this contribution is illustrated on several reproduceable examples. For compa-rative results modifications of the Haso-fer and Lind method, to which the algorithm is in certain sense related, are chosen.

1.1 The Hasofer and Lind method and its modifications

An iteration step of the method is done in a plane given by the origin, the starting po-int u^0 and the unit negative gradient vector $\alpha^1 = -\nabla g(u^0)/|\nabla g(u^0)|$. Following an in-terpretation shown in Fig.1 a new approxi-mation $u^1 = u^{HL}$ to the design point u^* is obtained by one step of Newton's iteration along the gradient line through u^0 towards the failure surface $g(u)=0$, thus, yielding u^{gN}, and the projection of u^{gN} onto α^1.

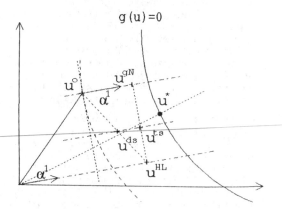

Fig.1 Interpretations of the Hasofer and Lind method and its modifications

Local convergence properties is advantageous to study in a suitable transformed
coordinate system. After rotation, u^* tends
upwards the u_n axis and the principal axes
of the failure surface at u^* coincide with
the coordinate axes u_i, $i=1,\ldots,n-1$. Now,
$u^* = (0,\ldots,\beta)$, where β is the reliability
index. We assume that u_i, $i=1,\ldots,n-1$
correspond to the ordering $\kappa_1 \geq \kappa_2 \geq \ldots$
$\geq \kappa_{n-1}$ of the principal curvatures. The
desired coordinate axes v_i, $i=1,\ldots,n$ are
obtained by the shift

$$v = u - u^*. \qquad (1)$$

Performing asymptotic analysis in an
ε-neighbourhood of the design point it can
be shown (Der Kiureghian and De Stefano
1991) that the new approximation of an
iteration step is given by equation

$$v^1 = (-\beta\kappa_1 v_1^0, \ldots,$$
$$- \beta\kappa_{n-1}v_{n-1}^0, 0) + 0(\varepsilon^2), \qquad (2)$$

where $v^0 = (v_1^0, \ldots, v_{n-1}^0, 0)$ denotes the
starting point.
    Let us introduce a constant

$$\kappa^* = 1/\beta > 0 . \qquad (3)$$

Obviously, for $|\kappa_i/\kappa^*| < 1$, $i=1,\ldots,n-1$ a
linear convergence of the algorithm takes
place. The i-th coordinate approaches zero
like a geometrical sequence with the convergence ratio $-\kappa_i/\kappa^*$ thus, the direction
of an iteration increment converges to that
of the axis with the most slowly converging
coordinate, i.e., v_1 or v_{n-1}. Employing
this feature of the method Der Kiureghian
and De Stefano (1991) suggested to
calculate the principal curvatures of the
failure surface at the design point by
repeated determinations of u^* in orthogonal
subspaces of u-space. The principal
curvatures are obtained in the order of
decreasing magnitude, which is their
importance for SORM analysis, with no need
to compute (and storage) the Hessian.

Since, for $\kappa_1 \geq \kappa^*$ the method fails to
converge Liu and Der Kiureghian (1991)
introduced a non-negative merit function in
order to achieve convergence by selecting
in each iteration a suitable value of a
step length parameter s along the direction
vector $u^{HL}-u^0$ (s=1 yielding u^{HL}). For this
contribution a constant s is applied to:
(i) the direction vector, (ii) the
tangential part of the direction vector,

thus obtaining the points u^{ds} and u^{ts},
respectively, see Figure 1. In both cases
the convergence rate is asymptotically
equivalent. It can be expressed by

$$v^1 = ((1 - s(1 + \kappa_1 / \kappa^*))v_1^0, \ldots,$$
$$(1 - s(1 + \kappa_{n-1} / \kappa^*))v_{n-1}^0, 0) + 0(\varepsilon^2) \qquad ($$

Considering $\kappa_{n-1}/\kappa^* > -1$, ($u^*$ cannot be
minimum-distance point for $\kappa_{n-1} < -\kappa^*$), w
easily deduce that for

$s = 2 / (2 + (\kappa_1 + \kappa_{n-1}) / \kappa^*)$ even an
optimal local convergence, in the sense o
an equal magnitude of the convergence rat
for v_1 and v_{n-1}, may be achieved. Note,
that there exist s values for which the
direction vectors of a converging sequen
approache the v_{n-1} axis.

2 THE CURVATURE EMPLOYING ALGORITHM

An iteration step of the proposed algorithm, visualized in Fig. 2, takes place
the plane of the Hasofer and Lind iteration. By taking into account curvature of
the failure function, profit on a better
new approximation is aimed. It is realize
on assuming that near the design point
contours of the failure function in the
plane can be approximated by circles havi
equal radii with centers on a common line
passing the origin.
    One iteration step of the algorithm pro
ceeds as follows

(1) Compute $\alpha^1 = -\nabla g(u^0)/|\nabla g(u^0)|$ at
    the initial point u^0 .
(2) Calculate $u^{gN} = u^0 + \alpha^1 g(u^0)/|\nabla g(u^0)|$
(3) Determine at u^0 the tangent vector
    $u^{1t} - u^0 = (u^0,\alpha^1)\alpha^1 - u^0$
    of $g(u)$ contour, and its unite vector
(4) Compute the curvature

$$\kappa = g_{rr}(u^0) / |\nabla g(u^0)|$$

    of $g(u)$ contour at u^0 .
(5) If $|\kappa|$ is small, set $u^1 = u^{HL}$ and go t
    (8).
(6) Calculate $u^{oc} = u^0 + \alpha^1/\kappa$ and its uni
    vector α^{oc} .
(7) Set $u^1 = u^q$, where u^q is the smaller o
    of the vectors

$$\alpha^{oc}[|u^{oc}|\pm 1 / |\kappa|$$
$$+ \text{sign } (u^{gN} - u^0, \alpha^{oc})|u^{gN} - u^0|].$$

(8) Compute $g(u^1)$, check the tolerance an
    decide.
    Asymptotic analysis of the algorithm gi

1400

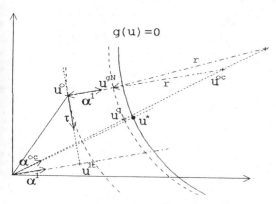

$$g(u) = 0$$

Fig. 2 Illustrations of the curvature employing algorithm

$$v^1 = (\frac{\kappa - \kappa_1}{\kappa + \kappa^*} v_1^0, \ldots,$$

$$\frac{\kappa - \kappa_{n-1}}{\kappa + \kappa^*} v_{n-1}^0, 0) + O(\varepsilon^2), \tag{5}$$

where

$$\kappa = \frac{\sum_{i=1}^{n-1} \kappa_i (1 + \kappa_i / \kappa^*)^2 (v_i^0)^2}{\sum_{i=1}^{n-1} (1 + \kappa_i / \kappa^*)^2 (v_i^0)^2}. \tag{6}$$

In the special case $\kappa_i = \text{const}, i=1,\ldots,n-1$, the algorithm converges superlinearly. Generally, if converging, the rate is linear. We see that the convergence ratio of the i-th coordinate depends on κ_i linearly attaining the smallest negative value for v_1 and the largest positive value for v_{n-1}. In the case n=3, the sequence generated by the algorithm converges for any allowed curvature values at u^* and any sufficiently small initial point, i.e., for

$$x = (\kappa_2 + \kappa^*) / (\kappa_1 + \kappa^*); \quad 0 < x \le 1,$$

$$y = (v_1^0)^2 / ((v_1^0)^2 + (v_2^0)^2); \quad 0 \le y \le 1,$$

$|v^0| < \varepsilon$. Moreover the product of two successive quotiens of v_1 and v_2 coordinates become asymptotically equal and tangent vectors of the iteration steps are asymptotically orthogonal. For n>3 the asymptotic orthogonality of successive tangent vectors can be shown, as well. Further, it has been observed, on numerical examples, that they

form a plane approaching the one given by the principal axes v_1, v_{n-1}, i.e., those with the largest and the smallest values of curvature. Obviously, we may anticipate that this feature of the algorithm may be utilized for even more efficient computations of the principal curvatures than by the Hasofer and Lind method.

3 NUMERICAL EXAMPLES

Four examples are presented. Exs.1,2,4 are taken from Der Kiureghian and De Stefano (1991), Ex.3 is the Example 5.6 of Madsen, Krenk and Lind (1986).
Ex.1: standard normal, n=10, tol=.00001,

$$g(u) = \beta - u_{10} + \frac{1}{2} \sum_{i=1}^{9} \kappa_i (u_i)^2,$$

$$\beta = 3, \quad \kappa_i = .3 - .5(i - 1), \quad i = 1,\ldots,9;$$

$u^{00} = [1,\ldots,1]$ (init . start . point).
Ex.2: as Ex.1 but $\beta = 4$.
Ex.3: log-normal, n=6, tol=.000001.
Ex.4: as Ex.1 but $\kappa_i = .3 - .01(i-1)$, $i=1,\ldots,9$.
In all the examples the gradient vector $\nabla g(u)$ is computed analytically. The HL method necessitates evaluations of $\nabla g(u^0)$ and $g(u^0)$ in each iteration step. The proposed algorithm needs moreover one evaluation of $g(u)$ for calculation of $g_{,n}(u^0)$ by a first-order backward finite difference scheme. Let us note that if computing the gradient by a first-order difference scheme of $O(\delta)$ error then the curvature should be calculated by a scheme of $O(\delta^2)$, thus, with additional one evaluation of $g(u)$.
The design point calculations are shown in Table 1. Since the performaces of the algorithms setting $u^1 = u^{ds}$ and $u^1 = u^{ts}$ are practically equivalent, only the results obtained for $u^1 = u^{ts}$ (algorithm HLts) are given. A finding of the design point is stopped when $|g(u)| \le \text{tol}$ and

$$|u_i^1 - u_i^0| / \max |u_j^0| \le \text{tol}, \quad \text{for} \quad i = 1,\ldots,n.$$

Table 2 shows computed principal curvatures of Ex.4 as obtained by the proposed algorithm (CA) in comparison with the results of Der Kiureghian and De Stefano (1991).

4 CONCLUSIONS

Based on a limited comparison, the proposed algorithm appears more efficient than the (modified) Hasofer and Lind method with respect to determinations of (i) the design

Table 1. Comparison of algorithms for determination of the design point.

Example	n	β	Algorithm	Curvature obtained	Number of iterations
1	10	3	HL (s=1)	.30000	100
			HLts - s=.9	.30000	32
			.8	.29976	18
			.75	-.00962	15
			.7	-.09992	16
			CA	.29998 } -.09996	14
2	10	4	HLts - s=.9		>200
			.8	.30000	40
			.7	.11390	19
			.65	-.09998	20
			CA	.30000 } -.09999	18
3	6	2.8825	HLts - s=.8	-.17431	24
			HL (s=1)	-.17431	18
			HLts - s=1.3	-.01719	14
			1.5	.00743	22
			CA	-.17386 } .00764	12

Table 2. Computed principal curvatures for example 4.

Exact curvature	HL (Der Kiureghian & De Stefano 1991)		CA (the author)	
	Computed κ_i in order found	Number of iterations	Computed κ_i in order found	Number of iterations
κ_1 = .30	.29996	83	.29994 } .22013	11
κ_2 = .29	.28881	37		
κ_3 = .28	.27863	28	.29000 } .22994	9
κ_4 = .27	.26897	25		
κ_5 = .26	.25897	22	.27999 } .24001	6
κ_6 = .25	.24921	20		
κ_7 = .24	.23970	18	.27001 } .24998	4
κ_8 = .23	.23156	16		
κ_9 = .22	.22539	14	.25991	0
		Σ = 263		Σ = 30

point - Table 1, (ii) the principal curvatures for SORM analysis without computation of the Hessian - Table 2. The algorithm can be yet characterized as simple, necessitating only one (possibly two) more evaluation of the failure function in each iteration than the Hasofer and Lind method.

REFERENCES

Der Kiureghian, A. & De Stefano, M. 1991. Efficient algorithm for second-order reliability analysis. *J.Eng.Mech. ASCE*, 117: 2904-2923.

Liu, P-L. & Der Kiureghian, A. 1991. Optimization algorithms for structural reliability. *Structural Safety*.

Madsen, H.O., Krenk, S. & Lind, N.C. 1986 *Methods of structural safety*. Englewood Cliffs: Prentice-Hall.

Structural Safety & Reliability, Schuëller, Shinozuka & Yao (eds) © 1994 Balkema, Rotterdam, ISBN 90 5410 357 4

Probabilistic model for tolerance synthesis: An analytical solution

E. Santoro

Istituto Costruzione di Macchine, Università di Napoli, Italy

ABSTRACT: The most common tolerance specification problem is the tolerance synthesis, which is the distribution of tolerance among the components of an assembly, to assure that design and manufacturing requirements will be met. This paper presents a probabilistic model for tolerance synthesis, where reliability indices are associated to either assembly conditions or component dimensions. For a frequent tolerance synthesis application an analytical solution is given, which can be easily computed, thus it is a step toward an automatic tolerance synthesis in a CAD system.

1 INTRODUCTION

One of the fundamental aspects of design and manufacturing is the specification of tolerances on each dimension of manufactured parts. As design variables, tight tolerances should be assigned for obtaining high performances of design; but as manufacturing variables they should be larger for reducing cost. Therefore, tolerance specification is a very critical link between design and manufacturing, where competing requirements will be met.

The most common tolerance specification problem is the tolerance synthesis, which is the distribution of tolerances among the components of the assembly, and it is formulated here as an optimization problem by treating cost minimization as the objective function and the assembly conditions as the constraints.

In these last years a great deal of research on tolerance synthesis have been devoted to the resolution of this trade-off between cost and performance. The tolerance synthesis can be performed on either a deterministic worst-case (Balling and all. 1986) approach or a probabilistic approach (Chase and all. 1990&1991, Dong and all. 1990, Kumar and all. 1992, Lee and all. 1989& 1990, Siddall 1982, Parkinson 1982&1984).The worst-case approach is very conservative and it may be useful only when the full satisfaction of assembly conditions is required and it will not be considered here.Vice versa, the probabilistic approach examines the probability of satisfying the assembly conditions, by permitting a fraction of the assemblies to be defective.It is very advantageous over the worst-case approach, since it allows larger tolerances to be assigned and consequently a considerable reduction of cost can be obtained.

In the probabilistic approach a random variable and its standard deviation are associated with a dimension and its tolerance; simbolically,if $\mathbf{X}$ is a random vector representing the dimensions $\mathbf{x} \in \Re^n$ with symmetric tolerance limits, it has a value in the tolerance region $R_T \subset \Re^n$ defined by

$$E[\mathbf{X}] - \alpha\sigma \le \mathbf{X} \le E[\mathbf{X}] + \alpha\sigma \qquad (1)$$

where $E[\mathbf{X}]$ is the expected value of $\mathbf{X}$, which may coincide with nominal dimension, and α and σ are the corresponding confidence coefficient and standard deviation vector, respectively. The bilateral tolerance of dimension x_i is thus equal to $t_i = \pm \alpha\sigma_i$, and it is determined by standard deviation and confidence coefficient.

The assembly conditions can be expressed mathematically as equations of the dimensions of the assembly components, and they can be viewed as the equations of the failure functions, which define two regions of the space, one corresponding to safe region the other to fail region. Let $\mathbf{F}(\mathbf{x}) : \Re^n \to \Re^m$ the assembly functions, the safe region is: $R_S = \{\mathbf{x} : \mathbf{F}(\mathbf{x}) \ge 0\} \subset \Re^n$. The intersection of R_T and R_S is referred to

as the reliable region $S=R_T \cap R_S$, which is bounded by $m+2n$ limit state functions : m assembly functions and $2n$ functions derived from upper and lower tolerance limits for the n dimensions (Eq. (1)). Therefore, if $M(X)$ represents the corresponding safety margins of above functions, then the probability of satisfying these safety margins is given by the following multiple integral

$$\gamma = P[M(X) \geq 0] = \int_S f(X) dX \qquad (2)$$

where $f(X)$ is a multivariate p.d.f. for X, and S is the reliable region. The γ parameter is called *yield* and it is an overall reliability measure (P_R), it is used to represent the probability that manufactured dimensions will be within a specified tolerance and satisfy also given assembly conditions.

Let $C(T):\Re^n \to \Re$ the manufacturing cost function, the tolerance synthesis can be formulated as

minimize $C(T)$

subject to $P[M(X) \geq 0] = \gamma$ $\qquad (3)$

where T is the tolerance vector with element t_i and γ is an assigned yield.
Problem (3) is computationally more complicated to solve and approximation solutions can be obtained by numerical integration or simulation techniques. Only in few cases it may be solved analytically, this occurs when $f(X)$ is a multivariate normal p.d.f. and the failure functions are hyperplanes.

In such a cases the reliable region is bounded by $k=m+2n$ hyperplanes, and the corresponding safety margins are defined by

$$M_i(X) = K_{i,0} + K_i^T X \qquad i=1,2,..,k \qquad (4)$$

where K_i^T is a row vector with element $K_{i,j}$ with $1 \leq j \leq n$.
The Hasofer-Lind reliability index for safety margin (4), which coincides with the Cornell reliability index, takes the value

$$\beta_{c,i} = \frac{K_{i,0} + K_i^T E[X]}{\sqrt{K_i^T C_X K_i}} \qquad (5)$$

where C_X is the matrix of covariances of X. Based on these indices the *yield* (2) thus

$$\gamma = \int_{-\infty}^{\beta_1} .. \int_{-\infty}^{\beta_k} \varphi_k(0;R) dy_1 ... dy_k \qquad (6)$$

where $\varphi_k(0;R)$ denotes a multivariate p.d.f. for a set of k normalized r.v., defined by

$$Y_i = \frac{E[M_i(X)] - M_i(X)}{(K_i^T C_X K_i)^{1/2}} \qquad (7)$$

having the correlation matrix R (Ditlevsen 1978).

2 APPLICATION

The most frequent tolerance synthesis problem is represented by the assembly equation $y=F(x):\Re^n \to \Re$ with lower and upper bounds, i.e.

$$y_{min} \leq y \leq y_{max} \qquad (8)$$

For examples, y can represent the assembly clearance, which is expressed as a function of the dimensions of the assembly components, while the lower and upper bounds define the limiting conditions where the minimum and maximum clearance is just attained.
For the sake of simplicity, considering symmetric tolerance about the expected value $E[Y]$ and let t the bilateral tolerance of y, i.e. $y_{max}-E[Y]=E[Y]-y_{min}=t$, the failure functions derived from (8) are

$$F_1(x) = F(x) - y_{min} = F(x) + t - E[Y]$$
$$F_2(x) = y_{max} - F(x) = -F(x) + t + E[Y] \qquad (9)$$

while the failure functions defined by tolerance limits (Eq.1) are

$$G_{i,1}(x_i) = x_i - E[X_i] + \alpha \sigma_i$$
$$G_{i,2}(x_i) = -x_i + E[X_i] + \alpha \sigma_i \quad \text{for } i=1,.,n \qquad (10)$$

In engineering practice, the tolerances are very small, therefore a linear approximation of assembly equation around the expected value $E[X]$ can be considered

$$F(x) = F(E[X]) + \Sigma \, Ki \, (xi - E[Xi])$$

where $Ki = |\partial F(x)/\partial xi|$ is the sensitivity of xi at $x=E[X]$.

Since, for a linear function $E[Y]=F(E[X])$, Eq. (9) becomes

$F_1(\mathbf{x}) = t + \sum K_i\,(x_i - E[X_i])$

(11)

$F_2(\mathbf{x}) = t - \sum K_i\,(x_i - E[X_i])$

The tolerance and the safe regions for two dimensions x1 and x2 with K1 =K2=1 are illustrated in Fig.1.

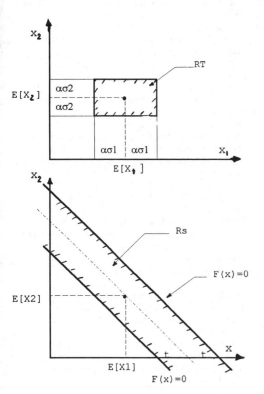

Fig. 1 : Tolerance and safe regions

Based on this linearized assembly equations (11) the Hasofer-Lind reliability indices (5) takes the value

$$\beta_c = \frac{t}{\sqrt{K^T C_X K}}$$

(12)

where K is the vector with element K_i; while for the failure functions (10) the corresponding reliability indices are all equal to α. Assuming that X_i are mutually independent normal r.v., Eq. (12) becomes

$$\beta_c = \frac{t}{\sqrt{\sum K_i^2 \sigma_i^2}} = \frac{t}{\sigma}$$

(13)

where σ_i and σ are the s.d. of X_i and Y, respectively.

For the above assumptions and considering only the failure functions (9), we get

$$P[x \in R_s] = P[y_{min} \le Y \le y_{max}] =$$

(14)

$$= \int_{E[Y]-t}^{E[Y]+t} \varphi(y; E[Y], \sigma)\,dy = 2\Phi(\beta_C) - 1$$

where $\Phi(.)$ is the cumulative p.d.f. of the standard normal variable; while considering the tolerance region we have

$P[x \in R_T] = [2\Phi(\alpha) - 1]^n$

The yield is less than or equal to either $P[x \in R_T]$ or $P[x \in R_S]$, but many practical applications suggeste use of the approximation $\gamma = P[x \in R_S]$.
Eq. (14) for Eq. (13), considering that bilateral tolerance is $t_i = \alpha\sigma_i$, becomes

$$t = \left[\frac{1}{\alpha}\Phi^{-1}\left(\frac{\gamma+1}{2}\right)\right]\left(\sum k_i^2 t_i^2\right)^{1/2}$$

(15)

Assuming the Bennett-Gupta tolerance-cost model (Bennett 1969), Problem (3) can be rewritten as

$$\text{minimize } \sum_{i=1}^{n} h_i t_i^a$$

$$\text{subject to } t^2 = \left[\frac{1}{\alpha}\Phi^{-1}\left(\frac{\gamma+1}{2}\right)\right]^2 \sum k_i^2 t_i^2$$

(16)

where h_i and a are manufacturing parameters depending on the manufacturing processes. Naturally, for $\alpha = 3$ we have the "natural" tolerance, which implies that 99.73% of the dimensions will be within tolerance range.
Problem (16) is an optimization problem with one equality constraint and since the constrain can be solved explicitly for one of the variable, then it may possible to utilize the variable-elimination method (Papalombros 1988) for solving it. After some rearrangements an explicit and analytical solution we get

$$t_i = \frac{\alpha}{\Phi^{-1}[(\gamma+1)/2]} t \left\{ \sum_{j=1}^{n} k_j^2 \left[\frac{h_i}{h_j} \left(\frac{k_j}{k_i} \right)^2 \right]^{2/(a-2)} \right\}^{-1/2}$$

for i=1,..,n　　　　　　　　　　　　　(17)

Tab. 1 shows the non increasing monotonicity between non dimensional tolerance t_i^* (respect to $\gamma=0.9973$) , yield and reliability index.

Tab.1 Reliability index and yield versus tolerances

γ	0.95	0.96	0.97	0.98	0.99	0.9973	0.999
β_C	1.96	2.06	2.17	2.33	2.58	3.0	3.3
t_i^*	1.53	1.46	1.38	1.29	1.16	1.0	0.91

For $\alpha=\beta C$ Eq. (17) reduces to classical statistical tolerance synthesis.

CONCLUSION

The probabilistic approach for tolerance synthesis permits a rational selection of tolerances for a given acceptable risk, since it takes into account either economy of production or design requirements.
There is no doubt that many revisions on tolerance analysis and synthesis problems will be carried out in the new development of CAD/CAM systems.
The solution obtained in this paper could be a step toward an automatic tolerance synthesis.

REFERENCES

Balling,R.J., Free J.C.& Parkinson A.R. 1986. Consideration of Worst-Case Manufacturing Tolerances in Design Optimization. ASME Jour. of Mechan.Aut. in Design. 108:438-441

Bennett G & Gupta L.C. 1969. Least Cost Tolerances:I and II. Intern. Jour. of Produc. Research. 8:64-74 & 169-181

Chase,K.W. & All. 1990. Least Cost Tolerance Allocation for Mechanical Assemblies with Automated Process Selection. Manufacturing Rewiew. 3:49-59

Chase,K.W. & Parkinson A.R. 1991. A Survey of Research in the Application of Tolerance Analysis to the Design of Mechanical Assemblies. Research in Eng. Design. 3:23-27

Ditlevsen O.D. 1979. Narrow Reliability Bounds for Structural Systems. ASCE Jour. of Struc. Mechanics. 7:453-472

Dong, Z. & Soon A. 1990. Automatic Optimal Tolerance Design for Related Dimension Chains. Manufacturing Review. 3:262-270

Kumar S. & Raman S. 1992. Computer-aided tolerancing:the past, the present and the future. Jour. of Design and Manufacturing. 2:24-41

Lee W-J. & Woo T.C. 1990. Tolerances:their Analysis and Synthesis. ASME Jour. of Engin. for Industry. 12:113-121

Lee W-J. & Woo T.C. 1989. Optimum Selection of Discrete Tolerances. ASME Jour. of Mechan. and Aut. in Design. 111:243-251

Lehtihet E.A. & Gunasena U.N. 1990. Statistical Models for the Relationship between Production Errors and the Position Tolerance of a Hole. Annals of the CIRP. 30:569-572

Madsen H.O,Krenk S. &Lind N.C. 1986. Method of Structural Safety. New Jersey: Prentice-Hall Inc.

Michael W. & Siddall J.N. 1982. The Optimal Tolerance Assignment with Less than Full Acceptance. ASME Jour. of Mechan. Design. 104:855-860

Papalambros P.Y. & Wilde D.J. 1988. Principles of Optimal Design. Cambridge:University Press

Parkinson D.B. 1984. Tolerancing of component dimensions in CAD. Computer Aided Design. 16:25-32

Parkinson D.B. 1982. The Application of Reliability Methods to Tolerancing. ASME Jour. of Mechan. Design. 104:612-618

Structural Safety & Reliability, Schuëller, Shinozuka & Yao (eds) © 1994 Balkema, Rotterdam, ISBN 90 5410 357 4

The reliability of elastic-plastic structures under time-dependent load

P.Śniady & S. Żukowski
Institute of Civil Engineering, Technical University of Wrocław, Poland

ABSTRACT: The problem of reliability of beam and frame structures subjected to time-varying load is considered. Taking into account the possible operation of the structure in the range of elastic-plastic strains, it is proposed to evaluate its reliability in terms of the shakedown theory for stochastic loading processes. The reliability problems relative to the incremental failure and to the fatigue failure are solved on the basis of the Neal's nonshakedown theorem and the Melan's shakedown theorem, respectively. The computer algorithm and a numerical example for the determination of the structure reliability are given.

1 INTRODUCTION

A considerable number of papers devoted to the evaluation of the reliability of structures, that take into account plastic strains have been published in recent years. In most cases, however, the problem of reliability of structures is considered in the aspect of the limit load capacity, or in other words with respect to generate of mechanisms of a rigid-plastic structure with probabilistic parameters (Augusti, Baratta, Casciati 1984, Melchers, Tang 1984, Murotsu 1985, Ranganathan, Deshande 1987). Very few papers look at it from the perspective of the theory of shakedown (Augusti, Baratta, Casciati 1984, Śniady, Żukowski 1991). When dealing with a load which is changing in time it is necessary to consider the problem in the terms of the theory of shakedown in order to obtain a correct assessment of the reliability.

In the paper the problem of reliability of beam and frame structures subjected to time varying load is considered. Taking into account the possible operation of the structure in the range of elastic-plastic strains, it is proposed to evaluate the reliability in terms of the shakedown theory for stochastic loading processes. The reliability problems relative to the incremental failure and to the fatigue failure are solved on the basis of the Neal's nonshakedown theorem and the Melan's shakedown theorem, respectively (Capurso 1979, Konig 1987).

The computer algorithm and a numerical example for the determination of the structure reliability are given.

2 PROBABILISTIC CHARACTERISTICS OF EXTREMUM BENDING MOMENTS

Maximum and minimum bending moments in i-th cross-section can be obtained in virtue of relationships

$$\max M_i = \sum_j M_i^j \cdot (\gamma_i^j \cdot P_j^\wedge + (1 - \gamma_i^j) \cdot P_j^\vee),$$

$$\min M_i = \sum_j M_i^j \cdot ((1 - \gamma_i^j) \cdot P_j^\wedge + \gamma_i^j \cdot P_j^\vee),$$

(1)

where $\gamma_i^j = \begin{Bmatrix} 0 \text{ for } M_i^j < 0 \\ 1 \text{ for } M_i^j > 0 \end{Bmatrix}$,

M_i^j is the bending moment in i-th cross-section due to the load $P_j = 1$,

$P_j^\vee$, $P_j^\wedge$ are the lower and the upper random bounds for loading parameter P_j.

It is assumed that the structure is subjected to some different transient stochastic loads.

Keeping in mind the randomness of loading, the average values and random fluctuations of these moments can be presented as

$$\max \dot{M}_i = \sum_j \dot{M}_i^j \cdot (\gamma_i^j \cdot \dot{P}_j^\wedge + (1-\gamma_i^j) \cdot \dot{P}_j^\vee),$$

$$\min \dot{M}_i = \sum_j \dot{M}_i^j \cdot ((1-\gamma_i^j) \cdot \dot{P}_j^\wedge + \gamma_i^j \cdot \dot{P}_j^\vee),$$

$$\max \tilde{M}_i = \sum_j \dot{M}_i^j \cdot (\gamma_i^j \cdot \tilde{P}_j^\wedge + (1-\gamma_i^j) \cdot \tilde{P}_j^\vee),$$

$$\min \tilde{M}_i = \sum_j \dot{M}_i^j \cdot ((1-\gamma_i^j) \cdot \tilde{P}_j^\wedge + \gamma_i^j \cdot \tilde{P}_j^\vee),$$

(2)

where the marks $\cdot$ and $\sim$ over symbols denote the mean value and the random fluctuation of the given item, respectively.

The covariances of these moments have the forms

$$Cov(\max M_i, \max M_s) = \sum_j \sum_k \dot{M}_i^j \cdot \dot{M}_s^k \cdot$$

$$[\gamma_i^j \cdot \gamma_s^k \cdot Cov(P_j^\wedge, P_k^\wedge) + (1-\gamma_i^j) \cdot (1-\gamma_s^k) \cdot$$

$$\cdot Cov(P_j^\vee, P_k^\vee) + (1-\gamma_i^j) \cdot \gamma_s^k \cdot Cov(P_j^\vee, P_k^\wedge) +$$

$$+\gamma_i^j \cdot (1-\gamma_s^k) \cdot Cov(P_j^\wedge, P_k^\vee)],$$

(3a)

$$Cov(\min M_i, \min M_s) = \sum_j \sum_k \dot{M}_i^j \cdot \dot{M}_s^k \cdot$$

$$[(1-\gamma_i^j) \cdot (1-\gamma_s^k) \cdot Cov(P_j^\wedge, P_k^\wedge) + \gamma_i^j \cdot \gamma_s^k \cdot$$

$$\cdot Cov(P_j^\vee, P_k^\vee) + \gamma_i^j \cdot (1-\gamma_s^k) \cdot Cov(P_j^\vee, P_k^\wedge) +$$

$$+(1-\gamma_i^j) \cdot \gamma_s^k \cdot Cov(P_j^\wedge, P_k^\vee)],$$

(3b)

$$Cov(\max M_i, \min M_s) = \sum_j \sum_k \dot{M}_i^j \cdot \dot{M}_s^k \cdot$$

$$[\gamma_i^j \cdot (1-\gamma_s^k) \cdot Cov(P_j^\wedge, P_k^\wedge) + (1-\gamma_i^j) \cdot \gamma_s^k \cdot$$

$$\cdot Cov(P_j^\vee, P_k^\vee) + (1-\gamma_i^j) \cdot (1-\gamma_s^k) \cdot Cov(P_j^\vee, P_k^\wedge) +$$

$$+\gamma_i^j \cdot \gamma_s^k \cdot Cov(P_j^\wedge, P_k^\vee)],$$

(3c)

$$Cov(\min M_i, \max M_s) = \sum_j \sum_k \dot{M}_i^j \cdot \dot{M}_s^k \cdot$$

$$[(1-\gamma_i^j) \cdot \gamma_s^k \cdot Cov(P_j^\wedge, P_k^\wedge) + \gamma_i^j \cdot (1-\gamma_s^k) \cdot$$

$$\cdot Cov(P_j^\vee, P_k^\vee) + \gamma_i^j \cdot \gamma_s^k \cdot Cov(P_j^\vee, P_k^\wedge) +$$

$$+(1-\gamma_i^j) \cdot (1-\gamma_s^k) \cdot Cov(P_j^\wedge, P_k^\vee)].$$

(3d)

3 SYSTEM SAFETY INDEX

The static and kinematic solutions with respect to shakedown theory can be formulated as dual linear programming problem from Neal's theory we have:

$$\text{Find} \quad \mu_p = \min\left[M_o^T \cdot (\Theta^+ + \Theta^-) \right]$$

under the constraints

$$m^T \cdot \Theta^+ - m^T \cdot \Theta^- = 0,$$

$$\max M^T \cdot \Theta^+ + \min M^T \cdot \Theta^- \geq 1,$$

$$\Theta^+ \geq 0, \qquad \Theta^- \geq 0,$$

(4)

where M_o is the vector of the plastic limit moments in critical cross-sections, Θ^+, Θ^- are the strains vectors in plastic hinges, μ_p is the parameter defining the limit load capacity in compliance with the incremental shakedown.

The optimal solution is connected with the equation

$$\mu_p \cdot \left[\max M^T \cdot \Theta^+ - \min M^T \cdot \Theta^- \right] =$$
$$= M_o^T \cdot \left[\Theta^+ + \Theta^- \right],$$

(5)

where the parameters Θ^+, Θ^- describe the plastic failure mode .

The structure limit load capacity due to the low-cycle fatigue shakedown is given by

$$\mu_z \cdot \left[\max M_k - \min M_k \right] = M_{e_k},$$

(6)

where M_{e_k} is the limit elastic moment in k-th cross-section, μ_z is the parameter defining the limit load capacity due to the low-cycle fatigue.

The elastic-state load capacity of the structure is determined by the relationships

$$\max M_k \cdot \mu_e = M_{e_k},$$
$$\min M_k \cdot \mu_e = -M_{e_k}.$$

(7)

The normalized plastic reserve of the structure load capacity with regard to the incremental and fatigue damage and elastic reserve can be expressed as

$Z_p = \mu_p - 1, \quad Z_z = \mu_z - 1, \quad Z_e = \mu_e - 1. \qquad (8)$

The expected values, fluctuations and variances of the plastic reserve are

$\dot{Z}_p = \dot{\mu}_p - 1, \quad \dot{Z}_z = \dot{\mu}_z - 1, \quad \dot{Z}_e = \dot{\mu}_e - 1, \qquad (9)$

$\tilde{Z}_p = \tilde{\mu}_p, \quad \tilde{Z}_z = \tilde{\mu}_z, \quad \tilde{Z}_e = \tilde{\mu}_e, \qquad (10)$

$\mathrm{Var}(Z_p) = \mathrm{Var}(\mu_p), \quad \mathrm{Var}(Z_z) = \mathrm{Var}(\mu_z),$
$\mathrm{Var}(Z_e) = \mathrm{Var}(\mu_e), \qquad (11)$

where

$\mathrm{Var}(\mu_p) =$

$= [\sum_i \sum_j \mathrm{Cov}(\max M_i, \max M_j) \cdot \Theta_i^+ \cdot \Theta_j^+ +$

$+ \sum_i \sum_j \mathrm{Cov}(\min M_i, \min M_j) \cdot \Theta_i^- \cdot \Theta_j^- +$

$- 2 \cdot \sum_i \sum_j \mathrm{Cov}(\max M_i, \max M_j) \cdot \Theta_i^+ \cdot \Theta_j^+] \cdot$

$\cdot \left[\dfrac{\dot{\mu}_p}{\sum_k (\max \dot{M}_k \cdot \Theta_k^+ - \min \dot{M}_k \cdot \Theta_k^-)} \right]^2,$

$\mathrm{Var}(\mu_z) = [\mathrm{Var}(\max M_k) +$
$+ \mathrm{Var}(\min M_k) - 2 \cdot \mathrm{Cov}(\max M_k, \min M_k] \cdot$

$\cdot \left[\dfrac{\dot{\mu}_z}{\max \dot{M}_k - \min \dot{M}_k} \right]^2,$

$\mathrm{Var}(\mu_e) = \mathrm{Var}(A) \cdot \left[\dfrac{\dot{\mu}_e}{\dot{A}} \right]^2,$

$A = \max[\max M_k, -\min M_k],$

$\mathrm{Cov}(\mu_p \mu_e) = \mathrm{Cov}(\mu_z \mu_e) = 0.$

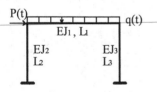

Fig. 1. Portal frame

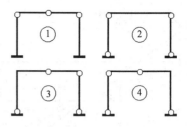

Fig 2. Collapse mechanisms

The reliability index is defined as

$\beta = \dfrac{\dot{Z}}{\sqrt{\mathrm{Var}(Z)}}, \qquad (12)$

where $Z = (Z_p, Z_z, Z_e).$

4 NUMERICAL EXAMPLE

Based on the algorithm presented, the computer program for evaluating the reliability of beams, plane frames and trusses in terms of the shakedown theory has been elaborated. The program has been applied to the reliability analysis of the portal frame shown in Figure 1.

The computations have been carried out for the following structures data:
spans length $\quad L_2 = L_3 = 0.5 \cdot L_1 = L,$
Young modulus $\quad E_1 = E_2 = E_3 = E,$
cross-sectional inertia moments
$J_2 = J_3 = 0.5 \cdot J_1 = J,$ limit plastic moments
$M_{o_2} = M_{o_3} = 0.6 \cdot M_{o_1} = M_o,$
limit elastic moments $M_{e_2} = M_{e_3} = 0.6 \cdot M_{e_1} =$
$= M_e = 0.85 \cdot M_o.$

The loading process is assumed to be a superposition of two stochastic processes (Wen 1977). The variation coefficients of the upper and lower boundaries of load parameters are $V_P^\wedge = V_P^\vee = V_q = 0.2.$ In the above formulas the quantities q, P, L, M_e, M_o, J and E have been regarded as the comparative values.
The collapse mechanisms due to the incremental unshakedown failure are presented in Figure 2.

In Figures 3 and 4 the relationships between the reliability index β and the load parameter

1409

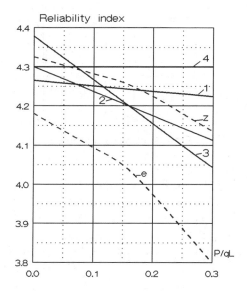

Fig.3. Reliability index vs load parameter
for $P^{\wedge}/P^{\vee} = 0$

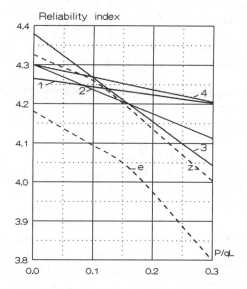

Fig.4. Reliability index vs load parameter
for $P^{\wedge}/P^{\vee} = 0.5$

capacity case (curve e) have been included for comparison purposes.

REFERENCES

Augusti, G., Baratta, A. & Casciati, F. 1984. *Probabilistic methods in structural engineering*. London, New York: Chapman and Hal.

Capurso, M. 1979. Some upper bound principles to plastic strains in dynamic shakedown of elastoplastic structures. *Jour. Struct. Mech.* 7: 1-20.

Konig, J.A. 1987. *Shakedown of elastic-plastic structures*. Warszawa: PWN.

Melchers, R.E. & Tang, L.K. 1984. Dominant failure modes in stochastic structural systems. *Structural Safety* 2: 127-143.

Murotsu, Y. 1985. Development in structural systems reliability theory. *Transactions of 8th Inter. Conf. on Struct. Mech. in Reactor Technology* M1: 7-14.

Ranganathan, R. & Deshande, A.G 1987. Generation of dominant modes and reliability analysis of frames. *Struct. Safety* 4: 217-288.

Śniady, P. & Żukowski, S. 1991. The reliability of bar structures in the light of shakedown theory. *Proc. 4th IFIP WG 7.5 Conference Munich*: 351-362. Berlin: Springer-Verlag.

Wen, Y.K. 1977. Statistical combination of extreme loads. *J. Struct. Div.*, ASCE 103(5): 1079-1093.

$P^{\wedge}/(q \cdot L)$ for different collapse mechanisms (curves 1, 2, 3 and 4) are shown, for $P^{\wedge}/P^{\vee} = 0$ and $P^{\wedge}/P^{\vee} = 0.5$, respectively. In Figures mentioned similar functions concerning the low-cycle fatigue case (curves z) and the elastic

Structural Safety & Reliability, Schuëller, Shinozuka & Yao (eds) © 1994 Balkema, Rotterdam, ISBN 90 5410 357 4

Homologous mode realization for reliability enhancement

N. Yoshikawa & S. Nakagiri
Institute of Industrial Science, University of Tokyo, Japan

ABSTRACT : An applicability of the homology design for reliability enhancement is investigated. Homology design is employed so as to let smart structures, for instance very flexible or intelligent structure equipped with control mechanism, maintain a certain geometric feature through its deformation in order to ensure reliable fulfillment of such demands given for the structure as mounting of precision instruments or reducing control cost. A formulation is proposed to synthesize the structural design for realizing homologous vibration mode of linear undamped eigenproblem. The formulation is based on the sensitivity analysis of the eigenproblem, and a governing matrix equation, whose coefficient matrix is rectangular, is derived for the design variables. The Moore-Penrose generalized inverse is used to determine the design variables. The validity of the proposed method is verified by the numerical example of the out-of-plane vibration of a planar lattice structure. The robustness required for the homology design to ensure the reliability enhancement is discussed through the numerical example.

1. INTRODUCTION

The concept of homologous deformation is to keep homology in a structural configuration, and was applied originally to the design of a large radio telescope so that the shape of the antenna is kept parabolic at any tilting angle regardless of the antenna deformation (von Hoerner 1967, Morimoto et al. 1982). Such a deformation is called homologous when a geometric relationship is maintained at the whole or a particular part of a structure before, during and after the deformation (von Hoerner 1967, Hangai and Guan 1989). The performance of the radio telescope was expected to be much improved at low cost once the homologous deformation of paraboloid was realized for the parabolic antenna. In the meantime, it has been found that stress state is mitigated by realizing adequate homology design, while the total weight is hardly increased (Yoshikawa and Nakagiri 1993). This means that homology design has possibility to afford remarkable enhancement of the reliability in the line of both safety and performance.

The concept of homologous deformation is extended to eigenproblem in this paper. As the first application to dynamic problems, this paper aims to develop a methodology for the homology design concerned with vibration mode shape of linear undamped eigenproblem. A mode shape is called homologous in this paper when a certain geometric relationship holds for the mode shape both during vibration and at standstill state. The finite element representation of structure is employed to deal with the eigenproblem. The sensitivity analysis of the eigenproblem with respect to the design variables adequately chosen is carried out to derive the governing equation of the design variables. The Moore-Penrose generalized inverse is employed for the determination of the design variables, as the coefficient matrix of the governing equation is not square but rectangular. The proposed formulation is examined and verified in problems of out-of-plane vibration of a planar lattice. How to keep the durability of the homology design itself is discussed briefly.

2. FORMULATION FOR HOMOLOGY DESIGN IN EIGENPROBLEM

The linear undamped eigenproblem concerning elastic structure is given in the form of eq.(1) by use of finite element discretization.

$$([K] - \lambda[M])\{\Phi\} = \{0\} \tag{1}$$

where $[K]$, $[M]$, λ and $\{\Phi\}$ denote the stiffness matrix, mass matrix, eigenvalue and eigenvector, respectively. Equation (1) indicates the eigenproblem after the incorporation of the boundary condition. The eigenvector consists of N ingredients. The dependent part of eigenvector denoted by $\{\Phi_h\}$ of J ingredients is prescribed by the independent part denoted by $\{\Phi_i\}$ of $N\text{-}J$ ingredients through the $J\text{x}(N\text{-}J)$ $[C]$ matrix which represents constraint for homologous mode as given in the following equation.

$$\{\Phi_h\} = [C]\{\Phi_i\} \tag{2}$$

$$\{\Phi\} = \begin{Bmatrix} \Phi_i \\ \Phi_h \end{Bmatrix} \tag{3}$$

The stiffness matrix and mass matrix in eq.(1) are partitioned in the form of eq.(4) according to the partition of the eigenvector.

$$\left(\begin{bmatrix} K_{ii} & K_{ih} \\ K_{hi} & K_{hh} \end{bmatrix} - \lambda \begin{bmatrix} M_{ii} & M_{ih} \\ M_{hi} & M_{hh} \end{bmatrix} \right) \begin{Bmatrix} \Phi_i \\ \Phi_h \end{Bmatrix} = \begin{Bmatrix} 0 \\ 0 \end{Bmatrix} \tag{4}$$

The above equation is rewritten in the separate form as given below by using the relation of eq.(2),

$$([K_s] - \lambda[M_s])\{\Phi_i\} = \{0\} \tag{5}$$

$$([K_r] - \lambda[M_r])\{\Phi_i\} = \{0\} \tag{6}$$

where $[K_s]$ and $[K_r]$ are defined as follows.

$$[K_s] = [K_{ii}] + [K_{ih}][C] \tag{7}$$

$$[K_r] = [K_{hi}] + [K_{hh}][C] \tag{8}$$

$[K_s]$ is $(N\text{-}J)\text{x}(N\text{-}J)$ square and asymmetric matrix, and $[K_r]$ $J\text{x}(N\text{-}J)$ rectangular one. $[M_s]$ and $[M_r]$ are defined similarly.

If the eigenpair of eq.(1) of the baseline design satisfies eqs.(5) and (6) simultaneously, homologous mode is realized. As eqs.(5) and (6) are not satisfied simultaneously in general, the baseline design is to be changed to form homologous mode by assigning M design variables α_m to structural parameters P_m as given below.

$$P_m = \overline{P}_m(1 + \alpha_m) \tag{9}$$

The upper bar means the value of baseline design hereafter. Homologous mode cannot be realized in all modes of eigenpair, so that design change is carried out with respect to eigenpair of a mode under interest. Equation (5) constitutes a different eigenproblem from eq.(1). The eigenpair of the asymmetric, that is, constrained eigenproblem is obtained as the solution of eq.(5). The governing equation for the structural parameters to make eqs.(5) and (6) hold simultaneously is formulated based on the first-order approximation of the structural behavior change with respect to the design variables. The change of the eigenpair of eq.(5) is expressed in the form of eq.(10) and (11) in the vicinity of the baseline design,

$$\lambda = \overline{\lambda} + \sum_{m=1}^{M} \lambda_m^I \alpha_m \tag{10}$$

$$\{\Phi_i\} = \{\overline{\Phi}_i\} + \sum_{m=1}^{M} \{\Phi_{im}^I\} \alpha_m \tag{11}$$

where superfix I and suffix m mean first-order sensitivity with respect to α_m. λ_m^I and $\{\Phi_{im}^I\}$ are derived by using both the left hand eigenvector and right hand one (Nelson 1976). The change of $[K_r]$ and $[M_r]$ is evaluated in the same form with the above equations by the first-order approximation.

$$[K_r] = [\overline{K}_r] + \sum_{m=1}^{M} [K_{rm}^I] \alpha_m \tag{12}$$

$$[M_r] = [\overline{M}_r] + \sum_{m=1}^{M} [M_{rm}^I] \alpha_m \tag{13}$$

Substituting eqs.(10), (11), (12) and (13) into eq.(6) and truncating the second-order term of α_m, the governing equation of α_m to make eqs.(5) and (6) hold simultaneously is finally given as follows.

$$\sum_{m=1}^{M} \left(\{[K_{rm}^I] - \overline{\lambda}[M_{rm}^I] - \lambda_{rm}^I[\overline{M}_r]\}\{\overline{\phi}_i\} \right.$$
$$\left. + \{[\overline{K}_r] - \overline{\lambda}[\overline{M}_r]\}\{\phi_{im}^I\} \right) \alpha_m \tag{14}$$
$$= -([\overline{K}_r] - \overline{\lambda}[\overline{M}_r])\{\overline{\phi}_i\}$$

Equation (14) is summarized in the following form,

$$[A]\{\alpha\} = \{b\} \qquad (15)$$

where $[A]$ is a $J \times M$ rectangular matrix, and $\{\alpha\}$ a design variable vector consisting of M unknowns.

We use the Moore-Penrose generalized inverse $[A]^-$ to determine the unknown design variable vector $\{\alpha\}$ (Rao and Mitra 1971). When the necessary and sufficient condition for the solution of eq.(15) to exist is satisfied, the general solution is expressed as the sum of two terms given as eq.(16),

$$\{\alpha\} = [A]^-\{b\} + ([I] - [A]^-[A])\{h\} \qquad (16)$$

where $[I]$ is an identity matrix, and $\{h\}$ the arbitrary vector. The first term on the right hand side of eq.(16) is called the particular solution, and the second one the complimentary solution. Aiming at minimum design change from current baseline design, we take only the particular solution, which minimizes the norm of the design variable vector, owing to the property of the Moore-Penrose generalized inverse, in this study.

The solution thus determined is affected by deficiency of the first-order approximation. The deficiency is overcome by renewing the baseline design iteratively, which consists of the eigenvalue analysis and sensitivity analysis, until the mode shape is judged to be homologous, until the right hand side of eq.(15) becomes equal to nil vector.

3. NUMERICAL EXAMPLE

Out-of-plane vibration of the planar lattice illustrated in Fig.1 is dealt with as numerical example to examine the validity of the proposed method. Each member of the lattice is represented by a finite beam element with torsion. Solid triangles indicate the points of simple support. The cross section of all members is circular with diameter $d = 50.0$ mm. Young's modulus, modulus of rigidity and mass density are taken equal to 70.0 GPa, 26.9 GPa and 3.0×10^3 Kg/m^3, respectively. Attention is paid to the first mode in this example. The first mode shape of baseline design, which is warped in saddle-shape, is shown in Fig.2. The first mode frequency is equal to 4.58Hz.

The homologous mode to be realized in this study is straight homologous mode, in which transverse members encircled by the broken line in Fig.1 is kept straight and parallel to those in standstill state. Hogging in the y direction is eliminated so that the stress state in the transverse members is mitigated

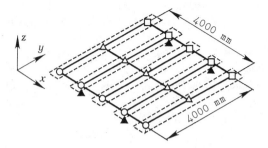

Fig.1 Finite element model of planar lattice

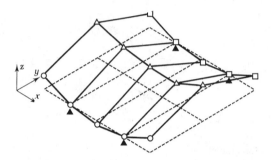

Fig.2 First mode shape of initial baseline design

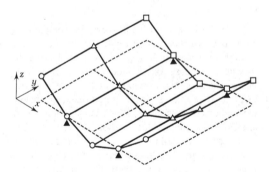

Fig.3 Homologous mode

very much. The straightened members afford reliable platform for mounting precision instrument. The straight homologous mode, once realized, is expected to be useful for the reliability enhancement form two aspects of safety and performance of the structure. The design variables are assigned to all the section diameters. The homologous mode obtained after the third renewal of the baseline design is shown in Fig.3. The corresponding frequency of the first mode is reduced to 3.44Hz.

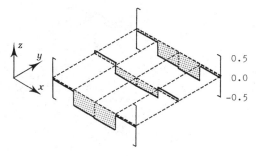

(a) *x* direction

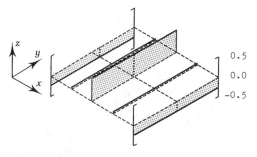

(b) *y* direction

Fig.4 Amount of design change to realize
homologous mode

A simple way to realize this straight homologous mode is to set the flexural rigidity of the transverse members sufficiently large. This way easily leads to inevitable increase of the total weight, which is undesirable from the view point of structural integrity in dynamic problems. The required diameter change for the homologous mode, which is shown in Fig.4 as ratio to the initial value, is about 0.5 at most. The total weight is reduced to 98.2% of the initial baseline design. This small change of weight arises from the property of the particular solution of the design variable vector, that is, the solution norm is minimum. In spite of the small reduction of total weight, the first mode frequency is lowered. This result implies that the structure is made so flexible by homology design that the homologous mode may be sensitive for uncertainties involved in the boundary conditions, material properties and structural dimensions.

4. CONCLUDING REMARKS

Homology design in eigenproblem is newly proposed. The validity of the proposed method is examined in problems of out-of-plane vibration of a planar lattice, in which the straight homologous mode is realized to enhance the reliability of the parts kept straight. It is expected from the result that reliability enhancement in dynamic problems can be incorporated in homology design which affords adequate homologous mode without increase of total weight of structure. The proposed homology design is confined within deterministic problems, however. The robustness of homology design against uncertainties in structural parameters is to be investigated in the future, aiming at the application of homology design for reliability enhancement in random vibration field.

REFERENCES

Hangai, Y. 1990. Shape analysis of structures, *Theoretical and applied mechanics, Vol.39, Proc. 39th Japan national congress for applied mechanics,* Ed. JNCTAM, Science council of Japan, Univ. of Tokyo Press, pp.11-28.
von Hoerner, S. 1967. Homologous deformation of tiltable telescope, *Journal of the structural division, Proceeding of the ASCE,* Vol.93, No.ST5, pp.461-485.
Morimoto, M. et al. 1982. Homology design of large antenna, *Mitubishi electric corp., Technical report,* Vol.56, No.7, pp.495-502, (in Japanese).
Nelson, R.B. 1976. Simplified calculation of eigenvector derivatives, *AIAA Journal,* Vol.14, No.9, pp.1201-1205.
Rao, C.R. and Mitra, S.K. 1971. *Generalized inverse of matrices and its applications,* John Wiley & Sons.
Yoshikawa N. and Nakagiri S. 1993. A note on finite element synthesis of structure (part 8), Formulation for homology design, *Monthly journal of Institute of Industrial Science, University of Tokyo,* Vol.45, No.7, pp.41-44.

Loads

Structural Safety & Reliability, Schuëller, Shinozuka & Yao (eds) © 1994 Balkema, Rotterdam, ISBN 90 5410 357 4

Multiple load combination analysis

C. Floris

Department of Structural Engineering, Politecnico, Milano, Italy

ABSTRACT: The paper reports the formulae for combining structural loads that are used by the Comité Eurointernational du Béton (CEB) and the Eurocode n.9. Wen's and Hasofer's methods for finding the extremes of a load combination are recalled. Then the results obtained by the use of Wen's or Hasofer's method for some load combinations are compared to the results deriving from CEB and Eurocode formulae: sometimes these are unconservative.

1 INTRODUCTORY REMARKS

The problem of combining structural loads and finding the extreme value in the structural life-time deserved attention in the seventies and in the early eighties, when many papers were published on this argument. Among others we cite Peir and Cornell (1973), Hasofer (1974), Wen (1977), Larrabee (1978), Der Kiureghian (1978), Madsen (1979), Turkstra and Madsen (1980), Larrabee and Cornell (1981), Breitung and Rackwitz (1982), Ditlevsen (1983) and Grigoriu (1984). In the successive years less contributions are found, but from a theoretical point of view any load combination could be studied by the use of the methods previously established.

Viceversa in the eighties and nineties the theoretical methods are not transferred in building codes, which were based on the semiprobabilistic or level I methods (CEB 1978, 1991) by now. Load combinations continue to be treated making use rather of experience and judgment than probabilistic tools. Loads are combined multiplying them by safety coefficients, while time fluctuations, load durations, load dispersions and coincidences are not accounted for in a quantitative manner: hence the risk associated with a design combination cannot be evaluated. In this way the potentially dangerous combinations, in which loads charac-

terized by large coefficients of variation (COV) and high rates of variability in time are preponderant, cannot be distinguished from the combinations that are less dangerous because of the predominance of dead load, which is constant in time and has a small COV.

In this paper two methods for combining stochastic loads are briefly recalled (a more complete examination of the methods for load combination is in Floris, 1993a). Successively both methods are used to derive the largest value fractiles of some load combinations involving dead, snow, wind, live and earthquake load; only combinations with three or more loads are considered. The exact results, which are found in this way, are compared to the results for the same combinations that are calculated according to the European codes.

2 LOAD COMBINATIONS ACCORDING TO SEMIPROBABILISTIC METHOD

The semiprobabilistic or level I methods apply a different factor to each random variable, that is defined by a single value, the so-called characteristic value. Thus, the following formula was proposed for structural load combinations (CEB 1978)

$$S_d = \gamma_G c_G G_k + \gamma_Q (c_1 Q_{1k} + \sum_{2i}^{n} c_i \psi_{0i} Q_{ik}) \quad (1)$$

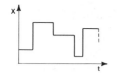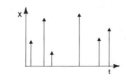

Fig. 1   Poisson square wave process (left);
Poisson spike process (right)

In Eq.1 S_d is a structural load effect; c_G and c_i are influence coefficients, γ_G and γ_Q are safety factors to be applied to dead load G_k and to variable loads Q_{ik} , respectively; and ψ_{0i} = companion load factors to take into account the reduced probability of all loads at their characteristic values, that are indicated with the index k. Eq.1 was confirmed in the model code for the nineties (CEB, 1991) and adopted by the Eurocode n. 9 (1990).

If the loads are unfavourable, the safety coefficients γ_G and γ_Q take the values 1.35 and 1.5 , respectively. The companion load factors should be determined using probabilistic methods, but in reality these are not used: hence, empirical values are suggested. With reference to the Eurocode ψ_0 is 0.7 for live load on floors (with the exception of floors subjected to possible overloads such as libraries and archives for which ψ_0 = 1) as well as for snow load; for wind load ψ_0 = 0.6.

A combination format similar to Eq. 1 is used in the Eurocode n.8 (1988) for the combinations involving earthquake load

$$S_d = \gamma_i \; E + G_k + P_k + \sum_j^n \psi_{2j} \; Q_j \qquad (2)$$

In Eq.2 the influence coefficients are included in load symbols; γ_i = importance factor of earthquake load (generally is one); E = earthquake load; P_k = prestressing effect. The coefficients ψ_{2j} are different from the coefficients ψ_{0i} of Eq.1 and they too do not derive from probabilistic tools; they are less or equal to one (and in some cases are zero).

Both Eqs.1,2 do not consider the rate of each load on total load, when the loads have different COV, arrival rates and durations. Thus, the design combined loads are characterized by different probabilities of exceedance, while the risk associated with a design format should be constant.

3 EXACT METHODS FOR COMBINING STRUCTURAL LOADS

If the variable loads are schematized as Poisson processes (Wen 1977), i.e. the number of load changes for always-on loads (Fig.1 left) and the number of load arrivals for spike loads (Fig.1 right) are both Poisson distributed, two methods are preferable for load combinations: Wen's load coincidence method (Wen 1977), if only Poisson spike processes are involved; and Hasofer's integral equation method (1974), when a Poisson square wave process is to be combined with a Poisson spike process. Earthquake, snow and wind load can be modelled as Poisson spike processes. The live load on floors is divided into two parts, the former of which is the sustained load process (Peir and Cornell 1973), that is schematized as a Poisson square wave process; the latter part is the extraordinary load process (Mc Guire and Cornell 1974), that is schematized as a Poisson spike process.

The largest value in time T CDF of the combination Q_T of many spike processes is according to Wen (1977)

$$F_{QT}(q) = \exp\{-T[\; \sum_i^n \nu_i G_{Xi}(q) + \sum_{i \neq j}^{n} \sum^{n} \nu_i \nu_j (d_i +$$
$$+ d_j) G_{ij}(q) + \sum_{i \neq j \neq k}^{n} \sum^{n} \sum^{n} \nu_i \nu_j \nu_k (d_i d_j + d_j d_k +$$
$$+ d_k d_i) G_{ijk}(q) \;]\} \qquad (3)$$

In Eq.3 ν_i are load arrival rates, d_i are load mean durations; G_i, G_{ij}, G_{ijk} = the complements of the distributions of X_i, X_i+ X_j, and X_i+ X_j+ X_k, respectively. The extreme load Q_T is combined with dead load G to find the load combination S_T in T years.

If a Poisson square wave process Q_S and a Poisson spike process Q_E act together, Hasofer's method (1974) is suitable. The CDF of the largest values L of the sum Q_S + Q_E in the generic dwelling time τ of Q_S is found firstly:

$$F_L(x) = \int_0^x f_{QS}(x - y) \cdot$$
$$\exp \{- \nu_E \tau[\; 1 - F_{QE}(y)]\} \cdot dy \qquad (4)$$

The distribution of the largest values in T is

$$F_{ST}(s) = \exp(- \nu_S T) \cdot \gamma(T, s) \qquad (5)$$

In Eqs. 4,5 ν_E and ν_S are the arrival rates of Q_E and Q_S, respectively. The function $\gamma(T, s)$ is found solving the following integral equation

$$F_L(T,s) + \nu_S \int_0^T F_L(T-u,s) \cdot \gamma(u,s) \cdot du = \gamma(T,s) \quad (6)$$

As Floris et Giommi (1991) pointed out, when there are more spike processes and one square wave process, Hasofer's method is still usable. In fact in this case the exponential function of Eq. 4, which is the distribution of the maximum of Q_E in τ, is replaced by Eq.3, which represents the maximum of the sum of the spike processes in $T = \tau$. Moreover, as a computational simplification, the PDF f_{QS} is substituted by the PDF of the sum $Q_S + G$. The results that are obtained agree well with the results from another method suggested by Wen (1977).

4 APPLICATIONS

The theoretical models presented in the third section are applied to some remarkable combinations: more extensive applications will be found in Floris (1993 a,b). For each combined load effect the values according to Eq.1 (or Eq.2) are also calculated: the characteristic values of all the loads are necessary to do that. The data of the loads are resumed below; it is recalled that climatic loads refer to the city of Milan, which has a well-behaved climate, and derive from an elaboration performed by the writer and his collaborators elsewhere. Thus, strictly speaking, the results would be valid only for Milan, where both wind and snow pressure are lognormal variates. The effects of the upper tail of the instantaneous PDF should be ascertained to determine at which extent these results apply to other cases in which distributions with larger upper tails - e.g. extreme type I CDF - are used. However, in the author's opinion these effects should be less important than the effects of the macrotime variability; moreover Milan should be representative of the temperate climates.
Dead load - It is a normal variate with a COV $\nu_G = 0.04$. The average μ_G is given by the parameter p, that may be equal to 20, 50 or 80. A change of scale is made in order to render the sum of the averages equal to 100: thus, it is possible to ascertain the effect of the dead load rate on the largest value distribution.
Live load - Assuming an office as utilization and an influence area of 50 m², the sustained live load has a mean value μ_{QS} = 555 N/m², a standard deviation σ_{QS} = 320.43 N/m², a change rate ν_S = 0.125 per year and is schematized as a Poisson square wave process. The extraordinary live load is a Poisson spike process, the amplitude of which has μ_{QE} = 427.2 N/m² and σ_{QE} = 246.64 N/m², while ν_E = 12 per year. Both parts of the live load are gamma variates. The duration of Q_E is 4.25 days.
Snow load - Snow load is a Poisson spike process with an arrival rate ν_{qs} of 4 per year; the duration, necessary for Eq.3, is 5.04 days. The instantaneous statistics have a lognormal PDF with μ_{qs} = 71.86 N/m² and σ_{qs} = 90.57 N/m²; yearly average μ_{qsy} is 143.80 N/m².
Wind load - Wind storms are considered so that a Poisson spike process is still used: ν_W is 13/year with a duration of 3 days. The instantaneous statistics have a lognormal PDF with μ_W = 85.84 N/m² and σ_W = 53.86 N/m²; yearly average μ_{Wy} is 199.87 N/m².
Earthquake load - The response acceleration of a simple oscillator with a vibration period T_0 = 0.50 s and a relative damping ζ_0 = 0.05 have been determined elsewhere (Floris 1993 c): the instantaneous average acceleration $\mu_{\ddot{x}}$ is 40.88 gals with $\sigma_{\ddot{x}}$ = 48.54 gals. It is intended that the change of scale transforms the acceleration into a load effect E, since a linear behavior is assumed.

The results are in Tables 1-5: the exact average μ_{ST} and the fractiles with a 5% (S_{Tk}), a 2% (S_{Tc}) and a 0.5% (S_{Td}) probability of exceedance are given for each combination; the last line or the last two lines report the values that are obtained according to the Eurocodes ns. 9 and 8. In all the combinations with snow load there are two sets of results according to the Eurocodes, the latter of which is derived increasing snow load by a 50%: the standard snow load has a 2% probability of exceedance (i.e. a return period of 50 years) and is relative to the statistics of yearly maxima; Italian standards (CNR 1985) multiply it by a return coefficient c_r e-

Table 1: combination of dead, snow and wind load

α	0.25			1.0			4.0		
p	20	50	80	20	50	80	20	50	80
μ_{ST}	169.50	143.46	117.98	222.73	179.63	132.11	336.61	252.22	161.89
S_{Tk}	234.60	184.14	135.11	389.55	281.00	172.59	610.62	417.88	227.73
S_{Tc}	266.90	204.34	143.02	486.32	341.48	196.78	766.00	516.02	266.61
S_{Td}	325.33	240.88	157.21	599.75	455.22	242.24	1057.3	698.33	339.43
EUROC	284.77	231.93	179.10	329.74	260.04	190.35	419.58	318.94	213.91
EUROC*	316.83	251.97	187.12	444.24	331.61	218.97	600.58	433.44	259.71

$\alpha = \mu_{qsy}/\mu_{Wy}$ $\mu_G + \mu_{Wy} + \mu_{qsy} = 100$ $^*c_r = 1.5$

qual to 1.5.

The combined load effects are obtained in this way: with reference to the combination of dead, snow and wind load and the case $p = 20$, $\alpha = \mu_{qsy}/\mu_{Wy} = 0.25$, the sum of the averages $\mu_G + \mu_{Wy} + \mu_{qsy}$ is put equal to 100. Then, μ_G is p, while two influence coefficients c_1 and c_2 are calculated for snow and wind load, respectively: $c_1 = 0.111266$, $c_2 = 0.320208$. In this way the yearly averages become $\mu_{qsy} = 16.0$, $\mu_{Wy} = 64.0$. The instantaneous transformed averages and standard deviation are, respectively: $\mu'_{qs} = 7.9956$, $\sigma'_{qs} = 10.0773$; $\mu'_W = 27.4867$, $\sigma'_W = 17.2464$ (the PDF are unchanged!). The CDF of the yearly maxima of snow load is according to Poisson model

$$F_{qsy}(q) = \exp\{-4 \cdot 1 \cdot [1 - \phi (\frac{\ln q - \lambda_{qs}}{\zeta_{qs}})]\} \quad (7)$$

in which λ_{qs} and ζ_{qs} are the parameters of the instantaneous PDF; and $\phi(.)$ = the standard normal CDF. Putting Eq.7 equal to 0.98, it is easy to find

$$q_{sk} = \exp[\lambda_{qs} + \zeta_{qs} \cdot \phi^{-1}(1 + \frac{\ln 0.98}{4})] =$$
$$= 61.0694 \quad (8)$$

Analogously for wind load it is found $q_{Wq} = 127.9102$. Being $G_k = 20(1 + 1.645 \cdot 0.04) = 21.316$, the combined load effect is

$$S_d = 1.35 \cdot 21.316 + 1.5(127.9102 + 0.7 \cdot 61.0694) = 284.765 \quad (9)$$

or $S_d = 316.826$, if snow load is multiplied by 1.5 . The other cases and combinations are treated in an analogous way.

Analyzing Tables 1-4 the different probabilities of exceedance inherent to the results of Eqs.1,2 are noteworthy: in some cases (e.g. combination of dead, live and wind load, Table 2) they are larger as far as a 60% than the exact fractiles with a 0.5% probability of exceedance; in some others - particularly when snow load is present: Tables 1, 3, 4 - they correspond to the fractiles with 5% probability of exceedance or are between these and the averages. Dead load rate p appears to have an important effect: when p is large, the results of Eqs. 1,2 are more conservative. Thus it is suitable to include the mutual rates of the various loads in load combination rules. However Tables 1, 3 and 4 prove also the inadequacy of the 2% fractile of snow load: in temperate climates snow load has a very large dispersion and fractiles from the CDF of the largest values in fifty years should be used. In fact, things improve when snow load is increased by a 50%.

Table 5 with earthquake load requires a separate comment: the combined load effects of Eq.2 appear to be very unconservative, as they are always lower than the 5% fractile S_{Tk} (and for ρ = 0.25 lower than the average). This lack of conservativism has two causes: 1) firstly the way with wich Eurocode n.8 prescribes the seismic load; this is the product of a dynamical amplification factor with a probability of exceedance as large as 25% and of the ground acceleration, which is not specified (here

Table 2: combination of dead, live and wind load

p	20		50		80	
ψ	0.25	0.75	0.25	0.75	0.25	0.75
μ_{ST}	173.55	170.56	148.37	144.23	118.96	118.61
S_{Tk}	232.07	210.15	191.40	169.03	132.81	129.05
S_{Td}	296.31	241.91	239.64	188.51	146.65	136.55
EUROC	293.85	382.24	237.61	292.86	180.19	203.47

$$\mu_{QE} + \mu_{QS} + \mu_{Wy} + \mu_{G} = 100 \qquad \psi = (\mu_{QE} + \mu_{QS})/(\mu_{QE} + \mu_{QS} + \mu_{Wy})$$

Table 3: combination of dead, live and snow load

p	20		50		80	
ψ	0.25	0.75	0.25	0.75	0.25	0.75
μ_{ST}	323.06	184.24	243.00	160.51	156.57	122.60
S_{Tk}	581.47	249.51	400.92	193.54	220.13	139.12
S_{Td}	916.70	339.60	610.28	249.50	304.44	159.93
EUROC	457.49	426.42	339.89	320.47	221.10	214.52
EUROC*	629.25	466.50	447.24	345.52	264.04	224.54

$$\mu_{QE} + \mu_{QS} + \mu_{qsy} + \mu_{G} = 100 \qquad \psi = (\mu_{QE} + \mu_{QS})/(\mu_{QE} + \mu_{QS} + \mu_{qsy})$$

Table 4: combination of dead, live, snow and wind load

p	20		50		80	
ψ	0.25	0.75	0.25	0.75	0.25	0.75
μ_{ST}	204.72	210.96	174.87	183.04	127.10	147.09
S_{Tk}	246.59	252.34	205.70	211.49	138.01	157.06
S_{Td}	316.48	288.50	240.54	236.09	149.18	162.77
EUROC	320.48	400.73	254.25	304.41	162.93	208.10
EUROC*	392.34	417.50	299.17	314.89	174.91	212.29

$$\mu_{QE} + \mu_{QS} + \mu_{qsy} + \mu_{Wy} + \mu_{G} = 100 \qquad \psi = (\mu_{QE} + \mu_{QS})/(\mu_{QE} + \mu_{QS} + \mu_{qsy} + \mu_{Wy})$$

the fractiles of the response acceleration with a 10% probability of exceedance are used, as in Floris 1993 c). It is perhaps suitable that design seismic load should have a smaller probability of exceedance. 2) The coefficient ψ_{2j} for snow load is 0.3 while for wind load it is null, which is very wrong since the first order terms in Eq. 3 are never negligible.

SUMMARY AND CONCLUSIONS

In this paper two methods for combining

Table 5: combination of dead, earthquake, snow and wind load

p	20			50			80		
ρ	0.25	0.50	0.75	0.25	0.50	0.75	0.25	0.50	0.75
μ_{ST}	171.38	132.41	112.59	149.83	120.37	108.10	117.92	108.19	103.40
S_{Tk}	296.99	248.35	288.72	223.16	192.77	217.97	149.57	137.41	147.33
S_{Tc}	370.16	320.02	413.37	268.89	237.64	295.88	167.82	155.31	178.49
S_{Td}	506.40	440.29	600.81	354.03	312.72	413.03	201.82	185.31	225.35
EUROC	113.25	159.78	206.11	110.87	139.83	168.79	108.30	119.87	131.45

$$\rho = \mu_E / (\mu_E + \mu_{qsy} + \mu_{Wy}) \qquad \mu_G + \mu_E + \mu_{qsy} + \mu_{Wy} = 100 \qquad \mu_{qsy} = \mu_{Wy}$$

stochastic loads are recalled, the former of which is due to Wen and is suitable to combine Poisson spike processes; Hasofer proposed the latter, that matches the combinations in which a renewal Poisson process is present. Some combinations involving three or more loads - dead, sustained, extraordinary, snow, wind and earthquake load - are considered: some fractiles of the maximum distribution are calculated by the use of the above cited methods. These fractiles are compared to the design values calculated according to the Eurocodes: it is found that these do not understand a constant level of reliability and sometimes are unconservative. The unconservativism is caused by many causes: a bad choice of the probability of exceedance of a load (snow and earthquake load), the wrong values of some companion load factors ψ_0 or ψ_2, and not taking dead load rate into account. Because of this fact heavy structures have a higher level of reliability since dead load, which has a small COV and is constant in time, is preponderant. It is concluded that standard rules for combining loads need many improvements.

Observing the Tables, all the combinations according to the Eurocodes that have a probability of exceedance larger than a 5% can be considered unconservative. Viceversa, when this probability is smaller than a 0.5%, there is an excess of conservativism. The Eurocodes give unconservative results when the snow load is preponderant: Table 1, $\alpha = 1 - 4$, Table 3, $\psi =$ 0.25; the results are always unconservative in the presence of earthquake load. However, as the amount of dead load increases (p = 80), the probabilities of exceedance decrease definitely. The relative rates of the various loads should influence the combination coefficients, while Eqs.1,2 do not vary them in any case. The author of this paper is studying an alternative rule for combining structural loads (Floris 1993a, 1993b). In this rule the basic values of loads are the average ones, which are summed up and multiplied by only a coefficient γ to transform the average sum into the fractile that is wanted. The coefficient γ depends on the type of combination, on p and other parameters. As an example, considering the combination of dead, live and snow load, if the fractiles of the largest value distribution are to be calculated, in the first place the instantaneous average of snow load is added to the averages of dead, sustained and extraordinary load. The result is multiplied by γ_m, γ_k, γ_c or γ_d, if the average, the fractiles with a 5%, a 2% or a 0.5% probability of exceedance are looked for, respectively. The general form of the coefficients γ is

$$\gamma = (a_0 + a_1\psi + a_2\psi^2) + (b_0 + b_1\psi + b_2\psi^2)p + (c_0 + c_1\psi + c_2\psi^2)p^2 \qquad (10)$$

being ψ defined in Table 3. In the case of γ_d the coefficient becomes:

$$\gamma_d = (14.6959 - 13.7919\ \psi - 0.3570\ \psi^2) +$$
$$(- 0.1366 + 0.1353\ \psi + 0.0062\ \psi^2)p +$$
$$(2.117 \cdot 10^{-7} + 0.1613\ \psi - 0.2178\ \psi^2) \cdot 10^{-4} \cdot p^2$$

(the constants in this and similar expressions for γ's are found by fitting the results of the exact methods). Eq. 10 accounts for the parameters ψ and p and, implicitly, for the influence area of live load: Eq.10 is valid for any influence area, but different sets of constants are effective for influence areas between 50 and 100 m^2 or equal to 17 m^2 (with linear interpolation between the two limits).

The principal advantage of this combination rule is in that it relies on load average values, that are much easier to be found and less affected by uncertainty than the so-called characteristic values. In fact, as a fractile has a smaller probability of exceedance q, or, in other words, a larger return period $T_r = \dfrac{1}{q}$, the sample of data for evaluating this fractile should become more numerous, which is difficult to obtain in many cases. It is recalled that the standard deviation of sampling error is inversely proportional to the square root of sample size (Simiu and Filliben 1976; Simiu et al. 1978).

REFERENCES

Breitung, K. & R. Rackwitz 1982. Nonlinear combination of load processes. J. Struct. Mech., 10(2): 145-166.
CEB 1978. Model code for concrete structures. Bull. 124/125f, Paris.
CEB 1991. Model code for concrete structures. St. Saphorin (Switzerland).
CNR-Consiglio Nazionale delle Ricerche 1985. Ipotesi per la Valutazione delle Azioni sulle Costruzioni. CNR 10012/85, Roma (in Italian).
Der Kiureghian, A. 1978. Second-moment combination of stochastic loads. J. Struct. Div., ASCE, 106(2): 1551-1567.
Ditlevsen, O. 1983. Level crossing of random processes. In Reliability Theory and its Applications: 57-83. Hague, Martinus Nijhoff Publ.
Eurocode n.8 1988. Structures in Seismic Regions: Design. Commission of European Communities, Brussels.
Eurocode n.9 1990. Eurocode for Actions on Structures. Commission of European Communities, Brussels.
Floris, C. 1993a. Load combinations for structural design. Part I. In preparation.
Floris, C. 1993b. Load combinations for structural design. Part II. In preparation.
Floris, C. 1993c. Seismic load for structural analysis and load combination. In preparation.
Floris, C. & C. Giommi 1991. On stochastic load combination modelling and algorithms for calculating probability of failure. Proc. of CERRA-ICASP 6 Conf., Vol. 2 : 630-635. Mexico City.
Grigoriu, M. 1984. Load combination analysis by translation processes. J. Struct. Engrg., ASCE, 110(8): 1725-1734.
Hasofer, A.M. 1974. Time dependent maximum of floors live loads. J. Engrg. Mech. Div., ASCE, 100(5): 111-121.
Larrabee, R.D. 1978. Approximate stochastic analysis of combined loading. Res. Rep. R78-28, MIT, Dept. of Civ. Engrg., Cambridge (Mass).
Larrabee, R.D. & C.A. Cornell 1981. Combination of various load processes. Jour. Struct. Div., ASCE, 107(1): 223-239.
Madsen, H.O. 1979. Load models and load combinations. Rep. No. R113, Technical University of Denmark, Lyngby.
Mc Guire, R.K. & C.A. Cornell 1974. Live load effects in office buildings. Jour. Struct. Div., ASCE, 100(7): 1351-1366.
Peir, J.C. & C.A. Cornell 1973. Spatial and temporal variability of live loads. J. Struct. Div., ASCE 99(5): 903-922.
Simiu, E., J. Biétry & J.J. Filliben 1978. Sampling errors in estimation of extreme winds. J. Struct. Div., ASCE, 104(3) : 491-501.
Simiu, E. & J.J. Filliben 1976. Probability distributions of extreme wind speeds. J. Struct. Div., ASCE, 102(9): 1861-77.
Turkstra, C.J. & H.O. Madsen 1980. Load combinations in codified structural design. J. Struct. Div., ASCE, 106(12) : 2527-2543.
Wen, Y.-K. 1977. Statistical combination of extreme loads. J. Struct. Div., ASCE, 103(5): 1079-1093.

Structural Safety & Reliability, Schuëller, Shinozuka & Yao (eds) © 1994 Balkema, Rotterdam, ISBN 90 5410 357 4

Stochastic live load model for multiple story columns

Hideki Idota
Department of Architecture, Aichi Sangyo University, Okazaki, Japan

Tetsuro Ono
Department of Architecture, Nagoya Institute of Technology, Japan

ABSTRACT :
   The purpose of this paper is to evaluate the live load statistically for a column to support plural floors. A stochastic live load model for this multiple story column is presented using the live load survey results for office buildings. Live load reduction factors for multiple story columns are calculated parametrically based on the stochastic live load model. The furniture concentration due to the moving and remodeling of a building is computer-simulated using the surveyed live loads of the office building. The design live load for the axial compression on the column at the time of furniture concentration is studied.

1. INTRODUCTION

Various design codes indicate that in calculating the axial load on a column produced by a live load, the design live load can be reduced according to the number of stories supported by the column. This live load reduction was specified in consideration of the fact that the variation of axial compression on a column caused by live loads gets smaller than the variation of axial compression on a column which supports only one floor as the number of floors supported increases. This live load reduction is affected by the variation of a live load extended over different floors and the variation of the live load on one floor. Therefore, statistical data on the position and weight of the furniture and the correlation between a different story is indispensable to evaluate the live load supported by a multiple-story column. Nevertheless, little statistical data on live loads related to the axial loads of multiple-story columns has been accumulating as yet. In order to evaluate a reduction factor of design live load for multiple-story columns stochastically, it is necessary to discuss the reduction factor parametrically based on the stochastic live load model for multiple-story columns.

   The purpose of this paper is to study the design live load reduction for multiple-story columns. First, statistical data on the sustained live load of multiple-story columns is shown by conducting a survey of

the live loads of three office buildings. This paper indicates that both the standard deviation of the column load and the 99 percentile value decreases with an increasing number of stories supported. Secondly, a stochastic live load model for multiple-story columns is presented based on the sustained live load survey results. In the live load model, the variation of multiple-story columns load is calculated by means of the NBS bay subdivision model. The design live load reduction factors are studied parametrically for different bay sizes based on the live load model. Lastly, the concentration of live loads due to the moving or remodeling of a building is computer-simulated using the surveyed live loads of the office building. The design live load for the axial compression on the column at the time of furniture concentration is studied.

2. LIVE LOAD SURVEY RESULTS OF OFFICE BUILDINGS

A live load survey was carried out for three office buildings shown in Table 1. Total floor area surveyed was 9,667m². Data was collected in terms of location and weight of every piece of furniture and personnel in office area. The surveyed data are samples of arbitrary point-in-time loads in ordinary situations, and can be considered as the data for the sustained live load model.

Table 2 shows statistical results of live load intensity in square units with a size varying from 1m×1m to 10m×10m. The mean value of live load intensity was approximately 700Pa irrespective of unit size, and the coefficient of variation decreases with increasing unit size. Table 2 also shows the 3rd and 4th moments which can represent the skewness and kurtosis of probability distribution (these moments will be 0 and 3 respectively for normal distribution). It is clear that the probability distribution of live load approaches a normal distribution gradually when the unit area increases. This is based on the fact that the 3rd and 4th moments approach 0 and 3 respectively. Fig. 1 shows the K-S test results of averaged load intensity in square units for four probability distribution types, normal, lognormal, gamma and extreme type I distribution. With a unit less than 5m×5m, gamma distribution fits well, and a normal distribution fits for those over 5m×5m. These results can explain the changing of the 3rd and 4th moment of averaged load intensity in square units shown in Table 2.

3. STOCHASTIC LIVE LOAD MODEL FOR MULTIPLE STORY COLUMN

Fig. 2 shows the relation of unit size with the coefficient of variation for averaged live load intensity in square units based on the surveyed data. The variation of unit load is affected considerably by the spacial correlation of live loads on a floor. The spacial correlation has been formulated as an exponential decay function by Pier(1971), or as a white noise process by McGuire, Cornell(1974) and Choi(1989) for simplicity. However, it is reasonable to consider that the spacial correlation of live load intensity on a floor has directional qualities considering a geometrical arrangement of the furniture in an office in ordinary situations. Fig. 3 shows the correlation coefficient $\rho(x,y)$ of live load intensity at an arbitrary point (x_0,y_0) and (x_0+x,y_0+y) on a floor. Though the correlation coefficient decreases rapidly when going away from a point $(0,0)$, a comparatively large correlation is observed along the x and y axes. A relatively highly

Table 1 Outline of office buildings surveyed

		Building A	Building B	Building C
Use		Office	Office	Office
The Number of Stories		15	12	10
Surveyed Floors		11(5th-15th)	4(1st-4th)	3(3rd-5th)
Area Surveyed	Each Floor	645.4m²	276.5m²	487.1m²
(Office)	Total	7,100m²	1,106m²	1,461m²
Total Floor Area Surveyed		9,667 m²		

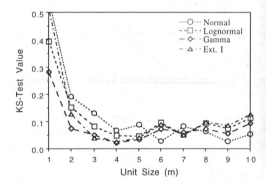

Fig. 1 K-S test results

Table 2 Results of unit analysis

Unit Size (m)	Unit Area (m²)	Points	Mean (Pa)	Standard Deviation (Pa)	Coefficient of Variation	3rd Moment	4th Moment
1×1	1	8,277	665	1069	1.61	2.23	11.03
2×2	4	1,887	704	605	0.86	1.36	6.67
3×3	9	818	708	460	0.65	1.21	7.16
4×4	16	441	713	361	0.51	0.83	4.23
5×5	25	254	719	340	0.47	0.90	4.94
6×6	36	186	703	278	0.40	0.34	4.81
7×7	49	95	700	281	0.40	0.58	4.47
8×8	64	72	704	241	0.34	-0.04	2.37
9×9	81	61	703	216	0.31	-0.18	2.51
10×10	100	57	698	203	0.29	-0.21	2.44

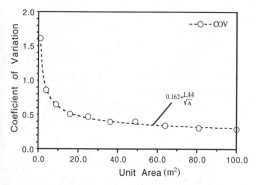

Fig. 2　Coefficient of variation of unit load

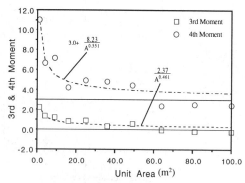

Fig. 4　3rd and 4th moment of unit load

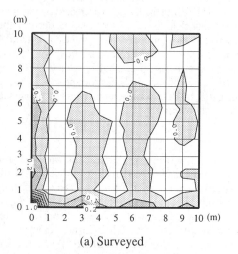

(a) Surveyed

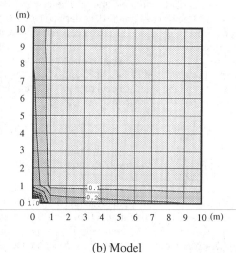

(b) Model

Fig. 3 Space correlation coefficients of surveyed data and statistical model

correlated point appears for the x direction in 3m intervals periodically corresponding to an interval of desk arrangements.　For simplicity, the periodicity of the spacial correlation is ignored, and this paper uses the next formula as a stochastic model for the spacial correlation coefficient $\rho(x,y)$ presented by Kanda(1985).

$$\rho(x,y) = e^{-C\{x^{r_1}+y^{r_1}\}^{r_2}} \qquad (1)$$

Where $C=1.20$, $r_1=0.12$, $r_2=1.0$.　The coefficient of variation for the unit live load intensity is modeled by the following equation based on a regression analysis.

$$\delta_u = 0.162+\frac{1.44}{A^{0.51}} \qquad (2)$$

A non-Gaussian property of the probability distribution type is considered by means of the 3rd and 4th moments shown in Table 2.　Fig. 4 shows the relationship between the 3rd and 4th moment of the averaged load intensity in square unit and the unit area.　The broken lines in Fig. 4 represent the regression lines.　Considering the 3rd and 4th moments approach 0 and 3 respectively, these higher order moments are modeled as the following equations using an exponential function.

$$\alpha_{3X} = \frac{2.37}{A^{0.461}} \qquad (3)$$

$$\alpha_{4X} = 3.0+\frac{8.23}{A^{0.551}} \qquad (4)$$

In case of axial load on a multiple story column, it is necessary to consider a statistical correlation of the live load vertically between different floors. This correlation represents the tendency for furniture to be placed on floors in a similar patterns, and is the so-called "stacking effect".　The correlation coefficients of column axial load

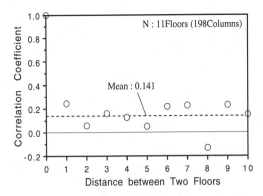

Fig. 5 Correlation coefficients between different Floors

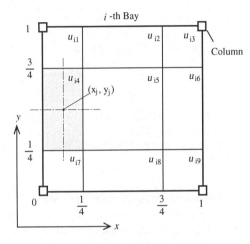

Fig. 6 Bay Subdivisions of NBS model

between two vertically different floors are presented arranged by the distance of the considered two floors in Fig.5. Since there is no significant influence in the changing of the distance between two floors, the correlation coefficient is determined to be 0.141 as a constant for the live load model.

4. LIVE LOAD REDUCTION FACTOR FOR MULTIPLE STORY COLUMN

Based on the stochastic live load model presented in a previous section, a multiple story column load is examined in this section. Assuming that live load intensity is $W(x,y)$ at a location (x,y), the axial load C of a column having influence area $A_i(=a{\times}b)$ can be calculated by;

$$C = \int_{A}^{a} \int_{0}^{b} I(x,y)W(x,y)dxdy \qquad (5)$$

where $I(x,y)$ denotes a influence function. Consequently, equivalent uniformly distributed load of a column L_C is expressed as follows.

$$L_C = \frac{C}{A_i} \qquad (6)$$

Where A_T is tributary area of a column. Calculation of statistics for L_C needs the covariance of $W(x,y)$, and it is very complicated to treat $W(x,y)$ as a continuous function considering the spacial correlation of live load intensity on a floor defined by Eq.(1). Since the live load is created by furniture or personnel, it is satisfactory to consider that the discrete live load model is used for the calculation of axial column load. In this paper, the NBS bay subdivision model presented by Corotis(1972) is used for calculating the axial column load. Using the NBS model, axial column load C is expressed by the following equation.

$$C = \sum_{i=1}^{n_F} \sum_{j=1}^{9} I_j(x_i,y_i)W_{ij} \qquad (7)$$

Where (x_i,y_i) and W_{ij} represent the x and y coordinates and live load intensity of j-th unit on i-th bay respectively. n_F in Eq.(7), that is the number of bays supported by the column, will be defined as 4 for an interior column, 2 for a perimeter column and 1 for a corner column. The statistical moments of W_{ij} are given as from Eqs.(2)(3) and (4) in a function of area A_{ij} of the divided unit. The correlation coefficient between W_{ij} and W_{kl} is defined by Eq.(1) using x and y components of the distance between the centers of two divided units. Eq.(7) shows that the column load C is defined as a linear function of random variables W_{ij}. Consequently, the mean value μ_C and the standard deviation σ_C of C can be calculated using the following equations.

$$\mu_C^2 = \sum_{i=1}^{n_F} \sum_{j=1}^{9} I_j(x_i,y_i)\mu_u A_{ij} \qquad (8)$$

$$\sigma_C^2 = \sum_{i=1}^{n_F} \sum_{j=1}^{9} \sum_{k=1}^{n_F} \sum_{l=1}^{9} I_j(x_i,y_i)I_l(x_k,y_k)\delta_u(A_{ij})\delta_u(A_{kl})\rho_{ijkl} \qquad (9)$$

Since the Eq.(7) shows that the axial column load C is expressed by a linear function, non-Gaussian properties of C can be considered with using the higher-order moments standardization technique, if necessary. A formulation of the standardization technique for the 3rd and 4th order moments is reported by the authors(1989).

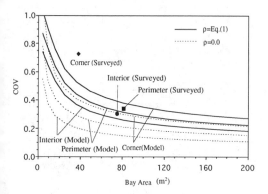

Fig. 7 Coefficient of variation of column Live Load

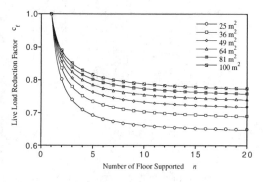

Fig. 8 Live load reduction factors and fitted curves

Table. 3 Parameters of Eq.(14)

Bay Area	Parameter	Interior Column	Perimeter Column	Corner Column
25 m²	a	0.635	0.613	0.592
	b	1.139	1.101	1.056
36 m²	a	0.674	0.647	0.622
	b	1.023	1.007	0.972
49 m²	a	0.701	0.673	0.647
	b	0.960	0.951	0.922
64 m²	a	0.723	0.694	0.670
	b	0.914	0.910	0.886
81 m²	a	0.741	0.712	0.684
	b	0.850	0.878	0.862
100 m²	a	0.755	0.726	0.698
	b	0.862	0.858	0.843

Fig. 7 shows the relation between the coefficient of variation for the column load $C\ (=\sigma_C/\mu_C)$ calculated by Eqs.(8)(9) and the bay area. The actual observed data points and the two kinds of curves are correlated as in Eq.(1) and non-correlated assumptions are given. The solid and broken curves represent correlated and non-correlated assumptions, respectively. These plots and curves are dividing into three groups; interior column, perimeter column and corner column. The curves

based on the live load model fits the actual observed data well except in the case of the corner column. The correlated assumption with Eq.(1), however, fits better than the curve for the non-correlated assumption.

The multiple story column load can be calculated by summing up the n number of single story column load; $L_{C1}, L_{C2}, ..., L_{Cn}$, which has the same statistics as L_C. Therefore, the mean value $\mu(n)$ and the standard deviation $\sigma(n)$ of the column live load supporting n number of stories are given as follows.

$$\mu(n) = n\mu_C \quad (10)$$

$$\sigma(n)^2 = n\sigma_C^2 + n(n-1)\sigma_C^2 \rho_F \quad (11)$$

Where ρ_F represents the correlation coefficient between two vertically different floors presented in Fig.5, and is defined as 0.141 for office buildings.

In order to evaluate the design live load for the column, percentile load L_e corresponding to non-excessing probability p_e must be calculated from the first four moments of the column live load. The higher-order moment method is used in calculating L_e as follows:

$$L_e(n) = \mu(n) + \beta_S \sigma(n) \quad (12)$$

Where β_S is the reliability index based on the higher-order moment standardization technique, and is defined as $\beta_S = S_X(\beta)$. $S_X(\cdot)$ denotes the higher-order moment standardization function (see reference (authors, 1989) for more detail). β is the reliability index based on the second-order moment method. β is defined as $\beta = \Phi^{-1}(1 - p_e)$, where $\Phi(\cdot)$ represents the probability distribution function of the standard normal variable.

Fig. 8 shows the live load reduction factor c_f. The reduction factor is calculated following the next equation.

$$c_f = \frac{L_{e99}(n)}{L_{e99}(1)} \quad (13)$$

Where $L_{e\,99}(n)$ indicates the 99 percentile load intensity for a column supporting n floors, and $L_{e99}(1)$ indicates the 99 percentile load intensity for a column supporting only one floor. Table 3 shows the parameters of the fitted curves approximated by the exponential function as shown;

$$c_f = a + (1-a)n^{-b} \quad (14)$$

Each fitted curve is shown by a broken line in Fig.8.

The statistics of L_{Cn} is defined by Eqs.(10) and (11). Therefore, $L_{e\,99}(n)$ is affected considerably by the correlation coefficient of live loads between two

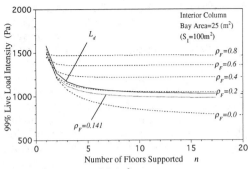

(a) Tributary area = 25 (m²)

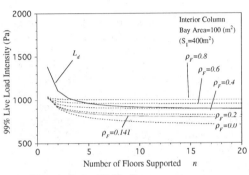

(b) Tributary area = 100(m²)

Fig. 9   99% live load intensity in changing of
tributary area and ρ_F

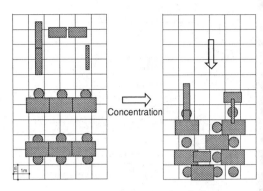

Fig. 10 Furniture concentration simulation

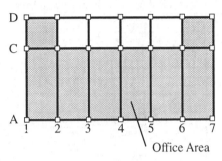

Office Area

Fig. 11 Floor plan of building A

different floors.   The effect of ρ_F on L_e has to be discussed for considering a different use in each floor.   Fig.17 shows L_{e99} of interior columns in changing ρ_F from 0.0 to 0.8 when the bay area is 25m² and 100m².   When the bay area is 25m² and ρ_F=0.0, L_{e99} in n=20 decreases to fifty percent of L_{e99} in n=1.   On the other hand, L_{e99} rarely changes in case of ρ_F=0.8.   Consequently, the influence of ρ_F is significant in changing L_e.   In that case the bay area is 100m², L_{e99} is smaller than that in a bay area of 25m², and L_{e99} in n=20 decreases to only seventy percent of L_{e99} in n=1.   When reducing the design live load for multiple story columns, the correlation coefficients between two different floors and bay area must be noticed for the other use than office.

5. REDUCTION FACTOR UNDER EXTRA-ORDINARY LIVE LOAD

Accompanying with the relocation or change of occupancy, it is often the case that the temporary concentration of furniture occurs.   In this chapter, the live load reduction factors for multiple story columns are investigated based on furniture concentration simulation using the live load survey results.   In this paper, the method for furniture concentration simulation is based on the method presented by Kanda and Kinoshita(1985).   The simulation method is simplified for calculation of the equivalently uniform distributed load for column axial force.   The concrete procedure is summarized below.

1) Furniture layout room plan is divided by the 1m × 1m grid, and the weight and occupied area of furnishings are calculated.   The weight of furnishings is treated as a concentrated load at the center of furnishings.

2) A grid unit with furnishings is gathered towards a wall side of a room gradually one by one from the nearest to the wall.

3) The weight and occupied area of furnishings are cumulated repeatedly in the grid units at the wall side.   If the occupied area reaches 0.9m², the furnishings are gathered at the next nearest grid unit to the wall.

4) The above procedures are repeated intended for

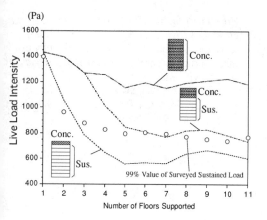

Fig.12 Load intensity of concentrated live load

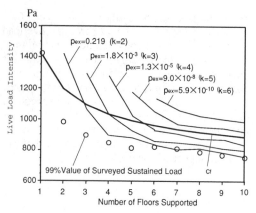

Fig.13 Load intensity and occurrence probability of concentrated live load

all furnishings in office area.

Fig.10 illustrates the furniture concentration, and the floor plan of Building A is shown in Fig. 11. After gathering, it is assumed that the weight of furnishings is loaded at the center of the unit.

Fig.12 shows the mean value of live load intensity for columns on axis 1 when furnishings are gathered towards a wall side on axis 1. In the figure, the circles indicate 99-percentile load intensity based on the sustained live load survey results. In the case that furniture concentration occurs on the tenth floor only, the concentrated column load is less than the 99 percentile column load based on the survey results at most of floors. On the other hands, when the furnishings concentrate on the top three floors, the concentrated column load is more than 99 percentile load at the top five floors. In the case that the concentration occurs at all ten floors, the concentrated load exceeds 99 percentile load at all floors.

Since the oncentration of furnishings occurs in random interval time, the extraordinary load can be expressed by the poisson square wave process having random duration time denoted by Δ. Dividing the lifetime of a structure, denoted by T, into n intervals of infinitesimal time denoted by λ, the probability that the extraordinary load doesn't exist at any point of time $t=0$, λ, 2λ, ..., $n\lambda$ is given by;

$$1-V_1 = (1-a)[1-c(\lambda)^n] = (1-a)[1-c(\lambda)]^{n/\lambda} \quad (15)$$

where a indicates a mean occurrence rate of an extraordinary load, and is expressed as follows.

$$a = \frac{\mu_\Delta}{\mu_I} \quad (16)$$

Where, μ_Δ and μ_I indicate the mean duration and the mean interval of an extraordinary load respectively. $c(\lambda)$ in Eq.(15) means the probability that the extraordinary load exists at a point of time of $k\lambda$ under the condition that no extraordinary load has existed at a point of time of $t=0$. Let λ approach to zero, and assuming $T>>\mu_\Delta$, the probability that the extraordinary load occurs once at least during T can be expressed as;

$$V = \frac{aT}{\mu_\Delta} \quad (17)$$

Since the occurrence probability of extraordinary loads at n floors simultaneously is a^n, the mean duration time μ_Δ of simultaneous occurrence of extraordinary loads at n floors is;

$$\mu_\Delta = \frac{\mu_{\Delta_1}}{n} \quad (18)$$

Consequently, the probability that n extraordinary loads occur simultaneously once in the lifetime T at least, can be expressed as follows.

$$V_n = \frac{a^n nT}{\mu_{\Delta_1}} \quad (19)$$

The thin solid lines in Fig.13 show expected live load intensity for a column in cases of that the number of simultaneous occurrence extraordinary load k changes from 2 to 6. $T=100$ years, $\mu_\Delta=10$ days and $\mu_I=5$ years are assumed as parameters for Eq.(19). The thick solid line in Fig.12 indicates the

design live load reduced by the reduction factor for a bay area of 25m² from the 99 percentile live load intensity of 1400Pa for a single story column. Although the reduced design load for $n \geq 6$ is greater than the extraordinary load for $k \leq 4$ (simultaneous occurrence probability p_{ex} is less than 1.3×10^{-5}), the extraordinary loads for all k are greater than reduced design load in the region of $n < 4$. Consequently, it is concluded that the live load reduction for multiple story column should be permitted when the number of floors supported n is $n \geq 3$.

6. CONCLUSIONS

Based on the actual surveyed data of office buildings, a stochastic live load model for multiple story column load was presented, and the multiple story column load was examined using the presented live load model. In order to evaluate the multiple story column load statistically, it is necessary to consider both the spacial correlation of a live load on a floor and the vertical correlation between two different floors. When reducing the design live load for multiple story columns, the correlation coefficients between two different floors and bay area must be noticed for the other use than office. Furthermore, the furniture concentration is computer-simulated using the surveyed live loads of the office building. The design live load for the axial compression on the column at the time of furniture concentration is studied. It is concluded that the live load reduction for multiple story column should be permitted when the number of floors supported n is $n \geq 3$. After this, it is hoped to accumulate and investigate the vertical correlation data of the live load between different floors in a building.

REFERENCES

Choi, E. C. C. , 1989. Live Load Model for Office Buildings, *The Structural Engineer*, Vol.67, No.24/19, December, 421-437.

Corotis, R. B., 1972. Statistical Analysis of Live Load in Column Design, *Journal of the Structural Division, ASCE*, August, 1803-1815.

Kanda, J. and K. Kinoshita, 1985. A Probabilistic Model for Live Load Extremes in Office Buildings, *ICOSSAR'85, Structural safety and reliability*, Vol.II, 287-296.

McGuire, R. K. and Cornell C. A., 1974. Live Load Effects in Office Buildings, *Journal of the Structural Division, ASCE*, July, 1351-1366.

Ono T. and H. Idota, 1989. System reliability using higher-order moments, *ICOSSAR'89, Structural safety and reliability*, Vol.II, 959-966.

Peir, J. C. , 1971. A Stochastic Live Load Model for Buildings, *Research Report R71-35,* School of Engineering MIT, Massachusetts, USA.

Structural Safety & Reliability, Schuëller, Shinozuka & Yao (eds) © 1994 Balkema, Rotterdam, ISBN 90 5410 357 4

Effects of probability distribution model of loads on optimum reliability

Jun Kanda
Department of Architecture, University of Tokyo, Japan

Takashi Inoue
Technical Research Institute, Hazama Corporation, Tsukuba, Japan

ABSTRACT: The optimum reliability concept based on the minimum total cost can be regarded as a basic principle to determine the target reliability of structural safety. One of the difficulties of optimum reliability formulation may arise due to uncertainty for the probabilistic behavior of random variables which are to be appropriately modeled. Optimum reliability is obtained for four distribution models of the load effect. The difference is significantly minor only except for the case of Kanda's distribution with very low upper bound limit. Results confirm that the effects of probability distribution models on the optimum reliability are not serious.

1 INTRODUCTION

The optimum reliability concept based on the minimum total cost or the maximum total utility can be regarded as a basic principle to determine the target reliability of structural safety. Quantification of the degree of safety together with cost estimate enables to obtain the design load factor based on the optimum reliability. One of difficulties of optimum reliability formulation may arise due to the uncertainty for the probabilistic behavior of random variables which are to be appropriately modeled. The effects of probability distribution model on the optimum reliability were evaluated by using four kinds of probability distribution model of loads.

2 OPTIMUM RELIABILITY CONCEPT

The optimum reliability can be defined as the reliability at which the total cost is minimum. The total cost C_T of structure is expressed as (Rosenblueth 1976, Lind 1976, Kanda et al. 1991)

$$C_T = C_I + P_f C_F \tag{1}$$

where C_I is the initial cost, P_f is the probability of failure and C_F is the cost associated with failure. The initial cost C_I is a function of the design load requirements. One simple form for C_I is a linear function, which can be written as

$$C_I = C_0 \{ 1 + k \left(\frac{r_d}{\mu_Q} - 1 \right) \} \tag{2}$$

where C_0 is the initial cost at $r_d = \mu_Q$, r_d is the design resistance which is assumed to be equal to the design load effect and μ_Q is the mean load effect (Kanda et al. 1991). Parameter k is a normalized cost ratio and may be obtained by a design study for a certain type of structure. The failure cost C_f is estimated at the time of construction as

$$C_F = g C_0 \tag{3}$$

in which g is the normalized failure cost.

As a simple case, suppose that both the strength and the load effects are described by normal distributions. The design point is written as the product of μ_Q and the load factor γ_Q, which is expressed in the well-known form by introducing the separation factor α_Q as

$$r_d = (1 + \alpha_Q \beta V_Q) \mu_Q \tag{4}$$

where V_Q is the c.o.v. of load effect, $\beta = (\mu_R - \mu_Q) / (\sigma_R^2 + \sigma_Q^2)^{1/2}$, μ_R is the mean of the

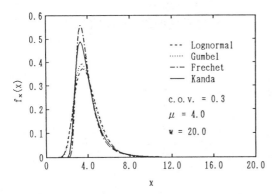

Fig. 1 Comparison of probability density models

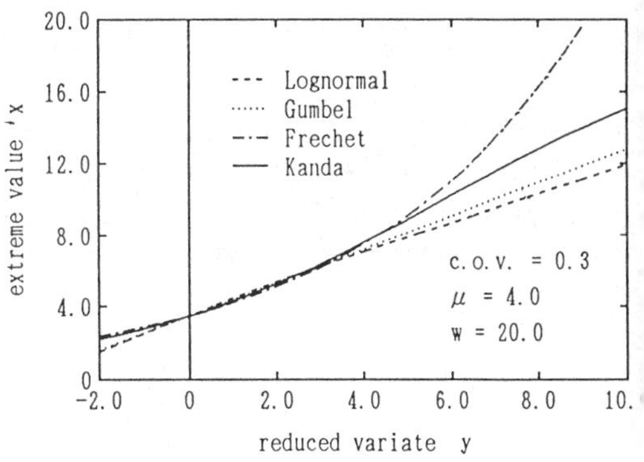

Fig. 2 Comparison of probabilistic distribution models

resistance, and σ_R and σ_Q are the standard deviation of the resistance and the load effect respectively. The probability of failure is given by the standard normal distribution function, i.e.,

$$P_f = \Phi(-\beta) \qquad (5)$$

The optimum reliability index β_{OPT} is obtained from the equation $dC_T / dr_d = 0$, after substituting eqs. (2),(3),(4) and (5) into eq. (1):

$$\beta_{OPT} = \sqrt{2\ln \frac{g}{\sqrt{2\pi}k\,\alpha_Q V_Q}} \qquad (6)$$

The log-normal distribution often has been used to represent both the resistance and the load effect. An simplified expression for the load factor is

$$\gamma_Q = \exp(\alpha_Q \beta V_Q) \qquad (7)$$

where approximations E $[\ln R] \approx \ln\mu_R$ and $V_R{}^2 \approx \ln(1+V_R{}^2)$ are employed and $\alpha_Q \approx V_Q / (V_R{}^2+V_Q{}^2)^{1/2}$ (separation factor). The solution for β_{OPT} is derived in a straightforward way from the equation $dC_T / dr_d = 0$ by substituting eqs. (2), (3), (5) and (7) into eq. (1) (Kanda et al. 1991):

$$\beta_{OPT} = -\alpha_Q V_Q + \sqrt{(\alpha_Q V_Q)^2 + 2\ln(\frac{g}{\sqrt{2\pi}k\,\alpha_Q V_Q})} \qquad (8)$$

A similar but slightly complicated form is also available (Sugiyama et al. 1982).

3 PROBABILITY DISTRIBUTION MODELS

The log-normal distribution is often used to represent random variables for the structural resistance and also for various load effects. However the tail of the probability distribution has not been well confirmed by any statistics particularly for random variables with a high coefficient of variation (c.o.v.) such as the load intensity caused by the natural environment.

The Gumbel distribution (Type I extreme value distribution) is conveniently applied to represent the probabilistic model of annual maximum load intensity, e.g., the annual maximum snow depth or the annual maximum wind speed (Galambos et al. 1982, A.I.J. 1986). The annual maximum earthquake intensity such as the peak ground acceleration (P.G.A.) or the peak velocity is sometimes modeled by Frechet distribution (Type II extreme value distribution, Galambos 1982), which may be improved by an empirical extreme value distribution with upper bound limit proposed by one of authors as follows (Kanda 1981):

$$F(x) = \exp[-\{\frac{w-x}{u(x-\varepsilon)}\}^\kappa] \qquad (9)$$

where u, κ, w, ε are scale parameter, shape parameter, upper bound limit and lower bound limit respectively (Kanda 1987).

The four probability distribution curves mentioned above are shown in Figs. 1 and 2. In Fig. 2, they are plotted on the double exponential probability paper where the Gumbel distribution is presented by a linear form. The c.o.v. is 0.3 and the mean is 4 for all the distributions. For Kanda's distribution $w = 20$ and $\varepsilon = 0$ are adopted. Differences of tail

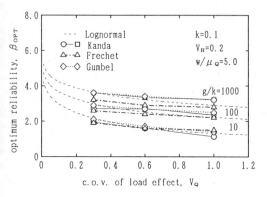

Fig. 3 Optimum reliability for four probabilistic distribution models

characteristics of probability distribution are quite significant.

4 OPTIMUM RELIABILITY FOR FOUR PROBABILITY DISTRIBUTION MODELS

The optimum second-moment reliability is given in a closed form for both the normal distribution and the log-normal distribution as described in previous section. The optimum reliability for the Kanda distribution model of load effect can be obtained numerically. The optimum reliability is obtained with the same parameter values for the initial cost, the failure cost, the coefficient of variation of resistance and load effect except for the probability distribution models for the load effect, i.e. the log-normal, Gumbel, Frechet and Kanda models. The log-normal distribution is used for the probability distribution model of resistance.

Failure probability, P_f, is evaluated by an iterative procedure using equivalent normal distribution at the failure point except for the case of log-normally distributed load effect, for which the closed form solution is given. The P_f is calculated for various mean values of resistance to get the minimum total cost C_T and corresponding β_{OPT}. In the case of Kanda distribution, the equivalent normal distribution can not be obtained when the failure point is close to or greater than the upper bound limit. Therefore, failure probability P_f is obtained by numerical integration of $f_Q(x)F_R(x)$, and β_{OPT}' is evaluated by using the central safety factor μ_R/μ_Q instead of the load factor r_d/μ_Q.

The optimum reliability indices calculated for the four probability distribution models of load effect are shown in Fig. 3. The values for log-

normal distribution are obtained by the closed form solution without approximations for the mean and the c.o.v. The ratio between the upper bound limit and the mean load effect $w/\mu_Q = 5$ and the lower bound limit $\varepsilon = 0$ are used in the case of Kanda distribution. The value of β_{OPT}' is shown by squares in the figure. The probability distribution model of resistance is the log-normal distribution with the c.o.v. $V_R = 0.2$. Although the values of β_{OPT} by the four probability distribution models of loads are a little different in the case of $g/k = 1000$, these values generally agree well in a wide range of V_Q.

5 EFFECTS OF UPPER BOUND LIMIT FOR LOADS

The values of the optimum reliability, β_{OPT}, and the optimum load factor, γ_{OPT}, for Kanda's distribution of load effect are numerically obtained for g/k's of 10, 100 and 1000, the ratio between the upper bound value and the mean, w/μ_Q, and the c.o.v. of load effect, V_Q. The value of γ_{OPT} is obtained by dividing the load effect at the design point, r_d, by μ_Q when β_{OPT} can be obtained, while $\gamma_{OPT}' = \phi \ \mu_R/\mu_Q$ assuming $\phi = 0.85$ when β_{OPT}' is evaluated. The results are listed in Tables 1-3 in comparison with the log-normal case. The values in the parenthesis are those of β_{OPT}' and γ_{OPT}'. The values of β_{OPT} and γ_{OPT} for the c.o.v. of load effect $V_Q = 0.6$ are shown in Figs. 4 and 5.

When $w/\mu_Q > 10$, β_{OPT} and γ_{OPT} are not affected by the values of w/μ_Q and stay fairly constant, and β_{OPT} agrees well with β_{OPT}'. When $w/\mu_Q < 10$, β_{OPT} tends to increase for a lower w/μ_Q except for $V_Q = 1.0$, and γ_{OPT} tends to saturate at about the value of w/μ_Q less than 4.

6 CONCLUSION

Effects of probability distribution models on optimum reliability are confirmed by using four probability distribution models of load effects. The differences among optimum reliability solutions due to various probability distribution models for loads is significantly minor except for the case of Kanda distribution with very low upper bound limit. Results confirm that the effect of probability distribution model on the optimum reliability is not serious and so in spite of a rather high uncertainty on the upper tail

probability the optimum reliability concept is highly applicable to design practices. However, these findings suggest that the sensitivity of the final design to the distribution tail assumption of load models does exist and could be taken into account by the decision principle of minimization of the total costs.

Table 1 The values of β_{OPT} and γ_{OPT} $(g/k=10)$

$\frac{w}{\mu_Q}$	V_Q					
	0.3		0.6		1.0	
	β_{OPT}	γ_{OPT}	β_{OPT}	γ_{OPT}	β_{OPT}	γ_{OPT}
2	(2.09)	(1.72)	(2.04)	(2.11)	–	–
5	1.94	1.63	1.60	2.06	1.13	2.07
10	1.85	1.56	1.54	1.87	1.26	1.93
20	1.84	1.55	1.56	1.83	1.34	1.89
40	1.84	1.54	1.60	1.54	1.40	1.87
L.N.	2.15	1.62	1.63	2.01	1.32	2.06

Table 2 The values of β_{OPT} and γ_{OPT} $(g/k=100)$

$\frac{w}{\mu_Q}$	V_Q					
	0.3		0.6		1.0	
	β_{OPT}	γ_{OPT}	β_{OPT}	γ_{OPT}	β_{OPT}	γ_{OPT}
2	(2.79)	(2.10)	(2.84)	(2.57)	–	–
5	2.71	2.18	2.69	3.50	(2.35)	(4.37)
10	2.66	2.25	2.36	3.29	2.16	4.54
20	2.63	2.26	2.35	3.26	2.13	4.12
40	2.63	2.25	2.38	3.27	2.11	3.72
L.N.	2.97	1.97	2.52	3.20	2.21	4.25

Table 3 The values of β_{OPT} and γ_{OPT} $(g/k=1000)$

$\frac{w}{\mu_Q}$	V_Q					
	0.3		0.6		1.0	
	β_{OPT}	γ_{OPT}	β_{OPT}	γ_{OPT}	β_{OPT}	γ_{OPT}
2	(3.54)	(2.56)	(3.70)	(3.14)	–	–
5	3.59	2.94	(3.34)	(4.82)	(3.22)	(5.90)
10	3.39	3.22	3.20	5.57	(3.03)	(7.88)
20	3.31	3.30	3.00	5.51	2.79	8.03
40	3.28	3.31	2.93	5.33	2.63	6.70
L.N.	3.62	2.31	3.20	4.56	2.90	7.39

Note: The values in the parenthesis are those of β_{OPT}' and γ_{OPT}'.

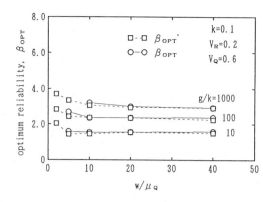

Fig. 4 Variation of optimum reliability with w/μ_Q

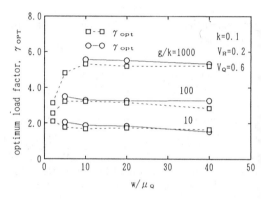

Fig. 5 Variation of optimum load factor with w/μ_Q

REFERENCES

A.I.J. 1986. *Load and resistance factor design for steel structures (proposal)*: 73-89. (in Japanese)

Galambos, T.V., Ellingwood, B., MacGregor, J.G. & Cornell, C.A. 1982. Probability based load criteria: Assessment of current design practice, *J. Struct. Div., ASCE, 108 (5)*: 959-977.

Kanda, J. & Ellingwood, B. 1991. Formulation of load factors based on optimum reliability. *Structural Safety, Vol. 9*: 197-210.

Kanda, J. 1981. A new extreme value distribution with lower and upper limits for earthquake motions and wind speeds. *Proc. of the 31th Japan National Congress for Applied Mechanics, Vol. 31*: 351-360.

Kanda, J. & Dan, K. 1987. Distribution of seismic hazard in Japan based on a empirical

extreme value distribution. *Structural Safety, Vol. 4*: 229-239.

Lind, N.C. 1976. Approximate analysis and economics of structures, *J. Struct. Div., ASCE, 102 (6)*: 1177-1196.

Rosenblueth, E. 1976. Towards optimum design through building codes, *J. Struct. Div., ASCE, 102 (3)*: 591-607.

Sugiyama, T., Sakai, Y., Fujino, Y. & Ito, M. 1982. Decisions on reliability level and safety factor for structural design, *Trans. Japan Assoc. Civil Eng., 327*: 21-28. (in Japanese)

Structural Safety & Reliability, Schuëller, Shinozuka & Yao (eds) © 1994 Balkema, Rotterdam, ISBN 90 5410 357 4

Design local ice pressures on ice breakers: The effect of exposure and load sharing on extreme values

M.A. Maes
Queen's University, Kingston, Ont., Canada

ABSTRACT : In this paper a probabilistic exposure model is developed which addresses the basic question of determining extreme values of local ice pressure on a fixed design area. The model takes into account partial coverage, load sharing, overlap, and time dependence during rams. It is based on assumptions regarding the creation, the growth and the movement, in time and in space, of so-called pressure hot-spots during impact. The paper also describes an application of the method to a data set recorded by the Canmar Kigoriak ice breaker during 1981 summer and fall ramming tests in the Canadian Arctic.

1. INTRODUCTION

The design of icebreaker plating and framing is governed by the intensity of local ice pressures on hull areas of specified size and location. Similarly, offshore structures operating in ice-infested waters are designed on the basis of assumed ice pressures on small areas. The current process of selecting design pressures is, to a large extent, empirical. It has typically been guided by extensive experience in the Baltic and in the Arctic. Recently, several ice ramming data sets have become available which should provide a strong incentive for rationalizing the design procedure. This question is addressed extensively in Jordaan et al. (1992). Possible codification of ice loading is also discussed in Maes (1989). The present paper describes a model of analysis that results in reliability-based design ice pressures on small design surface areas.

One of the most important aspects of ice loading is the development of a so-called pressure-area relationship. An instrumented panel on the hull of an icebreaker is subjected to different ice loading histories for each ram. This may include zero loading during a ram, if no direct contact occurs between the panel and the ice. An additional complication is the spatial spreading of the contact area; a large pressure panel may only be partially loaded, or it may share load with a contingent panel. Typically, data consist of pairs of peak pressures during a ram and the corresponding fraction (but not the location) of the area loaded within that panel. In this paper a probabilistic exposure model is developed in order to address the basic question of determining extreme values of local ice pressure on a fixed design area. The model takes into account partial coverage, load sharing, overlap, and time dependence during rams. It is based on assumptions regarding the creation, the growth and the movement, in time and in space, of so-called pressure hot-spots during impact.

The paper describes an application of the method to a data set recorded by the Canmar Kigoriak ice breaker during 1981 summer and fall ramming tests in the Canadian Arctic. Extreme values of local pressures on areas as small as 0.20m^2 may be determined using the exposure model.

2. KIGORIAK DATA SET

To illustrate the overall approach for determining design local ice pressure, a particular set of ice load data is considered. This data set originates from the August 1981 and October 1981 arctic field tests by the Canadian icebreaker CanMar Kigoriak. The bow of the Kigoriak was instrumented with two load panels of size 1.25m^2 and 6.0m^2. In this paper, only the data associated with the smaller panel will be analysed. A full analysis, together with a comparison with results obtained from different ice ramming tests, can be found in Maes and Hermans (1992). The ice loading data recorded on the panel with area $a = 1.25$m^2 consist of the following pairs of data

$$(y_k, a_k) \qquad (1)$$

where y_k represents the total load (not the pressure) on the panel; this load is defined as the peak load occurring during ram k (see Fig. 1). The value a_k is the associated loaded

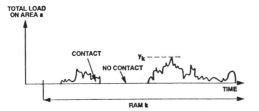

Fig. 1 : Peak Ice Load y_k on a Specified Panel During Ram k.

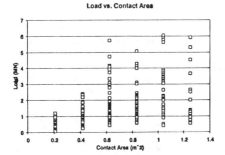

Fig. 2 : Load Versus Contact Area

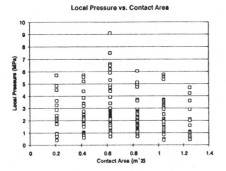

Fig. 3 : Pressure Versus Contact Area

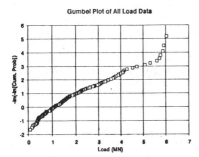

Fig. 4 : Empirical Distribution of Non-Zero Ice Loads

area within the panel ($a_k \leq a$). This loaded area was calculated using strain gauge information available for each loading event. The instrumentation layout was such that the loaded area could be determined to be a multiple of one-sixth of the total panel area, that is, a_k belongs to the discrete set of values :

$$a_k = \left\{ 0, \frac{a}{6}, \frac{2a}{6}, ..., a \right\} \qquad (2)$$

The exact location of the contact area a_k within the area a is, for all practical purposes, unknown. The value a_k merely indicates how many contingent subareas of size $a/6$ are loaded when the peak force y_k on the panel occurs.

A preliminary analysis of the ice load data, showed that, out of a total of 397 rams, 179 non-zero loading events were recorded on the panel of area a. The multi-year ice ramming tests were conducted at different impact speeds and different ice conditions. However, neither of these variables appeared to be statistically significant explanatory variables for the local ice pressure (Maes and Hermans, 1992). A scatterplot of the paired non-zero data (y_k, a_k) is given in Fig, 2; Fig. 3 shows the corresponding pressures y_k/a_k plotted versus contact area a_k .

A preliminary investigation of the load data is shown in Fig. 4. Here, all $n = 179$ force data are ordered and plotted on Gumbel probability paper; the ordinate shows the i-th force data point (ascending order) in the plotting position η , with :

$$\eta_i = -\ln (-\ln(i/n+1)) \qquad (3)$$

Lumping together all forces, irrespective of the associated contact areas, would be justified if we were to design an area on the bow, that is precisely equal to 1.25m²; in each ram k, the panel would either be fully loaded ($a_k = a$), partially loaded, or not loaded at all ($a_k = 0$). The Gumbel plot in Fig. 4 fails to show a linear trend in the tail; in fact, a bothersome positive curvature ("dog's tail") may be detected which, ceteris paribus, would be indicative of a Weibull type of tail convergence. Such a conclusion, however, would be hard to accept from an ice mechanics point of view; there seems to be no physical reason why impact force on an area would be bounded from above. This issue, together with the more general objective to be able to determine design ice pressures, not only on a window of size a, but on a window of any size less than a, have prompted the development of the subsequent exposure model.

3. AREA CLASSES

Following the observation that sizes of contact areas a_k associated with peak force events y_k can take on 6 non-zero values, it seems logical to partition the ice load data into 6 area classes. Class j, for instance, would contain values

of total load obtained when the size of the contact area would be found to be equal to $(j/6)a$. Table 1 lists the six area classes together with the number of load data belonging to each area class. For the sake of completeness, the table also shows the number of rams during which no contact (zero load) was observed on the pressure panel.

Clearly, for design purposes, it is necessary to derive a probability distribution of ice loads (or, pressures) on a *fixed* area. Let the random variable X_j represent the ice load that would be recorded on a fixed area of class j (i.e., having size $(j/6)a$); this random variable X_j is fundamentally different from the random variable Y_j which represents the load on an (arbitrary) *moving* area of the same size. The vector of observed force values $(y)_j$ (Table 1) constitutes a random sample of this random variable Y_j. The following exposure model attempts to establish a relationship between the random variables X_j and Y_j .

4. EXPOSURE MODEL

Fig. 5 illustrates what is referred to as a "loading pattern": a fixed area of, say, size $j = 3$, may be covered by a large number of loading patterns (indicated as shaded areas) of different shapes and different sizes. The "fixed" area could be *partially* covered by "moving" windows of a size less than or equal to the fixed area (case "a" in fig. 5.1) or by patterns with a size greater than it (case "b"). Alternatively, it could be wholly covered by patterns equal to or greater than the fixed area (case "c").

In order to transform the moving ice loads Y_j in classes j ($j = 1, ..., 6$) to ice loads X_j on fixed areas, it is therefore necessary to take into account the complicated interaction of loading patterns and their coverage properties. The following simplifying assumptions are made :

(1) Spatial coherence of the load : in considering loading patterns, only those combinations of subareas are allowed that are spatially contiguous. Experimental evidence from the size and the properties of so-called critical zones (relatively flaw-free zones in ice which can sustain high pressures, Jordaan et al. 1991) seems to support this assumption.
(2) Partial loading and load sharing : since a given coverage may arise as a result of many overlapping load patterns, it is assumed that the *pressure* in the overlapping area is the same as the pressure averaged over the entire contact area.
(3) Equally likely loading patterns : this assumption amounts to treating all possible patterns of the same total size as equally likely, irrespective of their shape.
(4) Peak loading events are stochastically independent.

Under these assumptions, a straightforward application of the theorem of total probability leads to the following relation :

Table 1   Ice Load Area Classes

Area Class #	Loaded Area (m^2)	# data
1	0.208	19
2	0.417	31
3	0.625	47
4	0.833	35
5	1.042	31
6	1.250	16
Total of Non-Zero Data		179
0	0	218
Total Number of Rams		397

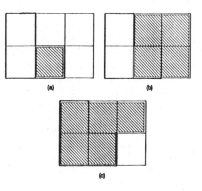

(a)   (b)

(c)

Fig. 5 : Loading Patterns

$$F_{X_j}(x) = 1 - \sum_{c=1}^{j} \sum_{i=c}^{n-j+c} P_{ji}(c) P_i(i) \left[1 - F_{Y_i}\left(\frac{i}{c}x\right)\right] \quad (j=1,...,n) \quad (4)$$

where F_{X_j} : the distribution function of ice loading on a fixed area of size class j.

F_{Y_i} : the distribution function of ice loading having a (moving) contact area belonging to class i.

$P_{ji}(c)$: the discrete distribution of all possible coverage values, arising from a combination of a fixed area of class j with a moving area of class i.

$P_i(i)$: the probability that the (moving) contact area belongs to class i.

The probabilities $P_i(i)$ are taken to be the relative frequencies of the area classes listed in Table 1; the probabilities $P_{ji}(c)$ are derived on a geometrical basis for each combination (i, j). As an example : $P_{11}(1) = 1/6$; $P_{34}(1) = 9/40$, $P_{34}(2) = 1/2$, and $P_{34}(3) = 11/40$. It can be seen that the transformation equation (4) amounts to mixing the tails of "moving" ice load distributions associated with each area class. The assumed load sharing mechanism accounts for the increased cut-off $(i/c)x$ for the exceedance probability of Y_i.

5. TAIL ESTIMATION

The principal objective is to be able to determine extreme values of all the fixed area ice loads X_j. Since their distributions can be obtained as mixtures of the tails of the distributions F_{Y_i}, it is important to select appropriate models for tail behaviour.

Many statistical estimation procedures focus on estimating the central part of a probability distribution. The use of a "central" method in conjunction with tail estimation amounts to estimating for a given parametric distribution family the values of the parameters (e.g. on the basis of maximum likelihood or a method of moments) and then using the tails of this estimated distribution in a subsequent analysis. As a result, the data in the central part of the distribution clearly have a dominating influence on the estimated distribution and the data in the tails which are usually few in number do not carry much weight. But, in order to get a good fit of the distribution tail, it is precisely these data that should have the most influence on the tail behaviour. In setting up a suitable fitting procedure for the different area classes, it was found to be desirable to accomplish the following objectives :

- to avoid the problem of choosing cut-offs and discarding large parts of the data set (this is a very common feature of many tail-fitting procedures)

- to assign greater weights to data in the tail of interest, and less weight to data in the central and opposite parts of the data range;

- to respect the principle of tail equivalence (Castillo, 1988);

- to produce a method which is consistent with risk criteria used in structural reliability analysis and which can be interpreted visually (Breitung and Maes, 1993);

- to produce estimates of the associated uncertainties.

It is shown in Breitung and Maes (1993) that these objectives can be met by selecting the following technique for tail-fitting ordered (ascending) univariate data y_i ($i = 1, ..., n$) :

1. use an appropriate extreme value distribution $F_y(y, \boldsymbol{\theta})$ (with a parameter vector $\boldsymbol{\theta}$) which is tail-equivalent with the tail of the (unknown) parent distribution; a suitable choice is the generalized extreme value (GEV) distribution for maxima or the i - dimensional GEV for the i highest order statistics (Smith, 1986).

2. Minimize the following expression $h(\boldsymbol{\theta})$ with respect to $\boldsymbol{\theta}$:

$$h(\boldsymbol{\theta}) = \sum_{i=1}^{n} w_i \, \epsilon_i^2 - \sum_{i=1}^{n-1} \sqrt{w_i w_{i+1}} \; \epsilon_i \, \epsilon_{i+1} \quad (5)$$

where

$$\epsilon_i(\boldsymbol{\theta}) = F_Y(y_i \mid \boldsymbol{\theta}) - p_i \quad (6)$$

$$w_i = (1 - p_i)^{-2} \quad (7)$$

and p_i is the expected value of the i-th order statistic of a uniform distribution :

$$p_i = E_{Y_i}\!\left(F_Y(y_i \mid \boldsymbol{\theta})\right) = i/(n+1) \quad (8)$$

In these expressions w_i represent weights associated which each residual ϵ_i ; those weights increase with i and they are consistent with several risk-based criteria (Breitung and Maes (1993); Castillo (1988)). The negative terms in (5) account for the covariance between the ordered residuals.

3. Step 2 is equivalent to a multivariate maximum likelihood problem for y_i , ..., y_n with a loglikelihood function proportional to

$$- 1/2 \;\; \boldsymbol{\epsilon}^T K^T V^{-1} K \, \boldsymbol{\epsilon}$$

where $\boldsymbol{\epsilon}$ is given by (6), and $K = \text{diag}\left(\sqrt{w_i}\right)$ and V^{-1} is a tridiagonal matrix with main diagonal elements equal to $2 / (n+1)$ and adjacent elements equal to $-1 / (n+1)$. Therefore, the observed information or the Fisher information method may be applied to determine the variance-covariance matrix of the vector $\hat{\boldsymbol{\theta}}$ which minimizes the expression (5); this procedure is illustrated in Maes and Breitung (1992).

4. Use the delta method (Taylor series approximation) to retrieve the approximate variance of the desired quantiles.

6. RESULTS

The procedure outlined above is applied to "moving" ice-loading data in each of the six area-classes. In Step 1,

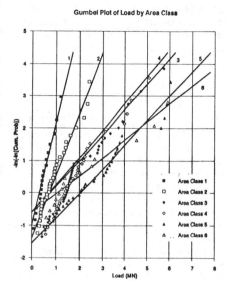

Fig. 6 : Distributions of Ice Loads on Moving Area

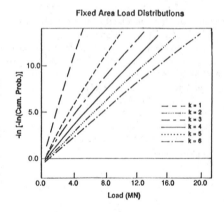

Fig. 7 : Distributions of Fixed Area Ice Loads

a Wilks' likelihood ratio test is used to compare the use of the GEV distribution with the use of the Gumbel distribution; since the latter is special case of the former, the two models are nested and Wilks' chi-squared test is applicable to determine whether the two models are significantly different. As a result of this comparison, the Gumbel distribution was selected in five of the six area-classes; only in area class 5, there appeared to be a strong tail curvature, substantial enough to disqualify the Gumbel model. Since this anomaly showed up only in one area class, it was decided to disregard the possibility of a better GEV fit in this particular area class on both statistical and physical grounds (i.e. (1) the compelling evidence of an

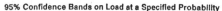

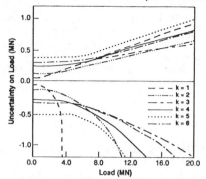

Fig. 8 : 95% Confidence Bands on Ice Load Quantiles

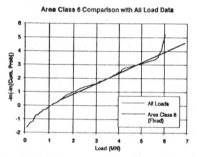

Fig. 9 : Comparison Between Area Class 6 Model Distribution and Empirical Distribution of All Load Data.

excellent Gumbel fit in all five other area classes, and (2) the aforementioned argument regarding upper bounds on ice pressures).

Fig. 6 shows a Gumbel plot of the distributions of Y_j obtained as a result of the above tail-fitting procedure. Subsequently, the exposure model (equation (3)) was used to transform the random variables Y_j into loads on fixed areas X_j ($j = 1, ..., 6$). The resulting distributions are not Gumbel as can be seen from the light nonlinearity in the Gumbel plot shown in Fig. 7. The uncertainties on Y_j may also be combined to determine the corresponding uncertainties on X_j. These uncertainties include the parameter uncertainties determined using the observed information method and subsequently transformed using a second-order approximation, as well as the model uncertainties resulting from the weighted least-squares procedure. As an example, Fig. 8 shows 95% confidence bands on a load quantile in area class j, plotted as a function of the estimated quantile.

A verification of the use of exposure model is possible by focusing on the distribution F_{X_6} obtained for area class

6. Since the random variable X_6 represents ice loading on a total fixed area of $a = 1.25m^2$, it may be compared with the empirical distribution of *all* readings on the pressure panel, regardless of their area class. The comparison is shown in Fig. 9; it provides a satisfactory explanation of the tail behaviour which would not have been possible exclusively on an empirical basis, i.e. without the use of an exposure model.

7. DESIGN LOCAL ICE PRESSURES

As discussed in Jordaan et al. (1992), there are three factors that determine the probability level at which quantiles of F_X should be determined in order to select reliability-based design ice loads and pressures. All three are linked with the exposure of the area to ice loading :

(1) n_r : the expected annual number of ice encounters (rams). In the case of a vessel, this factor is a direct function of the design arctic class; for instance Canadian Arctic Class I corresponds roughly to about 10,000 rams per year. In the case of a structure, the geographical location is relevant.

(2) β : the expected relative duration of an ice encounter (relative with respect to the average duration of rams used in the present analysis).

(3) r : the ratio of the expected number of hits on the design panel divided by the total number of rams. This aspect of exposure is related to position on the ship or structure (e.g. bow vs. side shell) and it is usually accounted for by the use of so-called "area-factors".

These three factors determine the required exceedance probability level α_m of X_j in order to determine a m-year return period load on an area of size j :

$$\alpha_m = 1 / (r n_r \beta_m) \qquad (9)$$

Equivalently, the annual extreme value distribution can be found on a Gumbel plot by shifting the F_{X_j} curve upwards by an amount equal to $+ \ln (r\, n_r\, \beta)$. As an illustration, the 100-year return ice pressure (together with a 95% confidence interval) on an area of $0.63m^2$ ($j = 3$) and $1.25m^2$ ($j = 6$) would be 13.2 *MPa* (12.5, 13.6) and 10.8 *MPa* (9.6, 11.4) respectively, using $m_r = 100$ per year and $\beta = 1$.

ACKNOWLEDGEMENTS

The data used in the analysis were provided by the Canadian Coast Guard (Northern). A study of local ice pressure analysis techniques was performed in the context of a review of the revised CASPPR regulations. The help of Ian Jordaan, Peter Brown and Ivar Hermans is appreciated. Support form Memorial University of Newfoundland, the Natural Sciences and Engineering Research Council of Canada, and Mobil Research and Development Corporation are gratefully acknowledged.

REFERENCES

Maes, M.A. (1989), "Uncertainties and Diffficulties in Codified Design for Ice Loading", Procedings, Eighth International Offshore Mechanics and Arctic Engineering Symposium, *OMAE 1989*, the Hague, the Netherlands, Vol. 4, pp. 97-103.

Breitung, K. and Maes, M.A. (1993), "Risk-Based Tail Estimation", To be presented at the *IUTAM* Symposium on Probabilistic Structural Mechanics (Advances in Structural Reliability Methods), San Antonio, Texas, June.

Jordaan, I.J., Maes, M.A., Brown, P.W. and Hermans, I.P. (1992), "Probabilistic Analysis of Local Ice Pressures", Proceedings of the Eleventh International Conference on Offshore Mechanics and Arctic Engineering, *OMAE 1992*, Calgary, June, Vol. II, pp. 7 -14.

Maes, M.A. and Hermans, I.P. (1991), "Review of Methods of Analysis of Data and Extreme Value Techniques of Ice Loads", Report submitted to Memorial University of Newfoundland, Ocean Engineering Research Centre, Kingston, Canada, 51 p.

Jordaan, I.J., Kennedy, K.P., McKenna, R.F. and Maes, M.A. (1991), "Loads and Vibration Induced by Compressive Failure of Ice", Cold Regions Engineering, Proc. Sixth International Specialty Conference, ASCE, Hanover, NH, pp. 638-649.

Castillo E. (1988). "Extreme Value Theory in Engineering", Academic Press, Inc., Boston.

Smith R.L. (1986), "Extreme Value Theory based on the *r* Largest Annual Events", *J. Hydrology*, 86, pp. 27 - 43.

Structural Safety & Reliability, Schuëller, Shinozuka & Yao (eds) © 1994 Balkema, Rotterdam, ISBN 90 5410 357 4

Poisson processes in load space reliability

Robert E. Melchers

Department of Civil Engineering and Surveying, The University of Newcastle, N.S.W., Australia

ABSTRACT: A Monte Carlo procedure for estimating the zero–time failure probability and the outcrossing rate for structural systems acted upon by loads modelled as mixed Poisson pulse processes is presented. The problem is formulated in the space of the load processes and the structural system resistance is represented by probabilistically defined region boundaries obtained from the (known) limit state functions. The results are compared to those obtained from a consideration of bounds on the outcrossing rate for mixed sparse Poisson processes. The theory is applied to an example system with multiple discrete load processes and a comparison to previous results is given.

KEYWORDS: Probability; Loads; Outcrossing Rate; Time–Dependence; Reliability; Structural Systems; Poisson Processes.

1. INTRODUCTION

For a structural system the probability of failure may be expressed as

$$p_f = \int_R p_f(\mathbf{r}) f_{\mathbf{R}}(\mathbf{r}) \, d\mathbf{r} \qquad (1)$$

where $p_f(\mathbf{r})$ is the structural failure probability conditional on the structural system capacity vector $\mathbf{R}$ having the vector value $\mathbf{r}$ and $f_{\mathbf{R}}(\ )$ is the joint probability density function (pdf) of $\mathbf{R}$. For any realisation of $\mathbf{R} = \mathbf{r}$, $p_f(\mathbf{r})$ represents the probability of structural failure, irrespective of the manner this is attained or the loading situation.

Consider now the situation where the structure is subject to a set of load processes $\mathbf{Q}(t)$ and let the integral in equation (1) governed by the dimension of the vector of load processes $\mathbf{Q}(t)$. It follows that $\mathbf{R} = \mathbf{R}(\mathbf{X})$ represents an n–vector of random variables such that each component of $\mathbf{R}$ corresponds to one component only of $\mathbf{Q}$. Further, in the n–dimensional load process space $\mathbf{Q}(t)$, the conventional limit state equations of structural reliability theory [e.g., Melchers, 1987], given by $G_i(\mathbf{q}, \mathbf{x}) = 0$ are represented as probabilistic boundaries, governed by the m–dimensional vector of random variables $\mathbf{X}$. The components of $\mathbf{X}$ describe the structure but may include also any uncertainties in the specification of the load processes $\mathbf{Q}(t)$. Similarly, $\mathbf{Q}(t)$ may include

processes other than load processes. However, it is useful to keep the simple separation of the load space and the random variables $\mathbf{X}$ representing structural resistance parameters as this allows a ready identification of the roles of each; this is a convenient but not a necessary distinction.

For this situation, the conditional failure probability $p_f(\mathbf{r})$ for $\mathbf{R} = \mathbf{r}$ is a function of the process vector $\mathbf{Q}(t)$ "crossing–out" of the safe domain D. This is a function only of the probability that $\mathbf{Q}(t)$ lies on the limit state function describing D and the expectation that the process will then move "out" of, rather than "in" to D. This is, conventionally, expressed through the generalised Rice formula. The actual path followed by $\mathbf{Q}(t)$ in arriving at the limit state is of no interest (provided it remains, until then, within D).

It will be convenient, as in Melchers [1992], to rewrite equation (1) in (hyper–) polar co–ordinate form with $\mathbf{R} = S \cdot \mathbf{A} + \mathbf{c}$, where $\mathbf{A}$ is a vector of direction cosines, S is a (scalar) radial distance and $\mathbf{c}$ is some point selected as the origin. As will be seen below, it will be extremely convenient (but not essential) to assume that $\mathbf{c} = 0$. Expression (1) may now be written as

$$p_f = \int_{\substack{\text{unit} \\ \text{sphere}}} f_{\mathbf{A}}(\mathbf{a}) \left[\int_S p_f(s|\mathbf{a}) \cdot f_{S|\mathbf{A}}(s|\mathbf{a}) \, ds \right] d\mathbf{a} \qquad (2)$$

where $f_{S|A}(\ )$ is the conditional pdf of the structural resistance (denoted S here) for a given radial direction $A = a$. It will be assumed that $f_{S|A}(\ )$ can be obtained [Melchers, 1992] Also $f_A(\ )$ is the pdf of A. In the absence of other information, $f_A(\ )$ would be taken as unbiased such that all directions A = a have equal likelihood of occurrence. The integral over the unit sphere in equation (2) may be replaced by the expected value E_A, which then may be used in a Monte Carlo estimate.

Since each load is modelled as a random process, the conditional failure probability $p_f(s|a)$ in the polar co-ordinate system defined above can be obtained, for a given structural life $(0 - t_L)$, from the well-known (upper) bound on the outcrossing rate v_D^+ of the vector process $Q(t)$ out of the safe domain D [e.g., Veneziano et al., 1977]. For the comparatively high reliability problems of interest and for wide-band processes, individual outcrossings may be considered approximately as independent Poisson events. For a given direction $A = a$, the failure probability for a deterministic limit state at A = s is then

$$p_f(s|a) \le p_f(0, s|a) + [1 - p_f(0, s|a)]$$

$$\left(1 - \exp\left(-\int_0^{t_L} v_D^+(\tau, s|a)\, d\tau \right) \right)$$

$$\le p_f(0, s|a) + v_D^+(s|a) \cdot t_L \qquad (3)$$

where the latter expression holds for $v_D^+(\ )$ independent of time and for $p_f(0)$ small relative to unity. Also $p_f(0, s|a)$ denotes the (conditional) failure probability at time t = 0. For a given load process model, both $p_f(0, s|a)$ and $v_D^+(s|a)$ may be evaluated in the (hyper–) polar co-ordinate system.

2. OUTCROSSING RATE FOR DISCRETE LOAD PROCESSES

Let the loading be rectangular pulse processes and renewal square wave load processes of the Poisson type. For these, the time between events or between changes of state is exponentially distributed. Let the mean pulse arrival rate (the "intensity") be v_m and the mean pulse duration be μ_d. Also, let the processes be sparse so that the proportion of time the process is active is rather low, i.e., $v_m \cdot \mu_d \ll 1$, so that pulses are assumed to occur only relatively rarely. The rate at which a renewal process of intensity v_i upcrosses the level "a" is

$$v_a^+ = v_i\, F(a)\{[1 - F(a)]\} \qquad (4)$$

Evidently the term F(a) represents the probability that the process is below level "a" and the square bracketted term the probability that it is above level "a". This may be extended immediately to obtain the local outcrossing rate out of a domain D at the point $(s \cdot a + c)$ in (hyper–) polar space. For a renewal load process vector $Q(t)$ for which the load components are independent, the local outcrossing rate applies to each Q_i direction and the sum of all such results is sought. For systems of high reliability and if the pulse height is Gaussian distributed the (conditional) outcrossing rate is [cf. Breitung and Rackwitz, 1982]

$$v_D^+\Big|_{\substack{A = a \\ S = s}} = \sum_{i=1}^n v_i\, f_{Q(i)}(q(i))\{\Phi(y_i) \cdot \Phi(-y_i)\}\, |n_i|$$

$$(5)$$

where $y_i = (q_i - \mu_{Qi})/D(Q_i)$ with $D(\ )$ = standard deviation. If the pulses always return to zero at the end of each pulse duration, the term $\Phi(y) \cdot \Phi(-y)$ in equation (5) reduces to $\Phi(+y_i)$ for $n_i < 0$; $\Phi(-y)$ for $n_i > 0$.

If there is a finite probability p of zero pulse height, expression (5) becomes

$$v_D^+\Big|_{\substack{A = a \\ S = s}} = \sum_{k=1}^{ncomb} \sum_{i=1}^n v_i\, f_{Q(i)}(q(i))$$

$$\{p_i + q_i\, \Phi(y_i) \cdot q_i \cdot \Phi(-y_i)\}\, |n_i| \qquad (6)$$

where

v_i = pulse rate when the ith process is "on" = $1/\mu_i$
μ_i = mean duration of ith process
q_i = probability of the ith process being "on"
   = $v_{mi} \cdot \mu_i$
v_{mi} = mean pulse rate for the ith process
   ("on" or "off") = $q_i\, v_i$

and the number of combinations of "on", "off" states

is given by $ncomb = \sum_{\ell=1}^n 2^{(\ell-1)}$. Further, $f_{Q(i)}(\ )$ is given by $f_{Q(j)}(u) = \delta(u = 0)\, p_j + q_j\, f_j(u)$ where $\delta(u = 0) = 1$ if u = 0, = 0 otherwise.

Expression (6) now may be made unconditional by substitution into equation (2) (with the outcrossing rate substituted for failure probability) thus

$$v_D^+ = \int_{\substack{unit \\ sphere}} f_A(a) \int_{s=0}^{\infty} \left[\sum_{k=1}^{ncomb} \sum_{i=1}^n \{v_i f_{Q(i)}(q(i))\} \right.$$

1446

TABLE 1
LOAD AND RESISTANCE PARAMETER

VARIABLE	LOAD $Q_1(t)$	LOAD $Q_2(t)$	RESISTANCE X_1	RESISTANCE X_2
Mean μ	100	50	360	480
Standard Deviation σ	10	15	54	72
Intensity of Occurrence ν_m	5/year	0.2/year	–	–
Pulse Duration μ_d	10^{-2} year	10^{-2} year	–	–

$$\{(p_i + q_i\,\Phi(y_i))\,q_i\,\Phi(-y_i)\}\,|n_i|$$

$$f_{S|A}(s|a)\,s^{(n_k-1)}\{a\cdot n\}^{-1}\Big]\;ds\,da \qquad (7)$$

where n_k = number of "active" components of the load process vector $Q(t)$ for the kth combination. The term $s^{(n_k-1)}\{a\cdot n\}^{-1}$ corrects the elemental (hyper–) surface area ΔS_D at radius s to the equivalent area on the unit sphere [Melchers, 1992].

3. SIMULATION

The procedure used in the directional simulation described herein follows essentially the approach outlined in Melchers [1992], although the modelling of the load processes as "mixed" processes requires allowance to be made for sampling in each sub–space defined by the pairs (p_i, q_i). Thus with the centre of the sampling co–ordinate system centred at $c = 0$, a radial integration is required in each co–ordinate axis direction to account for the cases of only that process being "on" whilst all others are "off". For sparse processes, such as considered here, these cases contribute significantly to the final result and cannot be left for random sampling. Sampling in all successively higher dimensional sub–spaces is also required to ensure that all load combinations are properly considered. Each such sub–space may be considered for separate sampling. Also, constraining the problem to one quadrant, as when all load processes can be positive–valued only, considerably simplifies the simulation, but this is not a necessary restriction.

When the sampling co–ordinate system is not centred at $c = 0$ no particular difficulties arise for all sub–spaces greater than the axis sub–spaces. These need still to be considered, however, in the same manner as before. For higher dimensional sub–spaces (i.e., ≥ 2) care must be taken to ensure that any directional simulation ray does not extend beyond the integration domain.

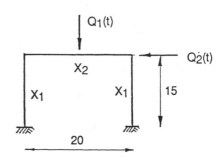

FIGURE 1 Rigid–Plastic Frame Example

4. EXAMPLE

The simple ideal rigid–plastic frame shown in Figure 1 is subjected to sparse Poisson load processes $Q_1(t)$ and $Q_2(t)$. It has been previously analysed for annual failure probabilities by Wen and Chen [1986, 1989]. Table 1 shows the parameters for the load processes (which have Gaussian pulse heights) and for the Gaussian resistance random variables. (There is a small inconsistency in this formulation. Gaussian pulse height implies a range $-\infty$ to $+\infty$, but the loading model suggests that all pulse heights are greater than zero. In the present example this may introduce a small error in comparison to earlier work which appears to have ignored the matter). The relevant plastic collapse mechanisms are shown in Table 2.

The outcrossing rates ν_D^+ obtained in the present study are shown in column 2 of Table 3. In each case the upper value corresponds to assuming that the pulses automatically return to zero prior to a change of load intensity – the lower value corresponds to the pulse value not doing so. All the results were verified by a "brute–force" Monte Carlo simulation in X space using expression (1) directly [but with X for R]. The results were found to be extremely stable with regard to sample sizes for

TABLE 2
RIGID–PLASTIC COLLAPSE MECHANISMS
(LIMIT STATE EQUATIONS)
AND OUTWARD UNIT NORMAL DIRECTION COSINES AT $Z_i = 0$

MECHANISM	COLLAPSE MODE EQUATION	$\alpha_1 \propto \partial Z/\partial Q_1$	$\alpha_2 \propto \partial Z/\partial Q_2$
1	$Z_1 = 4X_1 + 2X_2 - 10Q_1 - 15Q_2$	0.555	0.832
2	$Z_2 = 4X_1 \qquad\qquad - 15Q_2$	0	1.0
3	$Z_3 = 2X_1 + 4X_2 - 10Q_1 - 15Q_2$	0.555	0.832
4	$Z_4 = 3X_1 + X_2 \qquad - 15Q_2$	0	1.0
5	$Z_5 = \qquad 4X_2 - 10Q_1$	1.0	0
6	$Z_6 = 2X_1 + 2X_2 \qquad - 15Q_2$	0	1.0
7	$Z_7 = X_1 + 3X_2 - 10Q_1$	1.0	0
8	$Z_8 = 2X_1 + 2X_2 - 10Q_1$	1.0	0

directional simulation and number of integration steps in the radial direction.

Wen and Chen [1986, 1989] reported the results shown in columns 3, 4 and 5 of Table 3. Even without detailed inspection it is clear that the present results are considerably different.

In attempting to clarify this situation, two courses of action were followed:

i.   bounds to the theoretical outcrossing rate were derived, and

ii. a detailed investigation was performed into the assumptions underlying the results given in columns 3 – 5.

Details of this work will be given elsewhere. Suffice it to note that the simulation results (column 2) lie in all cases within the bounds (columns 6 and 7) and that the results in columns 3 – 4 do not all lie within the bounds. (Ignoring the slight error introduced by not allowing for the truncation of the distributions at zero load in columns 3 – 7). The reasons for the discrepancy between these latter results and those of column (2) will now be examined.

In Table 3, the "ensemble average" results (column 3) are based on the "load–coincidence" method [see Wen and Chen, 1989]. They: (i) give a higher estimate of the outcrossing rate results obtained by any other means, but (ii) coincide entirely with the "approximate upper bound" of column 7. The first observation is readily explained. For very sparse loading processes, such as exemplified by Load

Case 1 in Table 3, there would be very little difference between "ensemble average" results and more correctly determined estimate of v_D^+ since the individual processes would only rarely have "on" times coincident with those of the other processes. It is evident that the outcrossing rate for such process combination is essentially equal to the sum of the separate upcrossing rates for each of the processes. Further, for the individual (homogeneous and stationary) processes, there is no distinction between ensemble and correct upcrossing rates [e.g., Melchers, 1987]. This is confirmed by an analysis of the result given for Case 1 in column 2; the contribution to v^+ due to Load 1 is 0.00640, due to Load 2 is 0.00256, and that due to both loads is only 0.00032. Hence it is clear that for Load Case 1, at least, the outcrossing rate v^+ would be expected to be close to that estimated by the ensemble average estimate. For the other cases it follows that an ensemble estimate will overestimate the actual outcrossing rate – evidently considerably so when the "on" fraction is at all significant.

The second observation merely reflects the fact that the "load–coincidence" method is really an upper bound result obtained essentially in the same way that the upper bound given in column 7 has been derived. The socalled "exact" results of Wen and Chen [1989] are based on multiple integration. The procedure used to obtain these values was essentially the same as that of the "brute–force" method of Monte Carlo simulation used to verify the results given in column 2. These values are therefore subject to the same type of difficulties in computations and are likely to be underestimated

TABLE 3
OUTCROSSING RATE COMPARISON

	(1)	(2)	(3)	(4)	(5)	(6)	(7)
CASE	LOAD PARAMETERS	SIMU–LATION 1,2	ENSEMBLE ESTIMATE[3]	EXACT[4]	MONTE CARLO[5]	LOWER BOUND	APPROX UPPER BOUND
			[Source: Wen and Chen, 1986, 1989]				
1	$v_{m1} = 5$/yr $v_{m2} = 0.2$/yr $\mu_1 = \mu_2 = 0.01$ yr	0.0093 0.0091	0.00967	0.0073	0.0075	0.00894	0.00967
2	$v_{m1} = v_{m2} = 5$/yr $\mu_1 = \mu_2 = 0.01$ yr	0.079 0.077	0.088	0.0672	0.062	0.0702	0.088
3	$v_{m1} = v_{m2} = 20$/yr $\mu_1 = \mu_2 = 0.01$ yr	0.381 0.365	0.525	0.236	0.236	0.237	0.525
4	$v_{m1} = v_{m2} = 20$/yr $\mu_1 = \mu_2 = 0.025$ yr	0.510 0.440	0.87	0.315	0.329	0.148	0.87

NOTES

1. Directional sampling based on 500 directional samples and 200 integration steps in the radial direction.

2. Upper values assume non–automatic return to zero. Lower values assume automatic return to zero.

3. Valued obtained in the present study and in accord with Chen (private communication).

4. Converted from annual failure probabilities p_1 using $v^+ = -\ln(1 - p_1)$.

5. Based on 800 samples of the structural resistance vector $\mathbf{X}$.

unless considerable care is taken with the limits of integration and a sufficiently large number of samples is employed.

5.   CONCLUSIONS

In conclusion, the major advantage of the proposed approach is that it allows outcrossing rates to be determined as well as initial (time zero) failure probabilities for any probabilistic description of the load processes. Simulation in the total $(\mathbf{X} + \mathbf{Q})$ space does not allow this to occur very readily. Furthermore, the proposed approach is formulated in terms of load processes and therefore requires simulation only in a (generally) low–dimensional space. This means that Monte Carlo simulation can be economic in computer time. It also means that situations with complex limit state formulations (e.g., finite element analyses, dynamic analyses) can be accommodated readily and economically. Nevertheless, the full potential of the proposed approach, and its potential limitations, such as determining radial distributional properties such as for system strength and expressed in $f_{S|A}(\ )$, still needs to be more fully explored.

6.   REFERENCES

BREITUNG, K. and RACKWITZ, R., (1982), Non–Linear Combination of Load Processes, J. Struct. Mech., 10(2), 145–166.

MELCHERS, R.E., (1987), Structural Reliability Analysis and Prediction, John Wiley and Sons.

MELCHERS, R.E., (1992), Load Space Formulation for Time Dependent Structural Reliability, J. Engg. Mech., ASCE, 118(5), 853–870.

VENEZIANO, D., GRIGORIU, M. and CORNELL, C.A., (1977), Vector–Process Models for System Reliability, J. Engg. Mech. Div., ASCE, 103(EM3), 441–460.

WEN, Y.K. and CHEN, H.C., (1986), System Reliability Under Multiple Hazards, Report No. 526, Structural Research Series, Department of Civil Engineering, University of Illinois at Urbana–Champaign.

WEN, Y.K. and CHEN, H.C., (1989), System Reliability Under Time Varying Loads: I, J. Engrg. Mech., ASCE, 115(4), 808–823.

Structural Safety & Reliability, Schuëller, Shinozuka & Yao (eds) © 1994 Balkema, Rotterdam, ISBN 90 5410 357 4

Combination of non-stationary rectangular wave renewal processes

R. Rackwitz

Technical University of Munich, Germany

ABSTRACT:
Suitable, widely asymptotic formulae for the failure probability under combined non-stationary rectangular wave renewal processes are derived via the outcrossing approach. Non-stationarity can exist either in the limit state function or the parameters of the stochastic models. An importance sampling scheme for the treatment of non-ergodic variables is proposed. An example illustrates some theoretical findings.

1 INTRODUCTION

Rectangular wave processes are frequently used to model the time variations of occupancy loading. Such processes can also approximate other loading phenomena and, in particular, may be used to model the main characteristics in so called missions, e.g. the journey of a ship between two places, the different sea states it experiences during the journey or the loading environment of a processing plant between the shut down periods. In some cases another more rapidly fluctuating loading process then is superimposed upon the simple rectangular wave.

Stationary rectangular wave processes have been studied repeatedly. The special case of constant durations has been proposed first and has found the earliest solutions (Ferry Borges/Castanheta, 1971; Rackwitz/Fießler, 1978). The more general model of multivariate rectangular wave renewal process and its combination has been studied by Breitung/Rackwitz (1982) and (Rackwitz, 1985) via the outcrossing approach. Interesting and practically useful generalizations of the simple rectangular wave model have been proposed by Wen (1990), Shinozuka (1981) and Schrupp/Rackwitz (1988) and others. Considerable improvement and simplification was achieved by applying asymptotic concepts, i.e. when failure probabilities tend to zero (Breitung, 1984). Breitung did not only show that computation of outcrossing rates essentially reduces to simple volume integral evaluations but also indicated that under quite general conditions the optimal asymptotic expansion point of the limit state surface is, in fact, the same as in time-invariant analysis. The asymptotic Poissonian nature of the crossings into the failure domain has already been shown earlier.

In the following an approximate computation scheme for the non-stationary crossing rates and first passage times is proposed together with some indications how to deal consistently with non-ergodic variables. Those variables distort the asymptotic nature of crossings. Therefore integration with respect to these variables has to be performed separately. Special emphasis is given to a number of non-trivial numerical problems.

2 GENERAL CONCEPTS FOR TIME VARIANT RELIABILITY

Consider the general task of estimating the probability $P_f(t)$ that a realization $z(\tau)$ of a random state vector $Z(\tau)$ enters the failure domain $V = \{ z(\tau) \mid g(z(\tau),\tau) \leq 0, 0 \leq \tau \leq t \}$ for the first time given that $Z(\tau)$ is in the safe domain at $\tau = 0$. $g(\cdot)$ is the state function. The limit state is defined for $g(.) = 0$. $Z(\tau)$ may conveniently be separated into three components as

$$Z(\tau)^T = (R^T, Q(\tau)^T, S(\tau)^T) \qquad (1)$$

where R is a vector of random variables independent of time, $Q(\tau)$ is a slowly varying stationary and ergodic random vector sequence and $S(\tau)$ is a vector of not necessarily stationary, but sufficiently mixing random process variables having fast fluctuations as compared to $Q(t)$.

Consider first the case where only $S(\tau)$ is present. If it can be assumed that the stream of crossings of the vector $S(\tau)$ into the failure domain V is Poissonian it is well known that the failure probability $P_f(t)$ can be estimated from

$$P_f(t) \approx 1 - \exp(-\,\mathbb{E}[N_S^+(t)]) \leq \mathbb{E}[N_S^+(t)] \qquad (2)$$

with

$$\mathbb{E}[N_S^+(t)] = \int_0^t \nu_S^+(\tau)\, d\tau \qquad (3)$$

for high reliability problems. $\mathbb{E}[N_S^+(t)]$ is the expected number of crossings of $S(\tau)$ into the failure domain V in the considered time interval $[0,t]$ and $\nu_S^+(\tau)$ the outcrossing rate. It is assumed that there is negligible probability of failure at $\tau = 0$ and $\tau = t$, respectively. The upper bound in eq. (2) is a strict upper bound but close to the exact result only for rather small $P_f(t)$. The approximation in eq. (2) has found many applications in the past not only because of its relative simplicity but also because there has been no real practical alternative except in some special cases. It is already worth noting that for the upper bound solution there is no particular problem of integration because one needs not to distinguish between the different types of variables as introduced above (see below).

When both process variables $S(\tau)$ and time invariant random variables R are present the Poissonian nature of outcrossings is lost. Eq. (2) can furnish only conditional probabilities. The total failure probability must be obtained by integration over the probabilities of all possible realizations of R. Then the equivalent to eq. (2) is

$$\begin{aligned} P_f(t) \quad &\approx \mathbb{E}_R[1 - \exp(-\,\mathbb{E}[N_S^+(t\,|\,R)])] \\ &= 1 - \mathbb{E}_R[\exp(-\,\mathbb{E}[N_S^+(t\,|\,R)])] \\ &\le \mathbb{E}_R[\mathbb{E}[N_S^+(t\,|\,R)]] \end{aligned} \qquad (4)$$

In the general case where all the different types of random variables R, $Q(\tau)$ and $S(\tau)$ are present the failure probability $P_f(t)$ not only must be integrated up over the time in-variant variables R but an expectation operation must be performed over the slowly varying variables $Q(\tau)$. In Schall et al. (1991) the following formula has been established in part by making use of the ergodicity theorem

$$\begin{aligned} P_f(t) &\approx 1 - \mathbb{E}_R[\exp(-\,\mathbb{E}_Q[\mathbb{E}[N_S^+(t\,|\,R,Q)]])] \\ &\le \mathbb{E}_R[\mathbb{E}_Q[\mathbb{E}[N_S^+(t\,|\,R,Q)]]] \end{aligned} \qquad (5)$$

Eq. (5) is a rather good approximation for the stationary case but must be considered as a first approximation whenever $S(\tau)$ is non-stationary or the limit state function exhibits strong dependence on τ as shown in the mentioned reference. The approximation concerns the expectation operation with respect to Q in the non-stationary case. The bounds given in eqs. (2, 4 and 5) again are strict but close to the exact result only for even smaller failure probabilities. In fact, while the approximation with respect to the expectation operation inside the exponent in eq. (5) may be accepted also for the non-stationary case in most practical applications the expectation with respect to R must be taken outside the exponent because, depending on the relative magnitude of the variabilities of the R- and the S, Q-variables, errors up to several orders of magnitude can occur (see Schall et al. 1990). An exact evaluation of eq. (5) may also be necessary if the failure event in eq. (5) has to be conditioned on some other event, e.g. in inspection planning. Therefore, it is of particular interest to design effective computations schemes especially in view of the fact that the R-vector can be high-dimensional (e.g. in stochastic finite elements with hundreds of variables describing random system properties). To be complete it ought to be mentioned that consideration of the initial and final conditions of the processes, i.e. at $\tau = 0$ and $\tau = t$, respectively, sometimes can result in noticeable improvements of the results (Plantec/Rackwitz, 1988). In the following those effects are not considered.

3 CONDITIONAL OUTCROSSING RATES FOR NON-STATIONARY RECTANGULAR WAVE RENEWAL PROCESSES

If the components of a stationary rectangular wave renewal process are independent with marks S_k with distribution function $F_S(s;q,r)$ and renewal rates λ_i it has been shown that the mean number of exits into the failure domain is (Breitung/Rackwitz, 1982)

$$\begin{aligned} \mathbb{E}[N^+(t_1,t_2;q,r)] &= \\ = (t_2 - t_1) \sum_{i=1}^{n} \lambda_i\, &\mathbb{P}(\{\,S_i^- \in \bar{V};q,r\} \cap \{\,S_i^+ \in V;q,r\,\}) \end{aligned} \qquad (6)$$

where $\bar{V}$ and V are the safe and failure domain, respectively. S_i^+ is the total load vector when the i-th component of the renewal process had a renewal. S_i^- denotes the total load vector just before the renewal. Therefore, S_i^- and S_i^+ differ by the vector S_i which is to be introduced as an independent vector in the second set. Applying asymptotic concepts and using eq. (6) it can further be shown that asymptotically (Breitung 1984)

$$E[N^+(t_1,t_2;r,q)] \sim (t_2 - t_1) \sum_{i=1}^{n} \lambda_i\, \mathbb{P}(\{\,S \in V;q,r\}) \qquad (7)$$

with $\mathbb{P}(\{\,S \in V;r,q\})$ computed as a volume integral in the usual manner by SORM. Very rarely this formula is noticeably improved for not small probabilities $\mathbb{P}(\{\,S \in \bar{V};r,q\})$ by replacing the term $\mathbb{P}(\{\,S \in V;r,q\})$ by

$\mathbb{P}(\{\,S_{\bar{i}}\in \bar{V};q,r\} \cap \{\,S_i^+\in V;q,r\}) =$

$= \mathbb{P}(\{\,S\in V;r,q\}) - \mathbb{P}(\{\,S_{\bar{i}}\in V;r,q\} \cap \{\,S_i^+\in V;q,r\})$

as in eq. (6). Note that integration with respect to q is performed simultaneously with the integration with respect to s. If unconditional mean numbers of exits need to be computed as in eq. (5) integration is also over r (see below). If there is complete dependence of jump events for subsets of wave processes summation in eq. (6) or (7) is only over the independent components.

The non-stationary case is not substantially more difficult. The renewal rates $\lambda_k(\tau)$, $k = 1,2,...$, are assumed to vary slowly in time. The distribution function of S may contain distribution parameters $r(\tau)$ varying in time and the failure domain can be a function of time, i.e. $V = \{g(s,q,r,\tau) \le 0\}$. Then, eq. (6) needs to be modified as

$$\mathbb{E}[N^+(t_1,t_2|r)] \sim \int_{t_1}^{t_2} \sum_{i=1}^{m} \lambda_i(\tau)\, \mathbb{P}(\{\,S\in V)|q,r,\tau\})\, d\tau$$

$$= \int_{t_1}^{t_2} \int_V \sum_{i=1}^{n} \lambda_i(\tau)\, f_{S,Q}(s,q,\tau|r)\, ds\, dq\, d\tau \quad (8)$$

The time-volume integral (8) can be approximated using FORM/SORM concepts. For small $\mathbb{P}(\{\,S\in V|q,r,\tau\})$ the integrand is dominated by the probability term in the neighborhood of the most likely failure point (s^*,q^*,τ^*) in $\{\,S\in V|q,r,\tau\}$ to be determined by an appropriate algorithm. One such algorithm has been proposed by Abdo/Rackwitz (1991). Here and in the following integrations are best performed in the standard space after applying a suitable probability distribution transformation. Because $\lambda_k(\tau)$ is slowly varying it is drawn in front of the integral with value $\lambda_k(\tau^*)$ by virtue of the mean value theorem of integral theory. Classical FORM/SORM algorithms can then be applied according to (Hagen/Tvedt, 1991) after transforming the integral in eq. (8) into a simple probability integral by introducing an additional uniform density $f_T(\tau) = (t_2 - t_1)^{-1}$ into eq.(8) such that

$$\mathbb{E}[N^+(t_1,t_2|r)] = (t_2 - t_1) \sum_{k=1}^{m} \Big[\lambda_i(\tau^*) \times$$

$$\times \int_{\mathbb{R}^1} \int_V f_{S,Q}(s,q,\tau|r)\, f_T(\tau)\, ds\, dq\, d\tau \Big] \quad (9)$$

With the transformation $\tau = (t_2 - t_1)\,\Phi(u_\tau)$ the probability integral can be determined in the usual manner. $\Phi(.)$ is the standard normal integral. The results turn out to be quite accurate whenever τ^* lies within a large interval $[t_1,t_2]$ and the associated norm $\|u\|$ of the standard uncertainty vector is large and thus the asymptotic conditions are met.

For smaller intervals $[t_1,t_2]$ the results become less accurate. They are no more acceptable in general when τ^* is a boundary point.

Therefore, a slightly different scheme is advantageous in all cases. In the critical point (s^*,q^*,τ^*) the probability $\mathbb{P}(\{\,S\in V|q,r,\tau\})$ is estimated as

$$\mathbb{P}(\{\,S\in V|q,r,\tau\}) = \Phi(-\,\beta(\tau^*)) \times C(s^*,q^*,\tau^*|r) \quad (10)$$

with $C(s^*,q^*,\tau^*)$ the well known curvature correction term (in the s-q-space) in SORM. Then, one can write

$\mathbb{E}[N^+(t_1,t_2|r)]$

$$\sim \int_{t_1}^{t_2} \sum_{i=1}^{n} \lambda_i(\tau|r)\, \Phi(-\,\beta(\tau|r)) \times C(s^*,q^*,\tau|r)\, d\tau$$

$$\approx C(s^*,q^*,\tau^*|r) \sum_{i=1}^{n} \lambda_i(\tau^*|r) \int_{t_1}^{t_2} \Phi(-\,\beta(\tau|r))\, d\tau$$

$$= C(s^*,q^*,\tau^*|r) \times$$

$$\times \sum_{i=1}^{n} \lambda_i(\tau^*|r) \int_{t_1}^{t_2} \exp[\ln[\Phi(-\,\beta(\tau|r))]]\, d\tau$$

$$= C(s^*,q^*,\tau^*|r) \sum_{i=1}^{n} \lambda_i(\tau^*|r) \int_{t_1}^{t_2} \exp[f(\tau)]\, d\tau$$

$$(11)$$

where $f(\tau) = \ln[\Phi(-\,\beta(\tau))]$. The time integral in eq. (11) is perfectly suited for application of Laplace's integral approximation. Expanding $f(\tau)$ to first and second order with derivatives

$$f'(\tau) = -\frac{\varphi(-\,\beta(\tau))}{\Phi(-\,\beta(\tau))}\frac{\partial \beta(\tau)}{\partial \tau} \approx -\,\beta(\tau)\frac{\partial \beta(\tau)}{\partial \tau}$$

$$f''(\tau) = -\frac{\varphi(-\,\beta(\tau))}{\Phi(-\,\beta(\tau))}\Big[\Big[\frac{\partial \beta(\tau)}{\partial \tau}\Big]^2 \times$$

$$\times \Big[\beta(\tau) + \frac{\varphi(-\,\beta(\tau))}{\Phi(-\,\beta(\tau))}\Big] + \frac{\partial^2 \beta(\tau)}{\partial \tau^2}\Big]$$

yields integrals which have analytical solutions. While $\partial \beta / \partial \tau$ is directly obtained as a parametric sensitivity the second derivative $\partial^2 \beta(\tau)/\partial \tau^2$ must be determined numerically by a simple difference scheme. The results are

- $\tau^* = t_1$ or $\dfrac{\partial \beta(\tau)}{\partial \tau} > 0$:

$$\mathbb{E}[N^+(t_1,t_2|r)] \approx C(s^*,q^*,t_1|r) \sum_{i=1}^{n} \lambda_i(t_1|r)\, \Phi(-\,\beta(t_1|r))$$

$$\times \Big\{\frac{\exp[f'(t_1)\,t_2] - \exp[f'(t_1)\,t_1]}{\exp[f'(t_1)\,t_1]\; f'(t_1)}\Big\} \quad (12a)$$

$$- \quad \tau^* = t_2 \text{ or } \frac{\partial \beta(\tau)}{\partial \tau} < 0:$$

$$\mathbb{E}[N^+(t_1,t_2|\mathbf{r})]$$

$$\approx C(\mathbf{s}^*,\mathbf{q}^*,t_2|\mathbf{r}) \sum_{i=1}^{n} \lambda_i(t_2|\mathbf{r}) \, \Phi(-\beta(t_2|\mathbf{r})) \times$$

$$\times \left\{ \frac{\exp[f'(t_2)\,t_2] - \exp[f'(t_2)\,t_1]}{\exp[f'(t_2)\,t_2]\;f'(t_2)} \right\} \tag{12b}$$

$$- \quad t_1 < \tau^* < t_2 \text{ or } \frac{\partial \beta(\tau)}{\partial \tau} = 0 \text{ and } \frac{\partial^2 \beta(\tau)}{\partial \tau^2} > 0:$$

$$\mathbb{E}[N^+(t_1,t_2|\mathbf{r})]$$

$$\approx C(\mathbf{q}^*\mathbf{s}^*,\tau^*|\mathbf{r}) \sum_{i=1}^{n} \lambda_i(\tau^*|\mathbf{r}) \, \Phi(-\beta(t_1|\mathbf{r}))$$

$$\times \left[\frac{2\pi}{|f''(\tau^*)|} \right]^{1/2} \times$$

$$\times \left\{ \Phi(|f''(\tau^*)|^{1/2}(t_2 - \tau^*)) - \Phi(|f''(\tau^*)|^{1/2}(t_1 - \tau^*)) \right\} \tag{12c}$$

If the conditions for eq. (12c) are met but the critical point is at one of the boundaries the mean number of outcrossings is just one half of the value in eq. (12c) (see Bleistein/Handelsman, 1986). The modifications in eqs. (12) result in more accurate exit means than eq. (9) especially for smaller time distances $t_2 - t_1$ although the interaction between τ and the other variables is neglected. A genuine first order result does not exist because time integration always is an approximation in the second order sense. In many applications the computation of the correction factor $C(\mathbf{s}^*,\mathbf{q}^*,\tau^*|\mathbf{r})$ involving the second order derivatives in the s, q-space will yield only small improvements, however. Of course, some conditions must be met for the validity of the approach. In particular, there should be $\beta(\tau^*) > 1$ and $\mathbb{E}[N^+(t_1,t_2|\mathbf{r})] \ll 1$.

4 INTEGRATION WITH RESPECT TO TIME-INVARIANT VARIABLES R

If there are time-invariant random vectors **R** several possibilities exist the most straightforward being numerical integration. However, even for small dimensions of **R** the computational effort can be considerable. So called nested FORM/SORM has been proposed as an alternative. Unfortunately, it turned out to be rather time consuming and not reliable in the non-stationary case. Also a simple first order Taylor expansion of the exponent in eq. (1) or application of a standard (inner point) result of Laplace's integral approximation method has been found to be not sufficiently accurate at least if the expansion point $\mathbf{r}^*$ is not exactly the critical point. A first approximation for this point can be obtained from one of the upper bound solutions as in eq. (5). The exact location would

require some iteration which now is quite involved as the eqs. (12) must be solved in each iteration step. Even then Laplace's solution still contains an error which has been found to become quite large in extreme cases.

Alternatively and arbitrarily exact at increasing numerical effort, the expectation operation in eq. (5) can be performed either by crude Monte Carlo integration or, more efficiently, by importance sampling. For convenience it is assumed that R is an uncorrelated standardized Gaussian vector which can always be achieved by a suitable probability distribution transformation. There is

$$\mathbb{E}_{\mathbf{R}}[1 - \exp\{- \mathbb{E}_{\mathbf{Q}}[\mathbb{E}[N_{\mathbf{S}}^+(t|\mathbf{Q},\mathbf{R})]]\}]$$

$$= \int_{\mathbb{R}^{n_r}} \left[1 - \exp\{- \mathbb{E}_{\mathbf{Q}}[\mathbb{E}[N_{\mathbf{S}}^+(t|\mathbf{Q},\mathbf{r})]]\} \right] \frac{\varphi_{\mathbf{R}}(\mathbf{r})}{h_{\mathbf{R}}(\mathbf{r})} h_{\mathbf{R}}(\mathbf{r}) \, d\mathbf{r} \tag{13}$$

where $h_{\mathbf{R}}(\mathbf{r})$ is the sampling density. Then,

$$\mathbb{E}_{\mathbf{R}}[1 - \exp(- \mathbb{E}_{\mathbf{Q}}[\mathbb{E}[N_{\mathbf{S}}^+(t|\mathbf{Q},\mathbf{R})]])]$$

$$\approx \frac{1}{N} \sum_{i=1}^{N} \left[1 - \exp\{- \mathbb{E}_{\mathbf{Q}}[\mathbb{E}[N_{\mathbf{S}}^+(t|\mathbf{Q},\mathbf{r}_i)]]\} \right] \frac{\varphi_{\mathbf{R}}(\mathbf{r}_i)}{h_{\mathbf{R}}(\mathbf{r}_i)} \tag{14}$$

The sampling density (standard space) can conveniently be chosen as the standard normal density with mean $\mathbf{r}^*$ from the upper bound solution eq. (5) and covariance matrix **I**. A crude conservative estimate is already obtained for $\mathbf{r}_i = \mathbf{r}^*$. From extensive testing of examples it is concluded that the scheme according to eq. (14) is rather robust provided that an efficient and reliable algorithm is available to locate the critical point $(\mathbf{s}^*,\mathbf{q}^*,\tau^*)$ for every simulated **r**.

5 HAZARD RATES

The hazard function as an additional useful reliability characteristic for time-variant reliability problems can also be computed. By FORM/SORM the hazard function is computed simply from parametric sensitivities. It is

$$h(t) = \frac{f_T(t)}{1 - P_f(t)} = \frac{1}{1 - P_f(t)} \frac{\partial P_f(t)}{\partial t}$$

$$= \frac{\varphi(\beta(t))}{\Phi(\beta(t))} \frac{\partial \beta(t)}{\partial t} \tag{15}$$

by assuming that the considered time interval is $[0,t]$. $\beta(t)$ is to be interpreted as the equivalent reliability index $\beta(t) = \Phi^{-1}(P_f(t))$. The foregoing theory allows to compute these hazard rates rigorously if the parametric sensitivities $\partial \beta(t)/\partial t$

are available. Unfortunately, this can be done only at the expense of some additional numerical effort.

However, good approximations are possible if the computation is for the critical point τ^* only. For a critical inner point the hazard rate then remains essentially constant even if there are parameter variations. Rewriting eq. (5) as

$$P_f(t) = \mathbb{E}_R[1 - \exp(-\mathbb{E}_Q[\mathbb{E}[N_s^+(t\,|\,Q,R)]])]$$

$$= \int_{\mathbb{R}^{n_r}} \left[\varphi_R(r) - \frac{1}{(2\pi)^{n_r/2}} \right.$$

$$\left. \times \exp(-\mathbb{E}_Q[\mathbb{E}[N_s^+(t\,|\,Q,r)]] - \tfrac{1}{2}r^T r) \right] dr \quad (16)$$

and formal application eq. (15) yields

$$h(t) \approx \exp\left[-\tfrac{1}{2}r^{*T}r^*\right] \frac{\exp\left[-\mathbb{E}_Q[\mathbb{E}[N^+(t)]]\right]}{1 - P_f(t)} \times$$

$$\times \mathbb{E}_Q[\nu^+(\tau^*)] \quad (17)$$

An estimate of the (critical) conditional outcrossing rate, of course, is

$$\nu^+(\tau^*\,|\,r^*) = \mathbb{E}_Q[\nu^+(\tau^*\,|\,r^*)] \quad (18)$$

and the unconditional outcrossing rate can be obtained from

$$\nu^+(\tau^*) = \exp\left[-\tfrac{1}{2}r^{*T}r^*\right] \mathbb{E}_Q[\nu^+(\tau^*)] \quad (19)$$

Eqs. (17) to (19) rest on the assumption that the r-variables have only small variability as compared with the q- and s-variables. Therefore, the hazard rate as computed by eq. (17) or (19) must be considered as a crude approximation whenever the r-variables dominate. According to Krzykacs/Kersken-Bradley (1976) the hazard rate should decrease in the stationary case.

6 DISCUSSION

It should be noted first that the rectangular wave renewal process model always yields slightly larger failure probabilities than the simpler model of rectangular waves with constant durations and integer ratios between the so called repetition numbers. This is due to the partial or complete removal of jump dependencies in the latter case.

The above developments concentrated on various bounds and then on certain approximations. The motivation for the approximations was to keep the numerical effort as small as possible especially in high dimensional spaces. All approximations are based on some information from an upper bound solution. Its accuracy is practically always sufficient if the resulting failure probability is

small, say, smaller than 10^{-3} the reason simply being the fact that then $1 - \exp[-x] \approx x$ in very good approximation. This is true even when the non-ergodic variables dominate. It should be clear that any upper bound solution is an upper bound solution only under the condition that the mean number of upcrossings can be computed exactly. This is, in principle, possible by adding to FORM or even SORM a suitable importance sampling scheme in the usual manner, i.e. the scheme proposed by Hohenbichler/Rackwitz (1988). For small failure probabilities very accurate results are thus obtained. For larger failure probabilities an "exact" computation of the upper bound makes hardly sense. If the mean number of upcrossings is found only by FORM (and sometimes also by SORM) the upper bound is no more strict. Remember that the time integration is always a SORM solution according to the approximate scheme followed for the derivation of eqs. (12). Whenever the dimension of the r-vector is large a complete SORM solution may be out of reach due to the considerable effort required for determination of the full Hessian matrix. Therefore, it must be recommended to compute the "upper bound solution" with the simplest method possible in this case.

Eqs. (12) can be evaluated strictly by FORM or SORM in the q,s-space but it should be noted that asymptotic concepts are applied in both cases when time-variant jump rates are involved (see transition from eq. (8) to (9)).

In applications with large non-ergodic uncertainties the integration over r is most critical for the accuracy of the final result. Surprisingly, the point r* obtained for the upper bound solution according to eq. (5) can be remarkably away from the correct one in extreme cases (i.e. for large failure probabilities). It may, therefore, be advisable to use some adaptive importance sampling scheme in those cases. Experience with such schemes showed, however, that the gain in required sampling points is rather modest in the cases investigated. The reason appears to be a rather large domain in the r-space where considerable contributions to the integral are obtained. In this case it is clear that a sampling distribution with relatively large spread can also account for the inaccuracy in the central point.

7 ILLUSTRATION

For the purpose of illustration we take the following simple function of structural states.

$$g(t,p,R,Q,S) = p_1(1 + p_2 t + p_3 t^2)\tfrac{\pi}{4}R_1 R_2^2$$
$$- (p_4 S_1 + p_6 S_2 + p_7 R_3)$$

with the p_i certain deterministic constants and a fixed time window of $[0, 2]$. The stochastic model and the parameters are summarized in table 1.

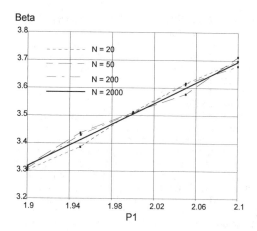

Figure 1: Convergence of equivalent reliability index with number of sampling points versus parameter p_1

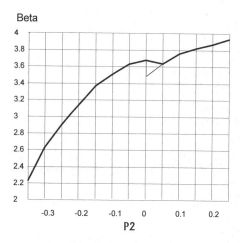

Figure 2: Transition from interior to boundary point solution with increasing parameter p_1

It is seen that this example includes almost all types of non-stationarities and dependencies which one can imagine to occur in practice. By variation of the parameters any kind degree of non-stationarity can be produced. Letting p_2 running between - 0.35 and + 0.25 will change the nature of the solution at $p_2 \approx 0.0$ from an interior point solution to a boundary point solution. It should be clear from the theoretical developments that there can be correlations between r-variables and q-variables, respectively, but no cross correlations. Also, there may be dependencies of each of the s-variable via its distribution parameters on q- and possibly r-variables (e.g. for taking account of statistical un-

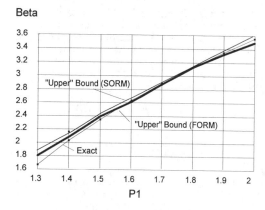

Figure 3: First order upper bound solution and simulation versus parameter p_1

certainties) but all q-variables must be conditionally independent of each other. Figure 1 gives an impression of the convergence of the simulations for the r-variables towards the correct solution for various p_1. It is seen that roughly 50 simulations are already sufficient in this case. Figure 2 with only N = 20 illustrates what happens when the interior point solution switches over to a boundary point solution for running parameter p_2. Typically, the failure probability jumps by roughly a factor of two near $p_2 \approx 0.0$. Around this point both solutions are not very reliable. Figure 3 finally compares the first order and second order upper bound solution with the second order solution in the q,s-space and with number of samples for the r-integration of N = 500, which brings the coefficient of variation for the resulting failure probability down to less than 3 %. It is seen that the differences are quite small. The first order upper bound is not a strict upper bound over the whole parameter domain.

8 SUMMARY AND CONCLUSIONS

Conditional non-stationary crossing rates for rectangular renewal wave processes are computed in part making use of asymptotic concepts. Hereby, the parameters of the wave processes may depend on slowly fluctuating random and ergodic sequences. Numerical analysis is performed by applying classical FORM/SORM concepts. For time integration a separate computational step is proposed. Some effort is spent to remove the conditioning by simple random variables by appropriate numerical schemes. It is found that a importance sampling scheme is the by far most satisfying numerical method in the stationary as well as the non-stationary case whenever failure probabilities are not small. For small failure probabilities the upper bound solution always is sufficiently accurate for practical purposes. It,

Table 1: Stochastic Model for Example

Name	Distribution	Mean Value	Std. Deviation	Jump Rates
R_1	Lognormal	2.0	0.2	-
R_2	Lognormal	3.0	0.3	-
R_3	Gumbel	6.0	0.6	-
Q_1	Lognormal	5.0	0.5	-
S_1	Normal	10.0	$Q_1 (1 + p_5 t)$	10.0
S_2	Gumbel	8.0	3.0	$p_8 (1 + p_9 t)$

p_1	2.000	Ultimate resistance multiplier
p_2	-0.100	Time multiplier for linear term
p_3	0.100	Time multiplier for quadratic term
p_4	0.330	Load multiplier for first life load
p_5	0.010	Time multiplier for standard deviation function $\sigma = Q_1 (1 + p_5 t)$ for Q_1
p_6	0.330	Load multiplier for second life load
p_7	0.330	Load multiplier for dead load
p_8	0.100	Constant parameter of jump rate function $\lambda_1 = p_8 (1 + p_9 t)$ for Q_2
p_9	1.000	Linear parameter of jump rate function $\lambda = p_8 (1 + p_9 t)$ for Q_2

nevertheless, must be admitted that time-variant reliabilities can be considerably more difficult and laborious to determine by FORM/SORM than time-invariant reliabilities.

ACKNOWLEDGEMENTS

T. Abdo and S. Gollwitzer performed the extensive programming work which is appreciated.

REFERENCES

Abdo, R., Rackwitz, R., A New Beta-Point Algorithm for Large Time-Invariant and Time-Variant Reliability Problems, *Proc. 3rd IFIP WG 7.5 Conference*, Berkeley, USA, 1990, Springer, pp. 1-12

Bleistein, N., Handelsman, R.A., *Asymptotic Expansions of Integrals*, Dover, 1986

Breitung, K., Asymptotic Approximations for Multinormal Integrals, *Journ.Eng. Mech.*, *ASCE*, 110, 3, 1984, 357-366

Breitung, K., Asymptotic Approximations for the Maximum of the Sum of Poisson Square Wave Processes, in: *Berichte zur Zuverlässigkeitstheorie der Bauwerke, Technische Universität München*, 69, 1984, pp. 59-82

Breitung, K., Rackwitz, R., Nonlinear Combination of Load Processes, *Journal of Struct. Mech.*, Vol. 1O, No.2, 1982, pp. 145-166

Ferry Borges, J., Castanheta, M., *Structural Safety*, Laboratorio Nacional de Engenharia Civil, Lisbon, 1971

Hagen, O., Tvedt, L., Parallel System Approach to Vector Outcrossings, *Proc. 10th OMAE Conf.*, Stavanger, 1991, Vol. II, pp.165-172

Hohenbichler, R., Rackwitz, R., Improvement of Second-order Reliability Estimates by Importance Sampling, *Journ. of Eng. Mech.*, *ASCE*, Vol.114, 12, 1988, pp. 2195-2199

Krzykacs, B., Kersken-Bradley, M., Wahrscheinlichkeitstheoretische Analyse der Lebensdauerverteilung nach Freudenthal et al., *Berichte zur Sicherheitstheorie der Bauwerke, Technische Universität München*, 12, 1976

Plantec, J.-Y., Rackwitz, R., Structural Reliability under Non-Stationary Gaussian Vector Process Loads, *Berichte zur Sicherheitstheorie der Bauwerke, Technische Universität München*, 85, 1989

Rackwitz, R., Reliability of Systems under Renewal Pulse Loading, *Journ. of Eng. Mech.*, *ASCE*, Vol. 111, 9, 1985, pp. 1175-1184

Rackwitz, R., and Fiessler, B., Structural Reliability under Combined Random Load Sequences, *Comp. & Struct.*, 1978, 9, 484-494

Schall, G., Faber, M., Rackwitz, R., The Ergodicity Assumption for Sea States in the Reliability Assessment of Offshore Structures, *Journ. Offshore Mechanics and Arctic Engineering, ASME*,1991, Vol. 113, No. 3, pp. 241-246

Schrupp, K., Rackwitz, R., Outcrossing Rates of Marked Poisson Cluster Processes in Structural Reliability, *Appl. Math. Modelling*, 12, 1988, Oct., 482-490

Shinozuka, M., Stochastic Characterization of Loads and Load Combinations, *Proc. 3rd ICOSSAR*, Elsevier, Amsterdam, 1981

Wen, Y.-K., *Structural Load Modeling and Combination for Performance and Safety Evaluation*, Elsevier, Amsterdam, 1990

Loads (ongoing research)

Structural Safety & Reliability, Schuëller, Shinozuka & Yao (eds) © 1994 Balkema, Rotterdam, ISBN 90 5410 357 4

Dynamic loads for continuous span bridges

E.-S. Hwang
Korea Institute of Construction Technology, Seoul, Korea

S. P. Chang
Seoul National University, Korea

ABSTRACT: The purpose of this paper is to develop the analytical model to calculate dynamic load factors and quantify the effect of various parameters for continuous span bridges. Analytical model including bridge, truck and road profile model is developed. Bridges are modeled by finite element method. Trucks are modeled as the masses with nonlinear springs representing the suspension systems and linear springs representing the tires. A series of 5 axle tractor-trailers are considered. Road roughness is considered as a gaussian random process. Analysis utilizes Newmark β time-integration method to calculate the maximum static and dynamic displacements and flexural moments at midspan. Parametric study is performed to quantify the effect of weight, speed, headway distance and number of trucks. Results are also compared with the case of the simple span bridge.

1 INTRODUCTION

Bridges are subjected to many loads including dead load, vehicle loads, earthquake load, and so on. The vehicle loads are due to the vehicles (especially trucks) running over the bridge and their effect is considered as static and dynamic. In the codes the dynamic load factor (impact factor) is specified as a function of the span length (Korea, United States, and Japan) or the fundamental frequency (Ontario, Canada and Switzerland). However, recent measurements and analysis show that the dynamic load factor is also affected by road roughness and dynamic characteristics of the vehicle.

The purpose of this paper is to develop an analytical model for dynamic loads and quantify the effect of various parameters for continuous span bridges. Since the continuous span bridge is usually longer than the simple span bridge, a series of trucks and various headway distances are considered in the analysis.

The analysis utilizes Newmark β time-integration method to calculate the maximum response at midspan. The dynamic load factors are calculated as the ratio of maximum dynamic response and maximum static response. Parametric analysis is performed to quantify the effect of weight, speed, headway distances, and number of trucks for simple and continuous span bridges.

2 DYNAMIC SYSTEM MODELS

2.1 truck model

Trucks used in the study are 5 axle semi tractor-trailers. Their models include masses with nonlinear springs representing the suspension systems and linear springs representing tires. Equations of motion are described in the reference(Hwang 1990). Characteristics of nonlinear springs are taken from Fancher(1980). A series of trucks are considered in the simulation.

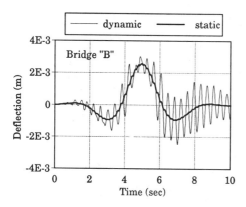

Fig. 1 : Time History of Midspan Deflection

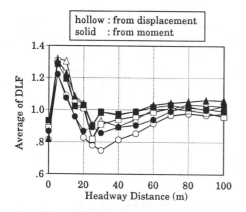

Fig. 2 : DLFs from Displacement and
Moment Response

Table 1. Summary of Bridge Properties

Bridge	"A"	"B"	"C"
No. of Spans	1	5	8
Total Bridge Length (m)	50	210	400
Maximum Span Length (m)	50	50	50
Girder Type	PC Box	PC Box	PC Box
Fundamental Frequency (Hz)	2.56	2.74	2.57

2.2 road profile model

Road roughness is considered as a gaussian random process. Road profiles representing the road roughness are generated by inverse Fourier transform of its power spectral density function. Characteristics of power spectral density function are taken from Fancher(1980). Details of generations are in the reference(Hwang 1990). To account for the random nature of road profiles, 10 road profiles are generated and their average values are used in the study. Effects of loss of contact between wheel and pavement and construction joints are not considered.

2.3 bridge model

Bridges are modeled by finite element method. In the study, three bridges with 50 m span lengths are selected: simple span, five and eight continuous span bridges. Their properties are summarized in Table 1. Even though the developed bridge model can consider three-dimensional behavior, this study uses two-dimensional frame elements to model the bridge. Therefore, effects of truck transverse positions are not included. Damping matrix is formulated by Rayleigh's method to be equivalent to 2% of critical damping.

3 SIMULATION

To calculate the maximum response, Newmark β time integration method is used. Since the suspension systems are nonlinear, predictor-corrector scheme is used. Relative tolerance limit is
taken as 0.01. Time step is taken as 0.005 sec. Typical time history of midspan deflection from developed computer program is shown in Fig. 1. In the study, the headway distance is defined as the distance from the rear axle of the trailer and the front axle of the tractor. Therefore minimum distances is assumed as 5 m(KICT 1991).

Simulations are performed to compare the DLFs (dynamic load factors) from displacement response and flexural moment response. Fig. 2 shows the results for Bridge "B". It shows little

difference between the DLFs calculated from displacement response and flexural moment response. Therefore, DLFs from displacement comparison are used throughout the study.

4 RESULTS

4.1 truck weight

To find the effect of the truck weight, simulations are performed for Bridge "B" with two trucks having the headway distance of 5 m. For truck weight ranging from 18.1 ton (40 kips) to 54.4 ton (120 kips), the DLFs decrease as the weight increases as shown in Fig. 3. This result is the same as in the simple span case (Hwang 1990).

4.2 truck speed

Fig. 4 shows the effect of the truck speed on the DLFs. For Bridge "A" and "B" with two trucks having the total weight of 36.3 ton and the headway distance of 5 m, the result shows the highest values for the speed of 20 m/sec. In the figure, the zero headway distance means one truck on the bridge.

4.3 headway distance

For each bridge, headway distances from 5 to 100 m are considered. DLFs are defined as maximum dynamic response for a particular headway distance divided by maximum static response from all headway distance cases.

The DLFs of 50 m span from three bridges are shown in Fig. 5 to 6. They show the highest values for headway distance of 5 m. Maximum DLFs for headway distance of 5 m from three bridges are shown in Fig. 7. Their average values show little difference even though values from continuous span bridges show some scattered values.

4.4 number of trucks

For Bridge "B", one to ten trucks with equal

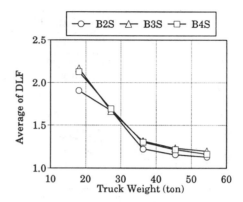

Fig. 3 : Effect of Truck Weight on DLF

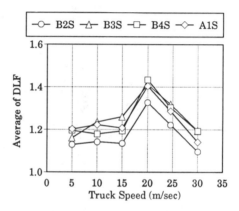

Fig. 4 : Effect of Truck Speed on DLF

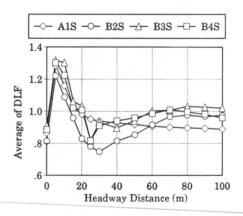

Fig. 5 : Effect of Headway Distance for Bridge "A" and "B"

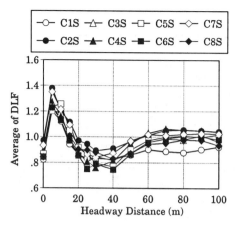

Fig. 6 : Effect of Headway Distance for Bridge "C"

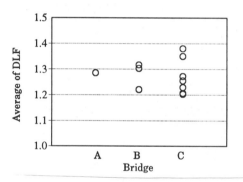

Fig. 7 : Maximum DLF for Bridge "A", "B", and "C"

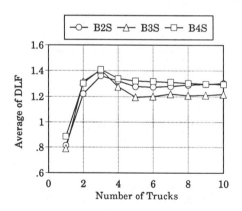

Fig. 8 : Effect of Number of Trucks on DLF for Bridge "B"

headway distances of 5 m are simulated and the results are shown in Fig. 8. It shows almost same values for two or more trucks and highest values for three trucks (Three trucks give the largest static values).

5. CONCLUSIONS

From the results of simulations, following conclusions may be stated:

1. DLFs calculated from displacement and flexural moment response are almost same,

2. DLFs decreases as the truck weight increases,

3. DLFs are high at the truck speed of 20 m/sec,

4. DLFs are largest for headway distance of 5 m and almost same regardless of simple or continuous span,

5. for 50 m span, two or more truck cases give almost same dynamic load factors.

Also, it may be concluded that maximum number of trucks on a span with minimum headway distances should be considered to find the DLF for the bridge. Further research should be carried out for negative moment and other span length.

REFERENCES

Hwang, E.-S. 1990. *Dynamic Loads for Girder Bridges.* Ph.D. Thesis, The University of Michigan, Ann Arbor, Michigan.

Fancher, P.S. and et. al. 1980. Measurement and Representation of the Mechanical Properties of Truck Leaf Springs: *Society of Automotive Engineers Technical Report 800905,* Warrendale, Pennsylvania.

KICT 1991. *Probabilistic Analysis of Bridge Design Loads:* Final Report, Korea Institute of Construction Technology, Seoul, Korea.

Structural Safety & Reliability, Schuëller, Shinozuka & Yao (eds) © 1994 Balkema, Rotterdam, ISBN 90 5410 357 4

Load and resistance index design: A new probabilistic format

J.W. Murzewski
Politechnika Krakowska, Kraków, Poland

ABSTRACT: Three coordinates define the state of a structural element: its resistance, load effect and model uncertainty. Their design values should not be exceeded with a specified hazard ratio and not necessarily - specified probability. Modal values are characteristic for the coordinates of state. The design values are proportional. A permanent load effect is asymptotically the Gauss-normal variable and a variable action maximum in a reference time is the Gumbel random variable. The resistance of a structural element and a professional error are log-normal. Applied loads are ordered according to the Ferry-Borges and Castanheta's rule and the load combinations are defined according to an extended probabilistic model. The hazard ratio and coefficients of variation are calibrated so that design results will be the same in simple cases as they are according to the Eurocode partial factor design.

1 RELIABILITY MEASURES

Random load effect S, resistance R and model uncertainty M are the coordinates in three-dimensional space of structural states. The overall probability of failure p_f depends on efficiency φ of quality control

$$p_f = (1-\varphi)p_{f0} + \varphi p_{f1} \qquad (1)$$

where

$p_{f0} = \text{Prob}(S<R/M)$, $p_{f1} = \text{Prob}(S<S_d, R>R_d, M<M_d)$,
S_d, R_d, M_d - limit values subject to control.

The conditional probability $1-p_{f0}$ has been indicated as the best measure of reliability and a conventional reliability index β has been defined as the basis of codified design

$$\beta = -\Phi^{-1}(p_{f0}) \,(\text{CEN}) \ \text{or} \ \beta = \frac{-\ln(\breve{S}\breve{M}/\breve{R})}{\sqrt{v_S^2 + v_R^2 + v_M^2}} \,(\text{ASCI}) \,(2)$$

where $\Phi(.)$ - normal probability function,
$\breve{S}, \breve{R}, \breve{M}$ - median values of random variables,
v_S, v_R, v_M - log. coefficients of variation.

The two definitions (2) are equivalent if the explicit coordinates of state S,R,M are independent and log-normal.

It has been also believed that the best design limit point $\mathbf{X_d}(S_d, R_d, M_d)$ is such that the probability density p_{f0} attains maximum on the plane of limit states $g(S,R,M)=0$.

However it has been found that it is impossible for practical reasons to keep the β index constant. Split indices have been proposed by the Eurocode (CEN, 1993):

$$\beta_S = 0,7\beta, \ \beta_R = 0,8\beta \ \text{and no explicit M} \qquad (3)$$

with $\beta=3,8$ for buildings and common works

and by the LRFD specification (ASCI, 1986):

$$\beta_S = \beta_R = \beta_M = \beta/\sqrt{3} \approx 0,55\beta \qquad (4)$$

with differentiated $\beta=2,5$, 3,0 or 3,7, 4,5 - for some load combinations and design cases.

There are no problems of β-value setting nor β-splitting in the new probabilistic design format (Murzewski,1974;1989). The overall probability (1) is taken under consideration and its minimum value is required

$$p_d(S_d, R_d, M_d) = \min \qquad (5)$$

with the limit state condition $S_d = R_d/M_d$.

The optimal solution is independent from the control efficiency φ. It satisfies equations

$$S_d h(S_d) = R_d h(R_d) = M_d h(M_d) = \text{const} \qquad (6)$$

where $h(X) = [d[\ln P(X)]/dX]$ - hazard function,
$P(S_d) = \text{Prob}(S<S_d) = F(S_d)$ cumulative
$P(R_d) = \text{Prob}(R>R_d) = 1-F(R_d)$ probability
$P(M_d) = \text{Prob}(M<M_d) = F(M_d)$ functions.

The constant (6) is called hazard ratio. Its value $1/k=1/1,8$ gives the same results in simple cases of design as the Eurocode rules do. Optimal values of the component indices β_S,β_R,β_M may be found after "normalization" of the probabilities $P(X)$ (Fig.1). The joint index β is not constant. It is linear combination of component indices

$$\beta = \alpha_S\beta_S + \alpha_R\beta_R + \alpha_M\beta_M \qquad (7)$$

where $\alpha_S,\alpha_R,\alpha_M$ - the sensitivity factors.

2 LOAD EFFECT INDEX

An internal force or moment $S_j=N,V,M...$ is a linear combination of independent random loads and other actions Q_i, $i=0, 1, 2.. n$, unless a nonlinear global analysis is used,

$$S_j = \sum_i a_{ij}Q_i . \qquad (8)$$

An equivalent load effect S has to be linearized for probabilistic design unless an interaction formula is given already by codemakers in a linear form

$$S = \sum_j b_jS_j . \qquad (9)$$

Decorrelation of the load effect S may be done by rearrangement of the components

$$S = \sum_j b_j \sum_i a_{ij}Q_i = \sum_i c_iQ_i \qquad (10)$$

where $c_i = \sum_j a_{ij}b_j$.

It is a simpler way than application of the Pugachev-Rosenblatt's theorem about principal directions in the space of basic variables.

The permanent loads G_l, $l=1,2..m$, may have any probability distributions and their effect $c_0G = \sum c_lG_l$ will have the Gauss-normal asymptotic distribution,

$$F(G) \rightarrow \Phi\left[(G - \overline{G})/\sigma_G\right] \quad \text{for} \quad m \rightarrow \infty. \qquad (11)$$

The modal values $\tilde{G}_l = \overline{G}_l$ are identified with the characteristic values specified for various materials.

The point-in-time probability distribution of a variable action Q_i' has not necessarily to be defined. The maximum Q_i of a sequence of r_i independent repetitions of the Q_i' actions has an asymptotic extreme value distribution. The Gumbel one is supposed

$$F(Q_i) \rightarrow \exp\left\{-\exp\left[-(Q_i - \tilde{Q}_i)/u_Q\right]\right\}\text{for } r_i \rightarrow \infty. \quad (12)$$

The modal values $\tilde{Q}_i$, $i=1,2.. n$, are identified with the characteristic values Q_{ik} specified for a reference period t_r. There is $t_r=50$ years for buildings (CEN, 1993).

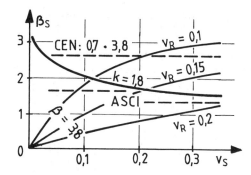

Fig.1 Load effect index versus the c.o.v.

A random value Q_i' remains constant during an elementary period $\theta_i = t_r/r_i$. If there are more simultaneous loads, $Q_1,Q_n,.. Q_n$, their order must be such that $r_1 \le r \le.. r_n$. It is the essential rule of the Ferry-Borges and Castanheta's combination model. The combination values Q_{ic} are less than the characteristic values except a dominant load Q_{kc}

$$Q_{ic} = \psi_{ic}Q_{ik} \qquad (13)$$

where $\psi_{ic} < 1$ if $i \neq k$ and $\psi_{kc} = 1$.

The Ferry-Borges and Castanheta's model is developed now. The combinations $c=1,2..2^{n-1}$ are ordered in a well defined sequence and combination factors ψ_{ic} are derived from the Gumbel distribution properties

$$\psi_{ic} = 1 - \left(u_Q/\tilde{Q}_{ik}\right)\ln r_{ic} \qquad (14)$$

with relation repetition numbers !
$r_{ic}=1$ for the dominant load, $c=2^i(j-1/2)$,
$r_{ic}=r_{i-1},r_{i-2}.. r_1$ for $2^j(j-1)<c<2^j(j-1/2)$,
$r_{ic}=r_i$ for $2^i(j-1/2)<c<2^ij+1$ and $j=1,2...$

The combination point X_k and the design point X_d are colinear

$$G_d = \gamma_SG_k \text{ and } Q_{id} = \gamma_SQ_{ik}, \text{ i=1,2..n.} \qquad (15)$$

This is a new uniqueness theorem which helps to select a linearization point on a limit load hypersurface or a relevant segment of a piece-wise linear interaction formula (2). It may replace cumbersome calculations of the β-point which should satisfy the Hasofer-Lind's theorem.

The calculations are as follows.

$$\overline{S} = c_0\overline{G} + \sum c_i\psi_{ic}\overline{Q}_i - \text{mean value,}$$

$$\sigma_S^2 = c_0^2\sigma_G^2 + \sum c_i^2\sigma_Q^2 - \text{variance, } v_S = \sigma_S/\overline{S},$$

$$\left(1/v_S + \beta_S\right)\phi(\beta_S) = 1/k \quad \rightarrow \quad \beta_S \qquad (16)$$

where $\phi(.)=\varphi(.)/\Phi(.)$ - Mills'function;
$S_d = \overline{S} + \beta_S\sigma_S$ - the design value.

The moments $\bar{Q}_i$, σ_Q^2 of the Gumbel distribution (12) are

$$\bar{Q}_i = \tilde{Q}_i + 0,45\sigma_Q \ , \qquad \sigma_Q^2 = u_Q^2 \, \pi^2/6 \ . \tag{17}$$

Normalization of the Gumbel distribution by means of collocation of the probability curves at the design point (Murzewski, 1989) gives more exact results in meaning of the first type definition (2) of the component safety index β_S.

3 RESISTANCE INDEX OF A STRUCTURAL ELEMENT

Non-negative mechanical properties f_y, f_u, E and section moduli A, W etc. are characterized by the log-normal probability distributions. The plastic resistance of a steel member under axial force N_S has also the log-normal distribution with parameters

$$\tilde{N}_{pl} = \tilde{f}_y \tilde{A} \ - \ \text{median value}, \tag{18}$$

$$v_{pl} = \sqrt{v_f^2 + v_A^2} \ - \ \text{log. coefficient of variation}$$

and the design value for a specified k

$$N_{pl,d} = \tilde{N}_{pl} \exp\left(-\beta_R v_{pl}\right) = \tilde{N}_{pl}/\gamma_R \tag{19}$$

where $\beta_R = \phi^{-1}\left(v_R/k\right)$ - resistance index.

E.g. $N_{pl,d}=23,5 \ \tilde{A}$ [kN] and $\gamma=1,21$ if k=1,8, $\tilde{f}_y=285$ MPa, $v_{pl}=0,10$ and $\tilde{A}$ in cm^2.

A compression member of a high slenderness $\tilde{\lambda} > 3$ subject to elastic buckling has resistance parameters

$$\tilde{N}_{cr} = \pi^2 \, \tilde{E}\tilde{I}/l^2 = \tilde{f}_y \, A/\tilde{\lambda}^2 \, , \ v_{cr} = \sqrt{v_E^2 + v_I^2 + 4v_l^2} \tag{20}$$

$$\tilde{\lambda} = \left(\lambda/\pi\right)\sqrt{\tilde{E}/\tilde{f}_y} \ - \ \text{relative slenderness ratio.}$$

The elastic-plastic buckling resistance N_R may be modelled as a minimum of two independent variables in the most frequent case of moderate slenderness $\tilde{\lambda} < 3$,

$$N_R = \min(N_{pl}, N_{cr}) \ . \tag{21}$$

The log-normal probability distributions,

$$\Phi\left(\ln \sqrt[v]{N_{pl}/\tilde{N}_{pl}}\right), \quad \Phi\left(\sqrt[v]{N_{cr}/\tilde{N}_{cr}}\right) \tag{22}$$

for $v = v_{pl} = v_{cr}$, are replaced by the Weibull distributions,

$$1 - \exp\left(-\sqrt[v]{N_{pl}/\tilde{N}_{pl}}\right), \quad 1 - \exp\left(-\sqrt[v]{N_{cr}/\tilde{N}_{cr}}\right) \tag{23}$$

and a simple relation is derived:

$$\left(1/\check{N}_R\right)^v = \left(1/\check{N}_{pl}\right)^v + \left(1/\check{N}_c\right)^v \tag{24}$$

where $\check{N} = \tilde{N}\exp(0,45v)$, $\upsilon = v\sqrt{6}/\pi$.

The method of moments is applied again and the design value of buckling resistance is

$$N_{Rd} = \check{N}_R \exp\left(-\beta_R v\right)\Big/\left(1 + \sqrt[v]{\tilde{\lambda}^2}\right)^\upsilon , \tag{25}$$

E.g. $\gamma_R = N_{Rd}/\check{N}_R = 1,30$ if $v = 0,10$, k = 1,8.

The same procedure may be applied to other multiple mode of failure cases: yielding and/or fracture of tension members with holes, low-cycle fatigue failures etc.

4 MODEL UNCERTAINTY INDEX

A ratio M of experimental and theoretical values of carrying capacity defines the random model factor. The log-normal distribution and the bias factor $\check{M}=1$ are assumed in a general case. The logarithmic coefficients of variation are estimated as follows:

$v_M=0,04$ for steel members in simple cases,
$v_M=0,12$ for connections. (26)

Relative model uncertainty index for the middle safety class is

$\beta_M=\phi^{-1}(0,04/1,8)=2,4$ for simple cases,
$\beta_M=\phi^{-1}(0,12/1,8)=1,9$ for connections (27)

and the model factor :

$\gamma_M=\exp(2,4\cdot0,04)=1,10$ for simple cases,
$\gamma_M=\exp(1,9\cdot0,12)=1,25$ for connections. (28)

The same values are recommended by the Eurocode 3 (CEN, 1992) but they are called material factors γ_M.

Variations of buckling resistance N_R are often much more than 10%. It is due to residual stresses and other imperfections. Their influence depends on shape and way of fabrication and it is different for compression members of different slenderness. The maximum variations, up to 30%, have been observed for middle slenderness $\tilde{\lambda} \approx 1$. An interpolation formula is proposed (Murzewski,1989) for any ratio $\tilde{\lambda}$

$$v_b = \sqrt{0,04^2 + 2\Delta v^2 \, \tilde{\lambda}^2\Big/\left(1 + \tilde{\lambda}^2\right)^2} \tag{29}$$

with $\Delta v = 0,12, \ 0,17, \ 0,22, \ 0,30$ for the European buckling curves a,b,c,d.

The "erosion" of buckling resistance is taken into account by both reduced median value (25) and increased coefficient of variation (29) when perfectly plastic and perfectly elastic resistances are equal or close.

5 CONCLUSIONS

Minimum probability of failure is required by the new design format (Murzewski, 1989) instead of maximum probability density that has been the world-wide accepted basis of probabilistic design (Madsen, Krenk and Lind, 1986; Thoft-Christensen and Murotsu, 1986, etc.).

New γ-points on the limit states surface are different from the conventional β-points in the space of random states. A new design point can be easily found thanks to colinearity rule which gives always unique results. It is in agreement with proportional loading rule supposed by designers in limit analysis and stability problems. No standardization of random variables and no iterative calculation of the design point are necessary. That is why the new design format is user-friendly.

New design values of actions do not depend on resistance of structural elements and the strength of a structural material does not depend on actions which may be applied. No artificial splitting of the β index is necessary in order to avoid such unacceptable coupling. Now the component indices come directly from the probability optimization condition: β_S - relative to load effect and β_R - relative to resistance.

A probabilistic definition of the third component index β_M enables to determine the model uncertainty factor γ_M not only from engineer's judgement but also from statistical tests.

New design values S_d, R_d depend on the coefficients of variation: v_S of load effect and v_R - of resistance of the structural element, respectively. The factors γ_S, γ_R increase with v_S, v_R not so immoderately as they do when the β=const theory is used to calibrate the design values (Fig.2).

The component indices β_S, β_R, β_M and component factors γ_S, γ_R, γ_M must not be mistaken with partial indices β_i and partial factors γ_i, i=1,2..n, of the semiprobabilistic methods (AISC,1986;CEN,1993). The former are associated with structural elements and the latter are associated with particular loads and materials. Resistance factors γ_R can be determined from ultimate loading tests and compared with anticipated values. It is not available when the partial factor analysis is applied.

Now, the codified design methods take into consideration random variations within a type of structures nor they do not consider building size effects and autocorrelations among structural elements. These points may be introduced to future developments (Murzewski, 1982).

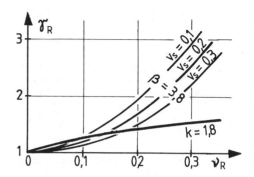

Fig.2 Resistance factor γ_R versus c.o.v. v_R

ACKNOWLEDGEMENT

The ongoing research on new rules of load combination and calibration of probabilistic parameters has been supported by the Polish Committee for Scientific Research under grant No 7.1035.91.01 (PB 619/7/91). Pilot design projects have been accomplished according to the new probabilistic method.

REFERENCES

AISC,1986. *Load and resistance factor design specification for steel building.* American Institute of Steel Construction, Chicago, ILL.

CEN,1992. *Eurocode 3: Design of steel structures. Part 1.1: General rules and rules for buildings.* European prestandard: ENV 1993-1-1.

CEN,1993. *Eurocode 1: Basis of design and actions on structures. Part 1: Basis of design,* CEN/TC250 Project team sixth draft of ENV 1991-1.

Ferry-Borges,J. and Castanheta,M. 1971. *Structural safety.* 3-rd edition. Lisbon: Laboratorio Nacional de Engenharia Civil.

Madsen,H.O., Krenk,S. and Lind,N.C. 1986. *Methods of structural safety.* Englewood Cliffs, NJ: Prentice-Hall Inc.

Murzewski,J. 1974. *Sicherheit der Baukonstruktionen,* Berlin: Verlag Bauwesen. (Translated from the Polish edition 1970, Warszawa: Arkady).

Murzewski,J. 1982. Discrete models of structural systems. Euromech 155: *Reliability theory of structural engineering systems.* DIALOG 6-82: 129-142.

Murzewski,J.1989. *Niezawodność konstrukcji inżynierskich,* Warszawa: Arkady.

Thoft-Christensen,P. and Murotsu,Y. 1986. *Application of structural systems reliability theory.* Berlin etc:Springer-Verlag.

Structural Safety & Reliability, Schuëller, Shinozuka & Yao (eds) © 1994 Balkema, Rotterdam, ISBN 90 5410 357 4

Modelling of crane loads and their combinations

H. Pasternak
Technical University of Braunschweig, Germany

B. Rozmarynowski
Technical University of Gdańsk, Poland

Y.-K. Wen
University of Illinois at Urbana-Champaign, Ill., USA

ABSTRACT: A random model for the vertical load on columns of single story steel buildings due to one or two moving cranes was developed. It was assumed that two cranes can move either on the same crane track or in neigbouring bays. Based on the defined pulse shapes that occur according to Poisson model, the mean upcrossing rate function and the probability distribution of the maximum value of a combined process were considered. For combined effect of loads only the occurrence dependence was included. Dependences between intensities and durations within each load and among loads were ignored in this study. Charakteristic values, patrial safety factors and combination factors were estimated on the basis of extensive numerical studies. The point crossing method and the load coincidence method with including "clustering" effects were utilized for developing numerical results.

1. INTRODUCTION

Several load models have been developed for needs in engineering structural systems, e.g. for wind, snow, earthquake etc. Up to now, there is no comprehensive model for loads caused by overhead travelling cranes. Only a few papers have been devoted to this subject giving individual results on the basis of statistical investigations.

For bridge cranes, the deterministic nominal wheel loads have been introduced as characteristic values. Moreover, the partial safety factor has been assumed a value a priori (i.e. without any probabilistic reasoning). According to Kikin 1966 this coefficient is about 1.2 to 1.4. It depends mainly on service intensity and unevenness of the runway. Following the new German DIN 18800/1 (1990) and the draft of EUROCODE 3 (1990) codes, forces caused by two cranes are to be treated as one action, consequently, no combination factors have been provided. In contrast to this, the Russian code (SNiP II-6-74 (1974)) contains a combination factor for the case of two cranes.

Depending on the load intensity, ψ is either 0.90 or 0.95. Belenja and Vasiljev 1976 present the combination factor ψ as a function of the span of the craneway girder.

2. PROBLEM STATEMENT

Considering the craneway as a system of simply supported beams, the vertical reaction R due to a moving crane can be described by a trapezoid pulse shape function as shown in Fig. 1.

The dashed line in Fig. 1 shows the actual influence line due to the superposition of two moving forces, whereas the solid line is a simplification of the actual one in order to facilitate subsequent analysis. The vertical forces F_1, F_2 acting on the column due to loading in different bays (Fig. 2) may be obtained from the following formulae

$$F_1 = \frac{Q_{KB}}{4} + \frac{Q + Q_K}{2}\left(\frac{L-y}{L}\right)$$

$$F_2 = \frac{Q_{KB}}{4} + \frac{Q + Q_K}{2}\left(\frac{y}{L}\right) \tag{1}$$

where F_1 and F_2 in Eq.1 are valid for bays L

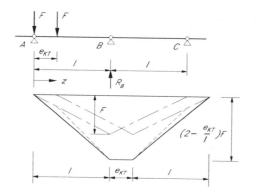

Fig.1   Influence lines and the pulse approximation

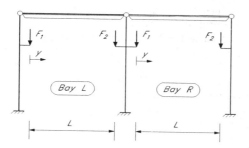

Fig.2   Forces acting on the column

and R respectively; Q_{KB}=the weight of the crane bridge; Q=the lifted load; Q_K=the weight of the trolley; L=the span of the crane and y=the variable describing the trolley position. The extreme value of the pulse can be obtained from the following relationship (Fig.1))

$$S_i = \left(2 - \frac{e_{KT}}{l}\right)F_i, \quad i = 1,2 \qquad (2)$$

in which e_{KT} is the wheel base and l denotes the girder span; F_i is given by Eq.1. Two random variables X_1^* and X_2^* are used to describe the uncertainty of forces given by Eq.1, i.e. X_1^*=Q (lifted load) and X_2^*=y (trolley position).

The probability density function (PDF) of each random variable is assumed to be known. For the first variable, the PDF is defined on the basis of typical empirical frequency diagrams (Kikin 1966), whereas the second random vari-

able is described by PDF of the trolley position which is proposed to be of two types: uniform or triangle. The condition of statistical independence between X_1^* and X_2^* is assumed to be fulfilled. In the analysis on load combination there are two functions: the mean upcrossing rate and the arbitrary-point-in- time (APIT) that are of special relevance (Larrabee and Cornell 1981).

The problem of vertical forces due to two cranes (or more) working separately or together on the same (or not) crane track is reduced to that of finding the mean upcrossing rate function $v_R^+(r)$ ("point crossing" method) and CDF, i.e. $F_{R_m}(r,T)$ (load coincidence method with use "clustering" effects) in which R_m is the maximum value of combined load effects in time T, i.e.

$$R(t) = C_1 X_1(t) + C_2 X_2(t), \quad t \in (0,T) \qquad (3)$$

X_1 and X_2 denote the vertical loads acting on a column caused by cranes, C_1 and C_2 are constant load effect coefficients. The mentioned methods are described in details by Wen 1990 and Larrabee and Cornell 1981. Process with "clustering" effects can be understood as a kind of correlated processes (Cox and Lewis 1972). Loading may be clustered around a common point in time that there is a much higher chance of coincidence. In case of crane loads this point may be given by the manufacturing cycle in the building. One of the most convenient ways for load combination analysis including the occurrence correlation among loads is through the use of conditional occurrence rate (COR) (Wen 1990). Two possible occurrences of moving cranes with the minimum distance between them are considered. Assuming the Poisson type of upcrossings of the maximum value of load effects over a time period T one can get an approximation of the probability distribution i.e. P[R < r in (0,T)]=F(r,T), where r is a level of load effects.

A load combination factor may be given as

$$\psi = \frac{Q_{R_m}}{Q_{X_1} + Q_{X_2}} \qquad (4)$$

where all magnitudes denote the 99% fractiles

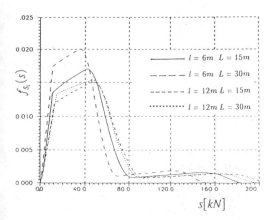

Fig.3   PDFs of the pulse intensity

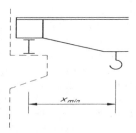

Fig.4   Definition of the x_{min} value

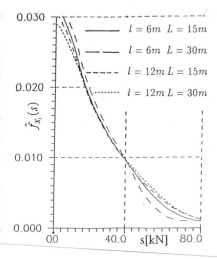

Fig.5   APIT distribution functions $\hat{f}_{X_i}(s)$

of the maximum value and two single load effects respectively.

3. NUMERICAL RESULTS

In Fig.3 the PDFs of the pulse extreme value S for one chosen frequency distribution (Scheer 1991) of the lifted load using triangular PDF of the trolley position and crane data from Tab.1 and Fig.4 are depicted.

The PDFs are presented for four cases of the girder and crane spans denoted by l and L respectively. Fig.5 shows the appropriate APIT functions.

The maximum of crane load combination effect increases with the period of time. This can be seen from Fig.6.

For sufficiently long periods the maximum values during these periods, however, are no longer statistically independent. Because of this one should not extrapolate to periods beyond 1000 - 3000 crane passages (Kikin 1966).

4. CONCLUSIONS

On the basis of extensive numerical analysis of combined load effects over a time period for two types of pulse shapes (the trapezoid and its rectangular approximation) the following conclusions can be drawn:

- In the meaning of GRUSIBAU 1981 the 99% fractiles are treated as characteristic values of crane loads. According to the lifted load and geometrical relations these fractiles result from 80% to 92% of the maximum vertical crane forces if the uniform PDF of the triolley position is used and the 85% to 94% if the triangular PDF is used. The values from 1.09 to 1.25 (for the uniform PDF) and from 1.06 to 1.18 (for the triangular PDF) give partial coefficients smaller than that suggested by DIN 18800/1 and Eurocode 3, i.e. $\gamma_f = 1.5$.

- The 99% fractile values of the vertical crane forces are more dependent on the distribution of the lifted load than the girder and crane spans.

- For many practical cases, especially for: (a) the independent operation of cranes (without

Table 1. Significant crane's data

Characteri-stic values	L=15m	L=30m
Q [kN]	196.2	196.2
Q_{KB} [kN]	49.776	156.96
Q_K [kN]	15.89	19.23
e_{KT} [m]	2.5	4.56
x_{min} [m]	0.95	0.95

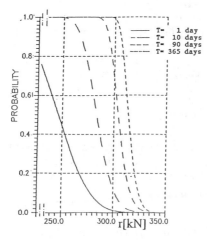

Fig.6   Influence of the period of time T on the Probability $P[R_m(t) > r, \ t \in (0,T)]$

"clustering"), (b) the rare occurrences (= 10% - 50% of the time) of the load process with "clustering" effects, (c) larger crane and girder spans - the combination factor ψ is smaller than 0.9.
• The dependency of the girder span on the factor ψ is weaker than suggested by Belenja 1976.

A comprehensive description of the investigations is given by Scheer and Pasternak and Rozmarynowski 1991.

Acknowledgement

The authors wish to thank the German Research Society (DFG) for financial support of the analysis.

5. REFERENCES

Belenija, E.E., Vasiljev, A. (1976). Osobiennosti dejstvitielnoj raboty podkranovych konstrukcyj. *Stavebnicky Casopis*, 24-886.

Cox, D.R., Lewis, P.A.W. (1972). Multivariate point processes. *Proc. 6th Berkeley Symposium on Mathematical Statistics*, Vol. 3, 401-448.

Grundlagen zur Festlegung von Sicherheitsanforderungen für bauliche Anlagen GRU-SIBAU (1981). Berlin, Beuth-Verlag.

Kikin, A.I. et al. (1966).*Povysenije dolgovecnosti metalliceskich konstrukcyj promyslennych zdanij*. Moskva, Izd. po Stroit.

Larrabee, R.D., Cornell, C.A. (1981). Combination of various load processes. *J. Struct. Div. ASCE*, 107(ST1), 223-239.

Scheer J., Pasternak H., Rozmarynowski B. (1991).*Modellierung von Kranlasten und deren Kombinationen*. Bericht 6201 des Institut für Stahlbau der TU Braunschweig.

Wen, Y.K. (1990). *Structural load modeling and combination for performance and safety evaluation*. Amsterdam, Elsevier.

Structural Safety & Reliability, Schuëller, Shinozuka & Yao (eds) © 1994 Balkema, Rotterdam, ISBN 90 5410 357 4

A stochastic model for human induced rhythmic loads

C. Rebelo
Civil Engineering Department, University of Coimbra, Portugal

R. J. Scherer
Institut für Massivbau und Baustofftechnologie, University of Karlsruhe, Germany

ABSTRACT: The load model presented in this paper is based on assumed functions for the auto-correlation and for the space correlation of the load field, which is considered homogeneous. The envelope of the loads' auto-correlation for individuals is modeled by a bilinear function defined by the total variance of the time series, the auto-correlation length and the deterministic part of the quasi-periodic dynamic loads. The modulus of the complex valued space correlation is related to the randomness of the envelope of the load time series and to the time-dependent fluctuation of the basis-frequency. The phase angle is related to the time lag between persons. To identify the model parameters measurements were carried out on a load platform and on one floor during real loading.

1 Introduction

The dynamic forces arising from rhythmic activities such as dancing or aerobics are characterized by narrow peaks in the frequency domain. For structures having natural frequencies near these peaks resonance phenomena must be checked.

To estimate the reliability of those structures information about the loads must be available and, therefore, the following topics have to be investigated:

1. the zone of the spectrum where those peaks will occur

2. the concentration of energy depending on the type of human activity

3. the increase of spectral energy with an increasing number of participants

Concerning the first and second topic some investigations [1, 2] are already known. The third topic, however, can only be clarified by developing a stochastic model for the load field [3, 4], as undertaken in this work.

The model is developed on the basis of individual load histories measured on a force platform and on the basis of the structural response of a dancing hall measured during two festivities.

2 Load model

The proposed load model reflects the following characteristics of these loads:

- they are almost periodic, showing, in the frequency domain, several narrow peaks at the harmonic frequencies, which are established by any rhythmic effect such as music.

- The whole energy a person can develop by a certain rhythmic movement can be mathematically represented by the integral of the variance of the load function also known as the signal energy. Because of the clear separation of the peaks in the spectrum this integral can be computed separately for each harmonic.

The ability of one person to maintain the rhythm, that is, to concentrate the energy around the harmonic frequencies, is given by the width of those peaks. This effect can be represented by a suitable function for the autocorrelation of the individual load function and must be a characteristic of each harmonic.

To take into account the interaction between individuals in a group, the space correlation of the load field has to be considered. The loads are

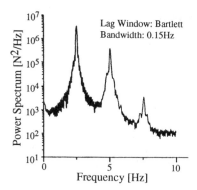

Figure 1: Power spectral density of the vertical loads produced by one person during rhythmic jumping.

$$a(\tau) = \begin{cases} \sigma^2[\eta_d + (1 - \eta_d)\frac{\Gamma - \tau}{\Gamma}] & \text{for } 0 \leq |\tau| \leq \Gamma \\ \sigma^2\eta_d & \text{for } |\tau| > \Gamma \end{cases}$$

$$(4)$$

where the index n is dropped for simplicity and σ^2 represents the part of the load variance, that is imputable to the harmonic n.

According to this model the envelope of the variance decreases linearly until $\tau = \Gamma$ and then remains constant at the level $\sigma^2\eta_d$. The parameter η_d represents, therefore, a deterministic portion in the load function, which is due to the rhythmic stimulus (for instance the music beat).

The Fourier transform of (4) smoothed by a Bartlett window is shown schematically in figure 2-b.

SPACE CORRELATION

The space correlation coefficient is defined, according to [5], in the same way as the cross-spectrum and is, therefore, a complex function of the frequency and of the distance between points in the load field:

$$\rho_n(\xi) = \frac{C(\xi, f_n)}{S(f_n)} \qquad (5)$$

Such a definition allows the identification of the time lag between persons as the phase of the complex valued coefficient ρ_n, as well as to relate its module to the randomness of the envelope of the load function and to the time-dependent fluctuation of the center frequency f_0.

The function for the module of the correlation coefficient must allow the existence of a deterministic portion, according to the time correlation model. The following function is assumed:

$$|\rho_n(\xi)| = k_d + (1 - k_d)e^{-k_a\xi} \qquad (6)$$

where k_d and k_a are the model parameters.

For the phase angle ϕ_n is assumed that it can be expressed independently of the distance between the persons in a group, so that:

$$\phi_n(\xi) = \begin{cases} 0 & \text{for } \xi = 0 \\ \Phi & \text{for } \xi \neq 0 \end{cases} \qquad (7)$$

3 Model identification

TIME CORRELATION MODEL

To estimate the model parameters, load functions from one individual at rhythmic jumping

therefore a stochastic function of time and space $P(x, y, t)$.

Under the presumption of stationarity in time and homogeneity and isotropy in space, the load's randomness is described by the cross spectrum

$$C(\xi, f) = R(\xi, f) - iQ(\xi, f) \qquad (1)$$

where ξ represents the distance between any two points on the load field.

TIME CORRELATION

The real valued power spectrum of the individual loads are represented by (1) for $\xi = 0$:

$$S(f) = C(0, f) = R(0, f) \qquad (2)$$

As pointed out before, this spectrum presents peaks at the harmonics of the base frequency, each of them having different height and width, and therefore different randomness. Fig.1 represents one typical spectrum of the individual's vertical load arising from rhythmic jumping at a frequency of 2.5 Hz.

When the time functions, corresponding to each of the peaks, are analyzed by bandpass filtering the original time series at the nth harmonic, the correlation function for this harmonic can be written in the form

$$R_n(\tau) = a_n(\tau)\cos(2n\pi f_0\tau) \qquad (3)$$

where f_0 represents the center frequency of the peak. For the envelope $a_n(\tau)$ the following model is proposed(s. Fig.2-a):

and bouncing were measured and the time correlation for each harmonic computed. The time series corresponding to each harmonic were built by filtering the measured load with a trapezoidal bandpass filter. The width of the filter was 1 Hz at the corner frequencies and 2Hz at the cut-off frequencies. Variations of these values have negligible influence on the results.

The values of the deterministic portion η_d for the first harmonic at the jumping frequency were very high (fig. 3) and decreased for higher harmonics as expected. This confirms the idea of different randomness for each harmonic, which is due to some random effects, such as the way the foot contacts the load platform. This becomes important when the load level decreases, as it happens for the secondary harmonics.

For a certain harmonic the tendency of η_d is to follow a concave curve with a maximum within the measured values, at a base frequency of about 2Hz.

The parameter Γ was about the same for both rhythmic movements and lay between 2 and 8 seconds for the 20 seconds time series.

The values of the variance σ^2 within a certain harmonic follows approximately the form of the curves for η_d. They are dependent on the harmonic under consideration and, otherwise than for η_d, strongly dependent on the type of human movement.

SPACE CORRELATION MODEL

The space correlation parameters were estimated by measuring the response acceleration at some points of a plate during a gala ball [3, 4]. The phase and coherency between the responses at the measurement points were computed and were plotted versus the distance between those points just like a correlation function.

A finite element model of the plate was then used to compute the responses of the plate under the load model. The parameters were varied until the measured cross correlations (coherencies and phases) could be approached (fig.4).

This could be done in this way because the boundaries of the loaded zone were well known. Despite of that, the position of the loads within those boundaries were slightly varied and, for each one, the curve built up by means of the spectral values at the considered harmonic using full spatial correlation was compared with the one built up with the measured values. It was noted

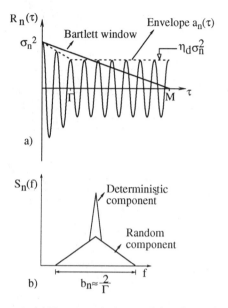

Figure 2: (a)Time correlation model and time lag window. (b)Schematic representation in the frequency domain.

that those small variations were not significant for the results of the model identification, which can be resumed as follows:

- The values of the parameter k_a were, in general, high, showing a short correlation length of the random portion corresponding to $(1 - k_d)$.

- The parameter k_d depends upon the length of the time series considered, just like the correlation in time. For small lengths (about 10-20 seconds) of an entire measurement (about 130 sec.), the assumption of homogeneity for the load field seems not to be adequate because the expected spectral values do not build up over the entire field. This could only be reached for time series of about 40 seconds, for which the values of k_d were approximately the same as those of η_d (table 1).

- The identification of the phase angle was, in general, more difficult to estimate and less precise than that of k_d. Some of the results are resumed in table 1.

Table 1: Parameter values for dancing (40 sec. time series)

Frequency [Hz]		η_d	k_d	Φ
1st harmonic	1.7	0.78	—	—
	2.0	0.84	0.85	10°
2nd harmonic	3.4	0.78	0.80	10°
	4.0	0.40	0.40	20°

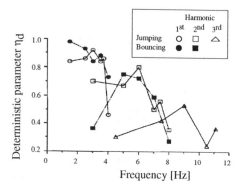

Figure 3: Values of η_d for jumping and bouncing (Time series=20 sec.)

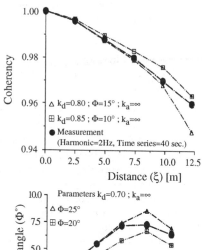

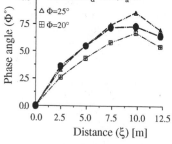

Figure 4: Parameter identification for space correlation. (a)Coherency. (b)Phase angle.

4 Conclusions

The principal advantage of the proposed model is in the physical meaning of the parameters, the most important being the deterministic portions of both correlations and the phase angle.

The deterministic portion η_d identified for each harmonic of the individual loads can be used also for the space correlation by making $k_d \approx \eta_d$. However, the model identification is dependent on the length of the time series used, that is, the shorter the time window used to build the correlation along the measurement the greater the maximum value of η_d obtained.

Concerning the space correlation the same problem can be expected, that is, for small groups of people greater maximum values of k_d may appear than for big groups, where the averaging effect through the structure will produce the same as the time average for η_d. In this work only big groups were investigated.

It was found that time series of about 40 seconds are suitable for big groups in dancing halls. The parameter k_d can be made equal to η_d, simplifying the determination of the space correlation. Values of about 0.85 and 0.40 for the first and second harmonic, respectively, were found. For the phase angle mean values of about 10° and 20° respectively can be used. The correlation length represented by k_a is negligible for practical uses.

References

[1] Allen,D.E., *et.al.* — Vibration criteria for assembly occupancies, Can. J. Civ. Eng.,12,617-623 (1985)

[2] Bachmann,H., Ammann,W. – Schwingungsprobleme bei Bauwerken, IABSE, Structural Engineering Documents,3d (1987)

[3] Rebelo,C. — Stochastische Modellierung menschenerzeugter Schwingungen, doctoral thesis at the University of Karlsruhe, Germany (1992).

[4] Rebelo,C., Scherer,R.J., Eibl,J. — Statistical modelling of dynamic loads imposed by occupancies, EURODYN'90, Bochum, R.F.G.(1990).

[5] Vanmarcke,E. — Random Fields, analysis and synthesis, MIT press (1984).

Structural Safety & Reliability, Schuëller, Shinozuka & Yao (eds) © 1994 Balkema, Rotterdam, ISBN 90 5410 357 4

Analysis on load combination rules

H. M. Sandi
INCERC Building Research Institute, Bucharest, Romania

ABSTRACT: Probabilistic analyses were performed in order to check the consistency of code provisions concerning load combination factors and to provide basic data for an improved safety control. Some analytical developments in this connection are followed by the presentation of a case study concerning the combination of earthquake and snow loading for single-storey industrial hall structures.

1. INTRODUCTION

Load combination rules are, as known, of direct importance for the safety level provided in design. Rules adopted in Romanian codes tend to provide very low design values for partial safety factors, while combination factors are conservative. Western codes, like Eurocodes or UBC, are better balanced from this viewpoint, yet there is still room for improvement there too.

According to current concepts, a satisfactory calibration of design parameters should rely on the use of appropriate probabilistic tools. Moreover, to cover the diversity of needs in this field, a third-level probabilistic approach (referred to as level P.3), that provides suficient generality of the field of application and flexibility as well, is suitable.

Level P.3 approaches to various structural safety problems were adopted in INCERC in the frame of several research projects. Parametric analyses like those of (Sandi et al 1986, Sandi et al 1987) were performed in this connection after 1980, and their outcome lay at the basis of improvement of design codes. The analysis presented in this paper, intended to contribute to a suitable calibration of load combination factors, was performed in 1991 (INCERC 1991), under the responsibility of the author.

2. CONCEPTUAL FRAMEWORK

For the case study presented, it is satisfactory to consider two DOF for the spatial distribution of loading, related to the axes (p) for gravity loads and (q) for horizontal loads respectively, with no apparent concern for the representation of the time-dependence of parameters referred to. The horizontal load is referred to as principal load, since it is critical for the risk affecting structures of the type dealt with, while the gravity load is referred to as secondary load, since it cannot lead to failure when applied isolated, and it can influence only to a limited extent the state of stress due to the combined application of the two kinds of loads.

The time dependence considered for the secondary load is continuous. The probability density $g(p)$ describing its distribution has in this case also the sense of time distribution characteristic. The time dependence of the principal load is idealized as a sequence of cases of instantaneous occurrence at various severity levels. Probabilistic independence of different cases of occurrence is assumed, such that the stochastic recurrence process is poissonian (macro level). Under these assumptions the main recurrence characteristic is the expected number of cases of occurrence at severity levels not lower than q, in a time interval of duration T, $N^{(h)}(q,T)$. This function has an expression

$$N^{(h)}(q,T) = T \int_q^\infty n^{(h)}(q') \, dq' \qquad (2.1)$$

where $n^{(h)}(q)$ is the density of the rate of cases of occurence at a severity level q (in a unit time interval).

The structural vulnerability is quantified by a system of probabilities $F_k(p,q)$, of exceedance of limit states L_k, conditional upon the couple of values

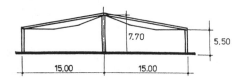

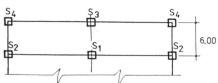

FIG. 1

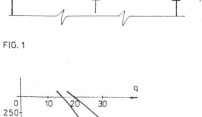

FIG. 2

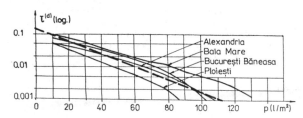

FIG. 3

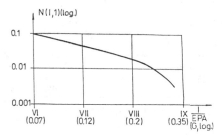

FIG. 4

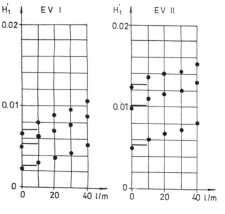

FIG 5

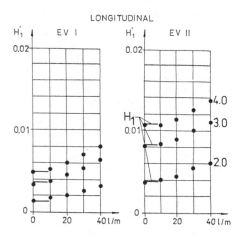

FIG. 6

1478

(p,q). This representation implicitly assumes that possible damage is due to the instantaneous state of loading, neglecting possible cumulative influences of the loading history.

The risk of failure is quantified under these assumptions by the expected number of cases of exceedance of a limit state L_k during a time interval of duration T, $N_k^{(r)}(T)$. In case one assumes that eventually damaged members are promptly and perfectly repaired after any case of damage occurrence, the sequence of cases of damage occurrence will be represented by a poissonian process too. The functions $N_k^{(r)}(T) = T H_{1k}$ will be determined on the basis of the expected (annual) rates of exceedance of limit states L_k, H_{1k},

$$H_{1k} = \int_0^\infty H'_{1k}(p) \, g(p) \, dp \qquad (2.2a)$$

$$H'_{1k}(p) = \int_0^\infty F_k(p,q) \, n^{(h)}(q) \, dq \qquad (2.2b)$$

The survival probabilities $H_{0k} = \exp(-T \cdot H_{1k})$ and the auxiliary safety characteristics $H_{2k} = -\ln H_{1k}$, which have a tendency of variation quite similar to that of classical safety factors, may be determined directly on this basis too.

3. CASE STUDY

The case study presented is related to a single-storey industrial structure, that is represented in fig. 1. The reinforced concrete columns are fixed at ground level and hinged at roof level. The permanent mass of the roof is of some 142 t (the distributed mass is of some 160 kg/m^2). There are four types of columns (front and current columns, for the lateral and central rows respectively).

A unique limit state, corresponding to the code definition of ultimate limit states, was considered. The analysis was conducted such, as to lead to results and comparisons expressed in terms of direct significance for code representation. (Axial force)-(bending moment) limit-state boundaries at the critical sections of the columns were converted to (gravity force)-(horizontal force) homologous boundaries, for the longitudinal and transverse directions of the structure respectively. The (axial force)-(bending moment) boundaries for individual columns, converted into homologous (gravity force)-(horizontal force) boundaries were combined by summing up the horizontal forces corresponding to the various columns. It was assumed in this connection that the columns have, if necessary, sufficient ductility in order to mobilize simultaneously their bearing capacities (expressed in terms of horizontal forces). The axial forces corresponding to various gravity loads, used in order to define the boundaries referred to, were determined assuming uniform plan distribution of the gravity loads (dead and meteorological both). The boundary referred to had to be determined only for a limited intensity interval of gravity loads, i.e. practically possible values. The resulting boundary (fig. 2) was fitted (for the domain of practical interest) to a standard analytical expression (a closed curve of the 4-th grade). It was assumed that the boundary is random, but its randomness is simple, depending upon a single factor of homothety with respect to the origin. A log-normal distribution with a low coefficient of variation (0.05) was assumed for this factor.

The mass concentrated at the upper end of columns consists of the mass of the roof, plus half of the mass of the columns, plus that of the snow. A log-normal distribution with low coefficient of variation (0.05) was assumed for the mass and the corresponding gravity load. For the mass of snow to be considered in the analysis, the results of statistical processing performed previously for several meteorological stations (INCERC 1988) were used. As an example, for the stations of Alexandria, Baia Mare, Bucharest-Baneasa and Ploiesti, the average fraction of time for which a certain water reserve of a snow layer is exceeded, $\tau^{(d)} = T^{(d)}(p) / T_{obs}$ ($T^{(d)}(p)$: average duration for which a value p is exceeded; T_{obs}: length of observation time interval), is presented in fig. 3. For the analysis performed an average relation $\lg \tau^{(d)}(p) = -0.8 - 0.02 \, p$, represented by a dotted line in fig. 3, was assumed.

The horizontal force was represented as a conventional static one. In agreement with the representation (2.1) and with the relations (2.2), the distribution of values q corresponded to extreme value laws (alternatively, of types I and II). The horizontal force variability was defined by the ratio q(50 years)/q(10 years), that varied up to 4.0 (according to Romanian data, this ratio ranges between 1.5 and 1.8 for wind pressure and between 3.5 and 4.0 for seismic acceleration. To illustrate this, a recurrence law for seismic intensity and for EPA (ATC 1978), that is realistic for Bucharest, is given in fig. 4. This corresponds (for the straight part) to a type II extreme value distribution. For the case of (conventional) seismic loading, which is dealt with here, an overall seismic coefficient $c_s = 0.08$ was adopted as a reference (the design spectrum is constant over a wide interval of oscillation periods, up to 1.5 s for Bucharest). Given the data of the Romanian code for earthquake resistant design (MLPAT 1992), the value referred to

corresponds to a return period of about 50 years.

The basic outcome of the parametric analysis was represented in terms of (discretized) values $H'_1(p_j)$ (2.2b). Besides that, the values H_1 (2.2a) were determined for the sums corresponding to the variation of p from 0 to p_j. The parameters H_2 and $H_0(T)$ (for T varying from 1 to 100 years) were determined too. To illustrate the outcome for the parameter H_1, the values $H'_1(p_j)$ (versus H_1, represented by a horizontal line) are represented in fig. 5 and fig. 6, for the two main directions respectively.

The results obtained can be summarized as follows:

1. The outcome is to some extent conventional, since conventional verification, as in usual design, was performed. Therefore, it is reasonable to look for results as expressed rather in relative, than in absolute, terms.

2. The risk of failure is strongly influenced by the random variability of the principal load (due to earthquake), expressed at its turn by the ratio q(50 years)/q(10 years), as well as by the type of extreme value distribution, referred to previously. The exceedance probabilities vary fivefold for the transverse direction and eightfold for the longitudinal one, as a function of assumptions accepted.

3. Totally neglecting the snow load has a very limited influence on exceedance probabilities as compared with the case of a rigorous approach. The deviation of probabilities is in the range of 5%, without exceeding a maximum of 10%. A rational snow design load in this combination should be less than 5% of the snow design load to be considered in checking the roof for gravity loads.

4. The Romanian regulations (as well as foreign regulations) are unduly conservative in prescribing load combination factors for snow, in cases of moderate climate and low altitude above sea level.

4. FINAL REMARKS

The use of a P.3 level probabilistic approach to structural safety problems, which is of interest when lower level approaches are no longer appropriate, represented a useful tool in this case. It is desirable to promote more frequent use of P.3 tools in this field.

The outcome of this study consisted of methodological developments, as well as of results of direct interest, expressed in qualitative and in quantitative terms.

Attempts to extend such an approach to cases of multi-DOF loading (variable distribution of silo padding in different cells, various positions of vehicles on bridges etc.) is in principle feasible, but would require a different approach, renouncing most likely at attempts of parametric analyses and using perhaps Monte-Carlo techniques.

Codes should be more open to such non-conventional analyses, allowing, besides conventional approaches, also more consistent ones, like the one presented. This would require flexibility and some freedom of choice in adoption of checking formats.

ACKNOWLEDGEMENTS

This study was developed in the frame of a contract with MLPAT (Ministry of Public Works of Romania). The author is indebted to his colleagues I. Borsaru (who contributed to the determination of (axial force)-(bending moment) boundaries) and I. Floricel (who performed the parametric analyses).

REFERENCES

Sandi, H., Floricel, I., Cazacu, D. 1986. Parametric analysis of risk and expected total cost for buildings. In Proc. 3-rd US Nat. Conf. on Earthquake Engineering, Charleston, N.C.

Sandi, H., Floricel, I. 1987. Parametric risk and cost-benefit analysis for structures subjected to snow loading. In Proc. ICASP-5, Vancouver.

INCERC (Building Research Institute, Bucharest) 1988. Analysis of meteorological actions (contract no. 509/88). Phase 1: Study on the long duration intensities of wind and snow actions (in Romanian).

INCERC (Building Research Institute, Bucharest) 1991. Probabilistic analysis of structural safety and durability (contract no. 137/90). Phase 3: Analysis of combination rules in case of variable secondary loading (in Romanian).

MLPAT (Romanian Ministry of Public Works and Land Use Planning) 1992. Code for the earthquake resistant design of buildings and industrial structures, P.100-92.

ATC 1978. Tentative provisions for the development of earthquake resistant regulations. Publ. ATC 3-06.

Structural Safety & Reliability, Schuëller, Shinozuka & Yao (eds) © 1994 Balkema, Rotterdam, ISBN 90 5410 357 4

System reliability under Poisson pulse loads

Lee Wang & Ying-Jun Chen
Civil Engineering Department, Northern Jiaotong University, Beijing, People's Republic of China

ABSTRACT: System reliability and load combination are two important aspects in structural safety analysis. When these two aspects are considered simultaneously, the situation becomes very complex. In this paper, a method based on the equivalent linear safety margin method and the existing load combination theory has been developed. The performance of the structure in time domain is treated by series system concept. To be compared the accuracies with the other methods, a numerical example on the plastic collapse of ductile frames is carried out, and it is found that the proposed method is simple and generally yields good results.

1. INTRODUCTION

Great progress has been made in recent years on the analysis of structural reliability, especially on the studies of load combination and system reliability.

However, the studies on load combination problem and system reliability have been quite isolated. Understandably, the theoretical problem involved in the time variant reliability of a structural system is extremely complex. The purpose of this study is to give a more systematic treatment of the problem based on the equivalent linear safety margin method and the existing load combination theory.

2. LOAD COMBINATION ANALYSIS IN SYSTEM RELIABILITY

2.1 Time domain analysis

We'll set up a series system model in time domain to handle the time variant property of the loads. Consider for exemple the case of two Poisson pulse processes (the principle is also

held in more than two load conditions). There are three possible load cases that may occur:

    Case 1: Only load 1 alone acting on the structure system;

    Case 2: Only load 2 alone;

    Case 3: Coincidence of load 1 and load 2.

Then the failure probability in given time period T can be estimated through a series system model(Fig.1). Each load case corresponds to a subsystem.

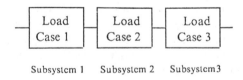

Subsystem 1   Subsystem 2   Subsystem 3

Fig.1 Subsystem concept in time domain

Let v_1, v_2, v_{12} are occurrence and coincidence rates of the loads in above cases (Appendix of Ref.[2]). In given time period T, the loads will independently occur v_iT times in case i. So, there will be v_iT elements in subsystem i (Fig.2), one element corresponds to one load

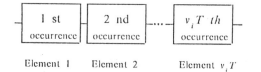

Fig.2   Element concept in time domain (Case i)

Table 1   Loading parameters

Variable	μ	σ	λ	μ_d	
M_1	360^{k-ft}	54^{k-ft}	---	---	
M_2	480^{k-ft}	72^{k-ft}	---	---	
S_1	100^k	10^k	5 / yr	10^{-2}	yr
S_2	50^k	15^k	0.2 / yr	10^{-2}	yr

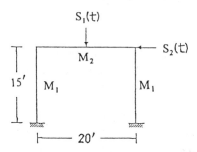

Fig.3   Simple frame under time varying loads

application. The subsystem is also a series system.

Since the resistance variables remain the same and the loads change independently from one application to the next, the elements are correlated between each other only through the resistances (Fig.2). And this kind of correlation is also held between the subsystems (Fig.1).

2.2 Mechanical system analysis

Consider a system with k potential mechanical failure modes. We must distinguish the mechanical failure modes from the elements in the time domain above. The different failure modes would have different performance functions

according to the mechanical properties of the structure. Each element may have several potential mechanical failure modes. On the basis of the load cases in Part 1 of this section, we can understand that the possible failure modes of the elements are part of that of the system.

The procedure and explanation are expressed as follows:

1. (Mechanical system—Time domain elements) Identify the potential failure modes of the elements according to the load case. By the method of equivalent linear safety margin, the reliability index of the elements in load case m β^e_m and the sensitivity factors (direction cosines) of the elements can be obtained.

2. (Elements—Subsystem) According to the independence occurrence of load and time invariant of the resistance assumptions, the correlation coefficient ρ^e_m between the elements in subsystem m can be obtained by

$$\rho^e_m = \sum_{i=1}^{l} \alpha^2_{ri,m}$$

in which $\alpha_{ri,m}$ is the sensitivity factor (direction cosine) of the ith resistance in the equivalent linear function of the elements. l is the number of the resistance.

It is obvious that the subsystem is a series system with equally correlated and equally reliability index elements. So, the reliability index of the subsystem m can be calculated by

$$\beta^{sub}_m = \Phi^{-1} \int_{-\infty}^{+\infty} \varphi(t) \left[\Phi\left(\frac{\beta^e_m - \sqrt{\rho^e_m}\, t}{\sqrt{1 - \rho^e_m}} \right) \right]^{v_m T} dt$$

3. (Subsystem) Again by the equivalent linear safety margin method, we can find that the sensitivity factors(direction cosines) of the resistances in the equivalent linear performance function of the subsystem m are equal to the sensitivity factors of the resistances in that of the elements.

$$\alpha^{sub}_{ri,m} = \alpha^e_{ri,m}$$

Because of the independence assumption of the loads, we are not concerned about the direction cosines of the loads $\alpha^{sub}_{si,m}$ or $\alpha^e_{si,m}$. Untill now, we have obtained the reliability in-

dex β_m^{sub} and the direction cosines $\alpha_{ri,m}^{sub}$ of the subsystem m.

The correlation coefficient $\rho_{m,n}^{sub}$ between the subsystem m and the subsystem n can be calculated by

$$\rho_{m,n}^{sub} = \sum_{i=1}^{l} \alpha_{ri,m}^{sub} \alpha_{ri,n}^{sub} = \sum_{i=1}^{l} \alpha_{ri,m}^{e} \alpha_{ri,n}^{e}$$

4. (Subsystems—System) By the system reliability methods, the failure probability of the system in Fig.1 can be calculated.

3.NUMERICAL EXAMPLE

Consider a simple frame under vertical and horizontal time variant loads as shown in Fig.3. The example was first mentioned in Ref.[2]. The loadings are sparse Poisson pulse processes representing transient loads. There are 8 possible mechanisms listed in Ref.[2]. For simplicity, only three failure modes were used in this analysis (also the same with Ref.[2]). They are:

(1)$4M_1+2M_2-10S_1-15S_2$
(2)$4M_1-15S_2$
(3)$4M_2-10S_1$

The statistics of the member and loading parameters are given in Table 1.

The reliability functions using the method proposed in this paper are compared with the methods of Wen(1989) in Fig.4 and Table 2. It is found that the results using the method in this paper are satisfactory.

4.SUMMARY AND CONCLUSIONS

In this paper, a method on the basis of equivalent linear safety margin method and the existing loads combination theory is proposed to treat the problem of the system reliability under time varying loads. Example shows that this method is simple and generally yields good results.

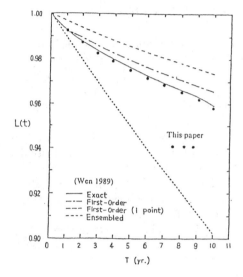

Fig.4 Comparison of reliability functions

Table 2. Failure probabilities in the first year

	EFRM	ES	TP
A	0.0097	0.0073	0.0073
B	0.084	0.065	0.072
C	0.37	0.21	0.26
D	0.51	0.27	0.32

Note:

EFRM: Ensemble failure rate method(Wen 1989)

ES: Exact solution(Wen 1989)

TP: This paper

A: $v_1 = 5 / \text{yr}, \ v_2 = 0.2 / \text{yr}$
$\qquad \mu_{d1} = \mu_{d2} = 0.01 \text{yr}$
B: $v_1 = v_2 = 5$
$\qquad \mu_{d1} = \mu_{d2} = 0.01$
C: $v_1 = v_2 = 20$
$\qquad \mu_{d1} = \mu_{d2} = 0.01$
D: $v_1 = v_2 = 20$
$\qquad \mu_{d1} = \mu_{d2} = 0.025$

5.ACKNOWLEDGEMENT

This study was supported by the National Science Foundation of China(NSFC).

REFERENCES

Thoft–Christensen, P. and Murotsu,Y., Application of structural systems reliability theory,Springer–Verlag, 1986.

Wen,Y.K. and Chen,H.C., System reliability under time varying loads, J. Eng. Mech., ASCE, 115(4), 1989.

Rackwitz,R., Reliability of systems under renewal pulse loading, J. Eng. Mech.,ASCE, 9(111), 1985.